Phosphate Minerals

Edited by
J. O. Nriagu and P. B. Moore

With Contributions by
F. Betts, N. C. Blumenthal, D. R. Bouldin, V. F. Buchwald
P. J. Cook, R. Flicoteaux, J. P. LeGeros, R. Z. LeGeros
J. Lucas, P. B. Moore, G. H. Nancollas, W. P. Nash, Y. Nathan
J. O. Nriagu, R. T. Oglesby, A. S. Posner, Y. Tardy
M. B. Tomson, P. Vieillard, L. Vignona

With 83 Figures

Springer-Verlag
Berlin Heidelberg New York Tokyo 1984

Dr. JEROME O. NRIAGU
National Water Research Institute;
Burlington, Ontario L7R 4A6, Canada

Professor Dr. PAUL B. MOORE
Department of Geophysical Sciences, University of Chicago,
Chicago, Illinois 60637, USA

ISBN 3-540-12757-7 Springer-Verlag Berlin Heidelberg New York Tokyo
ISBN 0-387-12757-7 Springer-Verlag New York Heidelberg Berlin Tokyo

Library of Congress Cataloging in Publication Data. Main entry under title: Phosphate minerals. Bibliography: p. Includes index. 1. Phosphates. I. Nriagu, Jerome O. II. Moore, Paul B. (Paul Brian), 1940– . III. Betts, F. QE389.64.P46 1983 549'.72 83-17160

Typesetting, printing and bookbinding: Brühlsche Universitätsdruckerei, Giessen
2132/3130-543210

Preface

The literature on the geology, chemistry, and biochemistry of phosphorus generally takes its mineralogy for granted. The incidental information on phosphate minerals given in these texts is often obsolescent and inaccurate. The few mineralogical texts that have dealt comprehensively with the phosphate minerals have now become outdated, and typically present the essential information in a manner unsuitable for nongeological readers. This volume is intended as a ready reference for workers who require good basic information on phosphate minerals or their synthetic equivalents. The topics covered should appeal to geologists and geochemists, lithologists, environmental scientists and engineers, chemists and biochemists who have any interest in the intricate world of phosphorus.

The hard tissues of many vertebrates and the many pathological calcifications consist mostly of phosphate minerals. The precipitation of these compounds also plays a major role in the ecological cycling of phosphorus, and occasionally even dominates the behavior of many trace metals in many geochemical and biological systems. Indeed, many pegmatitic phosphate minerals have acquired some notoriety because of the rarer trace metals which they tend to accumulate. With the commercialization of phosphate fertilizers since the early part of the 19th century, phosphate minerals have assumed an important role in industrial chemistry and agriculture. Clearly, the study of phosphate minerals is important from the economic, agricultural, environmental and (human and animal) health viewpoint.

Phosphate minerals are among the most complex and variegated in the entire mineral kingdom. This volume summarizes the determinative properties of the 300 or so known minerals which have been relatively well characterized. It reviews the current knowledge on the nucleation and growth of phosphate minerals and summarizes the available thermochemical data for these compounds. The volume deals comprehensively with the physical and chemical controls on the formation and weathering of phosphate minerals in various milieu, from the igneous and metamorphic rocks to biological systems. The recent influence of pollution on the precipitation of phosphate minerals in soils and sediments has also been highlighted.

Any success of this volume depends very much on the efforts of our distinguished group of contributors. We are also indebted to Springer-Verlag for invaluable editorial assistance.

Burlington, Ontario, Canada — Jerome O. Nriagu
Chicago, Illinois, U.S.A. — Paul B. Moore

Contents

Chapter 1 Phosphate Minerals: Their Properties and General Modes of Occurrence. J. O. NRIAGU 1

Chapter 2 The Nucleation and Growth of Phosphate Minerals. G. H. NANCOLLAS. (With 9 Figures) 137

Chapter 3 Crystallochemical Aspects of the Phosphate Minerals. P. B. MOORE. (With 6 Figures) 155

Chapter 4 Thermochemical Properties of Phosphates. P. VIEILLARD and Y. TARDY. (With 1 Figure) 171

Chapter 5 Phosphate Minerals in Meteorites and Lunar Rocks. V. F. BUCHWALD 199

Chapter 6 Phosphate Minerals in Terrestrial Igneous and Metamorphic Rocks. W. P. NASH. (With 5 Figures) 215

Chapter 7 Spatial and Temporal Controls on the Formation of Phosphate Deposits – A Review. P. J. COOK. (With 11 Figures) 242

Chapter 8 The Mineralogy and Geochemistry of Phosphorites. Y. NATHAN 275

Chapter 9 Weathering of Phosphate Minerals. R. FLICOTEAUX and J. LUCAS. (With 10 Figures) 292

Chapter 10 Formation and Stability of Base Metal Phosphates in Soils and Sediments. J. O. NRIAGU. (With 6 Figures) 318

Chapter 11 Chemistry and Structure of Precipitated Hydroxyapatites. A. S. POSNER, N. C. BLUMENTHAL and F. BETTS. (With 7 Figures) 330

Chapter 12 Phosphate Minerals in Human Tissues. R. Z. LEGEROS and J. P. LEGEROS. (With 21 Figures) 351

Chapter 13 Precipitation of Phosphate Minerals in Waste Water Treatment Systems. M. B. TOMSON and L. VIGNONA. (With 3 Figures) 386

Chapter 14 Phosphorus in the Environment. R. T. OGLESBY and D. R. BOULDIN. (With 4 Figures) 400

Subject Index 425

Author Index 435

List of Contributors

You will find the addresses at the beginning of the respective contributions

Betts, F. 330
Blumenthal, N.C. 330
Bouldin, D.R. 400
Buchwald, V.F. 199
Cook, P.J. 242
Flicoteaux, R. 292
LeGeros, J.P. 351
LeGeros, R.Z. 351
Lucas, J. 292
Moore, P.B. 155
Nancollas, G.H. 137
Nash, W.P. 215
Nathan, Y. 275
Nriagu, J.O. 1, 318
Oglesby, R.T. 400
Posner, A.S. 330
Tardy, Y. 171
Tomson, M.B. 386
Vieillard, P. 171
Vignona, L. 386

Chapter 1 Phosphate Minerals: Their Properties and General Modes of Occurrence

JEROME O. NRIAGU

Introduction

This chapter presents a review of the physical, chemical and crystallographic properties of phosphate minerals, and discusses their general modes of occurrence. It provides the necessary backdrop to the subsequent chapters which deal intensively with the various aspects of the important group of compounds.

I have relied heavily on the *Encyclopedia of Minerals* (by Roberts et al., 1974) for information on phosphate minerals published before 1973. Their mineral data have been checked and where necessary modified in the light of more recent findings. For the new minerals described since 1973, I have used the *New Minerals* section of the *American Mineralogist, Lists of New Mineral Names* of *Mineralogical Magazine* and the recent compilation, *A Manual of Mineral Names*, by Embrey and Fuller (1980). Other compendia consulted during the preparation of the report are listed below.

An attempt has been made to present what is considered to be the best information available on each mineral species. I have not endeavored to reconcile the conflicting data and opinions which certainly abound in the scientific literature. Also, no attempt has been made to present an exhaustive reference for each mineral; to do so could easily have doubled the size of the chapter. For easy reference, the minerals have been described alphabetically. Such a presentation strategy obviously obscures any important crystallo-chemical relationships which are described by Moore (Chap. 3 this Vol.).

Historical

Phosphate minerals have been known and prized since remote antiquity. Turquois has been found among the remains of many ancient civilizations, including Egypt, Mesopotamia, India and China. As early as 3,400 B.C., it was obtained from the Sinai peninsula, Egypt, in what was probably one of the earliest hard-rock mining operations in the world. A necklase with a turquois pendant rudely fashioned into the form of an ibex has been found which dates to the Old Empire, about 3,500 B.C. (Kunz, 1971). Its name means Turkish and was presumably acquired because the first supply of the mineral to Europe came from Persia by way of Turkey. The

National Water Research Institute, Box 5050, Burlington, Ontario L7R 4A6, Canada most famous ancient workings of turquois were located on the south slopes of the Ali-Mirsa-Kuh mountains near Nishapur, Iran and in the Kara Tube mountains near Samarkand (Eichholz 1962, p 253). Ancient turquois mines are also known in Siberia, Turkestan, Asia Minor, the Sinai Peninsula, Silesia, and Saxony in Germany. Numerous deposits in Arizona, California, Colorado, Nevada, New Mexico, and Utah were opened centuries ago by the American Indians and some are still being worked. Best known are the mines in the Cerrilos Hills, near Santa Fe, New Mexico (*Encyclopedia Britannica* 968, p 415).

Turquois was especially prized in Tibet, where it was used as one of the commonest decorative stones on all the ancient Tibetan works of art, taking a place almost equivalent to jade in China (Laufer 1913). The Tibetan word for it, *gyu*, is indigenous, suggesting that the mineral has been known since remote times. The Chinese term *sê–sê*, believed by some scholars to mean turquois, has been identified with the gem minerals of Badakshan (Laufer 1913). Turquois was mentioned frequently in the official histories of the Thang, and has been mined in China since the Yuan time (Needham 1959, p 672).

Theophrastus (ca. 4th century B.C.) referred to the blue turquois used by the Persians in inlay work (Eichholz 1962, p 215). Bolos of Mendes (ca. 200 B.C.), who wrote about *mirabilia* and on the counterfeiting of precious metals and gems, also spoke of the famous Therminaean and Persian green turquois (Eichholz 1962, p 219). Pliny (37:18) described various kinds of turquois under the heading of *smaragdis*. He described the various sources of green turquois (*callaina*) in ancient times, and then gives the following rather convoluted account of the prospecting and mining of this mineral (Pliny, 38:33): "It [*callaina,* green turquois] occurs in the hinterland beyond India among the inhabitants of the Caucasus, the Hyrcani, Sacae and Dahae. It is of exceptional size, but is porous and full of flaws. A far purer and finer stone is found in Carmania. In both localities, however, 'callaina' occurs amidst inaccessible icy crags, where it is seen as an eyeshaped swelling loosely adhering to the rocks, as though it had been attached to them, rather than formed upon them. Thus tribes accustomed to riding on horseback and too lazy to use their feet find it irksome to climb in search of the stones; and they are also deterred by the risks. They, therefore, shoot at them from a distance with their slings and dislodge them, moss and all. This is the article that pays their taxes, this they acknowledge to be the most beautiful thing that can be worn on neck or fingers, from this they derive their wealth, this is their pride and joy as they boast of the number that they have shot down since their childhood, an operation in which success varies, seeing that some win fine stones with their first shot, while many reach old age without obtaining one. Such, then, is the way in which they hunt the 'callaina'. Subsequently, the stone is shaped by the drill, being in other respects an easy stone to deal with. The best stones have the colour of 'smaragdus', so that it is obvious, after all, that their attractiveness is not their own. They are enhanced by being set in gold, and no gem sets off gold so well. The finer specimens lose their colour if they are touched by oil, unguents or even undiluted wine, whereas the less valuable ones preserve it more steadfastly. No gemstone is more easily counterfeited by means of imitations in glass. Some authorities say that 'callainae' are found in Arabia inside the nests of the birds known as 'melanchoryphi' or 'black caps'."

The accounts left by the Medieval writers indicate that turquois has continued to be highly sought-after as a gem and ornamental Stone in every age and culture for nearly 80 centuries (see Kunz, 1971 for detailed discussion).

Apatite has also been used since ancient times. Bone mineral was used in the cupellation of lead to recover the silver long before Roman times. The use of bone-ash is suggested in the following passage from Amos (2:1), "I will not turn away the punishment thereof, because he burned the bones of the king of Edom into lime". The white pigment of many ancient painters consisted of ground bone ash. The alchemists used bones and other phosphatic material, such as urine, in many of their operations and undoubtedly synthesized a number of metal orthophosphates of varying degrees of purity. With the commercialization of phosphate fertilizers during the early part of the 19th century, phosphate minerals assumed a major role in industrial chemistry, agriculture and global economics.

Beside turquois and apatite, only six other still valid phosphate minerals were characterized before 1800. These include lazulite, pyromorphite (in essence, lead apatite), torbernite and vivianite. Mandarino (1977) estimates that less than 100 valid mineral species were known prior to 1800, although many more names were used as synonyms. This implies that around 1800, the phosphate minerals constituted about 6% of all the known minerals, or roughly the same fraction as today. The historical pattern in the subsequent introduction of new phosphate minerals (still valid) into the scientific literature is as follows:

Period	New minerals reported
Pre –1800	7
1801–1820	7
1821–1840	16
1841–1860	11
1861–1880	32
1881–1900	16
1901–1910	11
1911–1920	16
1921–1930	21
1931–1940	13
1941–1950	16
1951–1960	41
1961–1970	35
1971–1980	65
1981	14

The figures show that between 1800 and 1900, the number of new phosphate minerals reported every 20 years averaged about 20. The unusually high number between 1861 and 1880 probably reflects the fact that determinative mineralogy attained the status of a scientific discipline during this period. Since 1900, the number of new phosphate minerals reported has increased quite dramatically. Twenty seven new minerals were reported between 1901 and 1920, 34 between 1921 and

1940, 57 between 1941 and 1960 and nearly 100 between 1961 and 1980. In 1981 alone, 14 new minerals were reported. Thus, about a third of all the 320 or so currently valid phosphate names were introduced during the last two decades; undoubtedly, some the most recent mineral names will eventually be invalidated and discarded.

Of the more than 2,000 phosphate mineral names in the older literature (Aballain et al. 1968; Palache et al. 1951), only about 350 are still regarded as valid mineral names. This compilation (see below) also retains a number of varietal names so that the total number of distinctive phosphate mineral species is believed to now number about 300.

Composition

The PO_4 radical combines with over 30 elements to form phosphate minerals. In addition, a large number and variety of substitution can occur among similar ions, and this makes it almost impossible to give satisfactory unique formulae for many of the phosphate minerals. For example, about 45% of the known phosphate minerals carry ferrous or ferric iron as a major component; 30%–35% contain substantial amounts of aluminum; 30%–35% feature calcium and roughly 25% have manganese as a major component. Many more phosphate minerals contain these metals only in minor amounts. In fact, virtually every known element has been found, at least in trace amounts, in a phosphate mineral. Phosphate minerals are particulary favored as hosts of uranium, thorium and many other rare elements.

Phosphate minerals are generally anisodesmic with the other XO_4^{3-} groups. Mutual substitution of P for As is very common and complete solid solution is known between some phosphates and arsenates. At ambient temperatures, the solid solution between phosphates and silicates is restricted, but the mutual solubility increases greatly with temperature. For example, complete solid solution has been demonstrated between tephroite $Mn_2[SiO_4]$ and lithiophilite $LiMn[PO_4]$ at about 950 °C and between the isostructural berlinite $AlPO_4$ and cristobalite above 1,450 °C, while almost 50 mol% $AlPO_4$ is soluble in β-eucryptite $LiAlSiO_4$ at 1,050 °C (Liebau and Koritnig 1970). Extensive replacement of P by B, V, Ge, and S in apatites and other phosphate minerals has also been reported.

Isotypism and isomorphism in phosphate minerals entail (a) the replacement of all or some of the associated cations, (b) the replacement of the phosphorus or vicarious atoms, (c) the replacement of the oxygen atoms bonded to the phosphorus or vicarious atoms, and (d) a combination of some of the above possibilities. The catholicity of the known phosphate mineral compositions clearly underscores the great reactivity of the phosphate ions under every conceivable geochemical environment.

Occurrence

Phosphorus ranks tenth in crustal abundance, and the phosphate minerals are formed as accessory minerals in virtually every known rock formation. The meteorites have yielded a number of important phosphate minerals (Buchwald, Chap. 5, this Vol.). Phosphate minerals, especially apatites, are a common accessory mineral group in most igneous and metamorphic rocks (Nash, Chap. 6, this Vol.). Pegmatites represent a major host of phosphate minerals (Moore 1982); these pegmatitic phosphates have attracted much commercial and scientific interest because of the rarer elements which tend to accumulate in them. The list of the secondary minerals which can be derived from the supergene alterations of the so-called primary phosphates is quite long (see Moore 1982). Likewise, a large number of species are generated in diverse subaqueous environments (Cook, Chap. 7, this Vol.; Flicoteaux and Lucas, Chap. 9, this Vol.). Even the discharge of human wastes can engender the formation of unique suites of phosphate minerals in soils (Nriagu, Chap. 10, this Vol.), sediments (Oglesby and Bouldin, Chap. 14, this Vol.), and wastewater treatment systems (Tomson and Vignona, Chap. 13, this Vol.). Biological systems generate their important suite of phosphate minerals (McConnell 1973); in general, however, most of the normal hard tissues of vertebrates are apatitic, as are many of the pathological calcifications (LeGeros and LeGeros, Chap. 12, this Vol., Posner et al., Chap. 11, this Vol.).

References

Aballain M, Chambolle P, Derec-Poussier F, Guillemin C, Mignon R, Pierrot R, Sarcia JA (1968) Index alphabetique de nomenclature mineralogique. Bureau Recher Geol Minier, Paris
Eichholz DE (1962) Introduction and notes in Pliny's natural history
Embrey PG, Fuller JP (1980) A manual of mineral names. University Press, Oxford
Kunz GF (1971) The curious lore of precious stones. Dover Publications, New York
Laufer B (1913) Notes on turquois in China. Field Mus Nat Hist (Chicago) Pub, Anthropol Ser 13 (1), Pub 169
Liebau F, Koritnig S (1970) Phosphorus, Part 15A. In: Wedephol KH (ed) Handbook of geochemistry. Springer, Berlin Heidelberg New York
Mandarino, JA (1977) Old mineralogical techniques. Can Miner 15:1–2
McConnell, D (1973) Biomineralogy of phosphates and physiological mineralization. p 425–449, In: Griffith EJ et al. (eds) Environmental phosphorus handbook. Wiley. New York
Moore PB (1982) Pegmatite minerals of P(V) and B(III). Mineral Assoc Can Handbook, vol 8. p 267–291
Needham J (1959) Science and Civilization in China, vol 3. University Press, Cambridge.
Palache C, Berman H, Frondel C (1951) Dana's system of mineralogy, vol 2. Wiley, New York
Pliny. Natural history, 1962, vol 10. The Loeb Classical Library. Harvard University Press. Cambridge MASS
Roberts WL, Rapp GR Jr, Weber J (1974) Encyclopedia of minerals. Van Nostrand. New York

Abukumalite = Britholite

Aldermanite
$Mg_5Al_{12}(PO_4)_8(OH)_{22} \cdot nH_2O$ (n ≃ 32)

Analyses (%): P_2O_5 25.9, MgO 8.4, CaO 1.2, Al_2O_3 28.4, H_2O 26.1
Color-Luster: Colorless
Crystal System: Orthorhombic
Density: 2.15 (Calc.)
First Reported: 1981
Habit: Minute, very thin, talc-like crystallites elongated in one direction with simple ensiform termination
Hardness: less than 2
Lattice Constants: a = 15.000, b = 8.330, c = 26.60 (Å)
Mode of Occurrence: Partly as an alteration product of fluellite, thinly coating cracks and cavities in a brecciated metamorphosed rock phosphate at Moculta, South Australia
Optical constants: Not certain; low birefringence
Space Group: Not given
Strongest Diffraction Lines: 13.40(100), 7.98(80), 5.55(60), 2.841(50)
Z: 2
Ref.: Harrowfield IR, Segnit ER, Watts JA (1981) Miner Mag 44:59–62

Alforsite (Barium chlorapatite)
$Ba_5(PO_4)_3Cl$

Analyses (%): BaO 67.7, CaO 4.6, SrO 2.7, PbO 0.8, P_2O_5 22.7, Cl 3.6, F 0.7
Color-Luster: Colorless, with distinctive reddish-violet cathodoluminescence
Crystal System: Hexagonal
Density: 4.73–4.83 (Calc.)
First Reported: 1981
Habit: Isolated small (0.05–0.2 mm) subhedral grains
Lattice constants: a = 10.25, c = 7.64 (Å)
Mode of Occurrence: Found as an accessory mineral in contact metamorphosed evaporitic rocks at the Incline, Mariposa County and Big Creek, Fresno County, California. The associated minerals include witherite, quartz, fluorapatite and sanbornite
Optical Constants: $\varepsilon = 1.694$, $\omega = 1.696$; Uniaxial negative with extremely low birefringence and high relief
Space Group: $P6_3/m$
Strongest Diffraction Lines: 3.06(100), 2.95(30), 2.13(30), 2.03(30), 1.928(30)
Ref.: Newberry NG, Essene EJ, Peacor DR (1981) Am Miner 66:1050–1053

Alluaudite Series
(includes *Ferroalluaudite* and *Maghagendorfite*)
$(Na, Ca)(Fe^{2+}, Mn^{2+})Fe_2^{3+}(PO_4)_3$

Analyses (%): MnO 17.9–28.2, Fe_2O_3 12.8–25.6, Na_2O 5.5–8.8, FeO 0–3.2, CaO 0–4.6, P_2O_5 41.2–46.7
Class: 2/m
Cleavage: {100} good, {010} good
Color-Luster: Dull greenish black; dirty yellow, brownish yellow to brownish black. Subtranslucent to opaque. Streak and powder dirty yellow to brownish
Crystal System: Monoclinic
Density: 3.52 (Meas.), 2.584 (Meas.)
First Reported: 1827
Habit: Massive, compact granular, radiating fibrous, and globular aggregates
Hardness: 5–5.5
Lattice Constants: a = 12.004, b = 12.533, c = 6.404 (Å), $\beta = 114.4°$
Mode of Occurrence: Occurs in large masses in pegmatite at several mines in the Custer-Pringle area, Custer County, and at the Etta mine, Pennington County, South Dakota. Also found in pegmatite at Chanteloube, France; Buranga, Ruanda; Varuträsk, Sweden; Sukula and Lemnäs, Finland; and in several other pegmatite deposits

Optical Constants: $\alpha = 1.760$, $\beta = 1.765$, $\gamma = 1.775$, (+)2V = moderate

Space Group: C2/c

Strongest Diffraction Lines: 2.73(100), 6.30(80), 3.07(65)

Z: 12

Ref.: Fisher DJ (1955) Am Miner 40:1100–1109, Fisher DJ (1965) Am Miner 50:1647–1669, Moore PB (1971) Am Miner 56:1955–1975, Moore PB, Ito J (1979) Miner Mag 43:227–235

Althausite

$Mg_4(PO_4)_2(OH,O,F)$

Analyses (%): MgO 45.4, CaO 4.50, Fe_2O_3 1.37, Al_2O_3 0.53, P_2O_5 43.6, F 2.86, H_2O 1.87

Cleavage: {001} perfect, {101} distinct

Color-Luster: Gray (from magnetite inclusions), luster vitreous

Crystal System: Orthorhombic

Density: 2.97 (Meas.), 2.91 (Calc.)

First Reported: 1975

Habit: Cleavable masses up to several hundred grams

Hardness: 3.5

Lattice Constants: a = 8.258, b = 14.383, c = 6.054 (Å)

Mode of Occurrence: Found in serpentine-magnesite deposits at Modum, Norway, where it is closely associated with apatite and Ca-Mg phosphates. It contains inclusions of magnetite and xenotime and is partly altered to apatite along cleavage planes

Optical Constants: $\alpha = 1.588$, $\beta = 1.592$, $\gamma = 1.598$; (+)2V 70° (Est.), 78.5° (Calc.)

Space Group: $Pna2_1$

Strongest Diffraction Lines: 3.593(100), 3.316(90), 3.024(80), 2.786(60)

Z: 8

Ref.: Raade G, Tysseland M (1975) Lithos 8:215–219, Rømming C, Raade G (1980) Am Miner 65:488–498

Amblygonite

(Amblygonite-Montebrasite Series)

$(Li,Na)Al(F,OH)(PO_4)$

Analyses (%): P_2O_5 46.7–49.3, Li_2O 8.8–10.2, Na_2O 0.04–2.2, Al_2O_3 33.9–35.8, F 0.3–7.85, H_2O 3.3–6.0

Class: $\bar{1}$

Cleavage: {100} perfect, {110} good, {0$\bar{1}$1} distinct, {001} imperfect. Fracture conchoidal to uneven. Brittle

Color-Luster: Usually white to grayish white; also colorless, yellowish, pinkish, tan, greenish, bluish. Transparent to translucent. Vitreous to greasy; pearly on cleavages

Crystal System: Triclinic

Density: 3.08 (Meas.), 3.065 (Calc.)

First Reported: 1817

Habit: Crystals equant to short prismatic; commonly with rough faces. Usually as large cleavable masses. Twinning on {$\bar{1}\bar{1}$1} common

Hardness: 5.5–6

Lattice Constants: a = 5.10, b = 7.17, c = 5.03 (Å), $\alpha = 112°$, $\beta = 97°$, $\gamma = 67°$

Mode of Occurrence: Occurs in granite pegmatites often in masses and crystals of very large size. Found at many places in the Black Hills, South Dakota, especially at the Beecher Lode, Custer County, as masses up to 200 tons in weight; as rounded masses as much as 20 feet long and 4–8 feet in thickness at the Giant-Volney mine at Tinton, Lawrence County; and as masses 20–40 feet long at the Hugo, Ingersoll, and Peerless mines, Keystone, Pennington County. It also is found in Arizona, New Mexico, California, New Hampshire, and at several localities in Maine, notably as fine transparent crystals at Newry, Oxford County. Among other occurrences, it is found in France, Germany, Czechoslovakia, Sweden, and Brazil

Optical Constants: $\alpha = 1.578$, $\beta = 1.593$ (vary with % F), $\gamma = 1.598$, (−)2V ≃ 50°

Space Group: $P\bar{1}$

Strongest Diffraction Lines: 4.64(100), 3.151(100), 2.925(100)

Z: 2

Ref.: Roberts WL, Rapp GR, Weber J (1974) Encyclopedia of minerals, New York, Van Nostrand Reinhold, p 17–18. Černá I, Černý P, Ferguson RB (1973) Am Miner 58:291–301

Anapaite
$Ca_2Fe^{2+}(PO_4)_2 \cdot 4H_2O$

Analyses (%): P_2O_5 34.4–35.5, CaO 26.4–28.3, FeO 17.5–20.0, H_2O 18.1–18.6

Class: $\bar{1}$

Cleavage: {001} perfect, {010} distinct

Color-Luster: Green to greenish white. Transparent. Vitreous. Streak white.

Crystal System: Triclinic

Density: 2.81 (Meas.), 2.80 (Calc.)

First Reported: 1890

Habit: Crystals tabular; as crusts of subparallel crystals and as rosette-like aggregates.

Hardness: 3.5

Lattice Constants: a = 6.41, b = 6.88, c = 5.86 (Å); $\alpha = 101.58°$, $\beta = 104.02°$, $\gamma = 71.01°$

Mode of Occurrence: Found as layers of pale green crystals at a depth of 500 feet in a core from the Lewis well, Kings County, California. It also occurs in phosphatic geodes in Miocene clay at Bellaver de Cerdena, Gerona Province, Spain; in a bituminous clay rock at Messel, Hesse, Germany; and in oolitic iron ore near Anapa, Taman peninsula, U.S.S.R.

Optical Constants: $\alpha = 1.602$, $\beta = 1.613$, $\gamma = 1.649$; (+)2V = 54°

Space Group: $P\bar{1}$

Strongest Diffraction Lines: 3.135(100), 2.866(70), 3.72(60)

Z: 1

Ref.: Roberts WL, Rapp GR, Weber J (1974) Encyclopedia of minerals, New York, Van Nostrand Reinhold, p 21

Andrewsite
$(Cu,Fe^{2+})_3Fe^{3+}_6(PO_4)_4(OH)_{12}$

Analyses (%): P_2O_5 25.8–26.1, FeO 7.1–8.6, Fe_2O_3 43.0–44.6, CuO 10.9–12.6, H_2O 8.8–9.8

Class: 2 2 2

Cleavage: In two directions parallel to fiber length

Color-Luster: Green to bluish green. Silky

Crystal System: Orthorhombic

Density: 3.475 (Meas.)

First Reported: 1871

Habit: Botryoidal aggregates with radial-fibrous structure

Hardness: 4

Lattice Constants: a = 14.16, b = 16.83, c = 5.18 (Å)

Mode of Occurrence: Occurs associated with limonite, chalcosiderite, and other minerals at the West Phoenix mine, near Liskeard, Cornwall, England

Optical Constants: $\alpha = 1.813$, $\beta = 1.820$, $\gamma = 1.830$; (+)2V = large

Space Group: $B2\,2_1\,2$

Strongest Diffraction Lines: 3.22(100), 2.12(80), 5.01(50)

Z: 4

Ref.: Roberts WL, Rapp GR, Weber J (1974) Encyclopedia of minerals, New York, Van Nostrand Reinhold, p 24

Apatite Series
$Me_{10}(XO_4)_6Z_2$; Me = Ca, Sr, Ba, Cd, Pb (typically); X = P, As, V, Mn, Cr; Z = OH, F, Cl, Br

1. Minerals of the apatite series are listed separately under:

Abukumalite	Francolite
Britholite	Hydroxyapatite
Carbonate apatite	Mimetite
Chlorapatite	Pyromorphite
Dahllite	Svabite
Ellestadite	Vanadinite
Fermorite	Wilkeite
Fluorapatite	

2. The following discredited apatitic names still appear occasionally in recent texts and reports:
   Collophane = Carbonate hydroxyapatite
   Deltaite = Crandallite + hydroxyapatite
   Glaubapatite = Monetite + apatite
   Grodnolite = Francolite
   Kurskite = Francolite
   Lessingite = Britholite
   Lewistonite = Potassium hydroxyapatite
   Monite = Carbonate hydroxyapatite
   Munkforssite = Manganoan apatite
   Nauruite = Francolite
   Ornithite = Carbonate hydroxyapatite
   Pyroguanite = Phosphate rock
   Rhodophosphite = Manganoan apatite
   Saamite = Strontium apatite
   Sombrerite = Phosphate rock
   Staffelite = Francolite
3. The cell parameters for natural and synthetic apatites and pyromorphites are summarized in the following table (Table 1)

Table 1. Lattice constants for synthetic and natural phosphatic apatites and pyromorphites

Compound	Unit cell data (Å)	
	a	c
$Ca_{10}F_2(PO_4)_6$ Fluorapatite	9.364	6.879
$Ca_{10}Cl_2(PO_4)_6$ Chlorapatite	9.52 10.43	6.85 7.79
$Ca_{10}Br_2(PO_4)_6$ Bromapatite	9.714	6.759
$Ca_{10}(OH)_2(PO_4)_6$ Hydroxyapatite	9.43_2	6.88
$Ca_{10}O(PO_4)_6$ Oxyapatite	9.38	6.93
$Ca_{10}SiP_4SO_{24}F_2$	9.45	6.96
$Ca_{10}SiP_4SO_{24}(OH)_2$	9.44	6.96
$Ba_{10}(OH)_2(PO_4)_6$	10.19	7.70
$Pb_{10}Cl_2(PO_4)_6$ Pyromorphite	9.95	7.31
$Pb_{10}O(PO_4)_6$	9.84	7.43
$Pb_{10}(OH)_2(PO_4)_6$	9.90	7.29
$Pb_5GeO_4(PO_4)_2$	9.88	7.32
$Pb_5(Si_{0.5}Ge_{0.5}O_4) \cdot (PO_4)_2$	9.81	7.30
$Pb_5(Si_{0.5}Ge_{0.5}O_4) \cdot (PO_4)(AsO_4)$	9.92	7.34
$Pb_5SiO_4.VO_4.PO_4$	9.88	7.33
$Pb_4Na(PO_4)_3$	9.71	7.18
$Pb_4K(PO_4)_3$	9.81	7.28
$Pb_4Tl(PO_4)_3$	9.82	7.37
$Pb_4Rh(PO_4)_3$	9.86	7.37
$Pb_4Bi(SiO_4)_2(PO_4)$	9.76	7.26
$Pb_3BiTl(SiO_4)(PO_4)_2$	9.78	7.34
$Pb_{4.5}Na_{0.5}(PO_4)_{2.5} \cdot (SiO_4)_{0.5}$	9.76	7.25
$Sr_{10}(OH)_2(PO_4)_6$	9.74	7.20
$Ca_9Cd.F_2(PO_4)_6$	9.36	6.88
$Ca_5Cd_5F_2(PO_4)_6$	9.12	6.71
$Ca_8Y_2Si_2P_4O_{24}(OH)_2$	9.41	6.87
$Ca_6Y_4Si_4P_2O_{24}(OH)_2$	9.39	6.83
$Ca_2Na_2La_6Si_4P_2O_{24} \cdot (OH)_2$	9.62	7.11
$Ca_4Na_2La_4Si_2P_4O_{24} \cdot (OH)_2$	9.51	7.00
$Pb_2^{4+}Pb_8^{2+}P_4Si_2O_{26}$	9.78	7.33
$Ca_8Y_2P_6O_{26}$	9.37	6.84
$Ca_6Y_4Si_2P_4O_{26}$	9.36	6.84
$Ca_4Y_6Si_4P_2O_{26}$	9.35	6.82
$Ca_8La_2P_6O_{26}$	9.48	6.95
$Ca_6La_4Si_2P_4O_{26}$	9.57	7.02
$Ca_4La_6Si_4P_2O_{26}$	9.62	7.07
$(Sr, Ca)_{10}(OH)_2(PO_4)_6$ Strontium apatite (11.6% SrO)	9.41	6.91
$(Ca, Mn)_6Ca_4F_2(PO_4)_6$ Mangano-apatite	9.33	6,80
$Ca_9(H_2O)_2(PO_4)_6$	9.25	6.88
$(Sr, Ce, Na, Ca)_{10}(OH, O)_2(PO_4)_6$ Belovite	9.62	7.12
$Ca_8Na_2P_4S_2O_{24}F_2$	9.52	6.90
$(CaSr)_6Ca_4(F, OH, O)_2[(P, As)O_4]_6$ Fermorite	9.60	7.00
Carbonate apatites:[c]		
Dahllite '$Ca_6(OH)_2(P, C)_6O_{24}(Ca, C)_4$'	9.41–9.43	6.89–6.90
Dehrnite '$(Ca, Na)_6(OH)_2 \cdot (PO_4)_6(Ca, C)_4$'	9.31	6.87

Table 1. (continued)

Compound	Unit cell data (Å)	
	a	c
Lewistonite '$(Ca, K, Na)_6(OH)_2 \cdot (PO_4)_6(Ca, C)_4$'	9.35	6.89
Francolite '$(Ca_3F)_2(P, C)_6 \cdot (O, OH, F)_{24}(Ca, C)_4$'	9.33–9.36	6.87–6.88
Wilkeite '$Ca_6(Cl, F, OH, O)_2 (P, S, Si, C)_6O_{24}$' $\cdot (Ca, C)_4$'	9.48	6.91
Ellestadite '$Ca_6(Cl, F, O, OH)_2 (S, Si, P, C)_6O_{24} \cdot (Ca, C)_4$'	9.53	6.91
Yttroapatite $(Ca, Y)_6(PO_4)_3(F, OH)$	9.397	6.860
Belovite $(Sr, Ce, Na, Ca)_{10} \cdot (PO_4)_6(O, OH)_2$	9.62	7.12
Britholite (Abukumalite, Fenghuangite, Lessingite, Beckelite) $(Ca, Th, Ce)_5 \cdot (PO_4, SiO_4)_3(OH, F)$	9.63	7.03
$Ba_5(PO_4)_3Cl$ Alforsite	10.25	7.64
$Sn_9Ca(PO_4)_6(OH, F, Cl)_2$	9.45	6.89
$Mn_{10}(PO_4)_6Cl_2$	9.54	6.20
$Cd_{10}(PO_4)_6F_2$	9.30	6.63
$Cd_{10}(PO_4)_6Cl_2$	9.67	6.50
$Ca_{10}(PO_4)_6F_2$	9.36	6.88
$Ca_{10}(PO_4)_6Cl_2$	9.63	6.78
$Ca_9Mg(PO_4)_6F_2$	9.355	6.867
$Ca_9Ni(PO_4)_6F_2$	9.364	6.870
$Sr_{10}(PO_4)_6F_2$	9.71	7.28
$Sr_{10}(PO_4)_6Cl_2$	9.87	7.19
$Pb_{10}(PO_4)_6F_2$	9.75	7.30
$Ba_{10}(PO_4)_6F_2$	10.16	7.69
$Ba_{10}(PO_4)_6Cl_2$	10.26	7.65

Archerite
$(K, NH_4)H_2PO_4$

Analysis (partial) (%): K_2O 10.8, NH_3 3.46

Cleavage: Indistinct

Color: White

Crystal System: Tetragonal

Density: 2.23

First Reported: 1977

Habit: Tetragonal crystals up to 2 mm long; water-soluble

Hardness: Relatively soft

Lattice Constants: a = 7.448, c = 6.977 (Å)

Mode of Occurrence: Occurs with biphosphammite and other phosphates and saline minerals as a constituent of stalactites and crusts in Petrogale Cave, Madura, W. Australia

Optical Constants: $\omega = 1.513$, $\varepsilon = 1.470$

Space Group: $I\bar{4}2d$

Z: 4

Ref.: Bridge PJ (1977) Miner Mag 41:33–35

Arctite
$Na_2(Ca, Ba)_4(PO_4)_3F$

Analyses (%): P_2O_5 33.8, SiO_2 3.60, CaO 31.8, BaO 12.0, Na_2O 9.72, K_2O 1.64, ZrO_2 3.25, F 5.25

Cleavage: {001} perfect

Color-Luster: Colorless, luster vitreous, pearly on the perfect cleavage

Crystal System: Tetragonal

Density: 3.13 (Meas.), 3.11 (Calc.)

First reported: 1981

Hardness: About 5

Lattice Constants: $a_{rh} = 14.32$ (Å). $\alpha_{rh} = 28°$; in hexagonal setting, a = 7.078, c = 41.203 (Å)

Mode of Occurrence: Found in a drill core in the valley of Vuonnemi River, Khibina alkalic massif, Kola Peninsula, U.S.S.R. in the pegmatitic veinlets; associated with rasvumite, villiaumite, aegirine and thenardite

Optical Constants: $\varepsilon = 1.577$, $\omega = 1.578$; Uniaxial neg., sometimes anomalously biaxial

Strongest Diffraction Lines: 2.746(100), 3.43(32), 13.80(25), 3.06(25), 2.804(25)

Z: 6

Ref.: Khomyakov AP, Bykova AV, Kurova TA (1981) Zapiski Vses Miner Obsh 110:506–508. Khomyakov AP, Bykova AV, Kurova TA (1982) Amer Miner 67:621

Ardealite
$CaHPO_4 \cdot CaSO_4 \cdot 4H_2O$

Analyses (%): P_2O_5 20.9–23.5, CaO 31.6–32.6, SO_3 21.3–23.1, H_2O 23.6–25.1

Cleavage: Not determined

Color-Luster: Pale yellow

Crystal System: Monoclinic

Density: 2.30 (Meas.), 2.38 (Calc.)

First Reported: 1931

Habit: As very fine grained powdery masses

Hardness: Not determined

Lattice Constants: a = 5.721, b = 30.992, c = 6.250 (Å); β = 117.3°

Mode of Occurrence: Occurs associated with gypsum and brushite in a phosphate deposit in a limestone cavern at Cioclovina, Transylvania, Rumania

Space Group: Cc

Z: 4

Ref.: Roberts WL, Rapp GR, Weber J (1974) Encyclopedia of minerals, New York, Van Nostrand Reinhold, p 32. Sakae T, Nagata H, Sudo T (1978) Am Miner 63:520–527

Arrojadite
(Dickinsonite – Arrojadite Series)
$KNa_4Ca(Mn^{2+}, Fe^{2+})_{14}$, $Al(OH, F)_2(PO_4)_{12}$

Analyses (%): P_2O_5 34.3–40.0, FeO 19.8–28.2, MnO 12.3–15.8, Na_2O 4.7–6.4, CaO 2.5–5.7, K_2O 1.4–1.7, Al_2O_3, F, H_2O 0.9–5.4

Class: 2/m

Cleavage: {001} distinct, {201} indistinct. Fracture subconchoidal to uneven

Color-Luster: Dark green; vitreous to greasy. Translucent

Crystal System: Monoclinic

Density: 3.55 (Meas.)

First Reported: 1891 (Unnamed); 1925

Habit: Large cleavable masses

Hardness: 5

Lattice Constants: a = 16.60, b = 10.02, c = 23.99 (Å); β = 93.62°

Mode of Occurrence: Occurs as large cleavable masses in granitic pegmatite associated with quartz, graftonite, sphalerite, cassiterite, beryl, muscovite, and spodumene at the Nickel Plate mine, Keystone, Pennington County, South Dakota; also as small cleavage masses in pegmatite at the Smith mine, Chandlers Mill, Newport, New Hampshire; and at the Serra Blanca pegmatite, Picuhy, Brazil.

Optical Constants: α – 1.664, β – 1.670, γ = 1.675; 2V = 86°

Space Group: C2/c

Strongest Diffraction Lines: 3.04(100), 2.72(80), 3.22(60)

Z: 12

Ref.: Lindberg, ML (1950) Am Min 35:59–76. Roberts WL, Rapp GR, Weber J (1974) Encyclopedia of minerals, New York, Van Nostrand Reinhold, p 35. Moore PB et al. (1981) Am Miner 66:1034–1049

Arsenocrandallite
(Member of the Beudantite-Crandallite group)
$(Ca, Sr)Al_3H[(As, P)O_4]_2(OH)_6$

Analyses (%): As_2O_5 22.9, P_2O_5 10.7, Al_2O_3 28.7, Fe_2O_3 1.20, CaO 6.9, SrO 6.0, BaO 4.3, CuO 1.8, ZnO 0.3, Bi_2O_3 2.4, SiO_2 3.2, H_2O 11.7

Cleavage: None; fracture conchoidal; may show triangular sectors under the microscope

Color-Luster: Blue to bluish-green; luster vitreous

Crystal System: Rhombohedral

Density: 3.25 (Meas.); 3.30 (Calc.)

First Reported: 1981

Habit: Reniform crusts and spherulitic aggregates up to 0.1 mm

Hardness: Approx. 5.5

Lattice Constants: a = 7.08, c = 17.27 (Å), Z = 3 for hexagonal setting.
a = 7.06 (Å), $\alpha = 60.2°$, Z = 1 for rhombohedral setting

Mode of Occurrence: Found on old dumps in the Neubulach mining district, Black Forest, Germany, associated with brochantite, chalcophyllite, parnauite, arseniosiderite, mansfieldite and corroded tennantite

Strongest Diffraction Lines: 2.99(100), 3.55(90), 1.769(60), 1.919(50)

Ref.: Walenta K (1981) Schweiz Miner Petrog Mitt 61:23–35

Attakolite (Attacolite)
$(Ca, Sr, Mn)_3(Al, Fe^{3+}, Mn)_6(PO_4, SiO_2)_{7-m}(OH)_{4m} \cdot 3H_2O$. A different formula is given as $Ca_2[Al_4(OH)_4(PO_4)_4]$ (P.B Moore and T. Araki (1975) Am Miner 60:460)

Analyses (%): P_2O_5 32.6–36.1, SrO 0–9.4, Al_2O_3 27.0–29.8, Fe_2O_3 0–4.0, MnO 7.1–8.0, CaO 11.4–13.2, H_2O 5.6–6.9

Color-Luster: Pale red

Crystal System: Orthorhombic

Density: 3.229 (Meas.)

First Reported: 1868

Habit: Massive, indistinctly crystalline

Hardness: Not determined

Lattice Constants: a = 11.38, b = 13.22, c = 14.08 (Å)

Mode of Occurrence: The mineral occurs associated with berlinite and lazulite in the iron mine at Västanå, southernmost Sweden

Optical Constants: $\alpha = 1.655$, $\beta = 1.664$, $\gamma = 1.675$; (+)2V = 84°

Strongest Diffraction Lines: 3.09(100), 3.13(80), 4.34(70)

Z: 4

Ref.: Roberts WL, Rapp GR, Weber J (1974) Encyclopedia of minerals, New York, Van Nostrand Reinhold, p 41. Geijer, P. and Gabrielson O. (1966) Am Miner 51:534

Auerlite = Variety of thorite high in PO_4

First Reported: 1888

Augelite
$Al_2PO_4(OH)_3$

Analyses (%): P_2O_5 34.6–35.5, Al_2O_3 49.1–51.4, CaO 0.9–1.1, H_2O 12.9–13.9

Class: 2/m

Cleavage: {110} perfect, {$\bar{2}$01} good, {001} imperfect, {$\bar{1}$01} imperfect. Fracture uneven. Brittle

Color-Luster: Colorless, white, yellowish, pale aquamarine-blue, pale rose. Transparent to translucent. Vitreous; pearly on perfect cleavage surface. Streak white

Crystal System: Monoclinic

Density: 2.696 (Meas.), 2.704 (Calc)

First Reported: 1868

Habit: Crystals thick tabular, also prismatic to acicular or as thin triangular plates. Also massive

Hardness: 4.5–5

Lattice Constants: a = 13.124, b = 7.988, c = 5.066 (Å); $\beta = 112.25°$

Mode of Occurrence: Occurs as transparent colorless crystals up to one inch size at the andalusite ore body at White Mountain, Mono County, California; as large cleavable masses and small crystals associated with morinite, wardite, apatite, and montebrasite at the Hugo mine, Keystone, South Dakota; as small crystals associated with lazulite, albite, and quartz at the Smith mine near Newport, and as small crystals with whitlockite at the Palermo mine, near North Groton, New Hampshire; and at localities in Bolivia and Sweden

Optical Constants: $\alpha = 1.5736$, $\beta = 1.5759$, $\gamma = 1.5877$; (+)2V = 55°

Space Group: C2/m

Strongest Diffraction Lines: 3.338(100), 3.506(90), 4.00(80)

Z: 4

Ref.: Roberts WL, Rapp GR, Weber J (1974) Encyclopedia of minerals, New York, Van Nostrand Reinhold, p 41. Araki T, Finney JJ, Zoltai T (1968) Am Miner 53:1096–1103. Wise WS and Loh SE (1976) Am Miner 61:409–413

Autunite

$Ca(UO_2)_2(PO_4)_2 \cdot 10\text{–}12H_2O$

Analyses (%): P_2O_5 13.4–15.2, CaO 5.3–6.6; UO_3 58.9–61.7, H_2O 14.9–20.3

Class: 4/m 2/m 2/m

Cleavage: {001} perfect, {100} indistinct. Flexible in thin plates

Color-Luster: Bright to pale sulfur yellow, lemon yellow, greenish yellow, pale green to dark green. Transparent to translucent. Vitreous, pearly on {001}. Earthy masses dull. Streak pale yellow. Fluoresces strong yellowish green in ultraviolet light

Crystal System: Tetragonal

Density: 3.05–3.2 (Meas.), 3.14 (Calc. for 10½H_2O)

First Reported: 1852 (Variously called *Uranglimmer*, *Urankalk*, or *Chalcolite* [Torbernite] prior to 1819)

Habit: Crystals very thin to thick tabular on {001} with rectangular or octagonal shape. Commonly in fan-like aggregates, as crusts, and as foliated or scaly aggregates. Also as small disseminated grains and as earthy masses

Hardness: 2–2.5

Lattice Constants: a = 6.989, c = 20.63 (Å)

Mode of Occurrence: Occurs as a secondary mineral formed by the alteration of uraninite or other uranium-containing minerals in pegmatites, granites, hydrothermal veins, and sedimentary deposits. Found in the United States as magnificent specimens composed of thick platy crystals more than an inch on edge from seams in granitic rock at the Daybreak mine, Mt. Spokane, Washington; also found widespread in small amounts in pegmatites in the Keystone and Custer districts, South Dakota; at numerous localities in Colorado, Utah, California, New Mexico, and other western states; and in pegmatites in North Carolina and in many pegmatites in the New England area. In Australia it is found in the Rum Jungle area, Northern Territory, and in thick masses at Mt. Painter in the Flinders Range. In Europe it occurs at several localities near Autun, Saône-et-Loire, France; at several mines in Saxony, Germany; at Sabugal and other places in Portugal; and in fine specimens at St. Austel and Redruth, Cornwall, England.

Optical Constants: $\omega = 1.577\text{–}1.578$, $\varepsilon = 1.553\text{–}1.555$. Usually biaxial (—), sometimes uniaxial (—), 2V variable ranging up to 53°

Space Group: I4/mmm

Strongest Diffraction Lines: 10.3(100), 4.96(80), 3.59(70)

Z: 2

Ref.: Frondel, Clifford (1958) USGA Bull 1064, p 160–170. Takano, Yukio (1961) Am Miner 46:812–822. Roberts WL, Rapp GR, Weber J (1974) Encyclopedia of minerals, New York, Van Nostrand Reinhold, p 43

Avelinoite = Cyrilovite

Babefphite

$BaBe(PO_4)(F, OH)$

Analyses (%): BeO 11.6, BaO 56.6, Fe_2O_3 0.3, P_2O_5 26.6, F 7.27, H_2O 0.64

Class: 4/m 2/m 2/m

Cleavage: None observed. Very brittle

Color-Luster: White; vitreous to greasy

Crystal System: Tetragonal

Density: 4.31 (Meas.), 4.44 (Calc.)

First Reported: 1966

Habit: Grains of isometric and rarely of flattened tabular form up to 1 mm × 1.5 mm in size

Hardness: Microhardness approx. 140–200 kg mm^{-2}

Lattice Constants: a=4.89, c=16.74 (Å)

Mode of Occurrence: Found in heavy concentrates along with zircon, ilmenorutile, fluorite, phenakite, and scheelite in "a rare-metal fluorite deposit in Siberia" genetically associated with subalkalic syenites, in the eluvial deposits located directly above the ore body

Optical Constants: $\omega = 1.629$, $\varepsilon = 1.632$;

Space Group: $I4_1/amd$

Strongest Diffraction Lines: 3.190(100), 2.163(100), 1.516(100)

Z: 4

Ref.: Nazarova AS, Kuznetsova NN, Shaskin DP (1966) Doklady Akad Nauk S.S.S.R. 167:895–897; Am Miner 51:1547. Roberts WL, Rapp GR, Weber J (1974) Encyclopedia of minerals, New York, Van Nostrand Reinhold, p 46

Baldaufite = Hureaulite

Barbosalite
$Fe^{2+}Fe_2^{3+}(PO_4)_2(OH)_2$

Analyses (%): FeO 13.1, MnO 2.82, Fe_2O_3 41.6, P_2O_5 37.5, H_2O 4.93

Class: 2/m

Cleavage: Not observed

Color-Luster: Greenish blue to almost black. Vitreous to dull and earthy

Crystal System: Monoclinic

Density: 3.60 (Meas.)

First Reported: 1955

Habit: Crystals prismatic, short, small. Massive; also as powdery crusts

Hardness: 5.5–6

Lattice Constants: a=7.25, b=7.46, c=7.49 (Å); $\beta = 120.25°$

Mode of Occurrence: Occurs in pegmatite in masses up to several hundred pounds in weight associated with pyrite, and often containing small vugs lined with small prismatic crystals of barbosalite, large crystals of metastrengite, strengite, and other phosphates at the Bull Moose mine, Custer County, South Dakota. Originally found as a very fine grained layer between heterosite and porous triphylite at the Sapucaia pegmatite mine, Minas Geraes, Brazil. Subsequently identified from several additional pegmatites in the Black Hills, South Dakota and other pegmatite districts in many parts of the world

Optical Constants: $\alpha = 1.77$, $\gamma = 1.835$

Space Group: $P2_1/c$

Strongest Diffraction Lines: 3.361(100), 3.313(80), 4.84(60)

Z: 2

Ref.: Roberts WL, Rapp GR, Weber J (1974) Encyclopedia of minerals, New York, Van Nostrand Reinhold, p 50

Baricite
(Magnesium analog of vivianite)
$(Mg_{1.64}Fe_{1.21}^{2+}Fe_{0.15}^{3+})$ $(PO_4)_2(OH)_{0.15}$ $\cdot 7.85H_2O$

Analyses (%): MgO 15.4, FeO 19.9, Fe_2O_3 2.77, P_2O_5 31.3, H_2O 29.0

Cleavage: {010} perfect and thin cleavage flakes are flexible

Color-Luster: Colorless to pale blue, the intensity of the blue color depending on the amount of the oxidation of the iron. The streak is white to pale blue, but the color of the powder changes to dark blue after several days. The luster is vitreous and is pearly on fresh surfaces parallel to {010}

Crystal System: Monoclinic

Density: 2.42 (Meas.), 2.448 (Calc.)

First Reported: 1976

Habit: Plates parallel to {010} up to 12 cm in maximum dimension and up to 5 cm thick

Hardness: 1.5–2

Lattice Constants: a=10.075, b=13.416, c=4.670 (Å); $\beta = 105°$

Mode of Occurrence: Found in fractures in a sideritic iron-formation in the Big Fish River – Blow River area, NE Yukon Territory, Canada. Quartz, siderite, vivianite and lazulite are the associated minerals

Optical Constants: $\alpha = 1.554$, $\beta = 1.564$, $\gamma = 1.595$; (+)2V = 59°

Space Group: C2/m

Strongest Diffraction Lines: 6.71(100), 2.699(70), 2.956(60), 2.526(50)

Z: 2

Ref.: Sturman BD, Mandarino JA (1976) Can Miner 14:403–406

Barrandite = Aluminous Strengite

Bassetite
$Fe^{2+}(UO_2)_2(PO_4)_2 \cdot 8H_2O$

Analyses (Stoichiometric) (%): P_2O_5 15.2, Fe_2O_3 8.4, UO_3 62.0, H_2O 14.5

Cleavage: {010} perfect

Color-Luster: Olive green, greenish brown, yellowish brown, bronze-yellow, yellow. Transparent. Vitreous, bronze on {010}

Crystal System: Monoclinic

Density: 3.4 (Meas.), 3.6 (Calc.)

First Reported: 1915

Habit: Crystals thin lozenge-shaped plates flattened on {010}. Forms {010}, {111}, {101}, {110}, and {001}. Often occurs as parallel growths with torbernite and uranospathite

Hardness: ~2.5

Lattice Constants: a = 7.01, b = 17.07, c = 6.98 (Å); $\beta = 90.53°$

Mode of Occurrence: Occurs in Utah at the Fuemrol mine, Temple Mountain, and at the Denise No. 1 mine on the Bowknot of the Green River, Emery County, Utah. It is also found as small crystals in oxidized vein material carrying pyrite and uraninite at the Wheal Basset, Redruth, Cornwall, England

Optical Constants: $\alpha = 1.603$, $\beta = 1.610$, $\gamma = 1.617$; 2V ~ 90°

Strongest Diffraction Lines: 4.89(100), 3.46(100), 8.59(60)

Z: 2

Ref.: Frondel, Clifford (1958) USGS Bull 1064, p 200–204

Bastinite = Lithium hureaulite

Beckelite = Britholite

Belovite (cf. Talmessite)
$(Na, Ca, Sr, Ce)_{10}(P, Si, SO_4)_6(OH)_2$
Strontium analog of hydroxylapatite

Analyses (%): P_2O_5 28.9–29.3, CaO 0.8–5.2, SrO 33.6–39.6, Na_2O 3.6–5.0, Ce_2O_3 11.3–24.0, La_2O_3 12.8, SiO_2 1.4, SO_3 1.1, H_2O 0.9

Class: 6/m

Cleavage: Prismatic and pinacoidal, imperfect. Fracture irregular. Brittle

Color-Luster: Honey yellow; vitreous, greasy on fracture

Crystal System: Hexagonal

Density: 4.19

First Reported: 1953

Habit: Prismatic crystals up to 2 cm in size with $(10\bar{1}0)$ and (0001) prominent

Hardness: 5

Lattice Constants: a = 9.62, c = 7.12 (Å)

Mode of Occurrence: Occurs in ussingite, formed by late-stage replacement of microcline, in the central part of a pegmatite in nepheline syenite at an unspecified locality, presumably on the Kola Peninsula

Optical Constants: $\omega = 1.660$, $\varepsilon = 1.640$; (−)

Strongest Diffraction Lines: 2.87(100), 1.998(80), 1.900(80)

Z: 2

Ref.: Borodin LS, Kazakova ME (1950) Doklady Acad Sci U.S.S.R. 71:925; *Idem* (1955) Am Miner 40:367–368. Roberts WL, Rapp GR, Weber J (1974) Encyclopedia of minerals, New York, Van Nostrand Reinhold, p 59

Belovite = Talmessite, arsenate-belovite

Beraunite
$Fe^{2+}Fe_5^{3+}(OH)_5(H_2O)_4(PO_4)_4 \cdot 2H_2O$

Analyses (%): P_2O_5 27.2–31.8, FeO 0–2.0, Fe_2O_3 49.6–56.0, H_2O 13.4–16.6

Class: 2/m

Cleavage: {100} good

Color-Luster: Reddish brown to dark red; also dull grayish green to dark greenish brown. Translucent. Vitreous to dull; pearly on cleavage. Streak yellow to greenish brown

Crystal System: Monoclinic

Density: 3.01 (Meas.), 2.962 (Calc.)

First Reported: 1841

Habit: Crystals small, rare, tabular and somewhat elongated {010}; vertically striated. Usually as radial-fibrous aggregates, foliated crusts and globules, or disk-like concretions. Twinning on {100} as interpenetration twins

Hardness: 3.5–4

Lattice Constants: a = 20.80–20.646, b = 5.156–5.129, c = 19.22–19.213 (Å); $\beta = 93.51°$

Mode of Occurrence: Occurs as a secondary mineral in iron ore deposits, and as an alteration product of primary phosphate minerals in granite pegmatites. Found as an alteration product of triphylite at the Big Chief and Hesnard mines, Keystone, Pennington County, South Dakota, and at the Palermo mine, North Groton, New Hampshire. In the United States it also is found in Arkansas, New Jersey, and Pennsylvania; other occurrences are found in Ireland, Germany, Czechoslovakia, and U.S.S.R.

Optical Constants: $\alpha = 1.775$, $\beta = 1.786$ (variable), $\gamma = 1.815$; usually (+)2V ~ medium to large

Space Group: C2/c

Strongest Diffraction Lines: 10.37(100), 4.825(60), 3.082(60)

Z: 4

Ref.: Roberts WL, Rapp GR, Weber J (1974) Encyclopedia of minerals, New York, Van Nostrand Reinhold, p 61. Moore PB (1970) Am Miner 55:135–169

Bergenite (cf. Phurcalite)
$(Ba,Ca)_2(UO_2)_3(PO_4)_2(OH)_4 \cdot 5\frac{1}{2}H_2O$

Analyses (%): P_2O_5 11.0, UO_3 62.5, BaO 14.0, CaO 2.44, H_2O 10.1

Cleavage: Not determined

Color-Luster: Yellow. Fluorescent weak orange-brown under short- and long-wave ultraviolet light

Crystal System: Monoclinic

Density: >4.1 (Meas.), 4.09 (Calc.)

First Reported: 1959

Habit: Crystals thin tabular

Hardness: Not determined

Lattice Constants: a = 22.32, b = 17.19, c = 20.63 (Å); $\beta = 93.0°$

Mode of Occurrence: Found associated with much uranocircite, some torbernite, renardite, autunite, barium uranophane, and unidentified uranium minerals on a mine dump at Streuberg near Bergen on the Treib, Vogtland, Saxony, Germany

Optical Constants: $\alpha = 1.660$, $\beta = 1.690$, $\gamma = 1.695$; (—)2V > 45°

Space Group: $P2_1/c$

Strongest Diffraction Lines: 7.73(100), 3.83(80), 3.054(60), 2.874(50), 2.826(50)

Z: 18

Ref.: Bültemann HW, Moh GH (1959) Neues Jahrb. Mineral Monatsh No. 10, 232–233. Ross V (1956) Am Miner 41:915–926. Piret P, Deliens M (1981) Bull Miner 104:16–18. Roberts WL, Rapp GR, Weber J (1974) Encyclopedia of minerals, New York, Van Nostrand Reinhold, p 61–62

Berlinite
$AlPO_4$

Analyses (%): Al_2O_3 40.3, Fe_2O_3 0.26, P_2O_5 54.8, H_2O 4.1

Class: 3 2

Cleavage: None. Fracture conchoidal

Color-Luster: Colorless, grayish, pale rose. Transparent to translucent. Vitreous

Crystal System: Hexagonal

Density: 2.64 (Meas.), 2.618 (Calc.)

First Reported: 1868

Habit: Massive, granular. Synthetic crystals similar in habit to quartz

Hardness: ~6.5

Lattice Constants: a = 4.92, c = 10.91 (Å)

Mode of Occurrence: Occurs at the Westanå iron mine, near Näsum, Sweden, associated with augelite and other phosphates

Optical Constants: $\omega = 1.5235$, $\varepsilon = 1.529$; (+)

Space Group: $P3_121$ or $P3_221$

Strongest Diffraction Lines: 3.369(100), 4.28(25), 1.835(16)

Z: 3

Ref.: Roberts WL, Rapp GR, Weber J (1974) Encyclopedia of minerals, New York, Van Nostrand Reinhold, p 62. Wise WS, Loh SE (1976) Am Miner 61:409–413

Bermanite

$Mn^{2+}Mn_2^{3+}(PO_4)_2(OH)_2 \cdot 4H_2O$

Analyses (%): P_2O_5 31.4, MnO 13.8, MgO 2.4, Fe_2O_3 3.0, Mn_2O_3 28.8, H_2O 19.3

Class: 2

Cleavage: {001} perfect, {110} imperfect, brittle

Color-Luster: Pale red to dark reddish brown; vitreous to slightly resinous

Crystal System: Monoclinic (pseudo-orthorhombic)

Density: 2.84 (Meas.), 2.840 (Calc.)

First Reported: 1936

Habit: Tabular crystals, often twinned, up to 0.5 mm in size. In subparallel, fan-shaped, or rosette-like aggregates of crystals, and as lamellar masses

Hardness: 3.5

Lattice Constants: a = 5.446, b = 19.25, c = 5.428 (Å); $\beta = 110.4°$

Mode of Occurrence: Occurs in pegmatite, associated with metastrengite, hureaulite, and other phosphates, coating seams in triplite, on the 7-U-7 Ranch, near Hillside, Arizona; in South Dakota as well-formed crystals in vugs in altered triphylite at the Tip-Top mine, Custer, and at several other pegmatite mines in the Custer and Keystone districts; also found at the Stewart Mine, Pala, California; Fletcher quarry, North Groton, New Hampshire; Williams prospect, Coosa County, Alabama; near Tanti, Córdoba, Argentina; Mangualde, Portugal; Sapucaia pegmatite, Minas Geraes, Brazil; and at unspecified locality in Madagascar

Optical Constants: $\alpha = 1.687$, $\beta = 1.725$, $\gamma = 1.748$; (—)2V = 74°

Space Group: $P2_1$

Strongest Diffraction Lines: 9.68(100), 5.08(65), 4.81(65)

Z: 2

Ref.: Hurlbut CS Jr, Aristarain LF (1968) Am Miner 53:416–431. Kampf AR, Moore PB (1976) Am Miner 61:1214–1248. Roberts WL, Rapp GR, Weber J (1974) Encyclopedia of minerals, New York, Van Nostrand Reinhold, p 62

Bertossaite

$(Li, Na)_2(Ca, Fe, Mn)Al_4(PO_4)_4(OH, F)_4$

Analyses: None reported

Class: 2/m 2/m 2/m or mm2

Cleavage: {100} good. Fracture uneven to subconchoidal

Color-Luster: Pale pink; vitreous

Crystal System: Orthorhombic

Density: 3.10 (Meas.), 3.10 (Calc.)

First Reported: 1966

Habit: Massive

Hardness: 6

Lattice Constants: a = 11.48, b = 15.73, c = 7.23 (Å)

Mode of Occurrence: Occurs associated with amblygonite, lazulite-scorzalite, augelite, brazilianite, apatite, crandallite, and quartz in the Buranga lithium pegmatite in Rwanda, Africa

Optical Constants: $\alpha = 1.624$, $\beta = 1.636$, $\gamma = 1.642$; (—)2V moderately large

Space Group: Imaa or I2aa

Strongest Diffraction Lines: 3.056(100), 3.286(70), 3.104(70)

Z: 4

Ref.: von Knorring O, Mrose ME (1966) Can Miner 8:668 (abstr.). Roberts WL, Rapp GR, Weber J (1974) Encyclopedia of minerals, New York, Van Nostrand Reinhold, p 63

Beryllonite
$NaBePO_4$

Analyses (%): Na_2O 23.3–24.4, BeO 19.1–19.8, P_2O_5 55.4–55.9

Class: 2/m

Cleavage: {010} perfect, {100} good, interrupted, {101} indistinct. Fracture conchoidal. Brittle

Color-Luster: Colorless, white, pale yellow. Transpartent to translucent. Vitreous; pearly on perfect cleavage

Crystal System: Monoclinic

Density: 2.84 (Meas.), 2.794 (Calc.)

First Reported: 1868

Habit: Crystals tabular to short prismatic; often highly modified. Crystal faces frequently etched or rough. Twinning on {101} common; also polysynthetic

Hardness: 5.5–6

Lattice Constants: a = 8.13 kX, b = 7.76, c = 14.17 (Å); $\beta = 90.0°$

Mode of Occurrence: Occurs sparingly in granite pegmatite at two localities in Maine. Found at Newry associated with albite, tourmaline, herderite, and eosphorite; and in a disintegrated pegmatite outcrop at McKean Mountain, near Stoneham, with muscovite, albite, smoky quartz crystals, and other minerals

Optical Constants: $\alpha = 1.5520$, $\beta = 1.5579$, $\gamma = 1.561$; (—)2V = 67.93°

Space Group: $P2_1/n$

Strongest Diffraction Lines: 2.84(100), 3.65(90), 2.28(70)

Z: 12

Ref.: Roberts WL, Rapp GR, Weber J (1974) Encyclopedia of minerals, New York, Van Nostrand Reinhold, p 65. Wehrenberg JP (1954) Am Miner 39:397

Beta-Lomonosovite
$Na_2Ti_2Si_2O_9 \cdot NaH_2PO_4$(?)

Analyses (%): P_2O_5 15.1–18.5, SiO_2 24.2–25.2, TiO_2 23.8–25.7, ZrO_2 1.89–2.50, Fe_2O_3 2.20, MnO 0.96–1.70, CaO 0.62–1.80, Na_2O 15.2–17.1, (Nb_2O_5 + Ta_2O_5) 3.05–4.78, H_2O 4.60–5.70

Cleavage: One perfect. Fracture uneven. Brittle

Color-Luster: Pale yellow-brown, brown, rose. Vitreous to pearly on cleavage. Vitreous to greasy on fracture

Crystal System: Triclinic

Density: 2.95–2.98

First Reported: 1961

Habit: Tabular to platy masses up to 5 × 4 × 0.3 cm in size

Hardness: ~4

Lattice Constants: a = 5.28, b = 7.05, c = 14.50 (Å); $\alpha = 102.4°$, $\beta = 96.85°$, $\gamma = 90°$

Mode of Occurrence: Occurs associated with microline, aegirine, arfvedsonite, eudialyte, sodalite, and nepheline in alkalic pegmatites of the Lovozero massif, Kola Peninsula

Optical Constants: $\alpha = 1.670$, $\beta = 1.770$, $\gamma = 1.779$; (—)2V = 10°–20°

Strongest Diffraction Lines: 2.77(100), 3.45(90), 13.53(80)

Ref.: Gerasimovskii VI, Kazakova ME (1962) Doklady Akad Nauk S.S.S.R., 142:670–673; *Idem* (1963) Am Miner

48:1413–1414. Roberts WL, Rapp GR, Weber J (1974) Encyclopedia of minerals, New York, Van Nostrand Reinhold, p 66

Beusite (Graftonite-Beusite Series)
$(Mn, Fe, Ca, Mg)_3(PO_4)_2$

Analyses (%): CaO 3.30–12.2, MgO 1.2–9.50, FeO 14.2–17.0, MnO 24.8–35.9, P_2O_5 40.0–43.8

Class: 2/m

Cleavage: {010} good, {100} fair

Color-Luster: Reddish brown. Translucent. Vitreous. Streak pale pink

Crystal System: Monoclinc

Density: 3.702 (Meas.), 3.715 (Calc.)

First Reported: 1968

Habit: Crystals prismatic, rough, up to 30 cm long; interlaminated with lithiophilite

Hardness: 5

Lattice Constants: a = 8.78, b = 11.52, c = 6.15 (Å); $\beta = 99.42°$

Mode of Occurrence: Found in granite pegmatites at Los Aleros, Amanda, and San Salvador, in Sal Luis Province, Argentina

Optical Constants: $\alpha = 1.702$, $\beta = 1.703$, $\gamma = 1.722$; (+)2V = 25°

Space Group: $P2_1/c$

Strongest Diffraction Lines: 3.49(100), 2.863(100), 2.708(60)

Z: 4

Ref.: Hurlbut CS Jr, Aristarain LF (1968) Am Miner 53:1799–1814. Roberts WL, Rapp GR, Weber J (1974) Encyclopedia of minerals, New York, Van Nostrand Reinhold, p 68

Biphosphammite
$NH_4H_2PO_4$

Analyses (%): K_2O 14.2, $(NH_4)_2O$ 12.3, SO_3 5.59, P_2O_5 51.1

Class: $\bar{4}$2m

Cleavage: None

Color-Luster: Colorless to light buff. Transparent to translucent. Soluble in water

Crystal System: Tetragonal

Density: 2.04

First Reported: 1972

Habit: Crystals prismatic, tapered, with pyramidal terminations. As rounded crusts with radiating groups of crystals (up to 2 × 0.2 mm)

Hardness: Very soft

Lattice Constants: a = 7.4935, c = 7.340 (Å)

Mode of Occurrence: Occurs, apparently as a crystalline product of the liquid fraction of bat guano, in Murra El Elevyn cave, Western Australia. Also reported to occur in the guano of Guañape Island

Optical Constants: $\omega = 1.525$, $\varepsilon = 1.480$

Space Group: $I\bar{4}2d$

Strongest Diffraction Lines: 3.75(100), 5.24(90), 3.02(90)

Z: 1

Ref.: Pryce MW (1972) Miner Mag 38:965–967

Bjarebyite Series
$(Ba, Sr)(Mn, Fe, Mg)_2Al_2(PO_4)_3(OH)_3$

Analyses (%): Ba 21.8, Sr 0.79, Mn 7.15, Fe 6.96, Mg 0.84, Al 7.01 (PO_4 and H_2O not determined)

Cleavage: {010} and {100} perfect

Color-Luster: Emerald-green with a faint bluish tinge, luster subadamantine, streak white

Crystal System: Monoclinic

Density: 3.95 (Meas.), 4.02 (Calc.)

First Reported: 1973

Habit: Highly faceted, complex spear-shaped crystals, usually pitted and etched, up to 3 mm in greatest dimension. Forms noted include *c* {001}, *a* {100}, *b* {010}, *m* {110}, *d* {120}, *r* {111}, *f* {121}, *e* {131}, *h* {141}, *w* {411}, *l* {211}, *t* {021}, *q* {011}, *v* {031}. All negative forms are missing

Hardness: 4+

Lattice Constants: a = 8.930, b = 12.073, c = 4.917 (Å); $\beta = 100.15°$

Mode of Occurrence: Found in the Palermo No. 1 pegmatite, near North Groton, New Hampshire, as rare crystals in open cavities along the contact between amblygonite-scorzalite remnants and Fe-Mn oxides. Associated minerals include amblygonite, augelite, childrenite, siderite, scorzalite, quartz and palermoite

Optical Constants: $\alpha = 1.692$, $\beta = 1.695$, $\gamma = 1.710$; (+)2V = 35°; weakly pleochroic from grayish tan to pale yellow-green

Space Group: $P2_1/m$

Strongest Diffraction Lines: 2.010(100), 8.81(70), 2.681(70), 1.495(50)

Z: 2

Ref.: Moore PB, Lund DH, Keester KL (1973) Miner Rec 4:282–285. Moore PB, Araki T (1974) Am Miner 59:567–572

Bobierrite

$Mg_3(PO_4)_2 \cdot 8H_2O$

Analyses (%): MgO 25.1–30.0, CaO 0–5.7, P_2O_5 34.5–34.9, H_2O 34.3–35.4

Class: 2/m

Cleavage: {010} perfect

Color-Luster: Colorless, white. Transparent. Weakly vitreous

Crystal System: Monoclinic

Density: 2.195 (Meas. Synthetic), 2.17 (Calc.)

First Reported: 1868

Habit: As minute acicular or fibrous crystals; also massive, as flattened aggregates, lamellar, and as minute crystalline aggregates

Hardness: 2–2.5

Lattice Constants: a = 9.946, b = 27.654, c = 4.639 (Å); $\beta = 104°$

Mode of Occurrence: Found in cavities in a fossil elephant tusk near Edgerton, Pipestone County, Minnesota; with apatite near Bamle, Norway; in guano on Mejillones island, Chile; and in New Zealand

Optical Constants: (Minn.) $\alpha = 1.510$, $\beta = 1.520$, $\gamma = 1.543$; (+)2V = 71°,53°. (N.Z.) $\alpha = 1.5468$, $\beta = 1.5533$, $\gamma = 1.5820$

Space Group: $P2_1/c$

Strongest Diffraction Lines: 6.96(100), 2.94(27), 8.04(18)

Z: 4

Ref.: Roberts WL, Rapp GR, Weber J (1974) Encyclopedia of minerals, New York, Van Nostrand Reinhold, p 77. Frazier AW, Lehr JR, Smith JP (1963) Am Miner 48:635–641

Bøggildite

$Na_2Sr_2Al_2PO_4F_9$

Analyses (Stoichiometric) (%): P_2O_5 13.3, SrO 37.6, Na_2O 11.6, Al_2O_3 19.4, F 31.7

Class: 2/m

Cleavage: Not determined

Color-Luster: Flesh red

Crystal System: Monoclinic

Density: 3.66

First Reported: 1952

Habit: Massive

Hardness: 4–5

Lattice Constants: a = 5.24, b = 10.48, c = 17.66 (variable, up to 18.52 reported) (Å); $\beta = 107.35°$

Mode of Occurrence: Occurs associated with siderite, fluorite, black cryolite, quartz, mica, and sphalerite at the contact of cryolite with greisen at Ivigtut, Greenland. Has also been found in uranium deposits of south-central British Columbia where it constitutes the major uranous phase present

Optical Constants: $\alpha = 1.462$, $\beta = 1.466$, $\gamma = 1.469$; (+)2V = 78°–80°

Space Group: $P2_1/c$

Strongest Diffraction Lines: 3.162(100), 3.893(80), 3.960(65)

Z: 4

Ref.: Pauly H, Moeller CK (1956) Am Miner 41:959. Bøgvad R (1954) Am Miner 39:848–849. Hawthorne FC (1982) Can Miner 20:263–270. Roberts WL, Rapp GR, Weber J (1974) Encyclopedia of minerals, New York, Van Nostrand Reinhold, p 77–78

Bolivarite

$Al_2(PO_4)(OH)_3 \cdot 4\text{–}5H_2O$

Analyses (%): Al_2O_3 44.1, P_2O_5 34.9, H_2O 20.6

Cleavage: None. Fracture conchoidal. Brittle

Color-Luster: Bright yellowish green to greenish white. Vitreous. Fluoresces strong bright green under ultraviolet light

Density: 1.97–2.05 (Meas.)

First Reported: 1921

Habit: Massive; as cryptocrystalline crusts, botryoidal masses, and veinlets

Hardness: 2.5–3.5

Mode of Occurrence: Occurs in crevices in granite near Pontevedra, Spain; also found in the weathering zone and phosphate-rich zone of the pegmatite of Kobokobo, Kivu, Zaire. Both occurrences contain abnormal amounts of uranium

Optical Constants: Biaxial to uniaxial (+); $\eta = 1.50\text{–}1.51$

Strongest Diffraction Lines: Amorphous to 1050 °C

Ref.: Van Tassel R (1960) Miner Mag 32:419–420. Van Wambeke L (1971) Miner Mag 38:418–423. Roberts WL, Rapp GR, Weber J (1974) Encyclopedia of minerals, New York, Van Nostrand Reinhold, p 78

Bořickite

$\sim CaFe_5(PO_4)_2(OH)_{11} \cdot 3H_2O$

Analyses (%): CaO 7.5–8.2, Fe_2O_3 52.3–53.2, P_2O_5 18.9–20.5, H_2O 19.1–20.4

Cleavage: Not determined

Color-Luster: Reddish-brown. Opaque. Weak waxy. Streak reddish-brown

Crystal System: Unknown

Density: ~2.70

First Reported: 1868

Habit: As compact reniform masses

Hardness: 3.5

Mode of Occurrence: Found at Nenačovik, south of Kladno, Czechoslovakia, and at Leoben, Styria, Austria. A similar poorly defined substance (foucherite) found at Fouchères, Champagne, France, and at localities in Czechoslovakia, may be bořickite

Optical Constant: Undetermined

Ref.: Roberts WL, Rapp GR, Weber J (1974) Encyclopedia of minerals, New York, Van Nostrand Reinhold, p 81

Bornemanite

$BaNa_4Ti_2NbSi_4O_{17}(F,OH) \cdot 1Na_3PO_4$ (Formula remains to be determined definitively)

Analyses (%): SiO_2 24.5, P_2O_5 6.69, Nb_2O_5 9.04, TiO_2 18.4, MnO 2.73, BaO 12.5, Na_2O 19.8, SrO 0.69, Ta_2O_5 0.66, K_2O 0.66

Cleavage: {001} very perfect

Color-Luster: Pale yellow, luster pearly, translucent to transparent

Crystal System: Orthorhombic

Density: 3.47–3.50 (Meas.), 3.49 (Calc.)

First Reported: 1975

Habit: Platy aggregates up to 10 × 8 × 0.2 mm of very fine leaflets along the cleavages

Hardness: 3.5–4; platy aggregates are brittle, fibers flexible

Lattice Constants: a = 5.48, b = 7.10, c = 48.2 (Å)

Mode of Occurence: Found on the surface of lomonosovite and in natrolite, in the natrolite zone of the "Jubilee" pegmatitic alkalic rocks of the Lovozero massif, Kola Peninsula, U.S.S.R.

Optical Constants: $\alpha = 1.682$, $\beta = 1.683$, $\lambda = 1.720$, 1.718; (+)2V = 66° (Calc.),

40° (Meas.); weakly pleochroic with X and Y colorless and Z brownish

Space Group: D_{2h}^{28}–Ibmm or C_{2v}^{2}–Ibm2

Strongest Diffraction Lines: 24.1(100), 8.04(100), 3.44(100), 3.02(100), 2.682(80)

Z: 4

Ref.: Men'shikov YP et al. (1975) Zapiski Vses Miner Obshch 104:322–326

Bosphorite = Oxidized Vivianite

Brabantite (Lingaitukuang, or **"Cathophorite"**; member of the monazite series)
$CaTh(PO_4)_2$

Analyses (%): CaO 11.9, MgO 0.56, ThO_2 52.6, P_2O_5 27.7, SiO_2 2.27

Cleavage: {100} and {001} are prominent

Color: Gray-brown, and altered to reddish-brown on the surfaces

Crystal System: Monoclinic

Density: 4,72 (Meas.), 5.26 (Calc.) for unheated specimen

First Reported: 1980

Habit: Elongated grains

Hardness: 5.5

Lattice Constants: a=6.726, b=6.933, c=6.442 (Å) for unheated specimen; a=6.718, b=6.916, c=6.442 (Å) for heated sample; $\beta = 103°$

Mode of Occurrence: Occurs in a zone of microcrystalline muscovite associated with thorite and uraninite in a pegmatite on the Brabant farm in the Karibib district, Namibia

Optical Constants: $\beta = 1.73$, $\alpha - \gamma = 0.05$

Space Group: $P2_1$

Strongest Diffraction Lines: 3.06(100), 2.85(75), 3.26(70), 4.15(30), 1.947(30)

Z: 2

Ref.: Rose D (1980) Neues Jahrb. Mineral Monatsh 6:247–257

Bradleyite
$Na_3PO_4 \cdot MgCO_3$

Analyses (%): Na_2O 37.5, MgO 15.4–16.2, P_2O_5 26.3–28.6, CO_2 17.7–18.4

Class: 2/m

Cleavage: Not determined

Color-Luster: Light gray due to clay impurities. Probably colorless or white

Crystal System: Monoclinic

Density: 2.734 (Meas.)

First Reported: 1941

Habit: As fine-grained masses

Hardness: Not determined

Lattice Constants: a=8.85, b=6,63, c=5,16 (Å); $\beta = 90.42°$

Mode of Occurrence: Occurs admixed with shortite and montmorillonite in a drill core from oil shale in the Green River formation, about 20 miles west of Green River, Sweetwater County, Wyoming

Optical Constants: $\alpha = 1.49$, $\beta = ?$, $\gamma = 1.56$; (−)2V = small

Space Group: $P2_1/m$

Strongest Diffraction Lines: 3.32(100), 2.66(100), 2.57(80)

Z: 2

Ref.: Roberts WL, Rapp GR, Weber J (1974) Encyclopedia of minerals, New York, Van Nostrand Reinhold, p 81

Brazilianite (Brasilianite)
$NaAl_3(PO_4)_2(OH)_4$

Analyses (%): P_2O_5 38.5–39.7, Na_2O 8.29–8.56 Al_2O_3 42.2–43.8, H_2O 9.6–10

Class: 2/m

Cleavage: {010} good. Fracture conchoidal. Brittle

Color-Luster: Colorless, pale yellowish to green; transparent; vitreous. Streak colorless

Crystal System: Monoclinic

Density: 2.983 (Meas.), 3.025 (Calc.)

First Reported: 1945

Habit: Crystals nearly equant to short prismatic [001], up to 12 cm long and 8 cm in width; also elongated [100], spear-shaped. Prism zone striated [001]. Also globular with radial fibrous structure

Hardness: 5.5

Lattice Constants: a = 11.19, b = 10.08, c = 7.06 (Å); $\beta = 97.37°$

Mode of Occurrence: Occurs as a hydrothermal mineral in cavities in pegmatite at the Palermo mine, near North Grafton, and also in pegmatite at the Smith mine near Newport, New Hampshire. Found as superb large gemmy crystals associated with muscovite, albite, apatite, and tourmaline in cavities in pegmatite near Conselheira Pena, and as spear-shaped crystals at Mantena in Minas Geraes; also as small (1–2 mm) grains and crystals associated with massive amblygonite, wardite, apatite, and other phosphates in the pegmatite of the Alto Patrimonio, at Piedras Lavradas, Paraíaba, Brazil

Optical Constants: $\alpha = 1.602$, $\beta = 1.609$, $\gamma = 1.621–1.623$; (+)2V = 71°, 75° (Calc.)

Space Group: $P2_1/n$

Strongest Diffraction Lines: 5.05(100), 2.99(80), 2.74(80)

Z: 4

Ref.: Roberts WL, Rapp GR, Weber J (1974) Encyclopedia of minerals, New York, Van Nostrand Reinhold, p 86–87

Brianite

$Na_2CaMg(PO_4)_2$

Analyses (%): Na_2O 22.1, MgO 12.6, CaO 18.8, FeO 0.5, P_2O_5 46.9

Cleavage: Not determined

Color-Luster: Colorless, transparent

Crystal System: Monoclinic

Density: 3.17 (Calc.), 3.10 (Meas. Synthetic)

First Reported: 1967

Habit: Grains up to 0.1–0.2 mm

Hardness: 4–5

Lattice Constants: a = 13.38, b = 5.23, c = 9.13 (Å); $\beta = 91.2°$

Mode of Occurrence: Occurs associated with whitlockite, panethite, albite, and enstatite in small pockets in the metallic phase of the Dayton octahedrite meteorite

Optical Constants: $\alpha = 1.598$, $\beta = 1.605$, $\gamma = 1.608$; (—)2V = 63–65°

Space Group: $P2_1/a$

Strongest Diffraction Lines: 2.625(100), 3.734(90), 2.679(90)

Ref.: Fuchs LH, Olson E, Henderson EP (1968) Am Miner 53:508–509; (1967) Geochim Cosmochim Acta 31:1711–1719. Moore PB (1975) Am Miner 60:717–718. Roberts WL, Rapp GR, Weber J (1974) Encyclopedia of minerals, New York, Van Nostrand Reinhold, p 88

Britholite

(Apatite Group, varieties include Fynchenite or Thorium-britholite, Alumobritholite, Hydrobritholite)

$(Ca, Th, Ce)_5(PO_4, SiO_4)_3(OH, F)$

Analyses (%): P_2O_5 1.9–17.0, Al_2O_3 0–14.9, CaO 11.5–28.8, ThO_2 3.2–9.8, SiO_2 12.3–22, Re_2O_3 27.6–57, F 1.4–2.1, H_2O 0–6.4

Class: 6/m

Cleavage: None. Fracture conchoidal. Brittle

Color-Luster: Yellow, brown, greenish brown, to nearly black. Translucent. Adamantine, resinous

Crystal System: Hexagonal

Density: 3.86 (Meas.), 3.95 (Calc.), Vlasov: 4.2–4.69 (Meas.), 4.65 (Calc.)

First Reported: 1900

Habit: Crystals prismatic, hexagonal in cross section. Usually compact massive or as disseminated grains

Hardness: 5

Lattice Constants: a = 9.63, c = 7.03 (Å)

Mode of Occurrence: Occurs chiefly in nepheline-syenites and in contact metasomatic deposits related to alkali syenites and granites. Found at Susan Lode, Custer County, South Dakota (thorium britholite); at Oka, Quebec, Canada; in the Julianehaab district, Greenland; at several places in the U.S.S.R.; in Tungpei, China (thorium britholite); and elsewhere

Optical Constants: Uniaxial (—) $\eta = 1.77–1.81$, $\omega–\varepsilon = .005–.008$, also (+) $\omega = 1.78$, $\varepsilon = 1.82$. Biaxial $\alpha = 1.772$, $\beta = 1.775$ 2V small (up to 44°), $\gamma = 1.777$

Space Group: $P6_3/m$

Strongest Diffraction Lines: 2.836(100), 3.48(80), 2.809(80) (Heated, partly metamict.)

Z: 2

Ref.: Vlasov KA (1966) Mineralogy of rare elements, vol II. Israel Program for Scientific Translations, p 297–300. Roberts WL, Rapp GR, Weber J (1974) Encyclopedia of minerals, New York, Van Nostrand Reinhold, p 89

Brockite

$(Ca, Th, Ce)\{(PO_4), (CO_3)\} \cdot H_2O$

Analyses (%): P_2O_5 23.6, ThO_2 42.7, CaO 9.7, Ce_2O_3 6.6, Fe_2O_3 4.6, BaO 1.1, SrO 1.3, CO_2 3.1, H_2O 7.5

Class: 622

Cleavage: Conchoidal fracture

Color-Luster: Deep red-brown; pale yellow on thin edges; translucent. Greasy to vitreous

Crystal System: Hexagonal

Density: 3.9 (Meas.), 4.0 (Calc.)

First Reported: 1962

Habit: Radial aggregates of fibroid grains. Also stubby, imperfect hexagonal prisms, usually with granulated margins. Average grain size 20 μm in length; rarely up to 50 μm long

Hardness: Not determined

Lattice Constants: a = 6.98, c = 6.40 (Å)

Mode of Occurrence: Occurs as very fine grained massive aggregates in nodules up to 35 mm in diameter associated with abundant inclusions of pyrite, and as earthy coatings in veins and altered granitic rocks near the Bassick mine, Querida, Custer County, Colorado. Also as yellow earthy coatings from the Hardwick mine and Nightingale shaft, Custer County, Colorado

Optical Constants: $\omega = 1.680$, $\varepsilon = 1.695$; (+)

Space Group: P622

Strongest Diffraction Lines: 3.03(100), 4.37(70), 2.83(70)

Z: 3

Ref.: Fisher, FG, Meyrowitz R (1962) Am Miner 47:1346–1355. Roberts WL, Rapp GR, Weber J (1974) Encyclopedia of minerals, New York, Van Nostrand Reinhold, p 89–90

Bröggerite = Variety of uraninite containing thorium

Brushite

$CaHPO_4 \cdot 2H_2O$

Analyses (%): CaO 30.8–32.7, P_2O_5 38.0–42.7, H_2O 26.2–30.9

Class: 2/m

Cleavage: {010} perfect, {001} perfect

Color-Luster: Colorless to ivory yellow. Transparent to translucent. Vitreous; pearly on cleavage

Crystal System: Monoclinic

Density: 2.328 (Meas.), 2.257 (Calc.)

First Reported: 1865 (Epiglaubite, 1856; Metabrushite, 1865; Stoffertite, 1901)

Habit: Crystals needle-like or prismatic to tabular {010}. Also earthy or powdery; foliated

Hardness: 2.5

Lattice Constants: a = 5.88, b = 15.15, c = 6.37 (Å); $\beta = 117.47°$

Mode of Occurrence: Occurs as nodular masses of platy crystals in the lower part of a bat guano and hair deposit in Pig Hole Cave, Giles County, Virginia. Also found as crystals up to 2 cm long in cavities in phosphorite at Quercy, near Limoges, France; in commercial quantities in a cave near Oran, Algeria; in guano from Aves Island in the Caribbean Sea; and widespread in small amounts in many other deposits throughout the world

Optical Constants: $\alpha = 1.539$, $\beta = 1.546$, $\gamma = 1.551$; (+)2V = 86°

Space Group: I2/a

Strongest Diffraction Lines: 7.62(100), 3.80(30), 1.90(10)

Z: 4

Ref.: Roberts WL, Rapp GR, Weber J (1974) Encyclopedia of minerals, New York, Van Nostrand Reinhold, p 93

Buchwaldite (Rhenanite)
$NaCaPO_4$

Analyses (%): CaO 35.6, MnO 0.06, Na_2O 19.5, P_2O_5 44.9

Cleavage: One prominent cleavage or parting

Color: Colorless in transmitted light

Crystal System: Orthorhombic

Density: 3.21 (Calc.)

First Reported: 1977

Habit: Minute inclusions (mostly < 10 μm across) in troilite nodules

Hardness: Fairly low, probably below 3

Lattice Constants: a = 5.167, b = 9.259, c = 6.737 (Å)

Mode of Occurrence: Found as compact, interlocking masses of fine needles within troilite nodules in the Cape York iron meteorite (specifically, the Agpalilik specimen)

Optical Constants: $\alpha = 1.607$, $\beta = 1.610$, $\gamma = 1.616$; (—)2V = 65° (Est.), 71° (Calc.); parallel extinction

Space Group: $Pmn2_1$

Strongest Diffraction Lines: 2.72(100), 3.79(90), 1.91(80), 2.63(60), 2.58(50)

Z: 4

Ref.: Olsen E, Erlichman J, Bunch TE, Moore PB (1977) Am Miner 62:362–364

Burangaite
$2(Na,Ca)_2(Fe,Mg)_2Al_{10}(PO_4)_8 \cdot (O,OH)_{12} \cdot 4H_2O$

Analyses (%): Al_2O_3 34.4, Fe_2O_3 1.14, FeO 6.26, MgO 2.00, CaO 1.88, Na_2O 2.93, P_2O_5 37.6, H_2O 11.6

Cleavage: {100} perfect

Color: Bluish to bluish-green; streak slightly bluish

Crystal System: Monoclinic

Density: 3.05

First Reported: 1977

Habit: Long prismatic crystals which show hourglass structure, with blue core and colorless margin. Forms noted include {301}, {102}, {104}, $\{\bar{1}01\}$, $\{\bar{2}23\}$, $\{3\bar{1}1\}$, $\{22\bar{3}\}$, $\{\bar{3}1\bar{1}\}$, and $\{40\bar{1}\}$

Hardness: 5

Lattice Constants: a = 25.09, b = 5.048, c = 13.45 (Å); $\beta = 110.9°$

Mode of Occurrence: Found in the Buranga pegmatite, Rwanda, associated with bertossaite, trolleite, scorzalite (which it replaces), apatite, bjarebyite and wardite

Optical Constants: $\alpha = 1.611$, $\beta = 1.635$, $\gamma = 1.643$; (−)2V = 58° (Calc., 60°). Pleochroic with X light blue, Y dark blue, Z colorless

Space Group: C2/c

Strongest Diffraction Lines: 11.71(100), 3.081(90), 3.115(70), 4.86(41)

Ref.: von Knorring O, Lehtinen M, Sahama ThG (1977) Bull Geol Soc Finland 49:33–36

Cacoxenite
$AlFe^{3+}_{24}(OH)_{12}O_6(PO_4)_{17}(H_2O)_{24} \cdot 51H_2O$

Analyses (%): P_2O_5 19.8–26.8, Fe_2O_3 40.4–48.6, H_2O 30.6–32.8

Class: 6/m

Cleavage: Not observed

Color-Luster: Golden yellow, yellow, brownish yellow, reddish yellow; rarely greenish

Crystal System: Hexagonal

Density: 2.26 (Meas.), 2.252 (Calc.)

First Reported: 1825

Habit: Crystals minute, rare, acicular, sometimes with hexagonal cross section. Usually as radial or tufted aggregates or fibrous crusts; spherulitic

Hardness: 3–4

Lattice Constants: a = 27.669, c = 10.655 (Å)

Mode of Occurrence: Occurs as a secondary mineral commonly associated with other phosphates and with limonite. Found at Shady, Polk County, Arkansas; in Cherokee County, Alabama; at the Vanleer mine, Iron City, Tennessee; at Hellertown and elsewhere in Pennsylvania; on hematite at Antwerp, New York; and at Tonapah, Nevada. It also occurs in France, Germany, Czechoslovakia, and Sweden

Optical Constants: $\omega = 1.580–1.585, 1.600$; $\varepsilon = 1.640–1.656, 1.680$; (+)

Space Group: $P6_3/m$

Strongest Diffraction Lines: 23.1(100), 11.9(100), 9.1(14)

Z: 12

Ref.: Roberts WL, Rapp GR, Weber J (1974) Encyclopedia of minerals, New York, Van Nostrand Reinhold, p 98. Fisher DJ (1966) Am Miner 51:1811–1814

Calcioferrite
$Ca_2Fe_2(PO_4)_3(OH) \cdot 7H_2O$

Analyses (%): CaO 14.8–19.5, Fe_2O_3 24.3–27.8, P_2O_5 32.9–34.0, H_2O 19.8–20.6

Cleavage: Perfect parallel to foliation. Brittle

Color-Luster: Yellowish white to yellow; also greenish yellow to yellowish green. Opaque; thin laminae translucent. Pearly

Crystal System: Monoclinic

Density: 2.53

First Reported: 1858

Habit: As foliated reniform or nodular masses

Hardness: 2.5

Mode of Occurence: Occurs as nodules in a clay bed at Battenberg, Germany

Optical Constants: $\omega = 1.57–1.58$ (variable); (−)

Ref.: Roberts WL, Rapp GR, Weber J (1974) Encyclopedia of minerals, New York, Van Nostrand Reinhold, p 100–101. Mead CW, Mrose ME (1968) U.S. Geol Surv Prof Pap 600-D, p D204–D206

Carbonate-Apatite (Dahllite, Francolite)
$Ca_5(PO_4, CO_3)_3(OH, F)$

Analyses (%): P_2O_5 34.6–38.6, CaO 51.3–56.5, CO_2 3.2–4.5, F 1.4–2.8, H_2O 1.2–4.4

Class: 6/m

Cleavage: {0001} indistinct, $\{10\bar{1}0\}$ trace. Fracture uneven to conchoidal. Brittle

Color-Luster: See fluorapatite

Crystal System: Hexagonal

Density: 2.9–3.1 (Meas.)

First Reported: 1906 (*Carbapatite*), 1916

Habit: See fluorapatite

Hardness: 5

Lattice Constants: a = 9.48, c = 6,89 (Å)

Mode of Occurrence: Occurs in granite pegmatites in Maine, New Hampshire, and the Black Hills, South Dakota. Among many other localities it is also found in Wyoming (dahllite), Arkansas, Greenland, France, Norway, and U.S.S.R.

Optical Constants: $\omega = 1.603, 1.628$; $\varepsilon = 1.598, 1.619$; (−)

Space Group: $P6_3/m$

Strongest Diffraction Lines: 2.822(100), 2.722(90), 3.451(70)

Z: 2

Ref.: Roberts WL, Rapp GR, Weber J (1974) Encyclopedia of minerals, New York, Van Nostrand Reinhold, p 107. McConnell D (1973) Apatites. Springer-Verlag, Wien, p 89–90

Cassidyite (Isostructural with collinsite)
$Ca_2(Ni, Mg)(PO_4)_2 \cdot 2H_2O$

Analyses (%): CaO 32.3, NiO 6.4–16.2, MgO 2.2–5.7, P_2O_5 39.2

Color-Luster: Pale green to bright green

Crystal System: Triclinic

Density: ~3.15

First Reported: 1967

Habit: Thin crusts and small spherules, finely fibrous

Hardness: Not determined

Lattice Constants: a=5.71, b=6,73, c=5.41 (Å); $\alpha=96.83°$, $\beta=107.37$, $\gamma=104.58°$

Mode of Occurrence: Occurs in cavities and cracks in weathered meteorites from the Wolf Creek crater in Western Australia

Optical Constants: $\alpha=1.64$–1.65, $\beta=?$, $\gamma=1.67$–1.68

Strongest Diffraction Lines: 2.70(100), 3.03(95), 2.67(79)

Z: 1

Ref.: White JS Jr, Henderson EP, Mason B (1967) Am Miner 52:1190–1197

Chalcosiderite (Turquoise Group)
$CuFe_6(PO_4)_4(OH)_8 \cdot 4H_2O$

Analyses (%): P_2O_5 28.6–33.8, Al_2O_3 4.5–20.8, Fe_2O_3 21.3–42.8, CuO 6.8–8.2, H_2O 11.9–16.5

Class: $\bar{1}$

Cleavage: {001} perfect, {010} good

Color-Luster: Dark green. Transparent to translucent. Vitreous

Crystal System: Triclinic

Density: 3.22 (Meas.), 3.26 (Calc.)

First Reported: 1824

Habit: Crystals short prismatic with forms {001}, {010}, {$1\bar{1}0$}, and {$\bar{1}\bar{1}1$}. As crusts or sheaf-like groups of distinct crystals

Hardness: 4.5

Lattice Constants: a=7.66, b=10.18, c=7.88 (Å); $\alpha=129.48°$, $\beta=115°18'$, $\gamma=69°$

Mode of Occurrence: Occurs as a secondary mineral in the oxidation zone of copper-bearing ore deposits. Found at Bisbee, Arizona; as fine specimens at the Wheal Phoenix mine, Cornwall, England; and at Siegen, Westphalia, and Schneckenstein, Saxony, Germany

Optical Constants: $\alpha=1.775$, $\beta=1.840$, $\gamma=1.844$; $(-)2V\simeq22°$

Space Group: $P\bar{1}$

Strongest Diffraction Lines: 3.77(100), 3.39(70), 3.02(60)

Z: 1

Ref.: Roberts WL, Rapp GR, Weber J (1974) Encyclopedia of minerals, New York, Van Nostrand Reinhold, p 120. Cid-Dresdner H (1965) Am Miner 50: 227–231

Chavesite
Hydrated Ca, Mn Phosphate

Analyses: None reported

Cleavage: Two good, in prism zone, nearly perpendicular to each other

Color-Luster: Colorless

Crystal System: Triclinic

First Reported: 1958

Habit: Crystalline coatings of tiny thin platy crystals

Hardness: ~3

Lattice Constants: a=5.79, b=13.07, c=5.49 (Å); $\alpha=99.73°$, $\beta=108.05°$, $\gamma=91.31°$

Mode of Occurrence: Occurs as a thin coating on lithiophilite, associated with hureaulite and tavorite, at the Boqueirão pegmatite near Parelhas, Rio Grande do Norte, Brazil

Optical Constants: $\alpha=1.60$, $\beta=1.62$, $\gamma=1.65$; (+2)2V large

Strongest Diffraction Lines: 3.35(100), 2.945(30), 2.23(30)

Ref.: Murdoch J (1958) Am Miner 43:1148–1156. Roberts WL, Rapp GR, Weber J (1974) Encyclopedia of minerals, New York, Van Nostrand Reinhold, p 122

Cheralite (Monazite Group)
(Th, Ca, Ce, La, U, Pb) (PO_4, SiO_4)

Analyses (%): P_2O_5 26.8, CaO 6.3, Ce_2O_3 14.2, La_2O_3 13.4, ThO_2 31.5, UO_3 4.1

Class: 2/m

Cleavage: {010} distinct, {100} difficult (tentative)
Parting on {001} poor (tentative)
Fracture uneven. Brittle

Color-Luster: Dark green to pale green; resinous to vitreous. Transparent

Crystal System: Monoclinic

Density: 5.3 (Meas.), 5.41 (Calc.)

First Reported: 1953

Habit: Masses up to 2 inches across

Hardness: 5

Lattice Constants: a = 6.717, b = 6.920, c = 6.434 (Å); $\beta=103.83°$

Mode of Occurrence: Occurs associated with black tourmaline, chrysoberyl, dark zircon, and smoky quartz in a kaolinized pegmatite dike at Kuttakuzhi in Halkulam taluk (parish), Travancore, southern India. It also occurs sparsely in the adjacent wall-rock (kaolinized granite-gneiss) and in surface wash

Optical Constants: $\alpha=1.779$, $\beta=1.780$, $\gamma=1.816$; 2V = 17.4–19°

Space Group: $P2_1/n$

Strongest Diffraction Lines: 3.07(100), 3.26(90), 2.86(90)

Z: 4

Ref.: Bowie SHU, Horne JET (1953) Min Mag 30:93–99. Finney JJ, Rao NN (1967) Am Miner 52:13–19. Roberts WL, Rapp GR, Weber J (1974) Encyclopedia of minerals, New York, Van Nostrand Reinhold, p 123

Childrenite (Childrenite-Eosphorite Series)
$(Fe^{2+}, Mn^{2+})AlPO_4(OH)_2 \cdot H_2O$

Analyses (%): P_2O_5 29.9–30.7, Al_2O_3 15.9–21.4, FeO 21.7–28.6, MnO 3.1–7.7, H_2O 15.8–17.1

Class: mm2

Cleavage: {100} poor. Fracture uneven to subconchoidal

Color-Luster: Brown and yellowish brown. Transparent to translucent. Vitreous to resinous

Crystal System: Orthorhombic

Density: 3.20 (Pure Fe end-member), 3.186 (Calc.)

First Reported: 1823

Habit: Crystals equant or pyramidal to short prismatic [001] and thick tabular {010}; also platy {100}. Often doubly terminated. Twinning by reflection on {100} and {001}

Hardness: 5

Lattice Constants: a = 10.38, b = 13.36, c = 6.911 (Å)

Mode of Occurrence: Occurs as fine crystals in hydrothermal vein deposits at several localities in Cornwall, England; and in the Tavistock area, Devon; in granite pegmatite at Greifenstein, Saxony, Germany; and at the Hugo mine, Keystone and Helen Beryl mine, Custer, South Dakota; and as exceptional short prismatic crystals up to several centimeters in length in Minas Geraes, Brazil

Optical Constants: $\alpha=1.63$–1.645, $\beta=1.65$–1.68, $\gamma=1.66$–1.685; (—)2V = 40–45°

Space Group: Bba2

Strongest Diffraction Lines: 2.81(100), 5.27(40), 2.42(40)

Z: 8

Ref.: Roberts WL, Rapp GR, Weber J (1974) Encyclopedia of minerals, New York, Van Nostrand Reinhold, p 124. Hurlbut, Cornelius S Jr (1950) Am Miner 35:793–805. Braithwaite RSW, Cooper BV (1982) Miner Mag 46:119–126

Chinoite = Libenthenite

Chlorapatite
$Ca_{10}(PO_4)_6Cl_2$

Analyses (%): CaO 53.0, P_2O_5 40.5, F 0.17, Cl 4.13, H_2O 0.56, Remainder 1.87

Cleavage: {0001} indistinct and {1010} in traces; fracture conchoidal to even; twin planes {1121} and {1013} are rare

Color-Luster: Sea-green, asparagus-green, bluish green; also violet-blue, violet, amethystine, sometimes colorless, gray, brown, rose-red, etc. Streak white, transparent to opaque in thin section. Often fluorescent in ultraviolet light or X-rays

Crystal System: Hexagonal, monoclinic variety also known

Density: 3.1–3.2

First Reported: 1860

Habit: Short to long prismatic crystals with {1010} and {1011} dominant, also thick tabular massive, coarse granular-to compact and sometimes globular or reniform

Hardness: 5 varying with orientation

Lattice Constants: a = 9.64, c = 6.78 (Å) (if hexagonal); a = 19,21, b = 6.785, c = 9.605 (Å); $\beta = 120°$ (if monoclinic)

Mode of Occurrence: In veins associated with basic rocks, iron ores of complex origin and silicified marbles

Optical Properties: $\gamma = \omega = 1.668$; $\alpha = \varepsilon = 1.665$, Biaxial negative

Space Group: $P6_3/m$ (Hexagonal); $P2_1/a$ (Monoclinic)

Strongest Diffraction Lines: 2.78(100), 2.86(50–60)

Ref.: Roberts WL, Rapp GR, Weber J (1974) Encyclopedia of minerals, New York, Van Nostrand Reinhold, p 126. McConnell D (1973) Apatites, Springer-Verlag, Wien, p 87

Churchite = Weinschenkite

Clinophosinaite
$Na_3CaPSiO_7$

Analyses (%): P_2O_5 25.4, SiO_2 20.1, CaO 16.0, SrO 2.50, Na_2O 32.6, MnO 0.57

Color-Luster: Pale lilac, vitreous luster

Crystal System: Monoclinic

Density: 2.85–2.88

First Reported: 1981

Fracture: Conchoidal

Hardness: 4

Lattice Constants: a = 7.303, b = 12.201, c = 14.715 (Å); $\beta = 92°$ for Koashva specimen; a = 7.30, b = 12.21, c = 14.81 (Å); $\beta = 92°$ for the Yukspor sample

Mode of Occurrence: Occurs in dumps from Mt. Yukspor, Lovozero massif, and in drill cores from Mt. Koashva, Khibina massif, Kola Peninsula, U.S.S.R. Typically intergrown with zirsinalite which appears to be a replacement product of eudialyte. It is analogous to phosinaite, except that the latter contains 13%–14% RE_2O_3

Optical Constants: $\alpha = 1.557$, $\beta = 1.561$, $\gamma = 1.567$; Biaxial pos., 2V = 80° (Koashva specimen), 75° (Yukspor specimen)

Space Group: P2/c

Strongest Diffraction Lines: 2.724(100), 2.639(57), 2.550(50), 2.031(50)

Z: 8

Ref.: Khomyakov AP, Pushcharovskii DY, Ronsbo JG (1981) Zapiski Vses Miner Obshch, 110:351–355; *Idem* (1982) Am Miner 67:414

Coconinoite
$Fe_2Al_2(UO_2)_2(PO_4)_4(SO_4)(OH)_2 \cdot 2OH_2O$

Analyses (%): Fe_2O_3 9.7, Al_2O_3 6.6, UO_3 34.9, P_2O_5 18.3, H_2O 24.0

Cleavage: Not determined

Color-Luster: Light creamy yellow

Crystal System: Monoclinic

Density: 2.70 (Meas.), 2.68 (Calc.)

First Reported: 1966

Habit: As aggregates of microcrystalline grains

Hardness: Soft, hardness cannot be measured accurately

Mode of Occurrence: Occurs in the oxidized zone of uranium deposits in Utah and Arizona, and occurrences have been documented from Wyoming and New Hampshire

Optical Constants: $\alpha = 1.550$, $\beta = 1.588$, $\gamma = 1.590$; (—)2V $\approx 40°$

Strongest Diffraction Lines: 11.05(100), 5.52(60), 5.61(40)

Ref.: Young EJ, Weeks AD, Meyrowitz R (1966) Am Miner 51:651–666. Roberts WL, Rapp GR, Weber J (1974) Encyclopedia of minerals, New York, Van Nostrand Reinhold, p 139–140

Coeruleolactite (Calcium analog of turquoise)
$(Ca, Cu)Al_6(PO_4)_4(OH)_8 \cdot 4\text{–}5H_2O$

Analyses (%): CaO 0–2.4, CuO 1.4–4.2, Al_2O_3 35.1–38.3, P_2O_5 36.3, H_2O 21.3–21.7

Class: $\bar{1}$

Cleavage: Fracture conchoidal to uneven

Color-Luster: Milk white to light blue

Crystal System: Triclinic

Density: 2.57–2.696 (Meas.)

First Reported: 1871

Habit: As minutely crystalline or fibrous crusts, veinlets, and botryoidal aggregates

Hardness: 5

Mode of Occurrence: Occurs associated with wavellite at General Trimble's mine, East Whiteland, Chester County, Pennsylvania, and with limonite at the Rindsberg mine, Nassau, Germany

Optical Constants: $\omega = 1.580$, $\varepsilon = 1.588$; (+)

Space Group: $P\bar{1}$

Strongest Diffraction Lines: 2.96(100) broad, 3.70(90), 3.48(50)

Ref.: Roberts WL, Rapp GR, Weber J (1974) Encyclopedia of minerals, New York, Van Nostrand Reinhold, p 140

Collinsite
$Ca_2(Mg, Fe)(PO_4)_2 \cdot 2H_2O$

Analyses (%): CaO 32.1, MgO 6.3–9.3, FeO 6.9–7.3, P_2O_5 39.8–41.1, H_2O 9.7–12.3

Class: $\bar{1}$

Cleavage: {001} perfect, {010} perfect. Brittle

Color-Luster: Colorless, white, light brown. Transparent to translucent. Vitreous or silky

Crystal System: Triclinic

Density: 2.99 (Meas.), 3.04 (Calc.)

First Reported: 1927

Habit: Crystals short prismatic to thin tabular; often in bundles, radiating outward in sheaf-like aggregates; sometimes doubly terminated. Usually as thick crusts with radial-fibrous structure

Hardness: 3.5

Lattice Constants: a = 5.71, b = 6.73, c = 5.39 (Å); $\alpha = 96.82°$, $\beta = 107.28°$, $\gamma = 104.53°$

Mode of occurrence: Occurs as fine crystals and radial aggregates lining cavities in altered phosphate nodules from granite pegmatite at the Tip Top mine, Custer County, South Dakota. Also found as alternate crusts with quercyite (carbonate-apatite), associated with small amounts of asphaltum, on fragments of andesite in a vein-like deposit at François Lake, British Columbia, Canada

Optical Constants: $\alpha = 1.632$, $\beta = 1.642$, $\gamma = 1.657$; (+)2V $= 80°$

Space Group: $P\bar{1}$

Strongest Diffraction Lines: 2.69(100), 3.04(80), 1.669(60)

Z: 1

Ref.: Roberts WL, Rapp GR, Weber J (1974) Encyclopedia of minerals, New York, Van Nostrand Reinhold, p 141–142

Collophane = Variety of apatite

Corkite
$PbFe_3PO_4SO_4(OH)_6$

Analyses (%): PbO 24.1–32.3, CuO 1.4–2.4, Fe_2O_3 34.6–40.7, SO_3 12.7–13.8, P_2O_5 9.0–9.4, H_2O 8.4–9.8

Class: 3m

Cleavage: {0001} perfect

Color-Luster: Dark green, yellowish green to pale yellow. Vitreous to resinous; fine-granular aggregates shiny to dull

Crystal System: Hexagonal

Density: 4.295 (Meas.), 4.423 (Calc.)

First Reported: 1869

Habit: Crystals, rhombohedral, commonly pseudocubic. Also as loosely coherent fine-granular aggregates

Hardness: 3.5–4.5

Lattice Constants: a = 7.22, c = 16.66 (Å)

Mode of Occurrence: Occurs in abundance associated with descloizite, vanadinite, pyromorphite, cerussite, and galena, at the Silver Queen mine, Galena, Lawrence County, South Dakota. It also is found at the Harrington-Hickory and Wild Bill mines, Beaver County, and at the Gold Hill and Honerine mines, Tooele County, Utah; in the Glendore iron mine, County Cork, Ireland; and at localities in Germany, Sardinia, Yugoslavia, and U.S.S.R.

Optical Constants: $\omega = 1.93$, ω–ε weak; uniaxial (–)

Space Group: R3m

Strongest Diffraction Lines: 3.03(100), 5.86(65), 2.237(55)

Z: 3

Ref.: Roberts WL, Rapp GR, Weber J (1974) Encyclopedia of minerals, New York, Van Nostrand Reinhold, p 146–147

Cornetite
$Cu_3PO_4(OH)_3$

Analyses (%): CuO 70.8, P_2O_5 19.8, H_2O 9.4

Class: 2/m 2/m 2/m

Cleavage: None. Brittle

Color-Luster: Dark blue to greenish-blue. Transparent to translucent. Vitreous

Crystal System: Orthorhombic

Density: 4.10 (Meas.), 4.10 (Calc.)

First Reported: 1917 (*Cesàro*, 1912)

Habit: Crystals short prismatic [001] to equant. {210} often rounded. As crusts of minute crystals

Hardness: ~4.5

Lattice Constants: a = 10.86, b = 14.07, c = 7.10 (Å)

Mode of Occurrence: Occurs as fine dark blue single crystals and crusts at the Blue Jay and Empire-Nevada mines, Yerington, Nevada; in association with libethenite, brochanite, pseudomalachite, malachite, atacamite, and chrysocolla at Saginaw Hill, about seven miles southwest of Tucson, Arizona; as incrustations of dark blue crystals on compact, gray argillaceous sandstone at Bwana Mkubwa, Zambia; and at the Etoile du Congo mine, Katanga, Zaire, associated with pseudomalachite on fine-grained argillaceous sandstone

Optical Constants: $\alpha = 1.765$, $\beta = 1.81$, $\gamma = 1.82$; (–)2V ~ 33°

Space Group: Pbca

Strongest Diffraction Lines: 3.04(100), 4.29(90), 3.17(80)

Z: 8

Ref.: Berry LG (1950) Am Miner 35:365–385. Roberts WL, Rapp GR, Weber J (1974) Encyclopedia of minerals, New York, Van Nostrand Reinhold, p 147

Crandallite (Pseudowavellite)
$CaAl_3(PO_4)_2(OH)_5 \cdot H_2O$

Analyses (%): CaO 7.5–16.9, Al_2O_3 28.2–38.8, P_2O_5 25.2–32.6, H_2O 17.4–20

Cleavage: {0001} perfect

Color-Luster: Yellow to yellowish white to white or gray. Transparent to subtranslucent. Vitreous; also dull and chalky

Crystal System: Triclinic, hexagonal

Density: 2.78–2.92 (Meas.)

First Reported: 1917 (*Kalkwavellit*, 1869)

Habit: Crystals minute trigonal prisms terminated by {0001}, or as rosettes of fibers. Commonly massive, as nodular masses or spherules with fibrous, fine-granular, or chalcedony-like structure

Hardness: 5

Lattice Constants: (Triclinic dimorph) a = 7.010, b = 9.819, c = 9.697 (Å); $\alpha = 103.17°$, $\beta = 91.23°$, $\gamma = 90.57°$. (Hexagonal dimorph) a = 7.005 (Å), c = 16.192 (Å)

Mode of Occurrence: Occurs abundantly associated with a wide variety of rare secondary phosphate minerals in variscite nodules near Fairfield, Utah County; with barite and quartz at the Brooklyn mine, Tintic District, Juab County; and at other places in Utah. It is also found in Esmeralda County, Nevada; at the Everly and Hugo mines, Keystone, Pennington County, South Dakota; in Lawrence County, Indiana; and in Brazil, Bolivia, Germany and Senegal. An unnamed apparently triclinic dimorph of crandallite occurs in the sediments of the Bajo de Sante Fé, El Patén, Guatemala

Optical Constants: (Hexagonal) $\omega = 1.60–1.622$, $\varepsilon = 1.61–1.631$; (+). (Triclinic) $\alpha = 1.602$, $\beta = 1.608$, $\gamma = 1.615$; (+)2V = 70°–75°

Space Group: $R\bar{3}m$

Strongest Diffraction Lines: (Hexagonal) 2.95(100), 2.98(50), 2.16(40). (Triclinic dimorph) 2.97(100), 2.18(45), 5.75(35)

Z: 3

Ref.: Roberts WL, Rapp GR, Weber J (1974) Encyclopedia of minerals, New York, Van Nostrand Reinhold, p 151. Blount AM (1974) Am Miner 59:41–47. Goreaud M, Raveau B (1980) Am Miner 65:953–956

Curetonite
$Ba_4Al_3Ti(PO_4)_4(O, OH)_6$

Analyses (%): P_2O_5 23.39, V_2O_5 1.27, BaO 53.09, Al_2O_3 12.81, TiO_2 6.84, H_2O 2.46

Cleavage: {011} good, parting on {010}

Color-Luster: Usually bright yellow-green, but nickel-green in some samples. White streak. Mostly colorless in section, but may show splotchy pleochroism in yellow with absorption X > Y = Z

Crystal System: Monoclinic

Density: 4.42 (Meas.), 4.31 (Calc.)

Habit: Crystals usually show forms {100} {010} {001} {011} and {$\bar{2}$01}. Polysynthetic twinning common on {100}

Hardness: 3.5, brittle

Lattice Constants: a = 6.957, b = 12.55, c = 5.22 (Å); $\beta = 102°$

Mode of Occurrence: Found in a barite mine near Golconda, Nevada

Optical Constants: $\alpha = 1.676$, $\beta = 1.680$, $\gamma = 1.693$; Biaxial pos., 2V = 60°

Space Group: $P2_1/m$

Strongest Diffraction Lines: 3.230(100), 3.286(80), 1.686(70), 2.991(60), 2.818(60)

Z: 1

Ref.: Williams SA (1979) Miner Rec 10:219–221

Cyrilovite (Avelinoite; cf Wardite)
$NaFe_3^{3+}(PO_4)_2(OH)_4 \cdot 2H_2O$

Analyses (%): P_2O_5 29.3, Na_2O 4.7, Al_2O_3 1.4, Fe_2O_3 48.3, MnO 1.0, H_2O 14.6

Class: 4 2 2

Color-Luster: Orange to brownish yellow. Vitreous. Powder is yellow

Crystal System: Tetragonal

Density: 3.081 (Meas.), 3.09 (Calc.)

First Reported: 1953

Habit: Crystals squat; dominant forms {001} and {113}; {012} occasionally present. Individual crystals usually less than 0.1 mm; often in intergrown aggregates

Hardness: Not determined

Lattice Constants: a = 7.32, c = 19.40 (Å)

Mode of Occurrence: The mineral occurs associated with metastrengite and leucophosphite in an altered zone surrounding frondelite at the Sapucaia pegmatite mine, Minas Geraes, Brazil. Also found as minute crystals in pegmatite at Cyrilov, near Velké Mežiřiči, West Moravia, and in Madagascar

Optical Constants: $\omega = 1.803$, $\varepsilon = 1.769$; (—)

Space Group: $P4_12_12$ or $P4_32_12$

Strongest Diffraction Lines: 4.85(100), 3.186(80), 2.658(80)

Z: 4

Ref.: Lindberg ML (1957) Am Miner 42:204–213. Roberts WL, Rapp GR, Weber J (1974) Encyclopedia of minerals, New York, Van Nostrand Reinhold, p 160–161

Dahllite = Carbonate apatite

Davisonite (Dennisonite)
(Inadequately described mineral)
$6CaO \cdot Al_2O_3 \cdot 2P_2O_5 \cdot 5H_2O$

Analyses (%): CaO 40.3–41.7, MgO 1.38–5.92, Al_2O_3 11.8–15.2, P_2O_5 30.3–33.1, H_2O 6.67–11.6

Cleavage: Basal, perfect

Color-Luster: White

Crystal System: Hexagonal (?)

Density: 2.85 (Meas.)

First Reported: As Dennisonite, 1930; corrected to Davisonite in 1951

Habit: As botryoidal or spherulitic crusts composed of stout fibers

Hardness: 4.5

Mode of Occurrence: Occurs lining cavities in crandallite in variscite modules at Fairfield, Utah County, Utah. It also is found at Damasio, near Diamantina, Minas Geraes, Brazil

Optical Constants: $\omega = 1.601$, $\varepsilon = 1.591$, (—)

Ref.: Roberts WL, Rapp GR, Weber J (1974) Encyclopedia of minerals, New York, Van Nostrand Reinhold, p 165–166

Deltaite = Crandallite plus hydroxylapatite

Delvauxite
(Inadequately described mineral)
$Fe_4(PO_4)_2(OH)_6 \cdot nH_2O$

Analyses (%): P_2O_5 16.0–24.5, Fe_2O_3 34.2–46.5, H_2O 28.0–*ca.* 50%

Cleavage: Not determined. Fracture conchoidal

Color-Luster: Yellowish brown to brown, reddish brown to brownish-black. Vitreous, greasy, weak waxy. Streak yellow

Cristal System: Unknown

Density: 1.85–2.83 (Meas.)

First Reported: 1845 (Delvauxene, 1839; Picite, 1877)

Habit: As concretionary masses, botryoidal crusts and coatings; stalactitic. Often gel-like

Hardness: 2.5–4

Mode of Occurrence: Occurs at Berneau, Liège, Belgium, and at localities in Austria, Germany, Czechoslovakia, and U.S.S.R.

Optical Constant: N = 1.72

Ref.: Roberts WL, Rapp GR, Weber J (1974) Encyclopedia of minerals, New York, Van Nostrand Reinhold, p 168

Dewindtite
(Perhaps a mixture), $8UO_3 \cdot 4PbO \cdot 3P_2O_5 \cdot 12H_2O$

Analyses (%): PbO 24.8–26.2, UO_3 54.8–55.8, P_2O_5 10.5, H_2O 6.71–7.93

Class: mm2 or 2/m 2/m 2/m

Cleavage: {100} perfect. Brittle

Color-Luster: Canary yellow; powder pale yellow. Translucent. Fluoresces green in ultraviolet light

Crystal System: Orthorhombic

Density: 5.03 (Meas.), 5.01 (Calc.)

First Reported: 1922 (called Stasite, 1922)

Habit: Microscopic tablets flattened {100} and terminated by {001}. {100} striated parallel to *c*-axis. Also very fine-grained and compact

Hardness: Not determined

Lattice Constants: a = 16.05, b = 17.50, c = 13.64 (Å)

Mode of Occurrence: Occurs as a secondary mineral associated with torbernite, dumontite, and kasolite at Shinkolobwe, Katanga, Zaire, and at Wölsendorf, Bavaria, Germany

Optical Constants: $\alpha = 1.760$–1.762, $\beta = 1.768$–1.767, $\gamma = 1.770$–1.768; (—)2V = moderate

Space Group: Bbm2 or Bmmb

Strongest Diffraction Lines: 8.01(100), 5.89(100), 3.14(90)

Z: $\simeq 12$(?)

Ref.: Frondel C (1958) USGS Bull 1064, p 230–232. Roberts WL, Rapp GR, Weber J (1974) Encyclopedia of minerals, New York, Van Nostrand Reinhold, p 170–171

Diadochite (Destinezite)
$Fe_2PO_4(SO_4)(OH) \cdot 5H_2O$

Analyses (%): P_2O_5 14.8–19.2, Fe_2O_3 35.1–39.7, SO_3 10.9–19.3, H_2O 25.7–34.5

Cleavage: Fracture conchoidal to uneven or earthy. Brittle to pulverulent

Color-Luster: Deep yellowish brown, dark amber; also reddish brown, greenish yellow, pale green, yellowish white. Translucent to opaque. Resinous, waxy to dull

Crystal System: Triclinic

Density: 2.0–2.4

First Reported: 1837

Habit: Usually gel-like and amorphous; as crusts or masses, botryoidal, stalactic. Also microcrystalline in masses composed of minute six-sided plates of various habits

Hardness: ~3–4

Lattice Constants: a = 9.61, b = 9.77, c = 7.36 (Å); $\alpha = 98.82°$, $\beta = 108°$, $\gamma = 64°$

Mode of Occurrence: Occurs as a recent deposit in mine workings, and as a near-surface secondary mineral. Found in the New Idria quicksilver mine, San Benito County, California; at Roberts Mountain, near Eureka, Nevada; in the Black Hills, South Dakota; at Shady, Polk County, Arkansas; and at numerous other localities in the United States and elsewhere throughout the world

Optical Constants: $\alpha = 1.62$, $\beta = 1.64$, $\gamma = 1.67$; (+)2V = small

Strongest Diffraction Lines: 4.35(100), 8.3(90), 8.7(80)

Z: 2

Ref.: Roberts WL, Rapp GR, Weber J (1974) Encyclopedia of minerals, New York, Van Nostrand Reinhold, p 171–172

Dickinsonite
(Dickinsonite-Arrojadite Series)
$KNa_4Ca(Mn, Fe)_{14}Al(OH, F)_2(PO_4)_{12}$

Analyses (%): P_2O_5 39.6–40.9, Na_2O 7.4, CaO 2.1, MgO 1.7, FeO 12.3–13.3, MnO 31.7, H_2O 1.6–1.8

Class: 2/m

Cleavage: {001} perfect, easy. Fracture uneven. Very brittle

Color-Luster: Yellowish green to olive green to brownish green. Transparent to translucent. Vitreous, pearly on cleavage. Streak white

Crystal System: Monoclinic

Density: 3.41 (Meas.), 3.42 (Calc.)

First Reported: 1878

Habit: Crystals tabular, often pseudorhombohedral; triangular striations on {001}. Commonly foliated to micaceous; also curved lamellar or radiated; as disseminated scales

Hardness: 3.5–4

Lattice Constants: a = 16.70, b = 9.95, c = 24.69 (Å); $\beta = 104.68°$

Mode of Occurrence: Occurs in granite pegmatites associated with lithiophilite, rhodochrosite, and a wide variety of secondary phosphate minerals. Found sparingly at Branchville and at Portland, Connecticut, and at the Berry quarry near Poland, Maine

Optical Constants: Undetermined; (+)2V

Space Group: C2/c

Strongest Diffraction Lines: 3.05(100), 2.72(90), 3.22(70)

Z: 4

Ref.: Roberts WL, Rapp GR, Weber J (1974) Encyclopedia of minerals, New York, Van Nostrand Reinhold, p 173. Fisher DJ (1965) Am Miner 50:1647–1669. Moore PB et al. (1981) Am Miner 66:1034–1049

Dittmarite

$(NH_4)MgPO_4 \cdot H_2O$

Analyses (%): MgO 25.7, $(NH_4)_2O$ 3.9, P_2O_5 46.5, H_2O 20.5

Class: mm2

Color-Luster: Colorless. Transparent

Crystal System: Orthorhombic

Density: 2.19 (Calc.)

First Reported: 1887

Habit: As minute crystals

Hardness: Low

Lattice Constants: a = 5.606, b = 8.758, c = 4.788 (Å)

Mode of Occurrence: The mineral occurs in bat guano, associated with newberyite, struvite, schertelite, and hannayite, in the Skipton Caves, southwest of Ballarat, Victoria, Australia

Optical Constants: $\alpha = 1.549$, $\beta = 1.569$, $\gamma = 1.571$; (—)2V = 40°

Space Group: $Pmn2_1$

Strongest Diffraction Lines: 8.77(100), 2.80(50), 2.92(40)

Z: 2

Ref.: Mrose ME (1971) U.S.G.S. Prof. Pap. 750-A, p. A115

Dorfmanite

$Na_2HPO_4 \cdot 2H_2O$

Analyses (%): P_2O_5 39.5–40.8, Na_2O 33.8–34.8, H_2O 25.4–25.6

Color: White

Crystal System: Orthorhombic

Density: 1.98 (Meas.), 2.06 (Calc.)

First Reported: 1980

Habit: Powdery aggregates

Hardness: 1–1.5

Lattice Constants: a = 10.34, b = 16.82, c = 6.01 (Å)

Mode of Occurrence: Formed by the alteration of lomonosovite in drill cores in alkalic pegmatites of the Khibina and Lovozero massifs, Kola Peninsula, U.S.S.R.

Optical Constants: $\alpha = 1.454$, $\beta = 1.461$, $\gamma = 1.461$; 2V = 65°

Space Group: Not given

Strongest Diffraction Lines: 3.35(100), 3.36(100), 4.67(90), 3.25(81)

Z: 8

Ref.: Kapustin YL, Pudovkina ZV, Bykova TE (1980) Zapiski Vses Mineral Obsh 109:211–216; *Idem* (1981) Am Miner 66:217

Drugmanite
$Pb_2(Fe^{3+}, Al)(PO_4)_2(OH) \cdot H_2O$

Analyses (%): Pb 59,6%, Fe 6.19, Al 0.85, P 8.89

Cleavage: {001} platelets bounded by {110}

Color: Colorless and transparent

Crystal System: Monoclinic

Density: 5.55 (Calc.)

First Reported: 1979

Habit: Platy crystals aggregated in bunches in vugs of a mineralized and silicified limestone

Lattice Constants: a = 11.100; b = 7.976; c = 4.644 (Å); β = 90.18°

Mode of Occurrence: Found at Richelle, Belgium in association with pyromorphite, anglesite, corkite and phosphosiderite

Optical Constants: ω = 1.87; 2V = 33°

Space Group: $P2_1/a$

Strongest Diffraction Lines: 4.63, 3.752, 3.35, 3.247, 2.912

Z: 2

Ref.: van Tassel R, Fransolet AM, Abraham K (1979) Miner Mag 43:463–467

Dufrenite
$CaFe_6^{3+}(PO_4)_4(OH)_6 \cdot 2H_2O$

Analyses (%): P_2O_5 30.3–31.8, Fe_2O_3 40.2–60.2, FeO 1.5–8.3, H_2O 8.4–14.4

Class: 2/m

Cleavage: {100} perfect. Brittle

Color-Luster: Dark green to greenish black, becoming greenish-brown to reddish-brown due to oxidation. Translucent to opaque. Vitreous to silky

Crystal System: Monoclinic

Density: 3.1–3.34 (Meas.)

First Reported: 1833 (called Strahlstein, 1803; Grüneisenstein, strahlicher, ochrichter, 1814)

Habit: Crystals rare, indistinct, rounded, in aggregates. Usually as botryoidal crusts or masses with radial-fibrous structure

Hardness: 3.5–4.5

Lattice Constants: a = 25.84, b = 5.126, c = 13.78 (Å); β = 111.2°

Mode of Occurrence: Occurs as a secondary mineral at Rock Run, Cherokee County, Alabama; at Wheal Phoenix, Cornwall, England; and at deposits in Saxony, Hesse, Westphalia, and Thuringia, Germany

Optical Constants: α = 1.820–1.842, β = 1.830–1.850, γ = 1.875–1.925; (+)2V = small

Space Group: C2/c

Strongest Diffraction Lines: 3.151(100), 12.00(90), 5.002(90)

Z: 4

Ref.: Roberts WL, Rapp GR, Weber J (1974) Encyclopedia of minerals, New York, Van Nostrand Reinhold, p 180. Moore PB (1970) Am Miner 55:135–169

Dumontite
$Pb_2(UO_2)_3(PO_4)_2(OH)_4 \cdot 5H_2O$

Analyses (%): PbO 27.2, UO_3 56.5, P_2O_5 8.65, H_2O 5.78

Class: 2/m

Cleavage: Not determined

Color-Luster: Pale yellow to golden yellow. Translucent. Powder pale yellow. Fluoresces weak green in ultraviolet light

Crystal System: Monoclinic

Density: 5.65 (Meas.)

First Reported: 1924

Habit: Minute crystals elongated on c-axis and flattened on {010}. Striated parallel to c-axis

Hardness: Not determined

Lattice Constants: a = 8.16, b = 16.73, c = 7.02 (Å)

Mode of Occurrence: Occurs as a secondary mineral with kasolite, autunite, and uranophane at the White Oak mine near Nogales, Arizona, and with kasolite at the Green Monster mine, Goodsprings, Nevada. It is also found associated with parsonsite and metatorbernite at Shinkolobwe, Katanga, Zaire

Optical Constants: $\alpha = 1.85$, $\beta = 1.87$, $\gamma \simeq 1.89$; (+2)V = large

Space Group: $P2_1/m$

Strongest Diffraction Lines: 4.27(100), 3.00(85), 2.95(85)

Z: 2

Ref.: Frondel C (1958) USGS Bull 1064, p 236–238. Roberts WL, Rapp GR, Weber J (1974) Encyclopedia of minerals, New York, Van Nostrand Reinhold, p 181

Egueiite
(Inadequately described mineral; possibly altered vivianite)
Near $CaFe^{3+}_{14}(PO_4)_{10}(OH)_{12} \cdot 21H_2O$

Analyses (%): CaO 2.28, Al_2O_3 1.50, Fe_2O_3 44.2, P_2O_5 30.3, H_2O 20.5

Cleavage: Very friable

Color Luster: Brownish yellow. Vitreous to slightly greasy. Streak yellow

Crystal System: Monoclinic (?)

Density: 2.60

First Reported: (Described but unnamed, 1909), 1910

Habit: As small nodules with fibrous-lamellar structure

Hardness: Not determined

Mode of Occurrence: Found associated with thenardite and trona in clay in the Eguéï region, Chad

Ref.: Roberts WL, Rapp GR, Weber J (1974) Encyclopedia of minerals, New York, Van Nostrand Reinhold, p 186

Ellestadite
(including hydroxyellestadite), Apatite Group
$Ca_5\{(Si, S, P, C)O_4\}_3(Cl, F, OH)$

Analyses (%): CaO 55.2, P_2O_5 3.06, SO_3 20.7, SiO_2 17.3, CO_2 0.61, Cl 1.64, F 0.57, H_2O 0.53

Class: 6/m

Cleavage: {0001} indistinct, $\{10\bar{1}0\}$ indistinct

Color-Luster: Pale rose. Translucent. Vitreous

Crystal System: Hexagonal

Density: 3.068 (Meas.)

First Reported: 1937

Habit: Massive, granular

Hardness: ~5

Lattice Constants: a = 9.53, c = 6.91 (Å). Synthetic $Ca_{10}(SiO_4)_3(SO_4)_3F_2$ has a = 9.43 and c = 6.93; the large dimension (a = 9.53) is for ellestadite containing a significant amount of structural water, but has been reduced to correct for Cl

Mode of Occurrence: Calc-silicic metamorphic rocks. For example, as veinlets in blue calcite associated with wilkeite, idocrase, diopside, and wollastonite in contact-metamorphes marble at Crestmore, Riverside County, California

Optical Constants: $\omega = 1.655$, $\varepsilon = 1.650$; (—)

Space Group: $P6_3/m$

Strongest Diffraction Lines: 2.86(100), 2.76(60), 1.97(60)

Z: 2

Ref.: Roberts WL, Rapp GR, Weber J (1974) Encyclopedia of minerals, New York, Van Nostrand Reinhold, p 187. McConnell D (1973) Apatites. Springer-Verlag, Wien, p 88

Embreyite
$Pb_5(CrO_4)_2(PO_4)_2 \cdot H_2O$

Analyses (%): PbO 74.4, CuO 1.70, Cr_2O_3 13.4, P_2O_5 9.09, H_2O 0.91

Class: 2/m

Cleavage: None. Fracture irregular. Brittle

Color-Luster: Various shades of orange. Transparent to translucent.

Dull to sparkling and resinous. Streak yellow

Crystal System: Monoclinic

Density: 6.45 (Meas.), 6.41 (Calc.)

First Reported: 1972

Habit: As drusy crystalline crusts composed of minute tabular crystals with the plane of flattening approximately normal to the surface of the crust. Crystals are composite and show sectored zoning and multiple twinning

Hardness: 3.5

Lattice Constants: a=9.755, b=5.636, c=7.135 (Å); $\beta=103.08°$

Mode of Occurrence: Occurs in association with crocoite, cerussite, phoenicochroite, and vauquelinite at Beresov, in the Urals, U.S.S.R.

Optical Constants: $\alpha=2.20$, $\beta=\gamma=2.36$. Birefringence high, 2V=0° to 11°

Space Group: $P2_1/m$

Strongest Diffraction Lines: 3.167(100), 4.751(60), 2.818(60)

Z: 1

Ref.: Williams SA (1972) Miner Mag 38:790–793

Englishite

(Inadequately described mineral)

$K_4Na_2Ca_9Al_{18}(OH)_{36}(PO_3OH)_{12} \cdot (PO_4)_6(H_2O)_8$

Analyses (%): K_2O 5.4, Na_2O 1.6, CaO 14.1, Al_2O_3 24.7, P_2O_5 37.8, H_2O 16.5

Cleavage: {001} perfect, micaceous

Color-Luster: Colorless. Transparent. Vitreous; pearly on cleavage

Crystal System: Monoclinic

Density: ~2.65

First Reported: 1930

Habit: As aggregates and layers of curved and composite plates

Hardness: ~3

Lattice Constants: a=38,43, b=11.86, c=20.67 (Å); $\beta=111.27°$

Mode of Occurrence: Occurs with crandallite, wardite, and other phosphate minerals in variscite nodules at Fairfield, Utah County, Utah

Optical Constants: $\alpha=1.570$, $\beta=?$, $\gamma=1.572$; (—)2V=small

Space Group: A2/a or Aa

Strongest Diffraction Lines: 9.3(100), 2.86(70), 1.72(60)

Z: 4

Ref.: Roberts WL, Rapp GR, Weber J (1974) Encyclopedia of minerals, New York, Van Nostrand Reinhold, p 191. Moore PB (1976) Miner Mag 40: 863–866

Eosphorite

(Eosphorite-Childrenite Series)

$(Mn, Fe)AlPO_4(OH)_2 \cdot H_2O$

Analyses (%): P_2O_5 29.9–31.0, FeO 1.4–7.4, MnO 23.5–29.9, Al_2O_3 20.5–22.4, H_2O 15.4

Cleavage: {100} poor. Fracture uneven to subconchoidal

Color-Luster: Colorless, pale pink, pale yellow, light brown, reddish-brown, black. Transparent to translucent. Vitreous to resinous

Crystal System: Monoclinic (pseudo-orthorhombic)

Density: 3.05 (Pure Mn end-member.)

First Reported: 1878

Habit: Crystals short to long prismatic; often flattened on {010}. Common forms a{100}, b{010}, m{110}, s{121}. Also as coarse radial aggregates. Twinning by reflection on {100}

Hardness: 5

Lattice Constants: a=10.45, b=13.49, c=6.93 (Å); $\beta=90°$

Mode of Occurrence: Occurs in granite pegmatite associated with manganese phosphates at Branchville, Connecticut; at Red Hill, Black Mountain, Newry, Hebron, Poland, Mt. Mica, and Buckfield, Maine; at the Hugo mine, Keystone, South Dakota; Palermo mine, North Groton, New Hampshire;

Hagendorf, Bavaria, Germany; and in Minas Geraes, Brazil as superb pink flattened long prismatic crystals as much as 1 cm wide and 4 cm in length

Optical Constants: $\alpha = 1.638$–1.639, $\beta = 1.660$–1.664; $\gamma = 1.667$–1.671; (—)$2V = 50°$

Space Group: Bbam (pseudocell)

Strongest Diffraction Lines: 2.81(100), 5.27(40), 2.42(40)

Z: 8

Ref.: Hurlbut C Jr (1950) Am Miner 35:793–805. Roberts WL, Rapp GR, Weber J (1974) Encyclopedia of minerals, New York, Van Nostrand Reinhold, p 191–192

Erikite = Rhabdophane or Monazite

Ernstite

$(Mn^{2+}_{1-x}Fe^{3+}_{x})Al(PO_4)(OH)_{2-x}O_x$ ($x = 0$–1)

Analyses (%): P_2O_5 29.7, Al_2O_3 24.2, Fe_2O_3 11.8, FeO 0.44, MgO 1.0, MnO 17.7, CaO 0.84, H_2O 13.4

Class: 2/m or m

Cleavage: {010} good, {100} good

Color-Luster: Yellow-brown

Crystal System: Monoclinic

Density: 3.07 (Meas.), 3.086 (Calc.)

First Reported: 1970

Habit: As radiating aggregates 10–15 mm long. Untwinned

Hardness: 3–3.5

Lattice Constants: $a = 13.32$, $b = 10.497$, $c = 6.969$ (Å); $\beta = 90.37°$

Mode of Occurrence: Occurs as an oxidation product of eosphorite (of which relicts remain) in granitic pegmatite near Karibib, South West Africa

Optical Constants: $\alpha = 1.678$, $\beta = 1.706$ (Na), $\gamma = 1.721$; (—)$2V = 74°$

Space Group: A2/a or Aa

Strongest Diffraction Lines: 2.829(100), 2.836(80), 2.438(50)

Z: 8

Ref.: Seeliger E, Mücke A (1970) Neues Jahrb Miner Monatsh, 289–298. Roberts WL, Rapp GR, Weber J (1974) Encyclopedia of minerals, New York, Van Nostrand Reinhold, p 196

Evansite

(Inadequately described mineral)

$Al_3PO_4(OH)_6 \cdot 6H_2O$(?)

Analyses (%): P_2O_5 18.0–21.7, Al_2O_3 34.5–40.2, Fe_2O_3 0–5.49, H_2O 37.0–41.5

Cleavage: Fracture conchoidal. Very brittle

Color-Luster: Colorless, white, or tinted yellow, green, or blue; sometimes red to brown due to impurities. Transparent to translucent. Vitreous to resinous or waxy. Streak whitish

Crystal System: Unknown

Density: 1.8–2.2

First Reported: 1864

Habit: Massive; stalactitic, botryoidal, or reniform. Opal-like

Hardness: 3–4

Mode of Occurrence: Occurs as a secondary mineral often associated with allophane and limonite. Found at Goldburg, Custer County, Idaho; at Coalville and Columbiana in the Coosa coal field, Alabama; and in England, France, Spain, Czechoslovakia, Hungary, Tasmania, and Madagascar

Optical Constant: $N \sim 1.445$–1.485

Ref.: Roberts WL, Rapp GR, Weber J (1974) Encyclopedia of minerals, New York, Van Nostrand Reinhold, p 201

Eylettersite (Crandallite Series)

$(Th, Pb)Al_3(PO_4, SiO_4)_2(OH)_6$(?)

Analyses (%): P_2O_5 18.3–19.7, ThO_2 18.4–20.1, UO_3 3.10, CaO 0.63, BaO 1.45–1.90, PbO 3.11–4.62, Al_2O_3 34.9–35.7, ZrO_2 1.0, SiO_2 1.20–1.65, H_2O 14.4–15.0

Class: $\bar{3}$ 2/m

Cleavage: None

Color-Luster: Creamy white

Crystal System: Hexagonal

Density: 3.38–3.44 (Meas.), 3.44–3.50 (Calc.)

First Reported: 1972

Habit: Massive; as pulverulent nodules

Hardness: Not determined

Lattice Constants: a = 6.98–6.99, c = 16.66–16.72 (Å)

Mode of Occurrence: The mineral occurs in the Kobokobo pegmatite, Kivu, Zaire. It is derived by the alteration of thorian crandallite

Optical Constants: Weakly anisotropic. N = 1.61–1.66

Space Group: R$\bar{3}$m

Strongest Diffraction Lines: 2.95(100), 3.51(60), 5.70(55)

Z: 1

Ref.: Van Wambeke L (1971) Am Miner 56:1366–1384. Roberts WL, Rapp GR, Weber J (1974) Encyclopedia of minerals, New York, Van Nostrand Reinhold, p 202

Faheyite

$(Mn, Mg, Na)Be_2Fe_2^{3+}(PO_4)_4 \cdot 6H_2O$

Analyses (%): P_2O_5 38.1, BeO 7.3, MnO 6.0, Na_2O 0.8, MgO 1.1, Fe_2O_3 21.4, H_2O 14.9

Class: 622

Cleavage: Perfect, parallel to c-axis

Color-Luster: White, bluish white, brownish white

Crystal System: Hexagonal

Density: 2.660 (Meas.), 2.670 (Calc.)

First Reported: 1952

Habit: Tufts, rosettes, and botryoidal masses of fibers. Fibers elongated parallel to c-axis. Individual fibers average about 0.08 mm in length and 0.01 mm in thickness

Hardness: Not determined

Lattice Constants: a = 9.43, c = 16.00 (Å)

Mode of Occurrence: As coating on other minerals, such as muscovite, quartz, variscite, and frondelite at the Sapucaia pegmatite mine, Minas Geraes, Brazil

Optical Constants: $\omega = 1.631$, $\varepsilon = 1.652$; (+)

Space Group: $P6_222$

Strongest Diffraction Lines: 5.72(100), 7.28(90), 3.244(60)

Z: 3

Ref.: Lindberg ML, Murata KJ (1953) Am Miner 38:263–270. Roberts WL, Rapp GR, Weber J (1974) Encyclopedia of minerals, New York, Van Nostrand Reinhold, p 203

Fairfieldite Series

$Ca_2(Mn^{2+}, Fe^{2+})(PO_4)_2 \cdot 2H_2O$

Analyses (%): CaO 28.8–30.8, MnO 12.4–19.7, FeO 3.4–7.1, P_2O_5 37.7–39.6, H_2O 9.7–10.0

Class: $\bar{1}$

Cleavage: {001} perfect, {010} good, {1$\bar{1}$0} distinct. Fracture uneven. Brittle

Color-Luster: White; also yellowish, greenish white. Transparent. Vitreous to pearly. Streak white

Crystal System: Triclinic

Density: 3.08 (Meas.), 3.09 (Calc.)

First Reported: 1879

Habit: Crystals prismatic to equant. Usually lamellar, foliated, or fibrous; in radiating masses

Hardness: 3.5

Lattice Constants: a = 5.77, b = 6.56, c = 5.47 (Å); $\alpha = 102.08°$, $\beta = 108.71°$, $\gamma = 90.09°$

Mode of Occurrence: The mineral occurs in granite pegmatite at Branchville, Fairfield County, Connecticut; at Buckfield and Poland, Maine; and at Hühnerkobel, Bavaria, Germany

Optical Constants: $\alpha = 1.636–1.640$; $\beta = 1.644–1.650$; $\gamma = 1.654–1.660$; (+)2V $\simeq 86°$

Space Group: $P\bar{1}$
Strongest Diffraction Lines: 3.23(100), 6.40(90), 3.03(80)
Z: 1
Ref.: Roberts WL, Rapp GR, Weber J (1974) Encyclopedia of minerals, New York, Van Nostrand Reinhold, p 204

Farringtonite
$Mg_3(PO_4)_2$

Analyses (%): P_2O_5 37.6–50, MgO 41.6–49, Fe_2O_3 5.1–8.3
Class: 2/m
Cleavage: {100} fair to good, {010} fair to good
Color-Luster: Wax-white to yellow
Crystal System: Monoclinic
Density: 2.76 (Calc.)
First Reported: 1961
Habit: Massive
Hardness: Not determined
Lattice Constants: a = 7.60, b = 8.23, c = 5.08 (Å); $\beta = 94.08°$
Mode of Occurrence: Occurs peripheral to olivine nodules in the Springwater pallasite meteorite, from near Springwater, Saskatchewan, Canada
Optical Constants: $\alpha = 1.540$, $\beta = 1.544$, $\gamma = 1.559$; (+)2V = 54°–55°
Space Group: $P2_1/n$
Strongest Diffraction Lines: 3.44(100), 3.85(90), 4.08(50)
Z: 2
Ref.: Dufresne ER, Roy SK (1961) Geochim Cosmochim Acta 24:198; *Idem* (1961) Am Miner 46:1513. Roberts WL, Rapp GR, Weber J (1974) Encyclopedia of minerals, New York, Van Nostrand Reinhold, p 205

Faustite
(Inadequately described mineral; the zinc analog of Turquoise)
$(Zn, Cu)Al_6(PO_4)_4(OH)_8 \cdot 5H_2O$

Analyses (%): P_2O_5 34.8, CuO 1.6, ZnO 7.7, Al_2O_3 35.3, H_2O 18.8
Cleavage: Fracture slightly conchoidal to smooth. Brittle
Color-Luster: Apple green; waxy to dull. Opaque. Streak white to pale yellow-green
Crystal System: Triclinic
Density: 2.92 (Meas.)
First Reported: 1953
Habit: Massive, compact
Hardness: 5.5
Mode of Occurrence: Occurs as a vein filling and as nodules in altered shales at Copper King mine, Maggie Creek district, Eureka County, Nevada
Strongest Diffraction Lines: 3.68(100), 2.89(80), 6.70(70)
Ref.: Erd RC, Foster MD, Proctor PD (1953) Am Miner 38:964–972. Roberts WL, Rapp GR, Weber J (1974) Encyclopedia of minerals, New York, Van Nostrand Reinhold, p 205–206

Fenghuangite = Britholite

Fermorite (Apatite Group)
$(Ca, Sr)_5\{(As, P)O_4\}_3(F, OH)$

Analyses (%): CaO 44.3, SrO 9.93, P_2O_5 20.1, As_2O_5 25.2, F 0.83
Class: 6/m
Cleavage: Fracture uneven
Color-Luster: Pale pinkish white. Translucent. Greasy. Streak white
Crystal System: Hexagonal
Density: 3.518
First Reported: 1910
Habit: Massive, granular
Hardness: 5
Lattice Constants: a = 9.55, c = 6.98 (Å)
Mode of Occurrence: Occurs in veinlets in manganese ore at Sitipár, Chindwara district, Central Provinces, India
Optical Constants: Mean index = 1.660; uniaxial (—)
Space Group: $P6_3/m$

Strongest Diffraction Lines: 2.86(100), 2.75(60), 3.49(50)

Z: 2

Ref.: Roberts WL, Rapp GR, Weber J (1974) Encyclopedia of minerals, New York, Van Nostrand Reinhold, p 208–209

Ferri-Sicklerite
(Intermediate Member of the Triphylite-Heterosite – [Sicklerite-Ferri-Sicklerite] Series)
$(Li, Fe^{3+}, Mn^{2+})PO_4$

Analyses (%): Li_2O 3.6, Na_2O 0.85, MnO 15.2–19.1, CaO 1.36–1.88, MgO 0.11–1.70, Fe_2O_3 27.2–29.1, H_2O 1.10, P_2O_5 44.7

Class: 2/m 2/m 2/m

Cleavage: {100} good

Color-Luster: Yellowish brown to dark brown. Opaque. Dull

Crystal System: Orthorhombic

Density: 3.2–3.4 (Meas.)

First Reported: 1937

Habit: Massive

Hardness: ~4

Lattice Constants: a = 5.939, b = 10.086, c = 4.787 (Å)

Mode of Occurrence: Occurs as an alteration of triphylite in the zone of weathering in granite pegmatites. Found at numerous places in the Keystone and Custer districts, Black Hills, South Dakota; in pegmatites in New Hampshire and Maine; at Varuträsk, Sweden; and in France, Germany, and Finland

Optical Constants: $\alpha = 1.750$, $\beta = 1.770$, $\gamma = 1.780$; (—)2V = medium large

Space Group: Pmnb

Strongest Diffraction Lines: 3.01(100), 2.53(100), 4.32(95)

Z: 4

Ref.: Roberts WL, Rapp GR, Weber J (1974) Encyclopedia of minerals, New York, Van Nostrand Reinhold, p 210

Fillowite Series
$Na_2Ca(Mn,Fe)^{2+}_7(PO_4)_6$

Analyses (%): P_2O_5 39.1–39.7, Na_2O 5.59, CaO 3.63–4.08, FeO 9.33–9.69, MnO 39.50, P_2O_5 39.1–39.7, H_2O 1.62

Class: $\bar{3}$

Cleavage: {001} perfect. Fracture uneven. Brittle

Color-Luster: Yellowish to reddish brown. Transparent to translucent. Subresinous to greasy. Streak white

Crystal System: Rhombohedral

Density: 3.43 (Meas.), 3.42 (Calc.)

First Reported: 1879

Habit: Crystals rhombohedral. As granular crystalline masses

Hardness: 4.5

Lattice Constants: a = 15.282, c = 43.507 (Å); $\alpha_{rh} = 16.91$, $\alpha = 53.52°$

Mode of Occurrence: Occurs in granite pegmatite associated with reddingite, fairfieldite, and triploidite at Branchville, Connecticut

Optical Constants: $\omega = 1.671$, $\varepsilon = 1.676$; (+)2V = small

Space Group: $R\bar{3}$

Strongest Diffraction Lines: 2.814(100), 3.017(70), 3.640(65)

Z: = 18

Ref.: Roberts WL, Rapp GR, Weber J (1974) Encyclopedia of minerals, New York, Van Nostrand Reinhold, p 214. Fisher DJ (1965) Am Miner 50:1647–1669. Araki T, Moore PB (1981) Am Miner 66:827–842

Finchenite = Variety of apatite

Florencite
$CeAl_3(PO_4)_2(OH)_6$

Analyses (%): P_2O_5 23.3–33.0, Al_2O_3 26.0–32.3, Re_2O_3 17.6–31.7, CaO 0.5–4.10, SrO 2.40–9.00, Fe_2O_3 1.4–6.62, H_2O 10.9–12.3

Class: $\bar{3}$ 2/m

Cleavage: {0001} good, {11$\bar{2}$0} in traces. Fracture subconchoidal to splintery

Color-Luster: Pink, pale yellow. Transparent to translucent. Resinous to greasy

Crystal System: Hexagonal

Density: 3.457–3.71 (Meas.)

First Reported: 1899

Habit: As small rhombohedral or pseudocubic crystals, and as rounded grains

Hardness: 5–6

Lattice Constants: a = 6.974, c = 16.36 (Å)

Mode of Occurrence: Reported occurrences of florencite are numerous; the reports describe the mineral as associated with a wide variety of parageneses. Florencite was first described as a rare constituent associated with monazite and xenotime in placer samples from Ouro Preto in Minas Gerais, Brazil, and as an accessory mineral in mica schists of the same area. Florencite was also observed associated with goyazite and gorceixite in diamond-bearing sands of the Jequitinhonha River near Diamantina, Brazil.
Other reported occurrences of florencite in alluvial placers are in Gabon and France, in the diamond placers of Ubangi, Central African Republic, and in the Ural Mountains, U.S.S.R. Florencite is abundant in the gold-sulfide deposits and associated placers of the gold-ore fields in the Donets Basin, U.S.S.R., and in the North Baikal Mountains, U.S.S.R. In these gold deposits, florencite is considered a hydrothermal mineral probably precipitated from rare-earth carbonate complexes. This origin is possibly similar to the gold-cinnabar-florencite mineralization of the Bondoukou region, Ivory Coast. Florencite has also been found in the Shituru (copper) deposit of Shaba in Zaire and in the fossiliferous shales of the Little Dal Group, Mackenzie Mountains, Northwest Territories, Canada.

Optical Constants: $\omega = 1.695, 1.680$, $\varepsilon = 1.705$; (+)

Space Group: R$\bar{3}$m

Strongest Diffraction Lines: 2.95(100), 5.71(70), 3.50(70)

Z: 3

Ref.: Roberts WL, Rapp GR, Weber J (1974) Encyclopedia of minerals, New York, Van Nostrand Reinhold, p 216. Lefebvre JJ, Gasparrini C (1980) Can Mineral 18:301–311

Fluellite (Kreuzbergite)
$Al_2PO_4F_2OH \cdot 7H_2O$

Analyses: None reported

Class: 2/m 2/m 2/m

Cleavage: {010} indistinct, {111} indistinct

Color-Luster: Colorless, pale yellow, white. Transparent. Vitreous

Crystal System: Orthorhombic

Density: 2.18 (Meas.), 2.16 (Calc.)

First Reported: 1824 (re-defined, 1920)

Habit: Crystals small, dipyramidal {111}, often modified by small {010}

Hardness: 3

Lattice Constants: a = 8.546, b = 11.222, c = 21.158 (Å)

Mode of Occurrence: Found as minute crystals on quartz, associated with fluorite, apatite (tavistockite), arsenopyrite, and torbernite at Stenna Gwyn, near St. Austell, Cornwall, England. It also occurs with secondary iron phosphate minerals in pegmatite at Hagendorf and Pleystein, Bavaria, and as an alteration of triplite at Königswart, Bohemia, Czechoslovakia

Optical Constants: $\alpha = 1.473$–1.490, $\beta = 1.490$–1.496, $\gamma = 1.506$–1.511; (+)2V = very large

Space Group: Fddd

Strongest Diffraction Lines: 6.48(100), 3.24(70), 3.09(60)

Z: 8

Ref.: Roberts WL, Rapp GR, Weber J (1974) Encyclopedia of minerals, New

York, Van Nostrand Reinhold, p 216–217. Guy B, Jeffrey GA (1966) Am Miner 51:1579–1592

Fluorapatite (Apatite Series)
$Ca_5(PO_4)_3F$

Analyses (%): P_2O_5 38.1–42.0, CaO 42.4–55.6, F 1.2–3.7, CO_2 0–3.4

Class: 6/m

Cleavage: {0001} indistinct, {10$\bar{1}$0} trace. Fracture conchoidal to uneven. Brittle

Color-Luster: Colorless, white, gray, yellow to yellowish green, pale to dark green, pale to dark bluish green, pale to very dark blue, violet-blue, violet, purple, various shades of red and brown. Transparent to opaque. Vitreous to subresinous, rarely silky. Often fluorescent, phosphorescent or thermoluminescent. Streak white

Crystal System: Hexagonal

Density: 3.1–3.2 (Meas.), 3.201 (Calc.)

First Reported: 1860, Antecedent names, Eupyrchroite, 1838; Francolite, 1850

Habit: Crystals short to long prismatic or thin to thick tabular, often complex. Also massive, compact to coarse granular; globular or reniform; stalactitic; fibrous, oolitic; earthy; as nodular concretions and as bedded deposits. Twinning on {11$\bar{2}$1} and {10$\bar{1}$3}, rare

Hardness: 5

Lattice Constants: a = 9.36, c = 6.88 (Å)

Mode of Occurrence: Occurs chiefly as a common and widespread mineral in many igneous rocks; in pegmatites; in hydrothermal and Alpine-type veins; as bedded deposits of marine origin; in metamorphic rocks; and as detrital deposits. Found as deep blue anhedral crystals two feet or more in diameter and several feet in length at the Hugo mine Keystone, South Dakota; and as superb crystals in Maine, New York, New Hampshire, Connecticut, South Dakota, and California. Fine specimens also are found at many localities in Ontario and Quebec, Canada; in Durango, Mexico; and in Bolivia, Brazil, Spain, Germany, Czechoslovakia, Switzerland, Austria, Italy, France, England, U.S.S.R., Sweden, Burma, and Japan. Recently found as magnificent crystals at Panasqueira, Portugal

Optical Constants: $\omega = 1.6357$, $\varepsilon = 1.6328$ (Na); (—)

Space Group: $P6_3/m$

Strongest Diffraction Lines: 2.800(100), 2.702(60), 2.772(55)

Z: 2

Ref.: Roberts WL, Rapp GR, Weber J (1974) Encyclopedia of minerals, New York, Van Nostrand Reinhold, p 217–218

Foggite
$CaAl(PO_4)(OH)_2 \cdot H_2O$

Analyses (%): CaO 25.1, Al_2O_3 22.9, P_2O_5 32.3

Cleavage: Perfect parallel to {010}, good parallel {100}

Color-Luster: Colorless to white

Crystal System: Orthorhombic

Density: 2.78 (Meas.), 2.771 (Calc.)

First Reported: 1975

Habit: Bunched platy aggregates hinged on a common axis like the pages of a partly opened book. Crystals are thin tabular parallel to b{010}, slightly elongated parallel to [001]. Crystals rarely exceed 0.2 mm in maximum dimension. Forms typically observed are c{001}, b{010}, and a{100}

Hardness: 4

Lattice Constants: a = 9.270, b = 21.324, c = 5.190 (Å)

Mode of Occurrence: Occurs as foliated plates upon quartz, childrenite and siderite in the Palermo No. 1 pegmatite, North Groton, Grafton County, New Hampshire

Optical Constants: $\alpha = 1.610$, $\beta = 1.610$, $\gamma = 1.611$; (+)2V = 40°–45°

Space Group: $A2_122$

Strongest Diffraction Lines: 4.24(100), 2.603(80), 4.20(50), 6.96(50)

Z: 8

Ref.: Moore PB, Irving AJ, Kampf AR (1975) Am Miner 60:957–964. Moore PB, Kampf AR, Araki T (1975) Am Miner 60:965–971

Foucherite = Bořickite?
(Inadequately described mineral)
Near $Ca(Fe, Al)_4(PO_4)_2(OH)_8 \cdot 7H_2O$

Analyses (%): CaO 7.71, Al_2O_3 4.5, Fe_2O_3 38.5, P_2O_5 19.5, H_2O 28.5

First Reported: 1910 (Incorrect analysis reported, 1836; called Fuchérite, 1867)

Francoanellite
$H_6K_3Al_5(PO_4)_8 \cdot 13H_2O$

Analyses (%): P_2O_5 45.0, Al_2O_3 18.2, Fe_2O_3 2.82, Na_2O 0.98, K_2O 9.75, H_2O 22.8

Color: Yellowish white

Crystal System: Rhombohedral

Density: 2.26

First Reported: 1976

Habit: Small nodular aggregates, 0.1 to 0.2 mm in diameter, or as earthy pulverulent masses

Hardness: Very soft

Lattice Constants: a = 8.71, c = 82.8 (Å)

Mode of Occurrence: Occurs at the contact of "terra rossa", with bat guano in the karst cave of 'Grotte di Castellana', Puglia, Italy

Optical Constants: $\omega = 1.510$, $\varepsilon = 1.515$; Uniaxial pos.

Space Group: R3c, or R$\bar{3}$c

Strongest Diffraction Lines: 13.83(100), 3.41(65), 2.799(62), 7.44(60)

Ref.: Balenzano F, Dell'Anna L, Dipiero M (1976) Neues Jahrb Miner Monatsh 49–57

Francolite = Carbonate fluorapatite

Frondelite
(Frondelite-Rockbridgeite Series)
$(Mg^{2+}, Fe^{2+})Fe_4^{3+}(PO_4)_3(OH)_5$

Analyses (%): MnO 4.1–7.74, Fe_2O_3 48.8–55.2, P_2O_5 31.4, H_2O 7.3

Class: 222

Cleavage: {100} perfect, {010} distinct, {001} distinct. Fracture uneven. Brittle.

Color-Luster: Dark olive green to greenish black, becoming brownish-green to reddish-brown on oxidation. Aggregates often exhibit concentric color banding. Subtranslucent. Vitreous to dull

Crystal System: Orthorhombic

Density: 3.476 (Meas.), 3.473 (Calc.)

First Reported: 1949

Habit: As radial-fibrous botryoidal crusts and masses of fine columnar structure

Hardness: 4.5

Lattice Constants: a = 13.89, b = 17.01, c = 5.21 (Å)

Mode of Occurrence: Occurs as a secondary mineral as an alteration product of triphylite or other manganese-iron phosphates in granite pegmatites. Found with metastrengite at the Fletcher mine, North Groton, New Hampshire, and in pegmatites in the Black Hills, South Dakota. It also occurs at the Sapucaia pegmatite, Minas Geraes, Brazil

Optical Constants: $\alpha = 1.860$, $\beta = 1.880$, $\gamma = 1.893$; (−)2V = moderate

Space Group: $B22_12$

Strongest Diffraction Lines: 3.20(100), 3.38(50), 1.598(50)

Z: 4

Ref.: Roberts WL, Rapp GR, Weber J (1974) Encyclopedia of minerals, New York, Van Nostrand Reinhold, p 224

Furongite
$Al_2(UO_2)(PO_4)_2(OH)_2 \cdot 8H_2O$

Analyses (%): UO_3 41.4, P_2O_5 18.5, SiO_2 0.80, Al_2O_3 13.7, Fe_2O_3 0.61, BaO 0.59, H_2O 14.5

Color-Luster: Bright yellow to lemon yellow, pale yellow under the microscope, translucent with vitreous luster

Crystal System: Triclinic

Density: 2.82–2.90 (Meas.), 2.85 (Calc.)

First Reported: 1976

Habit: Tabular crystals and dense cryptocrystalline aggregates; radioactive

Hardness: Brittle

Lattice Constants: a = 17.87, b = 14.18, c = 12.18 (Å); $\alpha = 67.8°$, $\beta = 77.5°$, $\gamma = 79.9°$

Mode of Occurrence: Found in the oxidized zone of an alluvial-type uranium deposit in the carbonaceous shale, western Hunan, China. Associated minerals include utahlite (variscite), evansite, opal, halloysite, limonite, and autunite

Optical Constants: $\alpha = 1.543–1.549$, $\beta = 1.564–1.567$, $\gamma = 1.570–1.575$; (−)2V = 65°; oblique extinction with small extinction angle. Strong pale yellowish-green fluorescence in UV light

Space Group: P1 or P$\bar{1}$

Strongest Diffraction Lines: 10.2(100), 8.62(80), 4.310(50), 3.639(40)

Z: 7

Ref.: Hunan 230 Laboratory, Hunan 305 Geological Team, and X-ray Laboratory, Wuhan Geological Institute (1976) Acta Geol Sinica (in Chinese with English abstract), 2:203–204

Gatumbaite

$CaAl_2(PO_4)_2(OH)_2 \cdot H_2O$

Analyses (%): Al_2O_3 28.1, Fe_2O_3 2.20, CaO 17.4, P_2O_5 41.4, H_2O 10.7

Cleavage: Brittle, giving fibrous asbestiform splinters with longitudinal cleavage and some cross-fractures

Color-Luster: Pure white, luster pearly

Crystal System: Monoclinic

Density: 2.92 (Meas.), 2.95 (Calc.)

First Reported: 1977

Habit: Sheaves and rosettes with radial fibrous structure up to 3–10 mm in diameter

Hardness: Less than 5

Lattice Constants: a = 6.907, b = 5.095, c = 10.76 (Å); $\beta = 91°$

Mode of Occurrence: Found in the Buranga pegmatite, near Gatumba, Gisenyi Province, Rwanda; in association with trolleite, scorzalite, apatite, bjarebyite and an undescribed Be phosphate

Optical Constants: $\alpha = 1.610$, $\beta = 1.63$, $\gamma = 1.639$; (—)2V = 65°; optical plane perpendicular to (010)

Space Group: P2/m, P2 or Pm

Strongest Diffraction Lines: 4.210(100), 2.241(100), 1.726(75), 2.772(70)

Z: 2

Ref.: von Knorring O, Fransolet AM (1977) Neues Jahrb. Mineral Monatsh, 561–568

Giniite

$Fe^{2+}(Fe^{3+})_4(PO_4)_4(OH)_2 \cdot 2H_2O$

Analyses (%): P_2O_5 36.99, Fe_2O_3 (total Fe) 46.07, Al_2O_3 1.20, MgO 0.68, MnO 0.63, H_2O 6.45

Cleavage: None. Fracture conchoidal

Color-Luster: Blackish-green to blackish-brown; streak olive; luster vitreous to greasy

Crystal System: Orthorhombic

Density: 3.42 (Calc.), 3.41 (Meas.)

First Reported: 1980

Habit: Idiomorphic crystals with {010}, {150}, {041} dominant, and with {310}, {001}, and {100} also present

Hardness: 3–4

Lattice Constants: a = 10.365, b = 26.582, c = 5.162 (Å)

Mode of Occurrence: Occurs in pegmatite at Sandamab near Usakos, Namibia, associated with hureaulite, tavorite, leucophosphite and an unknown phosphate, formed by the alteration of triphylite

Optical Constants: $\alpha = 1.775$, $\beta = 1.803$, $\gamma = 1.812$; (−)2V = 55°

Space Group: Cmm2, C222, or Cmmm
Strongest Diffraction Lines: 3.36, 2.04, 1.604
Z: 4
Ref.: Keller P (1980) Neues Jahrb Mineral. Monatsh, 49–56

Goedkenite
$(Sr, Ca)_2Al(PO_4)_2(OH)$

Analyses (%): SrO 32.7, CaO 11.4, Al_2O_3 13.7, P_2O_5 34.9, MgO 0.17
Cleavage: {100} fair
Color-Luster: Colorless to pale yellow, luster subadamantine
Crystal System: Monoclinic
Density: 3.83 (Calc.)
First Reported: 1975
Habit: Lozenge-shaped to spear-shaped crystals, tabular parallel to {001} and slightly elongated parallel to [100]. Crystals always warped, the c{001} face becoming increasingly bent toward the terminus, imparting "feathery" tips to the crystals. Observed forms include c{001}, l{011}, e{012}, d{$\bar{1}$01} and g{111}
Hardness: 5
Lattice Constants: a = 8.45, b = 5.74, c = 7.26 (Å); $\beta = 113.7°$
Mode of Occurrence: Found in close association with palermoite in the Palermo No. 1 pegmatite, North Groton, New Hampshire
Optical Constants: $\alpha = 1.669$, $\beta = 1.673$, $\gamma = 1.692$; (+)2V = 45°–50°
Space Group: $P2_1/m$
Strongest Diffraction Lines: 3.061(100), 2.585(70), 2.841(60), 7.76(50), 1.609(50)
Z: 2
Ref.: Moore PB, Irving AJ, Kampf AR (1975) Am Miner 60:957–964

Gormanite
(Gormanite-Souzalite Series)
$(Fe^{2+}, Mg)_3(Al, Fe^{3+})_4(PO_4)_4(OH)_6 \cdot 2H_2O$

Analyses (%): MgO 6.6–7.8, FeO 14.7–17.4, Al_2O_3 25.5–27.0, Fe_2O_3 3.82, P_2O_5 37.2–38.5, H_2O 11.4
Cleavage: Single poor cleavage parallel to {001}
Color-Luster: Blue-green, luster vitreous, pale green streak
Crystal System: Triclinic
Density: 3.12 (Meas.), 3.13 (Calc.)
First Reported: 1981
Habit: Elongate crystals (up to 3 × 0.5 × 0.1 mm) in radial aggregates and as blade-like crystals. The blades are elongated parallel to {010}, the plane of the blade being {001}. Other forms present include {$\bar{1}$00}, {102}, {$\bar{1}$02} and {010}
Hardness: 4–5
Lattice Constants: a = 11.79, b = 5.11, c = 13.61 (Å); $\alpha = 90°$, $\beta = 99°$, $\gamma = 90°$
Mode of Occurrence: Common in many outcrops of the phosphate-ironstone member in the Rapid Creek and Fish River areas, Yukon Territory, Canada. In thin section, it resembles a chlorite group mineral. It occurs with quartz, siderite, ludlamite, arrojadite, kryzhanovskite, oxidized vivianite and souzalite. Many crystals show strong chemical zoning with the mineral changing from gormanite to souzalite. Unanalyzed specimens should best be called gormanite-souzalite because the two species can only be differentiated by chemical analysis
Optical Constants: $\alpha = 1.619$, $\beta = 1.653$, $\gamma = 1.660$; Biaxial neg., dispersion very strong; Pleochroism, X colorless, Y blue, Z colorless
Space Group: P1 or P$\bar{1}$
Strongest Diffraction Lines: 3.395(100), 2.554(90), 2.925(80), 4.761(60), 3.154(60)
Z: 2
Ref.: Sturman BD, Mandarino JA, Mrose ME, Dunn PD (1981) Can Miner 19:381–387

Goyazite (Goyazite-Crandallite Series)
$(Sr, Ca)Al_3(PO_4)_2(OH)_5 \cdot H_2O$

Analyses (%): SrO 17.3–19.8, Al_2O_3 32.3–50.7, P_2O_5 14.9–28.9, CaO (?), H_2O 12.0–16.7

Class: $\bar{3}$ 2/m

Cleavage: {0001} perfect

Color-Luster: Colorless, yellowish, pink. Transparent. Resinous to greasy, pearly on base

Crystal System: Hexagonal

Density: 3.26 (Meas.), 3.29 (Calc.)

First Reported: 1884 (Hamlinite, 1890; Bowmanite, 1905)

Habit: Crystals small, rhombohedral, pseudocubic, or tabular. Rhombohedral faces commonly striated horizontally. Also as rounded grains or pebbles

Hardness: 4.5–5

Lattice Constants: a = 6.97, c = 16.51 (Å)

Mode of Occurrence: Occurs abundantly associated with ferberite throughout the Nederland tungsten belt of Boulder County, Colorado, especially at the Eagle Rock mine; found sparingly in pegmatites in Maine, New Hampshire, and South Dakota; near Diamantina, Minas Geraes, Brazil; at Lengenbach and in the Simplon tunnel, Switzerland; and in the U.S.S.R.

Optical Constants: $\omega = 1.620–1.653$, $\varepsilon = 1.630–1.661$; (+)

Space Group: $R\bar{3}m$

Strongest Diffraction Lines: 5.73(100), 2.96(100), 3.49(80)

Z: 3

Ref.: Roberts WL, Rapp GR, Weber J (1974) Encyclopedia of minerals, New York, Van Nostrand Reinhold, p 244–245

Graftonite
$(Fe^{2+}, Mn^{2+}, Ca)_3(PO_4)_2$

Analyses (%): P_2O_5 38.9–41.7, CaO 4.5–12.8, FeO 27.8–32.6, MnO 15.7–25.5

Class: 2/m

Cleavage: {010} distinct, {100} indistinct. Fracture subconchoidal to uneven

Color-Luster: Salmon pink to reddish brown; often dark brown from alteration. Vitreous to resinous. Translucent

Crystal System: Monoclinic

Density: 3.67–3.79 (Meas.), 3.72 (Calc.), also 3.66 (Calc.)

First Reported: 1900 (Repossite, 1935)

Habit: Massive, cleavable; rarely as rough composite crystals

Hardness: 5

Lattice Constants: a = 8.87, b = 11.57, c = 6.17 (Å); $\beta = 99.2°$

Mode of Occurrence: Occurs as a primary mineral in granite pegmatite as fresh masses as much as 3 feet in diameter associated with sarcopside and several secondary phosphates at the Bull Moose mine, near Custer, South Dakota. Also as clove brown masses as much as 18 inches in diameter at the Ross mine; in masses up to 6 inches across at the Victory mine near Custer; and in masses as much as 4 inches across associated with arrojadite at the Nickel Plate mine, Keystone, South Dakota. Also found in granite pegmatite at the Rice mine, Palermo mine, and Melvin Mountain, New Hampshire; at Greenwood, Maine; in the Kondakovo district, Eastern Siberia, U.S.S.R.; at Olgiasca, Lake Como, Italy; and at Brissago, Tessin, Switzerland

Optical Constants: $\alpha = 1.695–1.709$, $\beta = 1.699–1.714$, $\gamma = 1.719–1.736$; (+)2V = small, 43°, 60°

Space Group: $P2_1/c$

Strongest Diffraction Lines: 2.86(100), 3.50(90), 2.715(70)

Z: 4

Ref.: Lindberg ML (1950) Am Miner 35:59–76. Calvo C (1968) Am Miner 53:742–750. Roberts WL, Rapp GR Jr, Weber J (1974) Encyclopedia of Minerals, New York: Van Nostrand Reinhold

Grayite (cf. Brockite)
$(Th, Pb, Ca)PO_4 \cdot H_2O$

Analyses (%): (only rough approximations reported) Th >10, Si 7–10, Ca 3–10, Fe 1.5–10, Al 3–7, Mg 1.5–3, Ti 1.5–3, P 3–7

Class: 622

Cleavage: None. Fracture conchoidal

Color-Luster: Pale yellow; reddish brown. Resinous

Crystal System: Hexagonal

Density: 3.7–4.3

First Reported: 1961

Habit: Massive; very fine grained; also powdery

Hardness: 3–4

Mode of Occurrence: Occurs in a small vein in Precambrian rocks southwest of Gunnison in Gunnison County, and in the West Mountains of south-central Colorado; in the limestone member of the Pliocene Moonstone formation in Fremont County, Wyoming; and in the Mtoko district, Zimbabwe

Space Group: $P6_222$

Strongest Diffraction Lines: 3.04(100), 2.82(80), 2.14(80)

Z: 3

Ref.: Dooley JR Jr, Hathaway JC (1961) USGS Prof. Paper 424-C, p 339–341. Roberts WL, Rapp GR, Weber J (1974) Encyclopedia of minerals, New York, Van Nostrand Reinhold, p 247–248

Griphite
$(Mn, Na, Li)_6CaFeAl_2(PO_4)_6(F_2OH)_2$

Analyses (%): Na_2O 3.47–5.52, CaO 5.98–7.47, FeO 1.97–10.7, MnO 18.1–29.6, Al_2O_3 8.74–10.1, Fe_2O_3 2.36–6.95, P_2O_5 38.5–39.7, F 0–3.03, H_2O 2.62–4.29

Class: $2/m\,\bar{3}$

Cleavage: None. Fracture uneven to conchoidal. Brittle

Color-Luster: Dark brown to brownish black. Translucent. Resinous to vitreous

Crystal System: Cubic

Density: 3.40 (Meas.), 3.399 (Calc.)

First Reported: 1891

Habit: Massive, compact

Hardness: 5.5

Lattice Constants: a = 12.222 (Å)

Mode of Occurrence: The mineral occurs as masses ranging upwards of 50 pounds in weight in granitic pegmatite at the Riverton lode (Everly mine), and as nodular masses as much as 6 feet in diameter at the Sitting Bull mine, near Keystone, Pennington County, South Dakota. It also is found at Mt. Ida, Northern Territory, Australia, and in Turkestan, U.S.S.R.

Optical Constants: N = 1.63–1.66

Space Group: $P2_1/a\bar{3}$

Strongest Diffraction Lines: (Heated) 2.73(100), 3.04(30), 2.95(30). (Unheated), 2.75(100), 2.98(30), 2.00(30)

Z: 8

Ref.: Roberts WL, Rapp GR, Weber J (1974) Encyclopedia of minerals, New York, Van Nostrand Reinhold, p 248. McConnell D (1942) Am Miner 27: 452–461. Peacor DR, Simmons WB Jr (1972) Am Miner 57:269–272

Gutsevichite
(Inadequately described mineral)
$(Al, Fe)_3(PO_4, VO_4)_2(OH)_3 \cdot 8H_2O(?)$

Analyses (%): P_2O_5 16.0–17.6, CaO 2.7, Al_2O_3 24–26, Fe_2O_3 2.0–4.7, V_2O_4 2.3–2.9, V_2O_5 11.7–14.1, SiO_2 1–3, H_2O 30–33

Cleavage: Not determined

Color-Luster: Yellowish green to greenish brown to dark brown. Translucent. Waxy to dull

Crystal System: Unknown

Density: 1.90–2.00 (Meas.)

First Reported: 1959

Habit: Massive; as crusts, concretions, and cavity fillings

Hardness: 2.5

Mode of Occurrence: Occurs in the zone of oxidation of vanadium-containing shales of the Middle Cambrian in northwestern Kazakhstan

Optical Constants: N = 1.560–1.575

Strongest Diffraction Lines: 4.082(100), 2.506(90), 1.820(80)

Ref.: Ankinovich EA (1961) Memoir All-Union Min Soc 90–104; *Idem* (1961) Am Miner 46:1200. Roberts WL, Rapp GR, Weber J (1974) Encyclopedia of minerals, New York, Van Nostrand Reinhold, p 252

Hagendorfite

$(Na, Ca)(Fe^{2+}, Mn^{2+})_2(PO_4)_2$

Analyses (%): P_2O_5 42.0, Na_2O 8.5–9.7, CaO 1.4–1.7, MgO 0.7–2.6, FeO 15.4–18.2, MnO 9.8–18.5, Fe_2O_3 13.3–14.5

Class: 2/m

Cleavage: Three; one good, one less good, and the third poor

Color-Luster: Greenish black

Crystal System: Monoclinic

Density: 3.71

First Reported: 1954

Habit: Massive

Hardness: 3.5

Lattice Constants: a = 10.93, b = 12.59, c = 6.52 (Å); $\beta = 98°$

Mode of Occurrence: Occurs associated with triphylite, wolfeite, and hematite at the Hagendorf – South pegmatite, Bavaria; also from Norrö, Sweden

Optical Constants: $\alpha = 1.735$, $\beta = 1.742$, $\gamma = 1.745$; (—)2V

Space Group: $I2_1/a$

Strongest Diffraction Lines: 2.686(100), 2.593(75), 3.42(60)

Ref.: Fisher DJ (1956) Bull Geol Soc Am 67:1694–1695. Roberts WL, Rapp GR, Weber J (1974) Encyclopedia of minerals, New York, Van Nostrand Reinhold, p 254

Hannayite

$(NH_4)_2Mg_3H_4(PO_4)_4 \cdot 8H_2O$

Analyses (%): MgO 18.4–18.9, P_2O_5 44.6–45.7, $(NH_4)_2O$ 8.1, H_2O 28.4

Class: $\bar{1}$

Cleavage: {001} perfect, {110} poor, {1$\bar{1}$0} poor, {130} poor

Color-Luster: Yellowish. Transparent. Vitreous

Crystal System: Triclinic

Density: 2.03 (Meas.), 2.03 (Calc. synthetic)

First Reported: 1878

Habit: As small thin prismatic crystals elongated and striated parallel to c-axis. Synthetic crystals tabular {100}, elongate along c-axis, with prominent {01$\bar{1}$}, modified by narrow {110}

Hardness: Not determined. Soft

Lattice Constants: (synthetic) a = 7.70, b = 11.51, c = 6.70 (Å); $\alpha = 76.0°$, $\beta = 99.8°$, $\gamma = 115.8°$

Mode of Occurrence: Occurs in bat guano associated with struvite, newberyite, and brushite in the Skipton caves, Ballarat, Victoria, Australia

Optical Constants: $\alpha = 1.504–1.555$, $\beta = 1.522–1.572$, $\gamma = 1.539–1.579$; (—)2V = 45°–90°

Space Group: $P\bar{1}$

Strongest Diffraction Lines: 6.96(100), 3.46(75) (synthetic), 5.15(30)

Z: 1

Ref.: Roberts WL, Rapp GR, Weber J (1974) Encyclopedia of minerals, New York, Van Nostrand Reinhold, p 258. Frazier AW, Lehr JR, Smith JP (1963) Am Miner 48:635–641

Herderite

(Herderite-Väyrynenite Series), $(Ca, Mn)Be(PO_4)(F, OH)$

Analyses (%): CaO 30.3–34.8, BeO 15.0–16.5, P_2O_5 37.6–44.3, F 0.08–5.44, H_2O 0.61–7.97

Class: 2/m

Cleavage: {110} interrupted. Fracture subconchoidal

Color-Luster: Colorless to pale yellow or greenish white. Transparent to translucent. Vitreous

Crystal System: Monoclinic

Density: 2.95–3.01 (Meas.), 2.94 (Calc.)

First Reported: 1828

Habit: Crystals stout prismatic or thick tabular. Commonly pseudo-orthorhombic in appearance. Also as botryoidal crusts or aggregats with radial-fibrous structure. Commonly twinned on {001} or {100}

Hardness: 5–5.5

Lattice Constants: a = 4.80, b = 7.68, c = 9.80 (Å); $\beta = 90.10°$

Mode of Occurrence: Occurs as a late-stage hydrothermal mineral in granite pegmatites. Found at the Fletcher and Palermo mines, near North Groton, New Hamsphire; as fine crystals at several places in Maine, especially at Stoneham, Newry, Hebron, Paris, Greenwood, and Buckfield in Oxford County, and at Poland, Topsham, and Auburn, in Androscoggin County. It also occurs as superb crystals up to 5 × 12 cm in size in Minas Geraes, Brazil; at localities in Bavaria and Saxony, Germany; and at Mursinsk, Ural Mountains, U.S.S.R.

Optical Constants: $\alpha = 1.615–0.00059$ (%H), $\beta = 1.634–0.00056$ (%H), $\gamma = 1.644–0.00055$ (%H); (—)2V = 67° ~ 85°

Space Group: $P2_1/c$

Strongest Diffraction Lines: 3.14(100), 2.86(80), 2.20(70)

Z: 4

Ref.: Roberts WL, Rapp GR, Weber J (1974) Encyclopedia of minerals, New York, Van Nostrand Reinhold, p 269. Leavens PB, Dunn PJ, Gaines RV (1978) Am Miner 63:913–917. Moore PB (1982) MAC Short Course Handbook 8:267–291

Heterosite (Heterosite-Pupurite Series)
$(Fe^{3+}, Mn^{3+})PO_4$

Analyses (%): Na_2O 0.72–1.12, CaO 1.16–1.70, MnO 1.53–3.60, Mn_2O_3 12.1–24.4, Fe_2O_3 27.4–38.4, P_2O_5 38.7–43.8, H_2O 1.65–3.74

Class: 2/m 2/m 2/m

Cleavage: {100} good, {010} imperfect. Brittle. Fracture uneven

Color-Luster: Deep rose to reddish purple. Usually externally dark brown to brownish black due to alteration. Subtranslucent to opaque. Dull to satiny. Streak reddish purple

Crystal System: Orthorhombic

Density: 3.409 (Meas.), 3.702 (Calc.)

First Reported: 1825

Habit: Massive

Hardness: 4–4.5

Lattice Constants: a = 5.82, b = 9.68, c = 4.76 (Å)

Mode of Occurrence: Occurs as an alteration of triphylite in the zone of weathering in granite pegmatites. Found at numerous places in the Keystone and Custer districts, Black Hills, South Dakota; in pegmatites in New Hampshire, Maine, and Massachusetts; at Varuträsk, Sweden; and in Portugal, France, Germany, Australia, and Erongo, South West Africa

Optical Constants: $\alpha = 1.86$, $\beta = 1.89$, $\gamma = 1.91$; 2V = large

Space Group: Pmnb

Strongest Diffraction Lines: 3.48(100), 4.29(75), 2.73(75)

Z: 4

Ref.: Roberts WL, Rapp GR, Weber J (1974) Encyclopedia of minerals, New York, Van Nostrand Reinhold, p 271. Eventoff D, Martin R, Peacor DR (1972) Am Miner 57:45–51

Hinsdalite
$(Pb, Sr)Al_3PO_4SO_4(OH)_6$

Analyses (%): SrO 3.11, Al_2O_3 26.5, PbO 31.8, P_2O_5 14.5, SO_3 14.1, H_2O 10.2

Class: $\bar{3}$ 2/m

Cleavage: {0001} perfect

Color-Luster: Nearly colorless, with yellowish or greenish tint. Transparent to translucent. Vitreous to greasy

Crystal System: Hexagonal

Density: 3.65 (Meas.), 4.072 (Calc.)

First Reported: 1911

Habit: Crystals pseudocubic, rhombohedral, or tabular, sometimes with rough or dull faces. Also massive, granular

Hardness: 4.5

Lattice Constants: a = 6.99, c = 16.8 (Å)

Mode of Occurrence: Occurs as excellent microcrystals implanted on enargite or covellite crystals at Butte, Montana; and abundantly in a vein cutting volcanic rocks in the Golden Fleece mine, near Lake City, Hinsdale County, Colorado

Optical Constants: $\omega = 1.671$, $\varepsilon = 1.689$; (+)

Space Group: $R\bar{3}m$

Strongest Diffraction Lines: 2.78(100), 2.96(80), 5.59(65)

Z: 3

Ref.: Palache C, Berman H, Frondel C (1951) Dana's system of mineralogy, 7th edn, vol II. Wiley, New York, p 1004

Holtedahlite
(Mg analog of Satterlyite, and is dimorphous with Althausite):
$Mg_2(PO_4)(OH)$

Analyses (%): P_2O_5 41.19, CO_2 2.06, MgO 50.01, H_2O 6.22

Cleavage: None, fracture uneven

Color-Luster: Colorless with vitreous luster

Crystal System: Hexagonal

Density: 2.94 (Meas.), 2.936 (Calc.)

First reported: 1979

Hardness: 4.5–5

Lattice Constants: a = 11.188, c = 4.975 (Å)

Mode of Occurrence: In irregular patches associated with apatite, althausite, serpentine and talc. Found in serpentine-magnesite deposits, Modum, Norway

Optical Constants: Uniaxial (—), $\omega = 1.599$, $\varepsilon = 1.597$

Space Group: P321, P3ml, or $P\bar{3}ml$

Strongest Diffraction Lines: 2.438(100), 3.722(90), 3.475(50)

Z: 6

Ref.: Raade G, Mladeck MH (1979) 12:283–287

Hopeite
$Zn_3(PO_4)_2 \cdot 4H_2O$
Dimorphous with parahopeite

Analyses (%): ZnO 50.3–54.6, P_2O_5 30.5–32.3, H_2O 15.4–17.4

Class: 2/m 2/m 2/m

Cleavage: {010} perfect, {100}, {001} poor. Fracture uneven. Brittle

Color-Luster: Colorless, white, grayish white, pale yellow. Transparent to translucent. Vitreous, pearly on perfect cleavage. Streak white

Crystal System: Orthorhombic

Density: 3.05 (Meas.), 3.08 (Calc.)

First Reported: 1822

Habit: Crystals short to long prismatic or tabular. Crystal faces commonly irregular. As isolated individuals, aggregates, or crusts. Also compact massive

Hardness: 3.25

Lattice Constants: a = 10.66, b = 18.36, c = 5.04 (Å)

Mode of Occurrence: Occurs as a secondary mineral in zinc-bearing ore deposits. Found as superb crystal groups associated with tarbuttite and other secondary minerals at the Broken Hill mine, Zambia; with spencerite at the Hudson Bay mine, Salmo, British Columbia, Canada; and in small amounts at Altenberg, Belgium

Optical Constants: $\alpha = 1.574$, $\beta = 1.582$, $\gamma = 1.582$; (—)2V = small

Space Group: Pnma

Strongest Diffraction Lines: 9.04(100), 4.57(100), 2.857(100)

Z: 4

Ref.: Roberts WL, Rapp GR, Weber J (1974) Encyclopedia of minerals, New York, Van Nostrand Reinhold, p 279. Hill RJ, Jones JB (1976) Am Miner 61:987–995

Hühnerkobelite

$(Na_2, Ca)(Fe^{2+}, Mn^{2+})_2(PO_4)_2$

Analyses (%): Na_2O 3.73–9.67, CaO 1.35–9.70, MgO 0.68–2.55, FeO 7.09–18.15, Fe_2O_3 14.45–26.49, MnO 6.44–9.79, P_2O_5 39.40–41.93

Class: 2/m

Cleavage: {001} good, {010} distinct

Color-Luster: Greenish black, dull olive green. Translucent. Vitreous

Crystal System: Monoclinic

Density: 3.5–3.6 (Meas.)

First Reported: 1950 (described under the name of Arrojadite, 1942)

Habit: Crystals short prismatic, elongated along [010] or [001] with well-developed prism faces and w{211}, and poor f{132} predominating. Also as granular cleavable masses

Hardness: 5

Lattice Constants: a = 10.89, b = 12.54, c = 6.46 (Å); $\beta = 97.55°$

Mode of Occurrence: The mineral occurs as crystals up to 3.0 mm across, associated with dickinsonite and numerous other phosphates, at the Palermo No. 1 pegmatite, North Groton, New Hampshire. Also found at Norrö on the island of Ranö, Sweden

Optical Constants: $\alpha = 1.718–1.727$, $\gamma = 1.731–1.738$; (+)

Space Group: $I2_1/a$

Strongest Diffraction Lines: 2.703(100), 6.24(70), 3.085(55)

Z: 1

Ref.: Roberts WL, Rapp GR, Weber J (1974) Encyclopedia of minerals, New York, Van Nostrand Reinhold, p 283. Moore PB (1965) Am Miner 50:713–717

Hureaulite

$H_2(Mn, Fe)_5(PO_4)_4 \cdot 4H_2O$

Analyses (%): FeO 4.56–11.1, MnO 32.8–42.3, P_2O_5 38.1, H_2O 12.0–18.0

Class: 2/m

Cleavage: {100} good. Fracture uneven. Brittle

Color-Luster: Pale rose, violet-rose, red, yellowish, orange, orange-red, brownish orange, yellowish to reddish brown, gray, colorless. Transparent to translucent. Bright vitreous to greasy. Streak nearly white

Crystal System: Monoclinic

Density: 3.191 (Meas.), 3.23 (Calc.)

First Reported: 1826

Habit: Crystals short to long prismatic, up to 3.0 cm long; also tabular or equant. Crystals isolated or grouped. Also massive, compact, scaly, or imperfectly fibrous

Hardness: 3.5

Lattice Constants: a = 17.42, b = 9.12, c = 9.50 (Å); $\beta = 96.67°$

Mode of Occurrence: Occurs as a secondary mineral in granitc pegmatites associated with triphylite, lithiophilite, graftonite, and a wide variety of secondary phosphate minerals. Found widespread in the pegmatite districts of the Black Hills, South Dakota, especially at the Bull Moose and Tip Top mines in the Custer area, and at the Big Chief and Hesnard mines near Keystone. It also occurs at Branchville and at Portland, Connecticut; at the Palermo mine, North Groton, New Hampshire; and at Pala, San Diego County, California. Other localities include Huréaux and La Vilate, Haute Vienne, France; Mangualde, Portugal; Hagendorf, Bavaria, Germany; and Michelsdorf, Silesia, Poland

Optical Constants: $\alpha = 1.637$–1.652, $\beta = 1.645$–1.658; $\gamma = 1.649$–1.663; (—)$2V = 75°$

Space Group: C2/c

Strongest Diffraction Lines: 3.14(100), 8.01(70), 2.983(70)

Z: 4

Ref.: Roberts WL, Rapp GR, Weber J (1974) Encyclopedia of minerals, New York, Van Nostrand Reinhold, p 285. Moore PB, Araki T (1973) Am Miner 58:302–307

Hurlbutite
$CaBe_2(PO_4)_2$

Analyses (%): P_2O_5 52.0–56.2, Na_2O 1.5, BeO 19.4–21.3, CaO 21.7, H_2O 2.9

Class: 2/m 2/m 2/m

Cleavage: Fracture conchoidal. Brittle

Color-Luster: Colorless to greenish white; when stained, yellow. Vitreous to greasy. Transparent to translucent. Streak white

Crystal System: Orthorhombic

Density: 2.877 (Meas.), 2.88 (Calc.)

First Reported: 1951

Habit: Crystals stout prismatic [001], relatively large development of c{001} and m{110}. Crystals range in size from 4 to 25 mm or more along [110]. Faces delicately etched and exhibit striations on {110}

Hardness: 6

Lattice Constants: a = 8.29, b = 8.80, c = 7.81 (Å)

Mode of Occurrence: Occurs associated with muscovite, albite, triphylite, light-smoky quartz, and siderite in pegmatite at the Smith mine, Chandler's Mill, Newport, New Hampshire

Optical Constants: $\alpha = 1.595$, $\beta = 1.601$, $\gamma = 1.604$; (—)$2V = 70°$

Space Group: Pmmm

Strongest Diffraction Lines: 3.67(100), 3.03(90), 2.78(90)

Z: 4

Ref.: Mrose ME (1952) Am Miner 37:931–940. Roberts WL, Rapp GR, Weber J (1974) Encyclopedia of minerals, New York, Van Nostrand Reinhold, p 285

Hydroxylapatite (Hydroxyapatite)
$Ca_5(PO_4)_3OH$
Apatite Series

Analyses (%): CaO 44.1–55.8, P_2O_5 38.6–42.2, F 0–1.0, H_2O 1.2–2.9

Class: 6/m

Cleavage: {0001} indistinct, $\{10\bar{1}0\}$ trace. Fracture uneven to conchoidal. Brittle

Color-Luster: See fluorapatite

Crystal System: Hexagonal

Density: 3.08 (Meas.), 3.16 (Calc.)

First Reported: 1856

Habit: See fluorapatite

Hardness: 5

Lattice Constants: a = 9.418, c = 6.884 (Å)

Mode of Occurrence: Various metamorphic rocks, but probably always as an isomorphic variant containing F and/or Cl, and possibly CO_2. Occurs in talc-schist near Holly Springs, Cherokee County, Georgia; in talc-schist in the Hospenthal, Uri, Switzerland; and in diallage-serpentine rock at Rossa in the Val Devero, Ossola, Italy

Optical Constants: $\omega = 1.651$, $\varepsilon = 1.644$ (Na); (—)

Remarks: Isomorphic variants closely approaching $Ca_{10}(PO_4)_6(OH)_2$ are not known to occur naturally. All data given here are for synthetic preparations, and it is questionable whether a crystal of hydroxyapatite has ever been obtained with adequate size to permit single-crystal measurement. (The latter statement stands in contrast to some claims based on casual experimental methods.)

Space Group: $P6_3/m$

Strongest Diffraction Lines: 2.814(100), 2.778(60), 2.720(60)

Z: 2

Ref.: Roberts WL, Rapp GR, Weber J (1974) Encyclopedia of minerals, New York, Van Nostrand Reinhold, p 292. McConnell D (1973) Apatites. Springer-Verlag, Wien, p 89

Isokite

$CaMgPO_4F$

Analyses (%): CaO 29.6, MgO 21.2, P_2O_5 37.4, F 9.5

Class: 2/m

Cleavage: {010} very good

Color-Luster: Colorless to buff or pinkish; silky to slightly pearly. Fluoresces blue under long-wave ultraviolet light

Crystal System: Monoclinic

Density: 3.27 (Meas.), 3.29 (Calc.)

First Reported: 1955

Habit: As fibrous spherulites up to 3.0 mm in diameter; also as crystalline plates showing forms {001}, {100}, {101}, $\{10\bar{1}\}$, and $\{10\bar{2}\}$

Hardness: ~5

Lattice Constants: a = 6.52, b = 8.75; c = 7.51 (Å); $\beta = 121.47°$

Mode of Occurrence: Occurs associated with strontian fluorapatite, ilmenite, monazite, barite, pyrite, sellaite, and pyrochlore in ankeritic rocks of carbonatite plug at Nkumbwa Hill, 15 miles east of Isoka, Zambia

Optical Constants: $\alpha = 1.590$, $\beta = 1.594$, $\gamma = 1.614$; (+)2V = 51°

Space Group: C2/c

Strongest Diffraction Lines: 3.19(100), 3.02(100), 2.63(100)

Z: 4

Ref.: Deans T, McConnell JDC (1955) Miner Mag 30:691–690. Roberts WL, Rapp GR, Weber J (1974) Encyclopedia of minerals, New York, Van Nostrand Reinhold, p 305–306

Jagowerite

$BaAl_2(PO_4)_2(OH)_2$

Analyses (%): Al_2O_3 25.9, BaO 38.4, Fe_2O_3 0.26, P_2O_5 31.4 H_2O 4.09

Cleavage: {100} and {011} good, $\{0\bar{1}1\}$ fair

Color-Luster: light green, vitreous luster

Crystal System: Triclinic

Density: 4.01 (Meas.), 4.05 (Calc.)

First Reported: 1973

Habit: Crystalline masses up to 2.5 cm across

Hardness: 4.5

Lattice Constants: a = 6.049, b = 6.964, c = 4.971 (Å); $\alpha = 116.5°$, $\beta = 86.1°$, $\gamma = 112.6°$

Mode of Occurrence: Found in quartz veins in tension cracks in a carbonaceous argillite, 16 miles north of Hess River, Yukon Territory, Canada. Associated minerals are pyrite and hinsdalite

Optical Constants: $\alpha = 1.672$, $\beta = 1.693$, $\gamma = 1.710$; (—)2V = 80°; fluoresces greenish-white under long-wave UV

Space Group: $P\bar{1}$

Strongest Diffraction Lines: 3.00(100), 3.26(60), 2.94(55), 5.55(40)

Z: 1

Ref.: Meagher EP, Coates ME, Aho AE (1973) Can Miner 12:135–136. Meaher EP, Gibbons CS, Trotter J (1974) Am Miner 59:291–295

Jahnsite Series

$(Mg, Mn, Fe)_2CaMn^{2+}(Fe^{3+})_2(PO_4)_4(OH)_2 \cdot 8H_2O$

Analyses (%): MgO 9.4, CaO 6.6, MnO 8.0, Fe_2O_3 15.1, Al_2O_3 2.1, P_2O_5 32.2, H_2O 18.8

Class: 2/m

Cleavage: {001} good. Brittle

Color-Luster: Nut-brown, purplish-brown, yellow, yellow-orange and greenish yellow. Transparent to translucent. Vitreous to subadamantine

Crystal System: Monoclinic

Density: 2.706, 2.718 (Meas.), 2.715 (Calc.)

First Reported: 1974

Habit: Crystals commonly well developed and euhedral, striated parallel to [010], short to long prismatic parallel to [010], often tabular parallel to a{100}; usually 0.1–0.5 mm, rarely up to 5 mm in length. Frequently as parallel aggregates, twinned by reflection on c{001}

Hardness: 4

Lattice Constants: a = 14,94, b = 7.14, c = 9.93 (Å); $\beta = 110.168°$

Mode of Occurrence: Jahnsite occurs in moderate abundance in granite pegmatites as a late stage product in corroded triphylite-heterosite-ferrisicklerite-rockbridgeite masses, associated with leucophosphite, hureaulite, vivianite, laueite, collinsite, robertsite, and other secondary phosphate minerals. Found in South Dakota at the Tip Top, Bull Moose, White Elephant, and Linwood pegmatites in Custer County, and at the Big Chief, Gap Lode, and Hesnard mines in Pennington County. It also occurs with laueite and strunzite at the Palermo No. 1 mine, North Groton, New Hampshire

Optical Constants: $\alpha = 1.640$, $\beta = 1.658$, $\gamma = 1.670$; Biaxial (−), 2V large

Space Group: P2/a

Strongest Diffraction Lines: 9.27(100), 2.825(80), 4.91(60)

Z: 2

Ref.: Moore PB (1974) Am Miner 59:48–59. Moore, PB, Araki T (1974) Am Miner 59:964–973

Jezekite = Morinite

Johnsomervilleite
(Mg-Fe analog of Fillowite)
$Na_{10}Ca_6Mg_{18}(Fe^{2+}, Mn^{2+})_{25}(PO_4)_{36}$

Analyses (%): FeO 25.5–26.9, MnO 4.9–5.5, MgO 12.7–13.4, CaO 6.0–6.4, Na_2O 3.7–5.4, P_2O_5 44.3–45.0

Cleavage: Perfect basal {001} cleavage; subconchoidal fracture

Color-Luster: Very dark brown with vitreous luster; brown and non-pleochroic in transmitted light

Crystal System: Rhombohedral (hexagonal cell dimensions)

Density: 3.41 (Meas.), 3.35 (Calc.)

First Reported: 1980

Habit: Anhedral grains or pods measuring up to 1.5 mm across

Hardness: 4.5

Lattice Constants: a = 15.00, c = 42.75 (Å)

Mode of Occurrence: Occurs in phosphate-bearing metamorphic segregation pods in Moine gneiss at Loch Quoich area, Scotland. It is considered a primary phase together with green anhedral apatite, graftonite and an unidentified "mineral A"; the associated secondary phosphates include jahnsite, phosphosiderite, rockbridgeite, mitridatite and vivianite

Optical Constants: $\alpha = 1.655$, $\beta = 1.655$; Birefringence low; 2V ca. 10°

Strongest Diffraction Lines: 2.764(100), 3.70(70), 3.55(70), 11.20(50)

Z: 3

Ref.: Livingstone A (1980) Miner Mag 43:833–836

Jungite
$Ca_2Zn_4Fe_8^{3+}(PO_4)_9(OH)_9 \cdot 16H_2O$

Analyses (%): P_2O_5 31.0, Fe_2O_3 28.4, ZnO 14.6, CaO 5.89, MnO 2.80, H_2O 8.70–9.10

Cleavage: Perfect {010}

Color-Luster: Dark green; streak yellow; luster silky to vitreous

Crystal System: Orthorhombic

Density: 2.843 (Meas.), 2.849 (Calc.)

First Reported: 1980

Habit: Rosettes up to 1 cm in diameter of very thin, tabular, bent crystals

Hardness: 1

Lattice Constants: a = 11.98, b = 20.37, c = 9.95 (Å)

Mode of Occurrence: Occurs in association with mitridatite and manganese oxides at the Hagendorf South pegmatite, Bavaria, W.-Germany

Optical Constants: $\beta = 1.658$, $\gamma = 1.664$, Biaxial neg., 2V = 60°

Space Group: Pcmm, $Pcm2_1$, or Pc2m

Strongest Diffraction Lines: 9.96, 5.09, 3.79, 3.37, 3.30

Z: 2

Ref.: Moore PB, Ito J (1980) Aufschluss 31:56–61

Karnasurtite (Kozhanovite)
$(La, Ce, Th)(Ti, Nb)(Al, Fe)(Si, P)_2O_7(OH)_4 \cdot 3H_2O$(?)

Analyses (%): Al_2O_3 5.5–8.4, Fe_2O_3 0.8–3.5, Ce_2O_3 5.2–17.6, La_2O_3 8.6–10.6, ThO_2 2.6–6.2, TiO_2 2.8–12.3, SiO_2 20.5–30.3, Nb_2O_5 2.8–8.2, P_2O_5 2.4–6.8, H_2O 10.7–17.9

Cleavage: One good. Brittle

Color-Luster: Honey yellow when fresh to pale yellow when altered. Greasy. Streak yellowish

Density: 2.89–2.95

First Reported: 1959

Crystal System: Hexagonal (?)

Habit: Grains up to 1 cm in diameter; also aggregates of platy crystals up to 10 × 6 × 2 cm

Hardness: 2

Mode of Occurrence: Occurs associated with schizolite, natrolite, and epididymite, and is replaced by polylithionite, in a zoned pegmatitic stock at Mt. Karnasurt, Lovozero alkaline massif, Kola Peninsula

Optical Constants: $\omega = 1.617$, $\varepsilon = 1.595$; (−)

Strongest Diffraction Lines: 3.10(70), 2.88(70), 3.29(60), 3.49(50), 1.723(50). Amorphous to X-rays; after being heated at 900°, gives pattern close to huttonite

Ref.: Kuz'menko MV, Kozhanov SI (1960) Am Miner 45:1133–1134. Roberts WL, Rapp GR, Weber J (1974) Encyclopedia of minerals, New York, Van Nostrand Reinhold, p 319

Keckite
$(Ca, Mg)(Zn, Mn^{2+})_2(Fe^{3+})(PO_4)_4(OH)_3 \cdot 2H_2O$

Analyses (%): P_2O_5 35.8, Fe_2O_3 30.0, MnO 14.8, ZnO 2.24, CaO 5.88, MgO 0.66, H_2O 10.6

Cleavage: {001} and {100} good

Color-Luster: Brown, yellow-brown or dirty grayish-brown; luster dull

Crystal System: Monoclinic

Density: 2.6 (Meas.), 2.682 (Calc.)

First Reported: 1979

Habit: Aggregate of crystals up to 2 mm long

Hardness: 4.5

Lattice Constants: a = 15.02, b = 7.19, c = 19.74 (Å); $\beta = 110°$

Mode of Occurrence: Formed by the weathering of phosphophyllite or rockbridgeite in the Hagendorf pegmatite, Bavaria, W. Germany

Optical Constants: $\beta = 1.692$, $\gamma = 1.699$; Pleochroic with X reddish-brown, Y yellow and Z brighter yellow. Biaxial neg.

Space Group: $P2_1/a$

Strongest Diffraction Lines: 2.86(100), 9.3(80), 4.98(50), 3.51(50)

Z: 2

Ref.: Mücke A (1979) Neues Jahrb. Miner Abh 134:183–192

Kemmlitzite
(Sr, Ce)$Al_3(AsO_4)[(P, S)O_4](OH)_6$

Analyses (%): P_2O_5 5.69, As_2O_5 21.6, SO_3 5.99, SrO 8.27, RE_2O_3 12.6, CaO 0.9, MgO 0.9, Al_2O_3 27.2, Fe_2O_3 0.78, SiO_2 2.18, H_2O 12.0

Cleavage: Indistinct Basal

Color-Luster: Light grayish brown, partly colorless, and partly cloudy brownish with zonal structure

Crystal System: Hexagonal

Density: 3.63 (Meas.), 3.601 (Calc.)

First Reported: 1969

Habit: Pseudocubic rhombohedral crystals with average rhombohedron edge 0.1–0.15 mm. Basal pinacoid rarely present

Hardness: 5.5

Lattice Constants: (Hex.) $a=7.072$, $c=16.51$. (Rhomb.) $a_{rh}=6.837$ (Å), $\alpha=61.85°$

Mode of Occurrence: Found associated with apatite, anatase, and zircon in the heavy fraction separated from kaolinized quartz porphyry from the Kemmlitz deposit, Saxony, East Germany

Optical Constants: $\omega=1.701$, $\varepsilon=1.707$ (+)

Strongest Diffraction Lines: 2.959(100), 3.514(90), 1.903(90)

Z: 3

Ref.: Hak J, Johan Z, Kuacek M, Liebscher W (1969) Neues Jahrb. Miner Monatsh 201–212. Roberts WL, Rapp GR, Weber J (1974) Encyclopedia of minerals, New York, Van Nostrand Reinhold, p 321

Kidwellite
$Na(Fe^{3+})_9(PO_4)_6(OH)_{10} \cdot 5H_2O$

Analyses (%): P_2O_5 31.3, Fe_2O_3 52.3, Na_2O 1.60, Al_2O_3 0.98, H_2O 13.3

Cleavage: {100} perfect

Color-Luster: Lively pale chartreuse-green to greenish-yellow, greenish-white, and bright yellow with silky luster

Crystal System: Monoclinic

Density: 3.34 (Calc.)

First Reported: 1978

Habit: Mostly as mats or tufts made up of the finest feathery crystals. Elongation of crystals is parallel to [010]

Hardness: 3

Lattice Constants: $a=20.61$, $b=5.15$, $c=13.75$ (Å); $\beta=112°$

Mode of Occurrence: As late stage mineral associated with rockbridgeite, dufrenite, beraunite, and strengite in novaculite deposits. It has been found in several localities in the Ouachita Mountains, Arkansas, at Indian Mountain, Alabama, Irish Creek, Virginia and the Rotläufchen mine, Waldgirmes, Germany

Optical Constants: $\alpha=1.787$, $\beta=1.800$, $\gamma=1.805$; Biaxial neg., large 2V

Space Group: Not specified

Strongest Diffraction Lines: 9.41(100), 3.413(55), 3.813(40), 3.193(40)

Z: 2

Ref.: Moore PB, Ito J (1978) Miner Mag 42:137–140

Kingite
$Al_2O_3 \cdot Al(OH)_3 \cdot P_2O_5 \cdot 9H_2O$

Analyses (%): P_2O_5 28.2, Al_2O_3 31.9, F 0.8, H_2O 36.5

Color-Luster: White

Crystal System: Triclinic

Density: 2.21–2.30 (Meas.), 2.465 (Calc.)

First Reported: 1956

Habit: Nodular masses up to 5 cm in diameter; platelets or irregular thick fragments about 0.5 μm–1.0 μm in size

Hardness: Not determined

Lattice Constants: $a=9.15$, $b=10.00$, $c=7.24$ (Å); $\alpha=98.6°$, $\beta=93.6°$, $\gamma=93.2°$

Mode of Occurrence: Occurs coated and veined by brown powdery material consisting of talc, quartz, and halite, with possibly a little gypsum and hematite or goethite, at Robertstown, South Australia, and also in Precambrian sediments near the Clinton phosphate workings, 64 miles west-southwest of the first deposit

Optical Constants: Mean index ~1.514

Strongest Diffraction Lines: 9.1(100), 3.45(80), 3.48(65)

Z: 2

Ref.: Norrish K, Rogers LER, Shapter RE (1957) Miner Mag 31:351–357. Kato T (1970) Am Miner 55:515–517

Kingsmountite
(The Fe^{2+} analog of Montgomeryite), $\sim Ca_3Mn^{2+}Al_4(Fe^{2+}, Mn^{2+})(PO_4)_6(OH)_4 \cdot 12H_2O$

Analyses (%): P_2O_5 35.9, Al_2O_3 16.7, CaO 14.9, MnO 8.1, FeO (total Fe) 4.1, MgO 0.2, H_2O 20.6

Cleavage: Not discernible

Color-Luster: Very light brown, white streak

Crystal System: Monoclinic (isostructural with montgomeryite)

Density: 2.51 (Meas.), 2.58 (Calc.)

First Reported: 1979

Habit: Small fibers

Hardness: 2.5

Lattice Constants: a = 10.029, b = 24.46, c = 6.258 (Å); $\beta = 91.16°$

Mode of Occurrence: Found at the Foote Mineral Co. spodumene mine, near the town of Kings Mountain, North Carolina

Optical Constants: $\alpha = 1.575$, $\beta = 1.581$, $\gamma = 1.583$; Biaxial neg., 2V = 62°

Space Group: Probably C2

Strongest Diffraction Lines: 5.15(100), 2.624(60), 12.28(50), 2.950(40)

Ref.: Dunn PJ, Peacor DR, White JS, Ramik RA (1979) Can Miner 17:579–582

Kipushite = Veselyite

Kleemanite
$ZnAl_2(PO_4)_2(OH)_2 \cdot 3H_2O$

Analyses (%): ZnO 20.7, Al_2O_3 24.4, Fe_2O_3 1.10, Mn_2O_3, P_2O_5 35.7, H_2O 18.2

Color: Colorless in transmitted light

Crystal System: Monoclinic

Density: 2.76 (Calc.)

First Reported: 1979

Habit: Ochre-like coatings and thin veins 1–2 mm thick

Lattice Constants: a = 6.290, b = 7.194, c = 9.762 (Å); $\beta = 110°$

Mode of Occurrence: Occurs as coatings and thin fibers or crystallites on manganiferous iron ores at Iron Knob, South Australia

Optical Constants: $\alpha = 1.598$, $\gamma = 1.614$; inclined extinction up to 40°

Space Group: Not specified

Strongest Diffraction Lines: 4.76(100), 3.09(80), 9.09(60), 3.30(60)

Z: 2

Ref.: Pilkington ES, Segnit ER, Watts J (1979) Miner Mag 43:93–95

Kolbeckite (Sterrettite, Eggonite)
(Sc, Be, Ca)$PO_4 \cdot 2H_2O$

Analyses: None reported

Class: 2/m

Cleavage: {110} fair, {100} poor, {001} poor. Fracture conchoidal. Brittle

Color-Luster: Colorless, yellowish, bright blue to blue-gray. Transparent. Vitreous to pearly

Crystal System: Monoclinic

Density: 2.44 (Meas.), 2.47 (Calc.)

First Reported: 1926 (re-defined: 1959)

Habit: Crystals short prismatic, minute. Apparently twinned on {001} and perhaps {031} as a subordinate composition plane

Hardness: 5

Lattice Constants: a = 8.90, b = 10.20, c = 5.43 (Å); $\beta = 90.75°$

Mode of Occurrence: Occurs with wardite and crandallite in altered variscite nodules at Fairfield, Utah County, Utah; on specimens of silver ore from Felsöbánya, Roumania; and very sparingly in a quartz-wolframite vein in the Sadisdorf copper mine near Schmiedeberg, Saxony, Germany

Optical Constants: $\alpha = 1.572$, $\beta = 1.590$, $\gamma = 1.601$; (—)2V = 60°

Space Group: $P2_1/n$

Strongest Diffraction Lines: 4.88(100), 4.51(80), 2.90(80)

Z: 2

Ref.: Mrose ME, Wappner B (1959) Bull Geol Soc Am 70:1648–1649. Roberts WL, Rapp GR, Weber J (1974) Encyclopedia of minerals, New York, Van Nostrand Reinhold, p 329

Koninckite
(Strengite or Metastrengite?)
$(Fe,Al)PO_4 \cdot 3H_2O$

Analyses (%): P_2O_5 34.8, Al_2O_3 4.6, Fe_2O_3 34.4, H_2O 26.2

Cleavage: One; transverse to elongation

Color-Luster: Yellow. Transparent. Vitreous

Crystal System: Orthorhombic (?)

Density: 2.3

First Reported: 1883

Habit: As small spherical aggregates of radiating needles

Hardness: 3.5

Mode of Occurrence: Occurs associated with richellite, halloysite, and allophane at Richelle near Visé, Belgium

Optical Constants: $\alpha = 1.645$, $\beta = 1.65$, $\gamma = 1.656$, 2V = not very large

Ref.: Roberts WL, Rapp GR, Weber J (1974) Encyclopedia of minerals, New York, Van Nostrand Reinhold, p 330

Kribergite (Species status uncertain)
$2Al_2O_3 \cdot 2(P_2O_5, SO_3) \cdot 5H_2O$

Analyses (%): Al_2O_3 38.8, SO_3 13.1, P_2O_5 31.3, H_2O 17.2

Cleavage: Not determined

Color-Luster: White

Density: 1.92

First Reported: 1945

Habit: Massive, compact, chalk-like

Hardness: Not determined

Mode of Occurrence: Occurs as a cavity filling in copper-rich pyrite at the Kristineberg mine, Västerbotten, Sweden

Optical Constants: mean index = 1.484

Strongest Diffraction Lines: 11.6(100), 5.02(25), 6.62(20)

Ref.: Roberts WL, Rapp GR, Weber J (1974) Encyclopedia of minerals, New York, Van Nostrand Reinhold, p 333

Kryzhanovskite (Kruzhanovskite)
$Fe_3^{3+}(OH)_3(PO_4)_2$

Analyses (%): P_2O_5 35.3, CaO 1.5, MgO 1.3, MnO 16.4, Fe_2O_3 34.6, H_2O 9.50

Class: 2/m 2/m 2/m

Cleavage: {001} parting

Color-Luster: Deep red-brown to greenish brown, bronze on parting surface. Translucent. Vitreous to dull. Streak yellowish brown

Crystal System: Orthorhombic

Density: 3.31 (Meas.), 3.35 (Calc.)

First Reported: 1950

Habit: Crystals prismatic, poorly formed, up to 2–3 cm in diameter

Hardness: 3.5–4

Lattice Constants: a = 9.450, b = 10.013, c = 8.179 (Å)

Mode of Occurrence: Kryzhanovskite occurs intimately associated with sicklerite as the outer part of large nodules of altered triphylite in the Kalbinsk pegmatites, U.S.S.R.; also as an alteration of phosphoferrite at the Bull Moose, Big Chief, and Dan Patch pegmatites, Black Hills, South Dakota

Optical Constants: 2V = 40°–45°

Space Group: Pbna

Strongest Diffraction Lines: 3.156(100), 4.996(70), 4.701(50)

Z: 4

Ref.: Ginzburg AI (1951) Am Miner 36:382. Moore PB (1971) Am Miner 56:1–17. Moore PB, Araki T, Kampf AR (1980) Miner Mag 43:789–795

Kulanite
(Fe^{2+}-dominant analog of Bjarebyite)
$Ba(Fe, Mn, Mg, Ca)_2(Al, Fe)_2(PO_4)_3(OH)_3$

Analyses (%): MgO 3.08, CaO 0.77, MnO 6.32, FeO 10.5, BaO 23.7, Al_2O_3 14.3, Fe_2O_3 2.14, P_2O_5 33.3, H_2O 3.83

Cleavage: Fair to good parallel to {010} and {100}

Color-Luster: Blue to green in hand specimens, transparent to translucent in the thin sections; vitreous luster

Crystal System: Triclinic although crystals are pseudomonoclinic

Density: 3.91 (Meas.), 3.92 (Calc.)

First Reported: 1976

Habit: Tabular plates parallel to {$\bar{1}$01} and rosette-like aggregates. Forms (in monoclinic notation) present include {001}, {100}, {120}, {110}, {011}, {031}, {$\bar{1}$01}, {$\bar{3}$01}, {111}, {121}, {131}, and {161}

Hardness: 4

Lattice Constants: a = 9.032, b = 12.119, c = 4.936 (Å); $\alpha = 90°$, $\beta = 100°$, $\gamma \sim 90°$

Mode of Occurrence: Found in the sideritic iron-formation in the NE Yukon Territory, Canada. Associated minerals include baricite, maricite, penikisite, satterlyite and gormanite

Optical Constants: $\alpha = 1.703$, $\beta = 1.705$, $\gamma = 1.723$; 2V(+) = 32° (Meas.), 38° (Calc.); Pleochroic with X brown, Y green, Z pale brown; an optic axis is almost normal to {$\bar{1}$01}

Space Group: P$\bar{1}$

Strongest Diffraction Lines: 3.108(100), 2.927(80), 3.044(70), 2.659(70)

Z: 2

Ref.: Mandarino JA, Sturman BD (1976) Can Miner 14:127–131

Lacroixite
$NaAl(PO_4)(OH, F)$

Analyses (%): Na_2O 14.9, CaO 19.5, MnO 8.43. Al_2O_3 18.9, P_2O_5 28.8, F 6.53, H_2O 5.46

Class: 2/m

Cleavage: Parallel to {111} and {1$\bar{1}$1} indistinct

Color-Luster: Pale yellow to pale green. Vitreous to resinous

Crystal System: Monoclinic

Density: 3.126 (Meas.)

First Reported: 1914

Habit: As fragmentary crystals

Hardness: 4.5

Lattice Constants: a = 6.89, b = 8.22, c = 6.425 (Å); $\beta = 115.5°$

Mode of Occurrence: Found with morinite, apatite, childrenite, roscherite, and tourmaline in druses in granite at Greifenstein, Saxony, Germany; at Konigswart in Marienbad, Bohemia, and Ječlov, near Jihlava, Czechoslovakia, in association with natromontebrasite; and at Strickland quarry, Portland, Connecticut, in association with pink apatite, pink brazilianite, augelite, and natromontebrasite

Optical Constants: $\alpha = 1.545$, $\beta = 1.554$, $\gamma = 1.565$; $(-)2V \approx 90°$

Space Group: A2/a

Strongest Diffraction Lines: 2.900(100), 2.470(85), 3.159(71)

Z: 4

Ref.: Roberts WL, Rapp GR, Weber J (1974) Encyclopedia of minerals, New

York, Van Nostrand Reinhold, p 339. Mrose ME (1971) Program and Abstracts, 20th Clay Min Conf, p 10

Landesite
$Mn_2^{2+}Fe^{3+}(H_2O, OH)_3(PO_4)_2$

Analyses (%): CaO 1.40, MgO 3.05, MnO 32.1–33.6, Fe_2O_3 13.7, P_2O_5 31.9–35.7, H_2O 11.8–13.6

Class: 2/m 2/m 2/m

Cleavage: {010} distinct

Color-Luster: Brown

Crystal System: Orthorhombic

Density: 3.026 (Meas.)

First Reported: 1930

Habit: As octahedral-like crystals pseudomorphous after reddingite

Hardness: 3–3.5

Lattice Constants: a = 9.458, b = 10.185, c = 8.543 (Å)

Mode of Occurrence: Occurs in granite pegmatite associated with lithiophilite, apatite, eosphorite, fairfieldite, and rhodochrosite at the Berry quarry, Poland, Maine

Optical Constants: $\alpha = 1.720$, $\beta = 1.728$, $\gamma = 1.735$; (−)2V = large

Space Group: Pbcn

Strongest Diffraction Lines: 3.207(100), 5.096(80), 4.284(80)

Z: 4

Ref.: Roberts WL, Rapp GR, Weber J (1974) Encyclopedia of minerals, New York, Van Nostrand Reinhold, p 340. Moore PB, Araki T, Kampf AR (1980) Miner Mag 43:789–795

Laplandite (A mixture?)
$(Na, K, Ca)_4(Ce, Th)(Ti, Mg, Al, Nb)(P, Si)_8O_{22} \cdot 5H_2O$

Analyses (%): SiO_2 40.9, P_2O_5 9.62, TiO_2 4.08, Nb_2O_5 1.88, Al_2O_3 0.94, Fe_2O_3 0.64, MgO 1.02, CaO 0.56, ThO_2 1.32, RE_2O_3 16.8, Na_2O 9.81, K_2O 1.88, H_2O 8.96

Cleavage: None reported

Color-Luster: Light gray to yellowish, rarely bluish; luster vitreous to greasy

Crystal System: Orthorhombic

Density: 2.83

First Reported: 1974

Habit: Radiating fibrous deposits up to 0.5–1 cm thick and fan-shaped platy aggregates; plates about 0.01 mm thick often flattened on the c-axis. Crystals consist of prismatic individuals under the microscope. Fracture is splintery

Hardness: 2–3

Lattice Constants: a = 7.27, b = 14.38, c = 22.25 (Å)

Mode of Occurrence: Occurs in the natrolite zone of the Jubilee pegmatite, Mt. Karnasurt, Lovozero massif, Kola Peninsula, U.S.S.R., associated with belovite, nordite, serandite, thermonatrite, steenstrupine, leucosphenite, sphalerite, ilmajokite, raite and zorite

Optical Constants: $\alpha = 1.568$, $\beta = 1.584$, $\gamma = 1.585$; (—)2V = 28° (Calc.)

Space Group: Pmmm

Strongest Diffraction Lines: 3.79(90), 3.34(90), 2.82(90), 3.25(70), 3.01(70)

Z: 1

Ref.: Eskova EM et al (1974) Zapiski Vses Miner Obshch 103:571–575

Laubmannite
$Fe_3^{2+}Fe_6^{3+}(PO_4)_4(OH)_{12}$

Analyses (%): CaO 1.14, FeO 2.07–19.7, MnO 2.40, Fe_2O_3 43.0–57.9, P_2O_5 26.0–26.2, H_2O 9.07–10.5

Class: 2/m 2/m 2/m

Cleavage: Two probable cleavages parallel to fiber length. Brittle

Color-Luster: Bright yellowish green, grayish green to greenish brown to brown. Vitreous to silky. Subtranslucent to opaque

Crystal System: Orthorhombic

Density: 3.33 (Meas.)

First Reported: 1949

Habit: As botryoidal and mammillary aggregates with radial-fibrous or parallel-fibrous structure

Hardness: 3.5–4

Lattice Constants: a = 13.95, b = 30.77, c = 5.16 (Å)

Mode of Occurrence: Occurs filling crevices in novaculite at Buckeye Mountain, near Shady, Polk County, Arkansas; at the Nitzelbuch mine, Amberg-Auerbach district, Bavaria, Germany; and in vugs in a gossan zone surrounding magnetite ore at Leveäniemi, Norbotten Province, Sweden

Optical Constants: $\alpha = 1.840$, $\beta = 1.847$, $\gamma = 1.892$; (+)2V = moderate, variable

Space Group: Pbma

Strongest Diffraction Lines: 5.150(100), 15.30(90), 3.446(70)

Z: 4

Ref.: Moore PB (1970) Am Miner 55:135–169. Roberts WL, Rapp GR, Weber J (1974) Encyclopedia of minerals, New York, Van Nostrand Reinhold, p 344

Laueïte Series

$Mn^{2+}(Fe^{3+}, Al)_2(PO_4)_2(OH)_2 \cdot 8H_2O$

Analyses (%): MnO 11.1, FeO 1.34, MgO 0.52, CaO 0.23, Fe_2O_3 27.5, Al_2O_3 1.76, P_2O_5 26.5, H_2O 30.8

Class: $\bar{1}$

Cleavage: {010} perfect

Color-Luster: Honey-brown, yellowish-orange. Transparent to translucent. Vitreous

Crystal System: Triclinic

Density: 2.44–2.49 (Meas.), 2.56 (Calc.)

First Reported: 1954

Habit: Wedge-shaped crystals up to 2.0 mm in size. Most common forms: {100}, {010}, {110}, {1$\bar{1}$0}, {0$\bar{1}$1}, and {011}

Hardness: 3

Lattice Constants: a = 5.28, b = 10.66, c = 7.14 (Å); $\alpha = 107.92°$, $\beta = 111°$, $\gamma = 71.12°$

Mode of Occurrence: Occurs on rockbridgeite at the Hagendorf pegmatite, eastern Bavaria; associated with strunzite, stewartite, and pseudolaueite, all implanted upon siderite and ludlamite at the Palermo No. 1 quarry, North Groton, New Hampshire; and at several pegmatites in the Keystone and Custer districts, Black Hills, South Dakota especially at the Tip-Top mine near Custer, and at the Big Chief mine, Glendale

Optical Constants: $\alpha = 1.612$, $\beta = 1.658$, $\gamma = 1.682$; (—)2V = 50°

Space Group: P$\bar{1}$

Strongest Diffraction Lines: 9.91(100), 3.28(90), 4.95(80)

Z: 1

Ref.: Strunz H (1954) Naturwissenschaften 41:256; (1954) Am Miner 39:1038. Moore PB (1965) Am Miner 50:1884–1892. Bauer WH (1969) Am Miner 54:1312–1323. Roberts WL, Rapp GR, Weber J (1974) Encyclopedia of minerals, New York, Van Nostrand Reinhold, p 344

Lazulite Series

$(Mg, Fe^{2+})Al_2(PO_4)_2(OH)_2$

Analyses (%): P_2O_5 44.3–46.1, MgO 8.0–12.0, FeO 2.8–8.9, Al_2O_3 30.9–32.6, H_2O 5.8–6.5

Class: 2/m

Cleavage: {110} indistinct to good, {101} indistinct. Fracture uneven to splintery. Brittle

Color-Luster: Deep azure blue to light blue; also bluish green. Translucent to opaque; rarely transparent. Vitreous to dull

Crystal System: Monoclinic

Density: 3.08 (Extrapolated for Mg end member), 3.14 (Calc.)

First Reported: 1792

Habit: Crystals usually acute pyramidal, with large {111} and {$\bar{1}$11} and small {101}; also tabular on {101} or {$\bar{1}$11}. Massive, compact to granular.

Commonly twinned on composition plane a[100]

Hardness: 5.5–6

Lattice Constants: a = 7.15, b = 7.28, c = 7.25 (Å); $\beta = 120.58°$

Mode of Occurrence: Occurs in quartz veins, granitic pegmatites, and highly metamorphosed rocks, particularly quartzites. Often found associated with quartz, pyrophyllite, kyanite, muscovite, andalusite, sillimanite, rutile, corundum, and garnet. Found as crystals up to 6 inches in length at Horrsjöberg, Sweden, and as fine crystals at many European localities, also in Bolivia, Brazil, and in many depositis in Madagascar. Superb crystals are found in the United States especially at Graves Mountain, Lincoln County, Georgia, and large masses to 6 inches across occur at the Champion mine, Mono County, California

Optical Constants: $\alpha = 1.604–1.635$, $\beta = 1.633–1.663$, $\gamma = 1.642–1.673$; (—)2V = 68.9°

Space Group: $P2_1/c$

Strongest Diffraction Lines: 3.072(100), 3.136(95), 6.15(75)

Z: 2

Ref.: Roberts WL, Rapp GR, Weber J (1974) Encyclopedia of minerals, New York, Van Nostrand Reinhold, p 348–349. Pecora WT, Fahey JJ (1950) Am Miner 35:1–18

Lehiite
(Inadequately described mineral)
$(K, Na)_2Ca_5Al_8(PO_4)_8(OH)_{12} \cdot 12H_2O$

Analyses (%): Na_2O 3.08, K_2O 2.25, CaO 18.1, Al_2O_3 27.8, P_2O_5 34.6, H_2O 14.2

Cleavage: Not determined

Color-Luster: White to gray

Crystal System: Unknown

Density: 2.89

First Reported: 1930

Habit: As layers and crusts composed of subparallel fibrous aggregates

Hardness: 5.5

Mode of Occurrence: Occurs associated with wardite, crandallite, and other phosphate minerals in the outer shell of altered variscite nodules at Fairfeld, near Lehi, Utah County, Utah

Optical Constants: $\alpha = 1.600$, $\beta = 1.615$, $\gamma = 1.629$; (—)2V = large

Ref.: Roberts WL, Rapp GR, Weber J (1974) Encyclopedia of minerals, New York, Van Nostrand Reinhold, p 350

Lermontovite
(Inadequately described mineral)
$(U, Ca, Ce)_3(PO_4)_4 \cdot 6H_2O$

Analyses (%): BaO 1.0, CaO 1.0, UO_3 14.5, UO_2 36.5, RE_2O_3 1.67, P_2O_5 20.4, SiO_2 2.38, H_2O 13.4

Cleavage: Not determined. Very brittle

Color-Luster: Grayish green. Dull; silky in fractures

Crystal System: Unknown

Density: 4.50

First Reported: 1955

Habit: As botryoidal aggregates of radial fibrous needles

Hardness: Not determined

Mode of Occurrence: Occurs under sharply reducing conditions in the zone of cementation of hydrothermal deposits, associated with molybdenum sulfate, marcasite, hydrous silicates, and "thallium ocher" at an unspecified locality, presumably in the U.S.S.R.

Optical Constants: $\alpha = 1.562–1.574$; $1.562–1.702$, $\beta = ?$, $\gamma = 1.702–1.726$; $1.574–1.726$

Ref.: Anon. (1958) Mem Soc Russ Min 87:81; Am Min 43:379–380. Roberts WL, Rapp GR, Weber J (1974) Encyclopedia of minerals, New York, Van Nostrand Reinhold, p 353

Lessingite = Variety of britholite

Leucophosphite Series
$K_2(Fe^{3+}, Al)_4(PO_4)_4(OH)_2 \cdot 4H_2O$

Analyses (%): K_2O 7.88–10.9, Fe_2O_3 37.0–41.0, P_2O_5 27.0–34.7, H_2O 11.2

Class: 2/m

Cleavage: {100} excellent. Very brittle

Color-Luster: Pegmatites: Buff to chocolate brown, greenish, reddish, reddish-brown. Vitreous. Transparent to translucent. Sedimentary origin: White to greenish. Dull

Crystal System: Monoclinic (pseudo-orthorhombic)

Density: 2.948 (Meas.), 2.957 (Calc.)

First Reported: 1931

Habit: Crystals usually short prismatic with forms {$\bar{1}$01}, {$\bar{1}$11}, {011}, and {110} dominant. Also lamellar or as fine-grained chalk-like masses

Hardness: 3.5

Lattice Constants: a=9.782, b=9.658, c=9.751 (Å); β=102.24°

Mode of Occurrence: Occurs abundantly in superb crystals as much as 1 cm in length, and as lamellar masses several centimeters across, associated with rockbridgeite in altered triphylite nodules at the Tip-Top mine, Custer, South Dakota. Also in fine crystals at the White Elephant, Bull Moose, and Rock Ridge mines near Custer, and at the Gap Lode near Hill City, South Dakota. Found as minute crystals in vugs in frondelite at the Sapucaia pegmatite mine, Minas Geraes, Brazil; also as fine-grained chalk-like masses from two deposits, both of sedimentary origin, at Bomi Hill, western Liberia, and from Weelhamby Lake, Nighanboun Hills, Western Australia

Optical Constants: α=1.707, β=1.721, γ=1.739; (+)2V

Space Group: $P2_1/n$

Strongest Diffraction Lines: 6.79(100), 5.99(70), 3.061(70)

Z: 4

Ref.: Lindberg ML (1957) Am Miner 42:214–221. Moore PB (1972) Am Miner 57:397–410

Libethenite
Cu_2PO_4OH

Analyses (%): CuO 66.6, As_2O_3 0–2.3, P_2O_5 26.5–29.3, H_2O 3.74–4.04

Class: 2/m 2/m 2/m

Cleavage: {100} indistinct, {010} indistinct. Fracture conchoidal to uneven

Color-Luster: Light to dark green and blackish green; also light to dark olive green. Transparent to translucent. Vitreous to greasy

Crystal System: Orthorhombic

Density: 3.97 (Meas.), 3.93 (Calc.)

First Reported: 1823

Habit: Crystals equant or short prismatic; isolated crystals or as crusts

Hardness: 4

Lattice Constants: a=8.10, b=8.45, c=5.91 (Å)

Mode of Occurrence: Occurs chiefly as a secondary mineral in the oxidation zone of copper-bearing ore deposits. Found as excellent microcrystals at Inspiration and Castle Dome, Gila County, and in the Coronado mine, Greenlee County, Arizona; in the Santa Rita district, Grant County, New Mexico; at Yerington, Lyon County, Nevada; in glaucophane schist near Llanada, San Benito County, California; and at the Perkiomen mine, Montgomery County, Pennsylvania. It also occurs as crystals up to 6 mm in size at Libethen, near Neusohl, Rumania; and at localities in Chile, England, France, U.S.S.R., and Zaire

Optical Constants: α=1.701, β=1.743, γ=1.787; (−)2V=nearly 90°

Space Group: Pnnm

Strongest Diffraction Lines: 4.81(100), 2.63(100), 5.85(90)

Z: 4

Ref.: Roberts WL, Rapp GR, Weber J (1974) Encyclopedia of minerals, New York, Van Nostrand Reinhold, p 356

Likasite
$Cu_6PO_4(NO_3)_2(OH)_7$

Analyses (%): CuO 55.5, NO_3 15.1, P_2O_5 14.6, H_2O 16.5
Class: 2/m 2/m 2/m
Cleavage: {010} perfect and easy
Color-Luster: Blue
Crystal System: Orthorhombic
Density: 2.96–2.98 (Meas.)
First Reported: 1955
Habit: Crystals tabular with c{001} dominant; also r{101}, k{105}, g{108}, b{010}, d{012}, e{014}, and f{018}
Hardness: Not determined
Lattice Constants: a=6.72, b=21.65, c=5.79 (Å)
Mode of Occurrence: Occurs as crystals implanted on cuprite or as blue masses in cuprite, surrounded by a crust of malachite and by malachite pseudomorphs after likasite, at the Likasi copper mine, Zaire
Optical Constants: α=?, β=1.61, γ=1.69
Space Group: Pmcb
Z: 2
Ref.: Roberts WL, Rapp GR, Weber J (1974) Encyclopedia of minerals, New York, Van Nostrand Reinhold, p 357

Lingaitukuang ("Cathophorite") = Brabantite

Lipscombite
$(Fe^{2+}, Mn^{2+})Fe_2^{3+}(PO_4)_2(OH)_2$

Analyses (%): MnO 7.91, FeO 3.75, Fe_2O_3 50.4, P_2O_5 33.4, H_2O 4.45
Class: 422
Color-Luster: Olive green to black. Splendent. Opaque
Crystal System: Tetragonal
Density: 3.66 (Meas.), 3.706 (Calc.)
First Reported: 1953
Habit: Aggregates of microscopic crystals
Lattice Constants: a=7.40, c=12.81 (Å)
Mode of Occurrence: Found intimately intergrown with cyrilovite, leucophosphite, and metastrengite in vugs in the alkali-altered zone in frondelite at the Sapucaia pegmatite mine, Municipio of Counselheiro Pena, Minas Geraes, Brazil, and has been identified from other granitic pegmatites
Space Group: $P4_12_12$
Strongest Diffraction Lines: 3.314(100), 3.206(60), 1.656(40)
Z: 4
Ref.: Lindberg ML (1962) Am Miner 47:353–359; Moore PB (1970) Am Miner 55:135–169

Liroconite
$Cu_2Al(As, P)O_4(OH)_4 \cdot 4H_2O$

Analyses (%): CuO 36.4–37.2, Al_2O_3 9.7–11.8, P_2O_5 3.5–3.7, As_2O_5 22.2–26.5, H_2O 24.9–25.5
Class: 2/m
Cleavage: {110} indistinct, {011} indistinct. Fracture conchoidal to uneven. Brittle
Color-Luster: Sky blue to verdigris green. Transparent to translucent. Vitreous to resinous
Crystal System: Monoclinic
Density: 2.926–3.01 (Meas.), 2.95 (Calc. As:P=3:1)
First Reported: 1801; actually called Liroconite, 1832
Habit: Crystals flattened and wedge-shaped with forms m{110} and e{011} nearly equally developed; faces of both striated parallel to intersection edge. Crystals occur singly, up to 1 cm in length, and as subparallel groups; also coarsely granular
Hardness: 2–2.5
Lattice Constants: a=12.67, b=7.55, c=9.86 (Å); β=91.38°

Mode of Occurrence: Occurs very sparingly in the oxidized zone of copper deposits associated with limonite, azurite, malachite, cuprite, clinoclasite, olivenite, and chalcophyllite. Found at several mines in Cornwall and Devonshire, England; at several localities in Germany; Urals, U.S.S.R.; Herrengrund, Czechoslovakia; and with linarite and caledonite at the Cerro Gordo mine, Inyo County, California

Optical Constants: $\alpha=1.612$, $\beta=1.652$, $\gamma=1.675$; $(-)2V\sim72°$ (Calc.)

Space Group: I2/a

Strongest Diffraction Lines: 6.46(100), 3.01(100), 5.95(90)

Z: 4

Ref.: Roberts WL, Rapp GR, Weber J (1974) Encyclopedia of minerals, New York, Van Nostrand Reinhold, p 359. Berry LG (1951) Am Miner 36:484–503

Lithiophilite
(Lithiophilite-Triphylite Series)
$Li(Mn^{2+},Fe^{2+})PO_4$

Analyses (%): Li_2O 5.51–9.11, FeO 2.94–12.6, MnO 30.5–42.6, CaO 0–9.70, P_2O_5 43.4–45.1

Class: 2/m 2/m 2/m

Cleavage: {100} nearly perfect, {010} imperfect, {011} interrupted. Fracture subconchoidal to uneven

Color-Luster: Light reddish brown, yellowish brown, salmon; often externally brownish to blackish due to alteration. Translucent to transparent. Resinous to vitreous. Streak colorless to dirty white

Crystal System: Orthorhombic

Density: 3.34 (Meas. Mn:Fe=9:1), 3.54 (Calc. $LiMnPO_4$)

First Reported: 1879

Habit: As large anhedral or subhedral crystals with rough faces. Usually massive, cleavable

Hardness: 4–5

Lattice Constants: a=6.05, b=10.32, c=4.71 (Å)

Mode of Occurrence: Occurs as a primary mineral in granite pegmatites, often partly or completely altered to a wide variety of secondary phosphate minerals. Found at several localities in the Black Hills, South Dakota, especially at the Custer Mountain Lode, Custer district. It also occurs at Branchville and Portland, Connecticut; at several places in Oxford County, Maine; at Pala, San Diego County, California; in Canada; in the Karibib area, and in Namaqualand, South Africa; at Mangualde, Portugal; and in Argentina

Optical Constants: $\alpha=1.669$, $\beta=1.673$, $\gamma=1.682$; $(+)2V\sim65°$

Space Group: Pmnb

Strongest Diffraction Lines: 2.531(100), 3.01(90), 3.469(50)

Z: 4

Ref.: Roberts WL, Rapp GR, Weber J (1974) Encyclopedia of minerals, New York, Van Nostrand Reinhold, p 360

Lithiophosphate
Li_3PO_4

Analyses (%): Li_2O 37.1, P_2O_5 61.3

Class: mm2

Cleavage: {010} perfect, {110} distinct

Color-Luster: Colorless, white, light rose. Vitreous

Crystal System: Orthorhombic

Density: 2.478 (Meas.), 2.479 (Calc.)

First Reported: 1957

Habit: Crystals elongated parallel to b and superficially resembly topaz, up to 1 cm in length. Forms: {110}, {101}, {010}, {013}, and {001}. As masses up to 5.0 × 9.0 cm in size

Hardness: 4

Lattice Constants: a=6.1155, b=5.2340 (synthetic), c=4.8452 (Å)

Mode of Occurrence: Occurs as a hydrothermal replacement of montebrasite in the central zone of a pegmatite in amphibolite on the Kola Peninsula,

U.S.S.R.; also found as crystals and masses on quartz and albite crystal druses which line open fissures in massive albite-microcline-quartz-spodumene-muscovite pegmatite at Kings Mountain, Cleveland County, North Carolina

Optical Constants: $\alpha = 1.550–1.553$, $\beta = 1.557–1.558$, $\gamma = 1.566–1.567$; $(+)2V = 69°–80°$

Space Group: $Pmn2_1$

Strongest Diffraction Lines: 3.965(100), 2.635(100), 3.794(90)

Z: 2

Ref.: Matias VV, Bondareva AM (1957) Doklady Acad Sci U.S.S.R. 112:124; *Idem* (1957) Am Miner 42:585. White JS Jr (1969) Am Miner 54:1467–1469. Roberts WL, Rapp GR, Weber J (1974) Encyclopedia of minerals, New York, Van Nostrand Reinhold, p 361

Lomonosovite
$Na_5Ti_2(Si_2O_7)(PO_4)O_2$

Analyses (%): Na_2O 23.8–28.1, SiO_2 24.1, TiO_2 24.4–25.3, Fe_2O_3 2.4, P_2O_5 8.2–12.8

Class: 1

Cleavage: {100} perfect. Fracture uneven. Brittle

Color-Luster: Dark cinnamon brown to black, also rose-violet; vitreous to adamantine on cleavage, vitreous to greasy on fracture. Transparent in thin fragments

Crystal System: Triclinic

Density: 3.15 (Meas.)

First Reported: 1941

Habit: Laminated tabular crystals up to $7 \times 5 \times 0.6$ cm in size. Fine polysynthetic twinning

Hardness: 3–4

Lattice Constants: $a = 5.40$, $b = 7.03$, $c = 14.3$ (Å); $\alpha = 100°$, $\beta = 96°$, $\gamma = 90°$

Mode of Occurrence: Occurs associated with hackmanite, ussingite, lamprophyllite, eudialyte, arfvedsonite, microcline, ramsayite, and aegirine, in pegmatites in syenite, Lovozero Range, Kola Peninsula, U.S.S.R.

Optical Constants: $\alpha = 1.670$, $\beta = 1.750$, $\gamma = 1.778$; $(—)2V = 56°$

Space Group: P1

Strongest Diffraction Lines: 2.83(100), 1.778(90), 1.840(80)

Z: 1

Ref.: Gerasimovsky VI (1941) Compt Rend (Doklady) Acad Sci U.S.S.R. 32:498; *Idem* (1950) Am Miner 35:1092–1093

β-Lomonosovite (Metalomonosovite)
$Na_2Ti_2Si_2O_9 \cdot (Na, H)_3PO_4$ (variable)

Analyses (%): P_2O_5 15.1–18.5, SiO_2 24.2–25.2, $(Ta_2O_5 + Nb_2O_5)$ 3.05–4.78, TiO_2 23.8–25.7, ZrO_2 1.89–2.50, Fe_2O_3 2.13–2.38, MnO 0.96–1.70, CaO 0.62–1.80, Na_2O 15.2–17.1, K_2O 0-0.90, H_2O 5.48–6.0

Cleavage: Not given, fracture uneven

Color-Luster: Pale yellow-brown, rose, brown; luster vitreous to pearly on the cleavage, vitreous to greasy on fracture. Extinction inclined with respect to the perfect cleavage

Crystal System: Triclinic

Density: 2.95, 2.98

First Reported: 1962

Habit: Tabular to platy masses up to $5 \times 4 \times 0.3$ cm

Hardness: About 4. Brittle

Lattice Constants: $a = 5.28$, $b = 7.05$, $c = 14.50$ (Å); $\alpha = 102.4°$, $\beta = 96.8°$, $\gamma = 90°$

Mode of Occurrence: Occurs in alkalic pegmatites of the Lovozero massif, Kola Peninsula, U.S.S.R., in poikilitic sodalite syenite and in eudialyte lujavrites. Associated minerals include microcline, aegirine, arfvedsonite, eudialyte, sodalite and nepheline

Optical Constants: $\alpha = 1.670$, $\beta = 1.770$, $\gamma = 1.779$; Biaxial (—), $2V = 10°–24°$

Space Group: None specified

Strongest Diffraction Lines: 2.80, 2.77(100), 3.50, 3.45(90), 13.53(80), 3.11(50), 3.07(70)

Ref.: Gerasimovskii VI, Kazakova ME (1962) Doklady Akad Nauk U.S.S.R. 142:670–673. Semenov EI, Organova NI, Kukharchik MV (1961) Kristallografiya 6:925–932. Belov NV, Organova NI (1962) Geokhimiya 6–14 (translation in Geochemistry 4–13). See also (1963) Am Miner 48:1413–1415

Ludlamite
$Fe_3(PO_4)_2 \cdot 4H_2O$

Analyses (%): FeO 45.9–52.8, MgO 0–2.2, MnO 0–3.1, P_2O_5 30.1–33.9, H_2O 14.9–17.0

Class: 2/m

Cleavage: {001} perfect, {100} indistinct

Color-Luster: Bright green, apple green, pale green to greenish white to colorless. Transparent to translucent. Vitreous

Crystal System: Monoclinic

Density: 3.19 (Meas.), 3.21 (Calc.)

First Reported: 1877

Habit: Crystals thin to thick tabular, often wedge-shaped. Also massive, granular

Hardness: 3.5

Lattice Constants: a = 10.48, b = 4.63, c = 9.16 (Å); $\beta = 100.6°$

Mode of Occurrence: Occurs as a secondary mineral in the oxidation zone of ore deposits, and as an alteration product of primary iron phosphate minerals in granite pegmatites. Found as superb crystals up to 2 cm in size, often with vivianite, in the Blackbird district, Lemhi County, Idaho. In South Dakota as crystalline masses up to 30 cm in diameter, with vugs containing crystals up to 7 mm in length, at the Hesnard mine, Keystone, and as fine crystals in pegmatite at the Big Chief, Ferguson, and Dan Patch mines in the same district; and as excellent crystals in pegmatite at the Bull Moose and Tip Top mines, Custer district. It is also found at the Palermo pegmatite, North Groton, New Hampshire; at the Wheal Jane mine, Truro, Cornwall, England; and at Hagendorf, Bavaria, Germany

Optical Constants: $\alpha = 1.650–1.653$, $\beta = 1.667–1.675$; $\gamma = 1.688–1.697$; $(+)2V = 82°$

Space Group: $P2_1/a$

Strongest Diffraction Lines: 3.96(100), 2.765(100), 2.543(100)

Z: 2

Ref.: Roberts WL, Rapp GR, Weber J (1974) Encyclopedia of minerals, New York, Van Nostrand Reinhold, p 366

Lüneburgite (Lueneburgite)
$Mg_3(PO_4)_2B_2O_3 \cdot 8H_2O$

Analyses (%): P_2O_5 29.7, B_2O_3 12.8, MgO 25.2, H_2O 32.2

Class: Probably 2/m

Cleavage: {110} and {1$\bar{1}$0} fair

Color-Luster: White to brownish white; also green

Crystal System: Monoclinic

Density: 2.05 (Meas.)

First Reported: 1870

Habit: Crystals minute pseudohexagonal tablets. As flattened masses and nodules with fibrous to earthy structure

Hardness: ~2

Lattice Constants: a = 10.10, b = 7.62, c = 9.81 (Å); $\beta = 97.4°$

Mode of Occurrence: Occurs in clay associated with halite, sylvite, and polyhalite in the Permian salt basin in the vicinity of Carlsbad, New Mexico, and in adjacent parts of Texas. It also is found in Marl at Lüneburg, Hannover, Germany; and in a guano deposit at Mejillones, Peru

Optical Constants: $\alpha = 1.520–1.522$, $\beta = 1.54–1.541$, $\gamma = 1.545–1.548$; (—)2V = moderate

Space Group: Probably C2/m

Z: 2

Ref.: Roberts WL, Rapp GR, Weber J (1974) Encyclopedia of minerals, New York, Van Nostrand Reinhold, p 367–368

Lusungite

$(Sr, Pb)Fe_3(PO_4)_2(OH)_5 \cdot H_2O$

Analyses: None reported

Class: $\bar{3}$ 2/m

Cleavage: Not determined

Color-Luster: Dark brown

Crystal System: Hexagonal

Density: Not determined

First Reported: 1958

Habit: Massive, pulverulent

Hardness: Not determined

Lattice Constants: a = 7.04, c = 16.80; a_{rh} = 6.92 (Å), $\alpha = 61.2°$

Mode of Occurence: Occurs associated with goethite (limonite) and quartz in the phosphate zone of the Kobokobo pegmatite, Zaire

Optical Constants: n ~ 1.77–1.855, uniaxial, probably positive

Space Group: $R\bar{3}m$

Strongest Diffraction Lines: 2.98(100), 5.77(90), 3.53(61)

Z: 3 (hex.)

Ref: van Wambeke L (1958) Soc Belge Geol Paleontol et Hydrol Bull 67:162–169; *Idem* (1959) Am Miner 44:906–907. Roberts WL, Rapp GR, Weber J (1974) Encyclopedia of minerals, New York, Van Nostrand Reinhold, p 368

Marićite

$NaFePO_4$

Analyses (%): Na_2O 16.5, MnO 3.10, FeO 37.4, MgO 0.8, P_2O_5 42.5

Cleavage: None

Color-Luster: Colorless to gray and pale brown, vitreous, transparent to translucent

Crystal System: Orthorhombic

Density: 3.66 (Meas.), 3.64 (Calc.)

First Reported: 1977

Habit: Sub-parallel to radiating elongate grains, elongation is along [100]; forms recognized include {010}, {011}, {012}, and {032}

Hardness: 4–4.5

Lattice Constants: a = 6.867, b = 8.989, c = 5.049 (Å)

Mode of Occurrence: Found in shales along the Big Fish River, Yukon Territory, Canada. The associated minerals are quartz, ludlamite, vivianite, wolfeite, pyrite, apatite, varulite and satterlyite

Optical Constants: $\alpha = 1.676$, $\beta = 1.695$, $\gamma = 1.698$; Biaxial neg., 2V = 43°

Space Group. Pmnb

Strongest Diffraction Lines: 2.574(100), 2.729(90), 2.707(80), 1.853(60)

Z: 4

Ref.: Sturman BD, Mandarino JA, Corlett MI (1977) Can Miner 15:396–398. LePage Y, Donnay G (1977) Can Miner 15:518–521

Matulaite

$CaAl_{18}(PO_4)_{12}(OH)_{20} \cdot 28H_2O$

Analysis (%): P_2O_5 33.5, SiO_2 3.0, Al_2O_3 34.0, Fe_2O_3 1.05, CaO 1.59, H_2O 15.39

Cleavage: {100} very perfect

Color-Luster: Colorless to white; pearly luster

Crystal System: Monoclinic

Density: 2.330

First Reported: 1980

Habit: Small rosettes of thin tabular crystals and botroidal forms

Hardness: 1

Lattice Constants: a = 20.4, b = 16.7, c = 10.6 (Å); $\beta = 98.2°$

Mode of Occurrence: Occurs as encrustations on chert, the youngest mineral of the association beraunite, rockbridgeite, dufrenite, cacoxenite, strengite and wavellite. It has been found at the Bachman iron mine, Hellertown, Pennsylvania (type locality), from the Rotläufchen iron mine, Waldgirmes, West Germany, and at the LCA pegmatite, Gaston Co., North Carolina

Optical Constants: $\beta = 1.576$; $\gamma = 1.582$; Biaxial (−), 2V = 60°

Solubility: Soluble in hot H_2SO_4; slightly soluble in HCl

Space Group: Probably $P2_1/c$

Strongest Diffraction Lines: 9.96, 6.37, 4,43, 2.395

Z: 2

Ref.: Moore PB, Ito J (1980) Aufschluss 31:55–61

Melkovite
(Inadequately described mineral)
$CaFeH_6(MoO_4)_4(PO_4) \cdot 6H_2O$

Analyses (%): P_2O_5 7.86, MoO_3 57.2, Fe_2O_3 10.9, CaO 5.15, Na_2O 1.12, ZrO_2 0.97, H_2O 16.6

Cleavage: One perfect. Brittle

Color-Luster: Lemon yellow to brownish yellow; dull to waxy

Density: 2.971 (Meas.)

First Reported: 1969

Habit: Finely crystalline powdery films and veinlets consisting of platy crystals usually 0.001–0.002 mm in diameter

Hardness: ~3

Mode of Occurrence: Occurs associated with hydrous iron oxides, powellite, iriginite, ferrimolybdite, jarosite, and autunite in the oxidation zone of a molybdenite-fluorite deposit in the Shunak Mountains, central Kazakhstan

Optical Constants: Mean index = 1.838

Strongest Diffraction Lines: 2.92(100), 3.54(90), 8,42(80)

Ref.: Egorov BL, Dara AD, Senderova VM (1969) Zap Vses Miner Obshch 98:207–212; *Idem* (1970) Am Miner 55:320

Melonjosephite
$CaFe^{2+}Fe^{3+}(PO_4)_2(OH)$

Analyses (%): CaO 15.0, FeO 17.4, MgO 1.18, MnO 0.44, Fe_2O_3 21.8, Al_2O_3 0.17, P_2O_5 40.0, H_2O 2.5

Cleavage: Transverse imperfect cleavage

Color-Luster: Dark green nearly black; luster brilliant to slightly resinous

Crystal System: Orthorhombic

Density: 3.65 (Meas.), 3.61 (Calc.)

First Reported: 1973

Habit: Fibrous masses; brittle and breaks into splinters along a longitudinal cleavage

Hardness: Less than 5

Lattice Constants: a = 9.548, b = 10.85, c = 6.380 (Å)

Mode of Occurrence: Found in the Angarf-South pegmatite, Morocco, in the zone of surrounding triphylite

Optical Constants: $\alpha = 1.720$, $\beta = 1.770$, $\gamma = 1.800$; (−)2V = 80–85°; Dispersion strong, pleochroic with X deep brown to nearly opaque, Y greenish-brown, Z yellow to greenish-brown

Space Group: Pbam or Pba2

Strongest Diffraction Lines: 3.049(100), 5.42(90), 2.710(90), 2.624(60)

Z: 4

Ref.: Fransolet A-M (1973) Bull Soc Franc Mineral Cristallogr 96:135–142. Kampf AR, Moore PB (1977) Am Miner 62:60–66

Messelite (Neomesselite)
$(Ca, Fe, Mn, Mg)_3(PO_4)_2 \cdot 3H_2O$

Analyses (%): CaO 31.1, MnO 0–7.1, FeO 14.0–15.6, MgO 1.5, P_2O_5 37.7, H_2O 12.2

Class: $\bar{1}$

Cleavage: {001} perfect, {010} good, $\{1\bar{1}0\}$ distinct. Fracture uneven. Brittle

Color-Luster: Colorless, white to pale greenish white and greenish gray.

Transparent. Vitreous to pearly. Streak white

Crystal System: Triclinic

Density: 3.16 (Meas.)

First Reported: 1890

Habit: Crystals prismatic to equant. Usually lamellar, foliated, or fibrous; in radiating aggregates

Hardness: 3.5

Lattice Constants: $a \simeq 5.8$, $b \simeq 6.6$, $c \simeq 5.5$ (Å); $\alpha \simeq 102°$, $\beta \simeq 109°$, $\gamma \simeq 90°$

Mode of Occurrence: Messelite occurs as a late hydrothermal mineral in granite pegmatites. Found associated with graftonite, phosphoferrite, ludlamite, vivianite, and hureaulite at the Bull Moose mine, and with hureaulite in altered triphylite at the Tip Top mine, Custer County; as large crystals with siderite and ludlamite at the Big Chief mine, and with triphylite at the Ingersoll and Dan Patch mines near Keystone, Pennington County, South Dakota. It also occurs at the Palermo mine, North Groton, New Hampshire, and abundantly at King's Mountain, North Carolina

Optical Constants: $\alpha = 1.653$, $\beta = 1.659$, $\gamma = 1.676$; $(+)2V = 20°$ to $35°$

Space Group: $P\bar{1}$

Strongest Diffraction Lines: 6.34(100), 3.17(100), 3.02(80)

Z: 1

Ref.: Frondel C (1955) Am Miner 40:828–833. Roberts WL, Rapp GR, Weber J (1974) Encyclopedia of minerals, New York, Van Nostrand Reinhold, p 394–395

Meta-Ankoleite
$K_2(UO_2)_2(PO_4)_2 \cdot 6H_2O$

Analyses (%): UO_3 55.0–56.2, P_2O_5 14.3–16.0, K_2O 7.1–8.0, BaO 0.5–3.0, CaO 0–2.7, SiO_2 5.8, Al_2O_3 0.9, H_2O 11.0

Class: 4/m 2/m 2/m

Cleavage: {001} perfect, micaceous, {100} distinct

Color-Luster: Yellow. Fluoresces yellow-green in both short-wave and long-wave ultraviolet light

Crystal System: Tetragonal

Density: 3.54 (Calc.)

First Reported: 1966

Habit: As plates, usually about 0.5 mm and up to 1 mm in diameter

Hardness: Not determined

Lattice Constants: $a = 6.993$, $c = 8.891$ (Å)

Mode of Occurrence: Occurs intergrown with a phosphuranylite-type mineral associated with muscovite, quartz, and albite in the Mungenyi pegmatite, Ankole district, southwest Uganda. It also occurs associated with quartz, sericitized microcline, and accessory monazite, spinel, barite, and zircon in poorly cemented feldspathic quartzite, Sebungive district, Zimbabwe

Optical Constants: $n \sim 1.580(-)$

Space Group: P4/nmm

Strongest Diffraction Lines: 8.92(100), 3.73(65), 3.25(55)

Z: 1

Ref.: Roberts WL, Rapp GR, Weber J (1974) Encyclopedia of minerals, New York, Van Nostrand Reinhold, p 395

Meta-Autunite (Series)
$(Na, K, H_3O)UO_2(PO_4) \cdot 3H_2O$

Analyses (%): U 49.3, P 6.67, Na 2.70, K 2.37, H 1.47, C 0.17

Cleavage: {001} perfect, {100} indistinct

Color-Luster: Bright greenish-yellow, resinous luster

Crystal System: Tetragonal

Density: 3.7–4.7

First Reported: 1938

Habit: Aggregates of small crystallites up to 1.5 μm thick and 100 μm diameter (but mostly in the range 20–50 μm), with a micaceous habit. Strong, bright yellow fluorescence under UV light

Hardness: Soft, 2–2.5

Lattice Constants: a′ = 6.971, c′ = 8.545 (Å) [a ~ 19.71, c ~ 17.09 (Å)]

Mode of Occurrence: Occurs in a playa 105 km NNE of Kalgoorlie, Western Australia as void and fracture infillings in the near-surface gypsiferous muds; localities are given under autunite

Optical Constants: ω and ε are between 1.575 and 1.580; (—)2V near zero

Space Group: P4/nmm or $P4_22$

Strongest Diffraction Lines: 3.49(100), 4.94(75), 2.47(75), 1.559(50)

Ref.: Frondel C (1958) USGS Bull 1064, p 205–207. Butt CRM, Graham J (1981) Am Miner 66:1068–1072

Metaschoderite
$Al_2(PO_4)(VO_4) \cdot 6H_2O$

Analyses (%): P_2O_5 19.1, Al_2O_3 26.0, V_2O_5 27.0, H_2O 27.9

Class: 2/m

Cleavage: Not determined

Color-Luster: Yellowish orange

Crystal System: Monoclinic

Density: 1.610 (Calc.)

First Reported: 1960

Habit: Microscopic bladed tabular crystals, elongated parallel to (001)

Hardness: ~2

Lattice Constants: a = 11.4, b = 14.9, c = 9.2 (Å); β = 79°

Mode of Occurrence: Occurs sparsely as microcrystalline coatings associated with wavellite and vashegyite along fractures in phosphatic cherts of lower Paleozoic age near Eureka, Nevada

Optical Constants: α = 1.598, β = 1.604; γ = 1.626; (+)2V = 59° (Calc.)

Space Group: P2/m

Strongest Diffraction Lines: 7.5(100), 14.9(60), 11.1(40)

Z: 4

Ref.: Hausen DM (1962) Am Miner 47:637–648

Metatorbernite
$Cu(UO_2)_2(PO_4)_2 \cdot 8H_2O$

Analyses (%): UO_3 58.8–61.0, CuO 8.50–11.4, P_2O_5 13.4–14.0, H_2O 13.6–15.0

Class: 4/m

Cleavage: {001} perfect. Brittle

Color-Luster: Pale green to dark green; transparent to translucent. Vitreous to subadamantine, pearly on {001}. Also dull in pseudomorphs after torbernite. Not fluorescent

Crystal System: Tetragonal

Density: 3.7–3.8 (Meas.), 3.70 (Calc.)

First Reported: 1916 (Described as Metakupferuranit, 1901)

Habit: Crystals thin tablets flattened on {001}. Dominant forms {001} and {100}. As lamellar aggregates, rosettes, and sheaflike or subparallel aggregates. Also as dehydration pseudomorphs after torbernite

Hardness: 2.5

Lattice Constants: (Z = 1), a = 6.96, c = 8.62 (Å). (Z = 2) a = 6.969, c = 17.306 (Å)

Mode of Occurrence: Occurs chiefly as a secondary mineral in the oxidized portion of veins or other deposits containing uraninite together with copper-containing minerals, and possibly as a low-temperature hydrothermal mineral in vein deposits. Found widespread in small amounts in the copper-uranium deposits of the Colorado Plateau; in uraniferous lignite in the North Cave Hills, Harding County, South Dakota; and in quartzite in the Wilson Creek area, Gila County, Arizona. It is also found at the Nicholson mine, Lake Athabasca, Saskatchewan, Canada; in the Redruth area, Cornwall, England; in the Puy-de-Dôme district, France; at several localities in the Erzgebirge of Saxony and Bohemia; and as excellent specimens in the Katanga district, Zaire

Optical Constant: ω ≃ 1.626, ε ≃ 1.624; (+)

Space Group: P4/n

Strongest Diffraction Lines: 8.71(100), 3.68(100), 3.48(80)

Z: 1 or 2

Ref.: Frondel C (1958) USGS Bull 1064, p 208–211. Ross M, Evans HT Jr, Appleman DE (1964) Am Miner 49:1603–1621. Roberts WL, Rapp GR, Weber J (1974) Encyclopedia of minerals, New York, Van Nostrand Reinhold, p 400

Meta-Uranocircite I and II
I–$Ba(UO_2)_2(PO_4)_2 \cdot 8H_2O$
II–$Ba(UO_2)_2(PO_4)_2 \cdot 6H_2O$

Analyses (Phase I) (%): BaO 14.6, UO_3 56.9, P_2O_5 15.1

Class: 4/m 2/m 2/m or 4/m

Cleavage: {001} perfect, {100} distinct. Not brittle. Thin flakes flexible

Color-Luster: Yellow green; transparent to translucent. Pearly on {001}. Fluoresces green in ultraviolet light

Crystal System: Tetragonal

Density: 3.95 (Meas. I), 4.00 (Meas. II)

First Reported: 1877

Habit: Crystals thin plates flattened on {001} with rectangular outline. As subparallel aggregates or fan-like groups

Hardness: 2–2.5

Lattice Constants: a = 6.96, c = 16.90 (Å)

Mode of Occurrence: Found chiefly as a secondary mineral, but may be a primary low-temperature deposit in some vein occurrences. In the United States found at the Davier mine, Lawrence County, disseminated in uraniferous lignite in the Slim Buttes area, Harding County, and in a channel sandstone in the Chadron formation, Badlands, South Dakota; and with fluorite in the Honeycomb Hills, Utah. It also occurs in Germany at Menzenschwand (both I and II); at Schneeberg and at Bergen, Saxony; and at Wölsendorf, Bavaria. Also found in the Banat, Rumania; at Rosmaneira, Spain; and in an alluvial deposit near Antsirabe, Madagascar

Optical Constants: ε or $\alpha = 1.610$, $\beta = 1.623$, ω or $\gamma = 1.623$

Space Group: $P4_2/n$ (I), P4/nmm (II)

Strongest Diffraction Lines: I: 8.93(100), 3.73(80), 5.48(70). II: 8.55(100), 3.61(90), 5.39(70)

Z: 2

Ref.: Frondel C (1958) USGS Bull 1064, p 211–215. Walenta K (1966) Min Abs 17:695. Roberts WL, Rapp GR, Weber J (1974) Encyclopedia of minerals, New York, Van Nostrand Reinhold, p 401

Metavariscite (Lucinite)
$AlPO_4 \cdot 2H_2O$
Dimorphous with variscite

Analyses (%): Al_2O_3 32.4, P_2O_5 44.7, H_2O 22.7, V_2O_5 0.32

Class: 2/m

Cleavage: {010}. Brittle

Color-Luster: Pale green. Transparent to translucent. Vitreous. Streak white

Crystal System: Monoclinic

Density: 2.54 (Meas.), 2.535 (Calc.)

First Reported: 1925

Habit: Crystals minute, thin to thick tabular, slightly elongated or equant. Rarely long prismatic. Also granular, massive. Twinning on {102} as contact twins

Hardness: ~3.5

Lattice Constants: a = 8.47, b = 9.47, c = 5.16 (Å); $\beta \sim 90°$

Mode of Occurrence: Occurs as fine microcrystals in cavities in massive variscite near Lucin, Box Elder County, Utah; at Candelaria, Nevada; and on Malpelo Island, Pacific Ocean

Optical Constants: $\alpha = 1.551$, $\beta = 1.558$, $\gamma = 1.582$; (+)2V = 55°

Space Group: $P2_1/n$

Strongest Diffraction Lines: 2.71(100), 4.25(90), 4.575(80)

Z: 4

Ref.: Roberts WL, Rapp GR, Weber J (1974) Encyclopedia of minerals, New York, Van Nostrand Reinhold, p 402

Metavauxite
$Fe^{2+}Al_2(PO_4)_2(OH)_2 \cdot 8H_2O$

Analyses (%): FeO 1.46–17.0, Al_2O_3 17.4–21.0, CaO 0.27–0.80, MgO 0.56, P_2O_5 28.5–29.6, H_2O 14.3

Class: 2/m

Cleavage: Brittle

Color-Luster: Colorless, white, pale green. Transparent to translucent. Vitreous; fibrous aggregates, silky

Crystal System: Monoclinic

Density: 2.345 (Meas.), 2.35 (Calc.)

First Reported: 1927

Habit: Crystals long prismatic to acicular; as subparallel to radial aggregates

Hardness: 3

Lattice Constants: a = 10.21 kX, b = 9.57, c = 6.93 (Å); $\beta = 98.03°$

Mode of Occurrence: Occurs with paravauxite, vauxite, and wavellite, often encrusting quartz crystals, in the tin mines at Llallagua and Tasna, Bolivia

Optical Constants: $\alpha = 1.550$, $\beta = 1.561$ (Na), $\gamma = 1.577$; (+)2V = large

Space Group: $P2_1/c$

Strongest Diffraction Lines: 2.75(100), 4.67(90), 4.32(85)

Z: 2

Ref.: Roberts WL, Rapp GR, Weber J (1974) Encyclopedia of minerals, New York, Van Nostrand Reinhold, p 402

Metavivianite, = see Oxykerchenite

Millisite
(Ca, Na, K)$Al_3(OH, O)_4(PO_4)_2 \cdot 2H_2O$

Analyses (%): Na_2O 2.8, K_2O 1.7, CaO 7.1, Al_2O_3 36.3, P_2O_5 33.8, H_2O 18.0

Cleavage: {001}(?)

Color-Luster: White, light gray, greenish

Crystal System: Probably tetragonal

Density: 2.83 (Meas.), 2.87 (Calc.)

First Reported: 1930

Habit: As chalcedonic crusts or spherules with finely fibrous structure

Hardness: 5.5

Lattice Constants: a = 7.00, c = 19.07 (Å)

Mode of Occurrence: Occurs interlayered with wardite in variscite nodules at Fairfield, Utah County, Utah; as a microcrystalline intergrowth with crandallite as a major component of the aluminium phosphate zone of the Bone Valley formation of west-central Florida; and in the phosphate deposits of Thiès, Senegal, Africa

Optical Constants: $\alpha = 1.584$, $\beta = 1.598$, $\gamma = 1.602$; (+)2V = moderate

Strongest Diffraction Lines: 4.84(100vb), 2.98(80b), 2.81(80b)

Z: 4

Ref.: Roberts WL, Rapp GR, Weber J (1974) Encyclopedia of minerals, New York, Van Nostrand Reinhold, p 406–407

Minyulite
$KAl_2(PO_4)_2(OH, F) \cdot 4H_2O$

Analyses (%): K_2O 12.3, Na_2O 0.45, Al_2O_3 30.0, P_2O_5 35.6, H_2O 17.8

Class: mm2

Cleavage: {001} perfect. Brittle

Color-Luster: Colorless, white. Transparent. Silky

Crystal System: Orthorhombic

Density: 2.45

First Reported: 1933

Habit: As acicular radiating aggregates

Hardness: 3.5

Lattice Constants: a = 9.35, b = 9.74, c = 5.52 (Å)

Mode of Occurrence: Occurs at Noarlunga, South Australia, and in glauconitic phosphate deposits near Minyulo Well, Dandaragan, Western Australia

Optical Constants: $\alpha = 1.531$, $\beta = 1.534$, $\gamma = 1.538$; (+)2V large

Space Group: Pba2

Strongest Diffraction Lines: 5.6(100), 3.38(100), 6.8(60)

Z: 2

Ref.: Roberts WL, Rapp GR, Weber J (1974) Encyclopedia of minerals, New York, Van Nostrand Reinhold, p 408. Kampf AR (1977) Am Miner 62:256–262

Mitridatite

(Robertsite-Mitridatite Series)

$Ca_6(H_2O)_6[Fe_9O_6(PO_4)_9] \cdot 3H_2O$

Analyses (%): CaO 8.20–16.1, MgO 0.58–1.04, Fe_2O_3 29.0–43.3, P_2O_5 17.2–30.6, H_2O 7.56–26.8

Class: 2/m

Cleavage: {100} good. Fracture uneven. Brittle

Color-Luster: Crystals deep red to bronzy-red; translucent to nearly opaque; resinous to adamantine. Colloidal or fine-grained material greenish yellow to green and dark green; also brownish green, and yellowish brown to brownish black; dull and earthy to resinous. Streak dull olive green

Crystal System: Monoclinic

Density: 3.24 (Meas.), 3.25 (Calc.)

First Reported: 1910

Habit: Crystals thin tabular, pseudo-rhombohedral in aspect, minute. Forms: a{100}, c{001} (small), and v{$\bar{4}$23}. Also as colloidal or fine grained nodules, crusts, veinlets, and coatings

Hardness: 3.5

Lattice Constants: a = 17.52, b = 19.35, c = 11.25 (Å); $\beta = 95.92°$

Mode of Occurrence: Mitridatite crystals are found in association with jahnsite, collinsite, hureaulite, and hydroxylapatite lining cavities in altered triphylite at the White Elephant pegmatite, Custer County, and commonly as a stain or thin incrustation in numerous triphylite-lithiophilite-bearing pegmatites elsewhere in the Black Hills of South Dakota. It also occurs at the Yanysk-Tokila deposit and other localities on the Kerch and Taman peninsulas in the Soviet Union, and in several pegmatites in Central and East Africa

Optical Constants: $\alpha = 1.785$, $\beta = 1.85$, $\gamma = 1.85$; Biaxial (−), 2V 5°–10°

Space Group: Aa

Strongest Diffraction Lines: 8.64(100), 2.721(70), 5.55(60)

Z: 4

Ref.: Moore PB (1974) Am Miner 59:48–59. Moore PB, Araki T (1977) Inorg Chem 16:1096–1106

Monazite

(Ce, La, Y, Th)PO_4

Analyses (%): Ce_2O_3 21.1–34.0, Y_2O_3 1.15–4.66, (La, Nd)$_2O_3$ 27.9–41.8, Fe_2O_3 0.1–1.5, ThO_2 5.65–12.0, P_2O_5 24.9–29.7, (Pr, Y)$_2O_3$ 0.5–2.2

Class: 2/m

Cleavage: {100} distinct, {010} less distinct, {110} indistinct, {101} indistinct, {011} indistinct. Distinct parting often present on {001}; parting on {$\bar{1}$11} not common. Fracture conchoidal to uneven. Brittle

Color-Luster: Reddish-brown, brown, yellowish-brown, pink, yellow, greenish, grayish white, nearly white. Transparent to subtranslucent. Some crystals show "Alexandrite effect". Resinous, waxy, vitreous to subadamantine. Streak white or slightly colored

Crystal System: Monoclinic

Density: 4.6–5.4

First Reported: 1829

Habit: Crystals commonly small, thin to thick tabular on {100} or elongated along c-axis; sometimes equant, wedge-shaped or prismatic; also as large euhedral crystals weighing several pounds.

Crystal faces often uneven, rough, or striated. Also massive granular; as a detrital mineral. Twinning on {100} common

Hardness: 5–5.5

Lattice Constants: a = 6.77, b = 7.01, c = 6.43 (Å); $\beta = 103.17°$

Mode of Occurrence: Occurs widespread throughout the world as a detrital mineral in placer deposits and beach and river sands; in pegmatites, metamorphic rocks, and vein deposits. In the United States extensive detrital deposits are found in Idaho, Montana, Florida, North and South Carolina, and elsewhere in the Appalachian region. In Colorado it is produced in commercial amounts from the Climax molybdenum mine, and also occurs as fine crystals in many pegmatites throughout the state. Exceptional crystals weighing several pounds are found associated with euxenite near Encampment, Wyoming. It is also found in the Petaca district, New Mexico, and in pegmatite at Amelia, Virginia. Large commercial detrital deposits are found in Australia, Ceylon, India, Malaya, Nigeria, and Brazil. Fine crystals have been recovered from pegmatites at several localities in Norway, Finland, and Madagascar; in Switzerland fine crystals occur at several localities in Alpine vein deposits; and excellent twinned crystals are found associated with cassiterite at Callipampa, Bolivia

Optical Constants: $\alpha = 1.785$, $\beta = 1.787$, $\gamma = 1.840$; (+)2V ~ 12°

Space Group: $P2_1/n$

Strongest Diffraction Lines: 3.09(100), 2.87(70), 3.30(50)

Z: 4

Ref.: Roberts WL, Rapp GR, Weber J (1974) Encyclopedia of minerals, New York, Van Nostrand Reinhold, p 413–414. Frondel C (1958) USGS Bull 1064, p 150–160

Monetite
$CaHPO_4$

Analyses: Actual analysis lacking

Class: $\bar{1}$

Cleavage: Indistinct in three directions. Fracture uneven. Brittle

Color-Luster: Colorless, white, pale yellowish. Transparent to translucent. Vitreous

Crystal System: Triclinic

Density: 2.929 (Meas. synthetic), 2.92 (Calc.)

First Reported: 1882 (Described as Glaubapatite, 1856)

Habit: As massive aggregates composed of minute flattened crystals with rough faces and rhombohedral outline; also stalactitic or as crusts

Hardness: 3.5

Lattice Constants: a = 6.90, b = 7.00, c = 6.65 (Å); $\alpha = 96.35°$, $\beta = 91.27°$, $\gamma = 76.1°$

Mode of Occurrence: Occurs on the Islands of Moneta, Mona, and Los Monges in the Caribbean Sea; with newberyite and apatite on Ascension Island in the South Atlantic ocean; and in a cave deposit at Gunong Jerneh, Malaya

Optical Constants: $\alpha = 1.600$, 1.587, $\beta = 1.614$, $\gamma = 1.631$, 1.640; (+)2V = moderate to large

Space Group: $P\bar{1}$

Strongest Diffraction Lines: 2.96(100), 3.35(75), 3.37(70)

Z: 4

Ref.: Roberts WL, Rapp GR, Weber J (1974) Encyclopedia of minerals, New York, Van Nostrand Reinhold, p 414

Montebrasite
(Amblygonite-Montebrasite Series)
$(Li, Na)Al(PO_4)(OH, F)$

Analyses (%): Li_2O 8.80–9,24, Na_2O 0.66–2.17, Al_2O_3 33.4–35.8, P_2O_5 46.7–47.8, F 1.80–5.45, H_2O 4.54–5.99

Class: $\bar{1}$

Cleavage: {100} perfect, {110} good, {0$\bar{1}$1} distinct, {001} imperfect. Fracture conchoidal to uneven. Brittle

Color-Luster: Usually white to grayish white; also colorless, yellowish, pinkish, tan, greenish, bluish. Transparent to translucent. Vitreous to greasy; pearly on cleavages

Crystal System: Triclinic

Density: 3.027 (Meas.), 2.98 (Calc.)

First Reported: 1872

Habit: Crystals equant to short prismatic, commonly with rough faces. Usually as large cleavable masses. Twinning on $\{\bar{1}\bar{1}1\}$ common

Hardness: 5.5–6

Lattice Constants: a = 5.19, b = 7.12, c = 5.04 (Å); $\alpha = 112.04°$, $\beta = 97.82°$, $\gamma = 68.12°$

Mode of Occurrence: Occurs in granite pegmatites often in masses and crystals of very large size. Found widespread in the Black Hills, South Dakota, and at localities in Maine and Connecticut. It also occurs in Nova Scotia, Brazil, Western Australia, South West Africa, France, Spain, and Sweden. Many reported occurrences of amblygonite probably pertain to montebrasite

Optical Constants: $\alpha = 1.57–1.60$, $\beta = 1.61$, $\gamma = 1.62$; $(-)2V = 81.5°$

Space Group: $P\bar{1}$

Strongest Diffraction Lines: 2.968(100), 3.164(90), 4.672(70)

Z: 2

Ref.: Roberts WL, Rapp GR, Weber J (1974) Encyclopedia of minerals, New York, Van Nostrand Reinhold, p 415–416

Montgomeryite

$Ca_4MgAl_4(PO_4)_6(OH)_4 \cdot 12H_2O$

Analyses (%): CaO 19.0, Al_2O_3 21.4, P_2O_5 37.6, H_2O 21.7

Class: 2/m

Cleavage: {010} perfect, {100} poor

Color-Luster: Deep green, rarely pale green to white to colorless. Transparent to translucent. Vitreous

Crystal System: Monoclinic

Density: 2.46 (Meas. Etta mine), 2.53 (Meas. Fairfield, Utah), 2.51 (Calc.)

First Reported: 1940

Habit: Crystals minute, lath-like, often in subparallel growths. Also massive, as subparallel aggregates of coarse plates

Hardness: 4

Lattice Constants: a = 10.01, b = 24.15, c = 6.26 (Å); $\beta = 91.47°$

Mode of Occurrence: Occurs associated with crandallite and other phosphate minerals in variscite nodules at Fairfield, Utah County, Utah; and with mitridatite and frondelite-rockbridgeite in altered triphylite nodules at the Etta mine, Keystone, Pennington County, South Dakota

Optical Constants: $\alpha = 1.572$, $\beta = 1.578$, $\gamma = 1.582$; $(-)2V = 75°$

Space Group: C2/c

Strongest Diffraction Lines: 5.09(100), 12.0(90), 2.882(50)

Z: 2

Ref.: Mead CW, Mrose ME (1968) U.S. Geol Surv Prof Pap 600-D, p D204–D206. Moore PB, Araki T (1974) Am Miner 59:843–850. Fanfani L et al. (1976) Am Miner 61:12–14

Moraesite

$Be_2PO_4OH \cdot 4H_2O$

Analyses (%): BeO 25.3, P_2O_5 34.8, H_2O 39.8

Class: m or 2/m

Cleavage: Two excellent cleavages, the traces of which on {100} are parallel to the b- and c-axes

Color-Luster: White. Streak white

Crystal System: Monoclinic

Density: 1.805 (Meas.), 1.806 (Calc.)

First Reported: 1953

Habit: Spherulitic masses, as distinct crystals, and as crusts with a coarse fibrous structure. Crystals are singly terminated needles, acicular [001] with forms {100} and {130} and faces (131) and $(1\bar{3}1)$

Hardness: Not determined

Lattice Constants: a = 8.55, b = 36.90, c = 7.13 (Å); β = 97.68°

Mode of Occurrence: Occurs on the walls of vugs that have developed in or adjacent to beryl, and on surfaces of albite, quartz, and muscovite, often associated with frondelite and other secondary phosphates at the Sapucaia pegmatite mine, Minas Geraes, Brazil

Optical Constants: α = 1.462, β = 1.482, γ = 1.490; (—)2V = 65°

Space Group: Cc or C2/c

Strongest Diffraction Lines: 7.00(100). 3.278(90), 4.24(60)

Z: 12

Ref.: Lindberg ML, Pecora WT, Barbosa AL de M (1953) Am Miner 38:1126–1133. Roberts WL, Rapp GR, Weber J (1974) Encyclopedia of minerals, New York, Van Nostrand Reinhold, p 418–419

Morinite (Ježekite)
$Ca_4NaAl_2F_4(OH)(H_2O)_2(PO_4)_2$

Analyses (%): Na_2O 5.10, CaO 13.6, Al_2O_3 17.5, P_2O_5 33.0, F 13.0, H_2O 17.6

Class: 2/m

Cleavage: {100} perfect, {001} imperfect. Fracture subconchoidal to uneven. Brittle

Color-Luster: Colorless, white, pale pink. Translucent. Vitreous to slightly oily

Crystal System: Monoclinic

Density: 2.962 (Meas.), 2.911 (Calc.)

First Reported: 1891 (Ježekite, 1914)

Habit: Crystals long prismatic or tabular, vertically striated. Also columnar, radial-fibrous, or coarsely crystalline

Hardness: 4–4.5

Lattice Constants: a = 9.456, b = 10.690, c = 5.445 (Å); β = 105.46°

Mode of Occurrence: Occurs as large crystalline masses and as prismatic crystals up to 1 cm in length, associated with apatite, augelite, wardite, and montebrasite, at the Hugo mine, Keystone, South Dakota. It also is found in pegmatite at Montebras, Creuze, France; at Greifenstein, Saxony, Germany; and at Viitaniemi, Finland

Optical Constants: α = 1.551, β = 1.563, γ = 1.565; (—)2V = 43°

Space Group: $P2_1/m$

Strongest Diffraction Lines: 2.94(100), 3.47(80), 1.783(80)

Z: 2

Ref.: Fisher DJ, Runner JJ (1958) Am Miner 43:585–594. Hawthorne FC (1979) Can Miner 17:93–102. Roberts WL, Rapp GR, Weber J (1974) Encyclopedia of minerals, New York, Van Nostrand Reinhold, p 420

Mundite
$Al(UO_2)_3(PO_4)_2(OH)_3 \cdot 5\frac{1}{2}H_2O$

Analyses (%): P_2O_5 12.6, UO_3 72.2, Al_2O_3 4.45, H_2O 10.8

Cleavage: {010}, {100}, and {001} perfect

Color-Luster: Pale yellow, pleochroism very weak

Crystal System: Orthorhombic

Density: 4.295 (Calc.)

First Reported: 1981

Habit: Rectangular plates flattened on {010} and elongated on {001}

Hardness: Not given

Lattice Constants: a = 17.08, b = 30.98, c = 13.76 (Å); Structurally related to phosphuranylite

Mode of Occurrence: Found at Kobokobo, Kivu, Zaire, with phuralmite, upalite, and ranunculite

Optical Constants: α = 1.62, β = 1.682, γ = 1.688; Biaxial neg., 2V = 33°

Space Group: $P2_1cn$ or Pmcn

Strongest Diffraction Lines: 7.80(100), 3.37(70), 3.87(60), 3.06(50)

Z: 16

Ref.: Deliens M, Piret P (1981) Bull Miner (Soc Fr Miner Cristallogr) 104:669–671

Nacaphite
Na_2CaPO_4F

Analyses (%): P_2O_5 36.0, F 9.32, CaO 26.7, Na_2O 31.4, SrO 0.55, MnO 0.39

Color-Luster: Colorless, transparent; luster vitreous

Crystal System: Orthorhombic

Density: 2.85 (Meas.), 2.88 (Calc.)

First Reported: 1980

Fracture: Conchoidal

Habit: Inclusions of about 1 mm in diameter

Hardness: 3

Lattice Constants: a = 10.644, b = 24.423, c = 7.098 (Å); pseudo periods of a/2 and b/2 have been noted

Mode of Occurrence: As inclusions in thermonatrite, associated with apatite, aegirine and baryto-lamprophyllite in ijolite-urtite pegmatite in the subsurface apatite mines at Mt. Rasvumchorr, Khibina massif, Kola Peninsula, U.S.S.R.

Optical Constants: $\alpha = 1.508$, $\beta = 1.515$, $\gamma = 1.520$; 2V = 80°, biaxial neg.

Space Group: Cmma or C2ma

Strongest Diffraction Lines: 2.649(100), 3.054(45), 3.049(40), 2.005(40)

Z: 16

Ref.: Khomyakov AP, Kazakova ME, Pushcharovskii DY (1980) Zapiski Vses Miner Obsh 109:50–52; *Idem* (1981) Am Miner 66:218

Nagelschmidtite
$Ca_3(PO_4)_2 \cdot 2\alpha\text{-}Ca_2SiO_4$ [or more generally, $Ca_{3-4}(Si, P)_2O_8$, variable solid solution]

Analyses (%): SiO_2 27.9, P_2O_5 8.27, CaO 27.9, K_2O 0.84, Na_2O 0.82

Cleavage: In thin section, shows a complex set of lamellae intersecting at angles close to 60°. Cleavages {001} good, {110} fair

Color: Two polymorphs, one colorless and transparent, the other cloudy and slightly yellow or brown and turbid

Crystal Dimensions: Not reported

First Reported: 1977

Habit: Anhedral grains up to 150 μm

Mode of Occurrence: Found in the Hatrurim Formation ("Mottled Zone"), Israel. Associated minerals include gehlenite, rankinite, perovskite, titanian andradite and magnetite

Optical Constants: $\alpha = 1.680$, $\gamma = 1.698$ for colorless phase; $\alpha = 1.638$, $\gamma = 1.652$ for cloudy phase; Usually uniaxial pos., 2V small

Strongest Diffraction Lines: 2.66(vs), 2.80(s), 1.94(m) for the colorless, transparent phase; 3.45(m), 2.83(vs), 2.76(m), 1.94(m) for the cloudy phase

Ref.: Gross S (1977) Geol Survey Israel Bull 70, p 80

Nahpoite
Na_2HPO_4

Analyses (%): Na_2O 43.7, P_2O_5 49.5, H_2O 6.32

Color: Earthy white

Crystal System: Monoclinic

Density: 2.58 (Calc.)

First Reported: 1981

Habit: Fine-grained, somewhat elongated crystals that exhibit slightly irregular edges and have maximum dimension of 4 μm

Hardness: Very soft

Lattice Constants: a = 5.47, b = 6.84, c = 5.45 (Å); $\beta = 116°$

Mode of Occurrence: Occurs as a fine-grained, earthy white material in marićite nodules from the Big Fish River area, Yukon Territory, Canada. It is a stable alteration product of the parent marićite found in general association with wolfeite, satterlyite, barićite and vivianite

Optical Constants: Index of refraction, 1.490 (min), 1.505 (max)

Space Group: $P2_1/m$ or $P2_1$

Strongest Diffraction Lines: 2.803(100), 2.72(70), 3.84(55), 3.97(45)

Z: 2

Ref.: Coleman LC, Robertson BT (1981) Can Miner 19:373–376

Nastrophite
$Na(Ba, Sr)PO_4 \cdot 9H_2O$

Analyses (%): P_2O_5 19.0, SrO 22.2, BaO 8.17, CaO 0.41, Na_2O 8.21, K_2O 0.5, H_2O 40.5

Cleavage: None reported; conchoidal fracture

Color-Luster: Colorless, vitreous luster

Crystal System: Cubic

Density: 2.05 (Meas.), 2.12 (Calc.)

First Reported: 1981

Habit: Crystal generally 0.2–0.6, but up to 2–3 mm, and as deposits or irregular form up to 1 cm in diameter

Hardness: About 2, brittle

Lattice Constants: a = 10.559 (Å)

Mode of Occurrence: Found on Alluaiv mountain, Lovozero massif, Kola Peninsula and the Karnasurt mountain, Lovozero massif, U.S.S.R. Associated minerals include cancrinite, aegirine, vuonnemite, epistolite, mountainite, villiaumite, kogarkoite, and thermonatrite

Space Group: $P2_13$

Strongest Diffraction Lines: 2.54(100), 4.67(90), 5.21(80), 3.49(50), 1.953(50)

Z: 4

Ref.: Khomyakov AP, Kazakova ME, Popova GN, Malinovskii YA (1981) Zapiski Vses Miner Obshch 110:604–607

Natromontebrasite (Fremontite)
(Amblygonite Series)
$(Na, Li)Al(PO_4)(OH, F)$

Analyses (%): Li_2O 3.21, Na_2O 11.2, Al_2O_3 33.6, P_2O_5 44.4, F 5.63, H_2O 4.78

Class: $\bar{1}$

Cleavage: {100} perfect, {110} good, {0$\bar{1}$1} distinct, {001} imperfect. Fracture subconchoidal to uneven. Brittle

Color-Luster: Grayish white to white. Translucent. Vitreous

Crystal System: Triclinic

Density: 3.04–3.1 (Meas.)

First Reported: 1913

Habit: Crystals short prismatic, rough; also massive

Hardness: 5.5–6

Lattice Constants: a = 5.18 kX, b = 7.11, c = 5.03 (Å); $\alpha = 112.04°$, $\beta = 97.82°$, $\gamma = 68.12°$

Mode of Occurrence: Occurs in granite pegmatite associated with pink tourmaline and lepidolite at Eight Mile Park, Fremont County, Colorado. Also found at the Strickland quarry, Portland, Connecticut, and at localities in Czechoslovakia and Madagascar

Optical Constants: $\alpha = 1.594$, $\beta = 1.603$, $\gamma = 1.615$; (+)2V = very large

Space Group: $P\bar{1}$

Z: 2

Ref.: Roberts WL, Rapp GR, Weber J (1974) Encyclopedia of minerals, New York, Van Nostrand Reinhold, p 429

Natrophilite
$Na(Mn, Fe)PO_4$

Analysis (%): Na_2O 16.8, FeO 3.06, MnO 38.2, P_2O_5 41.0

Class: 2/m 2/m 2/m

Cleavage: {100} good, {010} indistinct, {021} interrupted. Fracture conchoidal

Color-Luster: Deep wine yellow. Transparent to translucent. Bright resinous, pearly on good cleavage

Crystal System: Orthorhombic

Density: 3.41 (Meas.), 3.47 (Calc.)

First Reported: 1890

Habit: Crystals rare, stout prismatic. Usually granular or as cleavable masses

Hardness: 4.5–5

Lattice Constants: a = 10.53. b = 5.00, c = 6.29 (Å)

Mode of Occurrence: Occurs sparingly in granite pegmatite at Branchville, Connecticut, associated with lithiophilite and secondary phosphate minerals

Optical Constants: $\alpha = 1.672$, $\beta = 1.674$, $\gamma = 1.684$; (+)2V ~ 75°

Space Group: Pnam

Strongest Diffraction Lines: (high natrophilite) 2.72(100), 2.60(80), 3.72(70)

Z: 4

Ref.: Roberts WL, Rapp GR, Weber J (1974) Encyclopedia of minerals, New York, Van Nostrand Reinhold, p 430. Fisher DJ (1965) Am Miner 50:1096–1097

Natrophosphate

$Na_6H(PO_4)_2F \cdot 17H_2O$

Analyses (%): Na_2O 28.4–29.2, P_2O_5 21.3–22.1, F 0.42–2.79, H_2O 41.8–49.2

Cleavage: {111} imperfect. Fracture conchoidal

Color-Luster: Colorless. Transparent. Vitreous to greasy. On exposure to air, surface alters to a secondary powdery coating. Fluoresces weak orange under ultraviolet radiation

Crystal System: Cubic

Density: 1.71–1.722 (Meas.)

First Reported: 1972

Habit: Massive; as dense monomineralic aggregates of irregular form, up to 5 × 3 cm in size

Hardness: ~2.5

Lattice Constant: a = 27.79 (Å)

Mode of Occurrence: Occurs in the central cavernous zone of a pegmatite of Yukspore Mt., Khibina massif, U.S.S.R. It is found in close association with fibrous aegirine, dehayelite, strontian apatite, natrolite, pectolite, and villiaumite in cavities, the walls of which are covered by druses of small crystals of albite, aegirine, and rarely villiaumite

Optical Constants: Constantly isotropic with *n* 1.460–1.462

Space Group: Fd3c

Strongest Diffraction Lines: 2.67(100), 2.42(90), 8.18(70)

Z: 56

Ref.: Kapustin YL, Bykova AV, Bukin VI (1972) Int Geol Rev 14:984–989. Roberts WL, Rapp GR, Weber J (1974) Encyclopedia of minerals, New York, Van Nostrand Reinhold, p 430–431

Newberyite

$MgHPO_4 \cdot 3H_2O$

Analyses (%): MgO 22.4, P_2O_5 40.7, H_2O 35.8

Class: 2/m 2/m 2/m

Cleavage: {010} perfect, {001} imperfect. Fracture uneven to subconchoidal. Brittle

Color-Luster: Colorless. Often grayish or brownish due to impurities. Transparent. Vitreous

Crystal System: Orthorhombic

Density: 2.10 (Meas.)

First Reported: 1879

Habit: Crystals short prismatic, tabular, or equant

Hardness: 3–3.5

Lattice Constants: a = 10.215, b = 10.681, c = 10.014 (Å)

Mode of Occurrence: Occurs as crystals up to 2.5 cm in size associated with struvite and hannayite in bat guano in the Skipton Caves, Ballarat, Victoria, Australia; at Mejillones, Chile; on Ascension Island in the South Atlantic; and on La Reunion Island, Indian Ocean

Optical Constants: $\alpha = 1.514$, $\beta = 1.517$, $\gamma = 1.533$; (+)2V = 44.77°

Space Group: Pbca

Strongest Diffraction Lines: (synthetic) 3.45(100), 3.05(80), 5.9(40); (natural) 4.71(100), 5.94(99), 3.46(94)

Z: 8

Ref.: Roberts WL, Rapp GR, Weber J (1974) Encyclopedia of minerals, New York, Van Nostrand Reinhold, p 434

Ningyoite
$(Ca, U, Ce)_2(PO_4)_2 \cdot 1\text{–}2H_2O$

Analyses (%): CaO 6.1–11.5, UO_3 24.8–50.8, Ce_2O_3(?), P_2O_5 16.8–29.4, H_2O 5.4–9.3

Class: 222

Cleavage: Not determined

Color-Luster: Brownish green to brown in transmitted light

Crystal System: Orthorhombic

Density: 4.75 (Calc.)

First Reported: 1959

Habit: Crystals acicular or elongated lozenge-shaped; extremely small, about 5 μm

Hardness: Not determined

Lattice Constants: a = 6.78, b = 12.10, c = 6.38 (Å)

Mode of Occurrence: Occurs as a coating on pyrite and other minerals or filling cracks and cavities in the ore in an oxidized zone of the Ningyo-toge mine, Tottori Prefecture, Japan

Optical Constants: N ~ 1.64–1.71

Space Group: P222

Strongest Diffraction Lines: 3.02(100), 2.81(80), 2.13(80)

Z: 3

Ref.: Muto T, Meyrowitz R, Pommer M, Murano T (1959) Am Miner 44:633–650. Roberts WL, Rapp GR, Weber J (1974) Encyclopedia of minerals, New York, Van Nostrand Reinhold, p 437

Nissonite
$Cu_2Mg_2(PO_4)_2(OH)_2 \cdot 5H_2O$

Analyses (%): P_2O_5 27.1, V_2O_5 0.2, CuO 32.2, MgO 15.8, H_2O 22.3

Class: 2/m or m

Cleavage: {100} fair

Color-Luster: Bluish green

Crystal System: Monoclinic

Density: 2.73 (Meas.), 2.74 (Calc.)

First Reported: 1966

Habit: Crystals tabular {100} and elongated [001]. Forms c{001}, a{100} and q{$\bar{1}$11}. As minute diamond-shaped crystals; more abundantly as thin crusts of crystal aggregates

Hardness: 2.5

Lattice Constants: a = 22.58, b = 5.027, c = 10.514 (Å); $\beta = 99.33°$

Mode of Occurrence: Occurs associated with malachite, azurite, libethenite, turquoise, chrysocolla, cuprite, barite, calcite, gypsum, riebeckite, and crossite in a small copper prospect in metamorphic rocks of the Franciscan Formation, Panoche Valley, California

Optical Constants: $\alpha = 1.584$, $\beta = 1.620$, $\gamma = 1.621$; (—)2V = 19° (Calc.)

Space Group: C2/c or Cc

Strongest Diffraction Lines: 11.14(100), 2.785(25), 4.374(21)

Z: 4

Ref.: Roberts WL, Rapp GR, Weber J (1974) Encyclopedia of minerals, New York, Van Nostrand Reinhold, p 438

Nováčekite (Autunite Group)
$Mg(UO_2)_2(AsO_4, PO_4)_2 \cdot 12H_2O$

Analyses (%): P_2O_5 6.9, MgO 3.4, As_2O_5 11.2, UO_3 59.7, H_2O 17.0

Class: 4/m

Cleavage: {001} perfect, {010} indistinct, {110} indistinct

Color-Luster: Pale yellow, yellow. Crystals lose luster and become opaque upon dehydration. Transparent to translucent. Fluoresces dull green in ultraviolet light

Crystal System: Tetragonal

Density: ~3.7

First Reported: 1951

Habit: Crystals rectangular plates flattened on {001} ranging in size up to 0.5

mm on an edge. As crusts or porous interlocking aggregates of thin plates and scales; also as lamellar aggregates filling tiny veinlets and cavities

Hardness: 2.5

Lattice Constants: a = 7.11, c = 20.06 (variable) (Å)

Mode of Occurrence: Occurs as a coating on sandstone in the Woodrow area, Laguna Reservation, Valencia County, New Mexico; also found near Aldama, Chihuahua, Mexico, and at Schneeberg, Saxony, Germany

Optical Constants: ε or $\alpha = 1.620$–1.625, $\beta = 1.620$–1.641, ω or $\gamma = 1.620$–1.641; $2V = 0°$–$20°$

Space Group: $P4_2/n$

Strongest Diffraction Lines: 10.15(100), 3.58(90), 5.06(80)

Z: 2

Ref.: Frondel C (1958) USGS Bull 1064, p 177–183. Roberts WL, Rapp GR, Weber J (1974) Encyclopedia of minerals, New York, Van Nostrand Reinhold, p 443

Olgite

$Na(Sr, Ba)PO_4$

Analyses (%): Na_2O 13.9–14.9, SrO 20.4–21.6, BaO 23.4–28.7, La_2O_3 2.3–6.1, Ce_2O_3 0.43–0.96, P_2O_5 31.6–32

Cleavage: Terminated crystals show the faces $(10\bar{1}0)$ and $(10\bar{1}1)$ dominant, and (0001), $(11\bar{2}0)$, $(11\bar{2}1)$ present

Color-Luster: Bright blue to bluish-green, luster vitreous

Crystal System: Hexagonal

Density: 3.94 (Meas.), 3.96 (Calc.)

First Reported: 1980

Habit: Grains up to 1 to 2 mm in diameter

Hardness: 4.5

Lattice Constants: a = 5.558, c = 7.037 (Å)

Mode of Occurrence: Occurs in analcime and natrosilite in nepheline syenite pegmatite, Mt. Karnasurt, Lovozero massif, Kola Peninsula, U.S.S.R.

Optical Constants: $\omega = 1.623$, $\varepsilon = 1.619$; Uniaxial neg.

Space Group: $P\bar{3}ml$, P3ml, or P321

Strongest Diffraction Lines: 2.842(100), 2.763(100), 1.982(63), 3.97(41)

Z: 2

Ref.: Khomyakov AP et al. (1980) Zapiski Vses Miner Obsh 109:347–351; *Idem* (1981) Am Miner 66, 438

Olmsteadite

$K_2(Fe^{2+})_4(Nb, Ta)_2(PO_4)_4O_4 \cdot 4H_2O$

Analyses (%): K_2O 7.8–9.4, MnO 2.9–5.0, FeO 23.2–28.6, Nb_2O_5 17.8–26.5, Ta_2O_5 0.2–12.2, P_2O_5 26.2–28.5, H_2O 0–7.2

Cleavage: Good parallel to {100} and {001}

Color-Luster: Deep brown, red brown to black with bronzy surfaces on some crystals; luster is subadamantine; streak and powder are olive green

Density: 3.36 (Hesnard), 3.31 (Big Chief)

First Reported: 1976

Habit: Well-developed crystals, thin tabular parallel to {001} (Big Chief); and thick tabular parallel to {100} to prismatic (Hesnard). Crystals elongated parallel to [010]. Forms shown generally include c{001}, a{100}, p{111}, p′$\{1\bar{1}1\}$, m{110}, and g{101}

Hardness: 4

Lattice Constants: a = 7.512, b = 10.00, c = 6.492 (Å)

Mode of Occurrence: Found at the Hesnard pegmatite and at the Big Chief pegmatite, Black Hills, South Dakota, where it is produced by hydrothermal leaching and corrosion of primary triphylite-lithiophilite, and presumably, columbite-tantalite. At Hesnard, it is embedded in dense aggregates of red-brown botroidal rockbridgeite

Optical Constants: $\alpha = 1.725$, $\beta = 1.755$, $\gamma = 1.815$; X blue-green, Y yellow, Z brown (Big Chief specimen). $\alpha = 1.765$, $\beta = 1.775$, $\gamma = 1.835$; X dark blue, Y light brown, Z dark brown (Hesnard specimen). (+)2V = 60°

Space Group: $Pb2_1m$

Strongest Diffraction Lines: 6.007(100), 2.990(70), 3.045(60), 2.856(60)

Z: 1

Ref.: Moore PB, Araki T, Kampf AR, Steele IM (1976) Am Miner 61:5–11

Olympite
Na_3PO_4

Analyses (%): P_2O_5 42.5, CO_2 2.30, Na_2O 54.5, MnO 0.50

Color-Luster: Colorless or translucent; luster vitreous

Crystal System: Orthorhombic

Density: 2.8 (Meas.), 2.85 (Calc.)

First Reported: 1980

Fracture: Conchoidal

Habit: Oval grains of diameter 1–3 mm

Hardness: About 4; brittle

Lattice Constants: a = 10.154, b = 14.819, c = 10.143 (Å)

Mode of Occurrence: Found in alkalic pegmatites of Mt. Rasvumchorr, Khibina massif, Kola Peninsula, U.S.S.R., in close association with sidorenkite, villiaumite, aegirine and an (un-named) mineral, $Na_2MnSi_2O_6 \cdot H_2O$

Optical Constants: $\alpha = 1.510$, $\beta = 1.510$, $\gamma = 1.512$; 2V = 46°, biaxial pos.

Space Group: Pnma, or $Pn2_1a$

Strongest Diffraction Lines: 2.582(100), 4.18(90), 2.531(70), 2.433(70), 1.472(70)

Z: 16

Ref.: Khomyakov AP, Bykova AV, Malinovskii YA (1980) Zapiski Vses Mineral Obsh 109:476–479; *Idem* (1981) Am Miner 66:438

Overite (Aluminium analog of segelerite)
$CaMg(H_2O)_4Al(OH)[PO_4]_2$

Analyses (%): CaO 11.6, Al_2O_3 28.0, P_2O_5 37.9, H_2O 22.0

Class: 2/m 2/m 2/m

Cleavage: {010} perfect, {100} poor. Brittle

Color-Luster: Light green to colorless. Transparent to translucent. Vitreous

Crystal System: Orthorhombic

Density: 2.53 (Meas.), 2.48 (Calc.)

First Reported: 1940

Habit: Crystals minute, platy to lath-like, often as subparallel growths on {010}. Also massive, as subparallel aggregates of coarse plates

Hardness: 3.5–4

Lattice Constants: a = 14.78, b = 18.78, c = 7.14 (Å)

Mode of Occurrence: Overite occurs associated with crandallite and other phosphate minerals in variscite nodules at Fairfield, Utah County, Utah

Optical Constants: $\alpha = 1.568$, $\beta = 1.574$, $\gamma - 1.580$; ( )2V = 75°

Space Group: Pcaa

Strongest Diffraction Lines: 2.832(100), 9.4(80), 5.29(60)

Z: 8

Ref.: Roberts WL, Rapp GR, Weber J (1974) Encyclopedia of minerals, New York, Van Nostrand Reinhold, p 453–454. Moore PB (1974) Am Miner 59:48–59

Oxykerchenite (Oxidized vivianite; Metavivianite, Bosphorite)
$(Fe^{3+}_x)(Fe^{2x}_{3-x})(PO_4)_2(OH)_x \cdot (8-x)H_2O$

Palermoite
(Palermoite-Bertossaite Series)
$(Li, Na)_2(Sr, Ca)Al_4(PO_4)_4(OH)_4$

Analyses (%): Li_2O 3.85, Na_2O 1.44, CaO 0.88–1.39, SrO 9.20–12.9, Al_2O_3 30.8–33.8, P_2O_5 42.9–44.6, H_2O 5.36–5.97

Class: 2/m 2/m 2/m

Cleavage: {100} perfect, {001} fair. Fracture fibrous to subconchoidal. Brittle

Color-Luster: Colorless to white. Transparent. Vitreous to subadamantine. Streak white. Fluoresces white in direct X-ray beam

Crystal System: Orthorhombic

Density: 3.22 (Meas.), 3.24 (Calc.)

First Reported: 1953

Habit: Crystals long prismatic, vertically striated

Hardness: 5.5

Lattice Constants: a = 11.556, b = 15.849, c = 7.315 (Å)

Mode of Occurrence: Occurs in association with siderite, childrenite-eosphorite, green fibrous beraunite, whitlockite, brazilianite, apatite, and quartz as a late hydrothermal mineral in open cavities at the Palermo pegmatite, North Groton, New Hampshire

Optical Constants: $\alpha = 1.627$, $\beta = 1.642$, $\gamma = 1.644$; (—)2V ~ 20°

Space Group: Imcb

Strongest Diffraction Lines: 3.089(100), 4.360(64), 3.129(60)

Z: 4

Ref.: Mrose ME (1952) Bull Geol Soc Am 63:1283. Mrose E (1953) Am Miner 38:354. Frondel C, Ito J (1965) Am Miner 50:777–779. Moore PB, Araki T (1975) Am Miner 60:460–465. Roberts WL, Rapp GR, Weber J (1974) Encyclopedia of minerals, New York, Van Nostrand Reinhold, p 456

Panasqueiraite
(The OH analog of Isokite)
$CaMgPO_4(OH, F)$

Analyses (%): CaO 31.0, FeO 0.4, MgO 22.9, P_2O_5 39.6, F 3.1, OH 6.7

Cleavage: {010} poor

Color-Luster: Pink, luster vitreous, streak white

Crystal System: Monoclinic

Density: 3.27 (Meas.), 3.22 (Calc.)

First Reported: 1981

Habit: Massive fine-grained crystals with few crystal faces

Hardness: 5

Lattice Constants: a = 6.535, b = 8.753, c = 6.919 (Å); $\beta = 112.3°$

Mode of Occurrence: Found in the veins at Panasqueira tin-tungsten deposits, Portugal. Like thadeuite, it occupies the vein selvages, sharing a localized phosphate-rich assembalge consisting largely of fluorapatite, wolfeite, topaz, vivianite and althausite

Optical Constants: $\alpha = 1.590$, $\beta = 1.596$, $\gamma = 1.616$; (+)2V = 51°. Blue cathodoluminescence in an electron beam at 12 kV potential

Space Group: Cc or C2/c

Strongest Diffraction Lines: 2.626(100), 3.02(86), 3.20(67), 2.584(45)

Ref.: Isaacs AM, Peacor DR (1981) Can Miner 19:389–392

Panethite
$(Na, Ca, K)_2(Mg, Fe, Mn)_2(PO_4)_2$

Analyses (%): Na_2O 15.2, K_2O 0.9, MgO 24.1, CaO 5.6, FeO 5.3, MnO 1.7, P_2O_5 44.7

Class: 2/m

Cleavage: Not determined

Color-Luster: Pale amber, transparent

Crystal System: Monoclinic

Density: 2.9–3.0 (Meas.), 2.99 (Calc.)

First Reported: 1967

Habit: Massive; as minute grains

Hardness: Not determined

Lattice Constants: a = 10.244, b = 14.715, c = 26.258 (Å); $\beta = 91.53°$

Mode of Occurrence: Occurs associated with whitlockite, brianite, albite, and enstatite in small pockets in the metallic phase of the Dayton octahedrite meteorite

Optical Constants: $\alpha = 1.567$, $\beta = 1.576$, $\gamma = 1.579$; (—)2V = 51°

Space Group: $P2_1/n$

Strongest Diffraction Lines: 3.007(100), 2.710(70), 5.10(60)

Z: 32

Ref.: Fuchs LH, Olson E, Henderson EP (1967) Geochim Cosmochim Acta 321:1711–1719

Parahopeite (Dimorphous with hopeite)
$Zn_3(PO_4)_2 \cdot 4H_2O$

Analyses (%): ZnO 53.0–54.7, P_2O_5 30.5–31.6, H_2O 15.3–15.6

Class: $\bar{1}$

Cleavage: {010} perfect

Color-Luster: Colorless. Transparent. Vitreous, pearly on cleavage

Crystal System: Triclinic

Density: 3.31 (Meas.), 3.304 (Calc.)

First Reported: 1908

Habit: Crystals tabular, elongated parallel to c-axis; often as subparallel aggregates or radial groups. Twinning on {100} common, polysynthetic

Hardness: 3.8

Lattice Constants: a = 5.755, b = 7.535, c = 5.292 (Å); $\alpha = 93.29°$, $\beta = 91.92°$, $\gamma = 91.32°$

Mode of Occurrence: Occurs as a secondary mineral in zinc-bearing ore deposits. Found as excellent crystal groups associated with tarbuttite, hemimorphite, pyromorphite, and limonite at the Broken Hill mine, Zambia; it also occurs associated with hopeite and other secondary zinc minerals at Salmo, British Columbia, Canada

Optical Constants: $\alpha = 1.614$, $\beta = 1.625$, $\gamma = 1.637$; (+)2V = nearly 90°

Space Group: $P\bar{1}$

Strongest Diffraction Lines: 7.56(100), 2.99(90), 4.48(70)

Z: 1

Ref.: Roberts WL, Rapp GR, Weber J (1974) Encyclopedia of minerals, New York, Van Nostrand Reinhold, p 460–461

Parascholzite
$CaZn_2(PO_4)_2 \cdot 2H_2O$

Analyses (%): ZnO 40.1, MnO 1.4, CaO 12.8, P_2O_5 35.7, H_2O 8.8

Cleavage: Not observable, but polysynthetically twinned crystals show a parting along the {100} composition plane

Color-Luster: White to colorless with white streak and vitreous luster

Crystal System: Monoclinic

Density: 3.12 (Meas.), 3.10 (Calc.)

First Reported: 1981

Habit: Tabular prismatic crystals twinned by reflection on {100}; crystals flattened on {100} and elongated parallel to {001}

Hardness: 4

Lattice Constants: a = 17.864, b = 7.422, c = 6.67 (Å); $\beta = 106°$ The unit cell corresponds to the subcell of scholzite, which has b/3 = 7.413A and symmetry Pbcn

Mode of Occurrence: Found in pegmatites at Hagendorf, Bavaria, West Germany where it is associated with vivianite, phosphosiderite and strengite. See also the scholzite paragenesis

Optical Constants: $\alpha = 1.587$, $\beta = 1.588$, $\gamma = 1.603$; 2V = 25°, biaxial pos.

Space Group: Cc or C2/c

Strongest Diffraction Lines: 8.55(100), 2.804(80), 4.158(50), 3.406(40), 2.779(40)

Ref.: Sturman BD, Rouse RC, Dunn PJ (1981) Am Miner 66:843–851

Paravauxite
$Fe^{2+}Al_2(PO_4)_2(OH)_2 \cdot 8H_2O$

Analyses (%): FeO 13.6, Al_2O_3 21.2, Fe_2O_3 0.47–1.52, P_2O_5 27.6–29.8, H_2O 34.1

Class: $\bar{1}$

Cleavage: {010} perfect. Fracture conchoidal. Brittle

Color-Luster: Colorless to pale greenish white. Transparent to translucent. Vitreous; pearly on cleavage. Streak white

Crystal System: Triclinic

Density: 2.36 (Meas.), 2.38 (Calc.)

First Reported: 1922

Habit: Crystals short prismatic to thick tabular. As randomly intergrown, subparallel, or radial aggregates

Hardness: 3

Lattice Constants: a = 5.23, b = 10.52, c = 6.96 (Å); $\alpha = 107°17'$, $\beta = 111.4°$, $\gamma = 72.48°$

Mode of Occurrence: Occurs as fine specimens in the tin deposits at Llallagua, Bolivia, often associated with vauxite, metavauxite, wavellite, and quartz

Optical constants: $\alpha = 1.552$, $\beta = 1.559$, $\gamma = 1.572$; (+)2V = 72°

Space Group: $P\bar{1}$

Strongest Diffraction Lines: 9.82(100), 6.38(90), 4.20(90)

Z: 1

Ref.: Roberts WL, Rapp GR, Weber J (1974) Encyclopedia of minerals, New York, Van Nostrand Reinhold, p 464. Hurlbut CS Jr, Honea RM (1962) Am Miner 47:1–8

Paravivianite (= oxidized vivianite?)

Parsonsite
$Pb_2UO_2(PO_4)_2 \cdot 2H_2O$

Analyses (%): PbO 44.7–47.4, UO_3 29.7–34.7, P_2O_5 14.5–15.1, H_2O 1.56–3.43

Class: $\bar{1}$

Cleavage: {010} indistinct. Fracture conchoidal

Color-Luster: Very pale yellow to yellowish amber, greenish brown, chocolate brown, rarely pale rose. Transparent to translucent. Subadamantine to greasy. Not fluorescent

Crystal System: Triclinic

Density: 5.72 (Meas.), 6.29 (Meas. Synthetic)

First Reported: 1923

Habit: Crystals prismatic, up to 4 mm long, elongated [001] and flattened on {010}; as acicular crystals without distinct form; as rudely radial clusters or tufts or crystals; also as radial fibrous masses and earthy crusts

Hardness: 2.5–3

Lattice Constants: a = 6.862, b = 10.425, c = 6.684 (Å); $\alpha = 101.43°$, $\beta = 98.25°$, $\gamma = 86.28°$

Mode of Occurrence: Occurs as a secondary mineral associated with autunite and phosphuranylite in pegmatite at the Ruggles mine, Grafton Center, New Hampshire. In France it is found associated with torbernite at Lachaux, Puy-de-Dôme, and with renardite and torbernite at Grury, Saône et Loire; at Wölsendorf, Bavaria, Germany, with uranocircite; and with metatorbernite at Shinkolobwe, Katanga, Zaire

Optical Constants: $\alpha = 1.85$, $\gamma = 1.86$; (—)2V ~ 11°–26°

Space Group: $P\bar{1}$

Strongest Diffraction Lines: 3.283(100), 3.253(100), 4.233(65)

Z: 2

Ref.: Frondel C (1958) USGS Bull 1064, p 233–236. Roberts WL, Rapp GR, Weber J (1974) Encyclopedia of minerals, New York, Van Nostrand Reinhold, p 465–466

Penikisite (Penikisite – Kulanite series)
$Ba(Mg, Fe)Al_2(PO_4)_3(OH)_3$

Analyses (%): MgO 6.1–6.5, CaO 1.4, FeO 9.5–10.1, BaO 24.8, Al_2O_3 17.8, P_2O_5 36.4–37.1, H_2O 3.9

Cleavage: Fair to good parallel to {010} and {100}

Color-Luster: Blue to green, transparent to translucent; vitreous luster, streak very pale green to white

Crystal System: Triclinic

Density: 3.79 (Meas.), 3.82 (Calc.)

First Reported: 1977

Habit: Very similar in appearance to kulanite; plates tabular parallel to {101} and rosette-like aggregates

Hardness: About 4

Lattice Constants: a = 8.999, b = 12.069, c = 4.921 (Å); α ~ 90°, β ~ 100°, γ ~ 90°

Mode of Occurrence: Found in the phosphate-ironstone occurrence in the northeastern corner of the Yukon Territory, Canada. It often grades into kulanite with the penikisite zones being near the edges of the kulanite-penikisite crystals

Optical Constants: α = 1.684, β = 1.688, γ = 1.705; Pleochroic with X grass green, Y blue-green, Z pale pink. Biaxial pos., 2V = 56°

Space Group: P1 or P$\bar{1}$

Strongest Diffraction Lines: 3.094(100), 2.915(80), 2.649(70), 8.81(60), 3.028(60)

Z: 2

Ref.: Mandarino JA, Sturman BD, Corlett MI (1977) Can Miner 15:393–395

Perhamite

$3CaO \cdot 3\frac{1}{2}Al_2O_3 \cdot 3SiO_2 \cdot 2P_2O_5 \cdot 18H_2O$

Analyses (%): SiO_2 13.6, P_2O_5 21.9, F 0.1, Al_2O_3 27.1, CaO 12.3, TiO_2 0.09, FeO 0.26, H_2O 24.6

Cleavage: Perfect {0001}

Color: Brown; vitreous luster; streak white

Crystal System: Hexagonal

Density: 2.64 (Meas.), 2.53 (Calc.)

First Reported: 1977

Habit: Isolated spherulitic masses (ca. 1 mm across) of platy crystals. Also occurs as very soft delicate white botroidal clusters

Lattice Constants: a = 7.02, c = 20.21 (Å)

Mode of Occurrence: As a rare mineral in a vuggy amblygonite-rich pegmatite zone at the Bell Pit, Newry Hill, Newry, Maine, where it is associated with siderite, colorless wardite, amblygonite, eosphorite and sphalerite. Has also been reported from Dunton Gem mine, atop Newry Hill, Maine

Optical Constants: ε = 1.577, ω = 1.564

Principal Diffraction Lines: 2.882(100), 5.80(70), 6.08(50), 3.51(50), 3.115(50)

Space Group: P6/mmm

Z: 1

Ref.: Dunn PJ, Appleman DE (1977) Miner Mag 41:437–442

Perloffite

(the Fe^{3+} analog of Bjarebyite)

$Ba(Mn, Fe)_2(Fe^{3+})_2(PO_4)_3(OH)_3$

Analyses (%): Ba 17.6, Ca 1.10, Mg 0.39, Mn 11.1, Fe 18.7, P 13.3 (probably in wt.%, but this is not stated)

Cleavage: {100} perfect

Color-Luster: Dark brown to greenish-brown in small crystals to black in larger ones; streak and powder greenish-yellow; luster vitreous to subadamantine

Crystal System: Monoclinic

Density: 3.996 (Calc.)

First Reported: 1977

Habit: Crystal forms c{001}, g{101}, t{021}, and x{131} result in spear-shaped crystals up to 1 mm long

Hardness: About 5

Lattice Constants: a = 9.223, b = 12.422, c = 4.995 (Å), β = 100°

Mode of Occurrence: Found at the Big Chief pegmatite, Glendale, S. Dakota as crystals perched on ludlamite, hureaulite and siderite in vugs

Optical Constants: α = 1.793, β = 1.803, γ = 1.808; (—)2V = 70°–80°; pleochroic with Y light greenish brown, X and Z dark greenish-brown

Space Group: $P2_1/m$

Strongest Diffraction Lines: 3.166(100), 2.979(60), 3.104(53), 2.742(43)

Z: 2

Ref: Kampf AR (1977) Miner Rec 8:112–114

Phosinaite

$H_2Na_3(Ca, Ce)(SiO_4)(PO_4) \cdot H_2O$

Analyses (%): P_2O_5 20.5–21.6, SiO_2 18.3–23.8, MnO 0–1.90, TR_2O_3 13.0–

13.8, CaO 5.39–12.2, Na_2O 28.0–28.8, K_2O 0.47–0.74, H_2O 4.41–6.07

Cleavage: Perfect on {100}, imperfect on {010} and {110}

Color-Luster: Colorless, pale rose to brownish-rose; luster vitreous

Crystal System: Orthorhombic

Density: 2.62 (Khibina); 3.00 (Lovozero)

First Reported: 1974

Habit: Columnar crystals up to 5×1 mm (Lovozero) or accumulations of irregular form (Khibina). Forms shown include {110} (predominant), also {100}, {010}, {001}

Hardness: 3.5

Lattice Constants: a=12.23, b=14.62, c=7.21 (Å) (Khibina specimen); a=12.24, b=14.59, c=7.20 (Å) (Lovozero specimen)

Mode of Occurrence: Found in alkalic pegmatite on the eastern slope of Mt. Koashva, Khibina where it fills interstices between large crystals of anorthoclase, associated with nepheline, aegirine, lomonosovite, and barian lamprophyllite. It also occurs in ussingite veinlets cutting alkalic rocks at Mt. Karnasurt, Lovozero, Kola Peninsula, U.S.S.R. where it is associated with belovite, nordite, neptunite, and vuonnemite

Optical Constants: $\alpha=1.567$, $\beta=1.569$, $\gamma=1.570$; (—)2V=68° (Khibina); $\alpha=1.570$, $\beta=1.570$, $\beta=1.572$, $\gamma=1.573$; (—)2V=70° (Lovozero)

Space Group: $P22_12$

Strongest Diffraction Lines (Khibina): 2.74(100), 7.44(55), 2.566(53), 6.92(50)

Ref.: Kapustin YI et al. (1974) Zapiski Vses Miner Obshch 103:567–570

Phosphoferrite
(Reddingite-Phosphoferrite Series)
$(Fe, Mn)_3(PO_4)_2 \cdot 3H_2O$

Analyses (%): CaO 1.20, FeO 37.5, MnO 13.6, P_2O_5 34.4, H_2O 13.3

Class: mmm

Cleavage: {010} poor. Fracture uneven. Brittle

Color-Luster: Colorless, pale green; often light to dark reddish brown from alteration. Transparent to translucent. Vitreous to resinous

Crystal System: Orthorhombic

Density: 3.29 (Meas.), 3.340 (Calc.)

First Reported: 1920

Habit: Crystals octahedral with {111}, or tabular {010}; often in parallel grouping. Also massive, granular, and coarse fibrous

Hardness: 3–3.5

Lattice Constants: a=9.41, b=10.02, c=8.66 (Å)

Mode of Occurrence: The mineral occurs as a hydrothermal alteration of triphylite or graftonite in granite pegmatites. Found in South Dakota as fine crystals with ludlamite, vivianite, messelite, and graftonite, at the Bull Moose mine, Custer County, and in association with ludlamite, vivianite, and triphylite at the Dan Patch and Big Chief mines, Pennington County. Originally found at Hagendorf, Bavaria, Germany, with ludlamite, vivianite, and triploidite

Optical Constants: $\alpha=1.672$, $\beta=1.680$, $\gamma=1.700$; (+)2V=68°

Space Group: Pbna

Strongest Diffraction Lines: 3.18(100), 2.724(80), 4.25(70)

Z: 4

Ref.: Roberts WL, Rapp GR, Weber J (1974) Encyclopedia of minerals, New York, Van Nostrand Reinhold, p 477. Moore PB and Araki T (1976) Inorg Chem 15:316–321

Phosphophyllite
$Zn_2(Fe, Mn)(PO_4)_2 \cdot 4H_2O$

Analyses (%): FeO 8.34–16.2, MnO 0.14–8.59, ZnO 34.2–36.0, P_2O_5 32.3–33.9, H_2O 16

Class: 2/m

Cleavage: {100} perfect, {010} distinct, {$\bar{1}$02} distinct. Brittle. Fracture uneven

Color-Luster: Colorless to vivid bluish green. Transparent. Vitreous

Crystal System: Monoclinic

Density: 3.13 (Meas.), 3.145 (Calc.)

First Reported: 1920

Habit: Crystals short to long prismatic or thick tabular. As isolated individuals or in groups. Twinning on {100} common; on {$\bar{1}$02} rare

Hardness: 3–3.5

Lattice Constants: a = 10.25, b = 5.09, c = 10.51 (Å); $\beta = 120.25°$

Mode of Occurrence: Occurs in vugs in massive sulfides at Potosi, Bolivia, as magnificent transparent single crystals up to 3 × 2 inches in size. It also is found in granite pegmatite as small crystals associated with triplite, triphylite, sphalerite, apatite, and a wide variety of secondary phosphate minerals at Hagendorf, Bavaria, Germany

Optical Constants: $\alpha = 1.595$, $\beta = 1.614$ (Na), $\gamma = 1.616$; $(-)2V \sim 45°$

Space Group: $P2_1/c$

Strongest Diffraction Lines: 2.825(100), 4.40(95), 8.84(80)

Z: 2

Ref.: Roberts WL, Rapp GR, Weber J (1974) Encyclopedia of minerals, New York, Van Nostrand Reinhold, p 477–478. Kleber et al. (1961) Acta Cryst 14:795. Hill RJ (1977) Am Miner 62:812–817. Dunn PJ, Nordberg J (1977) Am Miner 62:818

Phosphorroesslerite

$MgH(PO_4) \cdot 7H_2O$

Analyses (%): MgO 16.3, P_2O_5 28.1, H_2O 54.5

Class: 2/m

Cleavage: Fracture conchoidal

Color-Luster: Colorless; usually yellowish due to impurities. Transparent to translucent. Vitreous when fresh; usually dull

Crystal System: Monoclinic

Density: 1.725 (Meas.), 1.717 (Calc.)

First Reported: 1939

Habit: Crystals short prismatic or equant, also skeletal or crusted

Hardness: 2.5

Lattice Constants: a = 11.35, b = 25.36, c = 6.60 (Å); $\beta = 95°$

Mode of Occurrence: Found in abandoned mine workings near Schellgaden, Salzburg province, Austria

Optical Constants: $\alpha = 1.477$, $\beta = 1.485$, $\gamma = 1.486$; $(-)2V = 38.17°$

Space Group: A2/a

Z: 8

Ref.: Roberts WL, Rapp GR, Weber J (1974) Encyclopedia of minerals, New York, Van Nostrand Reinhold, p 478

Phosphosiderite

(Metastrengite; Clinostrengite)

$FePO_4 \cdot 2H_2O$ Dimorphous with strengite

Analyses (%): Fe_2O_3 42.7–44.4, P_2O_5 37.7–38.8, H_2O 17.3–19.3

Class: 2/m

Cleavage: {010} good, {001} indistinct. Fracture uneven. Brittle

Color-Luster: Reddish violet, peach-blossom red, purple, wine yellow, nearly colorless. Transparent to translucent. Vitreous to subresinous

Crystal System: Monoclinic

Density: 2.76 (Meas.), 2.76 (Calc.)

First Reported: 1890

Habit: Crystals tabular or stout prismatic. Also as botryoidal crusts and masses with radial-fibrous structure. Twinning on {101}, common as interpenetration twins

Hardness: 3.5–4

Lattice Constants: a = 5.32, b = 9.87, c = 8.69 (Å); $\beta = 90.6°$

Mode of Occurrence: Occurs as superb crystals up to 1 cm or more in length associated with strengite crystals in vugs in barbosalite at the Bull Moose mine, Custer County, and in other pegmatites in the Black Hills, South Dakota; in rockbridgeite with bermanite at the Williams prospect, Coosa County, Alabama; in pegmatite at Pala, San Diego County, California, and at the Fletcher and Palermo quarries near North Groton, New Hampshire; and with strengite at Manhattan, Nevada. It also is found near Chanteloube, France; in the Kalterborn mine near Eiserfeld, at Hagendorf, and at other places in Germany; in Sweden, Sardinia, and on Malpelo Island, Pacific Ocean

Optical Constants: $\alpha = 1.692$, $\beta = 1.725$, $\gamma = 1.738$; $(-)2V = 62°$

Space Group: $P2_1/n$

Strongest Diffraction Lines: 2.78(100), 4.69(80), 4.37(80)

Z: 4

Ref.: Palache C, Berman H, Frondel C (1951) Dana's system of mineralogy, 7th edn, vol II. Wiley, New York, p 769–771

Phosphuranylite
$Ca(UO_2)_4(PO_4)_2(OH)_4 \cdot 7H_2O$

Analyses (%): CaO 3.8, UO_3 76.7, P_2O_5 9.8–12.1, H_2O 10.8–11.2

Class: 2/m 2/m 2/m

Cleavage: {100} perfect, {010} indistinct. Not brittle

Color-Luster: Light golden yellow to deep yellow. Translucent. Not fluorescent

Crystal System: Orthorhombic

Density: ~4.1 (Meas.)

First Reported: 1879

Habit: Tiny scales, plates, or laths with rectangular outline. As dense, earthy aggregates or thin coatings and crusts

Hardness: ~2.5

Lattice Constants: a = 15.85, b = 17.42, c = 13.76 (Å)

Mode of Occurrence: Occurs widespread in small amounts as a secondary mineral in the weathered zone of pegmatites that carry uraninite, and also found sparingly in several sandstone-type uranium deposits on the Colorado Plateau. In the United States it is found in pegmatite at the Flat Rock and Buchanan mines, Mitchell County, North Carolina; in several pegmatites in the New England area, including the Ruggles and Palermo mines, Grafton County, New Hampshire; and at the Ferguson Lode claim near Keystone, Pennington County, South Dakota, associated with uraninite, meta-autunite, and beta-uranophane. Notable foreign localities include LaCrouzille and Margnac, Haute-Vienne, France; Carrasca and Urgeirica, Portugal; Wolsendorf, Bavaria, Germany; and Memoes, Rio Grande do Norte, Brazil

Optical Constants: $\alpha = 1.658–1.690$, $\beta = 1.699–1.724$, $\gamma = 1.699–1.724$; $(-)2V = 0°–25°, 51°$

Space Group: Bmmb

Strongest Diffraction Lines: 7.91(100), 3.96(60), 3.15(60)

Z: 6

Ref.: Frondel C (1958) USGS Bull 1064:222–227. Roberts WL, Rapp GR, Weber J (1974) Encyclopedia of minerals, New York, Van Nostrand Reinhold, p 478

Phuralumite
$Al_2(UO_2)_3(PO_4)_2(OH)_6 \cdot 10H_2O$

Analyses (%): P_2O_5 10.3, Al_2O_3 7.6, UO_3 65.9, H_2O 16.2

Color-Luster: Lemon-yellow; pleochroic from colorless to very pale yellow on Y and Z. Does not fluoresce in UV light

Crystal System: Monoclinic

Density: 3.5 (Meas.), 3.54 (Calc.)

First Reported: 1979

Habit: Prismatic crystals, max. length of 0.5 mm

Hardness: ~3

Lattice Constants: a = 13.87, b = 20.79, c = 9.38 (Å); $\beta = 112°$

Mode of Occurrence: Occurs in the berylcolumbite pegmatite at Kobokobo, Kivu, Zaire, associated with meta-autunite, phosphuranylite, threadgoldite, and other aluminum uranyl phosphates

Optical Constants: $\alpha = 1.559$ (Calc.), $\beta = 1.616$, $\gamma = 1.624$, X = b, Y near elongation; Biaxial neg., 2V = 40°

Space Group: $P2_1/a$

Strongest Diffraction Lines: 10.4(100), 3.08(80), 5.17(70), 3.40(50)

Z: 4

Ref.: Dalines M, Piret P (1979) Bull Miner 102:333–337

Phurcalite
$Ca_2(UO_2)_3(PO_4)_2(OH)_4 \cdot 4H_2O$

Analyses (%): CaO 8.3, UO_3 70.9, P_2O_5 11.5, H_2O 9.3

Cleavage: {001} and {010} perfect, {100} good

Color-Luster: Yellow, luster vitreous to nearly adamantine

Crystal System: Orthorhombic

Density: 4.14 (Calc.), over 4.03 (Meas.)

First Reported: 1978

Habit: Tablets flattened on [010] and elongated on [001]; crystals up to 1 mm long

Hardness: 3

Lattice Constants: a = 17.426, b = 16.062, c = 13.592 (Å)

Mode of Occurrence: Found at Bergen, Vogtland, Saxony where it is associated with specular hematite

Optical Constants: $\alpha = 1.690$ (Calc.), $\beta = 1.730$, $\gamma = 1.749$. Strongly pleochroic with X bright yellow, Y very pale yellow. Biaxial neg., 2V = 68°

Space Group: Pbca

Strongest Diffraction Lines: 8.05(100), 3.10(80), 3.09(80), 2.878(70)

Z: 8

Ref.: Deliens M, Piret P (1978) Bull Miner 101:356–358

Planerite = Intermediate phase in the Turquoise-Coeruleolactite Series

---

Plumbogummite
$PbAl_3(PO_4)_2(OH)_5 \cdot H_2O$

Analyses (%): PbO 34.4–38.9, Al_2O_3 21.0–29.5, P_2O_5 17.6–22.3, CO_2 2.77–4.66, H_2O 12.7–16.2

Class: $\bar{3}$ 2/m

Cleavage: Not determined. Fracture of masses uneven to subconchoidal. Brittle

Color-Luster: Grayish white, yellowish gray to yellowish or reddish brown; also yellow, greenish or bluish. Translucent. Dull to resinous, gum like. Streak colorless to white

Crystal System: Hexagonal

Density: 4.014 (Meas.), 4.08 (Calc.)

First Reported: 1779

Habit: Massive, compact; also as botryoidal or stalactitic crusts or masses with concentric structure. As minute crystals with hexagonal outline

Hardness: 4.5–5

Lattice Constants: a = 7.018, b = 16.787 (Å)

Mode of Occurrence: Occurs as a secondary mineral in the oxidation zone of lead-bearing ore deposits, and as rolled fragments in diamantiferous alluvial deposits. Found at the Cerro Gordo mine, Inyo County, California; at the Dallas mine, Gilpin County, Colorado; with marcasite at the Canton mine, Georgia; and at Mine la Motte, Missouri. It also occurs in the Diamantina district, Minas Geraes, Brazil; and at deposits in France and Cumberland, England

Optical Constants: $\omega = 1.653$, $\varepsilon = 1.675$; (+)

Space Group: $R\bar{3}m$

Strongest Diffraction Lines: 2.97(100), 5.73(80), 3.51(60)

Z: 3

Ref.: Roberts WL, Rapp GR, Weber J (1974) Encyclopedia of minerals, New York, Van Nostrand Reinhold, p 485–486

Polyquartz. Natural and artificial compounds, such as $AlPO_4$ (berlinite), $AlAsO_4$, $FePO_4$, BPO_4, which are isotypic with quartz, Si_2O_4. Such corresponding polymorphs have also been called Al-phosphorotridymite, and Al-phosphorocristobalite, etc.

Ref.: Embrey PG, Fuller JP (1980) A manual of new mineral names. Oxford University Press, p 284

Pseudoautunite

$(H_3O)_4Ca_2(UO_2)_2(PO_4)_4 \cdot 5H_2O$ (?)

Analyses (%): UO_3 53.1, CaO 7.78, P_2O_5 23.0, H_2O 15.6

Cleavage: {001} perfect, micaceous

Color-Luster: Pale yellow to white. Fluoresces intense greenish yellow in short-wave ultraviolet light; weakly in long-wave

Crystal System: Orthorhombic (?)

Density: 3.28 (Meas.), 3.29 (Calc.)

First Reported: 1964

Habit: Crystals hexagonal in aspect, platy, up to 0.1 mm in size; as crusts

Hardness: Not determined

Lattice Constants: a = b = 6.94, c = 12.85 (Å)

Mode of Occurrence: Occurs in fissures and cavities of albite-acmite veins in association with calcite, pyrochlore, oxidized sulfides, and sometimes apatite, in northern Karelia, U.S.S.R.

Optical Constants: $\alpha = 1.541$, $\beta = 1.568$ (Calc.), $\gamma = 1.570$; $(-)2V = 32°$

Strongest Diffraction Lines: 6.2(100), 3.25(100), 1.92(90)

Ref.: Sergeev AS (1964) Miner Geokhin. Univ Leningrad, Sbornik Statei 1:31–39. Sergeev AS (1965) Am Miner 50:1505–1506. Roberts WL, Rapp GR, Weber J (1974) Encyclopedia of minerals, New York, Van Nostrand Reinhold, p 494

Pseudolaueïte

$MnFe_2(PO_4)_2(OH)_2 \cdot 8H_2O$

Analyses (%): MnO 11.1, Fe_2O_3 28.0, P_2O_5 27.0, H_2O 31.0

Class: 2/m

Cleavage: Not determined. Brittle

Color-Luster: Orange-yellow; vitreous

Crystal System: Monoclinic

Density: 2.463 (Meas.), 2.51 (Calc.)

First Reported: 1956

Habit: Crystals prismatic to thick tabular with forms a{100}, c{001}, m{110}, and rare {011} and {$\bar{2}$01}

Hardness: 3

Lattice Constants: a = 9.647, b = 7,428, c = 10.194 (Å); $\beta = 104.63°$

Mode of Occurrence: Occurs as incrustations and as the core of stewartite crystals associated with oxides of manganese and iron in pegmatite at Hagendorf, Bavaria

Optical Constants: $\alpha = 1.626$, $\beta = 1.650$, $\gamma = 1.686$; $(+)2V = 80°$

Space Group: $P2_1/a$

Strongest Diffraction Lines: 9.926(100), 5.869(70), 3.472(40)

Z: 2

Ref.: Strunz H (1956) Naturwissenschaften 43:128. Strunz H (1956) Am Miner 41:815. Baur WH (1969) Am Miner 54:1312–1323. Roberts WL, Rapp GR, Weber J (1974) Encyclopedia of minerals, New York, Van Nostrand Reinhold, p 495

Pseudomalachite (Tagilite); Monoclinic polymorph has also been synthesized

$Cu_5(PO_4)_2(OH)_4 \cdot H_2O$

Analyses (%): CuO 67.0–71.4, P_2O_5 20.0–23.9, H_2O 6.76–8.51

Class: 2/m

Cleavage: {100} perfect and difficult. Fracture splintery

Color-Luster: Dark green to blackish green (crystals); Fibrous material bluish green to green. Translucent. Vitreous

Crystal System: Monoclinic

Density: 4.30–4.35 (Meas.-coarse crystalline material), 4.08–4.21 (Meas.-radiating and cryptocrystalline), 4.34 (Calc.)

First Reported: 1801

Habit: Crystals short prismatic [001]; often in subparallel aggregates; also reniform, botryoidal, massive with radial fibrous structure and concentric banding, foliated, microcrystalline or dense, colloform. Twinning on {100}

Hardness: 4.5–5

Lattice Constants: a = 17.06, b = 5,76, c = 4.49 (Å); $\beta = 91.03°$

Mode of Occurrence: Occurs as a secondary mineral associated with quartz, chalcedony, malachite, chrysocolla, tenorite, and limonite at numerous localities in France, Germany, Czechoslovakia, Roumania, U.S.S.R., Chile, England, Belgium, Zambia, Zaire, Western Australia, and United States

Optical Constants: $\alpha = 1.791$, $\beta = 1.856$, $\gamma = 1.867$; $(-)2V = 48°$

Space Group: $P2_1/a$

Strongest Diffraction Lines: 4.49(100), 2.386(70), 2.443(60)

Z: 2

Ref.: Roberts WL, Rapp GR, Weber J (1974) Encyclopedia of minerals, New York, Van Nostrand Reinhold, p 495–496. Berry LG (1950) Am Miner 35:365–385. Shoemaker GL, Anderson JB, Kostiner E (1977) Am Miner 62:1042–1048. Shoemaker GL, Anderson JB, Kostiner E (1981) Am Miner 66:109–115

Pseudowavellite = Crandallite

Purpurite (Heterosite-Purpurite Series)
$(Mn^{3+}, Fe^{3+})PO_4$

Analyses (%): Mn_2O_3 29.3, Fe_2O_3 15.9–24.6, CaO 0–1.48, P_2O_5 41.6–47.3, H_2O 5.26–7.62

Class: 2/m 2/m 2/m

Cleavage: {100} good, {010} imperfect. Fracture uneven. Brittle

Color-Luster: Deep rose to reddish purple. Usually externally dark brown to brownish black, due to alteration. Subtranslucent to opaque. Dull to satiny. Streak reddish purple

Crystal System: Orthorhombic

Density: 3.69 (Calc.)

First Reported: 1907

Habit: Massive

Hardness: 4–4.5

Lattice Constants: a = 5.83, b = 9.70, c = 4.77 (Å)

Mode of Occurrence: Occurs as an alteration of lithiophilite in the zone of weathering in granite pegmatites. Found at the Custer Mountain lode, Custer County, and at other places in the Black Hills, South Dakota; at Pala and Rincon, San Diego County, California; at Kings Mountains, Gaston County, North Carolina; at Chanteloube, France, at Feitcira, Portugal; and at Wodgina, Western Australia

Optical Constants: $\alpha = 1.85$, $\beta = 1.86$, $\gamma = 1.92$; 2V = moderate

Space Group: Pmnb

Strongest Diffraction Lines: 2.952(100), 2.448(100), 4.37(70)

Z: 4

Ref.: Roberts WL, Rapp GR, Weber J (1974) Encyclopedia of minerals, New York, Van Nostrand Reinhold, p 497

Pyromorphite (Pyromorphite Series)
$Pb_5(PO_4)_3Cl$

Analyses (%): PbO 77.3–81.2, CaO 0–12.3, P_2O_5 8.31–19.8, Cl 1.72–2.85, F(?), H_2O(?)

Class: 6/m

Cleavage: $\{10\bar{1}1\}$ trace. Fracture uneven to subconchoidal. Brittle

Color-Luster: Various shades of green, yellow, orange, brown, gray; rarely colorless or white. Transparent to translucent. Subadamantine to resinous. Streak white

Crystal System: Hexagonal

Density: 7.04 (Meas.), 7.10 (Calc.)

First Reported: 1748

Habit: Crystals short hexagonal prisms, commonly barrel-shaped and often cavernous; also equant, tabular, or pyramidal. As branching groups of prismatic crystals in nearly parallel position, tapering to slender point. Commonly globular, reniform, and botryoidal with subcolumnar structure; also granular, fibrous, earthy

Hardness: 3.5–4

Lattice Constants: a = 10.00, c = 7.33 (Å)

Mode of Occurrence: Occurs widespread as a secondary mineral in the oxidation zone of lead-bearing ore deposits. Found as fine specimens in the Coer d'Alene district, Shoshone County, Idaho; at the Wheatley mines, Phoenixville, Chester County, and the Ecton and Perkiomen mines, Montgomery County, Pennsylvania; at Galena and Carbonate, Lawrence County, South Dakota; and at deposits in Colorado, New Mexico, Utah, California, and other western states. Exceptional specimens also are found at the Society Girl claim, Moyie, British Columbia, Canada; at Bad Ems and other places in Germany; and at localities in Mexico; England, Scotland, France, Italy, Czechoslovakia, U.S.S.R., Zaire, Algeria, Burma, and at Broken Hill, New South Wales, Australia

Optical Constants: $\omega = 2.058$, $\varepsilon = 2.048$ (Na); (−)

Space Group: $P6_3/m$

Strongest Diffraction Lines: 2.957(100), 2.990(80), 2.065(70)

Z: 2

Ref.: Roberts WL, Rapp GR, Weber J (1974) Encyclopedia of minerals, New York, Van Nostrand Reinhold, p 500

Ranunculite

$AlH(UO_2)(PO_4)(OH)_3 \cdot 4H_2O$

Analyses (%): P_2O_5 13.2, UO_3 54.5, Al_2O_3 9.9, H_2O 20.3

Crystal System: Monoclinic (pseudo-orthorhombic)

Density: 3.4 (Meas.), 3.39 (Calc.)

First Reported: 1979

Habit: Gold-yellow (buttercup) nodules with mean diameter of 0.3 mm at type locality

Lattice Constants: a = 11.1, b = 17.7, c = 18.0 (Å); β = ca. 90°

Mode of Occurrence: With meta-autunite, phosphuranylite, and several other aluminum uranyl phosphates in a pegmatite at Kobokobo, Kivu, Zaire

Optical Constants: $\alpha = 1.643$, $\beta = 1.664$, $\gamma = 1.670$; Biaxial neg., $2V = 56°$

Space Group: not given

Strongest Diffraction Lines: 9.00(100), 3.133(80), 4.70(50)

Z: 14

Ref.: Daliens M, Piret P (1979) Miner Mag 43:321–323

Reddingite

(Reddingite-Phosphoferrite Series)

$(Mn, Fe)_3(PO_4)_2 \cdot 3H_2O$

Analyses (%): MnO 34.5–48.2, FeO 2.19–17.1, CaO 0.15–0.73, P_2O_5 34.5–35.2, H_2O 12.3–13.2

Class: 2/m 2/m 2/m

Cleavage: {010} poor. Fracture uneven. Brittle

Color-Luster: Pinkish to yellowish white, colorless; often reddish brown to dark brown from alteration. Transparent to translucent. Vitreous to resinous

Crystal System: Orthorhombic

Density: 3.23 (Meas.), 3.24 (Calc. Mn: Fe=3:1)

First Reported: 1878

Habit: Crystals octahedral with large {111}, or tabular {010}; often in parallel grouping. Also massive, granular, and coarse fibrous

Hardness: 3–3.5

Lattice Constants: a=9.49, b=10.08, c=8.70 (Å)

Mode of Occurrence: The mineral occurs as a hydrothermal alteration of lithiophilite in granite pegmatites. Found with fairfieldite and other secondary phosphate minerals at Branchville, Fairfield County, Connecticut; also at Buckfield and Poland, Maine

Optical Constants: $\alpha=1.643$–1.658, $\beta=1.648$–1.664, $\gamma=1.674$–1.685; $(+)2V=41°$–$65°$

Space Group: Pbna

Strongest Diffraction Lines: 3.20(100), 2.737(80), 4.28(70)

Z: 4

Ref.: Roberts WL, Rapp GR, Weber J (1974) Encyclopedia of minerals, New York, Van Nostrand Reinhold, p 512

Renardite

$Pb(UO_2)_4(PO_4)_2(OH)_4 \cdot 7H_2O$

Analyses (%): UO_3 64.8–69.0, PbO 12.3–13.0, P_2O_5 8.2–8.7, H_2O 8.7–9.3

Class: 2/m 2/m 2/m

Cleavage: {100} perfect. Brittle

Color-Luster: Lemon yellow, yellow, brownish yellow. Translucent to transparent; vitreous to adamantine. Not fluorescent

Crystal System: Orthorhombic

Density: 4.35 (Meas.), 4.34 (Calc.)

First Reported: 1928

Habit: Crystals, up to 2.0 mm long, usually rectangular plates flattened on {100} with {010}, {101}, and rarely {001}. Also lath-like by elongation along c-axis. As microcrystalline crusts, small crystal aggregates, fibrous radiated nodules, and as lamellar masses

Hardness: 3.5

Lattice Constants: a=16.01, b=17.5, c=13.7 (Å)

Mode of Occurrence: Occurs as a secondary mineral associated with torbernite, kasolite, and dumontite at Shinkolobwe, Katanga, Zaire; as an alteration of uraninite at Grabo, Ivory Coast, Africa; and at a number of localities in France including Kersegalec near Lignol, Morbihan; in the vicinity of Lachaux, Puy-de-Dôme; and at LaFaye, Grury, Saône et Loire

Optical Constants: $\alpha-1.715$–1.721, $\beta=1.736$–1.741, $\gamma=1.739$–1.745; $(-)2V\approx40°$–$45°$

Space Group: Bmmb

Strongest Diffraction Lines: 7.97(100), 3.99(90), 5.83(80)

Z: 6

Ref.: Frondel C (1958) USGS Bull 1064:227–230. Roberts WL, Rapp GR, Weber J (1974) Encyclopedia of minerals, New York, Van Nostrand Reinhold, p 514

Rhabdophane (also *Erikite* or titano-rhabdophane; silico-rhabdophane)

$(Ce, Y, La)(PO_4) \cdot H_2O$

Analyses (%): Ce_2O_3 0–61.7, La_2O_3 0–55.0, Y_2O_3 0–8.7, P_2O_5 24.8, H_2O 7.4

Class: 6 2 2

Cleavage: Fracture uneven

Color-Luster: Brown, pinkish, or yellowish white. Translucent. Greasy

Crystal System: Hexagonal

Density: 3.94–4.01 (Meas.)

First Reported: 1878

Habit: As stalactitic or botryoidal incrustations with radial-fibrous structure

Hardness: 3.5

Lattice Constants: a=6.98, c=6.39 (Å)

Mode of Occurrence: Occurs as thin crusts in the limonite deposit at Salisbury, Connecticut, and at a deposit in Cornwall, England

Optical Constants: $\omega = 1.654$, $\varepsilon = 1.703$; (+)

Space Group: $P6_222$

Strongest Diffraction Lines: 3.02(100), 4.40(80), 2.83(80)

Z: 3

Ref.: Roberts WL, Rapp GR, Weber J (1974) Encyclopedia of minerals, New York, Van Nostrand Reinhold, p 515–516

Richellite

$Ca_3Fe_{10}(PO_4)_8(OH, F)_{12} \cdot nH_2O$

Analyses (%): CaO 5.53–7.19, Al_2O_3 1.79–3.64, Fe_2O_3 28.7–29.7, P_2O_5 25.5–28.8, HF 0.96–6.11, H_2O 6.10–25.6

Class: 4(?)

Cleavage: Not determined

Color-Luster: Yellowish to reddish brown. Greasy to horn-like

Crystal System: Tetragonal

Density: ~2.0

First Reported: 1884

Habit: Massive; compact or foliated, also as globules with radial-fibrous structure

Hardness: 2–3

Lattice Constants: a = 5.18, c = 12.61 (Å), heated (500 °C, 30 min.)

Mode of Occurrence: Occurs associated with halloysite and allophane at Richelle near Visé, Belgium

Space Group: P4(?)

Strongest Diffraction Lines: 3.24(100), 1.590(80) (heated), 3.58(65)

Ref.: Roberts WL, Rapp GR, Weber J (1974) Encyclopedia of minerals, New York, Van Nostrand Reinhold, p 518

Robertsite

(Robertsite-Mitridatite Series)

$Ca_6(H_2O)_6Mn_9O_6(PO_4)_9 \cdot 3H_2O$

Analyses (%): None found in the literature

Class: 2/m

Cleavage: {100} good. Fracture uneven. Brittle

Color-Luster: Light to dark reddish-brown, deep red to bronzy, shiny black. Translucent to nearly opaque. Silky; resinous to adamantine. Streak chocolate-brown

Crystal System: Monoclinic

Density: 3.13, 3.17 (Meas.), 3.20 (Calc.)

First Reported: 1974

Habit: Crystals thick tabular, wedge-shaped, small; very thick crystals pseudo-rhombohedral in aspect. Forms: a {100}, c {001}, and p {031}. As bunched radial aggregates composed of plates up to 5 mm across; as botryoidal aggregates with fibrous structure; and as feathery aggregates. Frequently twinned by rotation normal to {100}

Hardness: 3.5

Lattice Constants: a = 17.36, b = 19.53, c = 11.30 (Å); $\beta = 96°$

Mode of Occurrence: Robertsite occurs in moderate abundance in granite pegmatites as a late stage product in corroded triphylite-heterosite-ferrisicklerite-rockbridgeite masses, usually associated with whitlockite, collinsite, hydroxylapatite, jahnsite, leucophosphite and other secondary phosphate minerals. Found at the Tip Top, Linwood, and White Elephant mines in Custer County, and at the Gap Lode mine, Pennington County, South Dakota

Optical Constants: $\alpha = 1.775$, $\beta = 1.82$, $\gamma = 1.82$; Biaxial (−), 2V ~ 8°

Space Group: Aa

Strongest Diffraction Lines: 8.63(100), 2.75(60), 5.61(50)

Z: 8

Ref.: Moore PB (1974) Am Miner 59:48–59

Rockbridgeite

(Frondelite-Rockbridgeite Series)

$(Fe^{2+}, Mn)Fe_4^{3+}(PO_4)_3(OH)_5$

Analyses (%): MnO 0.24–4.10, MgO 0–2.6, FeO 0.99–6.06, Fe_2O_3 50.8–55.8, Al_2O_3 0–1.31, P_2O_5 30.4–32.8, H_2O 7.96–8.35

Class: 2/m 2/m 2/m

Cleavage: {100} perfect, {010} distinct, {001} distinct. Fracture uneven. Brittle

Color-Luster: Light to dark olive green to nearly black, often becoming brownish green to reddish brown on oxidation. Aggregates often exhibit concentric color banding. Subtranslucent. Vitreous to dull

Crystal System: Orthorhombic

Density: 3.3–3.49 (Meas.)

First Reported: 1949

Habit: As minute eleongated slender prismatic crystals, often as isolated individuals or as radial aggregates. Usually as botryoidal crusts or as masses with radial-fibrous or fine columnar structure

Hardness: 4.5

Lattice Constants: a = 13.783, b = 16.805, c = 5.172 (Å)

Mode of Occurrence: Occurs in "limonite" beds, in novaculite deposits, and as an alteration product of triphylite or other iron-manganese phosphate minerals in granite pegmatites. Found widespread in the Black Hills, South Dakota, especially at the Bull Moose and Tip Top mines near Custer, and at the Big Chief mine near Keystone, as excellent microcrystals and large fibrous masses; in prospect pits in the Ouachita Mountains, Arkansas; in Cherokee and Coosa Counties, Alabama; at Greenbelt, Maryland; in Rockbridge County, Virginia; and at the Fletcher and Palermo mines, near North Groton, New Hampshire. It also is found in France, Germany, U.S.S.R., and Brazil

Optical Constants: $\alpha = 1.875$, $\beta = 1.880$, $\gamma = 1.897$; (+)2V = moderate

Space Group: Bbmm

Strongest Diffraction Lines: 3.196(100), 4.842(50), 3.573(50)

Z: 4

Ref.: Roberts WL, Rapp GR, Weber J (1974) Encyclopedia of minerals, New York, Van Nostrand Reinhold, p 522. Moore PB (1970) Am Miner 55:135–169

Roscherite
$(Ca, Mn, Fe)_3Be_3(PO_4)_3(OH)_3 \cdot 2H_2O$

Analyses (%): CaO 11.5, MnO 14.5, BeO (?), FeO 10.1, Al_2O_3 13.7, P_2O_5 38.0, H_2O 12.2

Class: 2/m

Cleavage: {001} good, {010} distinct

Color-Luster: Olive green, light brown, dark brown, reddish-brown

Crystal System: Monoclinic

Density: 2.934 (Meas.), 2.93 (Calc.)

First Reported: 1914

Habit: Crystals tabular parallel to (100); also short prismatic [001], with six- or eight-sided cross section; and as aggregates of thin plates or radiating fibrous masses

Hardness: 4.5

Lattice Constants: a = 15.95, b – 11.95, c = 6.62 (Å); $\beta = 94.83°$

Mode of Occurrence: Occurs in pegmatite at the Nevel quarry, Newry and at Black Mountain, Maine; as single crystals, crystal aggregates, and granular crusts in vugs in muscovite, associated with faheyite, variscite, frondelite, beryl, and euhedral crystals of quartz at the Sapucaia pegmatite mine, near Conselheiro Pena, Minas Geraes, Brazil. Also found with morinite, apatite, lacroixite, childrenite, and tourmaline in drusy cavities of a granite at Greifenstein, near Ehrenfriedersdorf, Saxony, Germany

Optical Constants: $\alpha = 1.636$, $\beta = 1.641$, $\gamma = 1.651$; (+)2V. $\alpha = 1.624$, $\beta = 1.639$, $\gamma = 1.643$; (−)2V = large

Space Group: C2/c

Strongest Diffraction Lines: 5.96(100), 9.58(90), 3.18(70)

Z: 4

Ref.: Roberts WL, Rapp GR, Weber J (1974) Encyclopedia of minerals, New York, Van Nostrand Reinhold, p 525–526. Lindberg ML (1958) Am Miner 43:824–838

Rosiérésite (Species status in doubt)
Hydrous phosphate of Pb, Cu, Al

Analyses (%): CuO 3.0, PbO 10.0, Al_2O_3 23.0, P_2O_5 25.5, H_2O 38.0

Cleavage: None

Color-Luster: Greenish yellow, yellow, pale brown

Crystal System: Unknown

Density: 2.2 (Meas.)

First Reported: 1910

Habit: As opaline stalactitic masses with concentric structure

Hardness: Not determined

Mode of Occurrence: Occurs as a recent deposit on mine walls at the copper mine at Rosiérés, Tarn, France

Optical Constant: N $\sim$ 1.50

Ref.: Roberts WL, Rapp GR, Weber J (1974) Encyclopedia of minerals, New York, Van Nostrand Reinhold, p 527

Rusakovite
(Inadequately described mineral)
$(Fe, Al)_5\{(V, P)O_4\}_2(OH)_9 \cdot 3H_2O$

Analyses (%): Al_2O_3 5.0, Fe_2O_3 45.0, V_2O_5 16.6, V_2O_4 5.0, P_2O_5 6.5, H_2O 18.3

Color-Luster: Yellow-orange to reddish yellow; dull. Streak ocher yellow

Crystal System: Unknown

Density: 2.73–2.80 (Meas.)

First Reported: 1960

Habit: As crusts, veinlets, and reniform concretions

Hardness: 1.5–2

Mode of Occurrence: Occurs in a surface, partially oxidized layer of carbonaceous shale, which contains apatite, colophane, vanadium-bearing mica, and sulfides of copper, zinc, lead, and vanadium in northwestern Kara-Tau, Kazakhstan

Optical Constant: n = 1.833

Strongest Diffraction Lines: 3.21(100), 2.945(90), 2.441(80)

Ref.: Ankinovich E (1960) Am Miner 45:1316. *Idem* (1960) Mem All-Union Miner Soc 89:440. Roberts WL, Rapp GR, Weber J (1974) Encyclopedia of minerals, New York, Van Nostrand Reinhold, p 529

Sabugalite
$HAl(UO_2)_4(PO_4)_4 \cdot 16H_2O$

Analyses (%): P_2O_5 16.1, UO_3 65.2, Al_2O_3 2.7, H_2O 15.9

Class: 4/m 2/m 2/m

Cleavage: {001} perfect

Color-Luster: Bright yellow to lemon yellow; transparent to translucent. Weakly vitreous. Fluoresces lemon yellow in ultraviolet light

Crystal System: Tetragonal

Density: 3.2 (Meas.), 3.15 (Calc.)

First Reported: 1951

Habit: Crystals thin plates up to 1.0 mm on edge, square or rectangular lathlike in shape. Plane of flattening uneven, warped or striated. As densely aggregated crusts

Hardness: 2.5

Lattice Constants: a = 6.96, c = 19.3 (Å)

Mode of Occurrence: Occurs in sandstone-type uranium deposits at the Huskon and Arrow Head claims near Cameron, Arizona, and at the Happy Jack mine, White Canyon, San Juan County, Utah. In Wyoming it is found at the Lucky Mc No. 20 mine, the Blue Buck claim, and in the Blarco Group, Fremont County; at the Del Linch claim, Johnson County; and at the Poison Creek claim, Crook County. It occurs in Portugal as a secondary mineral in the weathered outcrops of uraninite veins at Mina da Quarta Feira, Sabugal County, Beira Alta Province; at the Mina de Coitos, Bendada Parish; and at Kariz in Minko Province. It is also found at Margnac II, Haute-Vienne, France

Optical Constants: α or ε = 1.564–1.565, β = 1.581–1.583, γ or ω = 1.582–1.584; 2V = 0° to moderate

Space Group: I4/mmm

Strongest Diffraction Lines: 9.69(100), 4.86(90), 3.47(80)

Z: 1

Ref.: Frondel C (1958) USGS Bull 1064:196–200. Walenta K (1978) Am Miner 42:117–128. Roberts WL, Rapp GR, Weber J (1974) Encyclopedia of minerals, New York, Van Nostrand Reinhold, p 531

Saléeite (Autunite Group)
$Mg(UO_2)_2(PO_4)_2 \cdot 8\text{–}10H_2O$

Analyses (%): MgO 4.3–5.0, UO_3 60.3–64,1, P_2O_5 12.0–15.2, H_2O 16.5–19.3

Class: 4/m 2/m 2/m

Cleavage: {001} perfect, {010} indistinct, {110} indistinct

Color-Luster: Pale yellow, yellow, yellowish green. Crystals lose their luster and become opaque upon dehydration. Translucent to transparent. Fluoresces brilliant yellow-green in ultraviolet light

Crystal System: Tetragonal

Density: 3.27 (Meas.)

First Reported: 1932

Habit: Crystals rectangular plates flattened on {001} ranging in size up to 3.0 mm on edge, as flattened bipyramids. Forms are {001}, {010}, {012}, and {120}. In fanshaped groups, subparallel aggregates with torbernite, and as aggregates of thin plates and scales

Hardness: 2.5

Lattice Constants: a = 7.01, c = 19.84 (Å)

Mode of Occurrence: Occurs sparsely disseminated in carnotite-bearing sandstone at the Gull mine, Fall River County, and as a constituent of uraniferous lignite in the Cave Hills and Slim Buttes areas, Harding County, South Dakota. In Europe it occurs with autunite and limonite in quartz at Plessis, near Mortagne, Deux-Sèvres, France; with meta-autunite, phosphuranylite, and sabugalite in altered granite at Mina da Quarta Seira, Sabugal, Portugal; an arsenatian variety is found at Schneeberg, Saxony, Germany, associated with uranophane and zeunerite. Also found as superb crystals at Rum Jungle, Australia, and at Shinkolobwe, Katanga, Zaire

Optical Constants: ε or α = 1.544–1.565, β = 1.570–1.582, ω or γ = 1.571–1.585; 2V = 0°–10°, 61°–65°

Space Group: I4/mmm

Strongest Diffraction Lines: 9.85(100), 3.49(90), 4.95(80)

Z: 2

Ref.: Frondel C (1958) USGS Bull 1064:177–183. Roberts WL, Rapp GR, Weber J (1974) Encyclopedia of minerals, New York, Van Nostrand Reinhold, p 533–534

Salmonsite = Hureaulite + Jahnsite

Sampleite
$NaCaCu_5(PO_4)_4Cl \cdot 5H_2O$

Analyses (%): Na_2O 3.11, K_2O 1.49, CaO 5.83, MgO 0.52, CuO 41.1, Cl 4.00, P_2O_5 32.1, H_2O 9.74

Class: 2/m 2/m 2/m

Cleavage: {010} perfect, {100} good, {001} good

Color-Luster: Light blue to bluish green. Pearly on {010}

Crystal System: Orthorhombic

Density: 3.20 (Meas.), 3.274 (Calc.)

First Reported: 1942

Habit: Crystals very thin, lath-like, elongated along c-axis and flattened {010}

Hardness: ~4

Lattice Constants: a=9.70, b=38.40, c=9.65 (Å)

Mode of Occurrence: Occurs associated with gypsum, limonite, jarosite, and other minerals in highly oxidized ore at Chuquicamata, Chile

Optical Constants: $\alpha=1.629$, $\beta=1.677$ (Na), $\gamma=1.679$; (−)2V=22.51°

Strongest Diffraction Lines: 9.60(100), 3.04(100), 4.30(80)

Z: 8

Ref.: Roberts WL, Rapp GR, Weber J (1974) Encyclopedia of minerals, New York, Van Nostrand Reinhold, p 536

Samuelsonite

$(Ca, Ba)(Fe^{2+})_2(Mn^{2+})_2Ca_8Al_2(PO_4)_{10}(OH)_2$

Analyses (%): Na_2O 0.2–1.4, BaO 2.0–4.8, CaO 28.3–29.6, FeO 6.5–11.3, MnO 7.2–7.9, Al_2O_3 4.8–6.2, P_2O_5 43.1–44.0, H_2O 3.0

Cleavage: {001} fair

Color-Luster: Colorless, luster subadamantine

Crystal System: Monoclinic

Density: 3.353 (Meas.), 3.267 (Calc.)

First Reported: 1975

Habit: Striated flattened prismatic crystals parallel to [010]. Forms observed include *c*{001}, *a*{100}, *d*{101}, *f*{$\bar{1}$01}, *j*{$\bar{2}$01}, *e*{021}, *t*{$\bar{2}$11}, and *r*{$\bar{1}$12}

Hardness: 5+

Lattice Constants: a=18.495, b=6.805, c=14.00 (Å); $\beta=112.8°$

Mode of Occurrence: A rare constituent of the whitlockite-apatite paragenesis at the Palermo No. 1 pegmatite, North Groton, New Hampshire. Other associated phosphate minerals include childrenite and arrojadite

Optical Constants: $\alpha=1.645$, $\beta=1.650$, $\gamma=1.655$; (+)2V=70°–80°

Space Group: C2/m

Strongest Diffraction Lines: 3.058(100), 2.657(70), 3.032(60), 1.706(60)

Z: 2

Ref.: Moore PB, Irving AJ, Kampf AR (1975) Am Miner 60:957–964. Moore PB, Araki T (1977) Am Miner 62:229–245

Sarcopside

$(Fe, Mn, Mg)_3(PO_4)_2$

Analyses (%): CaO 3.40–5.23, MnO 10.8–20.6, FeO 30.5–39.9, Fe_2O_3 0.2–8.83, P_2O_5 33.3–40.5, H_2O 0.87–1.94, F 0–4.35

Class: 2/m

Cleavage: {100} good, {001} good, {010} poor. Fracture uneven to splintery

Color-Luster: Colorless when fresh; usually gray to reddish brown to brown due to slight alteration. Transparent to subtranslucent. Vitreous to silky and glistening when slightly altered

Crystal System: Monoclinic

Density: 3.79 (Meas.), 3.798 (Calc.)

First Reported: 1868

Habit: As irregular masses with fibrous structure, commonly as intergrowths with graftonite. Rarely in distorted six-sided plates. Twinning on {001}, polysynthetic

Hardness: 4

Lattice Constants: a=10.47, b=4.80, c=6.06 (Å); $\beta\sim90°$

Mode of Occurrence: Occurs in granite pegmatite as large masses intergrown with graftonite at the Bull Moose mine, Custer County, South Dakota; at the French King No. 2 quarry, East Alstead, and at Deering, New Hampshire; at Otav and at Domažlice, Bohemia, Czechoslovakia; near Tirschenreuth, Germany; and near Michelsdorf, Silesia, Poland

Optical Constants: $\alpha=1.670$, $\beta=1.728$, $\gamma=1.732$; (−)2V=28°

Space Group: $P2_1/a$

Strongest Diffraction Lines: 3.03(100), 3.54(80), 6.06(50)

Z: 2

Ref.: Hurlbut CS Jr (1965) Am Miner 50:1698–1707. Peacor DR (1969) Am Miner 54:969–972. Peacor DR, Garske D (1964) Am Miner 49:1149–1150. Moore PB (1972) Am Miner 57:24–35. Roberts WL, Rapp GR, Weber J (1974) Encyclopedia of minerals, New York, Van Nostrand Reinhold, p 538

Saryarkite
$(Ca, Y, Th)_2Al_4(SiO_4, PO_4)_4(OH) \cdot 9H_2O$

Analyses (%): CaO 6.3, Al_2O_3 26.6–28.7, ThO_2 7.8–9.1, RE_2O_3 10.9, Fe_2O_3 1.4–2.2, SiO_2 14.6, SO_3 2.9–3.4, P_2O_5 12.1, H_2O 14.0–14.8

Class: 422

Cleavage: None

Color-Luster: White; dull to greasy

Crystal System: Tetragonal

Density: 3.07–3.15 (Meas.), 3.35 (Calc.)

First Reported: 1964

Habit: Massive

Hardness: 3.5–4

Lattice Constants: a = 8.213, c = 6.55 (Å)

Mode of Occurrence: Occurs associated with barite, molybdenite, pyrite, hematite, hydrous iron oxides, and a thorium mineral in Devonian propylitized acid effusives and in altered granitic rocks, associated with secondary quartzites containing sericite, alunite, and andalusite from an unspecified locality, presumably in Kazakhstan, U.S.S.R.

Optical Constants: $\omega = 1.606$, $\varepsilon = 1.620$; (+)

Space Group: $P42_12$ or $P4_22_12$

Strongest Diffraction Lines: 3.014(100), 2.827(100), 1.854(100)

Z: 4

Ref.: Roberts WL, Rapp GR, Weber J (1974) Encyclopedia of minerals, New York, Van Nostrand Reinhold, p 539

Sasaite
$(Al, Fe^{3+})_{14}(PO_4)_{11}(OH)_7SO_4 \cdot 83H_2O$

Analyses (%): Al_2O_3 21.6, Fe_2O_3 1.05, SO_3 2.77, P_2O_5 24.2, H_2O 49.5

Color: White

Crystal System: Orthorhombic

Density: 1.747 (Calc.), 1.75 (Meas.)

First Reported: 1978

Habit: Nodular aggregation of rhombic plates 10–20 μm in greatest dimension. Dehydrates rapidly in ambient air

Lattice Constants: a = 21.50, b = 30.04, c = 92.06 (Å)

Mode of Occurrence: Found as white chalky nodules in the soil of a dolomitic cave, apparently derived from the action of bat guano on clay minerals. Type locality, West Driefontein Cave, Transvaal, South Africa

Optical Constants: $\alpha = 1.465$, $\beta = 1.473$, $\gamma = 1.477$

Strongest Diffraction Lines: 11.52(100), 2.901(42), 6.99(23), 7.51(22), 6.30(21)

Z: 10

Ref.: Martini J (1978) Miner Mag 42:401–404

Satterlyite
(Hexagonal polymorph of Wolfeite)
$(Fe^{2+}_{1.17}Mg_{0.35}Fe^{3+}_{0.18}H_{0.16}$-$Na_{0.10}Mn_{0.04})PO_4(OH)$

Analyses (%): MgO 7.1, FeO 43.1, Fe_2O_3 7.5, MnO 1.3, Na_2O 1.5, P_2O_5 34.8

Cleavage: None

Color-Luster: Transparent, pale yellow to pale brown, vitreous luster, streak pale yellow

Crystal System: Hexagonal

Density: 3.68 (Meas.), 3.60 (Calc.)

First Reported: 1978

Habit: Grains up to 1×1×40 mm, elongated parallel to [0001] in radiating aggregates

Hardness: 4.5–5

Lattice Constants: a = 11.36, c = 5.033 (Å)

Mode of Occurrence: Occurs as aggregates in nodules in shales along the Big Fish River in northeastern Yukon Territory, Canada. The associated minerals are quartz, pyrite, wolfeite and marcasite

Optical Constants: $\omega = 1.721$, $\varepsilon = 1.719$; Dichroic in thick grains with O pale and yellow and E brownish-yellow. Uniaxial neg.

Space Group: $P\bar{3}1m$, P31m, and P312

Strongest Diffraction Lines: 2.473(100), 2.840(80), 3.520(70), 1.447(60)

Z: 6

Ref.: Mandarino JA, Sturman BD, Corlett MI (1978) Can Miner 16:411–413

Schertelite

$(NH_4)_2MgH_2(PO_4)_2 \cdot 4H_2O$

Analyses (%): MgO 12.2, $(NH_4)_2O$ 16.2, P_2O_5 43.9, H_2O 27.6

Class: 2/m 2/m 2/m

Cleavage: None. Brittle

Color-Luster: Colorless, transparent. Vitreous. Soluble in water

Crystal System: Orthorhombic

Density: 1.83 (Calc.)

First Reported: 1887

Habit: As small indistinct flat crystals. Synthetic crystals tabular {100}, with nearly equally developed {010}, {001}, {111}, and {110}; also as single stout rods elongated along c-axis

Hardness: Not determined

Lattice Constants: a = 11.47, b = 23.63, c = 8.62 (Å)

Mode of Occurrence: Occurs in a bat guano deposit associated with struvite and newberyite in the Skipton Caves, Ballarat, Victoria, Australia

Optical Constants: $\alpha = 1.508$, $\beta = 1.515$, $\gamma = 1.523$; (+)2V = nearly 90°

Space Group: Pbca

Strongest Diffraction Lines: 5.94(100), 2.97(43), 5.21(37)

Z: 8

Ref.: Roberts WL, Rapp GR, Weber J (1974) Encyclopedia of minerals, New York, Van Nostrand Reinhold, p 543–544. Frazier A, Lehr JR, Smith JP (1963) Am Miner 48:635–641

Schoderite

$Al_2(PO_4)(VO_4) \cdot 8H_2O$

Analyses (%): Al_2O_3 23.8, V_2O_5 24.6, P_2O_5 17.4, H_2O 34.1

Class: 2/m

Color-Luster: Yellowish orange

Crystal System: Monoclinic

Density: 1.88 (Meas.)

First Reported: 1960

Habit: Microscopic bladed tabular crystals, elongated parallel to (001)

Hardness: ~2

Lattice Constants: a = 11.4, b = 15.8, c = 9.2 (Å); $\beta = 79°$

Mode of Occurrence: Occurs sparsely as yellowish orange microcrystalline coatings associated with wavellite and vashegyite along fractures in phosphatic cherts of lower Paleozoic age near Eureka, Nevada

Optical Constants: $\alpha = 1.542$, $\beta = 1.548$, $\gamma = 1.566$; (+)2V = 61° (Calc.)

Space Group: P2/m

Strongest Diffraction Lines: 7.9(100), 15.8(40), 11.1(20)

Z: 4

Ref.: Hausen DM (1962) Am Miner 47:637–648. Roberts WL, Rapp GR, Weber J (1974) Encyclopedia of minerals, New York, Van Nostrand Reinhold, p 544

Scholzite

$CaZn_2(PO_4)_2 \cdot 2H_2O$

Analyses (%): CaO 14.3, ZnO 35.7, MnO 1.4, MgO 1.0, P_2O_5 36.0, H_2O 10.4

Class: 2/m 2/m 2/m

Cleavage: {100} fair. Brittle

Color-Luster: Colorless to white. Transparent to translucent. Vitreous

Crystal System: Orthorhombic

Density: 3.11 (Meas.), 3.14 (Calc.)

First Reported: 1949

Habit: Crystals prismatic; platy; up to 5 mm long

Hardness: 3–3.5

Lattice Constants: a = 17.14, b = 22.19, c = 6.61 (Å)

Mode of Occurrence: Occurs associated with sphalerite, triplite, feldspar, and quartz in pegmatite at Hagendorf near Pleystein, Oberpfalz, Bavaria, Germany. Also found as crystals encrusting gossan about 40 miles southeast of Blinman, South Australia

Optical Constants: $\alpha = 1.581$, $\beta = 1.586$, $\gamma = 1.596$; (+)2V = 70°

Space Group: Pbmm

Strongest Diffraction Lines: 8.588(100), 2.788(90), 4.230(70)

Z: 12

Ref.: Strunz H (1951) Am Miner 36:382. Strunz H (1970) Miner Tab, 2nd edn. Akademische, Leipzig, p 164. Strunz H, Tennyson C (1961) Am Miner 46:1519. Taxer K (1975) Am Miner 60:1019–1022. Roberts WL, Rapp GR, Weber J (1974) Encyclopedia of minerals, New York, Van Nostrand Reinhold, p 545

Schoonerite
$ZnMn^{2+}(Fe^{2+})_2Fe^{3+}(PO_4)_3(OH)_2 \cdot 9H_2O$

Analyses (%): CaO 1.34, MgO 1.47, MnO 7.32–9.30, FeO 29.8–32.1, ZnO 7.95–10.6, P_2O_5 29.4–34.0, H_2O 23.4

Cleavage: Perfect to micaceous parallel to {010}, good parallel to {001}

Color-Luster: Pale tan to brown, reddish-brown to bronzy on exposed oxidized surfaces; transparent along [010]; streak is pale brown

Crystal System: Orthorhombic

Density: 2.87–2.92

First Reported: 1977

Habit: Thin laths elongated parallel to [100] and thin tabular grains parallel to {010}. Crystal aggregates occur as rosettes, laminae (parallel to {010}), sprays and mats; individual crystals which rarely exceed 1 mm in greatest dimension are generally curved and crinkled

Hardness: About 4

Lattice Constants: a = 11.119, b = 25.546, c = 6.437 (Å)

Mode of Occurrence: A late-stage low temperature mineral which occurs associated with siderite, mitridatite, jahnsite, whitmoreite, laueite, etc., in oxidized masses of ludlamite, messelite and vivianite derived from parent triphylite in the Palermo No. 1 pegmatite near North Groton, New Hampshire

Optical Constants: $\alpha = 1.618$, $\beta = 1.652$, $\gamma = 1.682$; (−)2V = 70°–80°; pleochroic with Z brown, Y pale brown, X pale yellow

Space Group: Pmab

Strongest Diffraction Lines: 12.77(100), 2.768(90), 8.35(70)

Z: 4

Ref.: Moore PB, Kampf AR (1977) Am Miner 62:246–249

Schreibersite (Rhabdite)
$(Fe, Ni)_3P$

Analyses (%): Fe 41.5–68.4, Ni 10.1–42.6, Co 0.05–1.32, P 12.8–15.3

Class: $\bar{4}$

Cleavage: {001} perfect, {010} or {110} imperfect. Very brittle

Color-Luster: Silver white to tin white; tarnishes to brass yellow or brown. Opaque. Metallic. Strongly magnetic

Crystal System: Tetragonal

Density: 7.0–7.8 (Meas.), 7.24 (Calc.)

First Reported: 1847

Habit: Crystals rare, often rounded. As plates, tablets, rods, or needles

Hardness: 6.5–7

Lattice Constants: a=9.013, c=4.424 (Å)

Mode of Occurrence: Occurs as a combustion product in the coal mines of Commentry and Cranzac, France. It is also found as a component in many iron meteorites

Optical Constants: Opaque

Space Group: $I\bar{4}$

Strongest Diffraction Lines: 2.19(100), 2.11(70), 1.972(70)

Z: 8

Ref.: Roberts WL, Rapp GR, Weber J (1974) Encyclopedia of minerals, New York, Van Nostrand Reinhold, p 546

Scorzalite (Lazulite-Scorzalite Series)
$(Fe^{2+}, Mg)Al_2(PO_4)_2(OH)_2$

Analyses (%): FeO 14.7–17.1, MgO 2.93–4.23, Fe_2O_3 0.13–0.54, Al_2O_3 30.8, P_2O_5 42.8, H_2O 6.0

Class: 2/m

Cleavage: {110} indistinct to good, {101} indistinct. Fracture uneven. Brittle

Color-Luster: Deep azure blue to bluish green. Subtranslucent to opaque. Vitreous to dull. Streak white

Crystal System: Monoclinic

Density: 3.38 (extrapolated for Fe end member), 3.39 (Calc.)

First Reported: 1947

Habit: Massive, compact to granular

Hardness: 5.5–6

Lattice Constants: a=7.15, b=7.32, c=7.14 (Å); β=119.0°

Mode of Occurrence: Occurs as irregular grains in the border zone of granitic pegmatite associated with quartz, muscovite, albite, apatite, brazilianite, and souzalite at the Corrego Frio mine, Minas Geraes, Brazil; also in granitic pegmatite in masses up to 5 cm in diameter, associated with quartz, tourmaline, wyllieite, muscovite, and albite at the Victory mine, Custer, South Dakota; and as irregular masses in a quartz dike at the White Mountain andalusite mine, Mono County, California

Optical Constants: α=1.633, β=1.663, γ=1.673; (−)2V=62°

Space Group: $P2_1/n$

Strongest Diffraction Lines: 3.24(100), 3.20(100), 3.14(80)

Z: 2

Ref.: Roberts WL, Rapp GR, Weber J (1974) Encyclopedia of minerals, New York, Van Nostrand Reinhold, p 549. Pecora WT, Fahey JJ (1950) Am Miner 35:1–18

Seamanite
$Mn_3^{2+}(OH)_2[B(OH)_4][PO_4]$

Analyses (%): MnO 56.4, (MgO+CaO) 1.33, P_2O_5 15.9, B_2O_3 9.94, H_2O 14.6

Class: 2/m 2/m 2/m

Cleavage: {001} distinct. Brittle

Color-Luster: Pale pink to yellow or yellowish brown. Transparent. Vitreous

Crystal System: Orthorhombic

Density: 3.128 (Meas.), 3.132 (Calc.)

First Reported: 1930

Habit: Crystals acicular, elongated along c-axis

Hardness: 4

Lattice Constants: a=7.811, b=15.114, c=6.691 (Å)

Mode of Occurrence: Seamanite occurs very sparingly as fine micro crystals associated with sussexite, calcite, and manganic oxyhydroxides in fractures cutting siliceous rock at the Chicagon mine, Iron River, Iron County, Michigan

Optical Constants: $\alpha = 1.640$, $\beta = 1.663$, $\gamma = 1.665$; (+)2V = 40°

Space Group: Pbnm

Strongest Diffraction Lines: 6.917(100), 2.843(80), 3.777(70)

Z: 4

Ref.: Roberts WL, Rapp GR, Weber J (1974) Encyclopedia of minerals, New York, Van Nostrand Reinhold, p 549. Moore PB, Ghose S (1971) Am Miner 56:1527–1538

Segelerite Fe^{3+} analog of Overite
$CaMg(H_2O)_4Fe^{3+}(OH)[PO_4]_2$

Analyses (%): CaO 13.6, MgO 9.5, Fe_2O_3 16.4, P_2O_5 33.1, H_2O 19.1

Class: 2/m 2/m 2/m

Cleavage: {010} perfect

Color-Luster: Pale yellow-green, chartreuse, inclining to colorless. Transparent to translucent. Vitreous

Crystal System: Orthorhombic

Density: 2.67 (Meas.), 2.610 (Calc.)

First Reported: 1974

Habit: Crystals long prismatic, vertically striated, nearly square in cross-section, and up to 0.5 mm in length. Forms: a{100}, b{010}, m{110}, and q{121}

Hardness: 4

Lattice Constants: a = 14.826, b = 18.751, c = 7.307 (Å)

Mode of Occurrence: Segelerite occurs sparsely along joint fractures in dense ferrisicklerite-rockbridgeite, usually in close association with collinsite, minor hydroxylapatite and globules of red-brown robertsite, at the Tip Top pegmatite, Custer County, South Dakota

Optical Constants: $\alpha = 1.618$, $\beta = 1.635$, $\gamma = 1.650$; Biaxial (−), 2V large

Space Group: Pcca

Strongest Diffraction Lines: 2.868(100), 9.31(90), 5.34(60)

Z: 8

Ref.: Moore PB (1974) Am Miner 59:48–59

Senegalite
$Al_2(OH)_3(PO_4) \cdot H_2O$

Analyses (%): P_2O_5 31.8, Al_2O_3 46.2, Fe_2O_3 0.28, H_2O 21.0

Cleavage: {100} imperfect, observed only under the microscope

Color-Luster: Colorless to pale yellow, luster vitreous

Crystal System: Orthorhombic

Density: 2.55 (Meas.), 2.55 (Calc.)

First Reported: 1976

Hardness: 5.5

Lattice Constants: a = 7.675, b = 9.711, c = 7.635 (Å)

Mode of Occurrence: Found in the Komondiako Iron Mine (magnetite ore) in the Faleme River Basin of east Senegal. It occurs in the oxidation zone of the deposit, associated with turquois, augelite, wavellite, and crandallite

Optical Constants: $\alpha = 1.562$, $\beta = 1.566$, $\gamma = 1.587$; Biaxial (−), 2V = 53° (Meas.), 48° (Calc.)

Space Group: $P2_1nb$

Strongest Diffraction Lines: 3.834(100), 4.089(90), 2.990(90), 3.610(80), 2.348(80)

Z: 4

Ref.: Johan Z (1976) Lithos 9:165–171. Keegan TD, Araki T, Moore PB (1979) Am Miner 64:1243–1247

Sicklerite
(Ferri-sicklerite-Sicklerite Series)
$Li(Mn^{2+}, Fe^{3+})PO_4$

Analyses (%): Li_2O 3.26–4.89, Na_2O 0.81–1.22, MnO 15.2–33.6, CaO 0.20–3.44, Fe_2O_3 4.72–29.1, Mn_2O_3 0–8.19, P_2O_5 43.1–44.8, H_2O 1.02–4.13

Class: 2/m 2/m 2/m

Cleavage: {100} good

Color-Luster: Yellowish brown to dark brown. Opaque. Dull

Crystal System: Orthorhombic

Density: 3.45 (Meas.)

First Reported: 1912 (As Pseudoheterosite, 1910)

Habit: Massive

Hardness: ~4

Lattice Constants: a = 5.95, b = 10.10, c = 4.80 (Å)

Mode of Occurrence: Occurs as an alteration of lithiophilite in the zone of weathering in granite pegmatites. Found at the Custer Mountain Lode, Custer County, South Dakota, and at other places in the Black Hills; at Pala, San Diego County, California; at Wodgina, Western Australia; at Varuträsk, Sweden; and in Finland

Optical Constants: $\alpha = 1.715$, $\beta = 1.735$, $\gamma = 1.745$; (−)2V = medium large

Space Group: Pmnb

Strongest Diffraction Lines: 3.013(100), 2.531(100), 4.32(95)

Z: 4

Ref.: Roberts WL, Rapp GR, Weber J (1974) Encyclopedia of minerals, New York, Van Nostrand Reinhold, p 558–559

Sidorenkite

$Na_3Mn(PO_4)(CO_3)$

Analyses (%): MnO 22.4, CaO 2.20, Na_2O 32.4, K_2O 0.44, P_2O_5 25.8, CO_2 15.7

Cleavage: {100}, {010} perfect, {001} imperfect

Color-Luster: Pale rose, luster vitreous to pearly on cleavages

Crystal System: Monoclinic, pseudo-orthorhombic

Density: 2.90 (Meas.), 2.98 (Calc.)

First Reported: 1979

Habit: Irregular grains and in crystals of match-box shape up to 2 cm; fracture stepped

Hardness: About 2, brittle

Lattice Constants: a = 8.979, b = 6.729, c = 5.150 (Å); $\beta = 90°$

Mode of Occurrence: Found in pegmatites cutting cancrinite syenites of the Mt. Alluaiv region, Lovozero massif, Kola Peninsula, U.S.S.R. It is associated with K-feldspar, nepheline, sodalite, cancrinite, and is cut by the aegirine

Optical Constants: $\alpha = 1.521$, $\beta = 1.563$, $\gamma = 1.585$; Biaxial neg., 2V = 68° (Meas.), 70° (Calc.)

Space Group: C_{2h}^2–$P2_1/m$ or C_2^2–$P2_1$

Strongest Diffraction Lines: 3.36(100), 8.97(20), 1.682(20), 2.69(15)

Z: 2

Ref.: Khomyakov AP, Semenov EI, Kazakova ME (1979) Zapiski Vses Miner Obshch 108:56–59. Khomyakov AP, Semenov EI, Kazakova ME (1979) Am Miner 64:1332

Sigloite

$(Fe^{2+}, Fe^{3+})\ Al_2(PO_4)_2\ (O, OH)_2 \cdot 8H_2O$

Analyses (%): Al_2O_3 21.1, Fe_2O_3 16.5, MgO 0.9, P_2O_5 27.5, H_2O 33.6

Class: $P\bar{1}$

Cleavage: {010} perfect, {001} good

Color-Luster: Pale straw yellow to light brown

Crystal System: Triclinic

Density: 2.35 (Meas.), 2.36 (Calc.)

First Reported: 1962

Habit: Crystals short prismatic [001] to thick tabular {010}. An oxidation pseudomorph after paravauxite

Hardness: 3

Lattice Constants: a = 5.26, b = 10.52, c = 7.06 (Å); $\alpha = 106.97°$, $\beta = 111.5°$, $\gamma = 69.5°$

Mode of Occurrence: Found on the 305 meter level of the Contacto vein, Siglo XX mine, Llallagua, Bolivia, in cavities filled with intergrowths of crystals and in other cavities with crystals perched on wavellite overgrowths on quartz

Optical Constants: $\alpha = 1.563$, $\beta = 1.586$, $\gamma = 1.619$; (+)2V = 76°

Space Group: $P\bar{1}$

Strongest Diffraction Lines: 9.69(100), 6.46(90), 4.86(90)

Z: 1

Ref.: Hurlbut CS Jr, Honea R (1962) Am Miner 47:1–8. Roberts WL, Rapp GR, Weber J (1974) Encyclopedia of minerals, New York, Van Nostrand Reinhold, p 560

Sincosite
$Ca(VO)_2(PO_4)_2 \cdot 5H_2O$

Analyses (%): CaO 12.1–13.3, V_2O_4 36.3–37.8, P_2O_5 31.4, H_2O 17.1–19.9

Cleavage: {001} perfect, {100} distinct, {110} distinct. Brittle

Color-Luster: Grass green, yellowish green to olive green or brownish green. Vitreous; altered crystals submetallic. Transparent to translucent. Streak green

Crystal System: Tetragonal

Density: ~2.84

First Reported: 1922

Habit: Crystals thin to thick tabular, usually square in outline; often in compact, irregular or scaly masses. Also as flattened, nearly circular rosettes, or as radial aggregates and small nodules admixed with carbonaceous material and gypsum. Twinning on {110}, rare

Hardness: Not determined. Soft.

Mode of Occurrence: Occurs in vugs, small cavities, and incrusting minute fissures in siliceous gold ore at the Ross Hannibal mine, near Lead, Lawrence County, South Dakota. Also found near Bloomington, Bear Lake County, Idaho, and as thin veinlets and irregular masses in carbonaceous shale near Sincos, Junin, Peru

Optical Constants: $\omega \sim 1.680$, $\varepsilon \sim 1.655$; (−) (sometimes biaxial)

Strongest Diffraction Lines: 6.28(100), 3.14(60), 3.35(20)

Ref.: Roberts WL, Rapp GR, Weber J (1974) Encyclopedia of minerals, New York, Van Nostrand Reinhold, p 562–563

Sodium Autunite (Autunite Group)
$Na(UO_2)(PO_4) \cdot 4H_2O$

Analyses (%): Na_2O 6.9, UO_3 62.5, P_2O_5 14.7, H_2O 14.8

Class: 4/m 2/m 2/m

Cleavage: {001} perfect, {100} less perfect. Brittle

Color-Luster: Lemon yellow to lettuce yellow, vitreous, pearly on basal plane. Fluoresces strong yellow-green under ultraviolet light

Crystal System: Tetragonal

Density: 3.584 (Meas.), 3.89 (Calc. for $8H_2O$)

First Reported: 1929

Habit: As thin plates; also in foliated and radiating masses

Hardness: 2–2.5

Lattice Constants: a = 6.97, c = 8.69 (Å)

Mode of Occurrence: Occurs in several lignite deposits in the Cave Hills and Slim Buttes areas in Harding County, South Dakota: also "in one of the granodiorite massifs of the U.S.S.R."

Optical Constants: $\omega = 1.578$, $\varepsilon = 1.559$ (−)

Space Group: P4/nmm

Strongest Diffraction Lines: 3.67(100), 2.675(80), 1.566(80)

Z: 2

Ref.: Chernikov AA, Krutetskaya OV, Organov NI (1958) Am Miner 43:383. Chernikov AA, Krutetskaya OV, Organov NI (1957) Atomaya Energ 3:135–140. Roberts WL, Rapp GR, Weber J (1974) Encyclopedia of minerals, New York, Van Nostrand Reinhold, p 568

Souzalite
$(Al, Mg, Fe)_6Al(PO_4)_4(OH)_6 \cdot 2H_2O$

Analyses (%): MgO 9.62, FeO 11.5, MnO 0.31, Al_2O_3 26.1, Fe_2O_3 2.65, P_2O_5 37.7, H_2O 12.0

Cleavage: One good, one fair, approximately at right angles

Color-Luster: Dark green

Crystal System: Probably monoclinic

Density: 3.087 (Meas.)

First Reported: 1949

Habit: As coarse fibrous masses. Twinning polysynthetic

Hardness: 5.5–6

Mode of Occurrence: Occurs as an alteration product of scorzalite at the Corrego Frio pegmatite, Minas Geraes, Brazil

Optical Constants: $\alpha = 1.618$, $\beta = 1.642$, $\gamma = 1.652$; $(-)2V = 68°$

Strongest Diffraction Lines: 2.690(100), 3.79(90), 5.35(80)

Ref.: Roberts WL, Rapp GR, Weber J (1974) Encyclopedia of minerals, New York, Van Nostrand Reinhold, p 571

Spencerite
$Zn_4(PO_4)_2(OH)_2\ 3H_2O$

Analyses (%): ZnO 60.3, P_2O_5 26.2, H_2O 13.3

Class: 2/m

Cleavage: {100} perfect, {010} good, {001} distinct

Color-Luster: White. Vitreous; pearly on cleavages

Crystal System: Monoclinic

Density: 3.145 (Meas.), 3.253 (Calc.)

First Reported: 1916

Habit: Crystals tabular, elongated [001]. Terminations blunt or lance-like. Commonly striated. Also massive, stalactitic, with platy to columnar structure. Twinning on {100}, polysynthetic

Hardness: 3

Lattice Constants: a = 10.441, b = 5.280, c = 11.196 (Å); $\beta = 116.83°$

Mode of Occurrence: Occurs in oxidized zinc ore associated with hemimorphite and hopeite at the Hudson Bay mine, Salmo, British Columbia, Canada

Optical Constants: $\alpha = 1.586$, $\beta = 1.602$, $\gamma = 1.606$; $(-)2V = 49°$

Space Group: P2/c

Strongest Diffraction Lines: 9.4(100), 3.501(55), 2.332(30)

Ref.: Roberts WL, Rapp GR, Weber J (1974) Encyclopedia of minerals, New York, Van Nostrand Reinhold, p 572

Stanfieldite
$Ca_4(Mg, Mn, Fe)_5(PO_4)_6$

Analyses (%): CaO 24.6, MgO 12.9, FeO 13.8, MnO 1.4, P_2O_5 46.6

Class: 2/m

Cleavage: Not determined

Color-Luster: Reddish to amber; under the microscope clear and transparent

Crystal System: Monoclinic

Density: 3.15 (Meas.), 3.15 (Calc. for Mg/Fe = 3:2)

First Reported: 1967

Habit: Massive; as irregular grains and thin veinlets

Hardness: 4–5, nearer to 5

Lattice Constants: a = 17.267, b = 10.002, c = 22.917 (Å); $\beta = 100.52°$

Mode of Occurrence: Occurs as grains up to 1 mm in diameter along the walls of cracks in the Estherville mesosiderite meteorite, and as veinlets less than 1 mm in width cutting olivine in the Santa Rosalia, Albin, Finnmarken, Imilac, Mt. Vernon, and Newport pallasites

Optical Constants: $\alpha = 1.619$, $\beta = 1.622$, $\gamma = 1.631$; $(+)2V = 50°$

Space Group: P2/c(?)

Strongest Diffraction Lines: 2.817(100), 3.747(80), 2.505(80)

Z: 8

Ref.: Fuchs LH (1967) Science 158:910–911

Stercorite
$Na(NH_4)H(PO_4) \cdot 4H_2O$

Analyses (%): Na_2O 14.5–15.8, $(NH_4)_2O$ 7.68–8.48, P_2O_5 34.4, H_2O 42.3

Cleavage: None

Color-Luster: Colorless, white, yellowish, brownish. Transparent. Vitreous

Soluble in water

Crystal System: Triclinic (pseudomonoclinic)

Density: 1.554 (Meas.)

First Reported: 1850

Habit: As crystalline masses and nodules. Synthetic crystals short prismatic. Twinning on {010} repeated, simulating monoclinic symmetry

Hardness: 2

Lattice Constants: a:b:c = 2.908:1:1.859; $\beta = 98.5°$

Mode of Occurrence: Occurs in guano deposits on the Guañape Islands, off the coast of Peru, and on Ichaboe Island, South West Africa

Optical Constants: $\alpha = 1.439$, $\beta = 1.442$, $\gamma - 1.469$; (+)2V − 35.57°

Strongest Diffraction Lines: 6.6(100), 2.89(100), 4.6(50)

Ref.: Roberts WL, Rapp GR, Weber J (1974) Encyclopedia of minerals, New York, Van Nostrand Reinhold, p 579

Stewartite
$MnFe_2(OH)_2(PO_4)_2 \cdot 8H_2O$

Analyses: None reported

Class: $\bar{1}$ or 1

Cleavage: One distinct

Color-Luster: Yellow to brownish yellow. Transparent to translucent. Vitreous to somewhat silky

Crystal System: Triclinic

Density: 2.94 (Meas.)

First Reported: 1912

Habit: As minute crystals and tufts of fibers.

Hardness: Not determined

Lattice Constants: a = 2 × 5.23, b = 10.77, c = 7.25 (Å); $\alpha = 90.58°$, $\beta = 109.97°$, $\gamma = 71.35°$

Mode of Occurrence: Occurs in granite pegmatite as minute yellowish crystals associated with rockbridgeite, metastrengite, and other secondary phosphate minerals in altered triphylite nodules at the Hesnard mine, Keystone, Pennington County, South Dakota; at the Stewart mine, Pala, San Diego County, California; at the Fletcher and Palermo pegmatites, North Groton, New Hampshire; and at Newry, Maine. It also is found at Alto Boqueiro, Rio Grande do Norte, Brazil, and at Chanteloube, Haute-Vienne, France

Optical Constants: $\alpha = 1.612$–1.63, $\beta = 1.653$–1.658, $\gamma = 1.660$–1.681; (−)2V = large

Space Group: $P\bar{1}$ or P1

Strongest Diffraction Lines: 9.98(100), 3.93(40), 2.99(40)

Z: 2

Ref.: Roberts WL, Rapp GR, Weber J (1974) Encyclopedia of minerals, New York, Van Nostrand Reinhold, p 580. Peacor DR (1963) Am Miner 48:913–914

Strengite (Variscite-Strengite Series – Dimorphous with metastrengite)
$FePO_4 \cdot 2H_2O$

Analyses (%): Fe_2O_3 41.4–43.4, P_2O_5 36.3–38.2, H_2O 18.9–21.6

Class: 2/m 2/m 2/m

Cleavage: {010} good, {001} poor. Fracture conchoidal. Brittle

Color-Luster: Colorless, pale to deep violet, red, carmine. Transparent to translucent. Vitreous. Streak white

Crystal System: Orthorhombic

Density: 2.87 (Meas.), 2.848 (Calc.)

First Reported: 1877 (as Barrandite, 1867)

Habit: Crystals octahedral, thin to thick tabular, stout prismatic. As small spherical aggregates with radial-fibrous structure and drusy surface, and as crusts. Twinning on {201} not common

Hardness: 3.5–4.5

Lattice Constants: a = 10.05, b = 9.92, c = 8.74 (Å)

Mode of Occurrence: Occurs as a secondary mineral formed by the alteration of iron-rich phosphate minerals. Found as superb crystals up to 4 mm in size associated with metastrengite in vugs in barbosalite at the Bull Moose mine, Custer County, South Dakota; as spherical aggregates and fine crystals with cacoxenite on limonite at Indian Mountain, Cherokee County, Alabama; in pegmatites in the Pala district, San Diego County, California; and at localities in Nevada, Virginia, Pennsylvania, and New Hampshire. It also occurs at several deposits in Germany, especially at Hagendorf, Bavaria; in Sweden; and altered triphylite at Mangualde, Portugal

Optical Constants: $\alpha = 1.707$, $\beta = 1.719$, $\gamma = 1.741$; (+)2V = moderate, sometimes small

Space Group: Pcab

Strongest Diffraction Lines: 4.38(100), 5.50(80), 3.11(80)

Z: 8

Ref.: Roberts WL, Rapp GR, Weber J (1974) Encyclopedia of minerals, New York, Van Nostrand Reinhold, p 586

Strontium-Apatite (Saamite)
(Apatite Group)
$(Sr, Ca)_5(PO_4)_3(OH, F)$

Analyses (%): BaO 2.7, SrO 46.1, Al_2O_3 0.4, MgO 1.6, CaO 10.8, SiO_2 0.9, P_2O_5 30.4, F 1.7, H_2O 1.6

Class: 6/m

Color-Luster: Pale green to yellowish green, colorless and transparent in small crystals; vitreous, on fracture greasy

Crystal System: Hexagonal

Density: 3.84 (Meas.)

First Reported: 1952

Habit: Poorly formed oval crystals 0.2 × 1.0 cm in longitudinal section and as small crystals with corroded faces. Forms: prism $\{10\bar{1}0\}$ and dypyramid $\{10\bar{1}1\}$

Hardness: 5

Lattice Constants: a = 9.66, c = 7.19 (Å)

Mode of Occurrence: Occurs associated with batisite, innelite, ramsayite and rare eudialyte in sugary albite filling interstices between crystals of aegirine and eckermannite in veins in alkalic pegmatites that cut the dunite core of the concentrically zoned Inagil massif of ultrabasic-alkalic rocks, Southern Yakutia

Optical Constants: $\omega = 1.651$, $\varepsilon = 1.637$; (−)

Space Group: $P6_3/m$

Strongest Diffraction Lines: 2.89(100), 3.167(70), 2.78(70)

Z: 2

Ref.: Efimov AS, Kravchenko SM, Vasil'eva ZV (1962) Am Miner 47:808. Efimov AS, Kravchenko SM, Vasil'eva ZV (1962) Compt Rend Acad Sci (URSS) 142:439. McConnell D (1973) Apatite. Springer, Berlin Heidelberg New York, p 90. Roberts WL, Rapp GR, Weber J (1974) Encyclopedia of minerals, New York, Van Nostrand Reinhold, p 588

Strunzite (Dimorphous with laueite)
$MnFe_2^{3+}(PO_4)_2(OH)_2 \cdot 8H_2O$

Analyses (%): MnO 9.1, Fe_2O_3 36.0, P_2O_5 33.0, H_2O 22.5%

Class: 2/m

Cleavage: Not determined

Color-Luster: Straw yellow to brownish yellow

Crystal System: Monoclinic

Density: 2.47–2.56 (Meas.)

First Reported: 1957

Habit: As divergent tufts and felted coatings of tiny hair-like or lath-like crystals. Laths flattened on (010). Twinning on (100)

Hardness: Not determined

Lattice Constants: a = 9.80, b = 18.06, c = 7.34 (Å); $\beta = 100.17°$

Mode of Occurrence: Occurs as a near-surface weathering product in pegmatites that contain triphylite at Hagendorf and other localities in Bavaria; Palermo and Fletcher, New Hampshire; Norway, Newry, Rumford, and Unity, Maine; at numerous localities in the Keystone and Custer districts, South Dakota; and as films in weathered outcrops of phosphate rock of the Phosphoria formation at Rasmussen Valley, Idaho

Optical Constants: $\alpha = 1.619$, $\beta = 1.670$, $\gamma = 1.720$; $(-)2V$ = moderate

Space Group. C2/c

Strongest Diffraction Lines: 9.02(100), 5.32(80), 4.35(60)

Z: 4

Ref.: Frondel C (1958) Am Miner 43:793–794. Roberts WL, Rapp GR, Weber J (1974) Encyclopedia of minerals, New York, Van Nostrand Reinhold, p 588

Struvite

$NH_4MgPO_4 \cdot 6H_2O$

Analyses (%): MgO 13.2–16.6, MnO 0–2.01, FeO 0.81–3.06, $[H_2O + (NH_4)_2O]$ 53.6–54.5, P_2O_5 28.1–28.8

Class: mm2

Cleavage: {001} good, {100} poor. Fracture uneven to subconchoidal. Brittle

Color-Luster: Colorless, becoming white on dehydration. Transparent to translucent. Vitreous; dehydrated material dull

Crystal System: Orthorhombic

Density: 1.711 (Meas.), 1.706 (Calc.)

First Reported: 1845

Habit: Crystals widely varied in habit; short prismatic, thick tabular, equant, coffin-shaped, wedge-shaped; distinctly hemimorphic. Crystals range up to 2.0 cm or more in size. Twinning on {001} common

Hardness: 2

Lattice Constants: a = 6.945, b = 11.208, c = 6.1355 (Å) (synthetic)

Mode of Occurrence: Occurs in bat guano associated with newberyite and hannayite in the Skipton Caves, Ballarat, Victoria, Australia; at Saldanha Bay, Cape Province, South Africa; on La Reunion Island, Indian Ocean; and among other localities, in Denmark and in Germany

Optical Constants: $\alpha = 1.495$, $\beta = 1.496$, $\gamma = 1.504$; $(+)2V = 37.37°$

Space Group: $Pm2_1n$

Strongest Diffraction Lines: 4.257(100), 5.601(60) (synthetic), 2.919(55)

Z: 2

Ref.: Roberts WL, Rapp GR, Weber J (1974) Encyclopedia of minerals, New York, Van Nostrand Reinhold, p 588–589

Svanbergite

$SrAl_3PO_4SO_4(OH)_6$

Analyses (%): CaO 1.75–3.25; Al_2O_3 32.7–36.9, P_2O_5 16.7–21.2, SO_3 11.5–17.3, H_2O 11.6–12.5

Class: $\bar{3}$ 2/m

Cleavage: {0001} distinct

Color-Luster: Colorless to yellow, rose, reddish brown. Translucent. Vitreous to adamantine

Crystal System: Hexagonal

Density: 3.22 (Meas.), 3.24 (Calc.)

First Reported: 1854

Habit: Crystals rhombohedral or pseudocubic. Also granular

Hardness: 5

Lattice Constants: a = 6.99, c = 16.75 (Å); $a_{rh} = 6.91$; $\alpha = 60.98°$

Mode of Occurrence: Occurs associated with pyrophyllite and diopside at Thorne, Mineral County, Nevada; at the Westanå iron mine in Skåne, and at Horrsjöberg, Sweden; with pyrophyllite near Chalmoux, Saône et Loire, France; and in the diamond gravels of Bahia, Brazil

Optical Constants: $\omega = 1.631$, $\varepsilon = 1.646$; (+)

Space Group: $R\bar{3}m$

Strongest Diffraction Lines: 2.98(100), 2.22(100), 5.74(90)

Z: 1 (rhomb.), 3 (hexag.)

Ref.: Roberts WL, Rapp GR, Weber J (1974) Encyclopedia of minerals, New York, Van Nostrand Reinhold, p 593–594

Switzerite

$(Mn, Fe)_3(PO_4)_2 \cdot 4H_2O$

Analyses (%): MnO 36.2, FeO 9.30, Fe_2O_3 3.84, P_2O_5 32.5, H_2O 17.7

Class: 2/m

Cleavage: {100} perfect, {010} fair

Color-Luster: Pale pink or light golden brown. Medium brown to chocolate brown when oxidized. Luster pearly to adamantine on the cleavage

Crystal System: Monoclinic

Density: 2.95 (Meas.), 3.18 (Calc.)

First Reported: 1967

Habit: As micaceous flakes or bladed crystals, up to 3 mm long, flattened on {100} and elongated [001]

Hardness: ~2.5

Lattice Constants: a = 17.099, b = 12.694, c = 8.282 (Å); $\beta = 95.92°$

Mode of Occurrence: Found at the Foote Mineral Company spodumene mine, Kings Mountain, North Carolina, in association with vivianite in seams in spodumene-rich pegmatite

Optical Constants: $\alpha = 1.602$ (Calc.), $\beta = 1.628$, $\gamma = 1.632$; (−)2V = 42°

Space Group: P2/a

Strongest Diffraction Lines: 8.550(100), 2.585(60), 7.128(40)

Z: 8

Ref.: Leavens PB, White JS Jr (1967) Am Miner 52:1595–1602. Roberts WL, Rapp GR, Weber J (1974) Encyclopedia of minerals, New York, Van Nostrand Reinhold, p 594–595

Tagilite = Pseudomalachite

Tancoite

$HNa_2LiAl(PO_4)_2(OH)$

Analyses (%): P_2O_5 50.1, Al_2O_3 18.3, Na_2O 20.0, Li_2O 5.2, CaO 0.5

Cleavage: {010}, {001} fair; fracture conchoidal

Color-Luster: Colorless to pale pink, vitreous luster

Crystal System: Orthorhombic

Density: 2.752 (Meas.), 2.724 (Calc.)

First Reported: 1980

Habit: Found as druses of columnar individuals or isolated crystals up to 1 mm long. Forms found are {100}, {010}, and {111} dominant, {021}, {001} minor

Hardness: 4–4.5

Lattice Constants: a = 7.041, b = 14.130, c = 6.975 (Å)

Mode of Occurrence: Occurs in the Tanco spodumene-bearing pegmatite, Bernic Lake, Manitoba, associated with lithiophosphate, pink apatite and a hydrous sodium phosphate

Optical Constants: $\alpha = 1.541$, $\beta = 1.563$, $\gamma = 1.564$; Biaxial neg., 2V = −23°

Space Group: Cmmb, Cm2a, or C2mb

Strongest Diffraction Lines: 4.672(100), 3.150(100), 3.413(90), 2.479(90)

Z: 4

Ref.: Ramik RA, Sturman BD, Dunn PJ, Povarennykh AS (1980) Can Miner 18:185–190

Tangeïte = Calciovolborthite

Taranakite

$H_6K_3Al_5(PO_4)_8 \cdot 18H_2O$, more likely $K_x[Al_{2-y}(H_3)_y](OH)_2[Al_xP_{4-x-z}(H)_z O_{10}]$

Analyses (%): K_2O 4.20–8.28, CaO 0–1.40, Fe_2O_3 0.83–1.17, Al_2O_3 18.6–22.9, P_2O_5 35.1–42.3, H_2O 28.2–33.1

Class: 3m or $\bar{3}$ 2/m

Color-Luster: White; also gray or yellowish white

Crystal System: Hexagonal

Density: 2.09 (Meas.), 2.11 (Calc.)

First Reported: 1865 (Described as Minervite 1893; Palmerite 1904)

Habit: As minute lath-like crystals; also massive, clay-like, pulverulent to compact

Hardness: Very soft

Lattice Constants: a = 8.71, c = 96.1 (Å)

Mode of Occurrence: Occurs as flour-like masses near contacts between guano and clay and along fractures within brecciated clay in Pig Hole Cave, Giles County, Virginia. Also found at widespread localities as a product of the reaction of phosphatic solutions derived from bat or sea-bird guano on clays or aluminous rocks. Described from the Sugarloaves, Taranaki, New Zealand; also occurs at Mt. Alburno, Salerno, Italy; in caves at Grotte de Minerve, Hérault, France; and on the island of Réunion, Indian Ocean

Space Group: R3c or R$\bar{3}$c

Strongest Diffraction Lines: 15.5(100), 7.6(80), 3.83(80)

Z: 6

Ref.: Roberts WL, Rapp GR, Weber J (1974) Encyclopedia of minerals, New York, Van Nostrand Reinhold, p 604. Smith JP, Brown WE (1959) Am Miner 44:138–142. McConnell D (1976) Am Miner 61:329–331. Sakae T, Sudo T (1975) Am Miner 60:331–334

Tarbuttite
$Zn_2PO_4(OH)$

Analyses (%): ZnO 66.6, P_2O_5 29.2, H_2O 3.8

Class: $\bar{1}$

Cleavage: {010} perfect. Fracture uneven. Brittle

Color-Luster: Colorless, white, or pale shades of yellow, brown, red, or green. Transparent to translucent. Vitreous; pearly on cleavage. Streak white

Crystal System: Triclinic

Density: 4.140 (Meas.), 4.21 (Calc.)

First Reported: 1907

Habit: Crystals short prismatic or equant, commonly rounded and deeply striated. As aggregates or crusts

Hardness: 3.8

Lattice Constants: a = 5.657, b = 6.432, c = 5.521 (Å); $\alpha = 102.45°$, $\beta = 87.67°$, $\gamma = 102.57°$

Mode of Occurrence: Occurs as a secondary mineral associated with limonite, hopeite, hemimorphite, pyromorphite, vanadinite, and other secondary minerals, in the zone of oxidation at the zinc mines of Broken Hill, Zambia

Optical Constants: $\alpha = 1.660$, $\beta = 1.705$, $\gamma = 1.713$; (−)2V = 50°

Space Group: P$\bar{1}$

Strongest Diffraction Lines: 2.78(100), 6.12(90), 3.70(90)

Z: 2

Ref.: Roberts WL, Rapp GR, Weber J (1974) Encyclopedia of minerals, New York, Van Nostrand Reinhold, p 604–605. Finney JJ (1966) Am Miner 51:1218–1220

Tavistockite = Fluorapatite

Tavorite
$LiFe^{3+}PO_4OH$

Analyses (%): FeO 2.39, Li_2O 7.64, MnO 1.47, Fe_2O_3 42.6, P_2O_5 39.8, H_2O 6.16

Cleavage: 3 distinct cleavages

Color-Luster: Grass green to yellowish green; transparent to translucent. Vitreous. Coatings earthy

Crystal System: Triclinic

Density: 3.288 (Meas.)

First Reported: 1954

Habit: Crystals equant or short prismatic; often in druses of interlocking crystals. Also in fine granular crystalline masses and coatings

Hardness: Not determined

Mode of occurrence: Occurs as minute yellow-green to grass green crystals in sheaf-like aggregates lining cavities in altered triphylite at the Bull Moose mine; and as superb grass green crystals up to 1 mm in size, associated with leucophosphite and hureaulite crystals in altered triphylite, at the Tip-Top mine, Custer County, South Dakota, and in several other pegmatites in the Custer and Keystone districts in the same state. Found originally sparsely disseminated in altered triphylite at the Sapucaia pegmatite mine, Minas Geraes, Brazil, and subsequently identified from pegmatites in Germany, Czechoslovakia, Madagascar, and the New England area

Optical Constants: Mean index = 1.807

Strongest Diffraction Lines: 3.045(100), 3.285(90), 4.99(50)

Ref.: Lindberg ML, Pecora WT (1955) Am Miner 40:952–966. Lindberg ML, Pecora WT (1954) Science 119:739. Roberts WL, Rapp GR, Weber J (1974) Encyclopedia of minerals, New York, Van Nostrand Reinhold, p 605

Thadeuite

$Mg(Ca, Mn)(Mg, Fe, Mn)_2(PO_4)_2(OH, F)_2$

Analyses (%): CaO 14.7, MnO 5.7, FeO 11.2, MgO 24.3, P_2O_5 38.7, F 2.4, OH 7.0

Cleavage: {010} good in hand specimens; two perpendicular cleavages observed in thin section

Color-Luster: Yellow-orange; vitreous luster

Crystal System: Orthorhombic

Density: 3.25 (Meas.), 3.21 (Calc.)

First Reported: 1979

Habit: Massive, coarse-grained cleavable material

Hardness: Slightly less than 4

Lattice Constants: a = 6.412, b = 13.563, c = 8.54 (Å)

Mode of Occurrence: Occurs locally in topaz- and muscovite-rich selvages along the borders of the tin-tungsten veins of the Panasqueira mining district in Central Portugal. Associated minerals include fluorapatite, isokite, wolfeite, althausite, and vivianite

Optical Constants: $\alpha = 1.568$, $\beta = 1.597$, $\gamma = 1.600$; Biaxial neg., $2V = 33°$

Space Group: C222

Strongest Diffraction Lines: 3.38(100), 3.00(31), 2.793(31), 2.626(23)

Z: 4

Ref.: Isaacs A, Peacor DR, Kelly WC (1979) Am Miner 64:359–361

Threadgoldite

$Al(UO_2)_2(PO_4)_2(OH) \cdot 8H_2O$

Analyses (%): P_2O_5 13.7, Al_2O_3 5.4, UO_3 63.5, H_2O 17.4

Cleavage: Cleavage planes present include {100}, {001}, {010}, and {012}

Color-Luster: Greenish-yellow, under the microscope, very pale yellow to colorless; fluoresces green in long-wave UV, pale green in short-wave UV light

Crystal System: Monoclinic

Density: 3.4 (Meas.), 3.32 (Calc.)

First Reported: 1979

Habit: Micaceous tabular crystals, elongated on *b*

Hardness: Not given

Lattice Constants: a = 20.25, b = 9.85, c = 19.75 (Å); $\beta = 114°$

Mode of Occurrence: With phuralumite and uplite in the beryl-columbite pegmatite at Kobokobo, Kivu, Zaire

Optical Constants: $\alpha = 1.573$ (Calc.), $\beta = 1.583$, $\gamma = 1.583$; Biaxial neg., $2V = 70°$

Space Group: Cc or C2/c

Strongest Diffraction Lines: 9.43(100), 4.474(80), 3.366(60), 2.197(60)

Z: 8

Ref.: Deliens M, Piret P (1979) Bull Miner 102:338–341

Tinticite

(Inadequately described mineral)

$2FePO_4 \cdot Fe(OH)_3 \cdot 3\frac{1}{2}H_2O$

Analyses (%): Fe_2O_3 49.1, P_2O_5 28.4–30.2, SO_3 0.7–1.07, H_2O 18.9–19.7

Cleavage: Not determined

Color-Luster: White with yellowish green tint

Crystal System: Orthorhombic (?)

Density: ~2.8

First Reported: 1946

Habit: As dense, porcelaneous to earthy masses

Hardness: ~2.5

Mode of Occurrence: Occurs in a limestone cavern near the Tintic Standard mine, Juab County, Utah, associated with jarosite and limonite

Optical Constants: Mean index ≈1.745

Strongest Diffraction Lines: 3.91(100), 3.28(95), 3.01(85)

Ref.: Stringham B (1946) Am Miner 31:395. Roberts WL, Rapp GR, Weber J (1974) Encyclopedia of minerals, New York, Van Nostrand Reinhold, p 619

Torbernite *(Chalcolite; Cuprouranite).* (The Torbernite Group include Autunite, Uranocirute, Saléeite, Zeunerite, and Uranospinite.)

$Cu(UO_2)_2(PO_4)_2 \cdot 8\text{–}12H_2O$

Analyses (%): CuO 7.15–8.58, PbO 0–2.95, UO_3 57.0–60.7, As_2O_5 0–3.24, P_2O_5 13.5–14.5, H_2O 14.3–20.3

Class: 4/m 2/m 2/m

Cleavage: {001} perfect, {100} indistinct. Brittle

Color-Luster: Emerald green to grass green; sometimes leek green, apple green, or siskin green. Vitreous to subadamantine; pearly on {001}. Transparent to translucent. Streak paler than the color

Crystal System: Tetragonal

Density: 3.22 (Meas.), 3.28 (Calc. for 12 H_2O)

First Reported: 1793 [Before then, described as Mica viridis crystals (from Joachimsthal), 1772; Grüner Glimmer (from Saxony), 1780]

Habit: Crystals thin to thick tabular on {001}, with rectangular or octagonal shape; rarely pyramidal and terminated by small faces of {001}. As scaly or lamellar aggregates; rarely as granular to almost earthy masses. Twinning on {110}, rare

Hardness: 2–2.5

Lattice Constants:
(Synthetic) a = 7.025, c = 20.63 (Å);
(Natural) a = 7.06, c = 20.5 (Å)

Mode of Occurrence: In the United States torbernite is widespread in small amounts in pegmatites, vein deposits, and several sedimentary rock types. Excellent crystal specimens occur at Chalk Mountain, near Spruce Pine, Mitchell County, North Carolina. It also is found in Sonora, Mexico; in large amounts in the oxidized uranium ores in the Katanga district, Zaire; at Mt. Painter, South Australia; associated with pitchblende in vein deposits in Czechoslovakia and Saxony, Germany; finely crystallized at several localities in Cornwall, England; in France at Cap Garonne and various places in the Puy-de-Dôme district; and at numerous other localities throughout the world

Optical Constants: $\omega = 1.590\text{–}1.592$, $\varepsilon = 1.581\text{–}1.582$; (−)

Space Group: I4/mmm

Strongest Diffraction Lines: 10.3(100), 4.94(90) (synthetic), 3.58(90)

Z: 2

Ref.: Berman R (1957) Am Miner 42:905–908. Frondel C (1958) USGS Bull 1064:170–177. Roberts WL, Rapp GR, Weber J (1974) Encyclopedia of minerals, New York, Van Nostrand Reinhold, p 622

Triphylite
(Triphylite-Lithiophilite Series)
$Li(Fe^{2+}, Mn^{2+})PO_4$

Analyses (%): Li_2O 5.51–9.23, Na_2O 0–2.71, FeO 2.94–32.9, MnO 3.11–42.6, CaO 0–9.7, MgO 0–7.38, P_2O_5 41.1–46.0

Class: 2/m 2/m 2/m

Cleavage: {100} nearly perfect, {010} imperfect, {011} interrupted. Fracture subconchoidal to uneven

Color-Luster: Greenish gray, bluish gray; often externally brownish to blackish due to alteration. Translucent to transparent. Resinous to vitreous. Streak colorless to dirty white

Crystal System: Orthorhombic

Density: 3.423 (Meas.), 3.527 (Calc. Fe:Mn = 1:1)

First Reported: 1834

Habit: Crystals rare; stout prismatic, often with rounded or rough faces. Usually massive, cleavable

Hardness: 4–5

Lattice Constants: a = 6.0285, b = 10.3586, c = 4.7031 (Å)

Mode of Occurrence: Occurs as a primary mineral in granite pegmatites, often partly or completely altered to a wide variety of secondary phosphate minerals. Found widespread in the Black Hills, South Dakota, often as anhedral or subhedral crystals up to 6 or more feet in size, especially at the Dan Patch, Big Chief, and Hesnard mines, Keystone district, and at the Bull Moose and Tip Top mines, Custer district. It also occurs as fine crystals at the Smith mine, Chandler's Mill, at Palermo, and at other places in New Hampshire; in Maine; and in the Pala district, California. Other occurrences are found in Canada, Brazil, France, Sweden, Germany, and Finland

Optical Constants: $\alpha = 1.689–1.694$, $\beta = 1.689–1.695$, $\gamma = 1.695–1.702$; $(+)2V = 0°–55°$; also (−)

Space Group: Pmnb

Strongest Diffraction Lines: 2.54(100), 4.29(90), 3.51(90)

Z: 4

Ref.: Roberts WL, Rapp GR, Weber J (1974) Encyclopedia of minerals, New York, Van Nostrand Reinhold, p 626–627. Finger LW, Rapp GR Jr (1970) Carnegie Inst Wash Year Book 69:290–292

Triplite
(Triplite-Zwieselite Series)
$(Mn^{2+}, Fe^{2+}, Mg, Ca)_2(PO_4)(F, OH)$
Var. Magnesian (Talktriplite)

Analyses (%): CaO 0.50–14.9, MgO 0–17.4, FeO 1.68–42.0, MnO 14.9–57.6, P_2O_5 30.6–33.3, F 6.02–9.09

Class: 2/m

Cleavage: {001} good, {010} fair, {100} poor. Fracture subconchoidal to uneven

Color-Luster: Reddish brown to dark brown; also salmon pink; often brownish black to black on alteration. Translucent to opaque. Vitreous to resinous. Streak brown to white

Crystal System: Monoclinic

Density: 3.55–3.87 (Meas.)

First Reported: 1802

Habit: As rough, poorly developed crystals. Commonly massive

Hardness: 5–5.5

Lattice Constants: a = 12.085, b = 6.536, c = 9.910 (Å); $\beta = 105.63°$

Mode of Occurrence: Occurs widespread as a primary mineral in granite pegmatites, and sparingly in high-temperature vein deposits. Found in the United States in California, Nevada, Arizona, Colorado, South Dakota, Virginia, Connecticut, and Maine. It also occurs in Argentina, France, Germany, Czechoslovakia, Portugal, Sweden, Finland, South West Africa, Rhodesia, and South Australia and the Megiliggar Rocks, Cornwall

Optical Constants: $\alpha = 1.643–1.696$, $\beta = 1.647–1.704$, $\gamma = 1.665–1.713$; $(+)2V = 25°–76°$

Space Group: I 2/m

Strongest Diffraction Lines: 2.87(100), 3.05(90), 3.26(70)

Z: 8

Ref.: Roberts WL, Rapp GR, Weber J (1974) Encyclopedia of minerals, New York, Van Nostrand Reinhold, p 627

Triploidite (Triploidite-Wolfeite Series)
$(Mn^{2+}, Fe^{2+})_2PO_4OH$

Analyses (%): CaO 0.19–2.00, MgO 0–4.74, MnO 13.1–48.4, FeO 14.9–44.4, Fe_2O_3 0.22–7.78, P_2O_5 30.9–34.2, H_2O 3.78–4.48

Class: 2/m

Cleavage: {010} good, {120} fair. Fracture subconchoidal to uneven

Color-Luster: Pinkish, yellowish to yellowish brown or reddish brown; rarely greenish. Transparent to translucent. Vitreous to greasy. Streak whitish

Crystal System: Monoclinic

Density: 3.697 (Meas.)

First Reported: 1878

Habit: Crystals prismatic, vertically striated. Usually as columnar to parallel-fibrous aggregates; also divergent fibrous; granular

Hardness: 4.5–5

Lattice Constants: a = 12.26, b = 13.38, c = 9.90 (Å); $\beta = 108.07°$

Mode of Occurrence: Occurs as pinkish masses in granite pegmatite, associated with alluaudite and loellingite, at the Ross mine, Custer County, and sparingly at the Peerless mine, Pennington County, South Dakota; with lithiophilite, rhodochrosite, and secondary phosphate minerals at Branchville, Fairfield County, Connecticut; and as an alteration of triplite at Wein, Czechoslovakia

Optical Constants: $\alpha = 1.725$, $\beta = 1.726$, $\gamma = 1.730$; (+)2V = moderate

Space Group: $P2_1/a$

Strongest Diffraction Lines: 2.94(100), 3.10(90), 3.19(80)

Z: 16

Ref.: Roberts WL, Rapp GR, Weber J (1974) Encyclopedia of minerals, New York, Van Nostrand Reinhold, p 627

Trolleite
$Al_4(PO_4)_3(OH)_3$

Analyses (%): CaO 0.97, Fe_2O_3 2.75, Al_2O_3 43.3, P_2O_5 46.7, H_2O 6.23

Cleavage: Indistinct in two directions at an angle of 110.87°. Fracture conchoidal to even

Color-Luster: Pale green. Vitreous

Crystal System: Monoclinic

Density: 3.09 (Meas.)

First Reported: 1868

Habit: Massive, lamellar

Hardness: ~8.5 (synthetic), 5.5–6 (natural)

Lattice Constants: a = 18.894, b = 7.161, c = 7.162 (Å); $\beta = 100°$

Mode of Occurrence: Occurs in association with berlinite and other phosphates at the iron mine of Westanå in Kristianstad, Sweden

Optical Constants: $\alpha = 1.619$, $\beta = 1.639$ (Na), $\gamma = 1.643$ (Na); (−)2V = 49°

Space Group: I2/c

Strongest Diffraction Lines: 3.20(100), 3.09(80), 2.51(45)

Ref.: Roberts WL, Rapp GR, Weber J (1974) Encyclopedia of minerals, New York, Van Nostrand Reinhold, p 629. Sclar CB, Carrison LC, Schwartz CM (1965) Am Miner 50:267. Moore PB, Araki T (1974) Am Miner 59:974–984

Tsumebite
$Pb_2Cu(PO_4)(SO_4)(OH)$

Analyses (%): PbO 63.8–65.1, CuO 11.8–12.1, P_2O_5 10.3–12.0, H_2O 12.3

Class: 2/m

Cleavage: Fracture uneven. Brittle

Color-Luster: Emerald green. Transparent. Brilliant, vitreous

Crystal System: Monoclinic

Density: 6.133 (Meas.)

First Reported: 1912

Habit: Crystals thick tabular, always twinned as trillings or more complex groups. As crusts of intergrown crystals. Twin plane $\{\bar{1}22\}$

Hardness: 3.5

Lattice Constants: a = 8.70, b = 5.80, c = 7.85 (Å); $\beta = 111.5°$

Mode of Occurrence: Occurs as a secondary mineral at Morenci, Greenlee County, Arizona; and with cerussite, smithsonite, and azurite at Tsumeb, South West Africa

Optical Constants: $\alpha = 1.885$, $\beta = 1.920$, $\gamma = 1.956$; (+)2V = near 90°

Space Group: $P2_1/m$

Z: 2

Ref.: Roberts WL, Rapp GR, Weber J (1974) Encyclopedia of minerals, New York, Van Nostrand Reinhold, p 631

Turquois(e) Group
(includes Chalcosiderite, Coeruleolactite, and Faustite)
$CuAl_6(PO_4)_4(OH)_8 \cdot 4H_2O$

Analyses (%): CuO 7.40–9.00, FeO 0–5.32, Al_2O_3 20.8–36.5, Fe_2O_3 0.21–21.3, P_2O_5 28.6–34.4, H_2O 11.9–20.1

Class: $\bar{1}$

Cleavage: {001} perfect, {010} good. Fracture conchoidal to smooth (massive)

Color-Luster: Crystals: Bright blue. Transparent. Vitreous. Massive: Pale blue to sky blue, bluish green to apple green to greenish gray. Waxy or dull. Subtranslucent to opaque

Crystal System: Triclinic

Density: 2.6–2.8 (massive), 2.84 (crystals), 2.91 (calc.)

First Reported: Antiquity, eg., *Callaina* in Pliny, Natural History, 37:56

Habit: Crystals rare, small, short prismatic. Usually massive, dense, cryptocrystalline to fine-granular; also concretionary, stalactitic, and as veinlets or crusts

Hardness: 5–6

Lattice Constants: a = 7.424, b = 7.629, c = 9.910 (Å); $\alpha = 68.61°$, $\beta = 69.71°$, $\gamma = 65.08°$

Mode of Occurrence: Occurs as a secondary mineral formed by the action of surface waters on aluminous rocks. Found as excellent microcrystals implanted on quartz or schist near Lynch Station, Campbell County, Virginia; as large deposits in the Los Cerillos Mountains, New Mexico, and elsewhere in the state; and in Arizona, Colorado, Nevada, and California. It also occurs in France, Germany, U.S.S.R., Egypt, and as fine gem -quality material in Iran

Optical Constants: $\alpha = 1.61$, $\beta = 1.62$ (Na), $\gamma = 1.65$; (+)2V ≃ 40°

Space Group: $P\bar{1}$

Strongest Diffraction Lines: 3.68(100), 2.91(80), 6.17(70)

Z: 1

Ref.: Cid-Dresdner H (1965) Am Miner 50:283. Roberts WL, Rapp GR, Weber J (1974) Encyclopedia of minerals, New York, Van Nostrand Reinhold, p 634

Upalite
$Al(UO_2)_3(PO_4)_2(OH)_3$

Analyses (%): P_2O_5 12.3, Al_2O_3 4.0, UO_3 80.2, H_2O 3.5

Color-Luster: Amber-yellow; strongly pleochroic, X colorless, Y and Z canary yellow

Crystal System: Orthorhombic

Density: 3.5 (Meas.), 3.58 (Calc.)

First Reported: 1979

Habit: Typically needles, with max. length of up to 0.33 mm

Hardness: Not given

Lattice Constants: a = 34.68, b = 16.81, c = 13.72 (Å)

Mode of Occurrence: Found in the beryl-columbite pegmatite at Kobokobo, Kivu Zaire, associated with meta-autunite, phospuranylite, threadgoldite, and other aluminum uranyl phosphates

Optical Constants: $\alpha = 1.649$, $\beta = 1.666$, $\gamma = 1.676$; Biaxial neg., 2V(calc.) = 74°

Space Group: Bbcm or Bba2

Strongest Diffraction Lines: 8.4(100), 4.18(80), 3.43(80), 2.903(75)

Z: 16

Ref.: Deliens M, Piret P (1979) Bull Miner 102:333–337

Uralolite
$CaBe_3(PO_4)_2(OH)_2 \cdot 4H_2O$

Analyses (%): BeO 18.9–19.6, CaO 16.5–19.1, P_2O_5 38.4–39.0, H_2O 21.9–23.6

Cleavage: Probable cleavage perpendicular to elongation and others along elongation

Color-Luster: Colorless to white, sometimes stained brown by iron oxides. Fibrous aggregates silky; needles vitreous

Crystal System: Monoclinic

Density: Probably 2.05–2.14, 2.042 (Calc.)

First Reported: 1964

Habit: As concretions composed of radiating fibrous spherulites 2–3 mm in diameter, and in sheaf-like growths

Hardness: 2.5

Lattice Constants: a = 8.43, b = 39.50, c = 7.12 (Å); $\beta = 94.97°$

Mode of Occurrence: Occurs in kaolin-hydromuscovite rocks (from the Urals) containing fluorite, beryl, apatite, crandallite, moraesite, and herderite

Optical Constants: $\alpha = 1.510$, $\beta = 1.525$, $\gamma = 1.536$; (−)2V

Strongest Diffraction Lines: 3.56(100), 3.04(80), 3.20(70)

Z: 8

Ref.: Grigor'ev NA (1964) Am Miner 49:1776. Grigor'ev NA (1964) Mem All-Union Miner Soc 87:156. Roberts WL, Rapp GR, Weber J (1974) Encyclopedia of minerals, New York, Van Nostrand Reinhold, p 639–640

Uramphite
(inadequately described mineral)
$NH_4UO_2PO_4 \cdot 3H_2O$

Analyses (%): $(NH_4)_2O$ 6.6, UO_3 68.7, P_2O_5 15.8, H_2O 11.0

Cleavage: Distinct in two directions

Color-Luster: Bottle green to pale green; vitreous. Fluoresces medium yellow-green under ultraviolet light

Crystal System: Unknown

Density: 3.7

First Reported: 1957

Habit: As square tablets up to 0.2 × 0.2 mm in size, in small rosettes, and as lichen-like deposits

Hardness: Not determined

Mode of Occurrence: Occurs in the oxidized zone of a uranium-coal deposit, in fractures in the coal 20–50 meters below the surface, at an unspecified locality in U.S.S.R.

Optical Constants: $\alpha = 1.564$, $\beta = \gamma = 1.585$; (−)2V = 0–3°

Strongest Diffraction Lines: 3.78(100), 2.22(90), 1.694(90)

Ref.: Nekrasova ZA (1957) Voprosy geologii urana, Atomizdat, p 67–72, Translated by Consultants Bureau, New York, 1958, p 56–60. Nekrasova ZA (1959) Am Miner 44:464. Roberts WL, Rapp GR, Weber J (1974) Encyclopedia of minerals, New York, Van Nostrand Reinhold, p 640

Uranite = General term for uranyl phosphates and arsenates of the autunite and meta-autunite groups

Uranocircite (Autunite Group)
Uranocircite I: $Ba(UO_2)_2(PO_4)_2 \cdot 12H_2O$
Uranocircite II: $Ba(UO_2)_2(PO_4)_2 \cdot 10H_2O$

Analyses (%): P_2O_5 15.1, UO_3 56.9, BaO 14.6, H_2O 12–16

Class: 4/m 2/m 2/m

Cleavage: {001} perfect, {100} distinct. Not brittle. Thin plates flexible

Color-Luster: Yellow. Transparent. Pearly on {001}. Fluoresces green under ultraviolet light

Crystal System: Probably tetragonal

Density: 3.46 (Calc. $10H_2O$)

First Reported: 1887

Habit: Crystals thin tabular with rectangular outline. Crystals usually as subparallel, composite, or fan-like aggregates

Hardness: 2–2.5

Lattice Constants: (I) a ~ 7, c = 22.59 (Å). (II) a = 7.01, c = 20.46 (Å)

Mode of Occurrence: Uranocircite I and II occur with other hydrated phases of $Ba(UO_2)_2(PO_4)_2$ at Farnwitte near Menzenschwand, Black Forest, Germany

Optical Constants: (I) $\alpha = 1.562$. (II) $\alpha = 1.574$, $\beta = 1.583$, $\gamma = 1.588$; $2V = 70°$

Space Group: P4/nnc (II)

Strongest Diffraction Lines: 5.10(100), 2.04(100) (uranocircite II), 10.1(60)

Z: 2 (II)

Ref.: Frondel C (1958) USGS Bull 1064:211–215, 177. Walenta K (1966) Min Abs 17:695. Roberts WL, Rapp GR, Weber J (1974) Encyclopedia of minerals, New York, Van Nostrand Reinhold, p 640–641

Uranospathite

(Member of the Torbernite Series)

$(HAl)_{0.5}(UO_2)_2(PO_4, AsO_4)_2 \cdot 20H_2O$

Analyses (%): P_2O_5 4.4, As_2O_5 13.8, UO_3 56.2, CuO 6.7, H_2O 18.0

Cleavage: {001} perfect, {100} good, {010} fibrous

Color-Luster: Yellow to pale green and bluish green

Density: 2.50 (Meas.)

First Reported: 1873

Habit: As fan-like groups of thin rectangular plates and laths flattened on {001}, elongated along *b*-axis and striated parallel to elongation. Sometimes twinned on {110}

Hardness: Not determined

Lattice Constants: a = 7.00, c = 30.02 (Å)

Mode of Occurrence: Occurs in parallel growths with bassetite at Redruth, Cornwall, England

Optical Constants: $\alpha \simeq 1.49$, $\beta = 1.510$, $\gamma = 1.521$; $(-)2V = 72°$

Space Group: $P4_2/n$

Strongest Diffraction Lines: 15.22(100), 7.60(100), 4.93(100), 3.50(80)

Z: 2

Ref.: Frondel C (1958) USGS Bull 1064:194–195. Walenta K (1978) Miner Mag 42:117–128. Roberts WL, Rapp GR, Weber J (1974) Encyclopedia of minerals, New York, Van Nostrand Reinhold, p 642

Variscite

(Variscite-Strengite Series, dimorphous with metavariscite)

$AlPO_4 \cdot 2H_2O$

Analyses (%): Al_2O_3 16.6–32.4, Fe_2O_3 0–14.4, CaO 0–0.85, Cr_2O_3 0–0.73, P_2O_5 41.2–44.7, H_2O 22.7–26.0

Class: 2/m 2/m 2/m

Cleavage: {010} good, {001} poor. Fracture splintery to uneven (massive), conchoidal (crystals)

Color-Luster: Pale green to emerald green, also bluish green to colorless. Transparent to translucent. Crystals vitreous; massive material waxy to dull

Crystal System: Orthorhombic

Density: 2.57 (Meas.), 2.61 (Calc.)

First Reported: 1837 (Described as Peganite, 1830)

Habit: Crystals octahedral (rare). Usually massive, as crusts, veinlets, or nodules. Also opal-like. Twinning on {201} not common

Hardness: 3.5–4.5

Lattice Constants: a = 9.87, b = 9.57, c = 8.52 (Å)

Mode of Occurrence: Occurs chiefly in deposits formed by the action of phosphatic meteoric waters on aluminous rocks. Found in Utah as nodules up to 25 cm in diameter near Fairfield, Utah County, associated with crandallite and other phosphate minerals; at Lucin, Box Elder County; and at several other places in the state. It also occurs as incrustations on rock at Bear, Montgomery County, Arkansas, and in California, Nevada, Arizona, and Pennsylvania. Other occurrences include localities in Czechoslovakia, Germany, Australia, Redonda Island, Antigua, Western Australia, and Rio Grande do Norte, Brazil

Optical Constants: $\alpha = 1.563$, $\beta = 1.588$, $\gamma = 1.594$; (−)2V = moderate

Space Group: Pcab

Strongest Diffraction Lines: 5.365(100), 4.257(100), 3.039(100)

Z: 8

Ref.: Roberts WL, Rapp GR, Weber J (1974) Encyclopedia of minerals, New York, Van Nostrand Reinhold, p 648

Varulite

$(Na_2, Ca)(Mn, Fe)_2(PO_4)_2$

Analyses (%): Na_2O 3.73–9.72, CaO 1.35–9.70, MgO 0.09–2.55, FeO 7.52–18.2, MnO 6.44–25.3, Fe_2O_3 5.32–26.5, P_2O_5 39.4–44.9, H_2O 0.64–4.49

Class: 2/m

Cleavage: {001} good, {010} distinct

Color-Luster: Dull yellowish green. Vitreous

Crystal System: Monoclinic

Density: 3.581 (Meas.)

First Reported: 1937

Habit: Massive, granular

Hardness: 5

Mode of Occurrence: Occurs in granite pegmatites at Varuträsk and at Skrumpetorp, Sweden

Optical Constants: $\alpha = 1.708, 1.720$, $\gamma = 1.722, 1.732$; (+)2V = large, 70°

Space Group: $I2_1/a$

Strongest Diffraction Lines: 2.74(100), 3.50(40), 2.56(40)

Ref.: Roberts WL, Rapp GR, Weber J (1974) Encyclopedia of minerals, New York, Van Nostrand Reinhold, p 649

Vashegyite

(Inadequately described mineral)

$Al_4(PO_4)_3(OH)_3 \cdot 13H_2O$ (?)

Analyses (%): Al_2O_3 26.5–28.3, Fe_2O_3 1.24, P_2O_5 26.6–31.3, H_2O 39.0–44.7

Cleavage: Not determined

Color-Luster: White, pale green to yellow and brownish. Subtranslucent to opaque. Dull

Crystal System: Unknown

Density: 1.96 (Meas.)

First Reported: 1909

Habit: Massive, compact to porous

Hardness: 2–3

Mode of Occurrence: Occurs with variscite in altered slate near Manhattan, Nye County, Nevada, and at localities in Germany, Czechoslovakia, and Hungary

Optical Constants: Mean index = 1.48

Strongest Diffraction Lines: 10.5(100), 7.2(70), 2.90(70)

Ref.: Roberts WL, Rapp GR, Weber J (1974) Encyclopedia of minerals, New York, Van Nostrand Reinhold, p 649

Vauquelinite (Laxmannite)

$(Cu, Pb)_3(CrO_4, PO_4)_2$

Analyses (%): PbO 61.1–70.6, CuO 4.57–12.4, CrO_3 10.1–26.8, Fe_2O_3 0–2.80, P_2O_5 3.55–9.94, H_2O 0.9–3.00

Class: 2/m

Cleavage: Fracture uneven. Brittle

Color-Luster: Green to brown to nearly black. Transparent to translucent. Adamantine to resinous. Streak greenish or brownish

Crystal System: Monoclinic

Density: 5.986–6.06 (Meas.), 6.03 (Calc. for $Pb_2Cu[CrO_4][PO_4]$)

First Reported: 1818

Habit: Crystals usually wedge-shaped, minute. As irregular aggregates; fibrous; compact; granular. Twinning on {102}

Hardness: 2.5–3

Lattice Constants: a = 13.68, b = 5.83, c = 9.53 (Å); $\beta = 93.97°$

Mode of Occurrence: Occurs as a secondary mineral associated with crocoite, pyromorphite, and mimetite at Beresov, Urals, U.S.S.R. Also reported from numerous unauthenticated localities

Optical Constants: $\alpha = 2.11$, $\beta = 2.22$, $\gamma = 2.22$; (−)2V ~ 0°

Space Group: $P2_1/n$

Strongest Diffraction Lines: 3.31(100), 4.73(70), 2.89(60)

Z: 4

Ref.: Roberts WL, Rapp GR, Weber J (1974) Encyclopedia of minerals, New York, Van Nostrand Reinhold, p 649–650

Vauxite
$Fe^{2+}_2Al_4(PO_4)_4(OH)_4 \cdot 12H_2O$

Analyses (%): FeO 15.0–15.5, CaO 0.08–0.77, Al_2O_3 21.4–23.8, P_2O_5 30.5, H_2O 30.6–31.4

Class: $\bar{1}$

Cleavage: None. Brittle

Color-Luster: Pale sky blue to dark blue. Transparent to translucent. Vitreous. Streak white

Crystal System: Triclinic

Density: 2.40 (Meas.), 2.41 (Calc.)

First Reported: 1922

Habit: Crystals minute, tabular, and somewhat elongated. As radial or subparallel aggregates. Twinning on {010}

Hardness: 3.5

Lattice Constants: a = 9.13, b = 11.59, c = 6.14 (Å); $\alpha = 98.3°$, $\beta = 92.0°$, $\gamma = 108.4°$

Mode of Occurrence: Occurs as excellent microcrystals associated with paravauxite, quartz, and wavellite in the tin deposits at Llallagua, Bolivia

Optical Constants: $\alpha = 1.551$, $\beta = 1.555$, $\gamma = 1.562$; (+)2V = 32°

Space Group: $P\bar{1}$

Strongest Diffraction Lines: 5.45(100), 5.97(80), 4.94(60)

Z: 2

Ref.: Gorden SG (1922) Science 56:56. Baur WH, Rao BR (1968) Am Miner 53:1025–1028. Roberts WL, Rapp GR, Weber J (1974) Encyclopedia of minerals, New York, Van Nostrand Reinhold, p 650

Väyrynenite
(Herderite-Väyrynenite Series)
$BeMnPO_4(OH, F)$

Analyses (%): FeO 4.59, MnO 30.6, CaO 1.82, Na_2O 1.42, K_2O 1.18, BeO 12.1, Al_2O_3 2.45, P_2O_5 40.4

Class: 2/m

Cleavage: {001} distinct. Brittle

Color-Luster: Pale pink to rose red; transparent; vitreous

Crystal System: Monoclinic

Density: 3.183–3.215 (Meas.), 3.23 (Calc.)

First Reported: 1954

Habit: Crystals prismatic; in drusy aggregates. Usually massive, granular

Hardness: 5

Lattice Constants: a = 5.411, b = 14.49, c = 4.73 (Å); $\beta = 102.75°$

Mode of Occurrence: Occurs associated with herderite, hurlbutite, beryllonite, microcline, morinite, and muscovite in lithium pegmatite at Viitaniemi, Eräjärvi, Finland

Optical Constants: $\alpha = 1.638$, $\beta = 1.658$, $\gamma = 1.664$; (−)2V = 54.13°

Space Group: $P2_1/a$

Strongest Diffraction Lines: 3.45(100), 7.25(85), 2.89(85)

Z: 4

Ref.: Volborth A (1954) Ann Acad Sci Fennicae Ser A. III-Geol Geogr 39:66. Volborth A (1956) Am Miner 41:371. Roberts WL, Rapp GR, Weber J (1974) Encyclopedia of minerals, New York, Van Nostrand Reinhold, p 650

Veszelyite (Arakawaite, Kipushite)
$(Cu, Zn)_3(PO_4)(OH)_3 \cdot 2H_2O$

Analyses (%): CuO 36.0–40.4, ZnO 23.6–28.9, P_2O_5 9.01–19.9, H_2O 14.3–17.1

Class: 2/m

Cleavage: {001}, {110}

Color-Luster: Green, greenish blue to dark blue. Translucent. Vitreous

Crystal System: Monoclinic

Density: 3.4 (Meas.), 3.42 (Calc. Cu:Zn = 3:2)

First Reported: 1874

Habit: Crystals short prismatic, thick tabular, equant, or octahedral. As granular aggregates

Hardness: 3.5–4

Lattice Constants: a = 9.84, b = 10.17, c = 7.48 (Å); $\beta = 103.42°$

Mode of Occurrence: Occurs at Moravicza (Vaskö), Banat, Rumania; at Kipushi, Katanga, Zaire (kipushite); at the Broken Hill mine, Kabwe, Zambia; and at the Arakawa mine, Ugo province, Japan (arakawaite)

Optical Constants: $\alpha = 1.640, 1.618$, $\beta = 1.658, 1.622$, $\gamma = 1.695, 1.658$; $(+)2V \sim 71°–38.5°$

Space Group: $P2_1/a$

Strongest Diffraction Lines: 3.642(100), 6.96(40), 4.489(30)

Z: 4

Ref.: Roberts WL, Rapp GR, Weber J (1974) Encyclopedia of minerals, New York, Van Nostrand Reinhold, p 652. Ghose S, Leo SR, Wan C (1974) Am Miner 59:573–581

Vibertite = Bassanite

Viitaniemiite (cf. Lacroixite)
$Na(Ca, Mn^{2+})Al(PO_4)(F, OH)_3$

Analyses (%): P_2O_5 28.3, Al_2O_3 22.4, FeO 0.70, MnO 10.5, CaO 14.7, Na_2O 11.6, F 12.3, H_2O 4.9

Cleavage: $\{10\bar{1}\}$ well developed

Color-Luster: Gray to white; vitreous luster. Colorless in section, optical axial plane is (010)

Crystal System: Monoclinic

Density: 3.245 (Meas.), 3.240 (Calc.)

First Reported: 1981

Habit: Fan-shaped tabular crystals 0.02 to 0.2 mm wide

Hardness: 5

Lattice Constants: a = 6.832, b = 7.143, c = 5.447 (Å); $\beta = 109°$

Mode of Occurrence: Found in the Viitaniemi granitic pegmatites of the Erajärvi area in Orivesi, southern Finland. Associated minerals include apatite, crandallite, and montebrasite. It occurs as an inclusion in eosphorite or a rim around grains of morinite

Optical Constants: $\alpha = 1.557$, $\beta = 1.565$, $\gamma = 1.571$; Biaxial neg., 2V = 81° (Meas.), 81.4° (Calc.)

Space Group: $P2_1$ or $P2_1/m$

Strongest Diffraction Lines: 2.883(100), 2.937(50), 3.223(46), 2.160(40)

Z: 2

Ref.: Lahti SI (1981) Geol Surv Fin Bull 314:51–56

Vilatéite = Metastrengite

Viséite (Zeolite Group)
$5CaO \cdot 6Al_2O_3 \cdot 3SiO_2 \cdot 3\frac{1}{2}P_2O_5 \cdot 1\frac{1}{2}F \cdot 36H_2O$

Analyses (%): P_2O_5 19.2–28.3, SiO_2 4.30–10.4, F 0.67–1.31, Al_2O_3 23.3–

28.4, CaO 12.2–17.2, Na_2O 0–1.77, FeO 0–1.43, H_2O 23.8–28.9

Cleavage: None. Brittle

Color-Luster: White, bluish white, yellowish white. Translucent; transparent in thin fragments. Vitreous

Crystal System: Cubic or pseudocubic

Density: 2.2 (Meas.), 2.17 (Calc.)

First Reported: 1943

Habit: Massive; as minute nodules

Hardness: 3–4

Lattice Constant: a = 13.65 (Å)

Mode of Occurrence: Occurs in association with delvauxite at Visé, Belgium

Optical Constant: N = 1.53

Strongest Diffraction Lines: 2.92(100), 1.74(60), 3.46(50)

Z: 2

Ref.: Mélon J (1943) Ann Soc Geol Belg 66:53. Deer WA, Howie RA, Zussman J (1963) Rock forming minerals, vol 4. Wiley, New York, p 342, 345, 408. McConnell D (1952) Am Miner 37: 609–617. Dunn PJ, Appleman DE (1977) Miner Mag 41:437–442. Roberts WL, Rapp GR, Weber J (1974) Encyclopedia of minerals, New York, Van Nostrand Reinhold, p 654

Vitusite

$Na_3(Ce, La)(PO_4)_2$

Analyses (%): P_2O_5 35.92–36.50, Na_2O 22.50–23.15, CaO 0.92–4.75, La_2O_3 9.25–15.56, Ce_2O_3 16.50–21.72, Nd_2O_3 4.33–7.69, Pr_2O_3 1.13–1.92

Cleavage: {100}, {010}, {001} good. Twinned after {160} and {120}

Color-Luster: Pale pink, white to pale green. Luster vitreous

Crystal System: Orthorhombic (pseudo-hexagonal)

Density: 3.60–3.70

First Reported: 1979

Hardness: 4.5

Lattice Constants: a = 5.342, b = 18.680, c = 14.062 (Å) (a = 58.75 Å for the superlattice)

Mode of Occurrence: Occurs in the natrolite zone of an alkaline pegmatite at Mt. Karnasurt (Greenland) associated with steenstrupine, belovite, neptunite, leucosphenite, and sazhinite. At the Ilimaussaq alkaline intrusions (south Greenland), it occurs in a melanocratic nepheline syenite, with arvedsonite, albite, microcline, nepheline, sodalite, steenstrupine and other unidentified minerals

Optical Constants: $\alpha = 1.604$, $\beta = 1.6465$, $\gamma = 1.654$; Biaxial neg., $(-)2V = 30^\circ$

Solubility: Readily dissolved in cold dilute HCl or H_2SO_4

Space Group: Pcmb or $Pc2_1b$

Strongest Diffraction Lines: 6.580, 4.634, 3.512, 2.811, 2.901, 2.690, 2.671

Ref.: Rønsbo JG et al. (1979) Neues Jahrb Miner Abh 137:42–53

Vivianite

$Fe_3(PO_4)_2 \cdot 8H_2O$

Analyses (%): FeO 32.7–44.1, MgO 0–1.92, MnO 0–2.01, Fe_2O_3 0–12.0, P_2O_5 27.0–30.0, H_2O 25.4–28.8

Class: 2/m

Cleavage: {010} perfect, easy; {106} trace; {100} trace. Fracture fibrous. Thin lamellae flexible. Sectile

Color-Luster: Colorless when fresh; darkens rapidly to shades of green or blue; on further exposure to light becoming dark green, dark bluish green, dark purplish, or bluish black. Streak colorless to bluish white, soon changing to dark blue or brown

Crystal System: Monoclinic

Density: 2.68 (Meas.), 2.71 (Calc.)

First Reported: 1758 (under various names; vivianite came into use in 1812)

Habit: Crystals short to long prismatic, elongated along c-axis, often flattened; also tabular or equant. Crystals often rounded. Commonly grouped in clusters or as radial and stellate groups.

Also massive bladed or fibrous, as crusts, concretionary, and earthy
Hardness: 1.5–2
Lattice Constants: a = 10.039 kX, b = 13.388, c = 4.687 (Å); $\beta = 104.3°$
Mode of Occurrence: Occurs as a secondary mineral in some metallic ore veins; as an alteration product of primary phosphate minerals in pegmatites: and as concretions or associated with organic matter in clays and other sedimentary deposits. Found as superb crystals with ludlamite in the Blackbird District, Lemhi County, Idaho; as fine crystals up to 5 inches long in Bingham Canyon, Salt Lake County, Utah; as magnificent crystal groups at Richmond, Virginia; in pegmatites in the Black Hills, South Dakota; and at localities in Colorado, California, New Jersey, Delaware, Maryland, Washington, D.C., and Florida. Excellent dark green crystals up to 6 inches or more in length occur at Llallagua and Poopo, Bolivia; spectacular crystals up to 4 feet long near N'gaoundere, Cameroons; and at places in Canada, England, Yugoslavia, Czechoslovakia, Roumania, France, Germany, U.S.S.R., Australia, and Japan

Optical Constants: $\alpha = 1.5788–1.616$, $\beta = 1.6024–1.656$, $\gamma = 1.6294–1.675$; (+)2V = 63.5–83.5°

Space Group: C2/m

Strongest Diffraction Lines: 6.80(100), 2.97(67) (synthetic), 2.71(67)

Z: 2

Ref.: Roberts WL, Rapp GR, Weber J (1974) Encyclopedia of minerals, New York, Van Nostrand Reinhold, p 654–655

Voelckerite (Oxyapatite)
An incompletely characterized mineral with probable composition of $Ca_9A(PO_4)_6O_2$ and/or $Ca_{10}(PO_4)_5(ZO_4)O_2$ where A and Z are elements with +3 and +6 charges respectively.

See McConnell D (1973) Apatites. Springer, Wien, p 91

Vuonnemite
$Na_4TiNb_2Si_4O_{17} \cdot 2Na_3PO_4$

Analyses (%): SiO_2 22.3, TiO_2 8.07, Nb_2O_5 23.2, MnO 0.53, Na_2O 30.2, Al_2O_3 0.46, P_2O_5 13.5

Cleavage: {001} very perfect, 2 other perfect cleavages

Color-Luster: Light yellow, luster vitreous

Crystal System: Triclinic

Density: 3.13

First Reported: 1973

Habit: Plates up to 1.8 × 1.5 × 0.5 cm

Hardness: 2–3, brittle

Lattice Constants: a = 7.02, b = 14.15, c = 5.38 (Å); $\alpha = 93°$, $\beta = 89°$, $\gamma = 87°$

Mode of Occurrence: Found in albitized alkalic rocks of the Khibina and Lovozero massifs, Kola Peninsula, U.S.S.R. Associated minerals include albite, microcline, nepheline, aegirine, lorenzenite, cancrinite, serandite, and villiaumite

Optical Constants: $\alpha = 1.639$, $\beta = 1.651$, $\gamma = 1.681$; (+)2V = 53°; plane of optic axis {001}

Space Group: P1 or $P\bar{1}$

Strongest Diffraction Lines: 2.869(100), 7.14(60), 3.582(60), 1.790(60)

Z: 1

Ref.: Bussen IV et al. (1973) Zapiski Vses Miner Obshch 102:423–426

Wagnerite
$(Mg, Fe, Mn, Ca)_2PO_4F$

Analyses (%): CaO 2.53–13.4, MgO 37.6–47.6, FeO 0–4.32, P_2O_5 40.2–43.4, F 5.06–11.5

Class: 2/m

Cleavage: {100} imperfect, {120} imperfect, {001} in traces. Fracture splintery to subconchoidal

Color-Luster: Various shades of yellow, grayish, greenish, yellowish red to brick brown. Translucent to opaque. Vitreous to resinous

Crystal System: Monoclinic

Density: 3.15 (Meas.), 3.15 (Calc.)

First Reported: 1821

Habit: Crystals prismatic, vertically striated or rounded, sometimes large and rough. Also massive

Hardness: 5–5.5

Lattice Constants: a = 11.90 kX, b = 12.51, c = 9.63 (Å); $\beta = 108.12°$

Mode of Occurrence: Occurs in quartz veins associated with chlorite, magnesite, and lazulite at Höllgraben and Radelgraben in Salzburg, Austria; at several places near Bamle, Norway; at Horrsjöberg, Sweden; in pegmatite near Mangualde, Portugal; and in lava on Mt. Vesuvius, Italy

Optical Constants: $\alpha = 1.5678$, $\beta = 1.5719$, $\gamma = 1.5824$; (+)2V = 28.41°

Space Group: $P2_1/a$

Strongest Diffraction Lines: 2.993(100), 2.839(100), 3.150(90)

Z: 16

Ref.: Roberts WL, Rapp GR, Weber J (1974) Encyclopedia of minerals, New York, Van Nostrand Reinhold, p 659

Wardite

$NaAl_3(PO_4)_2(OH)_4 \cdot 2H_2O$

Analyses (%): Na_2O 5.98–7.0, CaO 0–3.0, Al_2O_3 36.6–38.2, P_2O_5 34.6, H_2O 17.9

Class: 422

Cleavage: {001} perfect

Color-Luster: Colorless, white, pale green to bluish green; transparent to translucent. Vitreous

Crystal System: Tetragonal

Density: 2.81 (Meas.), 2.81 (Calc.)

First Reported: 1896

Habit: Crystals pyramidal, symmetrically developed with the same forms at opposite ends of the c-axis. Also as crusts and granular aggregates, as subparallel aggregates of coarse fibers, and as radially fibrous and concentrically banded spherulites. Crystals often striated horizontally

Hardness: 5

Lattice Constants: a = 7.04, c = 18.88 (Å)

Mode of Occurrence: Occurs associated with millisite and crandallite in variscite nodules at Fairfield, Utah County, and in the variscite deposits at Amatrice Hill and Lucin, Utah; as crystals up to 10 mm in maximum dimension coating quartz in pegmatite at Beryl Mountain near West Andover, New Hampshire; in narrow veins in massive amblygonite in the Stewart mine, Pala, San Diego County, California; and as small, well-formed crystals associated with silky-white apatite, morinite, augelite, and montebrasite at the Hugo mine, Keystone, South Dakota. Also found as an alteration product of amblygonite at Montebras, Creuse, France; and in the pegmatite of the alto Patrimonio, at Piedras Lavradas, Paraíba, Brazil

Optical Constants: $\omega = 1.586–1.594$, $\varepsilon = 1.595–1.604$; (+)

Space Group: $P4_12_12$ and $P4_32_12$

Strongest Diffraction Lines: 4.77(100), 4.73(100), 3.09(80)

Z: 4

Ref.: Davison JM (1896) Am J Sci 2:154. Hurlbut CS Jr (1952) Am Miner 37:849–852. Roberts WL, Rapp GR, Weber J (1974) Encyclopedia of minerals, New York, Van Nostrand Reinhold, p 662

Wavellite

$Al_3(OH)_3(PO_4)_2 \cdot 5H_2O$

Analyses (%): Al_2O_3 31.0–37.4, Fe_2O_3 0–3.18, P_2O_5 32.7–34.2, F 0.6–2.79, H_2O 26.4–28.6

Class: 2/m 2/m 2/m

Cleavage: {110} perfect, {101} good, {010} distinct. Fracture subconchoidal to uneven. Brittle

Color-Luster: White to greenish white to green, also yellowish green to yellow and yellowish brown, brown to brownish black, rarely bluish or colorless. Transparent to translucent. Vitreous to resinous or pearly. Streak white

Crystal System: Orthorhombic

Density: 2.36 (Meas.), 2.37 (Calc.)

First Reported: 1805

Habit: Crystals minute, rare, stout to long prismatic. Usually as acicular radiating aggregates, often distinctly spherical; also as crusts, stalactitic, or rarely opaline

Hardness: 3.25–4

Lattice Constants: a = 9.60, b = 17.31, c = 6.98 (Å)

Mode of Occurrence: Occurs as a secondary mineral in hydrothermal veins, in certain aluminous metamorphic rocks, and in phosphate-rock and limonite deposits. Found widespread as fine specimens in Garland, Hot Springs, and Montgomery counties, Arkansas; at East Whiteland, Chester County, and elsewhere in Pennsylvania; in St. Clair County, Alabama; at Dunellen, Marion County, Florida; in the King Turquois mine, Saguache County, and in the Cripple Creek district, Teller County, Colorado; and near Slate Mountain, El Dorado County, California. It also is found in Bolivia, England, Ireland, France, Portugal, Germany, Czechoslovakia, Bulgaria, Roumania, and Tasmania

Optical Constants: $\alpha = 1.520$–1.535, $\beta = 1.526$–1.543, $\gamma = 1.545$–1.561; (+)2V ~ 71°

Space Group: Pcmn

Strongest Diffraction Lines: 3.437(100), 3.263(100), 8.58(80)

Z: 4

Ref.: Roberts WL, Rapp GR, Weber J (1974) Encyclopedia of minerals, New York, Van Nostrand Reinhold, p 663

Waylandite
$(Bi, Ca)Al_3(PO_4, SiO_4)_2(OH)_6$

Analyses (%): CaO 2.90, Al_2O_3 29.3, Bi_2O_3 28.3, P_2O_5 22.2, SiO_2 4.7, H_2O 12.9

Class: $\bar{3}$ 2/m

Cleavage: Not determined. Fracture uneven

Color-Luster: White; vitreous to dull

Crystal System: Hexagonal

Density: 3.86 (Calc.)

First Reported: 1962

Habit: Massive; fine-grained, compact

Hardness: 4–5

Lattice Constants: a = 6.9649, c = 16.256 (Å)

Mode of Occurrence: Occurs as veinlets and marginal crusts replacing of bismutotantalite in the lithium pegmatite at Wampewo Hill, Busiro County of Buganda, Uganda

Optical Constants: Uniaxial (−)

Space Group: R$\bar{3}$m

Strongest Diffraction Lines: 2.93(100), 5.66(71), 3.483(50)

Z: 3

Ref.: Knorring O, Mrose ME (1962) Geol Soc Am Special Paper 73:256. Roberts WL, Rapp GR, Weber J (1974) Encyclopedia of minerals, New York, Van Nostrand Reinhold, p 663

Weinschenkite (Churchite)
$(Y, Er)PO_4 \cdot 2H_2O$

Analyses (%): Accurate analyses are lacking

Class: 2/m or m

Cleavage: {$\bar{1}$01} perfect. Fracture conchoidal

Color-Luster: Colorless, white, smoke gray. Transparent. Vitreous; pearly on cleavage. Also silky

Crystal System: Monoclinic

Density: 3.265 (Meas.)

First Reported: 1922

Habit: Crystals minute, lath-like, elongated along c-axis and flattened on {010}. As fan-shaped aggregates or rosettes, as crusts, and as spherulites with radial-fibrous structure

Hardness: 3

Lattice Constants: a = 5.61, b = 15.14, c = 6.19 (Å); $\beta = 115.3°$

Mode of Occurrence: Occurs in a manganese-rich limonite deposit at the Kelly Bank mine, Rockbridge County, Virginia. It also is found associated with wavellite, cacoxenite, and other phosphate minerals in a limonite deposit at Auerbach, Oberpfalz, Bavaria, Germany; and at a copper deposit in Cornwall, England

Optical Constants: $\alpha = 1.600–1.605$, $\beta = 1.608–1.612$, $\gamma = 1.645$; (+)2V = medium small

Space Group: A2/a or Aa

Strongest Diffraction Lines: 4.21(100), 7.50(90), 3.02(90)

Z: 4

Ref.: Roberts WL, Rapp GR, Weber J (1974) Encyclopedia of minerals, New York, Van Nostrand Reinhold, p 665–666

Wenzelite = Hureaulite

Whiteite (Al^{3+} analog of Jahnsite)
$Ca(Fe^{2+}, Mn^{2+})Mg_2Al_2(PO_4)_4(OH)_2 \cdot 8H_2O$

Analyses (%): CaO 5.98, MgO 12.55, FeO 9.57, Al_2O_3 11.5, P_2O_5 37.7, H_2O 10.4

Cleavage: None reported

Color: Tan

Crystal System: Monoclinic

Density: 2.58

First Reported: 1978

Hardness: Not given

Lattice Constants: a = 14.90, b = 6.98, c = 10.13 (Å); $\beta = 113°$ (Minas Gerais sample)

Mode of Occurrence: Occurs with other phosphates along joints and fractures in quartz and albite on various specimens from Ilha Taquaral, Minas Gerais, Brazil, in crystals up to 2 cm. Good specimens also found in the sideritic iron-formation in the Fish River area of NE Yukon Territory, Canada

Optical Constants: $\alpha = 1.580$, $\beta = 1.585$, $\gamma = 1.590$; (+)2V = 40°–50°

Space Group: $P2_1/a$

Strongest Diffraction Lines: 9.270(100), 2.781(80), 2.941(70), 4.822(65), 4.657(65)

Z: 2

Ref.: Moore PB, Ito J (1978) Miner Mag 42:309–323

Whitlockite
$Ca_9(Mg, Fe^{2+})(PO_4)_6(PO_3OH)$

Analyses (%): CaO 46.8–48.2, MgO 1.91–3.53, FeO 1.91–2.34, Fe_2O_3 0.05–1.73, P_2O_5 45.8

Class: $\bar{3}$ 2/m

Cleavage: None. Fracture subconchoidal to uneven. Brittle

Color-Luster: Colorless, white, gray, yellowish, pinkish. Transparent to translucent. Vitreous to resinous

Crystal System: Hexagonal

Density: 3.12 (Meas.), 3.123 (Calc.)

First Reported: 1941 (Described as Zeugite, 1865; Pyrophosphorite, 1878; Martinite, 1888)

Habit: Crystals rhombohedral, rarely tabular. Also coarse granular to earthy

Hardness: 5

Lattice Constants: (hexag.) a = 10.32, c = 36.9 (Å). (rhomb.) $a_{rh} = 13.67$; $\alpha = 44.35°$

Mode of Occurrence: Occurs as a secondary mineral in granite pegmatites. Found in Custer County, South Dakota, as fine crystals associated with quartz, carbonate-apatite, and a wide variety of secondary phosphate minerals at the Tip Top mine, and sparingly at the Bull Moose mine; also at the Palermo mine, North Groton, New

Hampshire, with siderite, quartz, triphylite, and secondary phosphate minerals. It also occurs in phosphate-rock deposits on several islands in the West Indies; as a cave deposit at Sebdou, Oran, Algeria; as a common accessory mineral in meteorites; and has been identified in lunar rocks

Optical Constants: $\omega = 1.629$, $\varepsilon = 1.626$; $(-)$

Space Group: R3c

Strongest Diffraction Lines: 2.837(100), 2.572(80), 1.701(80)

Z: 6 (hexag.), 2 (rhomb.)

Ref.: Roberts WL, Rapp GR, Weber J (1974) Encyclopedia of minerals, New York, Van Nostrand Reinhold, p 669

Whitmoreite

$(Fe, Mn)(Fe^{3+})_2(PO_4)_2(OH)_2 \cdot 4H_2O$

Analyses (%): FeO 4.90, MnO 4.20, Fe_2O_3 40.0, P_2O_5 31.2, H_2O 19.7

Cleavage: Fair parallel to {100}

Color-Luster: Pale tan to deep brown to greenish-brown; luster vitreous to subadamantine

Crystal System: Monoclinic

Density: 2.87 (Meas.), 2.85 (Calc.)

First Reported: 1974

Habit: Acicular crystals elongated parallel to [001] and twinned by reflection parallel to {100}. Forms observed are a{100}, m{110}, s{011}, t{021}, and u{112}. The crystals often possess a rhomboidal cross-section with a chisel-shaped termination

Hardness: 3

Lattice Constants: a = 10.00, b = 9.73, c = 5.471 (Å); $\beta = 93.8°$

OpticalConstants: $\alpha = 1.676$, $\beta = 1.725$, $\gamma = 1.745$; $(-)2V = 60°–70°$; pleochroic with X and Y light greenish-brown, Z dark greenish-brown

Space Group: $P2_1/c$

Strongest Diffraction Lines: 10.05(100), 7.01(70), 4.96(79), 4.21(70), 2.802(70)

Z: 2

Ref.: Moore PB, Kampf AR, Irving AJ (1974) Am Miner 59:900–905

Wicksite

$NaCa_2(Fe^{2+}, Mn^{2+})_4MgFe^{3+}(PO_4)_6 \cdot 2H_2O$

Analyses (%): Na_2O 3.08, FeO 22.7, MgO 3.77, MnO 4.72, CaO 11.1, Fe_2O_3 7.98, Al_2O_3 0.51, P_2O_5 41.6, H_2O 3.70

Cleavage: Good {010}

Color-Luster: Dark blue, almost black tending toward blue or green and transparent in thin fragments. Luster is submetallic and streak is green

Crystal System: Orthorhombic

Density: 3.54 (Meas.), 3.58 (Calc.)

First Reported: 1981

Habit: Platy parallel to {010}; some specimens are massive. Striations parallel to the *a* axis developed on many plates.

Hardness: 4.5–5

Lattice Constants: a = 12.890, b = 12.511, c = 11.634 (Å)

Mode of Occurrence: Formed syngenetically with wolfeite, satterlyite, pyrite and maricite in nodules in bedded ironstone and shale along the Big Fish River, northeastern Yukon Territory, Canada

Optical Constants: $\alpha = 1.713$, $\beta = 1.718$, $\gamma = 1.728$; biaxial pos. Strongly pleochroic with X blue, Y greenish blue, and Z pale yellowish brown. $2V = 66°$

Space Group: Pbca

Strongest Diffraction Lines: 2.753(100), 3.015(80), 2.910(80), 2.118(60)

Z: 4

Ref.: Sturman BD, Peacor DR, Dunn PJ (1981) Can Miner 19:377–380

Wilkeïte (Apatite Group)

$Ca_5(SiO_4, PO_4, SO_4)_3(O, OH, F)$

Analyses (%): CaO 54.4, P_2O_5 14.4–32.2, SO_3 12.3, SiO_2 2.9–11.2, CO_2 0–2.1, F 0.9–2.1, Cl 0.4–0.8

Class: 6/m

Cleavage: {0001} imperfect. Brittle

Color-Luster: Pale pinkish, yellowish. Translucent. Vitreous to resinous

Crystal System: Hexagonal

Density: 3.12–3.23 (Meas.)

First Reported: 1914

Habit: As indistinct rounded crystals; also massive granular

Hardness: ~5

Lattice Constants: a = 9.48, c = 6.91 (Å)

Mode of Occurrence: Occurs in blue calcite with idocrase, diopside, garnet, and tobermorite, in contact metamorphosed marble at Crestmore, Riverside County, California. It also is found at Kyshtym, Urals, U.S.S.R. and probably in the sanidinites of the Laacher See, Rhineland, Germany

Optical Constants: $\omega = 1.650–1.640$, $\varepsilon = 1.646–1.636$; (−)

Space Group: $P6_3/m$

Strongest Diffraction Lines: 2.80(100), 2.70(90), 2.24(80)

Z: 2

Ref.: Roberts WL, Rapp GR, Weber J (1974) Encyclopedia of minerals, New York, Van Nostrand Reinhold, p 670. McConnell D (1973) Apatites. Springer, Wien, New York, p 90

Wolfeite (Triploidite-Wolfeite Series)
$(Fe^{2+}, Mn^{2+})_2PO_4OH$

Analyses (%): FeO 33.4–44.4, MnO 13.1–21.7, CaO 0.19–2.00, Fe_2O_3 0.22–7.78, P_2O_5 32.4–34.2, H_2O 3.78–4.50, F 0.09–0.88

Class: 2/m

Cleavage: {010} good, {120} fair. Fracture subconchoidal to uneven

Color-Luster: Reddish brown to dark clove-brown; rarely greenish. Transparent to translucent. Vitreous to greasy. Streak whitish

Crystal System: Monoclinic

Density: 3.79 (Meas.)

First Reported: 1949

Habit: Crystals prismatic, vertically striated. Usually as columnar to parallel fibrous aggregates; also divergent fibrous; granular

Hardness: 4.5–5

Lattice Constants: a = 12.20, b = 13.17, c = 9.79 (Å); $\beta = 108.0°$

Mode of Occurrence: Occurs in granite pegmatite at the Bull Moose mine, Custer County, South Dakota; as an alteration of triphylite at the Palermo quarry, North Groton, New Hampshire; at Hagendorf, Bavaria, Germany; at Cyrilhof, Czechoslovakia; and with triplite at Skrumpetorp, Sweden

Optical Constants: $\alpha = 1.741$, $\beta = 1.742$, $\gamma = 1.746$; (+)2V = moderate

Space Group: $P2_1/a$

Strongest Diffraction Lines: 2.93(100), 3.09(90), 3.18(80)

Z: 16

Ref.: Frondel C (1949) Am Miner 34:692. Roberts WL, Rapp GR, Weber J (1974) Encyclopedia of minerals, New York, Van Nostrand Reinhold, p 674

Woodhouseite
$CaAl_3PO_4SO_4(OH)_6 \cdot 3H_2O$

Analyses (%): CaO 12.3, BaO 1.0, Al_2O_3 36.6, P_2O_5 18.1, SO_3 17.6, H_2O 13.4

Class: $\bar{3}$ 2/m

Cleavage: {0001} perfect

Color-Luster: White to flesh pink. Transparent to translucent. Vitreous; pearly on {0001}

Crystal System: Hexagonal

Density: 3.012 (Meas.), 3.001 (Calc.)

First Reported: 1937

Habit: As small crystals, usually pseudocubic, also tabular. Crystal faces commonly rough, curved, or striated

Hardness: 4.5

Lattice Constants: a = 6.961, c = 16.27; $a_{rh} = 6.75$ (Å); $\alpha = 62.07°$

Mode of Occurrence: Occurs in veins with quartz, lazulite, augelite, and topaz in the andalusite deposit on the western slope of the White Mountains, 7 miles east of Mocalno, north of Bishop, Mono County, California

Optical Constants: $\omega = 1.636$, $\varepsilon = 1.647$; (+)

Space Group: R $\bar{3}$m

Strongest Diffraction Lines: 2.94(100), 1.89(94), 2.19(86)

Z: 1 (rhomb.), 3 (hexag.)

Ref.: Lemmon DM (1937) Am Miner 22:939. Roberts WL, Rapp GR, Weber J (1974) Encyclopedia of minerals, New York, Van Nostrand Reinhold, p 675

Wyllieite

$Na_2Fe_2^{2+}Al(PO_4)_3$

Analyses (%): P_2O_5 43.8–45.3, Al_2O_3 7.9–8.4, FeO 29.6, MgO 1.90, MnO 3.6–4.3, CaO 2.35, Na_2O 8.0–9.75

Class: 2/m

Cleavage: {010} perfect, {$\bar{1}$01} distinct. Very brittle

Color-Luster: Deep bluish-green to deep oily green, greyish-green to greenish-black. Thin fragments pale green and transparent. Vitreous to submetallic. Streak and powder dirty olive-green

Crystal System: Monoclinic

Density: 3.601 (Meas.), 3.60 (Calc.)

First Reported: 1973

Habit: Crystals euhedral or subhedral, coarse with uneven surfaces, up to six inches across. Often as large masses of interlocking crude crystals weighing up to fifty pounds

Hardness: 4+

Lattice Constants: a = 11.868, b = 12.382, c = 6.354 (Å); $\beta = 114.52°$

Mode of Occurrence: Wyllieite occurs as a primary mineral in granite pegmatites. It is found in considerable abundance embedded in bull quartz, perthite and albitized perthite, frequently associated with muscovite, schorl, arrojadite, scorzalite, graftonite, and sarcopside, at the Victory mine near Custer, South Dakota; also with abundant arrojadite at the Nickel Plate mine near Keystone, Pennington County

Optical Constants: $\alpha = 1.688$, $\beta = 1.691$, $\gamma = 1.696$; Biaxial (+), 2V ~ 50°

Space Group: $P2_1/n$

Strongest Diffraction Lines: 2.693(100), 2.674(100), 6.15(60)

Z: 4

Ref.: Moore PB, Ito J (1973) Miner Rec 4:131–136. Moore PB, Molin-Case J (1974) Am Miner 59:280–290

Xanthoxenite

$Ca_4Fe_2(PO_4)_4(OH)_2 \cdot 3H_2O$

Analyses (%): CaO 28.0, MgO 0.48, Fe_2O_3 15.9, Al_2O_3 0.22, Mn_2O_3 3.94, P_2O_3 36.5, H_2O 9.73

Cleavage: One perfect

Color-Luster: Pale yellow to brownish yellow. Dull to waxy

Crystal System: Triclinic

Density: 2.8; 2.97 (Meas.)

First Reported: 1920

Habit: Rarely as distinct crystals; usually as masses and crusts composed of indistinct platy or lath-like crystals, or as fibrous, radiating, sheaf-like aggregates

Hardness: ~2.5

Lattice Constants: a = 6.70, b = 8.85, c = 6.54 (Å); $\alpha = 92.1°$, $\beta = 110.2°$, $\gamma = 93.2°$

Mode of Occurrence: Occurs as an alteration product of triphylite at the Palermo mine, North Groton, New Hampshire, and at Hühnerkobel, Bavaria, Germany

Optical Constants: $\alpha = 1.704$, $\beta = 1.715$, $\gamma = 1.724$; (−)2V = large

Space Group: P$\bar{1}$ or P1

Strongest Diffraction Lines: 3.05(100), 2.73(90), 3.22(80)

Z: 1

Ref.: Moore PB, Ito J (1978) Miner Mag 42:313–323

Xenotime (Isostructural with zircon)
YPO_4

Analyses (%): $(Y, Er)_2O_3$ 51.8–67.8, La_2O_3 0–2.14, Al_2O_3 0–4.80, ThO_2 0–2.47, UO_3 3.17–4.13, SiO_2 0.57–4.32, P_2O_5 25.4–36

Class: 4/m 2/m 2/m

Cleavage: {100} perfect. Fracture splintery to uneven. Brittle

Color-Luster: Yellowish brown to reddish brown; also pale gray, pale yellow, greenish, reddish. Vitreous to resinous. Translucent to opaque

Crystal Sytem: Tetragonal

Density: 4.4–5.1 (Meas.)

First Reported: 1824

Habit: Crystals short to long prismatic; also pyramidal or equant. As aggregates of rough crystals; as rosettes. Twinning on {111} rare

Hardness: 4–5

Lattice Constants: a = 6.88, c = 6.03 (Å)

Mode of Occurrence: Occurs widespread in small amounts in acidic and alkalic igneous rocks; in pegmatites, metamorphic rocks, Alpine-type vein deposits, and as a detrital mineral. Found in the United States in California, Colorado, North Carolina, Alabama, Georgia, and New York. It also occurs in Brazil, Norway, Sweden, Germany, Poland, Switzerland, Madagascar, South Africa, New Zealand, India, and Japan

Optical Constants: $\omega = 1.721, 1.720$, $\varepsilon = 1.816, 1.827$; (+)

Space Group: $I4_1/amd$

Strongest Diffraction Lines: 3.45(100), 2.56(50), 1.768(50)

Z: 4

Ref.: Roberts WL, Rapp GR, Weber J (1974) Encyclopedia of minerals, New York, Van Nostrand Reinhold, p 679

Xiangjiangite
$(Fe, Al)(UO_2)_4(PO_4)_{2-3}(SO_4)_{2-1}(OH)_{1-0} \cdot 24H_2O$

Analyses (%): UO_3 56.2–59.9, Al_2O_3 0.93, Fe_2O_3 2.10, P_2O_5 8.15–8.69, SO_3 5.65–6.02, SiO_2 4.42, H_2O 9.86–11.4

Color-Luster: Yellow to bright yellow; streak is pale yellow; luster silky

Crystal System: Tetragonal; absence of a four-fold axis suggests that the true symmetry is monoclinic, or more likely orthorhombic

Density: 2.775 (Calc.)

First Reported: 1978

Habit: Under electron microscope, most crystals appear as stretched six-sided plates with approx. 90° and 135° interfacial angles; some appear as eight-sided and rectangular plates

Hardness: 1–2

Lattice Constants: a = 7.17, c = 22.22 (Å)

Mode of Occurrence: Occurs with sabugalite, variscite, pyrite and quartz in the oxidized zone of a Permian uranium deposit in Hunan, China

Optical Constants: $\alpha = 1.558$, $\beta = 1.576$, $\gamma = 1.593$; pleochroism weak, extinction parallel or symmetrical; Biaxial neg.

Strongest Diffraction Lines: 11.11(100), 3.74(80), 3.294(80), 2.938(70); the strongest X-ray lines given in the English abstract are not consistent with those given the original Chinese text

Z: 1

Ref.: Hunan 230 Institute and X-ray Laboratory (1978) Wuhan Geol College, Sci Geol Sinica 2:183–188 (in Chinese with English abstract)

Yoshimuraite
$(Ba, Sr)_2TiMn_2(SiO_4)_2(PO_4, SO_4)(OH, Cl)$

Analyses (%): BaO 33.5, SrO 4.6, MnO 17.6, Fe_2O_3 1.8, TiO_2 10.0, SiO_2 18.3, SO_3 5.4, P_2O_5 4.0, H_2O 2.3

Class: $\bar{1}$

Cleavage: {010} perfect, {10$\bar{1}$} distinct, {101} distinct. Brittle

Color-Luster: Orange-brown

Crystal System: Triclinic

Density: 4.13 (Meas.), 4.21 (Calc.)

First Reported: 1959

Habit: Bladed, or in mica-like forms up to 5 cm long. Polysynthetic twinning on (010)

Hardness: 4.5

Lattice Constants: a = 7.00, b = 14.71, c = 5.39 (Å); $\alpha = 93.5°$, $\beta = 90.2°$, $\gamma = 95.3°$

Mode of Occurrence: Occurs in coarse-grained pegmatite along the boundary between massive manganese ore (tephroite with rhodonite, braunite, and hausmannite) and massive hornfels containing manganophyllite, richterite, and urbanite at the Noda-Tamagawa Mine, Iwate prefecture, Japan

Optical Constants: $\alpha = 1.763$, $\beta = 1.777$, $\gamma = 1.785$; (+)2V = 85°–90°

Space Group: P$\bar{1}$

Strongest Diffraction Lines: 2.91(100), 4.85(50), 3.21(50)

Z: 2

Ref.: Watanabe T, Takeuchi Y, Ito J (1961) Min J (Jpn) 3:156–167. Roberts WL, Rapp GR, Weber J (1974) Encyclopedia of minerals, New York, Van Nostrand Reinhold, p 682

Zairite

(Ferric analog of Waylandite; member of the Crandallite group)

$Bi(Fe^{3+}, Al)_3(PO_4)_2(OH)_6$

Analyses (%): Fe_2O_3 30.8, Al_2O_3 5.35, Bi_2O_3 28.7, BaO 1.98, CaO 0.72, CuO 0.82–2.0, P_2O_5 21.9, H_2O 9.05

Color: Greenish

Crystal System: Rhombohedral

Density: 4.37 (Meas.), 4.42 (Calc.)

First Reported: 1975

Hardness: 4.5

Lattice Constants: a = 7.015, c = 16.365 (Å)

Mode of Occurrence: In small masses associated with mica and other bismuth phosphates in the zone of weathering of quartz vein containing wolframite, at Eta-Etu, northern Kivu, Republic of Zaire

Optical Constants: ω and ε between 1.82 and 1.83; weak to moderate birefringence

Space Group: R$\bar{3}$m

Strongest Diffraction Lines: 2.945(100), 5.710(95), 3.503(40), 2.727(20)

Z: 3

Ref.: Van Wambeke L (1975) Bull Soc Fr Miner Cristallogr 98:351–353

Zapatalite

$Cu_3Al_4(PO_4)_3(OH)_9 \cdot 4H_2O$

Analyses (%): CuO 25.8–27.9, Al_2O_3 20.2, P_2O_5 20.7–22.5, H_2O 15.5–15.8, Fe_2O_3 1.16

Cleavage: {001} good. Sectile

Color-Luster: Pale blue (faience blue). Translucent. Streak pale blue

Crystal System: Tetragonal

Density: 3.016 (Meas.), 3.017 (Calc.)

First Reported: 1972

Habit: Massive; as poorly crystalline cavity fillings

Hardness: 1.5

Lattice Constants: a = 15.22, c = 11.52 (Å)

Mode of Occurrence: Occurs in association with libethenite, chenevixite, beaverite, alunite, and pseudomalachite in silicified and brecciated limestone at a small prospect northwest of Cerro Morita, 27 km southwest of Agua Prieta, Sonora, Mexico

Optical Constants: $\omega = 1.646$, $\varepsilon = 1.635$; (−) occasionally biaxial (−) with variable 2V

Strongest Diffraction Lines: 11.60(100), 7.62(100), 6.82(70)

Z: 6

Ref.: Williams S (1972) Miner Mag 38:541–544. Roberts WL, Rapp GR, Weber J (1974) Encyclopedia of minerals, New York, Van Nostrand Reinhold, p 684

Zwieselite (Triplite-Zwieselite Series) $(Fe^{2+}, Mn^{2+}, Mg, Ca)_2(PO_4)(F, OH)$

Analyses (%): FeO 1.68–42.0, MnO 14.9–57.6, CaO 0.5–14.9, MgO 0.31–11.9, P_2O_5 30.6–33.3, F 6.41–9.09

Class: 2/m

Cleavage: {001} good, {010} fair, {100} poor. Fracture subconchoidal to uneven

Color-Luster. Clove brown; brownish black to black on alteration. Translucent. Vitreous to resinous

Crystal System: Monoclinic

Density: 3.927–3.97 (Meas.)

First Reported: 1841

Habit: Crystals poorly developed; rough. Usually massive

Hardness: 5–5.5

Lattice Constants: a = 12.03, b = 6.46, c = 10.03 (Å); $\beta = 105.7°$

Mode or Occurrence: Occurs as a primary mineral in granite pegmatites. Found at Rabenstein, near Zwiesel, Bavaria, Germany; also at Kimoto, Finland

Optical Constants: $\alpha = 1.696$, $\beta = 1.704$ (variable), $\gamma = 1.713$; (+)2V = 87° (Calc.)

Space Group: I2/m

Z: 8

Ref.: Roberts WL, Rapp GR, Weber J (1974) Encyclopedia of minerals, New York, Van Nostrand Reinhold, p 692–693

General References

(The following monographs and texts were used in the compilation of the mineralogical data presented in the preceding section.)

Aballain M, Chambolle P, Derec-Poussier F, Guillemin C, Mignon R, Perrot R, Sarcia JA (1968) Index alphabetique de nomenclature mineralogique. Bureau Rech Geol Minièr, Paris

Chester AH (1896) A dictionary of mineral names. Wiley, New York

Embrey PG, Fuller JP (1980) A manual of mineral names. University Press, Oxford

Fleischer M (1975) Glossary of mineral species. Bowies (Maryland) Mineralogical Record, Inc.

Frondel C (1958) Systematic mineralogy of uranium and thorium. US Geol Survey Bull 1064, 400 p.

Kipfer A (1974) Mineralindex. Ott, Thun Munich

McConnell D (1973) Apatites. Springer, Wien New York

Mellor JW (1971) Treatise on inorganic and analytical chemistry, vol 8. [Suppl 3] Longman, London

Mitchell RS (1979) Mineral names. Van Nostrand, New York

Moore PB (1982) Pegmatite minerals of P(V) and B(III). Miner Assoc Can Handbook, vol 8, p 267–291

Palache C, Berman H, Frondel C (1951) Dana's system of mineralogy, vol 2. Wiley, New York

Povarennykh AS (1974) Crystal chemical classification of minerals, 2 vol. Plenum, New York

Roberts WL, Rapp GR Jr, Weber J (1974) Encyclopedia of minerals. Van Nostrand, New York

Strunz H (1970) Mineralogische Tabellen, 5. Aufl. Akademische Verlagsgesellschaft, Leipzig

Chapter 2 The Nucleation and Growth of Phosphate Minerals

GEORGE H. NANCOLLAS

Introduction

The precipitation and dissolution of phosphate salts in aqueous systems is involved in biological mineralization, in the environment, as well as in a large number of industrial processes. The resurgence of interest in the crystallization of inorganic salts from aqueous solution in recent years has evolved because of its involvement in areas such as the removal of phosphate from waste water, the fate of elements such as aluminum, calcium, iron, and other heavy metals in the formation of lake and ocean sediments, and industrial cooling-tower systems where the formation of scale on heated metal surfaces is a continuing problem. Consideration of phosphate balances in lakes and rivers near heavily populated areas indicates that despite a continuous entry of phosphate ions, the phosphate concentration does not increase proportionately, indicating that at least some phosphate is removed by precipitation (Brown 1973). Since the concentration of metal ions such as calcium is also relatively high through the use of lime additions for the removal of phosphates from sewage, calcium phosphate precipitation may be of particular importance in such systems. In the lime additon process, an induction period usually precedes the formation of precipitate, and as much as 85%–90% of the phosphate can be removed in a recycling reactor at a pH of 8.0 (Ferguson et al. 1971). The use of reclaimed sewage as cooling water in many industrial processes results in conditions favoring the precipitation of metal phosphates. In order to be able to control the precipitation of these salts, it is essential to elucidate the mechanism of the nucleation and growth reaction. However, the crystallization process is still one of the least well understood phenomena in chemistry.

Most solid/liquid studies of metal phosphate systems have been confined to solubility determinations. However, as a group of sparingly soluble salts, the phosphates have unique properties which make even solubility determinations very difficult to interpret. The solids may incorporate large amounts of impurities and lattice defects which are frequently characterized by a high degree of non-stoichiometry. Solid phases which may form through an incongruent dissolution process may be finely divided and persist for long periods despite the fact that other phases may be thermodynamically more stable. If the pH is allowed to change during the solubility determination, surface coatings of more acid metal phosphate phases may form. The apparent solubility behavior may therefore be quite different from that of the phase itself and this has probably been the source of the widely dis-

Chemistry Department, State University of New York at Buffalo, Buffalo, N.Y. 14214 crepant data in the literature for the solubilities of many metal phosphates. In constructing a solubility isotherm, it is essential to establish that the systems are at equilibrium in order to be able to use thermodynamic solubility products which refer to the free energies of single solid phases. It is now quite well established, however, that kinetic factors may be considerably more important in determining the nature of the solid phase present than are considerations based solely on equilibrium data.

Different phosphate phases may be stabilized or destabilized by the presence of various cations and anions which may not be significantly incorporated into the crystal lattice and so could not be detected by analysis of the solid phases in pseudo equilibrium with the aqueous solutions. For many of the heavy metal phosphates, these difficulties, combined with the formation of numerous basic salts, have confined research work to studies of apparent solubilities. It is only for the alkaline earth phosphate salts that attempts have been made to investigate the mechanism of the precipitation and dissolution processes. For the strontium phosphates, the formation of strontium hydroxyapatite is preceded by the precipitation of precursor phases amongst which, an amorphous phase, β-$SrHPO_4$, Collin's salt [$Sr_6H_3(PO_4)_5 \cdot 2H_2O$] and tri-strontium phosphate [$Sr_3(PO_4)_2$] have been proposed (Feenstra 1980; Collin 1966). In a recent pH-stat study, the precipitation of strontium phosphates at room temperatures in a solution containing 0.15 mol l^{-1} potassium nitrate was shown to be preceded by the precipitation of Collin's salt (Feenstra 1980). Upon the addition of fluoride, fluoridated strontium apatite was formed without the observed precipitation of this precursor phase. pH-stat studies of the spontaneous precipitation of strontium from phosphates revealed induction times for the formation of a fluoridated strontium apatite which showed a maximum value as the fluoride concentration in the solution was increased (Feenstra 1980). It was suggested that a change in interfacial tension of the precipitating fluoridated apatite or a change in the nature of the initially precipitating phase could account for this phenomenon. For the magnesium phosphates, an investigation of the morphology, habit and growth of newberyite ($MgHPO_4 \cdot 3H_2O$) was made by Boistelle and Abbona (1981). The aim of this study was to investigate the surface structure of the most important faces as a preliminary step to a subsequent kinetics study. It was shown that habit changes of the crystals occur in precipitation from pure aqueous solutions as a function of pH and supersaturation. When the supersaturation increased, the crystal habit changed successively from pseudo octahedral to flat and tabular, and finally to prismatic.

Of all the alkaline earth phosphates, those of the calcium ion have been most studied. These phases occur in all classes of rock and are found naturally in pegmatites and other veins of hydrothermal origin. As indicated above, the increase in phosphate concentration due to increased water reuse and the use of lower quality make-up water such as tertiary sewage plant effluent coupled with alkaline operating conditions has led to the formation of highly insoluble calcium phosphate scale deposits which are normally loosely attributed to the thermodynamically most stable phase, hydroxyapatite [$Ca_5(PO_4)_3OH$, hereafter HAP]. These hard insoluble and dense scale deposits on the heat exchanger surfaces lower cooling tower efficiencies by reducing the heat transfer coefficients. In some boiler water treatments, the addition of phosphate is made along with sludge conditioner in attempts

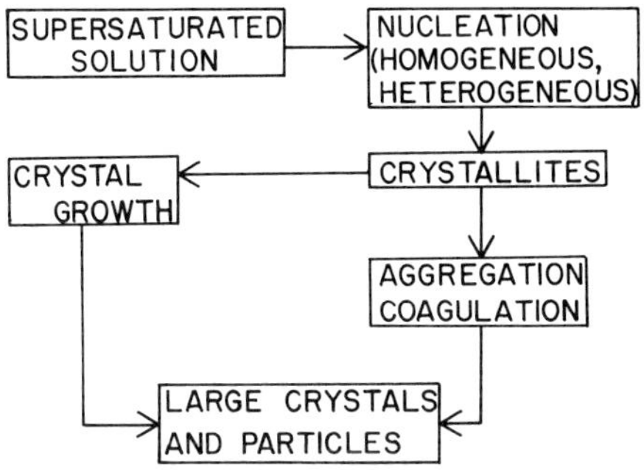

Fig. 1. The precipitation process

to change the morphology of the crystals comprising the scale deposits. Thus if the calcium phosphate precipitate can be maintained as a sludge, the crystals will not pack together as hard destructive scale, but will be swept away by the blow-down fluid flow. Since a great deal of recent work has been done on the precipitation and dissolution of calcium phosphate phases, the remainder of this chapter will be concerned with this system. It is quite likely that when more extensive studies have been made of the other metal phosphate systems, the pattern of precursor phases accompanying the precipitation reactions which has been found to hold in the case of calcium phosphates may also apply to the heavy metal salts. Research work on the precipitation and dissolution of these other phases using some of the newer physical chemical techniques outlined in this chapter is to be encouraged.

Although the rates of homogeneous reaction can often be predicted with considerable precision, for the heterogeneous precipitation process discrepancies amounting to orders of magnitude are not uncommon. In an attempt to predict the results of industrial precipitation processes, this may frequently lead to over-design of plants and expensive pilot plant studies. In general, crystallization in homogeneous solution can be considered to take place in two stages: nucleation, and the growth of these nuclei to macroscopic dimensions.

Nucleation

The overall precipitation process may be represented by the steps outlined in Fig. 1. The first stage, nucleation, corresponds to the production of new centers from which growth can occur. This process may be homogeneous, or, in the presence of foreign particles which can act as effective nucleating sites, heterogeneous nucleation will take place. In a moderately supersaturated homogeneous solution, crystals cannot form because the very small crystals or embryo are unstable. In terms of classical nucleation theories, statistical fluctuations in the solution result in the formation of ionic or molecular clusters through stepwise aggregation which can increase in size or dissociate according to the scheme:

$$
\begin{array}{lll}
A & +A \rightleftharpoons A_2 & \\
A_2 & +A \rightleftharpoons A_3 & \\
\vdots & \quad\vdots \qquad \vdots & \\
A_{n-1} & +A \rightleftharpoons A_n & \text{(critical cluster)},
\end{array}
$$

where A represents the single unit, A_2 the dimer, etc. These aggregates of varying size may be considered to be minute spherical droplets to which well-established theoretical considerations can be applied such as that involving the Kelvin-Gibbs equation relating the radius of the droplet to its surface tension. The creation of new surface requires energy which is provided by the tendency of the supersaturated solution to deplete itself. There is therefore an energy balance between the creation of solid surface and the deposition process. Statistical fluctuations within the supersaturated solutions will result in the formation of some clusters (A_n) which are of critical size and which will have an equal opportunity of growing or dissociating. Clusters larger than the critical size will tend to grow since the surface energy term becomes less important as the size increases. Once the critical size has been reached, by aggregation, the supersaturation collapses and the crystallites which have been formed will tend to grow by a process of crystal growth (Fig. 1).

Another mechanism for formation of large crystals and particles which precipitate from solution is that of aggregation and coagulation of crystallites which are formed initially in the supersaturated solution. In these processes, the charge at the solid/liquid interface plays an important part, since similarly charged particles will tend to remain dispersed in the aqueous phase due to electrostatic repulsion. Ions or molecules of opposite charge which may be adsorbed at the surface of these crystals may therefore reduce the surface charge and induce agglomeration or aggregation. In the overall precipitation process, it is important to separate these effects from the crystal growth reaction since both lead to the production of larger sized particles.

The frequency of nucleation was calculcated by Volmer (1945) by assuming a steady state flux of molecules through subcritical embryos as indicated in Fig. 2. The theory was developed by analogy with condensation of a vapour and, in accordance with current rate theory, the rate at which the critical nuclei form is proportional to [log (supersaturation)]$^{-2}$. Here, the supersaturation is expressed as m/m_0 where m is the concentration of electrolyte in the supersaturated solution and m_0 is the concentration at equilibrium (saturation). This inverse square relationship involving the logarithm of the supersaturation results in a nucleation rate curve with supersaturation which sharply increases at a critical value of the latter. This is illustrated in Fig. 3. It can be seen that as the supersaturation is increased, there is an appreciable metastable region within which the rate of nucleation is negligibly small. At the critical limit, however, the rate increases rapidly leading to the formation of crystals in the solution. This simple classical nucleation theory has been found to hold surprisingly well for vapour condensation and also nucleation of particles in solution for a number of systems. An imperfection in the theory, however, lies in its assumption of steady-state conditions. These may only be applied if the supersaturation is built up homogeneously and very slowly until the critical value is reached. The steady-state concentrations of clusters of various sizes will then always be present.

In many cases, there is a measurable time interval or induction period between the production of supersaturation (e.g., by mixing solutions containing the appropriate concentrations of crystal lattice ions) and the onset of precipitation even though immediate nucleation is predicted by the theory. It was proposed by Christiansen and Nielsen (1951) that the initial induction period was the time

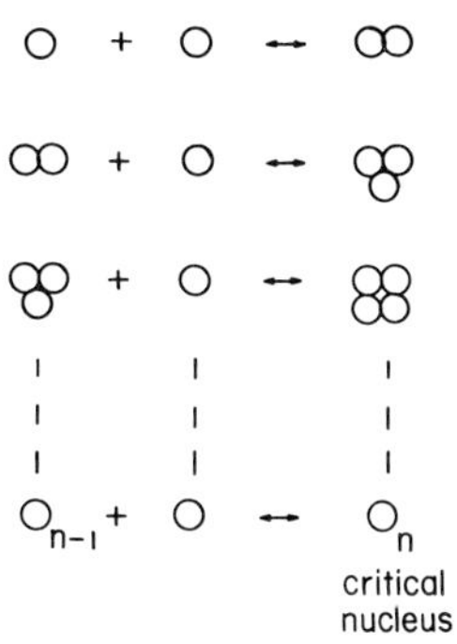

Fig. 2. Homogeneous nucleation

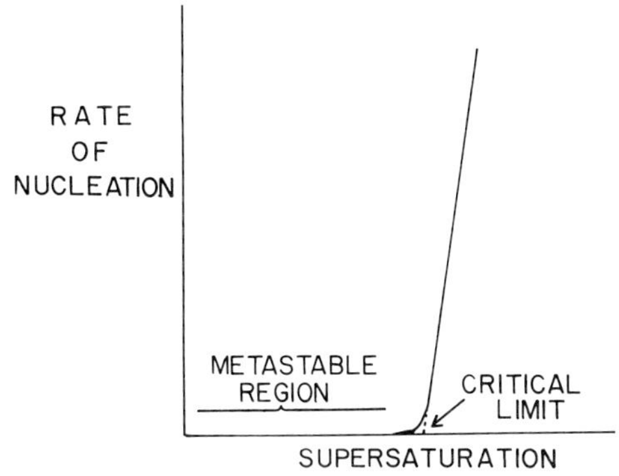

Fig. 3. Plot of rate of nucleation against time

required to build up clusters of critical size and that this could be related to the size of the critical cluster. Moreover, it was assumed that all the nucleation takes place during the initial period, the subsequent fall of concentration with time reflecting the growth of these nuclei. Using a rapid mixing technique, Christiansen and Nielsen were able to measure the times required for visible precipitation. The induction periods were also used to calculate the effective number of molecules in the critical nuclei and values of the order of three of four molecules per critical nucleus were found for a number of sparingly soluble salts.

Although this approach to the problem of critical nucleus size was very useful, it is essential to deal with situations in which nucleation is strictly homogeneous. When supersaturation is produced by the rapid mixing of two solutions it is possible that heterogeneous nucleation prevails, nucleation probably occurring in small volume elements where the supersaturation is larger than it is after complete homogenization. The preparation of metastable supersaturated solutions in the region illustrated in Fig. 3 is dependent upon careful experimental techniques in order to remove foreign particles which may act as sites for the initiation of precipitation. Thus in practice, when a series of solutions are prepared with increasing supersaturation, it is found that below the critical supersaturation value, the solutions are stable for appreciable periods extending for hours, days, or even months. Near the critical value, crystals develop irreproducibly and the results may depend much more on the care taken to prepare the solution and clean the apparatus than on the level of supersaturation. When the latter exceeds the critical value, however, the number of crystals formed increases very strongly (Fig. 3) typically increasing by several orders of magnitude when the supersaturation is increased by a factor of 10. This release of supersaturation will occur considerably below the critical limit if heterogeneous nucleation takes place.

Crystal Growth

In order to analyze kinetically the results of spontaneous precipitation of salts such as calcium phosphates, it is necessary to make assumptions about the extent of the nucleation preceding growth. Although it is usually assumed that all nuclei are formed in a short initial period, it is very difficult to determine the instant at which growth predominates. It is possible that for some time during the precipitation, new nuclei and the growth of existing ones will occur simultaneously, leading to a variation in the size distribution of particles. The irreproducibility of results of such experiments therefore makes it very difficult to undertake a rigorous kinetic analysis of the results. In order to avoid many of the problems associated with such chance nucleation events, it is possible to study the crystal growth of well-characterized seed materials in stable supersaturated solutions. Provided that special precautions are taken to avoid local concentration effects and the presence of foreign particles, it is possible to prepare such solutions of concentrations greater than the solubility value that are stable for as much as several days. Not only do such seeded crystallization studies avoid many of the problems associated with attempts to attain optimum conditions for "homogeneous" nucleation, but they are also very appropriate models for many of the applications outlined above. Thus there is little doubt that in scaling situations the growth of nuclei and the development of potential crystallizing sites is considerably more important than statistical fluctuation in the hypothetical "perfectly clean" solutions which lead to the formation of embryos of critical size.

Another process of importance in precipitation reactions under field conditions is the formation of secondary nuclei in which additional nucleation may be induced by the presence of seed crystals. Epitaxy or the growth of a phase different from that of the seed material upon its surface requires a suitable crystal lattice match between the phases. Epitaxial growth may play an important part, therefore, in the formation of scale and in the precipitation of phosphate minerals. Both secondary nucleation and epitaxial growth can be studied under highly reproducible conditions using the seeded growth technique.

In order to discuss the steps involved in the growth of the seed material in supersaturated solutions, we may represent the crystal surface as in Fig. 4. The first detailed theoretical treatment of crystal growth was given by Kossel (1934) and Stranski (1928) and later developed by Becker (1949). These theories of stepwise growth recognized the non-equivalence of sites on the crystal surface and proposed the formation of mono-molecular layers in a layer-by-layer growth mechanism. In the process of adsorption onto a crystal surface, the molecules or ions may be attached to the smooth surface, an edge or step or to a kink site (positions c, d, and e, respectively, in Fig. 4). Of these positions, the kink site, e, or repeatable step is most likely, since with three faces in contact with the crystal surface, it will have the highest binding energy. The first step in the crystal growth process, in terms of this model, is the formation of a two-dimensional nucleus (e.g., positions b, c, Fig. 4) from which layered growth proceeds. This layered growth has a particularly unfavorable energy requirement and the model predicts a stepwise energy barrier as each layer is completed and new surface nucleation is required. Such a suggestion was not supported by experiment, and many crystals grow at supersaturations below 1%, whereas the surface nucleation process would require values of 25%–50% in order to ensure two-dimensional nucleation on each completed lattice layer on the crystal surface. Frank (1949) pointed out that real crystals are imperfect and contain dislocations. He proposed the formation of screw dislocations on the sur-

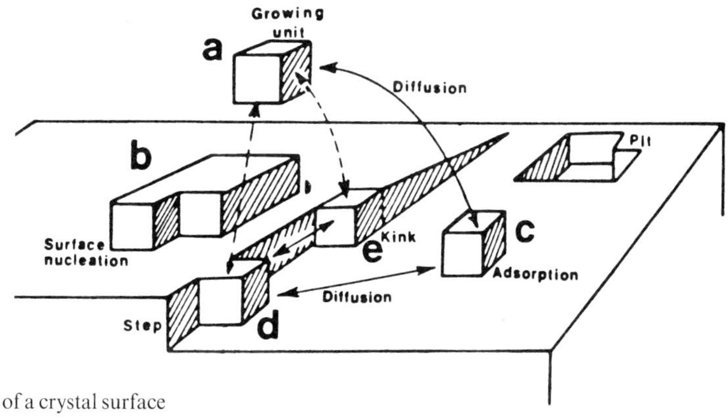

Fig. 4. Representation of a crystal surface

face which offer perpetual steps for crystal growth and avoid the necessity for two-dimensional nucleation. The emergence of a screw dislocation is shown in Fig. 4.

In terms of the surface dislocation model, the process of deposition of new material may be expected to take place through a number of simple step-wise events:

1. The transport, within the bulk solution phase, of ions, or molecules (position a) up to the crystal surface.
2. At the surface, the unit may become attached to the face (position c) by a process of adsorption or chemisorption and thereafter,
3. undergo two-dimensional diffusion to a neighbouring site (position d) offering an energetically more favourable position at an edge or step on the crystal surface.
4. A one-dimensional diffusion along the step to a kink, places the seed unit in an even more favourable energetic position (e), with three of its planes in contact with the crystal surface.
5. Attachment at the kink with concomitant dehydration completes the process of incorporation into the crystal lattice. The dehydration or partial dehydration might also take place in any of the steps (3–5).

In the crystal growth experiments, the observed reaction kinetics reflects the slow step or steps in the above model for crystal growth. If the equilibrium at the crystal surface is sufficiently rapid so as not to be involved in the rate-limiting step, the crystal growth is controlled by step (1), the bulk diffusion of material up to the surface. Under these circumstances the rate of crystal growth would be expected, through Fick's law of diffusion, to be proportional to the supersaturation. There is now considerable evidence from numerous electrolytes that the rate of crystallization is not simply a bulk diffusion reaction with a rate varying linearly with supersaturation. For many sparingly soluble salts $M_{\nu+}A_{\nu-}$ the rate of crystallization can be expressed by an equation of the form

$$\text{Rate} = d[M_{\nu+}A_{\nu-}]/dt = -ks\{([M^{m+}]^{\nu+}[A^{a-}]^{\nu-})^{1/\nu} - ([M^{m+}]_0^{\nu+}[A^{a-}]_0^{\nu-})^{1/\nu}\}^{n}. \quad (1)$$

In Eq. (1), $[M^{m+}]$, $[A^{a-}]$, and $[M^{m+}]_0$, $[A^{a-}]_0$, are the concentrations of crystal lattice ions in solution at time t and at equilibrium, respectively, k is the precipitation rate constant, s is proportional to the total number of available growth sites on the added seed material, and $\nu = (\nu_+ + \nu_-)$. The value of n, a constant, is typically 2 for the crystallization of a number of sparingly soluble salts (Nancollas 1979). The insensitivity of reaction rates to changes in fluid dynamics and the observed relatively high activation energies pointed to a surface-controlled crystallization. Thus, although diffusion is usually regarded as the principal mechanism controlling the solution of crystals, it appears to play little part in the crystallization process. The parabolic rate law [Eq. (1) with $n=2$] was proposed as long ago as 1908 by Marc (1908) and was also found in 1949 by Davies and Jones to hold for the seeded crystallization of silver chloride in supersaturated aqueous solution. Burton et al. (1951) explained the parabolic rate law at low supersaturations for the growth of non-electrolyte crystals in vapor and in solution by means of a surface diffusion mechanism with the spiral steps introduced in Fig. 4. Both the density of spiral steps and their velocity were suggested as being proportional to the supersaturation and thus the overall rate to the (supersaturation)^2 [i.e., $n=2$, Eq. (1)]. The significance of the parabolic rate law has been discussed recently by Nielsen (1981) who has shown that the Burton-Cabrera-Frank theory may be used to explain the kinetics of crystallization of electrolytes by assuming that the ions are adsorbed in electroneutral amounts on the crystal surface. Nielsen also suggested that the absolute rate was controlled by the integration process with an activation energy corresponding to the removal of one water molecule from the hydration shell of the cation at the crystal surface.

In some cases, the rate of crystal growth follows a kinetic equation similar to Eq. (1) with values of n greater than 2. These reactions may be explained by assuming that the step source is one of surface nucleation. Small crystals without dislocations grow very slowly at moderate supersaturation, since the probability that surface nucleation will occur is proportional to the surface area. With increasing supersaturation or size, the nucleation frequency increases and the growth rate accelerates. If the formation of a surface nucleus is rapid, a new layer may start before the development of the underlying layer is complete. This multiplicity leads to polynuclear growth and a number of nucleation theories are based mainly on this mechanism (Hillig 1966; Strickland-Constable 1968).

The Burton-Cabrera-Frank theory was derived for deposition from the vapor phase and it represented a real step forward in our understanding of the mechanisms of crystal growth. The dependence of growth rates upon the supersaturation is in quite good agreement with the theory, and the model has been extended and generalized by using computer simulation. In crystal growth from solution, the application of the theory is more complex. The lower diffusivity and its dependence on concentrations, the different hydration and dehydration characteristics of the ions and their surface diffusion in adsorbed layers on the crystal are some of the factors that still await satisfactory elucidation. Although the parabolic rate law has been shown to hold for many salts, the possible change to a linear dependence upon supersaturation at higher concentrations has never been observed. In aqueous systems, heterogeneous and secondary nucleation processes set in before such levels of supersaturation are achieved (Nancollas 1979).

As discussed above, the majority of slightly soluble salts, at accessible supersaturations, grow with a rate which is considerably smaller than the rate of simple mass transfer of electrolyte to the crystal surfaces. Moreover, there are numerous

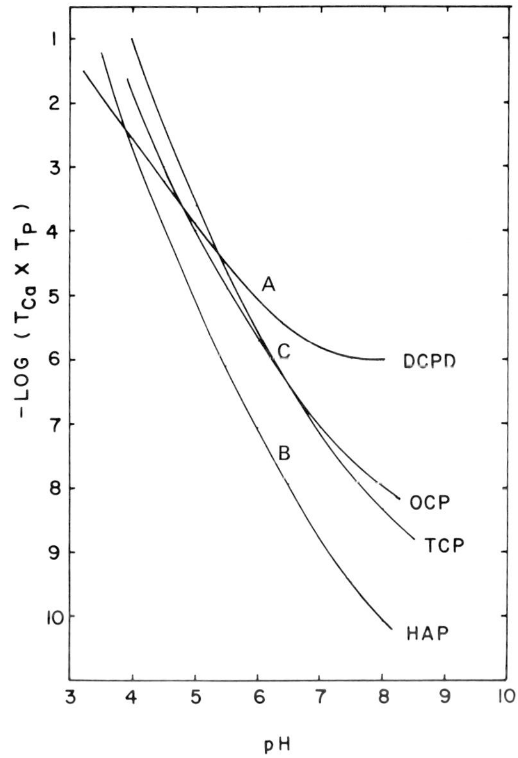

Fig. 5. Log of the product of calcium and phosphate concentration plotted against pH values of solutions at 37 °C, saturated with respect to each phase in the ternary system $Ca(OH)_2$–H_3PO_4–H_2O (Shyu L, Perez L, Zawacki S, Nancollas GH, 1983)

data pointing to activation energies which are larger than the value to be expected for simple diffusion (step 1, above) and the rates of reaction are normally independent of the rate of stirring. In this chapter, the nucleation and growth of calcium phosphate phases in aqueous solutions will be discussed in the light of some of the theories outlined above.

Calcium Phosphate Precipitation

In discussions of the precipitation of calcium phosphates, the phase which forms is usually loosely referred to as hydroxyapatite [$Ca_5(PO_4)_3OH$, hereafter, HAP], the thermodynamically most stable form. However, most calcium phosphate solutions in which precipitation experiments are made are initially supersaturated with respect to other calcium phosphate phases, tricalcium phosphate [$Ca_3(PO_4)_2$, hereafter TCP], octacalcium phosphate [$Ca_4H(PO_4)_3 \cdot 2 \cdot 5H_2O$, hereafter OCP], and dicalcium phosphate dihydrate ($CaHPO_4 \cdot 2H_2O$, hereafter DCPD). The solubility isotherms are shown in Fig. 5 expressed as products of total calcium (T_{Ca}) and total phosphate (T_P) concentrations plotted against pH of the solutions at 37 °C. For the constructions of these curves, it is necessary to calculate the free ionic concentration in the solutions and to take into account ion-pair and complex formation. Typical thermodynamic data are shown below for the protonation of phosphoric acid, calcium phosphate ion-pair formation, and the solubilities of the calcium phosphate phases.

$H^+ + H_2PO_4^- \rightleftharpoons H_3PO_4$, K = 164.1 l mol^{-1}, Bates and Acree (1943)
$H^+ + HPO_4^{2-} \rightleftharpoons H_2PO_4^-$, K = 1.58×10^7 l mol^{-1}, Bates and Acree (1943)
$H^+ + PO_4^{3-} \rightleftharpoons HPO_4^{2-}$, K = 2.33×10^{12} l mol^{-1}, Vanderzee and Quist (1961)
$Ca^{2+} + H_2PO_4^- \rightleftharpoons CaH_2PO_4^+$, K = 31.9 l mol^{-1}, Chughtai et al. (1968)
$Ca^{2+} + HPO_4^{2-} \rightleftharpoons CaHPO_4$, K = 681 l mol^{-1}, Chughtai et al. (1968)
$Ca^{2+} + PO_4^{3-} \rightleftharpoons CaPO_4^-$, K = 2.9×10^6 l mol^{-1}, Chughtai et al. (1968)
$Ca^{2+} + OH^- \rightleftharpoons CaOH^+$, K = 32.4 l mol^{-1}, Davies (1938)

The corresponding thermodynamic solubility products are:

for HAP, $[Ca^{2+}]^5[PO_4^{3-}]^3[OH]y_2^5y_3^3y_1 = 4.7 \times 10^{-59}(\text{mol l}^{-1})^9$, McDowell et al. (1977)

for TCP, $[Ca^{2+}]^3[PO_4^{3-}]^2y_2^3y_3^2 = 1.20 \times 10^{-29}(\text{mol l}^{-1})^5$, Gregory et al. (1974)

for OCP, $[Ca^{2+}]^4[PO_4^{3-}]^3[H^+]y_2^4y_3^3y_1 = 1.25 \times 10^{-47}(\text{mol l}^{-1})^8$, Moreno et al. (1960)

for DCPA, $[Ca^{2+}][HPO_4^{2-}]y_2^2 = 1.26 \times 10^{-7}(\text{mol l}^{-1})^2$, McDowell (1968)

and

for DCPD, $[Ca^{2+}][HPO_4^{2-}]y_2^2 = 2.49 \times 10^{-7}(\text{mol l}^{-1})^2$, Patel et al. (1974).

For the construction of Fig. 5, the concentrations of ionic species in the solutions must be determined from mass balance, electroneutrality, and the appropriate equilibrium constants as described previously (Nancollas 1966) by successive approximations for the ionic strength. For systems such as the phosphate salts containing multicharged ions, it is necessary to calculate the values for activity coefficients and, for this purpose, an extended form of the Debye-Hückel equation may be used such as that proposed by Davies (1962). This equation has been used satisfactorily to calculate activity coefficients of multiply charged ions up to ionic strengths of about 0.2 mol l^{-1} to within $\pm 1\%$. Neglect of activity coefficient corrections may introduce errors of several orders of magnitude in the calculated activity product for hydroxyapatite in calcium phosphate solutions of ionic strength 0.1 mol l^{-1}. It is important to note that in Fig. 5, the positions of the curves and the singular points will be dependent upon the ionic strength of the background electrolyte.

In Fig. 5 it can be seen that although HAP is the most stable salt under many conditions, it becomes less stable than DCPD if the solution is sufficiently acid. A calcium phosphate phase which is exposed to a solution more acid than the corresponding singular point may be expected to become covered by a surface coating of the more acid calcium phosphate phase. The apparent solubility behavior will then be quite different from that of the phase itself, and measurement of solution concentrations might lead to apparent high solubilities. Since calcium phosphate solutions are usually very dilute and solubilities are low, the surface phases may persist for long periods of time. This may account for many of the discrepant solubility data in the literature. Clearly when working with a diagram such as that in

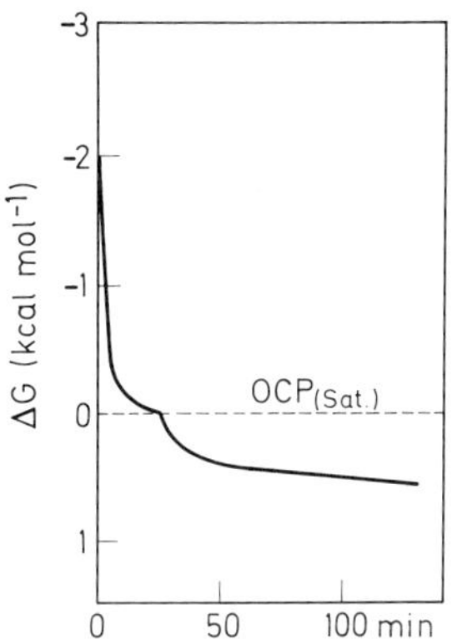

Fig. 6. Time variation of the degree of supersaturation with respect to OCP [$\Delta G = -RT \ln(IP/K_{so})$] for the spontaneous precipitation of calcium phosphate. [Reproduced from Eanes ED, Meyer JL (1977)]

Fig. 5, it is essential to establish equilibrium in order to be able to interpret the data in terms of free energies of single solid phases.

In typical precipitation experiments of calcium phosphates from solution supersaturated with respect to all phases in Fig. 5, the stoichiometric molar ratio, calcium/phosphate is frequently in the range 1.45 ± 0.05. In order to reach approximate HAP composition, extended solid/solution equilibration may be required. These stoichiometric molar ratios are usually calculated from measurements of calcium and phosphate concentrations as a function of time. Thus during a precipitation experiment, the ratio of the decrease in calcium concentration to that for phosphate ions yields the required stoichiometric ratio. However, it can be shown that an analytical error of only a few percent in these measured concentrations can preclude differentiation between the different calcium phosphate phases in Fig. 5 which may act as precursors to the formation of the thermodynamically stable HAP. Increasing the initial degree of supersaturation in order to improve the accuracy of these measurements may limit the significance of the errors but can result in a change in the mechanism of the precipitation process.

Much work has been done in attempts to identify the phases which form during the spontaneous precipitation of calcium phosphate from supersaturated solutions. It is usually regarded that the first formed phase is an unstable amorphous phase, ACP consisting of a highly hydrated cryptocrystalline calcium phosphate which appears as a precursor in the recipitation of OCP and HAP from neutral and slightly basic solutions (Eanes et al. 1966; Brecevic and Furedi-Milhofer 1972; Termine et al. 1970). ACP usually appears in the form of spherolytic monodispersed particles and it is still not clear whether this phase is a single stoichiometric unit. X-ray radial distribution measurements (Betts et al. 1981) on well-washed ACP with a Ca/P ratio of 1.5, suggested a basic structural unit consisting of spherical clusters with a stoichiometry $Ca_9(PO_4)_6$. The lifetime of this precursor phase was extremely sensitive to experimental conditions used for its preparation. Transformations into crystalline calcium phosphate phases were indicated by marked changes in pH and in calcium and phosphate concentrations.

A number of other precursor phases have been proposed in order to account for the apparent stoichiometry ($Ca/P = 1.45 \pm 0.05$) of the precipitating calcium phosphate phases. DCPD was proposed by Francis and Webb (1971) while Brown

(1966) and Brown et al. (1979) presented considerable X-ray and other evidence to show the existence of OCP. Eanes et al. (1966) proposed that a precursor TCP phase was converted to the thermodynamically stable HAP by an autocatalytic mechanism. The conversion process was studied as a function of solution environment and the presence of various stabilizing agents. Recently, there has been increased interest in attempts to follow concentration changes during the spontaneous precipitation process. Eanes and Meyer (1977) showed that the ripening of the crystals appears to be accompanied by inflexions in the time profiles of calcium and phosphate concentrations in the solution. Typical rate profiles are shown in Fig. 6. Calculation of the degrees of supersaturation with respect to the various calcium phosphate phases indicates that these inflexions occur in regions corresponding to the solubility of OCP. This provided added evidence that the precursor to the HAP precipitation was an OCP-like phase. Despotovic et al. (1975) and Feenstra and deBruyn (1979) have also provided evidence for OCP as an intermediate phase in the spontaneous precipitation of calcium phosphate.

As discussed previously, one of the problems associated with the use of apparent activity products of calcium and phosphate during the spontaneous precipitation reactions in order to identify the precursor phases is that the system is in a state of rapid kinetic change. Solubility products should be used only for solid phases which have a single free energy value, whereas in the precipitation reactions the solutions are metastable with respect to a variety of calcium phosphate phases during the reaction. Nancollas and Tomazic (1974) have shown that precursor phases may form and subsequently dissolve during the overall spontaneous precipitation reactions. The problems associated with the irreproducibility of the results of spontaneous precipitation studies were overcome with the development of seeded growth techniques (Davies and Jones 1949; Nancollas and Purdie 1964) which enabled the effects of factors such as temperature, supersaturation, and ionic strength to be studied quantitatively. The further development of a pH-stat method (Nancollas and Mohan 1970) enabled studies to be made of calcium phosphate crystal growth on well characterized seed material under conditions of constant hydrogen ion activity.

The simplest calcium phosphate phase for crystal growth studies is DCPD and, in Fig. 5, the use of supersaturated solutions at point A (pH approximately 6.0) and the addition of well characterized DCPD seed crystals resulted in rate plots shown in Fig. 7 in which calcium and phosphate concentrations were followed as a function of time. The kinetics of DCPD crystal growth in these metastable supersaturated solutions was shown to follow the rate Eq. (1) with $n = 2$. This is illustrated by a typical plot of the rate of crystallization as a function of supersaturation in Fig. 8. The observed independence of the rate of crystallization upon fluid dynamics and an activation energy of 48.9 ± 4 kJ mol^{-1} obtained from the temperature coefficient of the rate constant pointed to a surface-controlled process.

It can be seen in Fig. 5 that as the pH is increased, other calcium phosphate phases may be stabilized as precursors to the formation of apatite. Measurements of the specific surface area, SSA, of the products grown at various times indicate that the initial formation of a micro crystalline or amorphous precursor leads to a rapid increase in SSA. The development of these phases is also observed by scanning electron microscopy, and dissolution kinetic studies of the grown materials

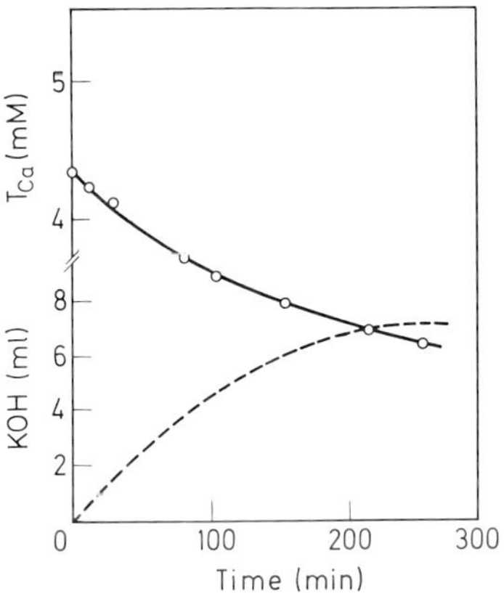

Fig. 7. Seeded growth of DCPD. *Dotted curve* volume of base (KOH) required to maintain the pH constant (5.60)

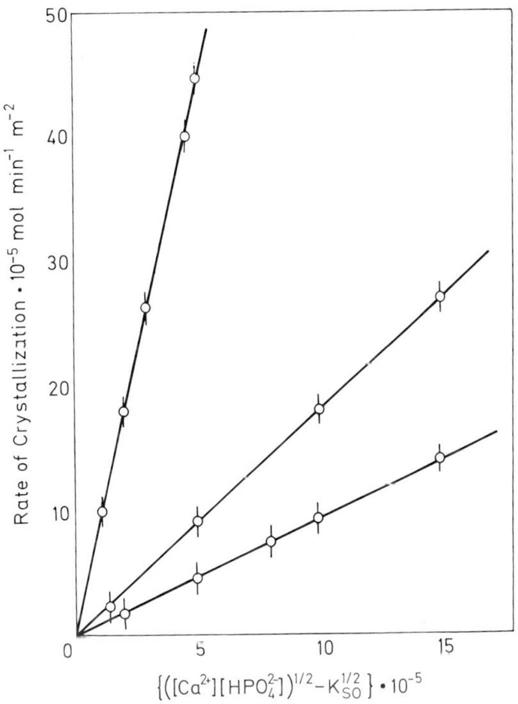

Fig. 8. Crystal growth of DCPD. Rate Eq. (1). Plot of rate of crystallization in 10^{-5} mol DCPD m^{-2} min^{-1} (y-axis) against $\{([Ca^{2+}][HPO_4^{2-}])^{1/2} - K_{SO}^{1/2}\}^2 \times 10^{-5}$ (x-axis)

have indicated the formation of OCP as a precursor phase (Nancollas and Tomazic 1974). The overall precipitation reaction appears to involve, therefore, not only the formation of different calcium phosphate phases, but also the concomitant dissolution of the thermodynamically unstable OCP formed rapidly in the initial stages of the reaction.

Although the seeded crystallization experiments yield highly reproducible results and, in the case of the pH-stat method, the hydrogen activity could be held constant during the reaction, they suffer from the disadvantage that the calcium and phosphate ionic concentrations decrease appreciably as the reaction proceeds towards solubility equilibrium. At each stage of the reaction, therefore, different calcium phosphate phases may form and subsequently redissolve as the concentrations in the supersaturated solutions decrease. To overcome this problem, a constant composition method was developed in which the chemical potentials of the solution species were maintained constant during the reaction (Tomson and Nancollas 1978). Following the addition of well-characterized seed material to metastable supersaturated solutions of calcium phosphate at the required pH, the concentrations of lattice ions were maintained constant by the simultaneous addition of reagent solutions containing calcium, phosphate, and hydroxide ions controlled by specific ion electrodes such as the glass electrode. The application of this method in OCP (Koutsoukos et al. 1980) and HAP (Amjad et al. 1978), crystallization has confirmed, kinetically, the formation of these phases during the precipitation reactions. These studies have provided direct evidence for OCP as a precursor to HAP formation, while studies at very low supersaturation have shown that macroscopic quantities of stoichiometric, highly crystalline, HAP can be formed. The method has enabled investigations of the influence of pH, supersaturation, seed-type and concentration, and traces of added substances with a precision hitherto unattainable. The constant composition method also enables the morphology of the HAP crystallites to be investigated as a function of the extent of crystallization. Ions such as lithium and chloride normally regarded as "inert" markedly affect the crystal shapes.

As indicated above, the addition of HAP seed crystals to supersaturated solutions of low concentration (corresponding to point B in Fig. 5) results in the exclusive formation of highly crystalline hydroxyapatite growth phase. The rates of crystallization are given by the rate at which titrant solutions are added in order to maintain the activities of ionic species constant. Extended experiments (Koutsoukos et al. 1980) have resulted in the growth of solid amounting to more than three times the initial seed used to inoculate the supersaturated solutions. Macroscopic amounts of highly crystalline HAP and, in the presence of fluoride ion, fluorapatite, over a wide range of pH from 5.6 to 8.5 have been confirmed without the involvement of precursor phases. Since the concentrations remain constant during the experiments, the analytical techniques used for their verification can be optimized with greatly improved precision. In addition, the method may be used with specific ion electrodes other than hydrogen and fluoride. The rates of crystallization of HAP in solutions of low supersaturation and pH 7.4 and 8.5 are plotted according to the kinetic Eq. (1) in Fig. 9. The slope of the line corresponds to an effective order of reaction, $n = 1.25 \pm 0.02$. As the supersaturation is raised, Ostwald's rule of stages predicts that other less stable and more soluble phases will form preferentially as precursors. Evidence has been advanced for the formation of OCP at points corresponding to C in Fig. 5. Indeed, the constant composition method may be used to prepare macroscopic quantities of highly crystalline OCP when OCP seed crystals are added to solutions at this supersaturation (Heughebaert and Nancollas unpublished results).

In the light of these data, we can conclude that the almost ubiquitous molar ratio of precipitating calcium phosphate phase, 1.45 ± 0.05, during the major part of the precipitation reactions may be interpreted either in terms of defect apatites or mixtures of discrete calcium phosphate phases. In terms of calcium-deficient arguments, when these phases are exposed to solutions containing calcium ions, these are taken up in an amount inversely proportional to the Ca/P ratio. The solid subsequently become less calcium deficient with Ca/P values approaching 1.67 for stoichiometric HAP. Unlike the discrete calcium phosphate phases, it is not possible to define solubilities in the thermodynamic sense, for such calcium-efficient apatites, even though this has been attempted by many workers. There is now considerable evidence (Brown et al. 1981; Young and Brown 1982; Koutsoukos et al. 1980) for the participation of OCP as a precursor phase. The conversion of OCP to HAP, readily interpretable in terms of the similar crystallographic interlayered structures is accelerated in the presence of fluoride ion (Newesley 1961; Brown 1962). It has

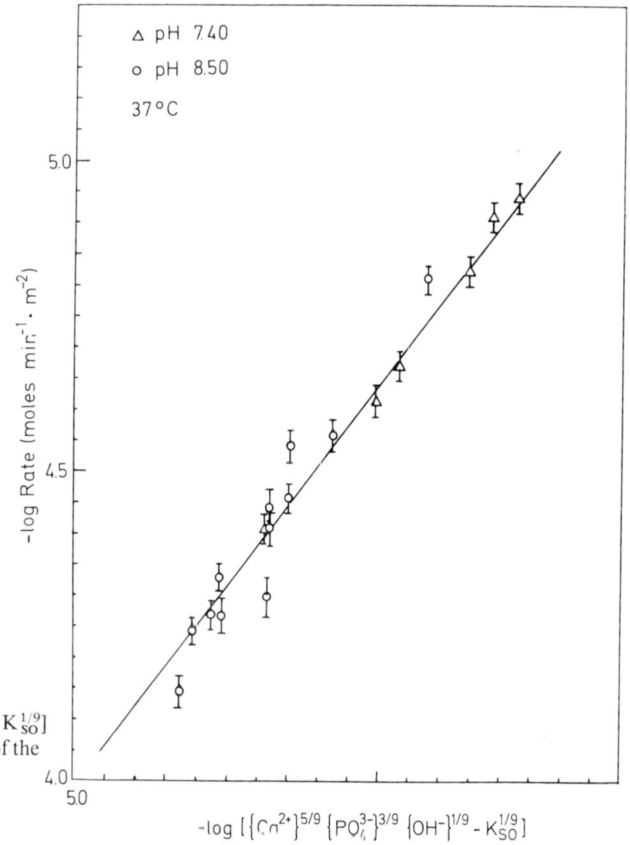

Fig. 9. Crystal growth of HAP. Plot of log (rate) against log $[\{Ca^{2+}\}^{5/9}\{PO_4^{3-}\}^{3/9}\{OH\}^{1/9}-K_{so}^{1/9}]$ where *braces* indicate activities of the species enclosed

also been shown that fluoride retards the formation of ACP which precedes the hydrolyses of OCP to HAP and inhibits or even eliminates the formation of OCP-like precursor phases. This is in contrast to the influence of magnesium ion during the precipitation of calcium phosphate which apparently stabilizes TCP-like precursor phases (Newesley 1961; Feenstra et al. 1981; Termine et al. 1970).

As indicated above, small errors in the determination of calcium and phosphate concentrations may lead to large errors in the calculated molar ratios of precipitating phases. X-ray powder diffraction evidence pointing to mixed phases containing HAP and TCP may not be convincing, however, because of the difficulty involved in distinguishing between these phases from the diffractograms. Crystallization of HAP at low constant supersaturation corresponding to point B in Fig. 5 and without the involvement of precursor phases indicates that this phase was formed in the presence of magnesium ion although the rate of formation was markedly reduced (Feenstra et al. 1981). The retarding effect could be interpreted in terms of a Langmuir adsorption isotherm representing the effective blocking of growth sites on the HAP seed crystal surface by the presence of magnesium ion (Koutsoukos and Nancollas in prep.). In contrast, the crystallization of HAP at low supersaturation in the presence of strontium ion is accompanied by the incorporation of this foreign ion into the apatite lattice to yield solid phases with considerably lower Sr/Ca molar ratios than those in the supersaturated solutions (Koutsoukos and Nancollas 1981). Changes in the unit cell lattice parameters and infra-red spectra were linearly related to the strontium content of the precipitated solid phases.

In addition to the alkali and alkaline earth cations, the calcium phosphates may take up numerous other metal ions and extensive studies have been made of the stoichiometry and crystal structures of these substituted apatites (LeGeros et al. 1981). Marked changes in lattice parameters accompany the substitution of many of these ions (LeGeros et al. 1977). Of the anions, the substitution of carbonate is probably the most interesting and at the same time most perplexing. Carbonate is known to be present in naturally occurring minerals and the carbonate-containing apatites have unit cell dimensions smaller than those of carbonate-free apatites. A great deal of work has been done in an attempt to locate the position of the carbonate ions by instrumental techniques (LeGeros et al. 1971). The substitution of two hydroxyl groups by one carbonate ion in HAP has been shown to increase the a axis length (Elliott 1964). This substitution of carbonate into the hydroxyl positions has been shown to take place, usually, at high temperatures (Bonel and Montel 1964). Carbonate ion has also been shown to replace phosphate, resulting in a decreased a axis and an increased c axis (LeGeros 1965). The two forms of substitution may be differentiated by infra-red measurements. Since the carbonate ion is smaller than the phosphate ion and is planar, it is likely that the former can occupy a face of the tetrahedral phosphate position. Recently, Chickerur et al. (1980) have studied the role of octacalcium phosphate in the process of incorporation of carbonate into apatite. It was shown that when OCP is hydrolyzed in the presence of sodium and carbonate ions in aqueous media, approximately one sodium and one carbonate ion substitute for a calcium and phosphate ion, respectively, in forming the apatite. Accompanying this substitution, the a axis is shortened and the evidence points to the carbonate occupying a phosphate site. The suggestion that carbonate is incorporated into HAP during hydrolysis of OCP is of particular importance and adds further weight to the participation of OCP as a precursor in the formation of the thermodynamically most stable calcium phosphate phase.

It is clear that the nature of the calcium phosphate phase which precipitates from solution is markedly dependent upon the supersaturation, pH and ionic strength of the solution. Indeed, even changes in the background "neutral" electrolyte may markedly influence both the rate of the crystallization reaction and the nature of the phases which form. In order to interpret the results for individual systems, it is important to calculate the activities of free ionic species in solution in order to obtain reliable estimates for the degree of supersaturation with respect to the different calcium phosphate phases. The presence of other divalent metal ions which may or may not be incorporated into the crystal lattices may also markedly influence the course of the precipitation reaction. Although it is not possible at this stage to decide between defect apatites formation and discrete precursor phase participation, there is now increasing evidence for the latter interpretation of the mechanism of precipitation in seeded systems.

Acknowledgments. This work was supported in part by grants from the National Institutes of Health, (Dental Research Grant No. DE03223) and Arthritic and Metabolic Diseases (No. AM-19048).

References

Amjad Z, Koutsoukos PG, Tomson MB, Nancollas GH (1978) The growth of hydroxyapatite from solution. A new constant composition method. J Dental Res 57:909–910

Bates RG, Acree SF (1943) ΔH values of certain phosphate chloride mixtures and the second dissociation constant of phosphoric acid from 0° to 60°. J Res Nat Bur Stand 30:129–155.

Becker R (1949) Kinetics of the formation of nuclei and statistical theory of condensation. Disc Farad Soc 5:56–61

Betts F, Blumenthal NC, Posner AS (1981) Bone mineralization. J Cryst Growth 53:63–73

Boistelle R, Abbona F (1981) Morphology, habit and growth of newberyite crystals ($MgHPO_4 \cdot 3H_2O$). J Cryst Growth 54:275–295

Bonel G, Montel G (1964) Synthetic carbonated apatite. CR Acad Sci [D] (Paris) 258:923–926

Brecevic Lj, Furedi Milhofer H (1972) Precipitation of calcium phosphates from electrolyte solutions II. The formation and transformation of the precipitates. Calc Tissue Res 10.82–90

Brown WE (1962) Crystal structure of octacalcium phosphate. Nature 196:1048–1050

Brown WE (1966) Crystal growth of bone mineral. Clin Orthop Relat Res 44:205–220

Brown WE (1973) Solubilities of phosphate and other sparingly soluble compounds. In: Griffith EJ, Beeton A, Spencer JM, Mitchell DT (eds) Environmental phosphorus handbook. Wiley & Sons, New York, pp 203

Brown WE, Mathew M, Tong MS (1981) Crystal chemistry of octacalcium phosphate. In: Pamplin J (ed) Prog crystal growth charact. Pergamon, London, pp 59–87

Brown WE, Schroeder LW, Ferris JS (1979) Interlayering of crystalline OCP and HAP. J Phys Chem 83:1385–1388

Burton WK, Cabrera N, Frank FC (1951) The growth of crystals and the equilibrium structure of their surfaces. Phil Trans R Soc A243:299–358

Chickerur NS, Tung MS, Brown WE (1980) A mechanism for incorporation of carbonate into apatite. Calcif Tissue Int 32:55–62

Christiansen JAC, Nielsen AE (1951) The kinetics of formation of precipitates of sparingly soluble salts. Acta Chem Scand 5:673–674

Chughtai A, Marshall R, Nancollas GH (1968) Complexes in calcium phosphate solutions. J Phys Chem 72:208–211

Collin RL (1966) Precipitate formation in the strontium phosphate system. Science 151:1386–1388

Davies CW (1938) The extent of dissociation of salts in water. Part VI. Some calcium salts of organic acids. J Chem Soc 277–281

Davies CW (1962) Ion association. Butterworths, London, p 39

Davies CW, Jones AL (1949) The precipitation of silver chloride from aqueous solutions. Part I. Disc Farad Soc 5:103–111

Despotovic R, Filipovic-Vincekovic N, Furedi-Milhofer H (1975) Precipitation of calcium phosphates from electrolyte solutions. III Radiometric studies of the kinetics of precipitation and aging of calcium phosphates. Calcif Tissue Res 18:13–26

Eanes ED, Meyer JL (1977) The maturation of crystalline calcium phosphates in aqueous suspensions at physiologic pH. Calcif Tissue Res 23:259–269

Eanes ED, Gilessen IH, Posner AS (1966) Mechanism of conversion of noncrystalline calcium phosphate to crystalline hydroxyapatite. In: Peiser HS (ed) Crystal growth. Pergamon, Oxford, p 373–376

Elliott JC (1964) The crystallographic structure of dental enamel and related apatites. PhD Thesis, University London

Feenstra TP, DeBruyn PL (1979) Formation of calcium phosphates in moderately supersaturated solutions. J Phys Chem 83:475–479

Feenstra TP, Van Straten HA, deBruyn PL (1981) Formation of strontium phosphates in supersaturated solutions. J Colloid Interface Sci 80:255–270

Feenstra TP (1980) The initial stages in the formation of calcium and strontium phosphates from supersaturated solutions. PhD Thesis, University Utrecht, p 66–89

Ferguson JF, Jenkins D, Eastman J (1971) Calcium phosphate precipitation. J Water Poll Control Fed Ser 45:620–631

Francis MD, Webb NC (1971) Hydroxyapatite formation from hydrated calcium monohydrogen phosphate precursor. Calcif Tissue Res 6:335–342

Frank FC (1949) The influence of dislocations in crystal growth. Disc Farad Soc 5:48–55
Gregory TM, Moreno EC, Patel JM, Brown WE (1974) Solubility of $Ca_3(PO_4)_2$ in the system $Ca(OH)_2$–H_3PO_4–H_2O at 5, 15, 25, and 37 °C. J Res Nat Bur Stand 78A:667–674
Hillig WB (1966) A derivation of classical two-dimensional nucleation kinetics and the associated crystal growth laws. Acta Met 14:1868–1869
Kossel W (1934) Theory of crystal growth. Ann Phys 21:455
Koutsoukos PG, Amjad Z, Tomson MB, Nancollas GH (1980) Crystallization of calcium phosphates. A constant composition study. J Am Chem Soc 102:1553–1557
Koutsoukos PG, Nancollas GH (1981) The influence of stontium ion on the crystallisation of hydroxyapatite. J Phys Chem 85:2403–2424
LeGeros RZ (1965) Effect of carbonate on the lattice parameters of apatite. Nature 206:403–404
LeGeros RZ, LeGeros JP, Tranz OR, Shirra WP (1971) Conversion of monetite, $CaHPO_4$, to apatites. Adv X-ray Anal 14:57–66
LeGeros RZ, Miravite MA, Quiroligico GB, Curzon MEJ (1977) The effect of some trace elements on the lattice parameters of human and synthetic apatite. Calc Tissue Res 22:362–367
LeGeros RZ, Taheri MH, Quiroligico GB, LeGeros JP (1981) Formation and stability of apatites: Effects of some cationic substituents. Proc Int Congr Phosphorus Comp IMPHOS, Rabat, p 89–103
Marc R (1908) Über die Kristallisation aus wässerigen Lösungen. Z Physik Chem 61:385–398
McDowell H (1968) Solubility of $CaHPO_4$ and ion-pair formation. PhD Thesis, Howard University, Washington DC
McDowell H, Gregory TM, Brown WE (1977) Solubility of $Ca_5(PO_4)_3OH$ in the system $Ca(OH)_2$–H_3PO_4–H_2O at 5, 15, 25, and 37 °C. J Res Nat Bur Stand 81A:273–281
Moreno EC, Brown WE, Osborn G (1960) Stability of dicalcium phosphate dihydrate in aqueous solutions and solubility of octacalcium phosphate. Soil Sci 24:99–102
Nancollas GH (1979) The growth of crystals from solution. Adv Colloid Interface Sci 10:215–252
Nancollas GH (1966) Interactions in electrolyte solutions. Elsevier, Amsterdam, p 73–90
Nancollas GH, Purdie N (1964) The kinetics of crystal growth. Quart Rev (Lond) Chem Soc 18:1–20
Nancollas GH, Tomazic BB (1974) Growth of calcium phosphate on hydroxyapatite crystals. Effect of supersaturation and ionic medium. J Phys Chem 78:2218–2225
Nancollas GH, Mohan MS (1970) The growth of hydroxyapatite crystals. Arch Oral Biol 15:731–745
Newesley H (1961) Changes in crystal types of low solubility calcium phosphates in the presence of accompanying ions. Arch Oral Biol 6:174–180
Nielsen AE (1981) Theory of electrolyte crystal growth. In: deJong, EJ, Jancic SJ (eds) Industrial crystallization. Elsevier North-Holland, Amsterdam, in press
Patel PR, Gregory TM, Brown WE (1974) Solubility of $CaHPO_4$–$2H_2O$ in the quaternary system $Ca(OH)_2$–H_3PO_4–NaCl–H_2O at 25 °. J Res Nat Bur Stand 78A:675–681
Shyu LJ, Perez L, Zawacki S, Heughebaert JC, Nancollas GH (1983) The solubility of octacalcium phosphate at 37 °C in the system $Ca(OH)_2$–H_3PO_4–KNO_3–H_2O. J Dental Res, 62:398–400
Stranski IN (1928) The theory of crystal growth. Z Phys Chem Abt A136:259–278
Strickland-Constable (1968) Kinetics and mechanism of crystallization. Academic Press, London, p 252–254
Termine JD, Peckauskas RA, Posner AS (1970) Calcium phosphate formation in vitro. II. Effect of environment on amorphous-crystalline transformation. Arch Biochem Biophys 140:318–325
Tomson MB, Nancollas GH (1978) Mineralization kinetics: A constant composition approach. Science 200:1059–1060
Vanderzee CE, Quist AS (1961) The third dissociation constant of orthophosphoric acid. J Phys Chem 65:118–123
Volmer M (1945) Kinetik der Phasenbildung, Bros. Ann Arbor, Michigan
Young RA, Brown WE (1982) Structures of biological minerals. In: Nancollas GH (ed) Biological mineralization and demineralization. Springer, Berlin Heidelberg New York, p 101–142

Chapter 3 Crystallochemical Aspects of the Phosphate Minerals

PAUL B. MOORE

Introduction

The details of phosphate mineral chemistry are intricate and voluminous. The great burst of knowledge, gained largely within the past 20 years, is due to the rapid increase of efficiency in artificial intelligence and the consequent practicability of systematic crystal structure analysis.

Since many points are touched only peripherally, I apologize for the rather loose jumble of facts which follow. Classificatory schemes are not to be taken as absolute irrefragables; rather they are devices of convenience. An entire chapter could be devoted to any crystallochemical theme presented. Rather this introduction is a prolegomenon to the subjects which follow. It seems that the major advances have been made through structure studies; let us hope thermochemical investigations will soon follow.

Incidentally, two rather detailed summaries of phosphate structure chemistry also appeared, one by Moore and the other by Corbridge. Both can be found in IMPHOS (1980).

A Cluster Hypothesis

The number of phosphate mineral species is large. Moore (1980) listed 118 species whose structures were known for the compounds involving M^{3+} octahedrally coordinating cations. Here, M corresponds to Fe^{3+} and Al^{3+} in the majority of cases. Among these, some isotypy with $[AsO_4]^{3-}$, $[VO_4]^{3-}$, and $[CrO_4]^{3-}$ is known. But if we narrow the list to distinct structure types, the number shrinks considerably (60 types). The question then remains: which structure types are intrinsically stable, and under what conditions? This is a big question, because if intrinsically stable structure types can be retrieved a priori (for example, through M.O. calculations), then phosphate crystal chemistry is inherently a solved problem. Unfortunately, this tantalizing question is not yet soluble, because it demands that the physically observed configurations be demonstrated as thermochemically more stable than all the *hypothetical arrangements which could geometrically exist.* Such an attack on this fundamental problem has so far been thwarted by the absence of combinatorial enumerative techniques to retrieve all such arrangements. To test their relative stabilities is a less difficult problem.

Department of the Geophysical Sciences, The University of Chicago, Chicago, Illinois 60637

Stepping back a bit from this fundamental question, Nature seems to be telling us something about intrinsically stable structural states. This has been gleaned largely from routine (and not so routine!) crystal structure analytical techniques, usually via single crystal X-ray diffraction. Presently, it appears evident that knowledge of the crystal structure is the first big step toward understanding a phase. Gross physical properties, bond distances and angles, structure topology, bonding type and many calculated vital statistics such as powder pattern can be rapidly assessed through knowledge of the atom positions and their point symmetries in the asymmetric unit of structure. Thermochemistry necessarily follows. Although we know a substantial amount about crystal structures of phosphates, their phase equilibria are far less completely understood (in comparison with, say, the rock-forming silicates) if for any other reason than that there has been no causal or more broadly defined impetus to define these phases thermochemically. For example, whitlockites are known from anhydrous meteoritic assemblages to low-temperature aqueous rest-liquid assemblages, to bone and tooth formation at ambient conditions, to pathological calcifications, yet the thermochemistry of whitlockite is poorly understood. Certain potentially important sedimentary phosphate phases beyond apatites and whitlockites, such as strengite, phosphosiderite and mitridatite are thermochemically barely understood at all! Since most phosphate species are phases formed at low temperature and since their solubility products tend to be low owing to the P–O bonds and links of high strength, detailed experimental determination of thermochemical parameters is relatively difficult to obtain. In addition, many of these phases probably form as metastable entities. Coupled with evident little interest in phosphates compared with rock-forming silicates, but little enthusiasm has been drawn to them.

From the considerable number of natural phosphate mineral structures now known, Nature seems to be telling us something. I prefer to phrase it like this in what I dub the "cluster hypothesis": that *among the great tribe of phosphate mineral species, the cluster principles which persist are relatively few*. Here as a cluster I mean a fundamental building block or aggregate of polyhedra which persists within limits of reasonable geometrical replication over a fairly extensive range of diverse structures. Hints of the cluster hypothesis appeared recently, for example in Moore (1970a, b) and Hawthorne (1979). Presumably, the modular units which are identified as these clusters are intrinsically thermochemically and structurally so stable that Nature exploits them repetitively, often over considerable temperature and composition interval, as building blocks of more complex arrangements.

At this point, it may be instructive to point out just which clusters have been found to appear repetitively in different structures. Here I may add that no distinction will be made among the great variety of combinatorial stereoisomerisms which may in fact exist. The reason for this loosening of the conditions is that many stereoisomers (such as trans-, cis-, cis-trans, etc.) are known to occur within a narrow paragenetic range and are evidently for that reason similarly stable on thermochemical grounds. The largest families I have singled out include the following:

I. The glaserite, $K_3Na(SO_4)_2$, derived structures and related structures
Ia. The lanthanide phosphate structures
II. The hexagonal and cubic close-packed structures
III. The 5 Å trimeric face-sharing structures

Table 1. The natural M^{3+} phosphates in categories III–VI[a]

III. The 5 Å trimeric face-sharing structures		
1.	Lazulite-scorzalite	$(Mg, Fe)Al_2(OH)_2(PO_4)_2$
	Barbosalite	$Fe^{2+}Fe_2^{3+}(OH)_2(PO_4)_2$
2.	Lipscombite	$Fe^{2+}Fe_2^{3+}(OH)_2(PO_4)_2$
3.	Trolleite	$Al_4(OH)_3(PO_4)_3$
4.	Rockbridgeite-frondelite	$(Fe, Mn)^{2+}Fe_4^{3+}(OH)_5(PO_4)_3$
5.	Laubmannite	$Fe_3^{2+}Fe_6^{3+}(OH)_{12}(PO_4)_4$
6.	Beraunite	$Fe^{2+}Fe_5^{3+}(OH)_5(H_2O)_4(PO_4)_4 \cdot 2H_2O$
7.	Souzalite-gormanite	$(Mg, Fe)_3Al_4(OH)_6(PO_4)_4 \cdot 2H_2O$
8.	Dufrenite	$CaFe_6^{3+}(OH)_6(H_2O)_2(PO_4)_4$
9.	Kidwellite	$NaFe_9^{3+}(OH)_{10}(PO_4)_6 \cdot 5H_2O$
IV. The 7 Å corner-sharing octahedral chain structures		
1.	Amblygonite-montebrasite	$(Li, Na)[Al(F, OH)(PO_4)]$
	Tavorite	$Li[Fe^{3+}(OH)(PO_4)]$
2.	Lacroixite	$Na[Al(OH, F)(PO_4)]$
3.	Strunzite	$Mn(H_2O)_4[Fe_2^{3+}(OH)_2(H_2O)_2(PO_4)_2]$
4.	Laueite	$Mn(H_2O)_4[Fe_2^{3+}(OH)_2(H_2O)_2(PO_4)_2] \cdot 2H_2O$
	Gordonite	$Mg(H_2O)_4[Al_2(OH)_2(H_2O)_2(PO_4)_2] \cdot 2H_2O$
	Paravauxite	$Fe(H_2O)_4[Al_2(OH)_2(H_2O)_2(PO_4)_2] \cdot 2H_2O$
	Sigloite	$Fe^{3+}(H_2O, OH)_4[Al_2(OH)_2(H_2O)_2(PO_4)_2] \cdot 2H_2O$
5.	Pseudolaueite	$Mn(H_2O)_4[Fe_2^{3+}(OH)_2(H_2O)_2(PO_4)_2] \cdot 2H_2O$
6.	Stewartite	$Mn(H_2O)_4[Fe_2^{3+}(OH)_2(H_2O)_2(PO_4)_2] \cdot 2H_2O$
7.	Wavellite	$[Al_3(OH)_3(H_2O)_4(PO_4)_2] \cdot H_2O$
8.	Kingite	$[Al_3(OH)_3(H_2O)_4(PO_4)_2] \cdot 5H_2O$
9.	Childrenite-eosphorite	$(Fe, Mn)[Al(OH)_2(H_2O)(PO_4)]$
	Oxychildrenite	$Fe^{3+}[Al(OH)_2(OH)(PO_4)]$
10.	Overite	$Mg(H_2O)_4[CaAl(OH)(PO_4)_2]$
	Segelerite	$Mg(H_2O)_4[CaFe^{3+}(OH)(PO_4)_2]$
11.	Jahnsite	$Mg_2(H_2O)_8[CaMn^{2+}Fe_2^{3+}(OH)_2(PO_4)_4]$
	Whiteite	$Mg_2(H_2O)_8[CaFe^{2+}Al_2(OH)_2(PO_4)_4]$
V. The octahedral corner-, edge-sharing sheet structures		
1.	Crandallite	$Ca[Al_3(OH)_6(PO_4)(PO_3OH)]$ + at least 24 other species
2.	Wardite	$Na[Al_3(OH)_4(H_2O)_2(PO_4)_2]$
	Millisite	$Ca[Al_3(OH)_4(H_2O, OH)_2(PO_4)_2]$
	Cyrilovite	$Na[Fe_3^{3+}(OH)_4(H_2O)_2(PO_4)_2]$
3.	Whitmoreite	$Fe(H_2O)_4[Fe_2^{3+}(OH)_2(PO_4)_2]$
4.	Kryzhanovskite	$[Fe_3^{3+}(OH)_3(PO_4)_2]$
5.	Mitridatite	$Ca_6(H_2O)_6[Fe_9^{3+}O_6(PO_4)_9] \cdot 3H_2O$
	Robertsite	$Ca_6(H_2O)_6[Mn_9^{3+}O_6(PO_4)_9] \cdot 3H_2O$
VI. The finite cluster structures		
1.	Strengite	$[Fe^{3+}(H_2O)_2(PO_4)]$
	Variscite	$[Al(H_2O)_2(PO_4)]$
2.	Phosphosiderite	$[Fe^{3+}(H_2O)_2(PO_4)]$
	Clinovariscite	$[Al(H_2O)_2(PO_4)]$
	Sterrettite	$[Sc(H_2O)_2(PO_4)]$
3.	Minyulite	$K[Al_2F(H_2O)_4(PO_4)_2]$
4.	Morinite	$Ca_2Na[Al_2F_4(OH)(H_2O)_2(PO_4)_2]$
5.	Jagowerite	$Ba[Al_2(OH)_2(PO_4)_2]$
6.	Leucophosphite	$K_2[Fe_4^{3+}(OH)_2(H_2O)_2(PO_4)_4] \cdot 2H_2O$
	Al-leucophosphite	$K_2[Al_4(OH)_2(H_2O)_2(PO_4)_4] \cdot 2H_2O$
7.	Augelite	$[Al_4(OH)_6(PO_4)_2]$

[a] Structure types arbitrarily labeled by numbers on left

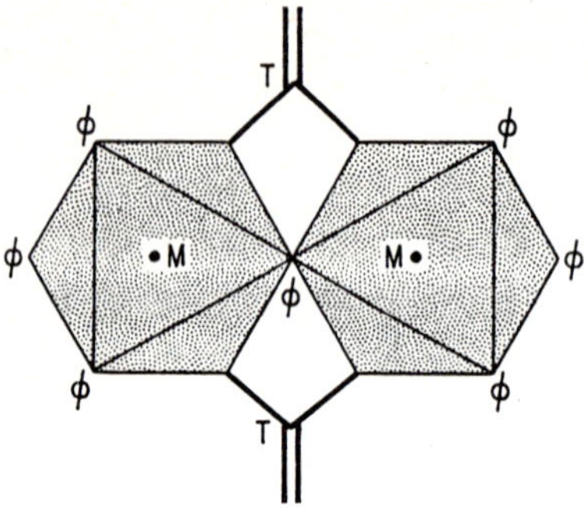

Fig. 1. The $M_2\phi_7(TO_4)_2$ cluster where M is an octahedrally coordinated cation and T is a tetrahedrally coordinated cation. The ϕ are assorted ligands (O^{2-}, OH^-, F^-, H_2O, etc.). The octahedra are *stippled* and the tetrahedra are drawn as T–O spokes

IV. The 7 Å corner-sharing octahedral chain structures
V. The octahedral corner-, edge-sharing sheet structures
Va. The actinide phosphate micaceous structures
VI. The finite cluster structures

Each family will be discussed in turn, but it is quite absurd to assume that the families are conveniently pigeonholed in the real world. On the contrary, representatives from several families may crystallize close to each other. They are roughly arranged according to diminishing temperature, categories I, Ia, and II essentially being the primary phases from which the others are derived through processes of hydrothermal reworking.

Categories Ia and Va are largely dictated by the behavior of the large lanthanide and actinide ions, and their bonding properties for certain ionic states. Categories II–VI, which doubtless elicit a greater fraction of our interest, embrace most of the phosphate minerals. In fact, of the 118 species in Table 1, about 65% are found within III–VI. Hawthorne's (1979) insight deserves special mention. Here, he extracted a single cluster type which persists in several seemingly unrelated structures. This cluster in Fig. 1 demonstrates the pitfalls which await the unwary in using such broad schemata as categories I–VI, since this fragment of a more complete structure may span several categories. It almost begs the following intriguing question: what is the hierarchy (from insular polyhedron on up) of clusters found in these crystal structures? This is a fascinating question indeed, for a solution would put all the species listed in a logical structural systematology. So far, this problem has met with failure, for it involves the more subtle and deeper reaches of combinatory and graph theory, and straightforward answers to this problem appear to be evanescent.

Certain problems can be answered with the aid of combinatory theory, however. For example, we may ask "what are all the edge-sharing octahedral clusters which can be found within the octahedron sharing all twelve of its edges with other octahedra"?

To address this problem more closely, the 144 geometrically distinct octahedral edge-sharing clusters ranging from $M\phi_6$ (central octahedron) to $M_{13}\phi_{38}$ (central octahedron with all twelve edges shared by other octahedra) listed in Moore (1974) were arranged and all center and corner coordinates were calculated for perfect octahedra. Consider a cluster $M_r^{n+}\phi_s^{q-}$. The $x = |n|/|q|$ is a ratio of the point charges on the cation and the anion. The s/r assesses the relative condensation of the

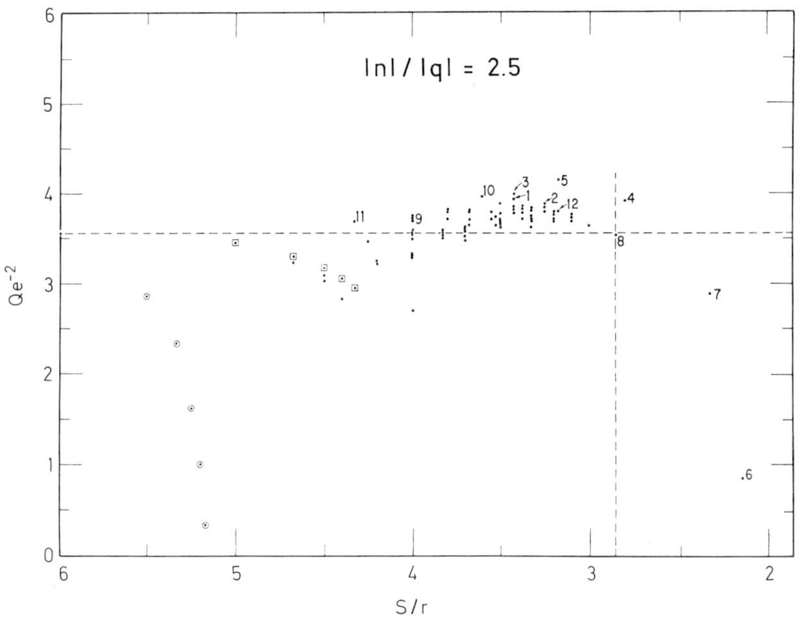

Fig. 2. Plot of Qe^{-2} (the electrostatic potential) vs cluster condensation s/r (or total anions divided by total cations) for the charge ratio $|n|/|q| = 2.5$. The cluster composition is $M_r^{n+}\phi_s^{q-}$. Points designated are *1* $[Mo_7O_{24}]^{6-}$; *2* $[Mo_8O_{26}]^{4-}$; *3* $M_7\phi_{24}$ (hypothetical, $\bar{3}$ 2/m); *4* $[V_{10}O_{28}]^{6-}$; *5* $[Nb_6O_{19}]^{8-}$; *6* $M_{43}\phi_{92}$ (hypothetical 4/m $\bar{3}$ 2/m); *7* $M_{19}\phi_{44}$ (hypothetical, 4/m $\bar{3}$ 2/m); *8* $M_{13}\phi_{38}$ (hypothetical, 4/m $\bar{3}$ 2/m); *9* $M_4\phi_{16}$ (hypothetical, 23m, mm2, and 2/m); *10* $M_5\phi_{18}$ (hypothetical, 4mm); *11* $M_3\phi_{13}$ (hypothetical, 3m); *12* $M_{12}\phi_{38}$ (hypothetical, 8 with a hole in the center)

clusters. For example, three isolated octahedra would have r = 3, s = 18 or s/r = 6.0. If these three octahedra shared edges to form a linear trimer, then four vertices would vanish and r = 3, s = 18 − 4 = 14, or s/r = 4.67.

The *electrostatic potential* $Qe^{-2} = -Ax^2 + Bx - C$ where e is the charge of the electron is the sum A over the M–M′ vectors in the cluster, B the sum of the M–ϕ vectors and C the sum over the ϕ–ϕ' vectors. This calculation for finite clusters is similar to electrostatic energy of crystals determined by diffraction data (Ewald 1973). A plot of Qe^{-2} vs. s/r appears in Fig. 2. The octahedral edge clusters occur as points, progressively condensed octahedral corner-chain fragments as open circles and progressively condensed octahedral edge-chain fragments as squares. The point labelled 8 marks the intersection between horizontal and vertical dashed lines and represents the cluster with all edges shared, $M_{13}\phi_{38}$. Since net attraction should increase with increasing Qe^{-2}, points above the horizontal dashed line should be relatively more stable than points below. In addition, points with higher Qe^{-2} than point 8 should represent configurations more stable relative to $M_{13}\phi_{38}$. Points 6 and 7 include the second and third shell edge enclosures beyond $M_{13}\phi_{38}$, with stoichiometries $M_{19}\phi_{44}$ and $M_{43}\phi_{92}$ respectively. Note that fewer than 144 points are represented on this diagram. This is because many arrangements have isomers and furthermore, many of these isomers are homometric so give identical results.

The above model, albeit simplistic, illustrates an interesting point. Although many clusters are represented, but few are stable (i.e., observed) entities. These occur along fixed s/r for arrangements with highest Qe^{-2}. Some example are (1) $[Mo_7O_{24}]^{6-}$, (2) $[Mo_8O_{26}]^{4-}$, (3) $[V_{10}O_{28}]^{6-}$, (5) $[Nb_6O_{19}]^{8-}$. As yet, no cluster has been found for fixed s/r which has a lower electrostatic potential relative to other isomers. These structures were selected since $2.5 \leq x \leq 3.0$. Note that further growth of $M_{13}\phi_{38}$ leads to decreased stability, that the growth of chain components decrease in stability as well. The many examples of infinitely extending chain structures for the phosphates would seem to contradict the decreasing stability of successive links in the chain components in Fig. 2. However, the simple model presented here involves only octahedral clusters, *not* octahedral-tetrahedral linkages such as $^{1}_{\infty}[M_2^{3+}(OH)_2(H_2O)_4(PO_4)_2]^{2-}$.

High Temperature Phases: Rod Structures and Hexagonal Close-Packed Structures

An intriguing family of structures includes some of the most complicated crystal chemistries in the mineral kingdom. These compounds are phosphates of intermediate to large-sized alkali and alkaline earth cations. Some phases, such as the apatite family, occur throughout a wide range of pressure and temperature conditions, from high pressure carbonatites, to alkali syenites, granitic pegmatites, regionally metamorphosed marbles, evaporite beds, as framework constituents in organisms and in pathological calcifications or metabolic disorders.

Many of these phases possess unusual physical properties, some of which have made them desirable for further study. One interesting crystallochemical property is the extremely efficient packing of the ions in space, often exceeding the packing of the classic hexagonal and cubic close-packed compounds. This property is fairly easy to obtain. Given the unit cell volume, one can calculate the *packing efficiency*, V_E, by dividing the unit cell volume by the total number of anions in that cell. Since this is usually used to assess densities of different polymorphs, the more efficiently packed structure has the smaller V_E. For this model, all the cations are assumed to be small spheres which pack interstitially within the close-packed framework of large anions. Effective ionic radii constitute the starting point in crystal chemistry. Goldschmidt (1927) explored the geometrical factors in calculating effective ionic radii from experimentally determined crystal cell parameters for simple oxides, sulphides, halides, and metals. The upper limit of fractional space coverage by identical hard spheres in cubic closest packing is $C = \sqrt{2}\pi/6 = 0.7405$. Shannon and Prewitt (1969) offer probably the most widely quoted effective ionic radii for oxides and fluorides. Since these were obtained from experimentally determined polyhedral interatomic distance averages, the anion radii were used initially with six-coordinate O^{2-} and six-coordinate F^- having radii $r_{O^{2-}} = 1.40$ Å and $r_{F^-} = 1.33$ Å. The volume of six-coordinate O^{2-} would be $V = 11.50$ Å^3, and the packing efficiency in cubic close-packing (c.c.p.) would be $V/C = V_E = 15.52$ Å^3. This value is close to some packing efficiencies of dense packed oxides, such as corundum, α-Al_2O_3, hexagonal close-packed, (h.c.p.) $V_E = 14.27$ Å^3; hematite, α-Fe_2O_3, h.c.p., $V_E = 16.83$ Å^3; and periclase, MgO, c.c.p., $V_E = 18.65$ Å^3. If ionic size were to have any meaning in the context of packing, clearly some consideration has to be made for size of the coordinating cations as well. In general, the larger the size of the cation, the greater the V_E. A large cation like Ca^{2+} forces a large V_E on the anion. For example in calcium oxide, CaO, c.c.p., it is $V_E = 27.82$ Å^3.

Even more dramatic relationships can be found among polymorphs. For low-temperature α-Ca_2SiO_4 which has the olivine structure, h.c.p., $V_E = 24.3$ Å^3. But for high-temperature "olivine" or silico-glaserite, γ-Ca_2SiO_4, $V_E = 21.8$ Å^3. Thus, γ-Ca_2SiO_4 is about 10% more efficiently packed than α-Ca_2SiO_4, the latter already having a close-packed structure! This "denser than dense-packed" structure of γ-Ca_2SiO_4 is a property which it shares with these dense, refractory phases among the phosphates, such as α-$Ca_3(PO_4)_2$, whitlockite, apatite, fillowite, brianite, etc. Related denser than dense-packed structures include wyllieite, dickinsonite-arrojadite, and garnet. All these structures appear to be based on the efficiently packed glaserite structure type, $KNa_3(SO_4)_2$.

Recently, Moore (1981) attempted to demonstrate their relationships through the vehicle of V_E. Structural diagrams emphasized the diverse order of polyhedra in such structures, from order 4 to order 12. Many polyhedra, particularly of order 8, are not familiar in the world of conventional dense packings. Some structures possess unusual complexity. Fillowite, for example, has 45 atoms in the asymmetric unit (Araki and Moore 1981) and dickinsonite-arrojadite has at least 49 (Moore et al. 1981). The α-$Ca_3(PO_4)_2$ has 78 atoms in the asymmetric unit, according to Mathew et al. (1977).

Recent evidence, much of it unpublished, suggests that a large family of compounds exists, which spans a diversity of radicals. How do these compounds tie together? One important principle is isomorphism, where an ion A may substitute for ion B in a crystal or molecule. For ions with inert gas configurations and very similar radii, the conceptual basis for this substitution is easy to grasp if the hypothesis of ionic radius is to have any validity. Thus, substitutions of the type (Na^+, Ca^{2+}), (Mg^{2+}, Al^{3+}), (Li^+, Mg^{2+}), (Al^{3+}, Si^{4+}) are well known. Ions with valence electrons present special problems, yet some form acceptable solid solutions with inert gas configuration ions, e.g., (Ca^{2+}, Ce^{3+}), (Mg^{2+}, Fe^{2+}), (Al^{3+}, Cr^{3+}). Ions with "bumps" on the surface or a lone pair of electrons tend not to form solid solutions, such as As^{3+}, Sb^{3+}, Tl^{1+}, Pb^{2+}, Bi^{3+}. From evidence largely based on structure refinement, several isomorphisms can be advanced:

$Na^{1+}P^{5+} \rightleftharpoons Ca^{2+}Si^{4+}$	example:	$Na_2CaMg(PO_4)_2$	brianite
		$Ca_2CaMg(SiO_4)_2$	merwinite
$Ca^{2+}P^{5+} \rightleftharpoons RE^{3+}Si^{4+}$	example:	$Ca_9Mg(PO_4)_6(PO_3OH)\square_3$	whitlockite
		$RE_9Fe^{3+}(SiO_4)_6(SiO_3OH)(OH)_3$	cerite
$RE^{3+}P^{5+} \rightleftharpoons AC^{4+}Si^{4+}$	example:	$Y(PO_4)$	xenotime
		$U(SiO_4)$	coffinite,

where RE = rare earth and AC = actinide.

More coupled relationships could also be written, and already demonstrate the importance of knowledge of the structure type. Within this family of compounds there appear to be yet further principles. Among phosphates and silicates of rare earths and calcium, a common thread occurs viz. $a \sim 10.7$ Å, $c \sim 3.7n$ Å, rhombohedral or hexagonal symmetry. So far, n=1 (cappelenite), n=3 (zirsinalite), n=4 (hilairite), n=7 (okanoganite), n=10 (cerite, whitlockite), n=12 (steenstrupine, fillowite). It is important to bear in mind that there is no evidence of polytypism among these compounds. Some structures with n>4 are very complex indeed. What is the vehicle for describing them?

The conventional approach would involve description of the entire structure. Drawing of such a structure as a clinographic projection, say, twelve layers in the repeat, would be a difficult task, and even more frustrating for the reader who would try to decipher the vast array of linked polyhedra. An alternative would involve breaking the structure down into components. One such means would break the structure into component slabs or two-sided planes. This vehicle often works well for classic hexagonal and cubic close-packed structures where slices are taken parallel to the close-packed layers or normal to the stacking axis. In fillowite, $18Na_2CaMn_7(PO_4)_6$, for example, twelve layers occur in the structure since n=12. On the other hand, a plan of the structure projected down the c-axis reveals a simple principle based on the packing of rods oriented parallel to the c-axis. These rods when projected and connected to each other reveal an hexagonal $\{6^3\}$ net where only three kinds of rods need to be described. One type of rod at xy=(00) and the second at $({}^1/_3\,{}^1/_3)$ are located at the centers of hexagons and each consists of twelve larger cation polyhedra. The third distinct rod is located at $({}^2/_9\,{}^1/_9)$ and consists of six non-equivalent (PO_4) tetrahedra, each tetrahedron alternating with four independent cations and two vacancies. In this manner, the entire structure can be described as a rod packing with central cation-oxygen polyhedra and circumjacent tetrahedral pinwheels. Fillowite is an extreme case and is therefore singled out as an example, with a=15.3 Å (${}^2/_3$ a ~ 10.2 Å), c=43.5 Å (c/12 ~ 3.6 Å), space group $R\bar{3}$. Space group, unit cell and asymmetric unit atom parameters automatically force the arrangement of rods according to the description above. This structure was discussed in detail by Araki and Moore (1981).

Closer examination of a substantial number of primary phosphate structures shows that a sizeable fraction of these compounds can be best considered as rod packings, related to the glaserite structure type. For example, Fig. 1 in Moore (1982) features a paragenetic outline for crystallization of pegmatite phosphates. Of the 20 highest temperature compounds listed, 10 are based on structures related to glaserite. All such compounds have some intermediate-sized cations such as Na^+, Ca^{2+}, Mn^{2+}. Five more compounds are based on hexagonal close-packing. Thus, ca. 75% of the high temperature compounds possess dense-packed or denser-than-dense-packed structures. These compounds are anhydrous or quite nearly so, refractory, brittle and – in the case of the glaserite-related structures – quite resistant to weathering.

For the glaserite-related arrangements, the structures are based on rod packings down the $\{6^3\}$ hexagonal net. These can be easily represented as the rod components in Roman numerals:

n			
2	Apatite	$2Ca_5(OH)(PO_4)_3$	I: $2[Ca_2]$
			II: $6[Ca(PO_4)]$
			III: $[(OH)_2]$

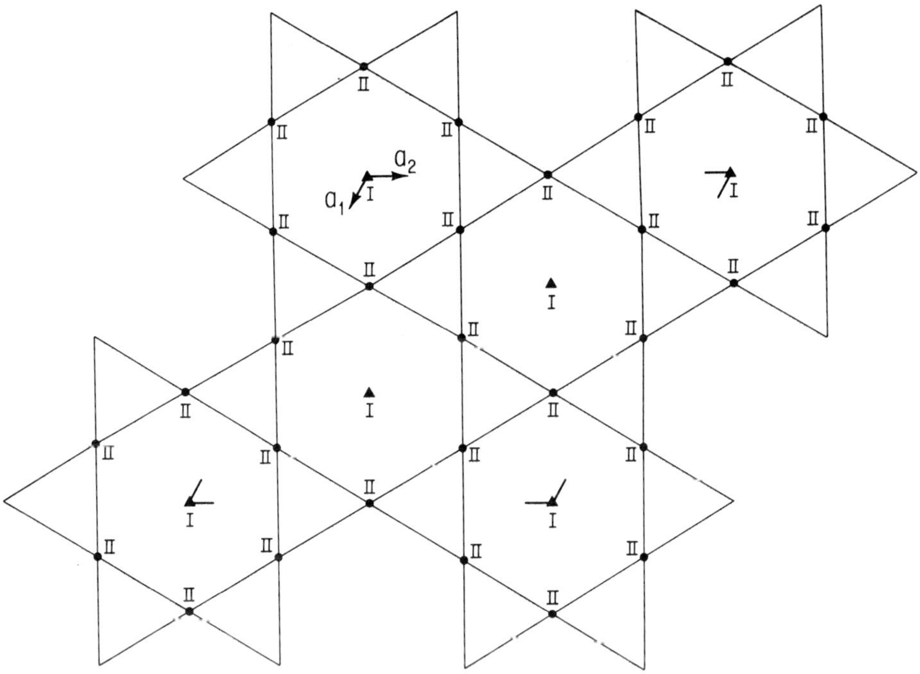

Fig. 3. The Kagomé semi-regular net $\{(6 \cdot 3 \cdot 6 \cdot 3)^4\}$ which is the basis of two kinds of rods in the whitlockite-related structures. *Rod I* is at the centers of hexagons and *rod II* is at the nodes

2 Brianite	$Na_2CaMg(PO_4)_2$	I:	[CaMg]
		II:	2[Na(PO_4)]
4 α-$Ca_3(PO_4)_2$	$6Ca_3(PO_4)_2$	I:	(=I–III): 3[Ca_4]
		II:	(IV–IX): 6[□Ca$(PO_4)_2$]
6 Griphite	$(Mn, Na)_6CaFeAl_2(PO_4)_6F_2$	I:	[□$_2$CaFeAl_2]
		II:	2[$Mn_3(PO_4)_3$]
12 Fillowite	$6Na_2CaMn_7(PO_4)_6$	I:	1[Na_6Mn_6]
		II:	6[□$_2$$Mn_4(PO_4)_6$]
		III:	2[$Ca_3Na_3Mn_6$]

Here the bracketed ions are the repeat components in a rod and the multipliers normalize the regions in the net so that the formula can be obtained by a simple additivity relationship. Also note that rods of type II contain the (PO_4) groups in the structure. Rods of type II occur at the nodes of $\{6^3\}$.

A related net occurs for the whitlockite structure and a family of complex rare earth borosilicates, such as cappelenite, $BaY_6B_6Si_3O_{25}$. It is the Kagomé semi-regular net $\{(6 \cdot 3 \cdot 6 \cdot 3)^4\}$ which states that around each node is an alternation of two hexagons and two triangles and the system is 4-connected.

The 4-connected Kagomé net is shown in Fig. 3. the rods are of two types, rod I which is situated in the centers of hexagons and rod II which occurs at the nodes of the Kagomé net. These loci approximate positions of projected atomic positions in a crystal structure. For hexagonal and rhombohedral crystals, it would correspond to a projection down the c-axis.

n

10 Whitlockite $2Ca_9Mg(PO_4)_6(PO_3OH)$
I: $1[\square_6Mg_2(PO_3OH)_2]$
II: $3[Ca_6(PO_4)_4]$

12 Steenstrupine $Na_{14}Mn_2^{2+}Fe_2^{3+}Th^{4+}RE_6^{3+}(OH)_2[Si_6O_{18}]_2[PO_4]_6[PO_3OH] \cdot 2H_2O$
I: $[Na_2Mn_2^{2+}Fe_2^{3+}Th^{4+}(OH)_2(PO_3OH)(H_2O)_2]$
II: $3[Na_4RE_2Si_4P_2]O_{24}O_{36}$

A truly elephantine structure is possessed by steenstrupine. It is based on twelve units in a rod as like fillowite. Unlike fillowite, it is derived from the Kagomé net. Interest is added to the phase because it is one of the few compounds which possesses both phosphate and silicate radicals. These two units do not polymerize to each other, but occur as insular $[PO_4]$ and $[Si_6O_{18}]$ radicals. In accord with Liebau (1966), the (PO_4) and (SiO_4) units do not polymerize to each other. Rather, the (SiO_4) tetrahedron polymerizes first, to form a $[Si_6O_{18}]$ ring. The relative infrequency of silicophosphates of phosphatosilicates still remains an intriguing problem. Rather, in natural systems phosphates and silicates tend to separate out to form their own distinct coexisting species.

Yet another net, neither regular nor semi-regular, accounts for two primary phosphate phases in pegmatites. One of these, arrojadite-dickinsonite, occasionally occurs as enormous crystals. The net can be constructed of triangles and hexagons but nodes are of two types: $(6 \cdot 6 \cdot 3 \cdot 3)$ and $(6 \cdot 3 \cdot 6 \cdot 3)$. Although the system is 4-connected, the nodes are inhomogeneous. A Schläfli symbol can be written, $\{(6 \cdot 6 \cdot 3 \cdot 3)^4 + (6 \cdot 3 \cdot 6 \cdot 3)^4\}$. Alternatively, the inhomogeneous net can be broken up into the triangular net $\{3^6\}$, or a system composed of hexagons and rhombi where each rhombus arises from the fusion of two triangles at a common edge. Selecting the net based on hexagons and rhombi, the net is still inhomogeneous since it is alternately 3- and 4-connected. Moore et al. (1981) deciphered its complex structure. At least 49 atoms occur in the asymmetric unit.

n

8 Arrojadite $2KNa_4CaFe_{14}Al(PO_4)_{12}F_2$
I: $2[Na_3AlFe_4]$
II: $1[Ca_2\square_6]$
III: $2[KNaFe_2(PO_4)_4]$
IV: $2[Fe_4(PO_4)_4]$
V: $2[Fe_4(PO_4)_4]$
VI: $1[\square_8]$

Stoichiometrically, the cell has 4I, 2II, 4III, 4IV, 4V, 2VI. Adding the contents of each rod and adjusting stoichiometry, the cation formula for arrojadite obtains. The phase wyllieite, $Na_2Fe_2Al(PO_4)_3$, is also a primary high-temperature phase and structurally allied to arrojadite-dickinsonite.

Other phases which are allied to the structures discussed above and which are also primary pegmatite phosphates include graftonite, $CaFe_2(PO_4)_2$; and triploidite, $Mn_2(OH)(PO_4)$. These phases were discussed with the others by Moore (1981) but the hexagonal plan of the net is quite distorted. However, they are exceedingly dense-packed structures, as calculations bore out.

Turning to structures based on the principles of dense-packing, it is opportune to further interject some comments on crystallochemical behavior where appropriate. A fairly complete discussion on phosphate species associated with pegmatites was offered by Moore (1973). Since about half – some 190 – of all known phosphate mineral species are from pegmatites, attention should be focused on these primary phases. The great majority of pegmatite phosphate species appear to be derived from the solid solution series triphylite-lithiophilite, Li(Fe, Mn)(PO_4). The field evidence is apparent: enormous single crystals of this series were subsequently attacked by aqueous rest liquid during consolidation of the pegmatite. The small alkali cation Li^+ is evidently quite leachable and the primary crystals easily suffer hydrothermal attack. (It is not known of this behavior is a consequence of the enormous hydrate radius of Li^+ vs. its small ionic radius.) Retrograde continued attack leads to pseudomorphs of heterosite after triphylite, $\square Fe^{3+}(PO_4)$. Heterosite is based on the same principle of h.c.p. as triphylite but ordered vacancies occur in the cation positions formerly occupied by lithium. Since Fe^{2+} and Mn^{2+} in solid solution possess different redox potentials and since other cations, particularly Mg^{2+}, can also occur in solid solution in the triphylite structure, a great variety of ions and liquid compositions can arise with diminishing temperature: Li^+, Fe^{2+}, Fe^{3+}, Mn^{2+}, Mn^{3+}, Mg^{2+}, $(PO_4)^{3-}$, O^{2-}, OH^-, H_2O°. Ions collected and transported by the aqueous rest liquid can contribute a substantial pool which depends on the particular pegmatite. Common examples in addition are Al^{3+}, Ba^{2+}, Be^{2+}, Ca^{2+}, Cu^{2+}, K^+, Na^+, Si^{4+}, Sr^{2+}, Zn^{2+}, and Zr^{4+}. So with a large total pool of ions, varying pH, bulk composition and diminishing T (from $\sim$800 °C to room temperature), a wide variety of species can arise with time. In addition, conditions are almost never dry, so that crystals of quality suitable for structure investigations are easily obtained. Indeed, the structure analyses of several species were only possible when pegmatite sources for their occurrence were found.

Another common feature is the occurrence of exsolution lamellae in host primary graftonite. Most frequently these are triphylite or sarcopside. Sarcopside, $\square FeFe_2(PO_4)_2$, is a h.c.p. structure with ordered vacancies; it can be easily derived from the structure of triphylite, $LiLiFe_2(PO_4)_2$. Recently, I drew the crystal structure of graftonite, solved and refined by Calvo (1968). I discovered that the graftonite structure possesses a site which corresponds to a hole or vacancy, with mean X–O = 2.13 Å distance. It corresponds to a distorted octahedral site. Mean distance and site shape are quite close to that expected for the Li–O mean distance in an octahedral environment. It is therefore possible that a rather complex series may exist whose site populations may be temperature-dependent:

$\square FeFe_2(PO_4)_2$–$\square FeFe_2(PO_4)_2$–$LiLiFe_2(PO_4)_2$
Fe-graftonite sarcopside triphylite

Calvo (1968) reported the synthetic $Fe_3(PO_4)_2$ with the graftonite structure.

Metasomatic Phases and Mixed-Valence Compounds

Another example, although of rare occurrence in a crystal, is the Li$\rightleftharpoons$Na metasomatic reaction. Perhaps the most interesting is natrophilite, $NaMnPO_4$, which is

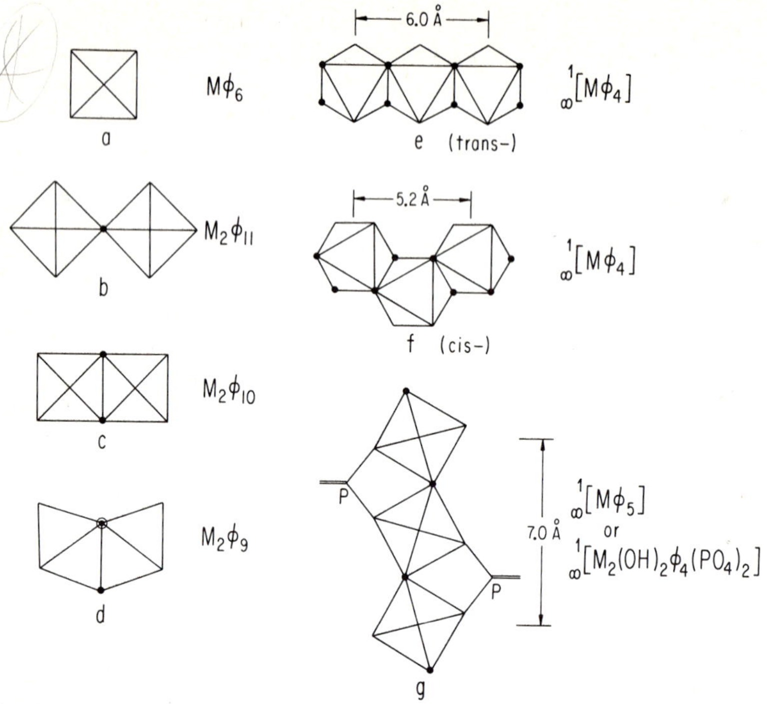

Fig. 4. Finite octahedral clusters and some octahedral chain structures which are the modules for many phosphate minerals. These, and a more extended list, appeared in Moore (1980)

isostructural to lithiophilite, $LiMnPO_4$. Another example is natromontebrasite, $NaAl(OH)(PO_4)$ which is a metasomatic product of montebrasite, $LiAl(OH)(PO_4)$. The amblygonite-montebrasite series, $LiAl(F, OH)(PO_4)$, is the remaining primary phosphate which can occur as large crystals. Its structural principle is based on the 7 Å octahedral corner-chain motif.

Mixed valence compounds form interesting series. One example is the low temperature phase phosphoferrite, $Fe_3^{2+}(H_2O)_3(PO_4)_2$. It can be easily oxidized to kryzhanovskite, $Fe_3^{3+}(OH)_3(PO_4)_2$. Both structures were refined from the same crystal by Moore and Araki (1976). In this structure, stability over the $Fe^{2+} \rightarrow Fe^{3+}$ valence change is evidently a result of each independent (H_2O) ligand being bound to two Fe^{2+} ions. Partial compensation of local change imbalance is achieved by loss of hydrogen. Vivianite, $Fe_3^{2+}(H_2O)_8(PO_4)_2$, is pale green when fresh but becomes progressively blue in color due to oxidation of Fe^{2+}. Homonuclear valence transfer-absorption accounts for this blue color where Fe^{2+}–Fe^{3+} across a shared octahedral edge occurs. The completely oxidized material is red-brown, $Fe_3^{3+}(OH)_3(H_2O)_5(PO_4)_2$. Unfortunately, it is amorphous to X-radiation. McCammon and Burns (1980), on the basis of extensive Mössbauer study, offered a mechanism for its oxidation which involved the destruction of some hydrogen bonds which are essential for structural integrity. It is interesting to note that, unlike phosphoferrite, vivianite has each (H_2O) ligand bound to only one Fe^{2+} ion. Ludlamite, $Fe_3^{2+}(H_2O)_4(PO_4)_2$, another member of the $Fe_3^{2+}(H_2O)_n(PO_4)_2$ homologous series, has a structural relation like phosphoferrite: each (H_2O) is bound to two Fe^{2+} ions. When the phase is completely oxidized, it becomes amorphous to X-radiation. Presumably, for the composition $Fe_3^{3+}(OH)_3(H_2O)(PO_4)_2$, one water molecule would be bound to two Fe^{3+} ions and thus would be seriously oversaturated.

Cluster Types

Turning to certain cluster types which seem to persist among the natural phosphates, Fig. 4 presents idealizations of octahedral clusters. These clusters occur as either insular octahedral units or as fragments of more complex structures. Most of them involve octahedrally coordinated cations $^{[6]}Fe^{3+}$ or $^{[6]}Al^{3+}$ and for this reason were extensively reviewed in Moore (1980). In fact, the greatest fraction of phosphate mineral species involve these and related clusters and account for most species in categories III, IV, V, and VI. For example, the insular octahedron, 4a, forms the basis for strengite and phosphosiderite; the corner-linked dimer 4b for minyulite and morinite; the edge-linked dimer 4c for jagowerite. These clusters occur as finite octahedral units in category VI. The linear edge-sharing chain, trans-${}^1_\infty[M\phi_4]$ and the bent edge-sharing chain, cis-${}^1_\infty[M\phi_4]$, 4e and 4f respectively, occur in bermanite, $Mn^{2+}(H_2O)_4[Mn_2^{3+}(OH)_2(PO_4)_2]$ and foggite, $Ca(H_2O)_2[CaAl_2(OH)_4(PO_4)_2]$. Figure 4g is the basis for the large family of 7 Å corner-chain structures which include category IV and which has 12 structure types as representatives.

Figure 5 illustrates how several categories can be spanned through successive condensations of octahedra. We would expect progressive octahedral condensation with increasing pH. As the concentration of the $(OH)^-$ anion increases, $M^{3+}-(OH)-M^{3+}$ links would be promoted. Indeed, Bandy (1938) observed a succession of progressively basic phases away from the primary sulfide ores. Relating the paragenetic sequence of phases away from the ores and outward toward increasing pH with knowledge of recently determined crystal structures suggests

$$\underset{\text{szomolnokite}}{{}^0_\infty[M(\phi H)_6]} \rightarrow \underset{\text{rhomboclase}}{{}^0_\infty[M\phi_6]} \rightarrow \underset{\text{copiapite}}{{}^0_\infty[M_2\phi_{11}]} \rightarrow \underset{\text{amarantite}}{{}^0_\infty[M_4\phi_{20}]} \rightarrow \underset{\text{botryogen}}{{}^1_\infty[M_2\phi_{10}]} \rightarrow \underset{\text{jarosite}}{{}^2_\infty[M_3\phi_{12}]} .$$

A similar sequence occurs in the phosphorites of west-central Florida and has been documented by Owens et al. (1960):

$$\underset{\text{wavellite}}{{}^1_\infty[M_3\phi_{15}]} \rightarrow \underset{\text{millisite}}{{}^2_\infty[M_3\phi_{14}]} \rightarrow \underset{\text{crandallite}}{{}^2_\infty[M_3\phi_{12}]} .$$

These paragenetic sequences show a progressive condensation of octahedra. In fact, VI (finite clusters)→IV (7 Å corner-sharing chains)→V (octahedral corner-sheets). Since a component can be identified in successive sequences of condensations, the proposition of intrinsically stable clusters is appealing.

Figure 6 represents a complex cluster consisting of three face-sharing octahedra, $M^{3+}M^{2+}M^{3+}$, four corner-sharing octahedra M^{3+} and four (PO_4) tetrahedral links. If such an insular cluster were to exist, it would have composition $[M^{2+}M_6^{3+}(OH)_4(PO_4)_4(H_2O)_{14}]^{4+}$. Homonuclear mixed-valence transfer leads to compounds which are deep greenish-black in color. If central M^{2+} is oxidized, they be-

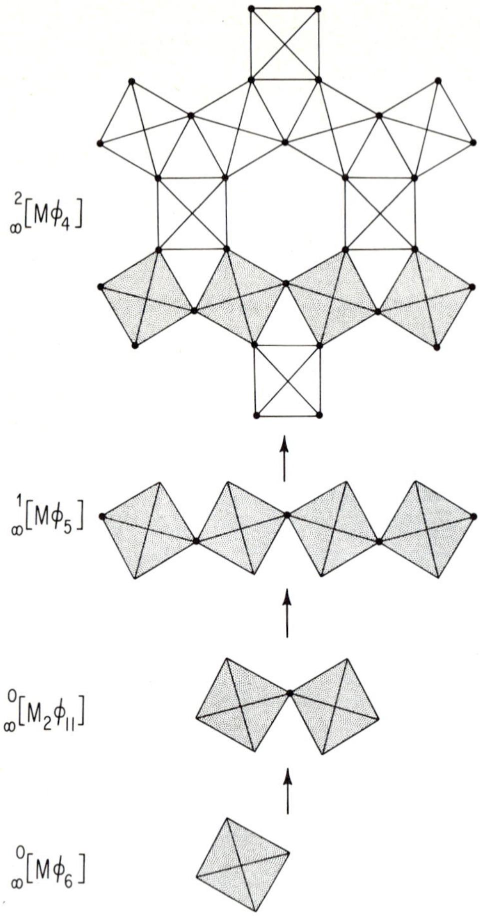

Fig. 5. The progressive polymerization of the $M\phi_6$ octahedron to the corner-dimer $M_2\phi_{11}$ (both in category VI), to the 7 Å corner-chain structure $^{1}_{\infty}[M\phi_5]$ (category IV), to the octahedral corner-sheet structure $^{2}_{\infty}[M\phi_4]$ as in crandallite (category V). Successive polymerized unit stressed by *stippling*

come orange in color. An infinite octahedral face-sharing chain along the horizontal arrows would lead to a thick slab of structure. As yet, no such face-sharing octahedral column has been described among these compounds but could conceivably exist in one of the polymorphs of lipscombite, $Fe^{3+}Fe^{2+}Fe^{3+}(OH)_2(PO_4)_2$. This unit in Fig. 6 can be found as a component in the so-called 5 Å structures, category III. The compounds listed in Table 1 often exist in close association and are often difficult to distinguish for this reason. Note that the inclined arrow indicates the direction of polymerization by corner-fusion of octahedra and 13.8 Å is the usual repeat in this direction for the structures.

Actinide and Lanthanide Phosphates

The uranyl phosphate salts span a considerable number of species and the structures are known for several of them. Most are products formed at low temperature,

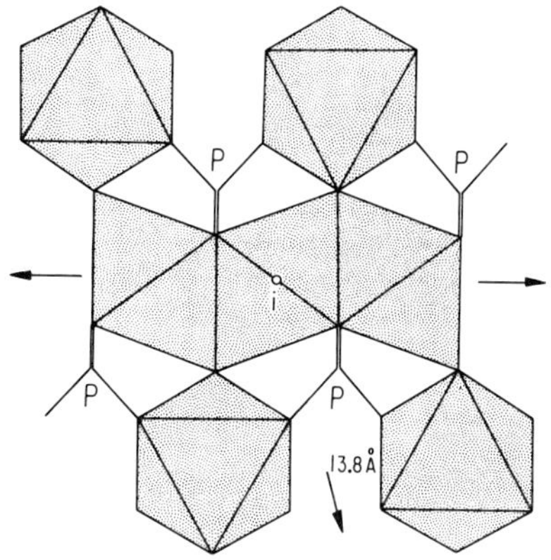

Fig. 6. The cluster involving three face-sharing octahedra M^{3+}–M^{2+}–M^{3+}, four additional corner-linking octahedra and four additional corner-linking tetrahedra drawn as spokes. The 13.8 Å repeat direction of polymerization is shown, as well as inversion center, i. This cluster can be found in the 5 Å structures, category III

and water, either as ligated groups or as channel or pore water of crystallization, plays an important role in structure stability. At least one step in dehydration appears for many of these compounds. Often such dehydration products are prefixed "meta-".

In a series of structural investigations, Ross et al. (1964) studied key sheet structures. For example, in meta-torbernite, $Cu(H_2O)_4[UO_2PO_4]_2^{1-}$, the underlying principle is a sheet of composition $^2_\infty[UO_2PO_4]^{1-}$. This is constructed by four nearly coplanar oxygens from four circumjacent (PO_4) tetrahedra coordinating to central U^{6+} and two additional oxygens above and below this plane linked to U^{6+} to form a compact $U^{6+}O_2O_4$ coordination. The result is a corner-linked sheet comprised of $(UO_2)^{2+}$ "dumbbells" and $(PO_4)^{3-}$ anions or $^2_\infty[UO_2PO_4]^{1-}$. Uranyl oxygens have U^{6+}–O ~ 1.80 Å and U^{6+}-phosphate oxygen U^{6+}–O ~ 2.32 Å. The uranyl phosphate sheets are linked in the third direction by the $Cu^{2+}O_2(H_2O)_4$ Jahn-Teller distorted octahedra where the links are to the uranyl oxygens.

The $(UO_2)^{2+}$ ion is a rather rigid entity and appears to be stabilized by the f atomic orbitals of type f_{z^3}. For this reason, it plays a central role in the stabilities and compositions of the uranyl phosphate structures.

The rare-earth phosphates constitute a small number of species. Several examples of near-isomorphism with Ca^{2+} compounds exist, for example:

Monazite	$CePO_4$	Huttonite	$ThSiO_4$
Xenotime	YPO_4	Zircon	$ZrSiO_4$
Whitlockite	$Ca_9Mg(PO_4)_6(PO_3OH)\square_3$	Cerite	$RE_9Fe^{3+}(SiO_4)_6 \cdot (SiO_3OH)(OH)_3$
Goedkenite	$Sr_2Al(OH)(PO_4)_2$	Törnebohmite	$RE_2Al(OH)(SiO_4)_2$.

Note that there exist isomorphous pairs $RE^{3+}P^{5+} \rightarrow X^{4+}Si^{4+}$ and $X^{2+}P^{5+} \rightarrow RE^{3+}Si^{4+}$ where X = inert gas configuration cation. The majority of natural lanthanide phosphate minerals are primary species.

Ionic size and ionic charge seem to play a major role in understanding their crystal chemistry. From the tables of Shannon and Prewitt (1969), the "lanthanide contraction" is pronounced when comparing the effective ionic radius of $^{[6]}Ce^{3+}$ 1.03 Å with $^{[6]}Lu^{3+}$ 0.85 Å. Thus the light rare earth ions often occur in solid solution with $^{[6]}Ca^{2+}$ 1.00 Å, and the heavy rare earth ions prefer to partition with $^{[6]}Y^{3+}$ 0.89 Å. Cerite presents another interesting effect resulting from increased bond strength, $\zeta = +^3/_8$ for RE^{3+} compared with whitlockite, with $\zeta = +^2/_8$ for Ca^{2+} in the phosphate structure. Although whitlockite and cerite are nearly isostructural, the coordination number of oxygens about RE ions is nine in cerite, a result of the insertion of one additional hydroxyl anion for a vacancy in whitlockite, e.g., $Ca^{2+}\square \rightarrow RE^{3+}(OH)$.

References

Araki T, Moore PB (1981) Fillowite, $Na_2Ca(Mn, Fe)^{2+}_7(PO_4)_6$: Its crystal structure. Am Miner 66:827–842

Bandy MC (1938) Mineralogy of three sulfate deposits of Northern Chile. Am Miner 23:669–760

Calvo C (1968) The crystal structure of graftonite. Am Miner 53:742–750

Ewald PP (1973) Diffraction data and electrostatic energy of a crystal. Am Crystall Assoc Prog Abst 2:122

Goldschmidt VM (1927) Geochemische Verteilungsgesetze der Elemente, vol 8. Untersuchungen über Bau und Eigenschaften von Kristallen. Norske Videnskaps-Akademi i Oslo: 1–156

Hawthorne FC (1979) The crystal structure of morinite. Can Miner 17:93–102

IMPHOS (1980) Second international congress on phosphorus compounds proceedings, April 21–25, Boston 1980. Institut Mondial du Phosphate, p 105–130, 145–163

Liebau F (1966) Die Kristallchemie der Phosphate. Fortschr Miner 42:266–302

Mathew M, Schroeder LW, Dickens B, Brown WE (1977) The crystal structure of α-$Ca_3(PO_4)_2$. Acta Crystallogr Sect B33:1325–1333

McCammon CA, Burns RG (1980) The oxidation mechanism of vivianite as studied by Mössbauer spectroscopy. Am Miner 65:361–366

Moore PB (1970a) Structural hierarchies among minerals containing octahedrally coordinating oxygen: I. Stereoisomerism among corner-sharing octahedral and tetrahedral chains. Neues Jahrb Miner Monatsh H1:163–173

Moore PB (1970b) A crystal-chemical basis for short transition series orthophosphate and orthoarsenate parageneses. Neues Jahrb Miner Monatsh H1:39–44

Moore PB (1973) Pegmatite phosphates: Descriptive mineralogy and crystal chemistry. Miner Rec, May–June 1973: 103–130

Moore PB (1974) Structural hierarchies among minerals containing octahedrally coordinating oxygen. II. Systematic retrieval and classification of octahedral edge-sharing clusters: An epistemological approach. Neues Jahrb Miner Abhand 120:205–227

Moore PB (1980) The natural phosphate minerals: Crystal chemistry. Second international congress on phosphorus compounds proceedings, April 21–25, Boston 1980. Institut Mondial du Phosphate: p 105–130

Moore PB (1981) Complex crystal structures related to glaserite, $K_3Na(SO_4)_2$: Evidence for very dense packings among oxysalts. Bull Soc Fr Minér Cristallogr 104:536–547

Moore PB (1982) Pegmatite minerals of P(V) and B(III). Miner Assoc Can Short Course Handb 8:267–291

Moore PB, Araki T (1976) A mixed-valence solid solution series: Crystal structures of phosphoferrite and kryzhanovskite. Inorg Chem 15:316–321

Moore PB, Araki T, Merlino S, Mellini M, Zanazzi PF (1981) The arrojadite-dickinsonite series, $KNa_4Ca(Fe, Mn)^{2+}_{14}Al(OH)_2(PO_4)_{12}$: Crystal structure and crystal chemistry. Am Miner 66:1034–1049

Owens JP, Altshuler ZS, Berman R (1960) Millisite in phosphorite from Homeland, Florida. Am Miner 45:547–561

Ross M, Evans HT Jr, Appleman DE (1964) Studies of the torbernite minerals (II): The crystal structure of meta-torbernite. Am Miner 49:1603–1621

Shannon RD, Prewitt CT (1969) Effective ionic radii in oxides and fluorides. Acta Crystallogr B25:925–946

Chapter 4 Thermochemical Properties of Phosphates

PHILIPPE VIEILLARD and YVES TARDY

Introduction

This chapter summarizes the thermochemical data of phosphate ions, aqueous complex species as well as minerals occurring in natural environments. The data are presented in such a way as to express all equilibrium reactions as a function of a set of reference aqueous ions. The sources of these data were selected, as far as possible, according to their internal consistency and the newness of the corresponding experimental works.

Thermochemical Data of Reference Aqueous Ions

The data listed in Table 1 are considered as reference for establishing an internal consistency among the thermodynamic data of phosphates. $\Delta H^0_{f\,298.15}$ and $\Delta G^0_{f\,298.15}$ stand respectively for the enthalpy and Gibbs free energy of formation of aqueous ions from the elements, at 298.15 K and 1 bar (10^5 Pascal), total pressure. They are given in kJ mol^{-1}. $S^0_{298.15}$ in J deg^{-1} mol^{-1} stands for the absolute entropy of the aqueous ions. It takes into account the entropy of aqueous hydrogen ion, recently measured and differs from the conventional entropy and the entropy of formation of the aqueous ions from the elements ($\Delta S^0_{f\,298.15}$). The corresponding relationships are the following:

$$S^0_{298.15}(\text{absolute}) = S^0_{298.15}(\text{conventional}) - 20.9\,Z\,,$$

where Z is the ionic charge and -20.9 J deg^{-1} mol^{-1} the value of the absolute entropy of aqueous hydrogen ion. Furthermore:

$$\Delta S^0_{f\,298.15} = S^0_{298.15}\text{ion (abs.)} + Z.S^0_{298.15}(e^-) - S^0_{298.15}\text{element},$$

where $S^0_{298.15}(e^-)$ is the absolute entropy of the electron and S^0 element is the absolute entropy of the corresponding element, with:

$$S^0_{298.15}(e^-) = \tfrac{1}{2}(S^0H_2(g) - 2S^0H^+_{(aq)})\,,$$

$$S^0_{298.15}(e^-) = 86.20 \text{ J deg}^{-1}\text{ mol}^{-1}\,.$$

The values of $S^0_{298.15}$ elements are listed in Table 6 and those of compounds in reference state are listed in Table 2.

Institut de Géologie, Centre de Sédimentologie et de Géochimie de la Surface, 1, rue Blessig, 67084 Strasbourg, France

Table 1. Thermochemical data of reference aqueous ions

Ions	ΔH_f^0 298° 15 kJ mol^{-1}	ΔG_f^0 298° 15 kJ mol^{-1}	S^0 298° 15 J deg^{-1} mol^{-1}	References[a]
H^+	0.0	0.0	− 20.90	(1)
Li^+	− 293.70	−278.60	− 6.60	(1)
Na^+	− 240.10	−261.90	+ 38.10	(2)
K^+	− 252.40	−283.26	+ 81.60	(2)
Rb^+	− 251.17	−283.97	+100.60	(2)
Ag^+	+ 105.58	+ 77.12	+ 51.78	(2)
NH_4^+	− 132.50	− 79.37	+ 92.45	(2)
Tl^+	+ 5.36	− 32.40	+104.60	(2)
Hg^{2+}	+ 172.40	+153.55	+ 42.80	(2)
Mg^{2+}	− 446,85	−454.80	−180.00	(2)
Ca^{2+}	− 542,83	−553.54	− 95.00	(2)
Sr^{2+}	− 545,80	−559.44	− 74.40	(2)
Ba^{2+}	− 537,64	−560.74	− 32.20	(2)
Zn^{2+}	− 153.90	−147.03	−153.80	(2)
Mn^{2+}	− 220.75	−228.00	−115.50	(2)
Cu^{2+}	+ 64.77	+ 65.52	−141.40	(2)
Cd^{2+}	− 75.90	− 77.58	−115.00	(2)
Fe^{2+}	− 89.10	− 78.87	−179.50	(2)
Co^{2+}	− 58.20	− 54.40	−154.90	(2)
Ni^{2+}	− 54.00	− 45.60	−170.70	(2)
Pb^{2+}	− 1.70	− 24.40	− 31.40	(2)
VO^{2+}	− 486.60	−446.40	−175.70	(2)
UO_2^{2+}	−1,018.80	−952.70	−138.90	(3), (4)
Al^{3+}	− 531.37	−485.34	−384.50	(2)
Ga^{3+}	− 211.70	−159.00	−393.30	(2)
In^{3+}	− 101.70	− 97.60	−214.00	(1)
Sc^{3+}	− 614.21	−586.60	−318.00	(2)
Y^{3+}	− 723.40	−693.70	−313.80	(2)
La^{3+}	− 707.10	−683.70	−280.30	(2)
Ce^{3+}	− 696.20	−671.90	−267.80	(2)
Pr^{3+}	− 704.60	−679.10	−272.00	(2)
Nd^{3+}	− 696.20	−671.50	−269.40	(2)
Sm^{3+}	− 691.60	−666.50	−274.50	(2)
Eu^{3+}	− 605.00	−574.00	−284.50	(2)
Gd^{3+}	− 684.50	−661.10	−268.60	(1)
Tb^{3+}	− 682.80	−651.90	−288.70	(2)
Dy^{3+}	− 698.60	−665.20	−293.70	(2)
Ho^{3+}	− 705.00	−673.60	−289.50	(2)
Er^{3+}	− 705.40	−669.00	−307.10	(2)
Tm^{3+}	− 697.90	−661.90	−305.40	(2)
Yb^{3+}	− 674.50	−643.90	−301.20	(2)
Lu^{3+}	− 665.00	−627.60	−326.30	(2)
Cr^{3+}	− 256.10	−215.50	−370.30	(1)
Fe^{3+}	− 48.50	− 4.60	−378.70	(2)
Bi^{3+}	–	+ 82.80	–	(2)
Am^{3+}	− 616.70	−599.10	−260.00	(3)
Ce^{4+}	− 537.20	−503.75	−384.90	(2)
Th^{4+}	− 769.00	−704.60	−507.55	(3)
U^{4+}	− 591.20	−530.90	−496.70	(3), (4)
Pu^{4+}	− 536.40	−481.60	−472.50	(3)
OH^-	− 230.00	−157.29	+ 10.14	(2)
F^-	− 332.63	−278.82	+ 7.11	(2)
Cl^-	− 167.16	−131.26	+ 77.35	(2)
Br^-	− 121.55	−103.97	+103.36	(2)

[a] (1) Barner and Scheuerman (1978), (2) Wagman et al. (1968, 1969, 1971, 1976), Parker et al. (1971), Schumm et al. (1973), (3) Fuger and Oetting (1976), (4) Langmuir (1978)

Table 2. Thermochemical data of H_2 (g), O_2 (g), F_2 (g), Cl_2 (g) and H_2O (1 and g)

Compounds	ΔH^0_f 298° 15 kJ mol^{-1}	ΔG^0_f 298° 15 kJ mol^{-1}	S^0 298° 15 J deg^{-1} mol^{-1}	Reference
O_2 (g)	0	0	205.03	(1)[a]
H_2 (g)	0	0	130.57	(1)
F_2 (g)	0	0	202.67	(1)
Cl_2 (g)	0	0	222.96	(1)
Br_2 (g)	0	0	152.23	(1)
H_2O (l)	−285.83	−237.18	69.91	(1)
H_2O (g)	−241.82	−228.59	188.72	(1)

[a] Wagman et al. (1968)

Finally, $\Delta S^0_{298.15}$ calculated from the above expression is related to enthalpy and Gibbs free energy of formation by the following expression:

$$\Delta G^0_{f298.15} = \Delta H^0_{f298.15} - 298.15\, \Delta S^0_{f298.15}\,.$$

Thermochemical Data of Complex Aqueous Species

The data listed in Table 3 correspond to the complex aqueous species involving orthophosphate ligand and various cations. The Gibbs free energies of formation presented are consistent with the dissociation constants and the Gibbs free energy of formation of reference aqueous ions, given in Table 1.

Additional $\Delta H^0_{f298.15}$ and $S^0_{298.15}$ data of complex aqueous species, when available, are listed, together with thermochemical data of reference ions, in Table 4.

Thermochemical Data of Phosphate Minerals

The enthalpy, Gibbs free energy of formation from the elements, entropy and solubility products of phosphate minerals are summarized in Table 5 and 6.

Solubility reactions and the corresponding solubility products (log Ksp) using the reference ions listed above, are presented in Table 5. These values are derived from solubility measurements and the ones indicated in italics are those which are selected for calculating the Gibbs free energies given in Table 6.

The values of enthalpies and entropies, listed in Table 6, come from calorimetric measurements. Most of the Gibbs free energies are derived from these data, but references noted (0) come from solubility measurements as already mentioned.

Heat Capacity

The heat capacities for the phosphates are listed in Table 7 as follows: $C_{p298.15}$ stands for the heat capacity at 298.15 K; a, b, and c are the Maier and Kelley

Table 3. Gibbs free energy of formation of complex aqueous species, dissociation reactions and logarithm of the dissociation constants

Aqueous Species	ΔG_f^0 298° 15 K kJ mol^{1-}	Reactions	Log K 298° 15	Reference[a]
$H_3PO_4^0$	−1,142.65	$H_3PO_4 \rightarrow H^+ + H_2PO_4^-$	− 2.148	(1), (6)
HPO_4^{2-}	−1,089.26	$HPO_4^{2-} + H^+ \rightarrow H_2PO_4^-$	+ 7.205	(1), (6)
PO_4^{3-}	−1,018.80	$PO_4^{3-} + 2H^+ \rightarrow H_2PO_4^-$	+19.55	(1), (6)
$AmH_2PO_4^{2+}$	−1,743.80	$AmH_2PO_4^{2+} \rightarrow Am^{3+} + H_2PO_4^-$	− 2.51	(3), (6)
$CaH_2PO_4^+$	−1,691.96	$CaH_2PO_4^+ \rightarrow Ca^{2+} + H_2PO_4^-$	− 1.407	(4)
$CaHPO_4^0$	−1,658.28	$CaHPO_4^0 + H^+ \rightarrow Ca^{2+} + H_2PO_4^-$	+ 4.494	(4)
$CaPO_4^-$	−1,609.20	$CaPO_4^- + 2H^+ \rightarrow Ca^{2+} + H_2PO_4^-$	+13.09	(4)
$CeH_2PO_4^{2+}$	−1,815.64	$CeH_2PO_4^{2+} \rightarrow Ce^{3+} + H_2PO_4^-$	− 2.33	(3), (6)
$CePO_4^0$	−1,796.53	$CePO_4^0 + 2H^+ \rightarrow Ce^{3+} + H_2PO_4^-$	+ 1.018	(5), (6)
$CrH_2PO_4^{2+}$	−1,360.50	$CrH_2PO_4^{2+} \rightarrow Cr^{3+} + H_2PO_4^-$	− 2.556	(8)
$CrHPO_4^+$	−1,358.50	$CrHPO_4^+ \rightarrow H^+ + H_2PO_4^-$	− 2.207	(9)
$FeH_2PO_4^+$	−1,224.70	$FeH_2PO_4^+ \rightarrow Fe^{2+} + H_2PO_4^-$	− 2.7	(10), (6)
$FeH_2PO_4^{2+}$	−1,166.00	$FeH_2PO_4^{2+} \rightarrow Fe^{3+} + H_2PO_4^-$	− 5.43	(11)
$FeHPO_4^0$	−1,188.70	$FeHPO_4^0 + H^+ \rightarrow Fe^{2+} + H_2PO_4^-$	− 3.601	(10), (6)
$FeHPO_4^+$	−1,156.00	$FeHPO_4^+ + H^+ \rightarrow Fe^{3+} + H_2PO_4^-$	− 3.681	(12)
$KHPO_4^-$	−1,378.74	$KHPO_4^- + H^+ \rightarrow K^+ + H_2PO_4^-$	+ 6.116	(13)
$MgH_2PO_4^+$	−1,593.80	$MgH_2PO_4^+ \rightarrow Mg^{2+} + H_2PO_4^-$	− 1.508	(13)
$MgHPO_4^0$	−1,560.67	$MgHPO_4^0 + H^+ \rightarrow Mg^{2+} + H_2PO_4^-$	+ 4.296	(14), (6)
$MgPO_4^-$	−1,511.20	$MgPO_4^- + 2H^+ \rightarrow Mg^{2+} + H_2PO_4^-$	−12.964	(13)
$NaHPO_4^-$	−1,358.00	$NaHPO_4^- + H^+ \rightarrow Na^+ + H_2PO_4^-$	+ 6.005	(13)
$PbH_2PO_4^+$	−1,163.35	$PbH_2PO_4^+ \rightarrow Pb^{2+} + H_2PO_4^-$	− 1.50	(15), (6)
$PbHPO_4^0$	−1,131.35	$PbHPO_4^0 + H^+ \rightarrow Pb^{2+} + H_2PO_4^-$	+ 4.11	(15), (6)
$UO_2H_2PO_4^+$	−2,100.40	$UO_2H_2PO_4^+ \rightarrow UO_2^{2+} + H_2PO_4^-$	− 3.03	(17), (16)
$UO_2(H_2PO_4)_2^0$	−3,244.70	$UO_2(H_2PO_4)_2^0 \rightarrow UO_2^{2+} + 2H_2PO_4^-$	− 5.47	(17), (16)
$UO_2(H_2PO_4)_3^-$	−4,384.85	$UO_2(H_2PO_4)_3^- \rightarrow UO_2^{2+} + 3H_2PO_4^-$	− 7.18	(17), (16)
$UO_2HPO_4^0$	−2,089.90	$UO_2HPO_4^0 + H^+ \rightarrow UO_2^{2+} + H_2PO_4^-$	− 1.19	(18), (16)
$UO_2(HPO_4)_2^{2-}$	−3,237.20	$UO_2(HPO_4)_2^{2-} + 2H^+ \rightarrow UO_2^{2+} + 2H_2PO_4^-$	− 4.15	(18), (16)
$YH_2PO_4^{2+}$	−1,839.23	$YH_2PO_4^{2+} \rightarrow Y^{3+} + H_2PO_4^-$	− 2.650	(3), (6)
$ZnH_2PO_4^+$	−1,286.55	$ZnH_2PO_4^+ \rightarrow Zn^{2+} + H_2PO_4^-$	− 1.60	(19)
$ZnHPO_4^0$	−1,255.13	$ZnHPO_4^0 + H^+ \rightarrow Zn^{2+} + H_2PO_4^-$	+ 3.905	(19)

[a] (1) Wagman et al. (1968, 1969, 1971, 1976), Parker et al. (1971), Schumm et al. (1973), (2) Fuger and Oetting (1976), (3) Borisov et al. (1966), (4) Chughtai et al. (1968), (5) Mayer and Schwartz (1950), (6) Smith and Martell (1976), (7) Barner and Scheuerman (1978), (8) Aleshchkina et al. (1976), (9) Sillen and Martell (1971), (10) Nriagu (1972b), (11) Nriagu (1972c), (12) Lahiri (1965), (13) Truesdell and Jones (1974), (14) Taylor et al. (1963b), (15) Nriagu (1972a), (16) Langmuir (1978), (17) Marcus (1958), (18) Moskvin et al. (1967), (19) Nriagu (1973b)

(1932) power function coefficients

$$C_p = a + bT + cT^{-2}$$

and

$$H_T - H_{T_0} = \int_{T_0}^{T} C_p dT,$$

$$S_T - S_{T_0} = \int_{T_0}^{T} \frac{C_p}{T} dT,$$

$$G_T - G_{T_0} = -S_{T_0}(T - T_0) + \int_{T_0}^{T} Cp\, dT - T \int_{T_0}^{T} \frac{Cp}{T} dT.$$

Table 4. Thermochemical data of reference ions and complex orthophosphate aqueous species

Aqueous species	ΔH_f^0 298° 15 kJ mol^{-1}	ΔG_f^0 298° 15 kJ mol^{-1}	S^0 298° 15 J deg^{-1} mol^{-1}	Reference[a]
H_3PO_4	−1,288.34	−1,142.65	158.20	(1)
$H_2PO_4^-$	−1,296.30	−1,130.40	111.30	(1)
HPO_4^{2-}	−1,292.14	−1,089.26	8.40	(1)
PO_4^{3-}	−1,277.38	−1,018.80	−159.00	(1)
Am^{3+}	− 616.70	− 599.10	−260.00	(2)
$AmH_2PO_4^+$	–	−1,743.80	–	(3), (6)
Ca^{2+}	− 542.83	− 553.54	− 95.00	(1)
$CaH_2PO_4^+$	−1,824.90	−1,691.96	91.10	(4)
$CaHPO_4^0$	−1,821.16	−1,658.30	11.55	(4)
$CaPO_4^-$	−1,807.24	−1,609.20	− 85.45	(4)
Ce^{3+}	− 696.20	− 671.95	−268.00	(1)
$Ce(H_2PO_4)^{2+}$	–	−1,815.64	–	(3)
$CePO_4^0$	–	−1,796.53	–	(5), (6)
Cr^{3+}	− 256.10	− 215.50	−370.30	(7)
$CrH_2PO_4^{2+}$	–	−1,360.50	–	(8)
$CrHPO_4^+$	–	−1,358.50	–	(9)
Fe^{2+}	− 89.10	− 78.87	−179.50	(1)
$FeH_2PO_4^+$	–	1,224.70	–	(10), (6)
$FeHPO_4^0$	–	−1,188.70	–	(10)
Fe^{3+}	− 48.53	− 4.60	−378.70	(1)
$FeH_2PO_4^{2+}$	–	−1,166.00	–	(11)
$FeHPO_4^+$	−1,356.44	−1.156.00	−215.10	(12)
K^+	− 252.40	− 283.26	81.60	(1)
$KHPO_4^-$	–	−1.378.74	–	(13)
Mg^{2+}	− 466.85	454.80	−180.00	(1)
$MgH_2PO_4^+$	–	−1,593.80	–	(13)
$MgHPO_4^0$	−1,746.44	−1,560.67	− 75.30	(14), (6)
$MgPO_4^-$	–	−1,511.20	–	(13)
Na^+	− 240.12	− 261.90	38.10	(1)
$NaHPO_4^-$	–	−1,358.00	–	(13)
Pb^{2+}	− 1.70	− 24.40	− 31.40	(1)
$PbH_2PO_4^+$	–	−1,163.35	–	(15)
$PbHPO_4^0$	–	−1,131.35	–	(15)
U^{3+}	− 489.10	− 480.74	−236.50	(2), (16)
U^{4+}	− 591.20	− 530.90	−496.70	(2), (16)
UO_2^{2+}	−1,018.80	− 952.70	−138.90	(2), (16)
$UO_2H_2PO_4^+$	–	−2,100.40	–	(17), (16)
$UO_2(H_2PO_4)_2^0$	–	−3,244.70	–	(17), (16)
$UO_2(H_2PO_4)_3^-$	–	−4,384.85	–	(17), (16)
$UO_2HPO_4^0$	–	−2,089.90	–	(18), (16)
$UO_2(HPO_4)_2^{2-}$	–	−3,237.20	–	(18), (16)
Y^{3+}	− 723.40	− 693.70	−313.80	(1)
$YH_2PO_4^{2+}$	–	−1,839,23	–	(3), (6)
Zn^{2+}	− 153.89	− 147.03	−153.80	(1)
$ZnH_2PO_4^+$	–	−1,286.55	–	(19)
$ZnHPO_4^0$	–	−1,255.13	–	(19)

[a] For references, see footnote to Table 3

Table 5. Log K (solubility products) of phosphate minerals

Compounds	Reactions	Log Ksp 298° 15 K	Reference[a]
Ag_3PO_4	$Ag_3PO_4(c)+2H^+_{(aq)} \rightarrow 3Ag^+_{(aq)}+H_2PO^-_{4(aq)}$	*2.00*	(1), (2)
		− 0.34	(3)
Hg_2HPO_4	$Hg_2HPO_4(c)+H^+_{(aq)} \rightarrow Hg^{2+}_{2(aq)}+H_2PO^-_{4(aq)}$	− *5.20*	(4), (1)
		− 7.32	(5)
$MgHPO_4$	$MgHPO_4(c)+H^+_{(aq)} \rightarrow Mg^{2+}_{(aq)}+H_2PO^-_{4(aq)}$	1.39	(6)
$MgHPO_4 \cdot 3H_2O$ Newberyite	$MgHPO_4 \cdot 3H_2O(c)+H^+_{(aq)} \rightarrow Mg^{2+}_{(aq)}+H_2PO^-_{4(aq)}+3H_2O(1)$	1.38	(7), (1)
$Mg_3PO_4)_2$ Farringtonite	$Mg_3(PO_4)_2(c)+4H^+_{(aq)} \rightarrow 3Mg^{2+}_{(aq)}+2H_2PO^-_{4(aq)}$	15.82	(7)
$Mg_3(PO_4)_2 \cdot 8H_2O$ Bobierrite	$Mg_3(PO_4)_2 \cdot 8H_2O(c)+4H^+_{(aq)} \rightarrow \rightarrow 3Mg^{2+}_{(aq)}+2H_2PO^-_{4(aq)}+8H_2O(1)$	13.90	(8), (1)
$Mg_3(PO_4)_2 \cdot 22H_2O$	$Mg_3(PO_4)_2 \cdot 22H_2O(c)+4H^+_{(aq)} \rightarrow 3Mg^{2+}_{(aq)}+2H_2PO^-_{4(aq)}+22H_2O(1)$	16.00	(8)
$CaHPO_4$ Monetite	$CaHPO_4(c)+H^+_{(aq)} \rightarrow Ca^{2+}_{(aq)}+H_2PO^-_{4(aq)}$	0.55	(9)
		0.31	(10)
$CaHPO_4 \cdot 2H_2O$ Brushite	$CaHPO_4 \cdot 2H_2O(c)+H^+_{(aq)} \rightarrow \rightarrow Ca^{2+}_{(aq)}+H_2PO^-_{4(aq)}+2H_2O(1)$	0.65	(11)
		0.61	(12)
$\beta Ca_3(PO_4)_2$ β Whitlockite	$\beta Ca_3(PO_4)_2+4H^+_{(aq)} \rightarrow 3Ca^{2+}_{(aq)}+2H_2PO^-_{4(aq)}$	10.18	(13)
$SrHPO_4$	$SrHPO_4(c)+H^+_{(aq)} \rightarrow Sr^{2+}_{(aq)}+H_2PO^-_{4(aq)}$	0.29	(14)
		0.24	(15)
$BaHPO_4$	$BaHPO_4(c)+H^+_{(aq)} \rightarrow Ba^{2+}_{(aq)}+H_2PO^-_{4(aq)}$	− 0.20	(14)
$Zn_3(PO_4)_2$	$Zn_3(PO_4)_2(c)+4H^+_{(aq)} \rightarrow 3Zn^{2+}_{(aq)}+2H_2PO^-_{4(aq)}$	7.06	(3)
$\alpha Zn_3(PO_4)_2 \cdot 4H_2O$ α-Hopeite	$\alpha Zn_3(PO_4)_2 \cdot 4H_2O(c)+4H^+_{(aq)} \rightarrow 3Zn^{2+}_{(aq)}+2H_2PO^-_{4(aq)}+4H_2O(1)$	*3.80*	(16), (1)
		8.10	(17)
$Cd_3(PO_4)_2$	$Cd_3(PO_4)_2(c)+4H^+_{(aq)} \rightarrow 3Cd^{2+}_{(aq)}+2H_2PO^-_{4(aq)}$	6.53	(18), (20)
$Cu_3(PO_4)_2$	$Cu_3(PO_4)_2(c)+4H^+_{(aq)} \rightarrow 3Cu^{2+}_{(aq)}+2H_2PO^-_{4(aq)}$	2.21	(18), (20)
$Cu_3(PO_4)_2 \cdot 3H_2O$	$Cu_3(PO_4)_2 \cdot 3H_2O(c)+4H^+_{(aq)} \rightarrow 3Cu^{2+}_{(aq)}+2H_2PO^-_{4(aq)}+3H_2O(1)$	3.98	(19), (20)
$Cu(UO_2)_2(PO_4)_2 \cdot 8H_2O$	$Cu(UO_2)_2(PO_4)_2 8H_2O(c)+4H^+ \rightarrow Cu^{2+}_{(aq)}+2UO^{2+}_{2(aq)}+2H_2PO^-_{4(aq)}+8H_2O(1)$	−12.8	(54)
$Fe_3(PO_4)_2$	$Fe_3(PO_4)_2(c)+4H^+_{(aq)} \rightarrow 3Fe^{2+}_{(aq)}+2H_2PO^-_{4(q)}$	9.20	(21)
$Fe_3(PO_4)_2 \cdot 8H_2O$	$Fe_3(PO_4)_2 \cdot 8H_2O(c)+4H^+_{(aq)} \rightarrow 3Fe^{2+}_{(aq)}+2H_2PO^-_{4(aq)}+8H_2O(1)$	3.70	(22)
		3.10	(23), (1), (53)
$CoHPO_4$	$CoHPO_4(c)+H^+_{(aq)} \rightarrow Co^{2+}_{(aq)}+H_2PO^-_{4(aq)}$	0.49	(18)
$Co_3(PO_4)_2$	$Co_3(PO_4)_2(c)+4H^+_{(aq)} \rightarrow 3Co^{2+}_{(aq)}+2H_2PO^-_{4(aq)}$	4.36	(18)
$Co(UO_2)_2(PO_4)_2 \cdot 7H_2O$	$Co(UO_2)_2(PO_4)_2 7H_2O(c)+4H^+ \rightarrow Co^{2+}_{(aq)}+2UO^{2+}_{2(aq)}+2H_2PO^-_{4(aq)}+7H_2O(1)$	− 9.9	(54)
$Ni_3(PO_4)_2$	$Ni_3(PO_4)_2(c)+4H^+_{(aq)} \rightarrow 3Ni^{2+}_{(aq)}+2H_2PO^-_{4(aq)}$	8.82	(18)
$Ni(UO_2)_2(PO_4)_2 \cdot 7H_2O$	$Ni(UO_2)_2(PO_4)_2 7H_2O(c)+4H^+ \rightarrow Ni^{2+}_{(aq)}+2UO^{2+}_{2(aq)}+2H_2PO^-_{4(aq)}+7H_2O$	− 9.5	(54)
$PbHPO_4$	$PbHPO_4(c)+H^+_{(aq)} \rightarrow Pb^{2+}_{(aq)}+H_2PO^-_{4(aq)}$	− 2.65	(3)
		− 4.16	(24)
		− *4.23*	(25), (1), (20)
$Pb_3(PO_4)_2$	$Pb_3(PO_4)_2(c)+4H^+_{(aq)} \rightarrow 3Pb^{2+}_{(aq)}+2H_2PO^-_{4(aq)}$	− 4.43	(24), (1), (20)
		− *5.30*	(25)
$H_2(UO_2)_2(PO_4)_2 \cdot 10H_2O$	$H_2(UO_2)_2(PO_4)_2 \cdot 10H_2O(c)+2H^+_{(aq)} \rightarrow 2UO^{2+}_{2(aq)}+2H_2PO^-_{4(aq)}+10H_2O(1)$	−11.85	(26)
UO_2HPO_4	$UO_2HPO_4(c)+H^+_{(aq)} \rightarrow UO^{2+}_{2(aq)}+H_2PO^-_{4(aq)}$	− 3.48	(27)
$UO_2HPO_4 \cdot 4H_2O$	$UO_2HPO_4 \cdot 4H_2O(c)+H^+_{(aq)} \rightarrow UO^{2+}_{2(aq)}+H_2PO^-_{4(aq)}+4H_2O(1)$	− 3.42	(28)
$(UO_2)_3(PO_4)_2$	$(UO_2)_3(PO_4)_2(c)+4H^+_{(aq)} \rightarrow 3UO^{2+}_{2(aq)}+2H_2PO^-_{4(aq)}$	− 7.20	(28)

Table 5 (continued)

Compounds	Reactions	Log Ksp 298° 15	Reference[a]
$(VO)_3(PO_4)_2$	$(VO)_3(PO_4)_2(c)+4H^+_{(aq)} \rightarrow 3VO^{2+}_{(aq)}+2H_2PO^-_{4(aq)}$	14.00	(29), (1)
$AlPO_4 \cdot 2H_2O$ Variscite	$AlPO_4 \cdot 2H_2O(c)+2H^+_{(aq)} \rightarrow Al^{3+}_{(aq)}+H_2PO^-_{4(aq}+2H_2O(1)$	− 2.52	(30)
$LaPO_4$	$LaPO_4(c)+2H^+_{(aq)} \rightarrow La^{3+}_{(aq)}+H_2PO^-_{4(aq)}$	− *2.62*	(31)
		− 2.75	(32)
$CePO_4$	$CePO_4(c)+2H^+_{(aq)} \rightarrow Ce^{3+}_{(aq)}+H_2PO^-_{4(aq)}$	− *2.69*	(31)
		− 2.83	(32)
		− 1.75	(33)
$PrPO_4$	$PrPO_4(c)+2H^+_{(aq)} \rightarrow Pr^{3+}_{(aq)}+H_2PO^-_{4(aq)}$	− *2.71*	(34), (31)
		− 2.85	(32)
$NdPO_4$	$NdPO_4(c)+2H^+_{(aq)} \rightarrow Nd^{3+}_{(aq)}+H_2PO^-_{4(aq)}$	− *2.77*	(31)
		− 2.87	(32)
$SmPO_4$	$SmPO_4(c)+2H^+_{(aq)} \rightarrow Sm^{3+}_{(aq)}+H_2PO^-_{4(aq)}$	− *2.83*	(31)
		− 2.95	(32)
$EuPO_4$	$EuPO_4(c)+2H^+_{(aq)} \rightarrow Eu^{3+}_{(aq)}+H_2PO^-_{4(aq)}$	− *2.89*	(31)
		− 2.97	(32)
$GdPO_4$	$GdPO_4(c)+2H^+_{(aq)} \rightarrow Gd^{3+}_{(aq)}+H_2PO^-_{4(aq)}$	− *2.85*	(31)
		− 3.03	(32)
$DyPO_4$	$DyPO_4(c)+2H^+_{(aq)} \rightarrow Dy^{3+}_{(aq)}+H_2PO^-_{4(aq)}$	− *2.87*	(31)
		− 3.01	(32)
$ErPO_4$	$ErPO_4(c)+2H^+_{(aq)} \rightarrow Er^{3+}_{(aq)}+H_2PO^-_{4(aq)}$	− *2.86*	(31)
		− 3.01	(32)
$YbPO_4$	$YbPO_4(c)+2H^+_{(aq)} \rightarrow Yb^{3+}_{(aq)}+H_2PO^-_{4(aq)}$	− *2.81*	(31)
		− 2.95	(32)
$LuPO_4$	$LuPO_4(c)+2H^+_{(aq)} \rightarrow Lu^{3+}_{(aq)}+H_2PO^-_{4(aq)}$	− *2.76*	(31)
		− 2.89	(32)
$CrPO_4$ (green)	$CrPO_4$ (green) $+2H^+_{(aq)} \rightarrow Cr^{3+}_{(aq)}+H_2PO^-_{4(aq)}$	− 3.07	(3)
$CrPO_4$ (purple)	$CrPO_4$ (purple) $+2H^+_{(aq)} \rightarrow Cr^{3+}_{(aq)}+H_2PO^-_{4(aq)}$	2.55	(3)
$FePO_4$	$FePO_4(c)+2H^+_{(aq)} \rightarrow Fe^{3+}_{(aq)}+H_2PO^-_{4(aq)}$	− 2.34	(3)
$FePO_4 \cdot 2H_2O$ Strengite	$FePO_4 \cdot 2H_2O(c)+2H^+_{(aq)} \rightarrow Fe^{3+}_{(aq)}+H_2PO^-_{4(aq)}+2H_2O(1)$	− 5.61, − 7.01	(35), (9)
		− 6.85	(36), (53)
		− 7.65	(37)
$BiPO_4$	$BiPO_4(c)+2H^+_{(aq)} \rightarrow Bi^{3+}_{(aq)}+H_2PO^-_{4(aq)}$	− 3.34	(3)
$Ce_3(PO_4)_4$	$Ce_3(PO_4)_4(c)+8H^+_{(aq)} \rightarrow 3Ce^{4+}_{(aq)}+4H_2PO^-_{4(aq)}$	−11.94	(33)
$Th(HPO_4)_2$	$Th(HPO_4)_2(c)+2H^+_{(aq)} \rightarrow Th^{4+}_{(aq)}+2H_2PO^-_{4(aq)}$	− 6.63	(38)
$Th_3(PO_4)_4$	$Th_3(PO_4)_4(c)+8H^+_{(aq)} \rightarrow 3Th^{4+}_{(aq)}+4H_2PO^-_{4(aq)}$	− 0.40	(27)
$U(HPO_4)_2$	$U(HPO_4)_2(c)+2H^+_{(aq)} \rightarrow U^{4+}_{(aq)}+2H_2PO^-_{4(aq)}$	−13.09	(38)
$Pu(HPO_4)_2$	$Pu(HPO_4)_2(c)+2H^+_{(aq)} \rightarrow Pu^{4+}_{(aq)}+2H_2PO^-_{4(aq)}$	−13.29	(38)
$Ca_8H_2(PO_4)_6 \cdot 5H_2O$ Octacalcium phosphate	$Ca_8H_2(PO_4)_6 \cdot 5H_2O(c)+10H^+_{(aq)} \rightarrow 8Ca^{2+}_{(aq)}+6H_2PO^-_{4(aq)}+5H_2O(1)$	23.48	(39)
		23.35	(40)
$CaU(PO_4)_2 \cdot 2H_2O$ Ningyoite	$CaU(PO_4)_2 \cdot 2H_2O(c)+4H^+_{(aq)} \rightarrow Ca^{2+}_{(aq)}+U^{4+}_{(aq)}+2H_2PO^-_{4(aq)}+2H_2O(1)$	−16.83	(26)
$Ca(UO_2)_2(PO_4)_2 \cdot 10H_2O$	$Ca(UO_2)_2(PO_4)_2 \cdot 10H_2O+4H^+_{(aq)} \rightarrow Ca^{2+}_{(aq)}+2UO^{2+}_{2(aq)}+2H_2PO^-_{4(aq)}+10H_2O(1)$	− 9.72	(26)
$KMgPO_4 \cdot 6H_2O$	$KMgPO_4 \cdot 6H_2O(c)+2H^+_{(aq)} \rightarrow K^+_{(aq)}+Mg^{2+}_{(aq)}+H_2PO^-_{4(aq)}+6H_2O(1)$	8.93	(41)
KUO_2PO_4	$KUO_2PO_4(c)+2H^+_{(aq)} \rightarrow K^+_{(aq)}+UO^{2+}_{2(aq)}+H_2PO^-_{4(aq)}$	− 3.55	(27)
$KUO_2PO_4 \cdot 3H_2O$	$KUO_2PO_4 \cdot 3H_2O(c)+2H^+_{(aq)} \rightarrow K^+_{(aq)}+UO^{2+}_{2(aq)}+H_2PO^-_{4(aq)}+3H_2O(1)$	− 3.56	(28)

Table 5 (continued)

Compounds	Reactions	Log Ksp 298° 15	Reference[a]
$MgNH_4PO_4$	$MgNH_4PO_4(c) + 2H^+_{(aq)} \rightarrow Mg^{2+}_{(aq)} + NH^+_{4(aq)} + H_2PO^-_{4(aq)}$	6.95	(3), (42)
$MgNH_4PO_4 \cdot 6H_2O$ Struvite	$MgNH_4PO_4 \cdot 6H_2O(c) + 2H^+_{(aq)} \rightarrow Mg^{2+}_{(aq)} + NH^+_{4(aq)} + H_2PO^-_{4(aq)} + 6H_2O(1)$	6.40	(41)
$NH_4UO_2PO_4$	$NH_4UO_2PO_4(c) + 2H^+_{(aq)} \rightarrow NH^+_{4(aq)} + UO^{2+}_{2(aq)} + H_2PO^-_{4(aq)}$	− 6.85	(27)
$NH_4UO_2PO_4 \cdot 3H_2O$	$NH_4UO_2PO_4 \cdot 3H_2O(c) + 2H^+_{(aq)} \rightarrow NH^+_{4(aq)} + UO^{2+}_{2(aq)} + H_2PO^-_{4(aq)} + 3H_2O(1)$	− 5.89	(43), (42)
$CaAlH(PO_4)_2 \cdot 6H_2O$ Calcium Aluminium phosphate	$CaAlH(PO_4)_2 \cdot 6H_2O(c) + 4H^+_{(aq)} \rightarrow Ca^{2+}_{(aq)} + Al^{3+}_{(aq)} + 2H_2PO^-_{4(aq)} + 6H_2O(1)$	0.10	(44)
$K_3Al_5(PO_4)_8H_6 \cdot 18H_2O$ Potassium taranakite	$K_3Al_5(PO_4)_8H_6 \cdot 18H_2O(c) + 10H^+_{(aq)} \rightarrow 3K^+_{(aq)} + 5Al^{3+}_{(aq)} + 8H_2PO^-_{4(aq)} + 18H_2O(1)$	−22.30	(45)
$(NH_4)_3Al_5(PO_4)_8H_6 \cdot 18H_2O$ Ammonium taranakite	$(NH_4)_3Al_5(PO_4)_8H_6 \cdot 18H_2O(c) + 10H^+_{(aq)} \rightarrow 3NH^+_{4(aq)} + 5Al^{3+}_{(aq)} + 8H_2PO^-_{4(aq)} + 18H_2O(1)$	−19.10	(45)
$Ca_{10}(PO_4)_6F_2$ Fluor apatite (synthetic)	$Ca_{10}(PO_4)_6F_2(c) + 12H^+_{(aq)} \rightarrow 10Ca^{2+}_{(aq)} + 6H_2PO^-_{4(aq)} + 2F^-_{(aq)}$	− 1.10	(9)
		− 3.57	(46)
		− 1.90	(47), (20)
$Ca_{10}(PO_4)_6(OH)_2$ Hydroxy-apatite (synthetic)	$Ca_{10}(PO_4)_6(OH)_2(c) + 12H^+_{(aq)} \rightarrow 10Ca^{2+}_{(aq)} + 6H_2PO^-_{4(aq)} + 2OH^-_{(aq)}$	*3.60*	(9)
		8.49	(48)
		1.80	(46)
		0.90	(49)
$Ca_{10}(PO_4)_6Cl_2$ Chlorapatite (synthetic)	$Ca_{10}(PO_4)_6Cl_2(c) + 12H^+_{(aq)} \rightarrow 10Ca^{2+}_{(aq)} + 6H_2PO^-_{4(aq)} + 2Cl^-_{(aq)}$	11.14	(46)
$Pb_5(PO_4)_3Cl$ Pyromorphite	$Pb_5(PO_4)_3Cl(c) + 6H^+_{(aq)} \rightarrow 5Pb^{2+}_{(aq)} + 3H_2PO^-_{4(aq)} + Cl^-_{(aq)}$	−20.47	(24)
		−10.35	(50)
		−25.75	(51), (20)
$Pb_5(PO_4)_3(OH)$ Hydroxy-pyromorphite	$Pb_5(PO_4)_3(OH)(c) + 6H^+_{(aq)} \rightarrow 5Pb^{2+}_{(aq)} + 3H_2PO^-_{4(aq)} + OH^-_{(aq)}$	−18.15	(25), (20)
$NH_4Al_2(PO_4)_2OH \cdot 2H_2O$	$NH_4Al_2(PO_4)_2OH \cdot 2H_2O(c) + 4H^+_{(aq)} \rightarrow NH^+_{4(aq)} + 2Al^{3+}_{(aq)} + 2H_2PO^-_{4(aq)} + OH^-_{(aq)} + 2H_2O(1)$	−17.90	(52)
$KAl_2(PO_4)_2OH \cdot 2H_2O$ Minyulite	$KAl_2(PO_4)_2OH \cdot 2H_2O(c) + 4H^+_{(aq)} \rightarrow K^+_{(aq)} + 2Al^{3+}_{(aq)} + 2H_2PO^-_{4(aq)} + OH^-_{(aq)} + 2H_2O(1)$	−15.90	(52)
$Cu_5O_2(PO_4)_2 \cdot 6H_2O$	$Cu_5O_2(PO_4)_2 \cdot 6H_2O(c) + 6H^+_{(aq)} \rightarrow 5Cu^{2+}_{(aq)} + 2H_2PO^-_{4(aq)} + 2OH^-_{(aq)} + 6H_2O(1)$	− 8.53	(17)
$Ca_{9.54}Na_{0.33}Mg_{0.13} \cdot (PO_4)_{4.8} (CO_3)_{1.2}F_{2.48}$ Carbonate-apatite (C.A.)	$(C.A)(c) + 4.8H^+_{(aq)} \rightarrow 9.54Ca^{2+}_{(aq)} + 0.33Na^+_{(aq)} + 0.13Mg^{2+}_{(aq)} + 1.2CO^{2-}_{3(aq)} + 4.8H_2PO^-_{4(aq)} + 2.48F^-_{(aq)}$	−20.56	(55)

[a] (1) Smith and Martell (1976), (2) Baldwin (1969), (3) Zharovskii (1951), (4) de Vries and Cohen (1949), (5) Brodsky (1929), (6) Webber and Racz (1970), (7) Racz and Soper (1968), (8) Taylor et al. (1963b), (9) Lindsay and Moreno (1960), (10) Mac Dowell et al. (1971), (11) Moreno et al. (1960a), (12) Patel et al. (1974), (13) Gregory et al. (1974), (14) Spitsyn et al. (1966), (15) Aia et al. (1964), (16) Nriagu (1973b), (17) Jurinak and Inouye (1962), (18) Chukhlantsev and Alyamovskaya (1961b), (19) Mills (1974), (20) Ball et al. (1980), (21) Singer (1972), (22) Rosenquist (1970), (23) Nriagu (1972b),

Table 6. Thermochemical data of phosphate minerals

Crystalline species	ΔH_f^0 298° 15 K kJ mol^{-1}	ΔG_f^0 298° 15 K kJ mol^{-1}	S^0 298° 15 K J deg^{-1} mol^{-1}	Original[a] References	Listed Reference
P (white)	0.0	0.0	41.10		(1)
P_4O_{10} (hex.)	−3,010.00	–	–	(8)	
	–	−2,723.60	228.80	(9)	
H_3PO_4	−1,279.00	−1,119.20	110.50		(1)
$2H_3PO_4 \cdot H_2O$	−2,862.70	−2,484.50	258.30		(1)
Li	0.0	0.0	29.08		(7)
Li_2O	− 598.52	− 561.90	37.90		(7)
LiH_2PO_4	−1,586.60	–	–	(10)	
Li_3PO_4 Lithiophosphatite	−2,092.00	–	–		(7)
Na	0.0	0.0	51.34		(7)
Na_2O	418.00	379.10	75.06		(7)
NaH_2PO_4	−1,543.90	–	–	(11)	
	–	–	127.50	(12)	
Na_2HPO_4	−1,754.80	–	–	(11)	
	–	–	150.50	(12)	
$Na_2HPO_4 \cdot 2H_2O$	–	–	221.30	(13)	
$Na_2HPO_4 \cdot 7H_2O$	–	–	434.60	(13)	
$Na_2HPO_4 \cdot 12H_2O$	–	–	621.60	(13)	
Na_3PO_4	−1,913.60	–	–	(11)	
	–	−1,785.00	173.80	(12)	
K	0.0	0.0	64.20		(6)
K_2O	− 361.50	–	–	(6)	
KH_2PO_4	−1,570.70	−1,418.30	134.85		(6)
K_2HPO_4	−1,779.90	–	–	(15)	
	–	–	179.10	(14)	
Rb	0.0	0.0	76.78		(6)
Rb_2O	− 338.90	–	–		(6)
RbH_2PO_4	−1,569.70	–	–	(16)	
	−1,562.90	–	–		(6)
Ag	0.0	0.0	42.50		(2)
Ag_2O	− 31.05	− 11.20	121.30		(2)
Ag_3PO_4	−1,005.00	–	–	(17)	
	–	− 887.60	–	(0)	
N_2 (g)	0.0	0.0	191.50		(1)
$NH_4H_2PO_4$	−1,445.10	−1,210.56	152.00		(1)

(24) Jowett and Price (1932), (25) Nriagu (1972a), (26) Muto (1965), (27) Chukhkantsev and Stepanov (1956), (28) Karpov (1961), (29) Zolotavin and Kuznetsova (1956), (30) Lindsay et al. (1959), (31) Marinova and Yaglov (1976), (32) Ousubaliev et al. (1975), (33) Chukhlantsev and Alyamovskaya (1961a), (34) Marinova and Rud'ko (1975), (35) Chang and Jackson (1957), (36) Nriagu (1972c), (37) Kharaka (1973), (38) Denotkina et al. (1960), (39) Moreno et al. (1960b), (40) Patel and Brown (1975), (41) Taylor et al. (1963a), (42) Sillen and Martell (1971), (43) Klygin et al. (1961), (44) Taylor and Gurney (1964a), (45) Taylor and Gurney (1961), (46) Valyashko et al. (1968), (47) McCann (1968), (48) Brown (1960), (49) Avnimelech et al. (1973), (50) Baker (1964), (51) Nriagu (1973a), (52) Taylor and Gurney (1964b), (53) Truesdell and Jones (1964), (54) Vochten et al. (1981), (55) Chien and Black (1976)

Table 6 (continued)

Crystalline species	ΔH_f^0 298° 15 K kJ mol^{-1}	ΔG_f^0 298° 15 K kJ mol^{-1}	S^0 298° 15 K J deg^{-1} mol^{-1}	Original[a] Refer-ences	Listed Refer-ence
$(NH_4)_2HPO_4$	−1,566.90	–	–		(1)
	–	–	177.00	(18)	
$(NH_4)_3PO_4$	−1,671.90	–	–		(1)
$(NH_4)_3PO_4 \cdot 3H_2O$	−2,555.60	–	–		(1)
Tl	0.0	0.0	64.20		(1)
Tl_2O	− 178.70	− 147.30	125.50		(1)
Tl_3PO_4	–	–	270.50	(19)	
Hg (1)	0.0	0.0	76.00		(2)
Hg_2O	–	–	–		
Hg_2HPO_4	–	−1,006.50	–	(0)	
Mg	0.0	0.0	32.68		(4)
MgO	− 601.70	− 569.45	26.94		(4)
$Mg(H_2PO_4)_2$	–	−2,730.10	–	(20)	
$Mg(H_2PO_4)_2 \cdot 2H_2O$	–	−3,200.30	–	(20)	
$Mg(H_2PO_4)_2 \cdot 4H_2O$	–	−3,674.40	–	(20)	
$MgHPO_4$	–	−1,577.30	–	(0)	
$MgHPO_4 \cdot 3H_2O$	–	−2,288.90	–	(0)	
Newberyite		−2,297.00		(20)	
$Mg_3(PO_4)_2$	−3,780.66	−3,538.80	189.20		(4)
Farringtonite		−3,534.90		(0)	
$Mg_3(PO_4)_2 \cdot 8H_2O$	–	−5,450.50	–	(20)	
Bobierrite		−5,443.40		(0)	
$Mg_3(PO_4)_2 \cdot 22H_2O$	–	−8,751.80	–	(0)	
Ca	0.0	0.0	41.42		(4)
CaO	− 635.10	− 604.04	39.75		(4)
$Ca(H_2PO_4)_2$	−3,104.10	–	–		(4)
	–	−2,824.80	–	(21)	
$Ca(H_2PO_4)_2 \cdot H_2O$	−3,409.70	−3,058.40	259.83		(4)
$CaHPO_4$ Monetite	−1,814.40	−1,681.30	111.40		(4)
$CaHPO_4 \cdot 2H_2O$ Brushite	−2,403.60	−2,154.75	189.45		(4)
$\alpha Ca_3(PO_4)_2$ High whitlockite	−4,109.95	−3,875.65	240.90		(4)
$\beta Ca_3(PO_4)_2$ Low whitlockite	−4,120.82	−3,884.82	236.00		(4)
Sr	0.0	0.0	52.30		(4)
SrO	− 592.00	− 561.90	54.40		(4)
$Sr(H_2PO_4)_2$	−3,134.60	–	–		(4)
$SrHPO_4$	−1,821.70	−1,688.70	121.30		(4)
$Sr_3(PO_4)_2$	−4,122.90	–	–		(4)
Ba	0.0	0.0	62.80		(4)
BaO	− 553.50	− 525.10	70.40		(4)
$Ba(H_2PO_4)_2$	−3,125.00	–	–		(4)
$BaHPO_4$	−1,814.60	–	–		(4)
	–	−1,692.30	–	(0)	
$Ba_3(PO_4)_2$ (am.)	−4,092.00	–	–		(4)
Zn	0.0	0.0	41.63		(1)
ZnO	− 348.28	− 318.30	43.64		(1)

Table 6 (continued)

Crystalline species	ΔH_f^0 298° 15 K kJ mol^{-1}	ΔG_f^0 298° 15 K kJ mol^{-1}	S^0 298° 15 K J deg^{-1} mol^{-1}	Original[a] References	Listed Reference
$Zn_3(PO_4)_2$	−2,899.60	−2,633.40	–	(23)	
$Zn_3(PO_4)_2 \cdot H_2O$	−3,211.70	−2,890.90	–	(24)	
$Zn_3(PO_4)_2 \cdot 2H_2O$	−3,516.30	−3,143.30	–	(25)	
para $Zn_3(PO_4)_2 \cdot 4H_2O$	−4,077.75	−3,606.30	–	(26)	
para Hopeite					
$\alpha Zn_3(PO_4)_2 \cdot 4H_2O$	–	−3,628.90	–	(0)	
α Hopeite	−4,102.00	–	–	(27)	
$\beta Zn_3(PO_4)_2 \cdot 4H_2O$	−4,091.50	–	–	(22)	
β Hopeite	–	−3,616.40	–	(28)	
Mn	0.0	0.0	32.00		(2)
MnO	− 385.20	− 362.90	59.70		(2)
$MnHPO_4$		−1,391.20	–		(2)
$Mn_3(PO_4)_2$(c)	−3,116.70	–	–		(2)
$Mn_3(PO_4)_2$(am)	−3,076.50	–	–	(29)	
$Mn_3(PO_4)_2 \cdot 3H_2O$	−3,979.00	–	–	(22)	
$Mn_3(PO_4)_2 \cdot 6H_2O$	−4,863.50	–	–	(22)	
Cd	0.0	0.0	51.76		(1)
CdO	− 258.20	− 228.40	54.80		(1)
$Cd_3(PO_4)_2$	–	−2,456.30	–	(0)	(1)
Cu	0.0	0.0	33.15		(2)
CuO	− 157.30	− 129.70	42.63		(2)
$Cu(H_2PO_4)_2$	−2,526.70	–	–	(30)	
$Cu_3(PO_4)_2$	–	−2,051.60	–	(0)	
	−2,231.40	–	–	(31)	
$Cu_3(PO_4)_2 \cdot 2H_2O$	−2,842.20	–	–	(22)	
	–	−2,564.80		(32)	
$Cu_3(PO_4)_2 \cdot 3H_2O$	−3,139.70	–	–	(22)	
	–	−2,753.05		(0)	
$Cu(UO_2)_2(PO_4)_2 \cdot 8H_2O$	–	−6,071.20	–	(0)	
Fe	0.0	0.0	27.30		(2)
FeO	− 272.00	–	–		(2)
$Fe_3(PO_4)_2$	–	−2,444.80	–	(0)	
$Fe_3(PO_4)_2 \cdot 8H_2O$ Vivianite	–	−4,377.20	–	(0)	(33)
Co	0.0	0.0	30.04		(2)
CoO	− 237.94	− 214.20	52.95		(2)
$CoHPO_4$	–	−1,182.00	–	(0)	(2)
$CoHPO_4 \cdot 1.5H_2O$	–	−1,575.10	–	(34)	
$Co_3(PO_4)_2$	–	−2,399.10	–	(0)	(2)
$Co(UO_2)_2(PO_4)_2 \cdot 7H_2O$	–	−5.937.40	–	(0)	
Ni	0.0	0.0	29.87		(2)
NiO	− 239.70	− 211.70	38.00		(2)
$Ni_3(PO_4)_2$	–	−2,347.30	–	(0)	
$Ni(UO_2)_2(PO_4)_2 \cdot 7H_2O$	–	−5,926.30	–	(0)	
Pb	0.0	0.0	64.80		(1)
PbO (red)	− 219.00	− 188.95	66.50		(1)
$PbHPO_4$	–	−1,178.90	–	(0)	
$Pb_3(PO_4)_2$	–	−2,364.00	–	(0)	
	–	–	353.34		(1)

Table 6 (continued)

Crystalline species	ΔH_f^0 298° 15 K kJ mol^{-1}	ΔG_f^0 298° 15 K kJ mol^{-1}	S^0 298° 15 K J deg^{-1} mol^{-1}	Original[a] References	Listed Reference
U	0.0	0.0	50.20	(35)	(36)
γUO_3	– 1,223.80	– 1,146.00	96.10		(36)
$H_2(UO_2)_2(PO_4)_2$	–	– 4,217.50	–	(37)	(36)
Hydrogen autunite					
$H_2(UO_2)_2(PO_4)_2 \cdot 10H_2O$	–	– 6,605.60	–	(0)	
UO_2HPO_4	–	– 2,102.90	–	(0)	
$UO_2HPO_4 \cdot 4H_2O$	–	– 3,051.30	–	(0)	
$(UO_2)_3(PO_4)_2$	–	– 5,160.00	–	(0)	
V	0	0	28.90		(3)
V_2O_4	– 1,427.20	– 1,318.40	102.50		(3)
$(VO)_3(PO_4)_2$	–	– 3,520.10	–	(0)	
Al	0.0	0.0	28.33		(1)
Al_2O_3	– 1,675.70	– 1,582.40	50.90		(1)
$AlPO_4$	– 1,733.85	– 1,601.20	90.80		(1)
Berlinite			–		
$AlPO_4 \cdot 2H_2O$	–	– 2,104.20	–	(0)	
Variscite	–	–	134.50	(38)	
Ga	0.0	0.0	40.88		(1)
Ga_2O_3	– 1,089.10	– 998.30	84.98		(1)
$GaPO_4$	–	– 1,297.50	–		(1)
	–	–	105.00	(39)	
Sc	0.0	0.0	34.64		(3)
Sc_2O_3	– 1,908.80	– 1,819.40	77.00		(3)
$ScPO_4$	–	–	–		
$ScPO_4 \cdot 2H_2O$	– 2,421.00	–	–	(40)	
Y	0.0	0.0	44.43		(3)
Y_2O_3	– 1,905.30	– 1,816.65	99.08		(3)
YPO_4	–	–	–		
La	0.0	0.0	56.90		(5)
La_2O_3	– 1,793.70	– 1,705.80	127.30		(5)
$LaPO_4$	–	– 1,829.00	–	(0)	
Ce	0.0	0.0	72.00		(5)
Ce_2O_3	– 1,796.20	– 1,706.20	150.60		(5)
$CePO_4$	–	– 1,817.70	–	(0)	
Monazite	– 1,942.60	–	–	(41)	
Pr	0.0	0.0	73.22		(5)
Pr_2O_3	– 1,809.60	–	–		(5)
$PrPO_4$	–	– 1,825.00	–	(0)	
Nd	0.0	0.0	71.55		(5)
Nd_2O_3	– 1,807.90	– 1,720.90	158.60		(5)
$NdPO_4$	–	– 1,817.70	–	(0)	
Sm	0.0	0.0	69.60		(5)
Sm_2O_3	– 1,823.00	– 1,734.70	151.00		(5)
$SmPO_4$	–	– 1,813.00	–	(0)	
Eu	0.0	0.0	77.78		(5)
Eu_2O_3 cub.	– 1,662.70	–	–		(5)
$EuPO_4$	–	– 1,720.90	–	(0)	

Table 6 (continued)

Crystalline species	ΔH_f^0 298° 15 K kJ mol^{-1}	ΔG_f^0 298° 15 K kJ mol^{-1}	S^0 298° 15 K J deg^{-1} mol^{-1}	Original[a] References	Listed Reference
Gd	0.0	0.0	68.07		(5)
Gd_2O_3	−1,819.60	–	–		(5)
$GdPO_4$	–	−1,807.80	–	(0)	
$GdPO_4 \cdot H_2O$	–	−2,050.16	–		(5)
Tb	0.0	0.0	73.20		(5)
Tb_2O_3	−1,865.20	–	–		(5)
$TbPO_4$	−1,924.64	–	–	(42)	
$TbPO_4 \cdot \frac{1}{2}H_2O$	−2,082.38	–	–	(42)	
Dy	0.0	0.0	75.30		(5)
Dy_2O_3	−1,863.10	−1,771.50	149.80		(5)
$DyPO_4$	–	−1,812.00	–	(0)	
Er	0.0	0.0	73.16		(5)
Er_2O_3	−1,897.90	−1,808.70	155.60		(5)
$ErPO_4$	–	−1,815.70	–	(0)	
Yb	0.0	0.0	59.87		(5)
Yb_2O_3	−1,814.60	−1,726.70	133.00		(5)
$YbPO_4$	–	−1,790.30	–	(0)	
Lu	0.0	0.0	50.95		(5)
Lu_2O_3	−1,878.20	−1,789.10	109.95		(5)
$LuPO_4$	–	−1,773.70	–	(0)	
Cr	0.0	0.0	23.77		(2)
Cr_2O_3	−1,139.70	−1,058.10	81.20		(2)
$CrPO_4$ (green)	–	−1,363.40	–	(0)	
$CrPO_4$ (purple)	–	−1,331.30	–	(0)	
$Cr(H_2O)_6PO_4$ (purple)	−3,233.40	–	–	(43)	
Fe	0.0	0.0	27.30		(2)
Fe_2O_3	− 824.20	− 742.20	87.40		(2)
$Fe(H_2PO_4)_3$	−3,875.20	–	–	(44)	
$Fe(H_2PO_4)_3 \cdot H_2O$	−4,162.70	–	–	(44)	
$FePO_4$	−1,297.50	–	–		(2)
Heterosite	–	−1,148.40	–	(0)	
$FePO_4 \cdot 2H_2O$ Strengite	−1,888.20	−1,657.70	171.25		(2)
V	0.0	0.0	28.90		(3)
V_2O_5	−1,550.60	−1,419.60	131.00		(3)
$VOPO_4$	−1,658.50	–	125.10	(45)	
$VOPO_4 \cdot H_2O$	−1,954.30	–	–	(46)	
$VOPO_4 \cdot 2H_2O$	−2,242.60	–	–	(45)	
Bi	0.0	0.0	56.74		(1)
Bi_2O_3	− 573.88	− 493.70	151.50		(1)
$BiPO_4$	–	−1,066.70	–	(0)	
Ce	0.0	0.0	71.96		(5)
CeO_2	−1,088.68	−1,024.66	62.30		(5)
$Ce_3(PO_4)_4$	−6,714.90	–	–	(47)	
Zr	0.0	0.0	39.00		(3)
ZrO_2	−1,100.56	−1,042.80	50.38		(3)
$Zr(HPO_4)_2$	−3,167.90	–	–	(48)	

Table 6 (continued)

Crystalline species	ΔH_f^0 298° 15 K kJ mol^{-1}	ΔG_f^0 298° 15 K kJ mol^{-1}	S^0 298° 15 K J deg^{-1} mol^{-1}	Original[a] References	Listed Reference
$Zr(HPO_4)_2 \cdot H_2O$	−3,463.70	–	–	(48)	
$Zr(HPO_4)_2 \cdot H_2O$ (am)	−3,435.80	–	–	(48)	
$Zr(HPO_4)_2 \cdot 2H_2O$	−3,741.20	–	–	(48)	
Th	0.0	0.0	53.40		(7)
ThO_2	–	−1,169.00	–		(7)
$Th(HPO_4)_2$	–	−3,003.20	–	(0)	
$Th_3(PO_4)_4$	–	−6,637.70	–	(0)	
U	0.0	0.0	50.21		(51)
UO_2	−1,084.91	−1,031.77	77.03		(52)
$U(HPO_4)_2$	–	−2,866.40	–	(0)	
Pu	0.0	0.0	56.15		(23)
PuO_2	–	–	–		
$Pu(HPO_4)_2$	–	−2,818.20	–	(0)	
$NaCaPO_4$	−2,004.14	–	–	(49)	
$Ca_8H_2(PO_4)_6 \cdot 5H_2O$	–	−12,263.30	–		(4)
Octacalcium phosphate	–	−12,261.20	–	(50)	
$CaU(PO_4)_2 \cdot 2H_2O$	−4,304.60	–	–	(51)	
Ningyoite	–	−3,915.60	–	(0)	
$Ca(UO_2)_2(PO_4)_2 \cdot 10H_2O$	–	−7,147.00		(0)	
Autunite					
$KMgPO_4 \cdot 6H_2O$	–	−3,241.00	–	(0)	
	−3,724.30	–	330.00	(52)	
KUO_2PO_4	–	−2,386.60	–	(0)	
$KUO_2PO_4 \cdot 3H_2O$	–	−3,098.20	–	(0)	
$MgNH_4 \cdot PO_4$	–	−1,624.90	–	(0)	
$MgNH_4 \cdot PO_4 \cdot 6H_2O$	−3,681.92	–	–		(4)
Struvite	–	−3,051.10	–	(0)	
$NH_4UO_2PO_4$	–	−2,201.60	–	(0)	
$NH_4UO_2PO_4 \cdot 3H_2O$	–	−2,907.60	–	(0)	
$Fe_3KH_8(PO_4)_6 \cdot 6H_2O$	–	–	929.60	(53)	(7)
$AlCaH(PO_4)_2 \cdot 6H_2O$	–	−4,722.20	–	(0)	
$K_3Al_5(PO_4)_8H_6 \cdot 18H_2O$	−18,908.20	−16,760.80	1,328.10	(54)	
Potassium Taranakite					
$(NH_4)_3Al_5(PO_4)_8H_6 \cdot 18H_2O$	−18,532.60	−16,129.15	1,347.50	(55)	
Ammonium Taranakite					
$Ca_{10}O(PO_4)_6$	–	−12,307.80	–	(56)	
Oxyapatite					
$Ca_{10}(PO_4)_6F_2$ (nat.)	−13,744.00	−12,983.00	775.70		(4)
Fluorapatite (synth.)	–	−12,886.00	–	(0)	
(pp)	–	−12,788.30	–	(57)	
$Ca_{10}(PO_4)_6(OH)_2$ (nat.)	−13,477.00	−12,677.00	780.70		(4)
Hydroxylapatite (synth.)	–	−12,611.80	–	(0)	
(pp)	–	−12,533.35	–	(57)	
$Ca_{10}(PO_4)_6(Cl_2)$ (synth.)	−13,272.00	−12,514.00	795.80	(0), (58)	
Chloroapatite (pp)	–	−12,447.50	–	(59)	
$Ca_{10}(PO_4)_6Br_2$ (pp)	–	−12,382.70	–	(60)	
Bromoapatite					

Table 6 (continued)

Crystalline species	ΔH_f^0 298° 15 K kJ mol^{-1}	ΔG_f^0 298° 15 K kJ mol^{-1}	S^0 298° 15 K J deg^{-1} mol^{-1}	Original[a] References	Listed Reference
$Mg_5(PO_4)_3F$	–	– 5,854.25	–	(20)	
$Mg_5(PO_4)_3(OH)$	–	– 5,758.00	–	(20)	
$Pb_5(PO_4)_3Cl$ Pyromorphite	–	– 3,791.50	– –	(0)	
$Pb_5(PO_4)_3OH$ Hydroxy-pyromorphite	–	– 3,774.00	–	(0)	
$Mg_4O(PO_4)_2$ Tetramagnesium phosphate	–	– 4,172.70		(20)	
$Ca_4O(PO_4)_2$ Hilgenstockite	–	– 4,588.00	–	(61)	
$Pb_4O(PO_4)_2$	–	– 2,582.80	–	(62)	
$NH_4Al_2(PO_4)_2(OH) \cdot 2H_2O$	–	– 4,044.70	–	(0)	
$KAl_2(PO_4)_2(OH) \cdot 2H_2O$	–	– 4,237.10	–	(0)	
Minyulite	–	–	294.68	(63)	(7)
Mg_2FPO_4 Wagnerite	–	– 2,305.40	–	(20)	
Ca_2FPO_4 Spodiosite	–	– 2,522.12	–	(64)	
Ca_2ClPO_4	–	– 2,371.70	–	(59)	
Ca_2BrPO_4	–	– 2,349.80	–	(60)	
$Al_4(PO_4)_3(OH)_3$ Trolleite	– 6,567.67	– 6,077.47	270.70	(65)	
$Al_2PO_4(OH)_3$ Augelite	–	– 2,809.56	–	(66)	
$Ca_{9.54}Na_{0.33}Mg_{0.13}(PO_4)_{4.8}(CO_3)_{1.2}F_{2.48}$ Carbonate-apatite	–	–12,294.5	–	(0)	

[a] (1) Wagman et al. (1968), (2) Wagman et al. (1969), (3) Wagman et al. (1971), (4) Parker et al. (1971), (5) Schumm et al. (1973), (6) Wagman et al. (1976), (7) Naumov et al. (1971), (8) Egan and Luff (1963), (9) Andon et al. (1963), (10) Rud'ko et al. (1972), (11) Irving and McKerrell (1967), (12) Andon et al. (1967), (13) Waterfield and Staveley (1967), (14) Luff and Reed (1978a), (15) Luff and Reed (1978b), (16) Rud'ko et al. (1973), (17) Cordfunke and Ouweltjes (1971), (18) Luff et al. (1976), (19) Coffy and Sunner (1980), (20) Duff (1971d), (21) Duff (1971c), (22) Volkov (1979), (23) ΔH_f^0 calculated form heat of dehydration of β Hopeite, $\Delta H = 224.7$ kJ mol^{-1} (Volkov et al. 1974a) for the reaction: $\beta Zn_3(PO_4)_2 \cdot 4H_2O$ (c) $\rightarrow$ $Zn_3(PO_4)_2$ (c) + $4H_2O$ (g); ΔG_f^0 calculated from the energy of dehydration of α Hopeite $\Delta G = 81.2$ kJ mol^{-1} (Yaglov 1978) for the reaction: $\alpha Zn_3(PO_4)_2 \cdot 4H_2O$ (c) $\rightarrow$ $Zn_3(PO_4)_2$ (c) + $4H_2O$ (g), (24) Calculated from heat and free energy of dehydration of β Hopeite, $\Delta H = 154.80$ kJ mol^{-1} and $\Delta G = 40.20$ kJ mol^{-1} (Yaglov 1978) for the reaction: $\beta Zn_3(PO_4)_2 \cdot 4H_2O$ (c) $\rightarrow$ $Zn_3(PO_4)_2 \cdot H_2O$ (c) + $3H_2O$ (g), (25) Calculated from heat and free energy of dehydration of α Hopeite, $\Delta H = 102.1$ kJ mol^{-1} and $\Delta G = 28.45$ kJ mol^{-1} (Yaglov 1978) for the reaction: $\alpha Zn_3(PO_4)_2 \cdot 4H_2O$ (c) $\rightarrow$ $Zn_3(PO_4)_2 \cdot 2H_2O$ (c) + $2H_2O$ (g), (26) Calculated from heat and free energy of dehydration of para hopeite $\Delta H = 210.9$ kJ mol^{-1} and $\Delta G = 58.60$ kJ mol^{-1} (Yaglov 1978) for the reaction: $Zn_3(PO_4)_2 \cdot 4H_2O$ (para) $\rightarrow$ $Zn_3(PO_4)_2$ (c) + $4H_2O$ (g), (27) Calculated from heat of dehydration of α Hopeite $\Delta H = 235.14$ kJ mol^{-1} (Yaglov 1978) for the reaction: $\alpha Zn_3(PO_4)_2 \cdot 4H_2O$ (c) $\rightarrow$ $Zn_3(PO_4)_2$ (c) + $4H_2O$ (g), (28) Calculated from free energy of dehydration of β Hopeite, $\Delta G = 68.60$ kJ mol^{-1} (Yaglov 1978) for the reaction: $\beta Zn_3(PO_4)_2 \cdot 4H_2O$ (c) $\rightarrow$ $Zn_3(PO_4)_2$ (c) + $4H_2O$ (g), (29) Calculated from heat of dehydration of $Mn_3(PO_4)_2 \cdot 3H_2O$, $\Delta H = 177.00$ kJ mol^{-1} (Yaglov

These functions are commonly used to calculate thermodynamic properties of minerals as a function of temperature.

Variations in the solubility of apatite as function of temperature is given as an example:

Fluorapatite

$Ca_{10}(PO_4)_6F_{2(c)} + 12H^+_{(aq)} \rightleftharpoons 10Ca^{2+}_{(aq)} + 6H_2PO^-_{4(aq)} + 2F^-_{(aq)}$

apatite

$\Delta G^0_{R,\,298.15} = 107.56\ kJ\ mol^{-1}$
$\Delta S^0_{R,\,298.15} = -792.64\ J\ deg^{-1}\ mol^{-1}$
$\Delta C_{p,\,298.15} = -1217.26\ J\ deg^{-1}\ mol^{-1}$
Δa -948.85

1978) for the reaction $Mn_3(PO_4)_2 \cdot 3H_2O$ (c) → $Mn_3(PO_4)_2$ (am) + $3H_2O$ (g), (30) Aleksandrov (1979), (31) Calculated from heat of dehydration of $Cu_3(PO_4)_2 \cdot 2H_2O$, $\Delta H = 127.20\ kJ\ mol^{-1}$ (Yaglov 1978) for the reaction $Cu_3(PO_4)_2 \cdot 2H_2O$ (c) → $Cu_3(PO_4)_2$ (c) + $2H_2O$ (g), (32) Calculated from free energy of dehydration of $Cu_3(PO_4)_2 \cdot 2H_2O$, $\Delta G = 56.10\ kJ\ mol^{-1}$ (Yaglov 1978) for the reaction: $Cu_3(PO_4)_2 \cdot 2H_2O$ (c) → $Cu_3(PO_4)_2$ (c) + $2H_2O$ (g), (33) Truesdell and Jones (1964), (34) Calculated from free energy of dehydration of $CoHPO_4 \cdot 1.5H_2O$, $\Delta G = 50.20\ kJ\ mol^{-1}$ (Yaglov et al. 1974) for the reaction $CoHPO_4 \cdot 1.5H_2O$(c) → $CoHPO_4$ (c) + $1.5H_2O$ (g), (35) Fuger and Oetting (1976), (36) Langmuir (1978), (37) Moskvin et al. (1967), (38) Egan and Wakefield (1966), (39) Ashuiko et al. (1975), (40) Volkov et al. (1975b), (41) Calculated from heat of reaction $\Delta H = 1{,}120.5\ kJ\ mol^{-1}$ (Guido et al. 1977) for the reaction $CePO_4$ (c) → CeO_2 (g) + PO_2 (g) with ΔH^0_f CeO_2 (g) = $-525\ kJ\ mol^{-1}$ and ΔH^0_f PO_2 (g) = $-297.1\ kJ\ mol^{-1}$ (Stull and Prophet 1971), (42) Yaglov et al. (1976), (43) Volkov et al. (1976), (44) Volkov et al. (1975a), (45) Volkov et al. (1974b), (46) Calculated from heat of dehydration of $VOPO_4 \cdot H_2O$, $\Delta H = 54.0\ kJ\ mol^{-1}$ (Yaglov 1978) for the reaction $VOPO_4 \cdot H_2O$ (c) → $VOPO_4$ (c) + H_2O (g), (47) Calculated from heat of reaction $\Delta H = -6.3\ kJ\ mol^{-1}$ (Afanasev and Starotin 1967) for the reaction $Ce_3(PO_4)_4$ (c) → $3Ce^{4+}$(aq) + $4PO_4^{3-}$(aq), (48) Allulli et al. (1979), (49) Bobrownicki and Jarmakowicz (1966), (50) Duff (1971b), (51) Calculated from heat of reaction $\Delta H = 22.2\ kJ\ mol^{-1}$ (Muto 1965) for the reaction $CaU(PO_4)_2 \cdot 2H_2O$ (c) + $6H^+$(aq) → Ca^{2+}(aq) + U^{4+}(aq) + $2H_3PO_4$(aq) + $2H_2O$ (l), (52) ΔH^0_f calculated from heat of solution $\Delta H = 60.79\ kJ\ mol^{-1}$ (Luff and Reed 1980) for the reaction ($MgCl_2 \cdot 10.05H_2O$) + KH_2PO_4 (c) → $MgKPO_4 \cdot 6H_2O$ (c) + 2(HCl $2.025H_2O$), S^0_f calculated from entropy of reaction $\Delta S^0 = 167.8\ J\ deg^{-1}\ mol^{-1}$ (Luff and Reed 1980) for the reaction $MgKPO_4 \cdot 6H_2O$ (c) → Mg^{2+}(aq) + K^+(aq) + PO_4^{3-}(aq) + $6H_2O$ (l), (53) Egan and Wakefield (1963), (54) Calculated from heat of reaction and entropy of reaction $\Delta H = 130.29\ kJ\ mol^{-1}$ and $\Delta S = -3{,}144.90\ J\ deg^{-1}\ mol^{-1}$ (Egan et al. 1961) for the reaction $H_6K_3Al_5(PO_4)_8 \cdot 18H_2O$ (c) → $6H^+$(aq) + $3K^+$(aq) + $5Al^{3+}$(aq) + $8PO_4^{3-}$(aq) + $18H_2O$ (l), (55) Calculated from heat of reaction and entropy of reaction $\Delta H = 114.30\ kJ\ mol^{-1}$ and $\Delta S = -3{,}131.70\ J\ deg^{-1}\ mol^{-1}$ (Egan et al. 1961) for the reaction $H_6(NH_4)_3Al_5(PO_4)_8 \cdot 18H_2O$ (c) → $6H^+$(aq) + $3NH_4^+$(aq) + $5Al^{3+}$(aq) + $8PO_4^{3-}$(aq) + $18H_2O$ (l), (56) Duff (1972a) has calculated the free energy of formation of oxyapatite from data of Garrels and Christ (1965) and obtained ΔG^0_f (oxyapatite) = $-12{,}272.1\ kJ\ mol^{-1}$, this value has been recomputed using these data, (57) The ΔG^0_f of fluor apatite and hydroxyapatite have been calculated from data of Duff (1971a) using new standards free energy of formation. These values obtained are usefull in consideration of the formation of phosphorites in sea water and in the mineralization of dental enamel. The ΔG^0_f of fluorapatite and hydroxyapatite quoted by Parker et al. (1971) are obtained from well crystallized materials prepared by sintering above 1,000 °C, (58) Valyashko et al. (1968), (59) The ΔG^0_f of this mineral has been recalculated from data of Duff (1972b) using new values of standard Gibbs free energies of formation, (60) Recalculated from data of Duff (1972c), (61) The ΔG^0_f of $Ca_4O(PO_4)_2$ has been recalculated from data of Duff (1971b), (62) Nriagu (1972a), (63) Egan et al. (1963), (64) Recalculated from Duff (1971c), (65) Bass and Sclar (1979), (66) Wise and Loh (1976).

Table 7. Heat capacity and Maier and Kelley power function of ions and phosphate minerals

Compounds and aqueous species	Cp, 298° 15 J deg^{-1} mol^{-1}	Heat content			Temperature range (K)	Reference[a]
		a	b, 10^{-3}	c, 10^{5}		
Aqueous Species						
H^+	0					
Na^+	46.44		155.8			(1)
K^+	21.80		72.80			(1)
Tl^+	4.20		13.80			(1)
NH_4^+	79.90		268.00			(1)
Mg^{2+}	38.50		129.10			(1)
Ca^{2+}	0.80		2.80			(1)
Pb^{2+}	− 52.70		− 176.80			(1)
Al^{3+}	38.10		128.00			(1)
Sc^{3+}	23.45		78.70			(1)
Y^{3+}	− 26.80		− 89.50			(1)
La^{3+}	− 56.90		− 190.85			(1)
Nd^{3+}	− 72.40		− 242.80			(1)
Eu^{3+}	15.10		50.65			(1)
Fe^{3+}	24.70		82.50			(1)
OH^-	−148.5		− 498.0			(1)
F^-	−105.0		− 352.3			(1)
Cl^-	−136.4		− 456.1			(1)
$H_3PO_4^0$	229.3		− 769.9			(1)
$H_2PO_4^-$	− 43.9		− 147.3			(1)
HPO_4^{2-}	−265.7		− 891.2			(1)
PO_4^{3-}	−334.7		−1,121.30			(1)
Compounds						
P_4O_{10}	211.7					(2)
		70.1	452.0		298– 631	(3)
H_3PO_4	106.1					(2)
$2H_3PO_4 \cdot H_2O$	252.0					(2), (3)
NaH_2PO_4	135.0	129.1	103.	− 22.	298– 500	(4)
Na_2HPO_4	116.0	29.0	282.	− 2.6	298– 460	(4)
$Na_2HPO_4 \cdot 2H_2O$	211.3					(5)
$Na_2HPO_4 \cdot 7H_2O$	438.4					(5)
$Na_2HPO_4 \cdot 12H_2O$	645.0					(5)
KH_2PO_4	116.57					(1)
K_2HPO_4	141.29					(6)
$NH_4H_2PO_4$	142.26					(1), (2)
$(NH_4)_2HPO_4$	179.16					(7)
TlH_2PO_4	124.14					(8)
Tl_3PO_4	163.76					(9)
$Mg_3(PO_4)_2$	213.47	261.54	76.36	− 63.0	298–1,626	(10)
$Ca(H_2PO_4)_2 \cdot H_2O$	258.82					(1), (2)
$CaHPO_4$	110.04					(1), (2)
Monetite		44.31	233.43	− 2.05	80– 310	(11)
$CaHPO_4 \cdot 2H_2O$	197.07					(1), (2)
Brushite		31.72	569.44	− 0.93	50– 310	(11)
$\alpha Ca_3(PO_4)_2$	231.58	201.84	164.18	− 21.	298–1,373	(3)
$\beta Ca_3(PO_4)_2$	227.82					(3)
$Ca_{10}(PO_4)_6F_2$ Fluorapatite	751.86	948.85	113.77	−205.3	298–1,600	

Table 7. (continued)

Compounds and aqueous species	Cp, 298° 15 J deg^{-1} mol^{-1}	Heat content a	b, 10^{-3}	c, 10^5	Temperature range (K)	Reference[a]
$Ca_{10}(PO_4)_6(OH)_2$ Hydroxyapatite	770.15	956.13	165.77	−209.20	298–1,500	(3)
$Ca_{10}(PO_4)_6Cl_2$	757.93					(12)
$Pb_3(PO_4)_2$	256.27					(1), (2), (3)
$AlPO_4$ Berlinite	93.18					(1), (2)
$AlPO_4 \cdot 2H_2O$ Variscite	165.98					(1)
$ScPO_4$	93.05	142.76	5.52	−45.65	298–1,600	(13)
YPO_4	103.81	124.85	25.02	−25.31	298–1,597 2	(14)
$LaPO_4$	101.55	118.03	35.15	−23.93	298–1,595 7	(14)
$NdPO_4$	104.39	122.21	36.74	−25.56	298–1,592 9	(14)
$EuPO_4$	109.41	141.67	17.49	−33.30	298–1,600	(13)
$FePO_4 \cdot 2H_2O$ Strengite	180.54					(1), (2)
$FePO_4 \cdot 2H_2O$ (am.)	189.41					(1)
$K_3Al_5(PO_4)_8H_6 \cdot 18H_2O$ (Potassium taranakite)	1,482.39					(1)
$(NH_4)_3Al_5(PO_4)_8H_6 \cdot 18H_2O$ (Ammonium taranakite)	1,592.01					(1), (2)

[a] (1) Naumov et al. (1971), (2) Wagman et al. (1968, 1969, 1971, 1976), Parker et al. (1971), (3) Kelley (1960), (4) Calculated from data of Ashcroft et al. (1969), (5) Waterfield and Staveley (1967), (6) Luff and Reed (1978a), (7) Luff et al. (1976), (8) Bousquet et al. (1978), (9) Coffy and Sunner (1980), (10) Oetting and McDonald (1963), (11) Calculated by Gregory et al. (1970) from the data of Egan and Wakefield (1964a, b), (12) Valyashko et al. (1968), (13) Tsagareishvili et al. (1975), (14) Tsagareishvili et al. (1972).

$$\Delta b \quad -1674.37 \times 10^{-3}$$
$$\Delta c \quad +205.3 \times 10^{5}$$
$$\log K_{R,\,298.15} = 10 \log[Ca^{2+}] + 6 \log[H_2PO_4^-] + 2 \log[F^-] + 12\,pH$$
$$= -15.87 \text{ at } 273.15\text{ K}$$
$$= -18.84 \text{ at } 298.15\text{ K}$$
$$= -20.22 \text{ at } 310.15\text{ K}$$
$$= -21.69 \text{ at } 323.15\text{ K}$$
$$= -27.13 \text{ at } 373.15\text{ K}$$
$$= -32.24 \text{ at } 423.15\text{ K}$$
$$= -37.39 \text{ at } 473.15\text{ K}.$$

Hydroxyapatite

$$Ca_{10}(PO_4)_6(OH)_2(c) + 12H^+_{(aq)} \rightleftharpoons 10Ca^{2+}_{(aq)} + 6H_2PO^-_{4\,(aq)} + 2OH^-_{(aq)}$$
$$\Delta G^0_{R,\,298.15} = 44.62\text{ kJ mol}^{-1}$$

Table 8. Values of Debye-Hückel $\mathring{a}_i$ parameter for some phosphate aqueous ions

Ions	Values (Å)
$H_2PO_4^-$	4.0 (1)[a], (2), 5.4 (3)
HPO_4^{2-}	4.0 (1), (2); 5.0 (3)
PO_4^{3-}	4.0 (1), (2); 5.0 (3)
$NaHPO_4^-$	4.0 (4), (2); 5.4 (3)
$KHPO_4^-$	4.0 (2); 5.4 (3)
$MgH_2PO_4^+$	5.4 (2), (3)
$MgPO_4^-$	4.0 (2); 5.4 (3)
$CaH_2PO_4^+$	5.1 (5); 5.4 (2), (3)
$CaPO_4^-$	4.0 (2); 5.4 (3)
$FeH_2PO_4^+$	5.4 (3)
$FeHPO_4^+$	5.4 (3)
$UO_2H_2PO_4^+$	6.0 (6)
$UO_2(HPO_4)_2^{2-}$	6.0 (6)

[a] (1) Kielland (1937), (2) Vieillard (1978), (3) Truesdell and Jones (1964), (4) Patel et al. (1974), (5) Gregory et al. (1970), (6) Dongarra and Langmuir (1980)

$$\Delta S^0_{R,\,298.15} = -791.58 \text{ J deg}^{-1} \text{ mol}^{-1}$$
$$\Delta C_{p,\,298.15} = -1322.55 \text{ J deg}^{-1} \text{ mol}^{-1}$$
$$\Delta a = -956.13$$
$$\Delta b \quad -2017.57 \times 10^{-3}$$
$$\Delta c \quad -209.20 \times 10^{5}$$
$$\log K_{R,\,298.15} = 10 \log [Ca^{2+}] + 6 \log [H_2PO_4^-] + 2 \log [OH^-] + 12 \text{ pH}$$
$$= -3.84 \text{ at } 273.15 \text{ K}$$
$$= -7.82 \text{ at } 298.15 \text{ K}$$
$$= -9.63 \text{ at } 310.15 \text{ K}$$
$$= -11.55 \text{ at } 323.15 \text{ K}$$
$$= -18.49 \text{ at } 373.15 \text{ K}$$
$$= -24.96 \text{ at } 423.15 \text{ K}$$
$$= -31.09 \text{ at } 473.15 \text{ K}$$

It is seen that apatite appears to be 1,000 times more soluble at 0 °C than at 25 °C. This fact illustrates the role played by the temperature in the mass transfer of phosphate accompanying the upwelling of deep oceanic waters.

Debye-Hückel å Coefficients

Activity coefficients (γ_i) in moderately dilute solutions are calculated from the Debye-Hückel expression:

$$\log \gamma_i = -\frac{AZ^2\, I}{1 + B\mathring{a}\, I}$$

Table 9. Total concentration of elements and the distribution of complex aqueous species in sea water (salinity 35⁰/₀₀, pH 8.1, $pCO_2 = 3 \times 10^{-4}$ atm, temperature 25° C

Aqueous species	Log (molality)	Log [activity]	Activity coefficient
K	− 2,000	–	–
K^+	− 2,007	− 2,210	0,626
KCl	− 4,731	− 4,661	1,174
KSO_4^-	− 3,882	− 4,069	0,649
$KHPO_4^-$	− 8,257	− 8,431	0,671
Na	− 0,301	–	–
Na^+	− 0,332	− 0,496	0,671
NaCl	− 1,867	− 1,197	1,174
$NaCO_3^-$	− 4,518	− 4,679	0,690
$NaHCO_3^0$	− 4,049	− 3,979	1,174
$NaSO_4^-$	− 1,969	− 2,135	0,682
$NaHPO_4^-$	− 6,433	− 6,607	0,671
Ca	− 1,989	–	–
Ca^{2+}	− 2,027	− 2,638	0,245
$CaCO_3^0$	− 4,960	− 4,891	1,174
$CaHCO_3^+$	− 4,464	− 4,611	0,714
$CaOH^+$	− 7,140	− 7.314	0,671
$CaPO_4^-$	− 7,209	− 7,382	0,671
$CaHPO_4^0$	− 6,928	− 6,858	1,174
$CaH_2PO_4^+$	− 8,943	− 9,085	0,720
$CaSO_4^0$	− 3,097	− 3,027	1,174
CaF^+	− 6,304	− 6,466	0,690
Mg	− 1,275	–	–
Mg^{2+}	− 1,317	− 1,819	0,315
$MgCO_3^0$	− 3,941	− 3,878	1,174
$MgHCO_3^+$	− 3,978	− 4,152	0,671
$MgOH^+$	− 4,973	− 5,147	0,671
$MgPO_4^-$	− 6,260	− 6,433	0,671
$MgHPO_4^0$	− 5,979	− 5,909	1,174
$MgH_2PO_4^+$	− 8,019	− 8,161	0,720
$MgSO_4^0$	− 2,338	− 2,268	1,174
MgF^+	− 4,845	− 5,007	0,690
Cl	− 0,262	–	–
Cl^-	− 0,248	− 0,451	0,630
HCl	− 14,721	− 14,651	1,174
S	− 1,550	–	–
SO_4^{2-}	− 1,923	− 2,699	0,167
HSO_4^-	− 8,636	− 8,809	0,671
C	− 2,934	–	–
CO_3^{2-}	− 4,725	− 5,453	0,187
HCO_3^-	− 3,072	− 3,233	0,690
$H_2CO_3^0$	− 5,053	− 4,983	1,174
P	− 5,547	–	–
PO_4^{3-}	− 9,422	− 11,204	0,017
HPO_4^{2-}	− 6,184	− 6,961	0,167
$H_2PO_4^-$	− 7,682	− 7,856	0,671
F	− 4,165	–	–
F^-	− 4,270	− 4,458	0,649
pH = 8, 10			

Table 10. Log activity product and log Ksp of some idealized apatite and β-whitlockite

Mineral species	Log activity product	Log Ksp	References[a]
$Ca_{10}(PO_4)_6F_2$ Fluorapatite	14.77	15.26	(1)
$Ca_{10}(PO_4)_{5.5}(CO_3)_{0.5}F_{2.5}$ Francolite	5.64	13.70	(2)
$Ca_3(PO_4)_2$	8.77	10.18	(3)
β Whitlockite		6.41	(4)

[a] (1) Solubility product calculated from ΔG_f^0 of fluorapatite given by Duff (1971a), (2) Vieillard (1978), (3) Gregory et al. (1974), (4) Solubility product calculated from ΔG_f^0 of β-Whitlockite of Parker et al. (1971)

in which A and B are constants depending on temperature and pressure, I, is the ionic strength and å is a coefficient characterizing the size of each aqueous species. The different values of $å_i$ for phosphate aqueous ions (i) are given in Table 8.

Aqueous Complex Species Distribution in Sea Water

The chemical distribution of aqueous complex species in sea water at 25 °C has been calculated by Vieillard (1978), using the computer programm DISSOL (Fritz 1976) and the thermochemical data given in Tables 3 and 8. The results are shown in Table 9.

The total amount of phosphorus in sea water (Turekian 1969, p. 309–311) is roughly 2.84 10^{-6} mol kg^{-1} H_2O. From Table 9, it can be calculated that phosphorus is mostly present as:

$$(P)_{tot} = (HPO_4^{2-}) + (MgHPO_4^{\circ}) + (MgPO_4^{-}) + (CaHPO_4^{\circ}) + (CaPO_4^{-}) + (NaHPO_4^{-})$$
$$= 2.80\ 10^{-6}\ \text{mol/kg}\ H_2O\ .$$

This distribution permits the calculation of degrees of saturation with respect to various minerals such as:

$$Ca_3(PO_4)_2 + 4H^+ \rightleftharpoons 2H_2PO_4^- + 3Ca^{2+}\ .$$

It comes to:

$$2\log[H_2PO_4^-] + 3\log[Ca^{2+}] - 4\log[H^+] = \log Ksp = 10.18 \text{ at } 25\ ^\circ C.$$

The corresponding log Q, the ionic product, in sea water is:

$$\log Q = 8.77\ .$$

It thus appears (Table 10) that sea water could be oversaturated with respect to most of the idealized apatites or calcium orthophosphates (Vieillard 1978).

Table 11. Gibbs free energies of formation estimated by Nriagu (1976) and Vieillard et al. (1979) of millisite-Na, crandallite, wavellite, variscite and augelite

Compounds	ΔG_f^0 298°15 estimated by Nriagu (1976) kJ mol^{-1}	ΔG_f^0 298°15 estimated by Vieillard et al. (1979) kJ mol^{-1}
$CaNaAl_6(PO_4)_9(OH)_9 \cdot 3H_2O$ Millisite-Na	–	–10,925.50
$Al_3(PO_4)_2(OH)_3 \cdot 5H_2O$ Wavellite	–5,602.40	– 5,565.00
$AlPO_4 \cdot 2H_2O$ Variscite	–2,104.50	– 2,087.80
$Al_2PO_4(OH)_3$ Augelite	–2,783.60	– 2,765.21
$CaAl_3(PO_4)_2(OH)_5 \cdot H_2O$ Crandallite	–5,614.90	– 5,592.75

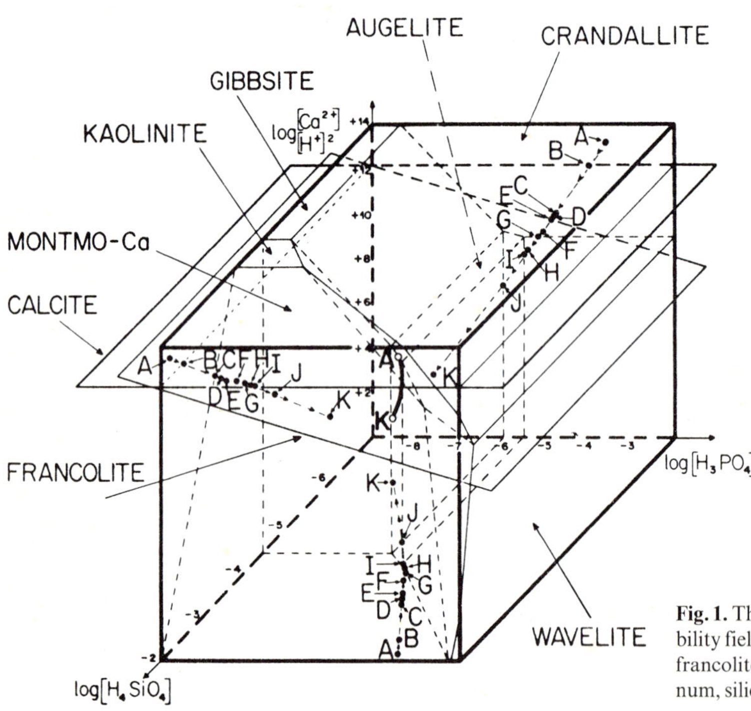

Fig. 1. Three-dimensional stability field diagram for calcite, francolite, calcium, aluminum, silicates, and phosphates

Estimation of Thermodynamic Properties of Aluminous Phosphates

In weathering lateritic profiles developed on phosphate deposits, clay minerals and apatite react together to form various calcium-aluminum-phosphates such as millisite, crandallite, wavellite, augelite, variscite and finally gibbsite. Thermochemical data for these minerals have been estimated by Nriagu (1976) and

Vieillard (1978). The latest ones seem to correspond better to the natural stability field and successive parageneses observed in nature. The Gibbs free energies of formation and the corresponding log Ksp of these minerals are given in Table 11.

In a three-dimensional diagram ($\log[Ca^{2+}]/[H^+]^2 - \log[H_3PO_4] - \log[H_4SiO_4]$) the clay-phosphate parageneses follow the general trend indicated in Fig. 1 (Vieillard et al. 1979).

Concluding Remarks

Phosphate minerals are numerous and their natural chemical composition is complex. The thermochemical data are currently available for relatively simple composition phases. Precise information on mineral composition, as well as more data on excess energies of mixing in solid solutions are needed for improving the discussion on the stability fields of phosphates in nature. Such data are especially required for (OH^-)–(F^-)–(Cl^-) and (HPO_4^{2-})–(CO_3^{2-}) apatites and for (Fe^{3+})–(Al^{3+}) millisites as examples of abundant minerals in nature.

References

Afanas'ev YA, Starotin AD (1967) Standard enthalpies of formation for orthophosphates of some rare earths. Izv Sib Otd Akad Nauk SSSR Ser Khim Nauk 12:104–106 (in russian)

Aia MA, Mathers JE, Mooney RW (1964) Thermodynamic solubility products of α and β $SrHPO_4$ from 25° to 90°. J Chem Eng Data 9:335–338

Aleksandrov YaI (1979) Nature of the adhesion of a galvanic deposit to aluminum through an anoxic oxide. Sov Electrochem 15:141–144

Aleshchkina AE, Masalovich VM, Agasyan PK, Sereda BP (1976) Chromium (III) phosphate-complexes. Russ J Inorg Chem 21:973–975

Allulli S, Massucci A, Tomassini N (1979) The standard enthalpy of formation of anhydrous monohydrated and dihydrated zirconium bis (mono-hydrogen phosphate). J Chem Thermod 11:613–618

Andon JL, Counsell JF, Mc Kerrell H, Martin JF (1963) Thermodynamic properties of phosphorus compounds. I. Entropy of phosphorus pentoxide. Trans Farad Soc 59:2702–2705

Andon JL, Counsell JF, Martin JF, Mash CJ (1967) Thermodynamic properties of phosphorus compounds. II. Low temperature, heat capacity and entropy of sodium mono-, di- and triphosphates. J Appl Chem 17:65–70

Ashcroft SJ, Keen E, Mortimer CT (1969) Thermochemistry of formation of sodium polyphosphate from sodium orthophosphates. Trans Farad Soc 65:2851–2855

Ashuiko VA, Lavrov AV, Rat'kovskii IA (1975) Mass spectrometric study of gallium and indium phosphates. Zh Fiz Khim 49:1958 (in russian)

Avnimelech Y, Moreno EC, Brown WE (1973) Solubility and surface properties of finely divided hydroxyapatite. J Res NBS 77A:149–155

Baker WE (1964) Mineral equilibrium studies of the pseudomorphism of pyromorphite by hindsalite. Am Miner 49:607–613

Baldwin WG (1969) Phosphate equilibria. II. Studies on the silver phosphate electrodes. Arkiv Kemi 31:407–414

Ball JW, Nordstrom DK, Everett AJ (1980) Additional and revised thermochemical data and computer code for WATEQ 2. A computerized chemical model for trace and major element speciation and mineral equilibria of natural waters. US Geol Surv Wat Res Invest 78–116

Barner HE, Scheuerman RV (1978) Handbook of thermochemical data for compound and aqueous species. Wiley, New York, p 156

Bass JD, Sclar CB (1979) The stability of trolleite and the Al_2O_3–$AlPO_4$–H_2O phase diagram. Am Miner 64:1175–1183

Bobrownicki W, Jarmakowicz J (1966) Energy balance of thomas superphosphate production. Estimation of enthalpy of formation of sodium calcium phosphate. Chem Stosow Ser A 10:365–367

Borisov MS, Elesin AA, Lebedev IA, Filomonov VT, Yakovlev GN (1966) Complex formation of the trivalent actinides and lanthanides in phosphoric acid solutions. Radiokhim 8:42–48 (in russian)

Bousquet J, Diot M, Tranquard A, Coffy G, Vignalou JR (1978) Heat capacities of dihydrogen and dideuterium thallium orthophosphates. Thermodynamic properties of solid phase transitions. J Chem Thermod 10:779–786

Brodsky A (1929) Electrochemistry of the mercurous ion. Z Elektrochem 35:833–837

Brown WE (1960) Behaviour of slightly soluble calcium phosphates as revealed by phase equilibrium calculation. Soil Sci 90:51–57

Chang SC, Jackson ML (1957) Solubility product of iron phosphate. Soil Sci Soc Am Proc 21:265–269

Chien SH, Black CA (1976) Free energies of formation of carbonate apatite in some phosphate rocks. Soil Sci Soc Am Proc 40:234–239

Chughtai A, Marshall R, Nancollas GH (1968) Complexes in calcium phosphate solutions. J Phys Chem 72:208–211

Chukhlantsev VG, Stepanov SI (1956) Solubility of uranyl and thorium phosphates. Zh Neorg Khim 1:478–484 (in russian)

Chukhlantsev VG, Alyamovskaya KV (1961a) Solubility product of uranyl, beryllium, and cerium phosphates. Izvest Vyss Uch Zaved Khim Khim Tekh 4:359–363 (in russian)

Chukhlantsev VG, Alyamovskaya KV (1961b) Solubility product of copper, cobalt, nickel, and cadmium phosphates. Izvest Vyss Uch Zaved Khim Khim Tekh 4:706–709 (in russian)

Coffy G, Sunner S (1980) Heat capacity and thermodynamic properties of trithallium phosphate from 5° to 320 °K. J Chem Termodyn 12:303–304

Cordfunke EHP, Ouweltjes W (1971) Heats of formation of silver phosphate and silver arsenate. A re-evalution. Rec Trav Chim Pays-Bas 90:1343–1344

Denotkina RG, Moskvin AI, Shevchenko VB (1960) The solubility product of plutonium. IV. Hydrogen phosphate and its solubility in various acids. Zh Neorg Khim 5:805–810 (in russian)

De Vries T, Cohen D (1949) Standard potential of the mercury-mercury I phosphate electrode. J Am Chem Soc 71:1114–1115

Dongarra G, Langmuir D (1980) The stability of UO_2OH^+ and $UO_2(HPO_4)_2^{2-}$ complexes at 25 °C. Geochim Cosmochim Acta 44:1747–1751

Duff EJ (1971a) Orthophosphates. II. The transformation Brushite→Fluorapatite and Monetite→Fluorapatite in aqueous potassium fluoride solution. J Chem Soc A:33–38

Duff EJ (1971b) Orthophosphates. III. The hydrolysis of secondary calcium orthophosphates. J Chem Soc A:917–921

Duff EJ (1971c) Orthophosphates. IV. Stability relationships of orthophosphates within the system CaO–P_2O_5–H_2O and CaF_2–CaO–P_2O_5–H_2O under aqueous conditions. J Chem Soc A:921–926

Duff EJ (1971d) Orthophosphates. VIII. The transformation of Newberyite into Biobierrite in aqueous alkaline solutions. J Chem Soc A:2736–2740

Duff EJ (1972a) Orthophosphates. VII. Thermodynamical considerations concerning the stability of oxyapatite $Ca_{10}O(PO_4)_6$ in aqueous media. J Inorg Nucl Chem 34:853–857

Duff EJ (1972b) Orthophosphates. IX. Chlorapatite: Phase relationships under aqueous conditions along $Ca_5F(PO_4)_3$–$Ca_5Cl(PO_4)_3$ and $Ca_5(OH)(PO_4)_3$–$Ca_5Cl(PO_4)_3$ joins of the system CaO–$CaCl_2$–CaF_2–P_2O_5–H_2O. J Inorg Nucl Chem 34:859–871

Duff EJ (1972c) Orthophosphates. XI. Bromoapatite: Stability of solid solutions of bromoapatite with other calcium apatites under aqueous conditions. J Inorg Nucl Chem 34:101–108

Egan EP, Luff BB (1963) Heat of formation of phosphorous oxides. Teness Valley Auth, p 44

Egan EP, Wakefield ZT (1963) Low temperature, heat capacity and entropy of the potassium, iron, phosphate $H_8KFe_3(PO_4)_6 \cdot 6H_2O$. J Chem Engl Data 8:182–184

Egan EP, Wakefield ZT (1964a) Low temperature, heat capacity and entropy of anhydrous dicalcium phosphate 10° to 310 °K. J Chem Engl Data 4:541–544

Egan EP, Wakefield ZT (1964b) Low temperature, heat capacity and entropy of dicalcium phosphate dihydrate 10° to 310 °K. J Chem Engl Data 9:544–545

Egan EP, Wakefield ZT (1966) Low temperature, heat capacity and entropy of variscite, $AlPO_4 \cdot 2H_2O$, 10° to 310 °K. J Chem Engl Data 11:610–611

Egan EP, Wakefield ZT, Luff BB (1961) Thermodynamic properties of potassium and ammonium taranakites. J Phys Chem 65:1609–1612

Egan EP, Wakefield ZT, Luff BB (1963) Low temperature, heat capacity and entropy of basic potassium aluminium phosphate. J Chem Engl Data 8:184–185

Fritz B (1976) Etude thermodynamique des réactions entre minéraux et solutions. Applications à la géochimie des altérations et des eaux continentales. Mem Sci Géol 41:p152

Fuger J, Oetting FL (1976) The chemical thermodynamic of actinide elements and compounds. II. The actinide aqueous ions. Int Atom Energ Agency Vienna, pp 16–60

Garrels RM, Christ CL (1965) Solutions, minerals and equilibria. Harper & Row, New York, pp 403–428

Gregory TM, Moreno EC, Brown WE (1970) Solubility of $CaHPO_42H_2O$ in the system $Ca(OH)_2$–H_3PO_4–H_2O at 5, 15, 25, and 37 °C. J Res NBS 74A:461–475

Gregory TM, Moreno EC, Patel JM, Brown WE (1974) Solubility of β $Ca_3(PO_4)_2$ in the system $Ca(OH)_2$–H_3PO_4–H_2O at 5, 15, 25, and 37 °C. J Res NBS 78A:667–674

Guido M, Balducci G, De Maria G, Gigli G (1977) Thermodynamic study of the vaporization of cerium orthophosphate. J Chem Soc Farad Trans I 73:121–127

Jowett M, Price III (1932) Solubilities of the phosphates of lead. Trans Farad Soc 28:568–581

Jurinak JJ, Inouye TS (1962) Some aspects of zinc and copper phosphate formation in aqueous systems. Soil Sci Soc Am Proc 26:144–147

Irving RJ, Mc Kerrell H (1967) Standard heats of formation of sodium dihydrogen phosphate, disodium monohydrogen phosphate and trisodium phosphate. Trans Farad Soc 63:2913–2916

Karpov VI (1961) The solubility of tri-uranyl phosphate. Zh Neorg Khim 6:531–533 (in russian)

Kelley KK (1960) High temperature, heat content, heat capacity and entropy data for the elements and inorganic compounds. Bur Mines Bull 584:232

Kharaka YK (1973) SOLMNEQ: Solution-mineral equilibrium computation. US Geol Surv comput contr, p 81

Kielland J (1937) Individual activity coefficients of ions in aqueous solutions. J Am Chem Soc 59:1675–1678

Klygin AE, Zavrazhnova DM, Nikol'skaya NA (1961) Isolation of uranium as ammonium uranyl phosphate and its gravimetric determination by calcination to $U_2O_3P_2O_7$. Zh Anal Khim 16:297–302 (in russian)

Lahiri SC (1965) Ferric-phosphoric acid system. J Ind Chem Soc 42:715–724

Langmuir O (1978) Uranium solutions-mineral equilibria at low temperatures with application to sedimentary ore deposits. Geochim Cosmochim Acta 42:547–569

Lindsay WL, Moreno EC (1960) Phosphate phase equilibria in soils. Soil Sci Am Proc 24:177–182

Lindsay WL, Peech M, Clark JS (1959) Solubility criteria for the existence of variscite in soils. Soil Sci Soc Am Proc 23:357–360

Luff BB, Reed RB (1978a) Low temperature, heat capacity and entropy of dipotassium orthophosphate. J Chem Engl Data 23:58–60

Luff BB, Reed RB (1978b) Standard enthalpies of monopotassium and dipotassium orthophosphate. J Chem Engl Data 23:60–62

Luff BB, Reed RB (1980) Thermodynamic properties of magnesium potassium orthophosphate hexahydrate. J Chem Engl Data 25:310–312

Luff BB, Reed RB, Nash RH (1976) Low temperature heat capacity and entropy of diammonium orthophosphate. J Chem Engl Data 21:418–419

Mac Cann HG (1968) The solubility of fluorapatite and its relationships to that of calcium fluoride. Arch Oral Biol 13:987–1001

Mac Dowell H, Brown WE, Sutter JR (1971) Solubility study of calcium hydrogen phosphate ion pair formation. Inorg Chem 10:1638–1643

Maier CG, Kelley KK (1932) An equation for the representation of high temperature heat content data. J Am Chem Soc 54:3243–3246

Marcus V (1958) Anion exchange of metal complexes: The uranyl phosphate system. Proc 2nd UN Inter Conf Peac Uses Atom Energy 3:465–471

Marinova LA, Rud'ko PK (1975) Thermodynamic study of praseodymium phosphate. Khim Khim Tekhnol (Minsk) 9:42–50 (in russian)

Marinova LA, Yaglov VN (1976) Thermodynamic characteristics of lanthanide phosphates. Zh Fiz Khim 50:802–803 (in russian)
Mayer SW, Schwartz SD (1950) The association of cerous ion with sulfite, phosphate and pyrophosphate ions. J Am Chem Soc 72:5106–5110
Mills KC (1974) Thermodynamic data for inorganic sulfides, selenides, and tellurides. Butterworths, London, p 845
Moreno EC; Brown WE, Osborn G (1960a) Solubility of dicalcium phosphate dihydrate in aqueous system. Soil Sci Soc Am Proc 24:94–98
Moreno EC, Brown WE, Osborn G (1960b) Stability of dicalcium phosphate dihydrate in aqueous solutions and solubility of octacalcium phosphate. Soil Sci Soc Am Proc 24:99–102
Moskvin AI, Shelyakina AM, Perminov PS (1967) Solubility product of uranyl phosphate and the composition and dissociation constants of uranyl phosphato complexes. Russ J Inorg Chem 12:1756–1760
Muto T (1965) Thermochemical study of ningyoite. Miner J 4:245–274
Naumov GB, Ryzenko B, Khodakovsky IL (1971) Handbook of thermochemical data. Moscou Atomizdat, p 239 (in russian)
Nriagu JO (1972a) Lead orthophosphates. I. Solubility and hydrolysis of secondary lead orthorphosphate. Inorg Chem 11:2499–2503
Nriagu JO (1972b) Stability of vivianite and ion-pair formation in the system $Fe_3(PO_4)_2$–H_3PO_4–H_2O. Geochim Cosmochim Acta 36:459–470
Nriagu JO (1972c) Solubility equilibrium constant of strengite. Am J Sci 272:476–484
Nriagu JO (1973a) Lead orthophosphate. II. Stability of chloropyromorphite at 25 °C. Geochim Cosmochim Acta 37:367–377
Nriagu JO (1973b) Solubility equilibrium constant of α-hopeite. Geochim Cosmochim Acta 37:2357–2361
Nriagu JO (1976) Phosphate-clay mineral relations in soils and sediments. Can J Earth Sci 13:717–736
Oetting FL, Mac Donald RA (1963) The thermodynamic properties of magnesium orthophosphate and magnesium pyrophosphate. J Phys Chem 63:2737–2743
Ousubaliev D, Batkibekóva M, Yousoupov V, Kydynov MK (1975) Thermochemical properties of phosphates and iodates of rare earth elements. 4th Conf Int Thermodyn Chem 1:217–223
Parker VB, Wagman DD, Evans WH (1971) Selected values of chemical thermodynamic properties. Natl Bur Stand Tech Note 270-6:119
Patel PR, Brown WE (1975) Thermodynamic solubility product of human tooth enamel: Powdered sample. J Dent Res 54:728–736
Patel PR, Gregory TM, Brown WE (1974) Solubility of $CaHPO_4 \cdot 2H_2O$ in the quaternary system $Ca(OH)_2$–H_3PO_4–NaCl–H_2O at 25 °C. J Res Nat Bur Stand 78A:675–681
Racz GJ, Soper RJ (1968) Solubility of dimagnesium phosphate trihydrate and trimagnesium phosphate. Can J Soil Sci 48:265–269
Rosenquist IT (1970) Formation of vivianite in Holocene clay sediments. Lithos 3:327–334
Rud'ko PK, Yaglov VN, Novikov GI (1972) Thermodynamics of the dehydration of lithium dihydrogen orthophosphate. Zh Fiz Khim 46:545 (in russian)
Rud'ko PK, Yaglov VN, Novikov GI (1973) Calorimetric study of rubidium dihydrogen orthophosphate. Vestsi Akad Navuk Belarus SSR Ser Khim Navuk 6:111–112 (in russian)
Schumm RH, Wagman DD, Bailey S, Evans WH, Parker VB (1973) Selected values of chemical thermodynamic properties. Natl Bur Stand Tech Note 270-7:76
Sillen LG, Martell AE (1971) Stability constants of metal ion complexes. Chem Soc Spec Publ [Suppl 1] 25:865
Singer PC (1972) Anaerobic control of phosphate by ferrous iron. J Water Poll Cont Fed 44:663–669
Smith RM, Martell AE (1976) Critical stability constants. vol 4: Inorganic complexes. Plenum NY, p 257
Spitsyn VI, Mikheev HB, Khermann A (1966) Thermodynamic study of the distribution of micro quantities of strontium between barium hydrogen phosphate and a solution. Dokl Akad Nauk SSSR 166:658–659 (in russian)
Stull DR, Prophet H (1971) JANAF thermochemical tables. Natl Stand Ref Data Ser Natl Bur Stand 37:1141
Taylor AW, Gurney EL (1961) Solubility of potassium taranakite and ammonium taranakite. J Phys Chem 65:1615–1617

Taylor AW, Gurney EL (1964a) The dissolution of calcium aluminum phosphate $CaAlH(PO_4)_2 \cdot 6H_2O$. Soil Sci Soc Am Proc 28:63–64

Taylor AW, Gurney EL (1964b) The dissolution of basic potassium and ammonium aluminum phosphates. Soil Sci Soc Am Proc 28:288–290

Taylor AW, Frazier AW, Gurney EL (1963a) Solubility products of magnesium ammonium and magnesium potassium phosphates. Trans Farad Soc 59:1580–1584

Taylor AW, Frazier AW, Gurney EL, Smith JP (1963b) Solubility product of di- and tri-magnesium phosphates and the dissociation of magnesium phosphate solutions. Trans Farad Soc 59:1585–1589

Truesdell AH, Jones BF (1964) WATEQ, a computer program for calculating chemical equilibria of natural waters. J Res US Geol Surv 2:233–248

Tsagareishvili DS, Gvelesiani GG, Orlovskii VP, Belyaevskaya TK, Rep'ko VP (1972) Enthalpies and specific heats of lanthanum, neodymium, and yttrium orthophosphate at high temperature. Izv Akad Nauk SSSR Neorg Mater 8:1790–1793 (in russian)

Tsagareishvili DS, Gvelesiani GG, Orlovskii VP, Belyaevskaya TV (1975) Enthalpy and heat capacity of scandium and europium orthophosphates at high temperatures. Izv Akad Nauk SSSR Neorg Mater 11:491–493 (in russian)

Turekian KK (1969) The oceans, streams and atmosphere. In: Wedepohl KH (ed) Handbook of Geochemistry. Springer, Berlin Heidelberg New York, p 297–323

Valyashko VM, Kogarko LN, Khodakovsky IL (1968) Stability of fluorapatite, chlorapatite, and hydroxylapatite in aqueous solutions at different temperature. Geoch Inter 5:21–30

Vieillard P (1978) Géochimie des phosphates. Etude thermodynamique. Application à la genèse et à l'altération des apatites. Mem Sci Géol 51:181

Vieillard P, Tardy Y, Nahon D (1979) Stability field of clays and aluminum phosphates: Parageneses in lateritic weathering of argillaceous phosphatic sediments. Am Miner 64:626–634

Vochten R, Piret P, Goeminne A (1981) Synthesis, crystallographic data, solubility and electrokinetic properties of copper, nickel, and cobalt uranylphosphate. Bull Miner 104:457–467

Volkov AI (1979) Thermochemical study of 3d element orthophosphates. Khim Khim Tekhnol (Minsk) 14:58–64 (in russian)

Volkov AI, Yaglov VN, Novikov GI (1974a) Heat of formation of trizinc diorthophosphate. Russ J Phys Chem 48:1697

Volkov AI, Yaglov VN, Novikov GI (1974b) Heat of formation of vanadyl phosphate and its hydrate. Russ J Phys Chem 48:1701

Volkov AI, Yaglov VN, Fakeev AA, Novikov GI (1975a) Thermodynamic study of iron III tris (dihydrogen) orthophosphate monohydrate. Khim Khim Tekhnol (Minsk) 9:29–35 (in russian)

Volkov AI, Yaglov VN, Glybin VP, Marinova LA (1975b) Determination of the standard heats of formation of scandium phosphates. Khim Khim Tekhnol (Minsk) 9:35–41 (in russian)

Volkov AI, Yaglov VN, Novikov GI (1976) Heat of formation of chromium orthophosphate hexahydrate. Vestsi Akad Navuk B SSR Ser Khim Navuk 120–121 (in russian)

Wagman DD, Evans WH, Parker VB, Halow I, Bailey SM, Schumm RH (1968) Selected values of chemical thermodynamic properties. Natl Bur Stand Tech Note 270-3:264

Wagman DD, Evans WH, Parker VB, Halow I, Bailey SM, Schumm RH (1969) Selected values of chemical thermodynamic properties. Natl Bur Stand Tech Note 270-4:152

Wagman DD, Evans WH, Parker VB, Halow I, Bailey SM, Schumm RH, Churney KL (1971) Selected values of chemical thermodynamic properties. Natl. Bur Stand Techn Note 270-5:49

Wagman DD, Evans WH, Parker VB, Schumm RH (1976) Chemical thermodynamic properties of compounds of sodium, potassium, and rubidium, an interim tabulation of selected values. Natl Bur Stand Techn Note 270:73

Waterfield CG, Staveley LAK (1967) Thermodynamic investigation of disorder in the hydrates of disodium hydrogen phosphate. Trans Farad Soc 63:2349–2356

Webber MD, Racz GJ (1970) Soluble complexes in the systems dicalcium phosphate dihydrate or dimagnesium phosphate trihydrate equilibrated with aqueous salt solutions. Can J Soil Sci 50:243–253

Wise WS, Loh SE (1976) Equilibria and origin of minerals in the system Al_2O_3–$AlPO_4$–H_2O. Am Miner 61:409–413

Yaglov VN (1978) Some characteristics of dehydration of hydrates of 3d elements orthophosphates. Khim Khim Tekhnol (Minsk) 13:7–14 (in russian)

Yaglov VN, Marinova LA, Novikov GI (1974) Thermodynamic features of the dehydration of cobalt and nickel phosphate crystal hydrates. Dokl Akad Nauk B SSR 18:624–627 (in russian)
Yaglov VN, Rud'ko PK, Bondar LA, Gerassimenko SV (1976) Thermodynamic characteristics of the crystal hydrates and anhydrous terbium phosphates. Khim Khim Ţekhnol (Minsk) 11:13–19 (in russian)
Zharovskii FG (1951) The solubility of phosphates. Trudy Kom Anal Khim Akad Nauk SSSR 3:101–115 (in russian)
Zolotavin VL, Kuznetsova VK (1956) Determination of the solubility product for vanadyl phosphate. Russ J Inorg Chem 1:2257–2259

Chapter 5 Phosphate Minerals in Meteorites and Lunar Rocks

VAGN FABRITIUS BUCHWALD

Introduction

Phosphorus is in meteorites and lunar rocks a minor, but in many respects important element. It occurs chemically bound as phosphates and phosphides, and it occurs in solid solution in the iron-nickel alloys kamacite and taenite. Broadly speaking, the phosphates are characteristic of the stone meteorites and of the inclusion-rich parts of the iron meteorites, while the phosphides, i.e., schreibersite and rhabdite, $(Fe, Ni)_3P$, are mainly present in iron meteorites. In pallasites, phosphates, and phosphides occur together in significant amounts.

From studies of the Solar photosphere and of carbonaceous chondrites the cosmic abundance of phosphorus atoms relative to silicon atoms, or relative to iron atoms, is found to be about 1:120, so one would perhaps not expect phosphorus compounds to be very important. However, their mere presence can be used to estimate the oxygen fugacity of the unknown location where the meteorites previously resided; this has, for example, been done by Olsen and Fuchs (1967). And in iron meteorites phosphorus has a pronounced influence upon the structural development below about 1,000 °C (Buchwald 1966; Romig and Goldstein 1980).

In meteorites and lunar rocks about a dozen phosphates are known, occurring as minor accessories, usually constituting less than 1 vol.-%. They are not all equally well defined, partly because they are rare and occur in rare rocks, partly because they are difficult to recognize and identify in small grains. It is characteristic that most of the extraterrestrial phosphates have first been found after the advent of the powerful electron microprobe. This means, however, that many of the phosphates are now known not only as to the ideal formulas, but in detail as to which minor elements occupy substitutional positions.

The list in Table 1 shows that five of the phosphates are peculiar to the extraterrestrial scenario, while the rest also are known on Earth. Except for chlorapatite and hydroxyapatite, the extraterrestrial phosphates are simply orthophosphates of the general type $M_3(PO_4)_2$, M representing varying amounts of major Ca, Mg, Fe, Mn or Na.

In the following we shall first treat the various meteorite classes, which have been studied for generations, and then turn to the new data deriving from the study of lunar rocks. It will be assumed that the reader is somewhat familiar with the phraseology within meteoritics and lunar work, otherwise the paper might have be-

Department of Metallurgy, Building 204, The Technical University of Denmark, 2800 Lyngby, Denmark

Table 1. List of named phosphates in extraterrestrial rocks

Whitlockite	$Ca_9(Mg, Fe)Na(PO_4)_7$	Ditrigonal	Frondel (1941), Fuchs (1962)
Chlorapatite	$Ca_5(PO_4)_3Cl$	Hexagonal	Fuchs and Olsen (1965)
Hydroxyapatite	$Ca_5(PO_4)_3OH$	Hexagonal	Fuchs (1969)
Farringtonite[a]	$(Mg, Fe)_3(PO_4)_2$	Monoclinic	DuFresne and Roy (1961)
Sarcopside	$(Fe, Mn)_3(PO_4)_2$	Monoclinic	Olsen and Fredriksson (1966)
Graftonite	$(Fe, Mn)_3(PO_4)_2$	Monoclinic	Olsen and Fredriksson (1966)
Brianite[a]	$Na_2CaMg(PO_4)_2$	Orthorhombic	Fuchs et al. (1967)
Panethite[a]	$Na_2(Mg, Fe)_2(PO_4)_2$	Monoclinic	Fuchs et al. (1967)
Stanfieldite[a]	$Ca_4Mg_3Fe_2(PO_4)_6$	Monoclinic	Fuchs (1967)
Buchwaldite[a]	$NaCaPO_4$	Orthorhombic	Olsen et al. (1977)

[a] Not reported on Earth

Table 2. Concentration of phosphorus, wt.%, in stone meteorites. Based on analytical data from Wiik (1956), Mason and Wiik (1964), Michaelis et al. (1969) and numerous individual papers mentioned in the text

Class	Name	Number analyzed	Range	Mean	Meteorite example
Chondrites					
E	Enstatite chondrites	9	0.08 –0.23	0.15	Hvittis
H	Olivine-bronzite-chondrites	34	0.02 –0.17	0.107	Pultusk
L	Olivine-hypersthene-chondrites	41	0.05 –0.15	0.105	Holbrook
LL	Amphoterites	12	0.08 –0.21	0.11	Soko-Banja
C1	Carbonaceous chondrites Type 1	4	0.05 –0.18	0.11	Orgueil
C2	Carbonaceous chondrites Type 2	11	0.07 –0.14	0.11	Murchison
C3	Carbonaceous chondrites Type 3	19	0.10 –0.17	0.12	Allende
Achondrites					
Ae	Aubrites	3	0.008	0.008	Norton County
Ah	Diogenites	2	0.005–0.006	0.0055	Johnstown
Au	Ureilites	3	0.03 –0.04	0.034	Haverö
Aho	Howardites	3	0.004–0.06	0.034	Malvern
Aeu	Eucrites	3	0.026–0.048	0.038	Pasamonte

come too voluminous. For a general introduction to meteoritics the reader may consult Mason (1962), Wasson (1974), Buchwald (1975), and Sears (1978), while the lunar work has been adequately summarized by Mason and Melson (1970); Taylor (1975), and King (1976).

Stone Meteorites

Phosphorus is a minor but ubiquitous element in all types of meteorites. It is determined either by wet chemical colorimetric methods (Moss et al. 1967) or by X-ray fluorescence methods (Michaelis et al. 1969). In stone meteorites phosphorus is reported as P_2O_5, but usually recalculated as elemental phosphorus, as has also been done here in presenting some selected data in Table 2. In many cases, when meteorites have been analyzed, the phosphates have not been separately studied, but have been assumed present and reported in the norm as unspecified "apatite".

In the common chondrites of type H, L, and LL the average weight percent phosphorus is 0.106. The range is, however, quite large, from about 0.02 in Menow (Hutchison et al. 1981) to 0.17 in Gorlovka (Kirova et al. 1977), and 0.21 in Alta'a-meem (Al-Bassam 1978). The range is apparently real and not due to inadequate chemical methods of identification. It reflects rather an uneven distribution of phosphates throughout the samples under investigation; the stone meteorites are known to be rather non-uniform in texture, but due to the rarity of specimens it is common practice to restrict the analytical sample to the order of 5–15 g.

All or almost all of the phosphorus in the H and L chondrites is present in two minerals, chlorapatite, and whitlockite. These minerals are best studied in polished sections, but the low abundance and uneven distribution may require quite extensive searches. The grains are usually 10–500 μm across, of irregular shape, and lacking crystal form (Fuchs 1969). However, chlorapatite has occasionally been reported as euhedral, colorless prisms, 300 μm long (Adib and Liou 1979), as subhedral crystals rimmed with whitlockite (Yagi et al. 1978), and as rare 1-mm-long grains (Graham et al. 1976). Typical analyses are presented in Table 3.

Whitlockite is apparently a problem in itself. At a time when analytical chemists were restricted to very small samples of phosphate minerals separated by tedious methods from crushed chondrites, the first meteoritic phosphate was reported and named merrillite (Shannon and Larsen 1925). Merrillite was supposed to be magnesium-free 3 CaO, Na_2O, P_2O_5. When Fuchs (1962) reexamined the type material, the two chondrites Allegan and New Concord, he was unable to confirm the merrillite, but supplied an analysis and X-ray diffraction patterns corresponding to whitlockite. This meteoritic whitlockite was composed of major Ca and P and minor, but significant, amounts of Na and Mg. Fuchs (1969) and later workers have thus found it difficult to sustain the magnesium-free, sodium-rich merrillite, and Strunz (1970) has discarded it in favor of whitlockite. Quite recently, however, Prewitt and Rothbard (1975) have reopened the question, suggesting that merrillite and whitlockite are distinct but closely related species, being compositional and structural variants of β-$Ca_3(PO_4)_2$. Unfortunately, the observations have only been presented in an abstract which is not readily accessible.

Meteoritic whitlockite is very homogeneous in its composition, not only in grains within the same meteorite, but also from meteorite to meteorite, and from class to class. The composition of the grains analyzed in Table 4 correspond well to the formula $Ca_9(Mg, Fe)Na(PO_4)_7$, proposed for terrestrial whitlockite by Calvo and Gopal (1975). Only in some pallasites and mesosiderites, exemplified by the meteorites Imilac, Marjalahti and Vaca Muerta, is there a significant depletion in sodium and a slight enrichment in iron (Buseck and Holdsworth 1977).

While the major elements in phosphates have been determined with the electron microprobe, minor and trace elements have been studied by spark source mass spectrometry. Mason and Graham (1970) examined nine different minerals, among them an unspecified phosphate fraction from various meteorites, for a large number of elements. It was found that the phosphate minerals (chlorapatite and whitlockite) accounted for most of the lanthanides, yttrium, uranium, and thorium in the chondrites. When apatite was present, it contained essential amounts of chlo-

Table 3. Electron microprobe analyses of chlorapatite, $Ca_5(PO_4)_3Cl$ (wt. %)

Meteorite	Type	Ca	P	Na	Fe	C1	F	OH[a]	Reference
New Concord	L6	37.9	18.2	0.4	0.2	4.8	–	–	Fuchs (1969)
Djati-Pengilon	H6	38.7	18.7	0.3	0.4	5.4	0.4[a]	0.4	van Schmus and Ribbe (1969)
Forksville	L6	38.3	18.6	0.4	0.4	5.3	0.4	0.5	van Schmus and Ribbe (1969)
Manbhoom	LL6	37.8	17.4	0.3	0.5	5.4	0.4	0.3	van Schmus and Ribbe (1969)
Vilna	L5	37.9	17.8	0.19	0.47	4.6	0.51	–	Smith et al. (1973)
Naragh	H6	38.5	18.0	0.32	0.32	5.5	–	0.56	Adib and Liou (1979)
Odessa (Iron)	IA	38.4	18.3	0.21	0.17	3.6	1.1	–	Bunch et al. (1970)
Landes (Iron)	IA	37.9	18.0	0.36	0.20	6.1	<0.05	–	Bunch et al. (1972)

[a] Estimated

Table 4. Electron microprobe analyses of whitlockite, $Ca_9(Mg, Fe)Na(PO_4)_7$ (wt. %)

Meteorite	Type	Ca	P	Na	Fe	Mg	Others	Reference
Allegan	H5	33.2	19.8	2.3	0.2	2.2		Fuchs (1969)
New Concord	L6	33.2	19.8	2.3	0.5	2.1		Fuchs (1969)
Djati-Pengilon	H6	33.9	21.0	2.1	0.3	2.1		van Schmus and Ribbe (1969)
Forksville	L6	33.1	20.3	2.1	0.4	2.1		van Schmus and Ribbe (1969)
Vilna	L5	34.0	20.1	1.9	0.41	2.0		Smith et al. (1973)
Naragh	H6	33.4	20.3	2.1	0.35	2.0		Adib and Liou (1979)
Dayton	IIID	33.0	20.0	2.5	0.2	2.2		Fuchs et al. (1967)
Imilac	Pall.	33.9	19.8	0.4	0.9	3.2	0.4 Si	Buseck and Holdsworth (1977)
Marjalahti	Pall.	33.8	20.2	<0.1	0.8	2.6		Buseck and Holdsworth (1977)
Springwater	Pall.	32.1	19.8	2.2	1.3	2.3	0.2 Si	Buseck and Holdsworth (1977)
Vaca Muerta	Mesos.	33.6	20.3	n.d.	2.4	2.3	0.2 Mn	Malissa (1974)
Landes (Iron)	Ia	33.4	20.0	2.2	0.3	2.3		Bunch et al. (1972)

rine and fluorine. Other elements enriched in the phosphates included Sc, Sr, Zr, Ba, and Hf.

Pellas (1981) and Pellas and Storzer (1981) studied the fission-track densities of whitlockite in a suite of H and L chondrites. They found that 99% of the tracks were due to fission of now extinct ^{244}Pu. The plutonium content was estimated to be 10–30 ppb (parts per billion), the uranium content to be below 200 ppb. The uranium content of apatites was about 10 times higher. From the calibrated fission-track density, it was possible to deduce cooling rates for the chondrites of the order of 2°–20 °C per million years.

The carbonaceous chondrites in Table 2 display almost the same phosphorus abundance as H and L chondrites, however, it has proven quite difficult to assign the phosphorus to specific minerals. Fuchs (1969) reported two chlorapatite grains in Karoonda, but found none in Warrenton and Lance. Clarke et al. (1970) found none in Allende, and Fuchs et al. (1973) none in Murchison. In general, while phosphorus is present on the 0.1 wt.-% level, all authors agree that both phosphates and phosphides are rare, so it appears that the P-atoms are in substitutional positions in other minerals.

The enstatite chondrites (Table 2) have 50% more phosphorus than other chondrites. Phosphates are, however, not present, but instead schreibersite in association with the abundant metallic grains. This observation is in accord with the general opinion that the enstatite chondrites have been subjected to a higher degree of reduction than other chondrites.

The achondrites (Table 2) are characterized by a very low concentration of phosphorus. Phosphate grains are only occasionally reported, for example recently in an Antarctic meteorite, Yamato (e), a howardite (Yagi et al. 1978), where (chlor) apatite grains were rimmed by whitlockite. Likewise, Steele and Smith (1976) analyzed a small amount of whitlockite in the eucrite Ibitira, and Smith and Hervig (1979) a few grains of whitlockite in Shergotty.

Hydroxyapatite has only been reported once, in the anomalous stone meteorite Shergotty (Fuchs 1969), which is characterized by significantly shock-altered minerals, such as maskelynite. Fuchs performed an experiment to support the idea that the hydroxyapatite of Shergotty had been formed by a shock reaction between whitlockite and traces of water.

Pallasites

Phosphates appear to be essential, albeit minor constituents of pallasitic meteorites. In a recent study, Buseck and Holdsworth (1977) observed phosphate minerals in 31 out of 41 studied pallasites. In decreasing order of abundance whitlockite, stanfieldite and farringtonite were found. Whitlockite was mentioned above (Table 4) as slightly more variable than the chondritic whitlockite.

Stanfieldite was first described by Fuchs (1967), who found it as irregular to subhedral grains or as veinlets, that penetrated cracks in olivine. The mineral was transparent with an amber tint and usually 0.1–1 mm in size.

Farringtonite was first described by DuFresne and Roy (1961) in the pallasite Springwater, where the phosphate formed mm^2-sized areas, filling interstices be-

Table 5. Electron microprobe analyses of stanfieldite and farringtonite from the Springwater pallasite (wt.%). (Buseck and Holdsworth 1977)

	Ca	P	Na	Fe	Mg	Mn	Si
Stanfieldite	18.3	21.8	0.08	2.9	12.7	0.43	0.01
Farringtonite	0.06	22.9	<0.1	3.2	26.2	0.19	0.06

tween olivine grains and metal. Farringtonite is essentially $Mg_3(PO_4)_2$, with some Fe^{2+} substitution for Mg, but with only traces of silicon (Olsen and Fredriksson 1966; Fuchs et al. 1973a). Typical analyses of stanfieldite and farringtonite from the same meteorite, Springwater, are given in Table 5.

In a comprehensive study of pallasites, Davis (1977) analyzed unspecified phosphates concentrated from the Springwater pallasite (as we may note from consulting Tables 4 and 5, Springwater contains three different phosphates). Davis found that the rare earth elements were strongly concentrated in the phosphate minerals relative to olivine, lanthanum and cerium alone adding up to about 50 ppm.

Most of the phosphates in pallasites apparently participated in the bulk development of the complex textures. Buseck and Holdsworth (1977) thus found that they were liquid after the olivine had crystallized and acted as a residual melt, filling interstices between olivine crystals and metal. Since phosphides and phosphates coexist in most, if not all, pallasites, relatively uniform conditions with respect to oxygen pressure must have prevailed when the pallasites developed. This is in contrast to what will be shown to be the case for iron meteorites.

Mesosiderites

This meteorite class displays complex, heterogeneous textures and variable mineral compositions. The metallic fraction constitutes about 50% by weight, while orthopyroxene, plagioclase, troilite, and accessories constitute the other 50%. In some mesosiderites, like Bondoc and Pinnaroo, the metal may occur as inch-sized globules, while in others it forms more discrete networks or individual grains. The phosphorus content of mesosiderites is not so well known as for stone and iron meteorites, because of this heterogeneity, but it may be estimated to be about 0.2 wt.-%, averaging over both stony and metallic fractions of the mesosiderites. The stony fraction probably accounts for 2/3 of the total phosphorus.

While schreibersite is responsible for phosphorus in the metallic fraction, it appears that four of the phosphates already discussed account for all of the P of the stony fraction: whitlockite, chlorapatite, farringtonite, and stanfieldite. Whitlockite coexists with chlorapatite in Vaca Muerta (Marvin and Klein 1964), with stanfieldite in Estherville (Fuchs 1967), and with farringtonite in Mincy (Malissa 1974). Whitlockite seems to be the only phosphate mineral in Emery, Barea, and Dyarrl Island (Mason and Jarosewich 1973), and in Patwar and Morristown (Fuchs (1969).

Prinz et al. (1980) examined all 21 mesosiderites and presented modal analyses of the silicate fractions. They found that phosphates in the silicate fraction ranged from nil to 3.7 vol.-% with a mean of 2.0%, and furthermore observed that the presence of phosphates was a distinctive characteristic of mesosiderites. Already Powell (1971) had suspected that whitlockite was an accessory component of all mesosiderites.

The composition of typical whitlockite from mesosiderites is entered in Table 4; there seems to be more iron in this whitlockite than in others, and Malissa (1974) also found a small amount (~0.2–0.5%) of manganese in mesosiderite phosphates.

Bull and Durrani (1980) examined the distribution of fossil tracks from the decay of ^{238}U in whitlockite from the Bondoc mesosiderite. By comparison with terrestrial apatite standards of well-defined age, they deduced an age of 2.68×10^9 years for Bondoc.

Iron Meteorites

There is more phosphorus in iron meteorites than in stone meteorites, cf. Table 6 with Table 2. The range in iron meteorites is from 0.01 to 2.1 wt.-%, but within individual groups the range is significantly smaller, in the most important group, IIIA, thus only from 0.08% to 0.37%. For the purpose of Table 6 the average phosphorus concentration has been calculated in two ways. The arithmetic mean of all 486 analyzed iron meteorites is 0.28 wt.-%. If, on the other hand, it is calculated on the basis of only the 32 most massive iron meteorites, each weighing more than 1,000 kg, a similar concentration, 0.26 wt.-%, is obtained. Both values are probably a little on the low side, because the larger schreibersite inclusions often escape sampling, when milling material for wet chemical analysis. In the author's opinion, a good bulk average value for the phosphorus content of iron meteorites is 0.30 wt.-%.

Virtually all phosphorus is in solid solution in the metal or occurs as phosphide, in two forms of the same mineral $(Fe, Ni)_3P$, called schreibersite and rhabdite. Traditionally the term rhabdite is restricted to mostly microscopic prismatic needles and platelets, while schreibersite covers macroscopic inclusions, rims on other minerals, and the microscopic, irregular grains in plessite and on grain boundaries.

Phosphates are rare and usually only present in small grains. Only lately, after the introduction of the electron microprobe, the first phosphate, chlorapatite, was reported from Odessa by Marshall and Keil (1965). Then Olsen and Fredriksson (1966) and Olsen and Fuchs (1967) found mm-sized euhedral grains of sarcopside in Bella Roca, Chupaderos, Sam's Valley, Verkhne Dnieprovsk, and Grant. Coexisting with sarcopside were found minor amounts of graftonite, both minerals calcium-free, in contrast to the terrestrial occurrences (Table 7).

Fuchs (1969), Bunch et al. (1970), and Bunch et al. (1972) presented a number of microprobe analyses on chlorapatite and whitlockite from iron meteorites; typical ones have been included in Tables 3 and 4. Buchwald (1971, 1975) reported small sarcopside crystals as nucleation sites for schreibersite in Wiley, and similar phosphates were also identified by Bild (1974). Fuchs et al. (1967) reported two entirely new minerals, brianite and panethite (Table 8) from Dayton, and Bild (1974) added San Cristobal as a brianite occurrence. Buchwald (1975, Fig. 176) and Kracher et al. (1977) discussed a number of difficultly identifiable phosphates in

Table 6. Concentration of phosphorus, weight percent, in 486 iron meteorites, group for group (Esbensen and Buchwald 1979). Iron meteorites known to contain phosphates in the last column

Group Structure	Number analyzed	Range	Mean	Meteorites with phosphates, number and names
IA Coarse octahedrites	75	0.12–0.40	0.25	13 Annaheim, Campo del Cielo, Copiapo, Goose Lake, Landes, Linwood, Odessa, Pine River, Shrewsbury, Toluca, Udei Station, Wichita County, Youndegin
IB Various	8	0.13–0.47	0.23	2 San Cristobal, Woodbine
IC Various	10	0.17–0.39	0.28	1 Arispe
IIA Hexahedrites	42	0.20–0.34	0.27	
IIB Coarsest octahedrites	14	0.25–0.90	0.48	
IIC Plessitic octahedrites	7	0.30–0.50	0.40	1 Wiley
IID Medium octahedrites	14	0.24–0.90	0.45	
IIIA Medium octahedrites	120	0.08–0.37	0.16	3 Avoca, Cape York, View Hill
IIIB Medium octahedrites	38	0.24–1.00	0.60	13 Augustinovka, Bear Creek, Bella Roca, Chupaderos, Delegate, Grant, Ilinskaya Stanitza, Mt. Edith, Sams Valley, Sanderson, Smiths Mt., Thurlow, Turtle River
IIICD Fine-finest octahedrites	15	0.15–0.60	0.29	1 Dayton
IIIE Coarse octahedrites	10	0.18–0.43	0.28	
IIIF Various	5	0.18–0.28	0.24	
IVA Fine octahedrites	38	0.02–0.18	0.04	2 Duchesne, Duel Hill (1854)
IVB Ataxites	10	0.05–0.19	0.07	
Ungrouped Various	80	0.01–2.1	0.20	8 Barranca Blanca, Britstown, Colomera, De Hoek, Kodaikanal La Primitiva, Netschaevo, Weekeroo Station
Total	486	0.01–2.1	0.28	

Table 7. Electron microprobe analyses of sarcopside, $(Fe, Mn)_3(PO_4)_2$ (wt. %)

Iron meteorite	Group	Ca	P	Na	Fe	Mg	Mn	Reference
Bella Roca	IIIB				46.3		1.3	Olsen and Fredriksson (1966)
Bear Creek	IIIB	<0.02	17.8		45.1	<0.01	1.4	Bild (1974)
Duel Hill (1854)	IVA	0.04	17.4	<0.01	47.7	<0.01	<0.01	Bild (1974)

Table 8. Electron microprobe analyses of three rare phosphate minerals (wt. %)

sphate	Meteorite	Group	Ca	P	Na	Fe	Mg	Mn	K	Reference
nite	Dayton	IIICD	13.5	20.4	16.4	0.4	7.6			Fuchs et al. (1967)
nite	San Cristobal	IB	13.6	20.9	15.4	0.02	8.1			Bild (1974)
ethite	Dayton	IIICD	4.0	20.8	11.3	4.1	14.5	1.3	0.7	Fuchs et al. (1967)
hwaldite	Cape York	IIIA	25.5	19.6	14.5					Olsen et al. (1977)

the typical iron meteorite, Cape York, and recently Olsen et al. (1977) identified one of these phosphates, $CaNaPO_4$, and named it buchwaldite. Bild (1977) found silicates and three phosphates, (chlor) apatite, whitlockite, and brianite in Britstown. A brief overview of phosphates was given by Buchwald (1977), who surveyed the mineralogy of iron meteorites in general.

Until the early 1960's potassium, sodium, magnesium, calcium, and manganese in meteorites were considered to be exclusively present in the various silicate minerals. From what has been said above it will, however, be noted that the situation has changed significantly. It appears that the phosphates, besides a few sulfides such as alabandite, djerfisherite, and sphalerite, have concentrated these elements rather actively, while the metal phase and troilite have become depleted in them. Furthermore, especially whitlockite has concentrated the rare earth elements.

There are controversial opinions with regard to the occurrence of chlorine in iron meteorites. In the author's opinion, chlorine is on a very low level, about 10 ppm or less, and this chlorine is mainly found in scattered chlorapatite inclusions. Others maintain that iron meteorites contain much more chlorine, and that this is present in the cosmic mineral lawrencite, $FeCl_2$ (see e.g., Taylor and Burton 1976). However, Buchwald (1977, p 484) has given good reasons that this is a misinterpretation of chloride formed by the action of aggressive terrestrial groundwater while the iron meteorite was buried in the soil for countless generations. Lawrencite has thus never been seen, nor suspected in a freshly fallen iron meteorite.

The observations quoted above, together with the author's own observations, are collected in Table 6, right column. A total of 44 iron meteorites, out of 486 examined, have now been shown to contain phosphates.

The phosphates are nowhere common, but they do occur with significant frequency in group IA and IIIB. In IA they are usually found in polymineralic troilite nodules, where silicates are frequently present; in IIIB they are concentrated in the residual liquid, the troilite nodules that may contain chromite in addition, but rarely (if ever) silicates.

Table 9. Mineral differentiation in iron meteorites. Within each iron meteorite group, phosphates and chromite only occur in the nickel-rich members

Group	Meteorite	% Ni	Minerals	Troilite
IIA	Coahuila	5.59	Daubreelite, cohenite	+
IIB	Sikhote-Alin	5.94	Chromite, olivine; phosphates?	++
IIC	Unter-Mässing	9.9	Daubreelite	+
IIC	Wiley	11.6	Phosphates	+
IIIA	Kenton County	7.45	Daubreelite	+
IIIA	Sacramento Mts.	7.93	Chromite, some daubreelite	+
IIIA	View Hill	8.87	Phosphates	++
IIIB	Baquedano	8.8	Daubreelite	+
IIIB	Bear Creek	9.9	Chromite, phosphates	++
IVA	Gibeon	7.93	Daubreelite, some chromite	+
IVA	Altonah	8.54	Chromite, some daubreelite	+
IVA	Duchesne	9.37	Chromite, phosphates	++
IVB	Tlacotepec	16.2	Daubreelite	+
IVB	Weaver Mts.	17.8	Unidentified oxides	+

In organizing the information now available about phosphates, it becomes clear that there is a trend within the individual chemical groups, Table 9. In the nickel-poor end of group IIIA, there are no phosphates and no chromite, Fe_2CrO_4, but instead daubreelite, Fe_2CrS_4, and phosphides. In the nickel-rich end there are chromite and phosphates, but no daubreelite. In between, members may be found, such as the Sacramento Mountains, that exhibit daubreelite and chromite side by side. Similar trends are well displayed in groups IIIB, IVA, IIC and, probably, IIAB. In the last mentioned, only chromite and olivine have been reported so far, but phosphates may be suspected in the nickel-rich end.

It appears that each of these groups, which for many reasons are all believed to have been entirely molten at one time, represents a redox system, where the oxygen partial pressure was highest in the nickel-rich end. Here, part of the phosphorus and chromium and all silicon were converted into oxygen-bearing minerals, probably already when the system was molten. Part of the iron was also removed, probably to form olivine and pyroxene, leading to an increased nickel concentration in the residue. Many of the new minerals disappeared as slag (to form other meteorites, possibly of pallasite type), but some were trapped and included with the low-melting, sulfide-rich residual melt. Finally, in the solid metal some chromite and phosphate precipitated out of solid solution to form scattered chromite and phosphate crystals.

Although the number of observations is still small, the systematic presence of chromite, phosphates, and silicates in only the nickel-enriched, i.e., iron-depleted meteorites, clearly points to a geologic-planetary scenario where oxygen is much more significant in iron meteorites than previously believed. For the mineralogist interested in hunting phosphates, it also indicates that the chances of finding them are best in the nickel-rich members of each group.

Lunar Rocks

Between 1969 and 1972, eight successful lunar missions brought back samples of the lunar surface for analysis. Apollo 11, 12, 15, and 17, and Luna 16 probed various Mare sites, while Apollo 14 and 16, and Luna 20 probed the highlands. The mare rocks proved to be generally of a basaltic nature, but the classification of them presents a challenge to petrologists, and this is still more true of the highlands rocks, which have few terrestrial analogs. All lunar rocks have been greatly altered by characteristic intense, repeated impact bombardment, the influx of "meteoritic" projectiles having been assimilated in the rocks, breccias, soils, and glasses.

However, the mineralogy is relatively simple on the Moon, because of the lack of water and chemical weathering. Only about 100 minerals have been reported compared to the more than 2,000 known on the Earth. A good survey of lunar mineralogy has been given by Frondel (1975), but anyone interested in the petrology and mineralogy of the lunar rocks must consult the extensive collection of papers that are annually edited by the Lunar and Planetary Institute and published by Pergamon Press as Proceedings of the Lunar and Planetary Science Conferences, in Houston. Twelve conferences have been held to date, resulting in 36 thick volumes of information. An index volume to the first nine conferences, 1970–1978, appeared in 1979.

The very large number of electron microprobe and wet chemical analyses performed up to now show that phosphorus is always present in lunar material, albeit at a rather low level of about 0.05 wt.-% on the average. Apatite and whitlockite have occasionally been reported, while farringtonite is only known from one rock. The rare earths (REE) and yttrium have ionic radii similar to calcium and are concentrated in the calcium-rich minerals, on the Moon especially whitlockite, apatite, and zirconolite ($CaZrTi_2O_7$). Apatite is presumably the major repository for fluorine and chlorine, and uranium and thorium seem also to be concentrated in the phosphates (Jovanovic and Reed 1980).

The phosphate grains are small, usually less than 5 μm across. For example, French et al. (1970) found only trace amounts of apatite in the Apollo 11 basalt 10017; the apatite formed very small crystals in the mesostasis, i.e., the fine-grained residual liquid which fills the interstices between major pyroxene and plagioclase crystals. In the apatite, they found 0.20 La, 0.80 Ce, 0.09 Pr, 1.90 Nd, and 0.09 Gd (wt.-%), a significant fraction of all the REE of this rock.

Keil et al. (1970) reported apatite from the basaltic rock 10045, but found the crystals too small for accurate probe work. Cl and F were searched for and Cl was detected in one grain, with a value of 0.34%. No F was detected. Keil et al. (1971) analyzed yttrian-cerian whitlockite and chlorine-bearing fluorapatite from two Apollo 12 igneous rocks, 12036 and 12039. Again, the grains were few and small, but the REE concentration was high, especially in the whitlockite, Table 10.

Dowty et al. (1973) analyzed 28 Mare basalts from the Apollo 15 site; phosphorus ranged from 0.01 to 0.08 wt.-%, and apatite and whitlockite were found to be present in most rocks.

McCallum and Mathez (1975) reported whitlockite and chlorapatite in chips from the 0.5 m norite boulder, sampled in the Apollo 17 mission (78235). From Table 10 it is seen that the whitlockite is rich in rare earth elements, while the

Table 10. Electron microprobe analyses (wt.%) of phosphates in lunar rocks

	Whitlockite			Apatite			Farringtonite
	1	2	3	4	5	6	7
SiO_2	0.54	2.29			0.71		
TiO_2	0.04						
Al_2O_3	0.25	1.26		0.49	0.06		0.09
FeO	2.13	0.89	0.30		0.18		3.5
MgO	3.8	3.28	3.21	0.05	0.25		40.8
CaO	42.3	39.94	41.13	52.8	53.56	53.87	0.59
Na_2O	0.33	0.47		0.20	0.01		÷
P_2O_5	42.5	40.28	42.92	39.6	39.52	40.97	51.0
Cl	0.02			0.30	0.74	1.20	
F	0.05			2.5	3.10	2.75	
Y_2O_3	2.28		2.02				
Ce_2O_3	1.69		2.06	2.6			÷
Dy_2O_3	0.37		0.36				
Ho_2O_3	0.17						
Er_2O_3	0.18	10.71^{+}	0.16		2.55^{+}		
La_2O_3	0.50		0.83	0.9			÷
Gd_2O_3	0.50		0.64				
Lu_2O_3	0.10						
Pr_2O_3	0.30		0.47	0.5			
Nd_2O_3			1.55	1.4			
Sm_2O_3			0.39				
Total	98.05	99.18	96.04	100.26	100.68	98.79	

1 and 4 from Keil et al. (1971), 2 and 5 from Albee et al. (1973), 3 and 6 from McCallum and Mathez (1975), and 7 from Dowty et al. (1974). Open spaces: not analyzed for these elements, ÷ looked for, but not detected; + sum of REE

apatite has accommodated fluorine and chlorine, the general picture for these phosphates when they are the last to crystallize from a trapped, highly fractionated liquid.

Tarasov et al. (1973), reporting on Luna 20, noted that phosphates were frequently found in Mare and in anorthositic rocks, especially interstitially with potassium-rich glasses in basalt of Mare type. However, the grains did not exceed 2–3 μm and were not investigated in detail.

Simonds et al. (1973) discussed the petrology of shock-generated rocks from the Apollo 16 site. Common to these was a rather high phosphorus content, ranging from 0.1 to 0.2 wt.-%; most of the small phosphate grains were located within the residual melt regions.

Albee et al. (1973) studied metamorphic rocks from the Apollo 16 and 17 missions. In one sample, 65015, whitlockite was abundant, while apatite was not found. The whitlockite formed needles up to 40 μm in length. In other samples, the phosphates formed REE-poor reaction rims around iron metal globules. Analyses are shown in Table 10. The isolated whitlockite grains are probably more enriched in REE than any other phosphate reported from the lunar material.

Dowty et al. (1974) analyzed an igneous rock, a so-called spinel troctolite, from the Apollo 16 site and concluded that this was a good candidate for the primary igneous cumulate formed during the original differentiation of the lunar crust. The bulk analysis of this rock showed only 0.02 wt.-% P, however both whitlockite and farringtonite were present (Table 10).

The Apollo 14 material appears to be identical in composition to the foreign fractions in the Apollo 12 regolith that were termed KREEP, i.e., a basalt enriched in potassium, rare earth elements and phosphorus.

It is speculated that this material is related to the Imbrium Basin or its throwout. The KREEP material is of a brecciated or glassy nature and has been well examined by, for example, Meyer et al. (1971). Phosphorus ranges from 0.1% to 0.7%, with an average of 0.4 wt.-%, and the minerals are apatite and whitlockite.

Vugs and vesicles are well known in basalts from the Apollo missions, and many of them have been examined for their mineral contents. McKay et al. (1972) studied a number of vugs in some of the Apollo 14 recrystallized breccias, and they found, among other minerals, perfect crystals of apatite. Whitlockite was rare. The apatite contained about 2% chlorine and formed both doubly terminated prisms, up to 50 µm long, and more stubby, micron-sized clusters, always on a substrate of plagioclase or pyroxene crystals. The crystals in vugs were shown to have been vapor-derived.

The lunar surface has been intensely reworked by impacts and this has crushed, ground and mixed the original surface components to an extent which is very difficult to comprehend. But it has also created glasses as matrix, as agglutinates and chondrules, and hot ejecta-blankets from large impacts have modified underlying material by recrystallization and mobilization of gases. Impact melts, created from the existing soils and rocks, have solidified as new igneous rocks, and these have in turn been comminuted by renewed impacts.

In order to bracket the conditions of formation, many experiments have been carried out on both synthetic and original lunar rocks and soils, only a few papers, which are concerned with phosphates, will be mentioned here. Taylor et al. (1976) examined the redistribution of phosphorus during long-time annealing. They found, for example, that in a basalt, rich in mesostasis and glass, phosphorus could diffuse out of the metal grains at 875 °C and form whitlockite. The experimental evidence supported the proposal by Agrell et al. (1976) that the phosphates commonly found adjacent to metal grains may be the result of a reaction in which phosphorus is oxidized out of the metal, either from solid solution or from schreibersite. Fuchs (1969) had already noted that the whitlockite in the mesosiderite Morristown was closely associated with the kamacite metal, wherefore he suggested that the phosphate originated by reaction of phosphorus in solid solution in the metal with the silicate minerals, above 750 °C.

On the other hand, Friel and Goldstein (1977) showed that such a redox reaction can go either way depending on three major factors: oxygen fugacity, temperature, and cooling rate. They also found that there was a difference in whitlockite composition according to location relative to iron-nickel grains; independent whitlockites contained nearly 8 wt.-% REE, while those adjacent to metal grains rarely had any. This is consistent with the formation of REE-free whitlockite by oxidation of phosphorus out of the metal, while REE-rich whitlockite formed from residual liquid, since the whitlockite structure, as noted above, is one of the few which is compatible with rare-earth ions.

Conclusion

Phosphates are minor accessories in meteorites and lunar rocks. They are water-free, except for one occurrence of hydroxyapatite in a rare type of meteorite, Shergotty. They are rather simple in structure, belonging to only three groups within Strunz's classification (1970): water-free, no foreign anions, medium-sized cations (graftonite, sarcopside, farringtonite); water-free, no foreign anions, large cations (whitlockite; stanfieldite, panethite, brianite, buchwaldite), and water-free, foreign anions, large cations (chlorapatite, hydroxyapatite).

The phosphates are small and inconspicuous, and they occur scattered, often in association with residual liquids. Especially the whitlockite is enriched in the rare earth elements, and sometimes also in such elements as scandium, thorium and uranium. Chlorapatite is the major repository for chlorine in iron meteorites, and possibly has a similar role in other meteorites and in lunar rocks.

The iron meteorites are, on the average, richest in phosphorus, with about 0.3 wt.-%, and display the best developed and most diversified phosphates. Most of the phosphorus is, however, present as phosphides and in solid solution in the metal. The lunar rocks and the achondrites are poorest in phosphorus, 0.05 wt.-% or below, and contain mainly micron-sized phosphates, usually whitlockite and fluor- or chlorapatite. There is certainly no basis for an economic recovery of phosphate rocks from the Moon!

References

Adib D, Liou JG (1979) The Naragh meteorite: A new olivine bronzite chondrite fall. Meteoritics 14:257–272

Agrell SO, Axon HJ, Goldstein JI (1976) A metallographic and petrological study of metal-silicate fragments in the >1 mm size range from lunar soil 14162,80. Miner Mag 40:565–577

Al-Bassam KS (1978) The mineralogy and chemistry of the Alta'ameen meteorite. Meteoritics 13:257–265

Albee AL, Gancarz AJ, Chodos AA (1973) Metamorphism of Apollo 16 and 17 and Luna-20 metaclastic rocks at about 3.95 AE. Proc Fourth Lunar Sci Conf, p 569–595

Bild RW (1974) New occurrences of phosphates in iron meteorites. Contrib. Miner Petrol 45:91–98

Bild RW (1977) Compositions of silicate inclusions as an aid in the classification of iron meteorites. Meteoritics 12:177

Buchwald VF (1966) The iron-nickel-phosphorus system and the structure of iron meteorites. Acta Polytech Scan 51:1–46

Buchwald VF (1971) Tritium loss resulting from cosmic annealing, compared with the microstructure and microhardness of six iron meteorites. Chem Erde 30:33–57

Buchwald VF (1975) Handbook of iron meteorites. Their history, distribution, composition, and structure. Univ California Press, Berkeley, vol 1–3

Buchwald VF (1977) The mineralogy of iron meteorites. Philos Trans R Soc Lond Math Phys Sci A286:453–491

Bull RK, Durrani SA (1980) Fission-track retention age of the Bondoc mesosiderite. Earth Planet Sci Let 49:181–187

Bunch TE, Keil K, Olsen E (1970) Mineralogy and petrology of silicate inclusions in iron meteorites. Contrib Miner Petrol 25:297–340

Bunch TE, Keil K, Huss GI (1972) The Landes meteorite. Meteoritics 7:31–38

Buseck PR, Holdsworth E (1977) Phosphate minerals in pallasite meteorites. Miner Mag 41:91–102

Calvo C, Gopal R (1975) The crystal structure of whitlockite from the Palermo quarry. Am Miner 60:120–133

Clarke RS Jr, Jarosewich E, Mason B, Nelen J, Gomez M, Hyde JR (1970) The Allende, Mexico, meteorite shower. Smithson Contrib Earth Sci 5:1–53
Davis AM (1977) The cosmochemical history of the pallasites. Diss NGR-07-004-166, Yale University
Dowty E, Prinz M, Keil K (1973) Composition, mineralogy, and petrology of 28 mare basalts from Apollo 15 rake samples. Proc Fourth Lunar Sci Conf, p 423–444
Dowty E, Keil K, Prinz M (1974) Igneous rocks from Apollo 16 rake samples. Proc Fifth Lunar Sci Conf, p 431–445
DuFresne ER, Roy SK (1961) A new phosphate mineral from the Springwater pallasite. Geochim Cosmochim Acta 24:198–205
Esbensen K, Buchwald VF (1979) A database for iron meteorite data. Statistical analysis of the global iron meteorite population. Meteoritics 14:573–576
French BM, Walter LS, Heinrich KJF (1970) Quantitative mineralogy of an Apollo 11 lunar sample. Proc First Lunar Sci Conf, p 433–444
Friel JJ, Goldstein JI (1977) The relationship between lunar metal particles and phosphate minerals. Proc Eight Lunar Sci Conf, p 3955–3965
Frondel C (1941) Whitlockite, a new calcium phosphate, $Ca_3(PO_4)_2$. Am Miner 26:145–152
Frondel JW (1975) Lunar mineralogy. Wiley & Sons, New York
Fuchs LH (1962) Occurrence of whitlockite in chondritic meteorites. Science 137:425–426
Fuchs LH (1967) Stanfieldite, a new phosphate mineral from stony-iron meteorites. Science 158:910–911
Fuchs LH (1969) The phosphate mineralogy of meteorites. In: Millman PM (ed) Meteorite research. Reidel, Dordrecht, Holland, p 683–695
Fuchs LH, Olsen E (1965) The occurrence of chlorapatite in the Mount Stirling octahedrite. Trans Am Geophys Un 46:122
Fuchs LH, Olsen E, Henderson EP (1967) On the occurrence of brianite and panethite, two new phosphate minerals from the Dayton meteorite. Geochim Cosmochim Acta 31:1711–1719
Fuchs LH, Olsen E, Gebert E (1973a) New X-ray and compositional data for farringtonite. $Mg_3(PO_4)_2$. Am Miner 58:949–951
Fuchs LH, Olsen E, Jensen KJ (1973b) Mineralogy, mineral-chemistry, and composition of the Murchison (C2) meteorite. Smithson Contrib Earth Sci 10:1–39
Graham AL, Easton AJ, Hutchison R, Jérome DY (1976) The Bovedy meteorite, mineral chemistry and origin of its Ca-rich glass inclusions. Geochim Cosmochim Acta 40:529–535
Hutchison R, Bevan AWR, Easton AJ, Agrell SO (1981) Mineral chemistry and genetic relations among H-group chondrites. Proc R Soc Lond A374:159–178
Jovanovic S, Reed GW Jr (1980) Cl, P_2O_5, U, and Br associated with mineral separates from a low and a high Ti mare basalt. Proc Eleventh Lunar Planet Sci Conf, p 125–134
Keil K, Bunch TE, Prinz M (1970) Mineralogy and composition of Apollo 11 Lunar samples. Proc First Lunar Sci Conf, p 561–598
Keil K, Prinz M, Bunch TE (1971) Mineralogy, petrology, and chemistry of some Apollo 12 samples. Proc Sec Lunar Sci Conf, p 319–341
King EA (1976) Space geology. An introduction. Wiley & Sons, New York
Kirova OA, Dyakonova MI, Charitonova WJ, Levskij LK (1977) A study of the chondrite Gorlovka. Meteoritika 36:46–52 (in russian)
Kracher A, Kurat G, Buchwald VF (1977) Cape York: The extraordinary mineralogy of an ordinary iron meteorite and its implication for the genesis of IIIAB irons. Geochem J 11:207–217
Malissa H (1974) Electron microprobe analysis of some minor and accessory components in mesosiderites. Tschermaks Miner Petrol Mitt 21:233–245
Marshall RR, Keil K (1965) Polymineralic inclusions in the Odessa iron meteorite. Icarus 4:461–479
Marvin UB, Klein C (1964) Meteoritic zircon. Science 158:910
Mason B (1962) Meteorites. Wiley & Sons, New York
Mason B, Graham AL (1970) Minor and trace elements in meteoritic minerals. Smithson Contrib Earth Sci 3:1–17
Mason B, Melson WG (1970) The lunar rocks. Wiley Intersci, New York
Mason B, Jarosewich E (1973) The Barea, Dyarll Island, and Emery meteorites, and a review of mesosiderites. Min Mag 39:204–215
Mason B, Wiik HB (1964) The amphoterites and meteorites of similar composition. Geochim Cosmochim Acta 28:533–538

McCallum IS, Mathez EA (1975) Petrology of noritic cumulates and a partial melting model for the genesis of Fra Mauro basalts. Proc Sixth Lunar Sci Conf, p 395–414
McKay DS, Clanton US, Morrison DA, Ladle GH (1972) Vapor phase crystallization in Apollo 14 breccias. Proc Third Lunar Sci Conf, p 739–752
Meyer C, Brett R, Hubbard NJ et al. (1971) Mineralogy, chemistry, and origin of the KREEP component in soil samples from the Ocean of Storms. Proc Sec Lunar Sci Conf, p 393–411
Michaelis H von, Willis JP, Erlank AJ, Ahrens LH (1969) The composition of stony meteorites. I. Analytical techniques. Earth Plan Sci Let 5:383–386
Moss AA, Hey MH, Elliott CJ, Easton AJ (1967) Methods for the chemical analysis of meteorites. Miner Mag 36:101–119
Olsen E, Fredriksson K (1966) Phosphates in iron and pallasite meteorites. Geochim Cosmochim Acta 30:459–470
Olsen E, Fuchs LH (1967) The state of oxidation of some iron meteorites. Icarus 6:242–253
Olsen E, Erlichman J, Bunch TE, Moore PB (1977) Buchwaldite, a new meteorite phosphate mineral. Am Miner 62:362–364
Pellas P (1981) Early thermal histories of L chondrites. Abstract, Twelfth Lunar Planet Sci Conf, March 1981, Houston, p 825–827
Pellas P, Storzer D (1981) ^{244}Pu fission track thermometry and its application to stony meteorites. Proc R Soc Lond A374:253–270
Powell BN (1971) Petrology and chemistry of mesosiderites II. Geochim Cosmochim Acta 35:5–34
Prewitt CT, Rothbard DR (1975) Crystal structures of meteoritic and lunar whitlockites. Abstr Lunar Sci VI 2:646–648
Prinz M, Nehru CE, Delaney JS, Harlow GE, Bedell RL (1980) Modal studies of mesosiderites and related achondrites, including the new mesosiderite ALHA 77219. Proc Eleventh Lunar Planet Sci Conf, p 1055–1071
Romig AD, Goldstein JI (1980) Determination of the Fe–Ni and Fe–Ni–P phase diagrams at low temperatures (700 to 300 °C). Metall Transact A11:1151–1159
Sears DW (1978) The nature and origin of meteorites. Hilger Bristol
Shannon EV, Larsen ES (1925) Merrillite and chlorapatite from stony meteorites. Am J Sci 9:250
Simonds CH, Warner JL, Phinney WC (1973) Petrology of Apollo 16 poikilitic rocks. Proc Fourth Lunar Sci Conf, p 613–632
Smith DGW, Folinsbee RE, Hall-Beyer M (1973) A note on the mineralogy and classification of the Vilna meteorite. Meteoritics 8:197–199
Smith JV, Hervig RL (1979) Shergotty meteorite: Mineralogy, petrography and minor elements. Meteoritics 14:121–142
Steele IM, Smith JV (1976) Mineralogy of the Ibitira eucrite and comparison with other eucrites and lunar samples. Earth Planet Sci Let 33:67–78
Strunz H (1970) Mineralogische Tabellen. Akademische Verlagsgesellschaft, Leipzig
Tarasov LS, Nasarov MA, Shevaleevsky ID, Marakov ES, Ivanov VI (1973) Mineralogy of anorthositic rocks from the region of the crater Apollonius C (Luna-20). Proc Fourth Lunar Sci Conf, p 333–349
Taylor LA, Burton JC (1976) Experiments on the stability of FeOOH on the surface of the Moon. Meteoritics 11:225–230
Taylor LA, Misra KC, Walker BM (1976) Subsolidus reequilibration, grain growth, and compositional changes of native FeNi metal in lunar rocks. Proc Seventh Lunar Sci Conf, p 837–856
Taylor SR (1975) Lunar science: A post – Apollo view. Pergamon, New York
Van Schmus WR, Ribbe PH (1969) Composition of phosphate minerals in ordinary chondrites. Geochim Cosmochim Acta 33:637–640
Wasson JT (1974) Meteorites. Classification and properties. Springer, Berlin Heidelberg New York
Wiik HB (1956) The chemical composition of some stony meteorites. Geochim Cosmochim Acta 9:279–289
Yagi K, Lovering JF, Shima M, Okada A (1978) Petrology of the Yamato meteorites (j), (k), (l), and (m) from Antarctica. Meteoritics 13:23–45

Chapter 6 Phosphate Minerals in Terrestrial Igneous and Metamorphic Rocks

WILLIAM P. NASH

Introduction

Phosphorus is a ubiquitous although minor component of most igneous and metamorphic rocks. Under the normal oxidative conditions of the Earth's crust and upper mantle, phosphorus is present as phosphates, and Koritnig (1978) lists over 200 phosphate minerals. Nonetheless, in common igneous and metamorphic rocks phosphate mineral diversity is extremely limited and is dominated by the occurrence of apatite. This chapter is devoted to phosphates in terrestrial igneous and metamorphic rocks in the strictest sense. Metasomatic, hydrothermal, ore-forming and pegmatite associations are not treated; the latter has been described in particular detail by Moore (1973). Accordingly, considerable attention is paid to apatite; other phosphates discussed include monazite, xenotime and whitlockite. Also omitted is any discussion of crystallography which is provided by Moore (this Vol.).

A number of workers have previously summarized the occurrence and chemistry of phosphate minerals in igneous and metamorphic rocks (Kind 1939; Taborsky 1962; Cruft et al. 1965; Deer et al. 1962; McConnell 1973; Koritnig 1978). This chapter focuses on the more recent literature on the occurrence and chemistry of phosphate minerals and applications to problems of petrogenesis.

Phosphorus in Terrestrial Igneous and Metamorphic Rocks

Based upon cosmochemical arguments Smith (1981) estimated that the phosphorus content of the whole earth is between 11×10^{21} and 13×10^{21} kg. Assuming that apatite on the average contains 3% fluorine, and accounting for the uptake of some fluorine in mica and amphibole, the limited fluorine reservoir of the earth would require 27×10^{19} kg of phosphorus as apatite which accounts for only 2% of the Earth's total phosphorus (Smith 1981). The remainder must be contained either in silicate minerals in the lower mantle or in the core, as suggested by Sun and Nesbitt (1977). The chondrite abundance of 0.2% P_2O_5 is equivalent to about 0.5% apatite. However, analyses of spinel and garnet lherzolites (Maaloe and Aoki 1977) indicate that upper mantle P_2O_5 is less than 0.06%, or an equivalent of 0.15% apatite. Accordingly, if the Earth has a chondrite abundance of P_2O_5, the conclusion is again reached that most phosphorus must reside in the lower mantle

Department of Geology and Geophysics, University of Utah, Salt Lake City, Utah 84112 – 1183 or core. The concentration of phosphorus in the inaccessible mantle and core remains speculative, although phosphides such as schreibersite should be anticipated.

As with other minor elements, the concentration of phosphorus in igneous and metamorphic rocks may vary by several orders of magnitude from a few percent to a few parts per million. Koritnig (1978) has compiled and summarized data on the abundance of phosphorus in igneous and metamorphic rocks. Highest values (1.5% to 2.0% as P_2O_5) occur in strongly alkaline, low-silica igneous rocks. With decreasing alkalinity and increasing silica, P_2O_5 contents decline to almost nothing in some granites and rhyolites. The phosphorus contents of most igneous rocks usually lie between 0.01 and 0.5% P_2O_5. Assuming that all phosphorus is contained in apatite, igneous rocks in general contain between 0.02% and 1.2% apatite.

Phosphorus concentrations in metamorphic rocks range from 0.01% to 1.3%. The data base for phosphorus in metamorphic rocks is much sparser than for igneous rocks, and there is every reason to expect the upper value to be increased. In most cases, phosphorus appears to be relatively immobile during metamorphism, and the phosphorus content of metamorphic rocks reflects the abundance in the unmetamorphosed precursors. Elliott (1973), in a study of the transition from gabbro to amphibolite, indicates that amphibolites have slightly greater P_2O_5 contents than their unmetamorphosed equivalents (0.169% versus 0.146%, respectively). Koritnig (1978) suggests that there is a general increase in phosphorus content with increasing metamorphic grade, but the data are sparse and no rigorous study exists.

Apatite in Igneous Rocks

Because virtually all igneous rocks contain phosphorus, and because the solubility of phosphorus in silicates is small (see below), almost all igneous rocks contain a phosphate, and it is almost always apatite. With the exception of rare phosphorus-rich rocks, apatite is an accessory mineral. Its abundance varies directly with the phosphorus content of the rock which tends to decline with increasing silica content (cf. Koritnig 1978). Thus in phosphorus-rich silica-poor alkaline lavas, apatite abundances may reach several percent, whereas in phosphorus-poor rhyolites (0.01% P_2O_5) apatite is rare. Hildreth (1977) reports that microscopically resolvable apatite constitutes only 10 ppm by volume of rhyolite from the Bishop Tuff of Long Valley, California.

Apatite saturation in magmas usually occurs in the mid to late stages of crystallization, and as a result apatite is commonly found in the groundmass or mesostasis of igneous rocks. For example, apatite began to crystallize well below the liquidus at about 1,000 °C in the tholeiitic basalt of Alae lava lake, Hawaii (Peck et al. 1966). Exceptions exist in ferrobasalts of the Snake River Plain (Leeman et al. 1976) and the chilled margin of the Shonkin Sag laccolith (Nash 1972b) where apatite phenocrysts indicate saturation near the liquidus. Apatite phenocrysts occur in trachyte from Reunion (Zielinski 1975), in trachyandesite, trachyte and aegerine-augite trachyte from Gough Island (Zielinski and Frey 1970) and with phenocrysts of melilite and nepheline in nephelinite obsidian from Nyiragongo. Hawaiite from New South Wales contains apatite megacrysts up to 1.5 mm in length. Apatite was produced experimentally near the liquidus in this hawaiite at 5 and 6.5 kb. and 1,040° to 1,060 °C with 2% H_2O (Knutson and Green 1975). Apatite may also crystallize from the vapor phase, and Birch (1980) has described apatite in vesicles in olivine leucitite from Australia. In rare instances, often associated with iron and titanium ore accumulations, carbonatites, pyroxenites and anorthosites, very high concentrations of apatite occur. Such accumulations may be due to liquid immiscibility.

Chemistry of Apatite from Igneous Rocks

Chemical analyses of apatites from a representative variety of igneous rocks are presented in Tables 1–4. Prior compilations have been made by Kind (1939), Cruft et al. (1965), Cruft (1966), and Deer et al. (1962). Earlier "classical" chemical analyses of apatite were fraught with potential analytical difficulties, some of which may not have been recognized by the analyst (Cruft et al. 1965). In particular, rare earth elements may have inadvertently been reported as Al_2O_3, MgO or CaO, and precise fluorine numbers were difficult to obtain. Accuracy is further complicated by problems of incomplete phase separation; apatites have been observed to contain a variety of solid and fluid inclusions (see below). A number of these problems can be obviated by the use of the electron microprobe, the method used for most of the major element analyses reported here. Precise trace element analyses may be obtained by a variety of instrumental techniques including X-ray fluorescence, isotope dilution mass spectrometry, and neutron activation analysis, all of which still require proper sample preparation.

Major Elements. Examination of Tables 1 and 2 reveals several features about the chemistry of igneous apatite. SiO_2 varies from 0.1% to almost 2% with highest values typically found in alkaline igneous rocks. Silicon substitutes, most likely, for phosphorus (Koritnig 1965), and the two elements are inversely correlated in apatites from carbonatites at Iron Hill, Colorado (Nash 1972). The difference in ionic radius for tetrahedral P^{5+}(0.24 Å) and Si^{4+}(0.34 Å) is probably sufficiently large to prohibit extensive phosphate-silicate substitution. For example the mineral britholite $Ce_3Ca_2(SiO_4)_3OH$ is readily obtained from apatite by the simple substitution of $REE^{3+}+Si^{4+}$ for $Ca^{2+}+P^{5+}$. Britholite from the Shonkin Sag laccolith contains 7.0% P_2O_5 and 16.2% SiO_2 (Nash 1972b); it does not coexist with apatite. Electron microprobe analyses demonstrate that titanium and aluminum contents of apatite are typically very low. However, significant aluminum occurs in apatite inclusions in an almandine garnet (analysis number 18, Table 2). Similarly, manganese and magnesium values rarely exceed 0.2% although manganoan apatites are not uncommon in pegmatites (cf. Deer et al. 1962). Iron contents are variable with moderately high values in iron-rich rocks from the Sandwich Horizon of the Skaergaard intrusion and in low temperature rhyolite from the Bishop Tuff (0.9%). Highest values of iron (up to 2.7% as Fe_2O_3) occur in apatites included in iron rich garnet and fayalite. Na_2O is low, reaching highest values in strongly

Table 1. Representative major element analysis of apatites from igneous rocks. (For key to analyses see Table

	1	2	3	4	5	6	7	8	9
SiO_2	0.47	0.10	0.73	0.34	0.65	0.78	1.93	0.18	0.8
TiO_2	–	–	–	–	–	–	0.06	0.02	–
Al_2O_3	0.06	–	0.01	–	0.04	0.05	n. d.	0.08	0.0
FeO[a]	0.45	0.38	0.71	0.32	0.29	0.17	0.03	0.09	0.9
MnO	0.05	0.03	0.05	0.04	0.07	0.09	n. d.	0.13	0.1
MgO	0.09	0.05	0.02	0.01	0.19	0.11	n. d.	0.01	0.1
CaO	54.0	54.91	54.6	54.52	54.0	53.2	49.08	54.87	52.5
Na_2O	0.05	n. d.	0.09	0.13	0.35	0.34	0.52	0.06	0.1
K_2O	–	–	–	–	0.06	0.23	n. d.	0.01	–
P_2O_5	41.8	42.12	41.5	42.30	40.6	39.8	41.59	41.97	39.5
CO_2	–	–	–	–	–	–	–	–	–
SO_3	0.04	–	0.03	–	0.20	0.17	–	–	–
F	2.63	3.64	3.37	3.54	2.4	3.2	–	3.36	2.[illegible]
Cl	0.15	0.05	0.09	0.11	0.18	0.19	n. d.	0.02	0.4
H_2O	0.51	0.04	0.18	0.09	–	–	–	–	–
sum[b]	101.2	101.37	103.0	102.13	100.9	101.2	101.0	101.2	99.6
–O=F, Cl	1.1	1.54	1.4	1.52	1.0	1.3	–	1.4	1.0
Total[b]	100.1	99.83	101.6	100.61	99.9	99.9	101.0	99.8	98.[illegible]

[a] Total Fe as FeO
[b] Includes trace elements
n. d. Not detected

alkaline rocks. Larson (1979) reports up to 3% Na_2O in rare earth-rich apatite containing 10.7% cerium, and suggests that in peralkaline magmas there may be moderate solid solution between apatite and vitusite $Na_3(La, Ce)(PO_4)_2$. K_2O contents are usually very low. When sought, sulfur is usually found in levels less than 0.2% as SO_3. Igneous apatites are typically fluorapatite; values in Tables 1 and 2 range from 1.4% to 3.4% F. Chlorine contents are generally low, ranging from 0.0% to 0.2% Cl; however, higher values do occur (see, for example, analysis 17). Halogens in apatites are discussed in greater detail below. Although CO_2 contents shown in Table 1 are small, carbonate apatite with substantial carbon has been reported from carbonatites (von Eckermann 1948; Davies 1956; Prins 1973). Several analyses of carbonate apatite are given by Deer et al. (1962) and the subject is discussed in detail by McConnell (1973).

Trace Elements. Strontium contents are extremely variable, typically falling between 500 to 8,000 ppm, but may reach major element proportions. Larsen et al. (1952) describe a strontian apatite in a syenite dike with 11.60% SrO. Yttrium is generally present at the several thousand ppm level; lesser amounts of niobium and barium are generally present, together with minor amounts of lead, thorium and uranium. Apatites contain significant amounts of rare earth elements (REE) substituting for Ca^{2+}. Absolute abundances are variable; chrondrite normalized values for apatites which were analyzed by neutron activation techniques are shown in Fig. 1. The REE distribution displays a typical negative slope indicating substantial relative enrichment in the light REE. Negative anomalies are evident

10	11	12	13	14	15	16	17	18
1.00	0.36	0.23	1.7	0.2	0.26	0.27	0.6	0.6
–	0.01	0.01	–	0.1	–	–	0.06	0.04
0.01	–	–	0.3	0.2	–	–	–	–
0.82	0.04	0.04	0.4	0.1	0.29	0.36	0.23	0.23
0.18	0.06	0.05	–	–	–	–	–	–
0.06	n.d.	0.03	0.9	0.7	0.15	0.22	0.01	0.31
52.94	55.2	55.0	52.3	53.2	53.3	53.8	–	–
0.09	0.08	0.23	–	0.9	–	–	0.02	0.03
–	n.d.	0.02	0.15	–	–	–	0.11	0.09
39.40	40.9	41.0	42.0	42.4	40.5	41.2	–	–
–	0.1	0.2	–	–	–	–	–	–
–	0.08	0.05	–	–	–	–	–	–
2.78	2.2	3.3	–	–	1.42	2.00	–	–
0.22	0.03	0.03	–	–	0.23	0.18		–
–	–	–	–	–	–	–	–	–
100.33	101.2	102.5	98.8	98.7	96.2	98.0	–	–
1.22	0.9	1.4	–	–	0.7	0.9	–	–
99.11	100.3	101.1	98.8	98.7	95.5	97.1	–	–

Table 2. Apatite inclusions in minerals (for key to analyses see Table 4)

	19	20	21	22	23	24	25
SiO_2	0.9	0.80	0.4	0.6	–	–	–
TiO_2	–	0.01	–	–	–	–	–
Al_2O_3	–	0.49		–	–	–	–
FeO[a]	–	1.93	2.7	1.9	–	–	–
MnO	–	0.12	–	–	0.12	0.18	0.11
MgO	–	0.27	–	–	0.37	0.22	0.59
CaO	52.1	54.1	53.2	53.1	53.3	52.1	53.1
Na_2O	–	0.15	–	–	–	–	–
K_2O	–	–	–	–	–	–	–
P_2O_5	39.1	41.5	41.0	41.2	42.0	41.8	42.6
F	1.41	2.04	–	–	1.3	1.3	1.5
Cl	0.87	0.01	–	–	0.28	0.31	0.82
Sum[b]	98.68	101.7	98.26	97.86	97.4	95.9	98.7
–O=F, Cl	0.79	0.9	–	–	–	–	–
Total[b]	95.89	100.8	98.26	97.86	–	–	–

[a] Total Fe as FeO
[b] Includes trace elements

in some apatite for europium and cerium, although the cerium anomalies are restricted to analyses by Puchelt and Emmermann (1976). REE concentrations can be substantial in apatite, particularly in strongly alkaline rocks. Larson (1979) reports typical values of 1.54% La and 3.51% Ce in apatite from foyaite from Ilimaussaq, Greenland, and a maximum content of 10.7% Ce.

Table 3. Trace element analyses of apatite from igneous rocks (values in ppm). For key to analyses see Table

	1	3	5	6	7	9	10	11	12	14	15	16	1'
Sr	500	1,100	8,200	13,500	–	850	1,000	7,000	6,300	–	–	–	2,
Y	1,300	2,220	–	–	–	1,760	4,670	1,072	1,070	–	–	–	–
Nb	840	770	630	500	–	–	–	900	900	–	–	–	–
Ba	–	–	450	600	–	–	–	300	200	–	–	–	4,
La	340	1,200	1,400	2,300	18,900	4,100	2,400	1,000	1,300	4,200	770	442	5,
Ce	1,100	2,900	2,100	4,300	32,200	9,500	7,800	1,400	2,600	8,000	1,860	930	1,
Pr	500	700	–	–	–	1,700	1,500	850	1,000	–	–	–	3,
Nd	1,500	2,600	1,900	2,800	2,800	5,800	6,300	1,100	2,100	3,400	796	416	4,
Sm	600	950	–	–	1,600	–	–	600	850	480	145	86	–
Eu	–	–	–	–	–	–	–	–	–	130	37.3	21.0	–
Gd	600	870	–	–	–	–	–	800	1,100	380	–	– –	1,
Tb	–	–	–	–	–	–	–	–	–	48	13.0	8.68	–
Dy	–	–	–	–	–	–	–	1,400	1,300	–	–	–	–
Ho	–	–	–	–	–	–	–	–	–	0	–	–	–
Er	–	–	–	–	–	–	–	800	900	–	–	–	–
Tm	–	–	–	–	–	–	–	–	–	–	–	–	–
Yb	–	–	–	–	–	–	–	–	–	25	9.8	6.23	–
Lu	–	–	–	–	–	–	–	–	–	2.5	1.10	0.95	–
Pb	–	–	600	400	–	–	–	–	–	–	–	–	–
Th	–	–	–	–	–	–	–	–	–	–	–	–	–
U	–	–	–	–	–	–	–	–	–	32	–	–	–

Concentrations of REE are a function of the distribution between apatite and host liquids during crystallization. There are limited natural data on partition coefficients for REE in apatite. Nagasawa (1970) has analyzed apatites and host matrices, and partition coefficients are similar to those estimated for granites by Exley (1980). Irving (1978) presents data on REE content of an apatite megacryst in a basanite lava. Pastor et al. (1974) and Roelandts and Duschene (1977) have calculated partition coefficients for apatite in layered intrusions, and Wass et al. (1980) have calculated coefficients for apatite from an ultramafic inclusion in basanite. Partition coefficients are estimated to range from 15–24 for La and 19–27 for Ce in strongly alkaline rocks and they increase with increasing peralkalinity (Larson 1979). The data in Fig. 2 illustrate the large variation in distribution coefficients with highest values in the most silicic rocks. The 50-fold variation in distribution coefficients may be in part due to increased polymerization in the more silicic liquids which reduces the available REE sites. Temperature may also have a significant effect as illustrated by two rhyolite samples from the Bishop Tuff (Hildreth 1977) where major element differences are not large; as expected, partition coefficients increase with falling temperature. Recently, Watson, and Green (1981) experimentally determined apatite/liquid REE and Sr partition coefficients for a variety of rock types from basanite to granite. Their observations are similar to the natural data. Melt composition, especially SiO_2 content, strongly affects partitioning with a factor of 4 increase between basaltic and silicic melts at constant temperature. Water content appears to not be important. At constant composition a decrease in temperature of 130 °C produces an increase in the partition coefficient by about a factor of two.

	19	26	27	28	29	30	31	32	33	34
0	7,300	–	–	–	–	–	–	–	–	–
	–	1,170	–	–	–	–	–	–	–	–
	–	–	–	–	–	–	–	–	–	–
0	–	–	–	–	–	–	–	–	–	–
0	3,000	10,600	3,000	–	–	8,200	903	–	–	–
0	5,500	18,000	1,700	2,200	965	4,700	2,649	344	–	–
0	1,400	1,930	–	–	–	–	–	947	–	–
0	1,600	6,480	1,400	1,220	757	3,100	2,108	–	1,490	733
	250	830	123	235	190	171	465	642	–	911
	–	130	40	20.8	30.8	34	45	180	1,360	–
0	700	340	–	218	224	–	423	31.4	543	2,200
	–	37	11	–	–	9.58	71.1	144	3.30	490
	–	170	–	168	177	–	300	26.7	–	10
	–	49	–	–	–	–	49.1	99.8	–	–
		50	–	90.6	81.8	–	142	14.8	615	70
	–	–	–	–	–	–	–	49.3	–	–
	–	–	–	73.9	55.8	25	88.8	–	256	–
	–	–	1.28	11.8	8.32	3.7	10.4	25.0	–	–
	–	–	–	–	–	–	–	2.4	188	152
	–	–	15	–	–	0.10	5.0	–	24.9	17
	–	–	22	–		193	–	–	–	–

Partition coefficients are shown in Fig. 2 as a function of deviation of the REE in question from the ionic radius of Ca^{2+}. REE closest to Ca^{2+} in radii have the highest partition coefficients; maximum values are for samarium whose radius is slightly smaller than Ca^{2+}. This deviation of the maximum from the Ca^{2+} radius may be due to complications arising from differing Ca-O and Ca-F bond lengths in apatite; values for ionic radii are for silicates (Shannon and Prewitt 1969). The distribution coefficient patterns are similar in shape to the crystal/liquid distribution coefficients of REE in $CaWO_4$ (Nassau and Loiacono 1966) and $CaMoO_4$ (Brixner 1967), demonstrating that the REE reside in the Ca^{2+} site.

Apatites in Igneous Processes

Differentiated Intrusions. There are few systematic studies of the behavior of apatite during magmatic differentation. In the Skaergaard intrusion (Nash 1976; Brown and Peckett 1977) and the Kiglapait intrusion (Huntington 1979) fluorine contents increase stratigraphically upward whereas chlorine contents decrease. Total REE contents are enriched in residual liquids in the Skaergaard intrusion, however there are no systematic variations in a number of other trace elements during differentiation. Brown and Peckett (1977) attribute this behavior in part to the crystallization of intercumulus apatite from pore liquids not in communication with the main differentiating magma mass. In contrast, apatites from the alkaline and silica-undersaturated Shonkin Sag intrusion exhibit systematic and continuous changes in morphology and chemistry with differentiation (Nash 1972b).

Table 4. Key to analyses

1. Chilled margin, Skaergaard intrusion (Nash 1976)
2. Upper Zone b, Skaergaard intrusion (Brown and Peckett 1977)
3. Sandwich Horizon, Skaergaard intrusion (Nash 1976)
4. Upper Border Group, Skaergaard intrusion (Brown and Peckett 1977)
5. Shonkinite, Shonkin Sag laccolith (Nash 1972b)
6. Soda syenite, Shonkin Sag laccolith (Nash 1972b)
7. Pulaskite, Kangerdlugssuaq intrusion (Henderson 1980)
8. Granitoid, Snake Range, Nevada (Lee et al. 1973)
9. Rhyolite, Bishop Tuff (Hildreth 1977, sample 77)
10. Rhyolite, Bishop Tuff (Hildreth 1977, sample 98)
11. Calcite carbonatite, Iron Hill, Colorado (Nash 1976a)
12. Dolomite carbonatite, Iron Hill, Colorado (Nash 1976a)
13. Lherzolite inclusion in basanite (Frey and Green 1974, sample 2700)
14. Lherzolite inclusion in basanite (Frey and Green 1974, sample 2638)
15. Apatite-amphibole inclusion in basanite (Wass et al. 1980, sample K 24)
16. Apatite-amphibole inclusion in basanite (Wass et al. 1980, sample K 41)
17. Woldigite, West Kimberly, Australia (Carmichael 1967, sample WK.3)
18. Wyomingite, Leucite Hills, Wyoming (Carmichael 1967, sample L.H.1).
19. Apatite in chrome diopside in kimberlite (Hervig and Smith 1981)
20. Apatite in almandine garnet (McCandless 1982, sample 47–6)
21. Apatite in fayalite phenocryst in rhyolite (Bacon and Duffield 1981)
22. Apatite in garnet phenocryst in rhyolite (Bacon and Duffield 1981)
23. Apatite in ferrokaersutite megacryst in basalt (Irving 1974, sample 2104d)
24. Apatite in anorthoclase megacryst in basalt (Irving 1974, sample 2104e)
25. Apatite in vein cutting phlogopite in kaersutite megacryst (Irving 1974, sample 8015)
26. Carbonatite (Eby 1975, sample 5-34-2)
27. Carbonatite (Puchelt and Emmermann 1976; includes 0.11 ppm Sc)
28. Dacite (Nagasawa 1970 sample 1A)
29. Dacite (Nagasawa 1970, sample 6A)
30. Phonolite (Puchelt and Emmermann 1976; includes .25 ppm Sc)
31. Nelsonite (Brunfelt and Roelandts 1974)
32. Norite (Roelandts and Duschene 1977, sample JCD 6460)
33. Granite (Nagasawa 1970, sample 10A)
34. Granite (Puchelt and Emmermann 1976, sample 5)

Microphenocrysts in the mafic shonkinite of the chilled margin average 0.15 mm in diameter with a length/breadth ratio of approximately 3:1. With differentiation of the host magma, apatites become elongate, ultimately reaching a maximum of 300:1 (0.05×15 mm) in soda syenite. As differentiation proceeds, apatites reflect changes in the magma composition and become enriched Si, Na, K, Sr, and REE, whereas Fe and Mg decrease in concentration. As in the Skaergaard and Kiglapait intrusion, F increases and Cl decreases with differentiation.

Occurrence in Mantle-Derived Rocks. Despite the fact that Thompson (1975) synthesized garnets with up to 0.6% P_2O_5, the phosphorus contents of natural silicate minerals are quite low (Table 5) and are insufficient to account for the phosphorus contents of peridotites (Bishop et al. 1978); accordingly, the presence of a phosphate mineral in the mantle is indicated. But apatite is not common in ultramafic rocks. Apatite has been reported in spinel peridotite inclusions in basanite from Victoria (Kleeman et al. 1969; Frey and Green 1974), and xenoliths rich in apatite

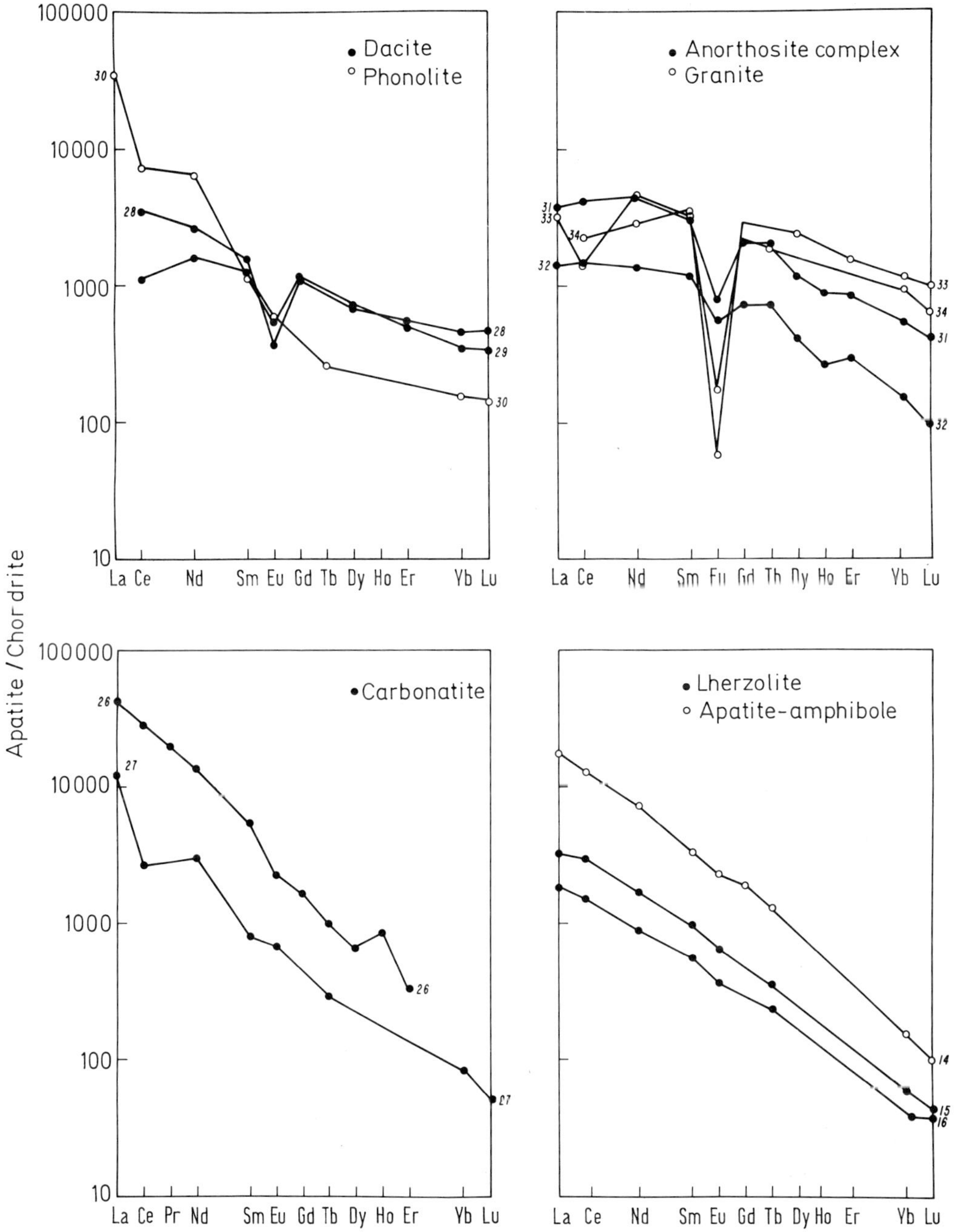

Fig. 1. Rare earth abundances in igneous apatite normalized to average chondrite (Evensen et al. 1978). *Numbers* refer to analyses in Table 3

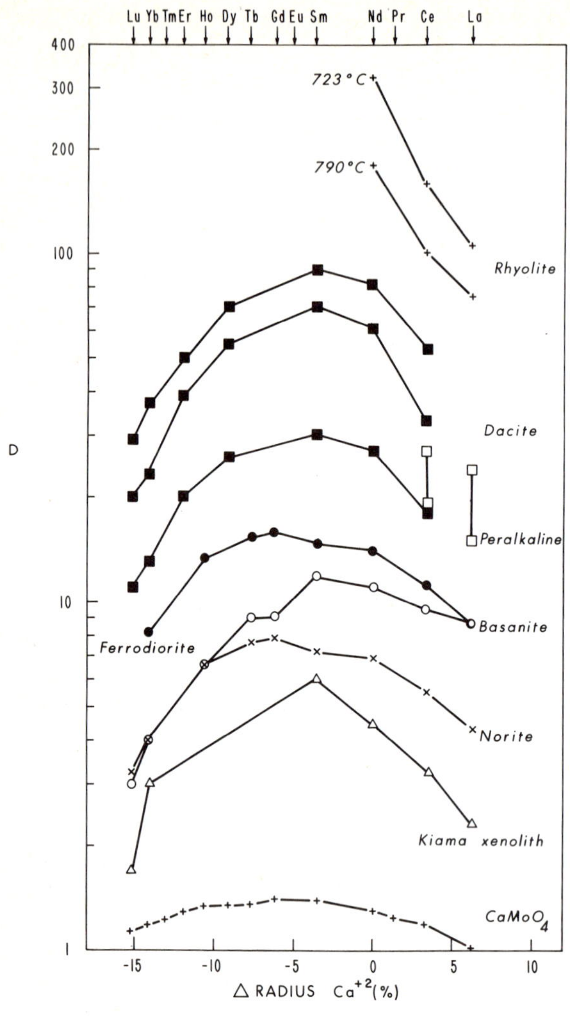

Fig. 2. Rare earth element partition coefficients for apatite from a variety of rock types plotted against REE deviation from the ionic radius of Ca^{2+}

occur in nepheline basanite from Kiama, New South Wales, where apatite coexists with abundant kaersutite (Wass 1980). Small grains of secondary apatite occur in xenoliths from kimberlite (Smith et al. 1981). Whitlockite has not been reported in mantle-derived rocks. In the absence of thorough experimental data on the stabilities of apatite and whitlockite under mantle conditions, the mineralogical locus of phosphorus remains speculative. Beswick and Carmichael (1978) calculated equilibria for apatite and whitlockite and concluded that apatite was the most likely phase in the source region of basalts. Smith et al. (1981) have documented a strong correlation between whole rock phosphorus and fluorine in oceanic basalts and gabbros with a P/F ratio near that of fluorapatite, and Bergman (1979) concluded that the whole rock REE contents of alkali basalts, nephelinites and kimberlite could be accounted for simply by their apatite content. These data suggest that apatite is present in the source regions of basaltic magmas and is con-

Table 5. Phosphorus contents of igneous silicates and oxides (values in ppm)

	1	2	3
Olivine	81	38	50– 200 (130)
Clinopyroxene	30–77	21–24	90– 630 (350)
Orthopyroxene	–	12	0– 90 (50)
Garnet	–	–	200–1,040 (460)
Plagioclase	46–1,400	20–37	–
Magnetite	8– 9	–	–
Ilmenite	20–22	–	–

1. Skaergaard intrusion (Henderson 1968)
2. Bushveld intrusion (Henderson 1968)
3. Lherzolite (Bishop et al. 1978) (values in parentheses are means)

sumed at least in part in the fusion process. Whitlockite does not contain fluorine, and is thus an unlikely candidate for the source of phosphorus in the upper mantle. Phosphorus also correlates well with the abundance of light REE (LREE) in basalts (Schilling 1973; Sun and Hanson 1975, 1976; Beswick and Carmichael 1978; Frey et al. 1980). However, because whitlockite also contains substantial amounts of REE (Griffin et al. 1972; Nash and Hausel 1973), the correlation does not serve to distinguish between apatite and whitlockite in the mantle.

The presumed presence of mantle apatite and the phosphorus content of basalts have been used to place constraints on partial fusion processes in the mantle. Frey et al. (1978) calculated the degree of melting to produce individual basalts by assuming that P_2O_5 is completely partitioned into the melt during partial fusion; that is, apatite is entirely consumed in the initial stages of magma formation, and the P_2O_5 concentration in the magma is inversely proportional to the degree of melting due to the effect of dilution. If, however, a phosphate mineral (either apatite or whitlockite) were refractory, then the REE characteristics of basalts, especially the LREE, may be controlled by the residual phase. For example, Mertzman (1977) attributed the REE and P_2O_5 behavior in lavas from Northern California to the presence of apatite and possibly priderite in the mantle or lower crustal source region. Beswick and Carmichael (1978, 1980) argue that the LREE-P_2O_5 correlation is due to residual apatite. However, experimental data indicate that fluorapatite is highly soluble in basaltic magmas, and that it is unlikely that apatite would survive a melting episode in the upper mantle (Watson 1979, 1980).

The fractionation of apatite may be recorded in the subsequent differentiation of magmas derived from the mantle. At Gough Island the increase in P_2O_5 from 0.29% in basalt to 0.36% in trachyte and the decrease to 0.04% in more evolved rocks was attributed to the saturation with apatite and its subsequent removal by Zielinski and Frey (1970). They successfully modeled REE abundances with the fractionation of apatite. Similarly, Zielinski (1975) modeled the role of apatite fractionation and its effect on the decline in P_2O_5 and REE in the differentiated lavas of Reunion. Price and Taylor (1973) attribute the same mechanism for the decline in P_2O_5 and slight depletion in REE in benmoreite compared to lesser evolved mugearite at Dunedin Volcano, New Zealand. Systematic chemical variations in apatite led Le Bas and Handley (1979) to conclude that carbonatites arise by immiscibility of carbonate and silicate melts as suggested previously by Le Bas (1977), Koster van Groos and Wyllie (1973), Nash (1972a) and others.

Halogens

Apatite has long been known to contain substantial amounts of halogens in addition to hydroxyl. Stoichiometric fluorapatite contains 3.7% F, and stoichiometric chlorapatite contains 6.8% chlorine. Halogen contents of apatite have been used in thermometry, in assessment of the effects of magmatic and post-magmatic fluids, and in evaluation of the total halogen budget of the Earth.

Table 6 provides halogen contents of apatites from single localities, general lithologic types, and mantle-derived material. Fluorine dominates in abundance, and in many instances stoichiometric fluorapatite is approached, particularly in the plutonic environment. Fluorine and chlorine contents vary considerably and systematically with magmatic differentiation (Fig. 3). In mafic plutons F/Cl ratios in apatite show as much as a 30-fold increase from early-formed cumulates to late-stage differentiates. Similarly, in the reconstructed, zoned rhyolitic magma chamber of the Bishop Tuff, F/Cl ratios in apatite increase upwards in the chamber, associated with a decrease in temperature. Total halogen contents increase in apatite despite a strong enrichment in H_2O upwards in the magma chamber (Hildreth 1977).

Although the sample population is small, it appears that apatites in mantle-derived xenoliths, as well as megacrysts in kimberlites and basanitic lavas, contain less fluorine than apatites from other lavas or crustal plutonic environments. This may be due to contrasting fluorine abundances in the crust and mantle or fluorine enrichment during differentiation, or both.

In general, chlorine contents of apatite fall between 0.0 and 1.0 weight percent. There appears to be no systematic variation between rock types with the exception that apatites from carbonatites are uniformly low in their chlorine content (<0.10 wt.-%). The chlorine content of an apatite megacryst from a kimberlite is similarly low (0.1 wt.-%). Chlorine exhibits antithetic behavior to fluorine and is relatively enriched in the least differentiated rocks of individual plutonic associations.

Stormer and Carmichael (1971) evaluated the distribution of fluorine and hydroxyl between apatite and coexisting biotite as a geothermometer using the following exchange reaction:

$$\underset{\text{fluorphlogopite}}{(KMg_3AlSi_3O_{10})_{1/2}F} + \underset{\text{hydroxypatite}}{Ca_5(PO_4)_3OH}$$

$$= \underset{\text{fluorapatite}}{Ca_5(PO_4)_3F} + \underset{\text{phlogopite.}}{(KMg_3AlSi_3O_{10})_{1/2}OH}$$

The necessary free energy data for the reaction was estimated from simple hydroxides and fluorides. Subsequently, Ludington (1978) refined the geothermometer with additional data on Fe-Mg substitution in mica, thermodynamic data on apatite endmembers (Duff 1971), and revised data on F-OH exchange in biotite.

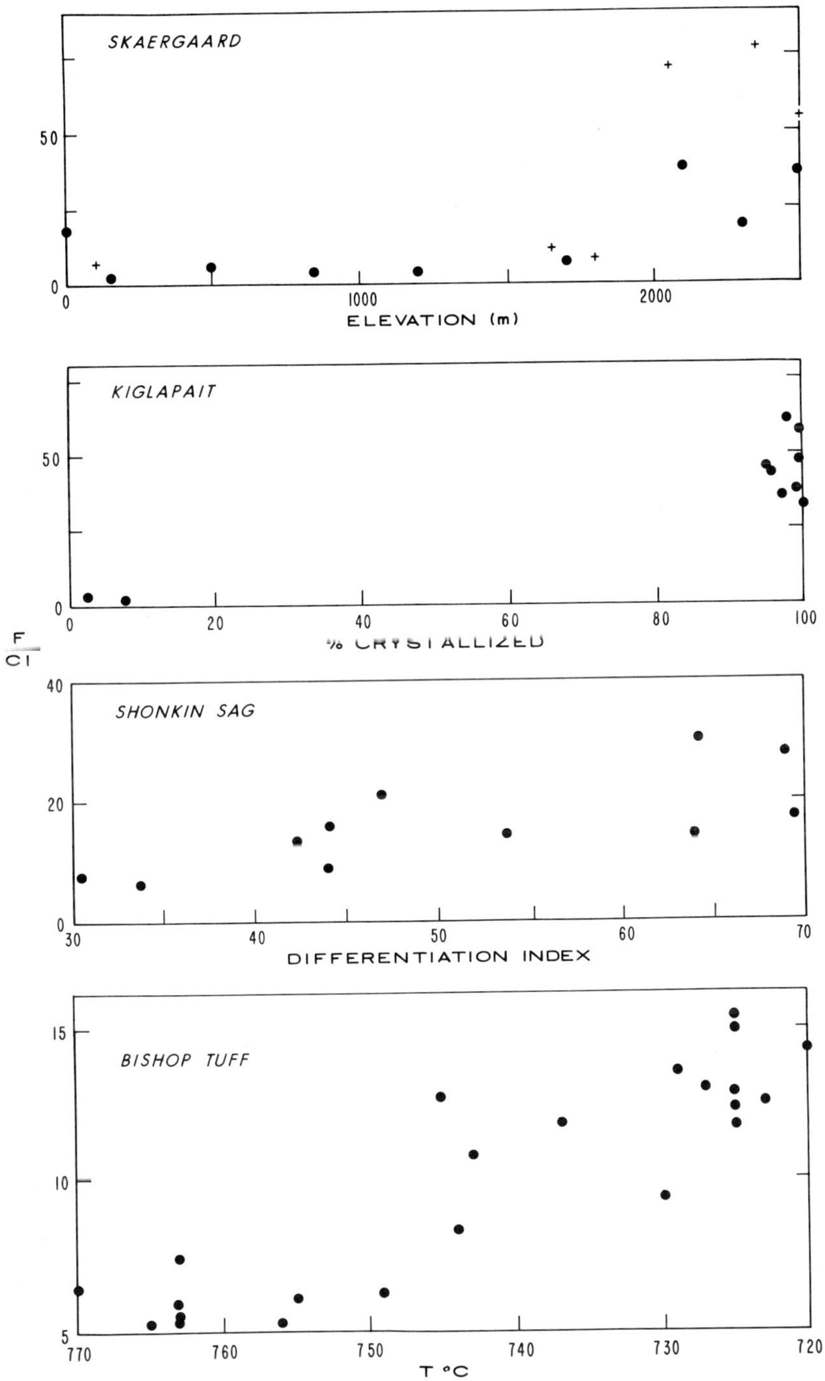

Fig. 3. F/Cl in apatite as a function of various indices of differentiation. Skaergaard + (Brown and Pekett 1977), ● (Nash (1976); Kiglapait (Huntington 1979); Shonkin Sag (Nash 1972 b); Bishop Tuff (Hildreth 1977)

Table 6. Halogen contents of igneous apatite

Host	No. of analyses	F(%)	Cl(%)	Br(ppm)	Reference
Plutonic					
Kiglapait	10	2.18–3.75	0.06–1.72		Huntington (1979)
Skaergaard	7	2.58–3.64	0.04–0.42		Brown and Peckett (1977)
Skaergaard	9	1.92–3.37	0.08–0.85		Nash (1976)
Diorite	3	2.1 –3.6	0.38–0.74		Roegge et al. (1974)
Quartz monzonite	32	1.0 –3.6	0.15–1.15		Roegge et al. (1974)
Quartz monzonite	5	1.73–3.21	0.04		Parry and Jacobs (1975)
Granite	25	2.19±0.50	0.00		Parry and Jacobs (1975)
Shonkin Sag	11	2.1 –3.3	0.11–0.34		Nash (1972b)
Ijolite series	3	2.07–3.16	–		Le Bas and Handley (1979)
Ijolite	2	2.1 –2.8	0.01–0.04		Nash (1972a)
Alkali pyroxenite	1	0.62	0.07		Dawson and Fuge (1980)
Carbonatite	8	1.98–3.72	0.00–0.10		Smith et al. (1981)
Carbonatite	3	2.2 –3.3	0.01–0.03		Nash (1972a)
Jumilla, Spain	1	–	0.56	5.5	Behne (1953)
Floitental, Aust.	1	–	0.05	1.3	Behne (1953)
Meteorite	1	–	4.3	5.3	Reed and Allen (1966)
Volcanic					
Leucite Hills	5	3.16–3.50	0.04–0.26		Stormer and Carmichael (1971)
Leucite basanite	2	3.3	0.06–0.75		Stormer and Carmichael (1971)
Trachyte	3	1.94–3.18	0.31–0.94		Stormer and Carmichael (1971)
Dacite	3	1.55–1.90	0.96–1.33		Stormer and Carmichael (1971)
Rhyolite	12	2.23–3.00	0.20–0.47		Hildreth (1977, 1979)
Mantle xenoliths					
Lherzolite	2	–	0.9 –1.0		Frey and Green (1974)
Harzburgite	1	1.80	0.58		Smith et al. (1981)
Websterite	1	0.71	0.54		Irving (1974)
Amphibole/apatite	5	1.42–2.52	0.15–0.23		Wass et al. (1980)
Mantle xenocryst hosts					
Diopside/dolomite	1	1.41	0.87		Smith et al. (1981)
Anorthoclase	1	1.3	0.31		Irving (1974)
Kaersutite	1	1.5	0.82		Irving (1974)
Ferrokaersutite	1	1.3	0.28		Irving (1974)
Apatite megacrysts					
Kimberlite	1	0.45	0.01		Smith et al. (1981)
Kimberlite	1	0.44	–		Aoki and Kanizawa (1979)
Basalt	3	1.3 –1.6	0.28–0.33		Irving (1974)

Some biotite-apatite pairs provide reasonable magmatic temperatures in lavas; others do not and yield anomalously low temperatures. In most cases, these low temperatures can be attributed to disequilibrium due to preferential loss of fluorine from biotite to hydrothermal fluids. Apatite is less susceptible than biotite to the loss of fluorine as shown by oxygen isotopes in the Skaergaard intrusion. Lowest fluorine contents in biotite occur in rocks which show substantial ^{18}O depletion; the fluorine contents of coexisting apatites appear unaffected (Nash 1976). Apatites from three Upper Zone gabbros show a uniform $\delta^{18}O$ of 4.8–4.9, whereas coexisting plagioclases exhibit varying degrees of ^{18}O depletion (Taylor and Forester 1979).

Bromine contents of igneous apatites are poorly known. Two analyses are given in Table 6 together with a chlorapatite from a meteorite for comparison. The similarity in ionic radii of chlorine and bromine (0.99 and 1.14 Å, respectively) may account for the substitution of bromine which is roughly correlated with chlorine concentration. Cl/Br ratios in apatite are 1025, 810, and 410 in these analyzed samples. Smith (1981) estimated the bromine reservoir of the mantle assuming apatite with 1% Cl and a Cl/Br ratio of 300, a representative whole rock value for mid-ocean ridge basalts. Because 1% Cl is probably a maximum value in the mantle apatite (Table 6), and because the Cl/Br ratio in apatite is apparently higher than 300, the bromine reservoir of the upper mantle could be substantially less than estimated by Smith (1981). On the basis of his evaluation of the budgets of F, Cl and Br in the Earth, Smith (1981) suggested that during accretion apatite was not distributed homogenously throughout the mantle, and that in time apatite may have become more refractory, i.e., fluorine-rich, due to preferential loss of OH, Cl and Br to magmas.

Iodine, with an ionic radius of 2.2 Å is unlikely to substitute effectively in the OH, F, Cl site in apatite. Dawson and Fuge (1980) report 0.24 ppm iodine in apatite with 6,230 ppm F and 742 ppm Cl occurring in an alkali pyroxenite. By comparison, chlorapatite (4.3%) from the Mt. Sterling meteorite contains 2.3 ppm iodine (Reed and Allen 1966).

Experimental Results

Skinner (1973) investigated the phase relations in the system CaO-P_2O_5-H_2O at 2 kb and up to 600 °C. Kogarko et al. (1977) examined the phase equilibria of the system fluorapatite-nepheline-diopside at magmatic temperatures. Although they observed liquation in the system, it occurs at very high temperatures and involves a silicophosphate not present in rocks. Accordingly, they concluded that apatite-rich rocks associated with the ijolite-urtite series do not arise by immiscibility between aluminosilicate and phosphate melts.

Watson (1979, 1980) determined that fluorapatite is stable over the range 1,275–1,350 °C and 8–25 kb. The solubility of apatite is a function of the silica content of the melt, decreasing with increased silica. Basaltic melts with 50% SiO_2 under mantle pressures can contain up to 3–4 wt.-% P_2O_5 without apatite saturation. Because of its high solubility, apatite is not expected to be a residual phase during partial melting in the mantle, especially if melting exceeds 3.5%. Accordingly apatite will not affect REE distribution patterns in initial mafic melts.

As described earlier, apatite phenocrysts occur more often in intermediate and silicic igneous rocks. Apatite is present near the liquidus in hawaiite with 1.16% P_2O_5 (Knutson and Green 1975). Watson and Capobianco (1981) demonstrated experimentally that apatite saturation is achieved at 0.14 wt.-% P_2O_5 in peralkaline felsic melts. The saturation concentration is insensitive to variations from 72% to 79% SiO_2. Moreover, there is no strong temperature effect. The implications for anatexis are several. Felsic magmas cannot contain more than 0.14% P_2O_5 unless apatite crystals are retained in the melt. Furthermore, felsic melts arising from anatexis must have 0.14% P_2O_5 if residual apatite is present in the refractory residuum. Finally, primary felsic melts with less than 0.14% P_2O_5 arose from source regions with less than 0.14% P_2O_5 in which apatite was completely consumed in the melting process, and P_2O_5 contents were subsequently diluted by any additional melt fractions. Saturation with apatite is an inevitable consequence of fractional crystallization of andesite magmas toward felsic compositions and results in depletion of REE in derivative melts. Watson and Capobianco (1981) modeled the effect of apatite consumption on the REE patterns of melts and concluded that apatite will have an effect on LREE enrichment or depletion, but that the effects would be secondary in the presence of residual amphibole or garnet.

Analytical studies by Price and Taylor (1977) support these experimental results by demonstrating that small amounts of residual apatite (1%–2%) could have a significant effect on the LREE abundances of melts modeled for the western metamorphic belt of south-eastern Australia. Similarly, Simmons and Hedge (1978) concluded that REE patterns in felsic magmas in the Colorado mineral belt could have been substantially affected by early fractionation of apatite and sphene.

Phosphorus Fugacity

It is possible to calculate the fugacity of phosphorus for a variety of rock types by constructing different reactions involving apatite. For example, Carmichael et al. (1974) proposed the following reaction for olivine-bearing rocks such as basalts and fayalite rhyolites:

$$\underset{\text{magnetite}}{\frac{15}{2}Fe_3O_4} + \underset{\text{forsterite}}{\frac{1}{2}Mg_2SiO_4} + \underset{\text{diopside}}{2CaMgSi_2O_6} + \underset{\text{anorthite}}{3CaAl_2Si_2O_8} + \underset{\text{glass}}{\frac{3}{4}SiO_2}$$

$$+ \underset{\text{steam}}{\frac{1}{2}H_2O} + \underset{\text{gas}}{3P} = \underset{\text{apatite}}{Ca_5(PO_4)_3OH} + \underset{\text{spinel}}{3MgAl_2O_4} + \underset{\text{fayalite}}{\frac{45}{4}Fe_2SiO_4}.$$

Using appropriate values for the compositions of mineral phases, the fugacity of phosphorus may be estimated as a function of temperature (Carmichael et al. 1974, Fig. 6–5). Fayalite rhyolites have f_p values two orders of magnitude higher than basalts (10^{-14} and 10^{-16}b, respectively). These values are several orders of magnitude less than the fugacity of phosphorus of 10^{-11}b at 750 °C calculated for the volcanic gases at Showa-shinzan (White and Waring 1963). Alkalic magmas with low activities of silica such as shonkinite and carbonatite have phosphorus fugacities substantially below that of basalt (Nash 1972a, b, 1973). Phosphorus fugacities were higher in lunar lavas where apatite and whitlockite coexist (Nash and Hausel 1973).

Apatite in Metamorphic Rocks

Although apatite is a common accessory in metamorphic rocks, little is known about the details of its chemical composition and paragenesis. Apatite occurs in

Table 7. Major and minor element concentrations in metamorphic apatite (values in wt.%)

	1	2	3	4	5	6	7	8	9
SiO_2	–	0.11	–	1.14	0.72	0.32	0.18	0.29	< 0.05
Al_2O_3	–	–	–	0.03	0.06	0.03	0.01	0.02	–
FeO[a]	0.21	0.08	0.18	0.26	0.11	0.11	0.01	0.07	–
MnO	tr	–	0.15	0.05	0.04	0.05	0.08	0.03	< 0.1
MgO	n.d.	–	0.57	0.20	0.21	0.21	tr	n.d.	–
CaO	51.60	55.09	55.31	–	–	–	–	–	0.7
BaO	0.28	–	–	–	–	–	–	–	71.3
Na_2O	0.5	–	–	–	–	–	–	–	–
K_2O	0.65	–	–	–	–	–	–	–	–
P_2O_5	40.70	–	42.14	–	–	–	–	–	21.0
H_2O^+	0.41	–	–	–	–	–	–	–	–
F	0.18	2.92	–	–	–	–	–	–	< 0.1
Cl	5.70	–	tr	–	–	–	–	–	3.5
ΣREE_2O_3	0.61	0.39	–	–	–	–	–	–	–
Sum	101.51	–	98.35	–	–	–	–	–	98.9
$-O=F, Cl$	1.80	–	–	–	–	–	–	–	0.8
Total	99.71[b]	–	98.35	–	–	–	–	–	98.1[c]

Key to analyses:
1. Chlorapatite in hornblende metagabbro: Taborszky (1972, analysis no. 1, Table 1)
2. Precambrian biotite gneiss, Colorado: Lee et al. (1973, sample I-CC)
3. Apatite inclusion in cordierite, Madagascar: Graziani and Guidi (1978)
4–8. Grenville gneiss: Cruft (1966)
9. Alforsite in metaevaporites, Mariposa Co., California: Newberry et al. (1981)
10. Dioritic gneiss: Puchelt and Emmermann (1976)
11. Migmatite: Puchelt and Emmermann (1976)
12. Anatectic gneiss: Puchelt and Emmermann (1976)
13. Blastic gneiss: Puchelt and Emmermann (1976)

[a] Total Fe as FeO
[b] Includes trace elements (see Table 8)
[c] Includes 2.0 wt.% SrO
n.d. = Not detected

regionally metamorphosed rocks ranging from low grade talc and chlorite schists (Burri et al. 1935; Mitchell et al. 1943) to high-grade biotite hornblende gneiss (Puchelt and Emmermann 1976). In cordierite-biotite gneiss in south-eastern Australia, apatite occurs as inclusions in biotite and as discrete accessory grains (Price and Taylor 1977). Apatite occurs in garnet granulite, garnet websterite and eclogite xenoliths of lower crustal origin in kimberlites (Griffin et al. 1979). Apatite has been described from contact metamorphism of carbonates such as Magnet Cove (McConnell and Gruner 1940) and Crestmore (Eakle and Rogers 1914) and from contact metamorphism of phosphate-rich sediments (Lowell 1955).

Chemical analyses of apatite from metamorphic rocks are given in Tables 7 and 8. Overall, metamorphic apatites are similar in composition to igneous apatites. Generalizations must be made with caution because of the limited analytical data. The K_2O content of 0.65 weight percent in analysis 1 (Table 7) exceeds that of reported igneous apatites, including those from the highly potassic West Kimberly and Leucite Hills lavas (Table 1). The same apatite has considerable chlorine (5.70

Table 8. Trace elements in metamorphic apatite (values in ppm). For key to analyses see Table 7

	1	2	4	5	6	7	8	10	11	12
Sc	–	–	–	–	–	–	–	0.76	1.17	0.73
Ti	–	20	–	–	–	–	–	–	–	–
V	100	7	n.d.	32	n.d.	tr	n.d.	–	–	–
Mn	23	–	406	347	378	639	225	–	–	–
Co	–	–	6,459	1,979	866	2,845	3,194	–	–	–
Ni	–	–	tr	n.d.	tr	tr	tr	–	–	–
Cu	–	7	10.2	1.3	2.9	1.7	2.8	–	–	–
Sr	170	300	5,768	7,645	5,700	9,839	8	–	–	–
Y	5,250	–	1,231	346	184	645	1,380	–	–	–
Zr	–	30	n.d.	2	3	n.d.	n.d.	–	–	–
Ba	2,500	2	8	8	1.7	4	10	–	–	–
La	–	–	5,297	1,119	481	3,255	1,680	553	393	385
Ce	–	–	–	–	–	–	–	302	220	408
Nd	–	–	–	–	–	–	–	296	267	792
Sm	–	–	–	–	–	–	–	28	26	124
Eu	–	–	–	–	–	–	–	4.62	5.78	17
Tb	–	–	–	–	–	–	–	2.30	3.44	17
Yb	–	–	–	–	–	–	–	1.8	12	41
Lu	–	–	–	–	–	–	–	0.22	1.90	4.90
Pb	–	10	–	–	–	–	–	–	–	–
Th	40	–	–	–	–	–	–	14	31	47
U	–	–	–	–	–	–	–	5.56	11	36

n.d. Not detected

wt.-%); it is from a hornblende metagabbro and coexists with phlogopite, scapolite, and talc pseudomorphs after enstatite (Taborszky 1972). Dunn (1975) described almandine with apatite inclusions containing 2.17% F and 0.21% Cl. Strontium contents range from 8 to 9,800 ppm, with several values lower than typical igneous apatites. Yttrium contents range from lower than in igneous apatite to comparable values. Most barium contents are very low (2–10 ppm) compared to igneous apatites (200–600 ppm). Alforsite, ideally $Ba_5(PO_4)_3Cl$, the barium analog of chlorapatite, occurs in metamorphosed evaporitic rocks in California (Newberry et al. 1981). An analysis is given in Table 7. Because alforsite coexists with nearly pure fluorapatite, these authors suggest there may be a solvus gap between Ba–Cl and Ca–F apatites. REE concentrations in metamorphic apatites are generally lower than in igneous apatites although some values fall within the igneous range. REE concentrations, normalized to an average chondrite, are shown in Fig. 4. The limited analytical data suggest that thorium and uranium concentrations are generally comparable in metamorphic and igneous apatites.

Fluid and Solid Inclusions in Apatite

Fluid inclusions have been described in greatest detail in apatites from ijolite and carbonatite (Girault 1967; Girault and Chaigneau 1967; Le Bas et al. 1972; Rankin and Le Bas 1973, 1974; Rankin 1975, 1977; Aspden 1980). Inclusions consist of

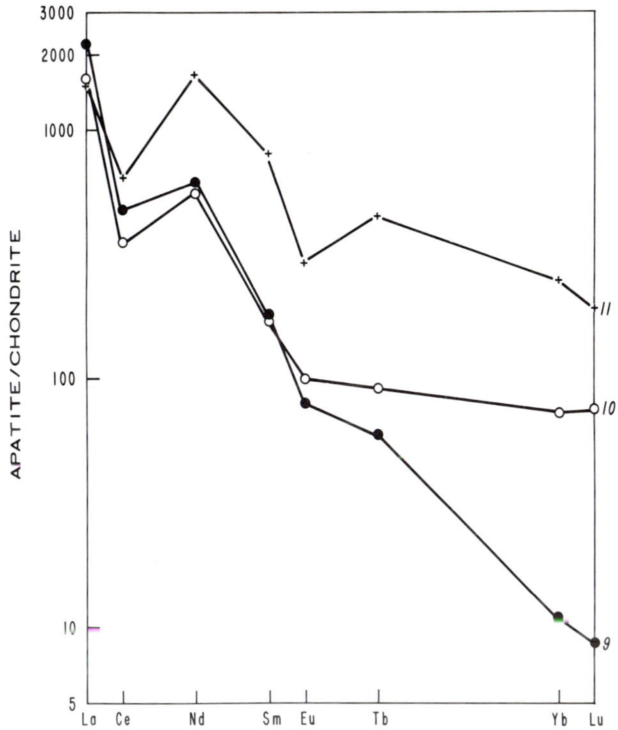

Fig. 4. Rare earth abundances in metamorphic apatite normalized to average chondrite. *Numbers* refer to analyses in Table 7

H_2O and CO_2 fluids, a vapor bubble and various daughter minerals including nahcolite ($NaHCO_3$), kalcine (?) ($KHCO_3$), halite, sylvite, alkali sulfate (probably a member of the aphthitalite group $(K, Na)_4(SO_4)_2$, Aspden 1980), hanksite (?), and opaques, magnetite (?) or pyrrhotite (?). Fluids from apatite in the Oka complex, Quebec consisted predominantly of CO_2 with lesser amounts of CO, H and N. HCl and HF were sometimes found, and H_2O was extremely scarce (Girault and Chaigneau 1967).

Solid inclusions reported in apatite include calcite, pyrrhotite, biotite, hematite, graphite (?), hornblende, fluorite, tourmaline (?), muscovite (?), feldspar (?), zircon (?), sphene, pyroxene, scapolite and molybdenite (Aspden 1980; Cruft 1966). Le Bas et al. (1977) describe glass inclusions of two distinct compositions in apatite from ijolite from Kenya. A K_2O-rich (6%) glass with low Na_2O (0.3%), and oversaturated with respect to SiO_2, coexists with a Na_2O-rich (6–14%) glass poor in K_2O (0.2%) which is undersaturated with respect to SiO_2.

Monazite, Xenotime and Whitlockite

Monazite, a lanthanide phosphate (REE) PO_4, occurs in minor amounts in silicic igneous rocks and in regional metamorphic terranes. Overstreet (1960, 1967) and Mertie (1960) provide considerable data on the occurrence and composition of monazite. In plutonic rocks monazite is an accessory in diorite, quartz monzonite,

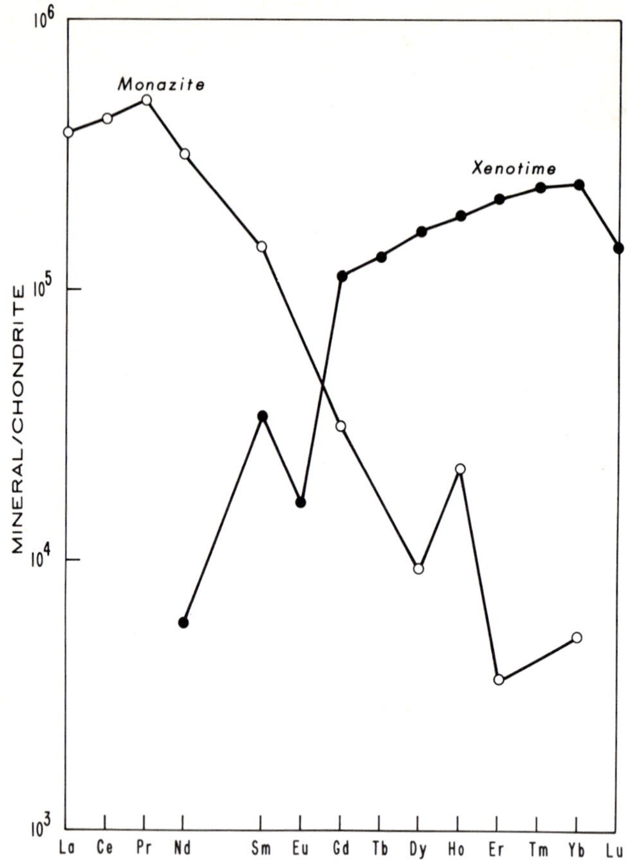

Fig. 5. Rare earth abundances in coexisting monazite and xenotime (Table 9)

muscovite and cassiterite granites, as well as pegmatites and veins. It occurs less commonly in syenite and carbonatite and has not been recognized in silicic lavas, plutonic mafic rocks or their metamorphosed equivalents. Monazite is most abundant in synkinematic granites associated with upper amphibolite or granulite facies rocks.

In metamorphic rocks, monazite is often derived from argillaceous sediments. Monazite occurs in Precambrian gneisses, schists and migmatites. It is rare in rocks derived from areanaceous and carbonate precursors. Overstreet (1960, 1967) concluded that monazite is not stable at low metamorphic grade; at medium to high grade monazite becomes more abundant as well as richer in thorium. Average ThO_2 contents of monazite range from 0.4 wt.-% in the greenschist facies to 4.9% in the amphibolite facies and ultimately 8.9% in the granulite facies (Overstreet 1967).

The chemical composition of monazite has been described by numerous authors (Wylie 1950; Murata et al. 1953, 1957; Vainshtein et al. 1956). Unfortunately for the purpose of this chapter, most analyzed monazites are from detrital, pegmatite or hydrothermal associations. Moreover, the majority of the analytical data are not complete and are expressed as weight fractions of the REE extracted

Table 9. Monazite, xenotime and apatite from pegmatite

	Monazite			Xenotime	Apatite	
	1	2	3	4	5	6
SiO_2	–	0.16	0.68	0.18	–	–
CaO	0.4	4.45	n.d.	n.d.	55.52	54.2
P_2O_5	27.6	31.02	28.66	34.49	42.01	41.4
Y_2O_3	–	1.01	0.96	43.78	n.d.	–
La_2O_3	12.4	13.89	11.05	n.d.	n.d.	0.07
Ce_2O_3	36.6	16.31	32.19	n.d.	n.d.	0.12
Pr_2O_3	5.8	1.64	5.71	n.d.	n.d.	–
Nd_2O_3	10.8	2.34	17.73	0.32	0.06	0.04
Sm_2O_3	–	–	2.61	0.61	n.d.	–
Eu_2O_3	–	–	n.d.	0.11	n.d.	–
Gd_2O_3	–	–	0.75	2.69	0.06	–
Tb_2O_3	–	–	n.d.	0.57	n.d.	–
Dy_2O_3	–	0.05	0.27	4.80	n.d.	–
Ho_2O_3	–	–	0.16	1.24	n.d.	–
Er_2O_3	–	–	0.07	4.17	n.d.	–
Tm_2O_3	–	–	n.d.	0.71	n.d.	–
Yb_2O_3	–	–	0.10	4.70	0.06	–
Lu_2O_3	–	–	n.d.	0.90	n.d.	–
ThO_2	3.9	11.34	–	–	–	0.06
UO_2	–	15.64	–	–	–	–
F	0.9	–	–	–	3.89	2.9
Cl	–	–	–	–	–	0.6
Sum	98.4	97.85	100.94	99.27	101.70	99.39
$-O=F, Cl$	0.4	–	–	–	1.63	1.3
Total	98.0	97.85	100.94	99.27	100.07	98.09

1. Monazite in pegmatitic granite, Sasketchewan, coexisting with apatite No. 6: Watkinson and Mainwaring (1975, sample K42)
2. Monazite in alpine pegmatite, Italy: Gramaccioli and Segalstad (1978)
3. Monazite in granite pegmatite, Norway, coexisting with xenotime No. 4 and apatite No. 5: Amli (1975, analysis 20)
4. Xenotime in granite pegmatite, Norway, coexisting with monazite No. 3 and apatite No. 5: Amli (1975, analysis 19)
5. Apatite in granite pegmatite, Norway, coexisting with monazite No. 3 and xenotime No. 4: Amli (1975, analysis 21)
6. Apatite in pegmatitic gneiss, Sasketchewan, coexisting with monazite No. 1: Watkinson and Mainwaring (1975, sample K42)

or as elemental ratios. In order to illustrate the general chemical features of monazite, Table 9 presents three moderately complete recent analyses from pegmatite associations.

Monazite has a strong preference for the light rare earths in contrast to xenotime, YPO_4, which preferentially concentrates heavy rare earths in the yttrium site. These characteristics are illustrated in Fig. 5 for coexisting monazite and xenotime. The scatter in the monazite data for heavy rare earths may reflect the difficulty in obtaining precise analyses of heavy rare earths in low concentration with the electron microprobe. Monazite may also contain substantial amounts of thorium and uranium. Several percent ThO_2 is common, and although UO_2 is usually less than

Table 10. Whitlockite and coexisting apatite

	Apatite	Whitlockite	
	1	2	3
SiO_2	0.20	< 0.01	1.34
TiO_2	–	–	0.20
Al_2O_3	–	–	0.24
FeO	0.05	0.87	3.03
MnO	< 0.01	< 0.01	–
MgO	0.04	3.20	2.15
CaO	53.1	47.1	40.4
Na_2O	< 0.01	0.22	0.29
K_2O	–	–	0.29
P_2O_5	41.4	46.7	39.8
BaO	–	–	0.40
SrO	–	–	< 0.02
Nb_2O_3	–	–	0.08
Y_2O_3	0.27	0.02	2.68
La_2O_3	0.01	< 0.01	0.98
Ce_2O_3	0.24	0.02	2.91
Pr_2O_3	0.04	< 0.01	–
Nd_2O_3	0.14	0.01	1.74
Sm_2O_3	0.04	< 0.01	0.81
Gd_2O_3	0.08	< 0.01	–
Dy_2O_3	0.03	< 0.01	–
Er_2O_3	–	–	0.23
Cl	1.60	< 0.01	0.06
F	0.29	0.29	0.1
Sum	97.43	98.74	97.7
–O=F, Cl	0.48	0.12	0.1
Total	96.95	98.62	97.6

1. Apatite, Ödegården, Norway: Griffin et al. (1972)
2. Whitlockite, ibid
3. Whitlockite, Lunar sample 14310,213: Nash and Hausel (1973)

1%, concentrations often reach 2.5%. An unusual monazite with 11.34% ThO_2 and 15.64% UO_2 is given in Table 9. Gulson and Krogh (1973) report 106,457 ppm U in monazite in granodiorite of the Bergell Massif, Swiss Alps; coexisting apatites contain 40 and 78 ppm U.

Xenotime occurs in much the same parageneses as monazite, but typically in lesser amounts. It also has been reported as an inclusion in diamond (Harris 1969). Y_2O_3 contents exceed 40 wt.-% (Table 9), and heavy REE are preferentially enriched over the light REE (Fig. 4). Figure 4 also reveals a moderate negative europium anomaly as is observed in some apatites (Fig. 1).

Whitlockite, $Ca_9(Mg, Fe)(PO_4)_6(PO_3OH)$, has not been observed in common igneous and metamorphic rocks on the earth. It is a well documented accessory in meteorites and in lunar lavas (Buchwald, this Vol.) where it may coexist with apatite and the phosphide schreibersite, $(Fe, Ni)_3P$. Griffin et al. (1972) describe whitlockite coexisiting with apatite in a "magmatic (?)" phosphate deposit in Norway. They suggest the whitlockite formed either from residual liquids after crystallization of apatite or as result of metamorphism during the emplacement of pegmatite dikes. Analyses of that whitlockite and apatite are given in Table 10 along with a lunar sample for comparison. In lunar lavas, whitlockite contains substantially greater amounts of REE than coexisting apatite (Gancarz et al. 1971; Griffin et al. 1972; Nash and Hausel 1973). The opposite is true for the Norwegian occurrence, where the REE contents of the whitlockite are less than in lunar whitlockites and lower than coexisting apatite (Griffin et al. 1972).

The apparent lack of whitlockite in common terrestrial igneous and metamorphic rocks may be due in part to its misidentification as apatite which it closely resembles in its physical properties. Thermochemical calculations, however, suggest that fluorine fugacities are sufficiently high in the Earth's upper mantle and crust to stabilize fluorapatite over whitlockite (Nash an Hausel 1973). Similarly, Beswick and Carmichael (1978) evaluated the breakdown of fluorapatite to whitlockite for a mantle assemblage, and concluded that whitlockite could be stable with fluorphlogopite at considerable depth, but in the source regions for most magmas fluorapatite should be the stable phase.

References

Amli R (1975) Mineralogy and rare earth geochemistry of apatite and xenotime from the Gloserheia granite pegmatite, Froland, southern Norway. Am Miner 60:607–620

Aoki K, Kanizawa S (1979) Fluorine contents of some hydrous minerals derived from upper mantle and lower crust. Lithos 12:167–171

Aspden JA (1980) The mineralogy of primary inclusions in apatite crystals extracted from Alno ijolite. Lithos 13:263–268

Bacon CR, Duffield WA (1981) Late Cenozoic rhyolite from the Kern Plateau, southern Sierra Nevada, California. Am J Sci 279:1–34

Behne W (1953) Untersuchungen zur Geochemie des Chlor und Boran. Geochim Cosmochim Acta 3:186–215

Bergman SC (1979) The significance of accessory apatite in the REE modelling of magma genesis. Trans Am Geophys Union 60:412

Beswick AE, Carmichael ISE (1978) Constraints on mantle source compositions imposed by phosphorus and the rare-earth elements. Contrib Miner Petrol 67:317–330

Beswick AE, Carmichael ISE (1980) Critical comments of Frey FA, Roden MF, Zindler A. A reply Contrib Miner Petrol 75:175–178

Birch WD (1980) Mineralogy of vesicles in an olivine leucitite at Cosgrove, Victoria, Australia. Miner Mag 43:597–603

Bishop FC, Smith JV, Dawson JB (1978) Na, K, P, and Ti in garnet, pyroxene and olivine from peridotite and eclogite xenoliths from African kimberlites. Lithos 11:155–173

Brixner LH (1967) Segregation coefficients of the rare earth niobates in $CaMoO_4$. J Electrochem Soc 114:108–110

Brown GM, Peckett A (1977) Fluorapatites from the Skaergaard intrusion, east Greenland. Miner Mag 41:227–232

Brunfelt AO, Roelandts I (1974) Determination of rare earths and thorium in apatites by thermal and epithermal neutron activation analysis. Talanta 21:513–521

Burri C, Jacob J, Parker R, Strunz H (1935) Über Hydroxyl-apatite von der Kemmleten bei Hospenthal. Sweiz Miner Petrol Mitt 15:327–335

Carmichael ISE (1967) The mineralogy and petrology of the volcanic rocks from the Leucite Hills, Wyoming. Contrib Miner Petrol 15:24–66

Carmichael ISE, Turner FJ, Verhoogen J (1974) Igneous petrology. McGraw-Hill, New York, p 739

Cruft EF (1966) Minor elements in igneous and metamorphic apatite. Geochim Cosmochim Acta 30:375–398

Cruft EF, Ingamells CO, Muysson J (1965) Chemical analysis and the stoichiometry of apatite. Geochim Cosmochim Acta 29:581–597

Davies KA (1956) The geology of part of southeast Uganda. Mem Geol Surv Uganda, p 8

Dawson JB, Fuge R (1980) Halogen content of some African primary carbonatites. Lithos 13:139–143

Deer WA, Howie RA, Zussman J (1962) Rock-forming minerals, vol 5: Non-silicates. Wiley & Sons, New York, p 323–346

Duff EJ (1971) Orthophosphates, part V. Phase equilibria in the system calcium oxide-phosphorus pentoxide – calcium fluoride – water along the fluorapatite-hydroxyapatite join under aqueous conditions. J Chem Soc (A) 1895–1898

Dunn PJ (1975) Inclusions in gem almandine from Idaho and New York. J Gemology 14:273–280

Eakle AS, Rogers AF (1914) Wilkeite, a new mineral of the apatite group, and okenite, its alteration product, from S. California. Am J Sci 37:262–267

Eby GN (1975) Abundance and distribution of the rare-earth elements and yttrium in the rocks and minerals of the Oka carbonatite complex, Quebec. Geochim Cosmochim Acta 39:597–620

Elliott RB (1973) The chemistry of gabbro/amphibolite transitions in south Norway. Contrib Miner Petrol 38:71–79

Evensen NM, Hamilton PJ, O'Nions RK (1978) Rare-earth abundances in chondritic meteorites. Geochim Cosmochim Acta 42:1199–1212

Exley RA (1980) Microprobe studies of REE-rich accessory minerals: Implications for Skye granite petrogenesis and REE mobility in hydrothermal systems. Earth Planet Sci Lett 48:97–110

Frey FA, Green DH (1974) The mineralogy, geochemistry and origin of lherzolite inclusions in Victorian basanites. Geochim Cosmochim Acta 38:1023–1059

Frey FA, Green DH, Roy SD (1978) Integrated models of basalt petrogenesis: A study of quartz tholeiites to olivine melilitites from southeastern Australia utilizing geochemical and experimental petrological data. J Petrol 19:463–513

Frey FA, Roden MF, Zindler A (1980) Constraints on mantle source compositions imposed by phosphorus and the rare-earth elements. Critical comment on the paper by Beswick AE, Carmichael ISE. Contrib Miner Petrol 75:165–173

Gancarz AJ, Albee AL, Chodos SA (1971) Petrologic and mineralogic investigation of some crystalline rocks returned by the Apollo 11 mission. Earth Planet Sci Lett 12:1–18

Girault J (1967) Fluid contents of apatites from the Oka complex, Quebec, and possible genetic implications. Can Miner 9:290–291

Girault J, Chaigneau M (1967) Sur les inclusions fluides presents dans les cristaux d'apatite des roches de la region d'Oka (Canada). Acad Sci (Paris) Compt Rend, Ser D 264:529–532

Gramaccioli CM, Segalstad TV (1978) A uranium- and thorium-rich monazite from a south alpine pegmatite at Piona, Italy. Am Miner 63:757–761

Graziani G, Guidi G (1978) Hydrous gem magnesian cordierite with inclusions of hydroxyapatite, dolomite, and rutile. Miner Mag 42:481–5

Griffin WL, Amli R, Heier KS (1972) Whitlockite and apatite from lunar rock 14310 and from Odegarden, Norway. Earth Planet Sci Lett 15:53–58

Griffin WL, Carswell DA, Nixon PH (1979) Lower-crustal granulites and eclogites from Lesotho, South Africa. In: Boyd FR, Meyer HOA (eds) The Mantle Sample: Inclusions in kimberlites and other volcanics. Am Geophys Union, p 59–86

Gulson BL, Krogh TE (1973) Old lead components in the young Bergell Massif, south-east Swiss Alps. Contrib Miner Petrol 40:239–252

Harris JW (1969) The recognition of diamond inclusions – Pt. 2: Epigenetic mineral inclusions. Diamond Rev 28:458–461

Henderson P (1968) The distribution of phosphorus in the early and middle stages of fractionation of some basic layered intrusions. Geochim Cosmochim Acta 32:897–911

Henderson P (1980) Rare earth element partition between sphene, apatite and other coexisting minerals of the Kangerdlugssuaq intrusion, E. Greenland. Contrib Miner Petrol 72:81–85

Hervig RL, Smith JV (1981) Dolomite-apatite inclusion in chrome-diopside crystal, Bellstank kimberlite, South Africa. Am Miner 66:346–349

Hildreth EW (1977) Magma chamber of the Bishop Tuff: Gradients in temperature, pressure and composition. Ph D Thesis, Univ California, Berkeley, p 328

Hildreth EW (1979) The Bishop Tuff: Evidence for the origin of compositional zonation in silicic magma chambers. In: Chapin CE, Elston WE (eds) Ash flow tuffs. Geol Soc Am Spec Paper 180:43–75
Huntington HD (1979) Kiglapait mineralogy I: Apatite, biotite and volatiles. J Petrol 20:625–652
Irving AJ (1974) Megacrysts from the Newer Basalts and other basaltic rocks of southeastern Australia. Bull Geol Soc Am 85:1503–1514
Irving AJ (1978) A review of experimental studies of crystal/liquid trace element partitioning. Geochim Cosmochim Acta 42:743–770
Kind A (1939) Der magmatische Apatit, seine chemische Zusammensetzung und seine physikalischen Eigenschaften. Chem Erde 12:50–81
Kleeman JD, Green DH, Lovering JF (1969) Uranium distribution in ultramafic inclusions from Victorian basalts. Earth Planet Sci Lett 5:449–458
Knutson J, Green TH (1975) Experimental duplication of a high-pressure megacryst/cumulate assemblage in a near-saturated hawaiite. Contrib Miner Petrol 52:121–132
Kogarko LN, Krigman LD, Petrova YeN, Solovova IP (1977) Phase equilibria in the fluorapatite-nepheline-diopside system and the origin of the Khibiny apatite deposits. Geochim Int 14:27–38
Koritnig S (1965) Geochemistry of phosphorus I. The replacement of Si^{4+} by P^{5+} in rock-forming silicate minerals. Geochim Cosmochim Acta 29:361–371
Koritnig S (1978) Phosphorus. In: Wedepohl KH (ed) Handbook of geochemistry, vol II-2. Springer, Berlin, Heidelberg, New York, p B-1–O-19
Koster van Groos AF, Wyllie PJ (1973) Liquid immiscibility in the join $NaAlSi_3O_8$–$CaAl_2Si_2O_8$–Na_2CO_3–H_2O. Am J Sci 273:465–487
Larsen ES, Fletcher MH, Cisney EA (1952) Strontian apatite. Am Miner 37:656–658
Larson LM (1979) Distribution of REE and other trace elements between phenocrysts and peralkaline undersaturated magmas, exemplified by rocks from the Gardar igneous province, south Greenland. Lithos 12:303–315
Le Bas MJ (1977) Carbonatite – nephelinite volcanism. Wiley, London, p 387
Le Bas MJ, Aspden J, Woolley AR (1977) Contrasting sodic and potassic glassy inclusions in apatite crystals from an ijolite. J Petrol 18:247–262
Le Bas MJ, Handley CD (1979) Variation in apatite composition in ijolitic and carbonatitic igneous rocks. Nature 279:54–56
Le Bas MJ, Mills AA, Rankin AH (1972) Preliminary evidence on the nature and composition of carbonatite magma. Nature 239:215
Lee DE, Brandt ELM, Van Loenen RE, Rose HJ Jr (1973) Chemistry of five accessory rock-forming apatites. J Res US Geol Surv 1:267–272
Leeman WP, Vitaliano CJ, Prinz M (1976) Evolved lavas from the Snake River Plain: Craters of the Moon National Monument, Idaho. Contrib Miner Petrol 56:35–60
Lowell WR (1955) Igneous intrusions and metamorphism in some phosphatic rocks of southwestern Montana. Econ Geol 50:715–737
Ludington S (1978) The biotite-apatite geothermometer revisited. Am Miner 63:551–553
Maaloe S, Aoki K (1977) The major element composition of the upper mantle estimated from the composition of lherzolites. Contrib Miner Petrol 63:161–173
McCandless T (1982) Detrital minerals in the Green River Basin, Wyoming. Unpub MS Thesis, Univ Utah, p 107
McConnell D (1973) Apatite; its crystal chemistry, mineralogy, utilization, and geologic and biologic occurrences. Springer, Wien, New York, p 111
McConnell D, Gruner JW (1940) The problem of the carbonate-apatites. III. Carbonate-apatite from Magnet Cove, Arkansas. Am Miner 25:157–167
Mertzman S (1977) Recent volcanism at Sonchin and Cinder Buttes, northern California. Contrib Miner Petrol 61:231–243
Mertie JB Jr (1960) Monazite and related minerals. In: Gillson JL (ed) Industrial minerals and rocks (nonmetallics other than fuels). Am Inst Min Metall Petrol Eng, p 623–629
Mitchell L, Faust GT, Hendricks SB, Reynolds DS (1943) The mineralogy and genesis of hydroxylapatite. Am Miner 28:356–371
Moore PB (1973) Pegmatite phosphates: Descriptive mineralogy and crystal chemistry. Miner Rec 4:103
Murata KJ, Rose HJ Jr, Carron MK (1953) Systematic variation of rare earths in monazite. Geochim Cosmochim Acta 4:292–300

Murata KJ, Rose HJ Jr, Carron MK, Glass JJ (1957) Systematic variation of rare-earth elements in cerium-rich minerals. Geochim Cosmochim Acta 11:141–161
Nagasawa H (1970) Rare earth concentration in zircon and apatite and their host dacites and granite. Earth Planet Sci Lett 9:359–364
Nash WP (1972a) Apatite-calcite equilibria in carbonatites: Chemistry of apatite from Iron Hill, Colorado. Geochim Cosmochim Acta 36:1313–1319
Nash WP (1972b) Apatite chemistry and phosphorus fugacity in a differentiated igneous intrusion. Am Miner 57:877–886
Nash WP (1973) Apatite chemistry and phosphorus fugacity in a differentiated igneous intrusion: correction. Am Miner 58:345
Nash WP (1976) Fluorine, chlorine, and OH-bearing minerals in the Skaergaard intrusion. Am J Sci 276:546–557
Nash WP, Hausel WD (1973) Partial pressures of oxygen, phosphorus and fluorine in some lunar lavas. Earth Planet Sci Lett 20:13–27
Nassau K, Loiacono GM (1966) Calcium tungstate – III. Trivalent rare earth substitution. J Phys Chem Solids 24:1503–1510
Newberry NG, Essene EJ, Peacor DR (1981) Alforsite, a new member of the apatite group: the barium analogue of chlorapatite. Am Miner 66:1050–1053
Overstreet W (1960) Metamorphic grade and the abundance of ThO_2 in monazite. US Geol Surv Prof Paper 400-B: 55–57
Overstreet W (1967) The geologic occurrence of monazite. US Geol Surv Prof Paper 530:327
Parry WT, Jacobs DC (1975) Fluorine and chlorine in biotite from Basin and Range plutons. Econ Geol 70:554–558
Pastor TP, Schauwecker DS, Haskin LA (1974) The behavior of some trace elements during solidification of the Skaergaard layered series. Geochim Cosmochim Acta 38:1549–1577
Peck DL, Wright TL, Moore JG (1966) Crystallization of tholeiitic basalt in Alae lava lake, Hawaii. Bull Volc 29:629–656
Price RC, Taylor SR (1973) The geochemistry of the Dunedin Volcano, east Otago, New Zealand: Rare earth elements. Contrib Miner Petrol 40:195–205
Price RC, Taylor SR (1977) The rare earth element geochemistry of granite, gneiss and migmatite from the western metamorphic belt of south-eastern Australia. Contrib Miner Petrol 62:249–263
Prins P (1973) Apatite from African carbonatites. Lithos 6:133–144
Puchelt H, Emmermann R (1976) Bearing of rare earth patterns of apatites from igneous and metamorphic rocks. Earth Planet Sci Lett 31:279–286
Rankin AH (1975) Fluid inclusion studies in apatite from carbonatites of the Wasaki area of western Kenya. Lithos 8:123–136
Rankin AH (1977) Fluid-inclusion evidence for the formation conditions of apatite from the Tororo carbonatite complex of eastern Uganda. Miner Mag 41:155–64
Rankin AH, Le Bas MJ (1973) A study of fluid inclusions in alkaline rocks with special reference to critical phenomena. J Geol Soc 129:319
Rankin AH, Le Bas MJ (1974) Nahcolite ($NaHCO_3$) in inclusions in apatites from some E. African ijolites and carbonatites. Miner Mag 39:564–70
Reed GW, Allen RO (1966) Halogens in chondrites. Geochim Cosmochim Acta 30:779–800
Roegge JS, Logsden MJ, Young HS, Borr AB, Borcsik M, Holland HD (1974) Halogens in apatite from the Providencia area, Mexico. Econ Geol 69:229–240
Roelandts I, Duschene JC (1977) Rare-earth elements in apatite from layered norites and iron-titanium oxide ore-bodies related to anorthosites, Rogaland, SW Norway. In: Ahrens LH (ed) Origin and distribution of the elements. Phys Chem Earth 11:199–212
Schilling JG (1973) Iceland mantle plume: Geochemical study of the Reykjanes ridge. Nature 242:565–571
Shannon RD, Prewitt CT (1969) Revised values of effective ionic radii. Acta Cryst B25:946–960
Simmons EC, Hedge CE (1978) Minor elements and Sr-isotope geochemistry of Tertiary stocks, Colorado mineral belt. Contrib Miner Petrol 67:379–396
Skinner HCW (1973) Phase relations in the CaO–P_2O_5–H_2O system from 300° to 600 °C at 2 kb H_2O pressure. Am J Sci 273:545–560
Smith JV (1981) Halogen and phosphorus storage in the earth. Nature 289:762–765
Smith JV, Delaney JS, Hervig RL, Dawson JB (1981) Storage of F and Cl in the upper mantle: Geochemical implications. Lithos 14:133–147

Stormer JC, Carmichael ISE (1971) Fluorine-hydroxyl exchange in apatite and biotite: A potential igneous geothermometer. Contrib Miner Petrol 31:121–131

Sun SS, Hanson GN (1975) Origin of Ross Island basanitoids and limitations upon the heterogeneity of mantle sources for alkali basalts and nephelinites. Contrib Miner Petrol 52:77–106

Sun SS, Hanson GN (1976) Rare earth element evidence for differentiation of McMurdo volcanics, Ross Island, Antarctica. Contrib Miner Petrol 54:139–155

Sun SS, Nesbitt RW (1977) Chemical heterogeneity of the Archean mantle, composition of the earth and mantle evolution. Earth Planet Sci Lett 35:429–448

Taborsky FK (1962) Geochemie des Apatits in Tiefengesteinen am Beispiel des Odenwaldes. Beitr Miner Petrogr 8:354

Taborszky FK (1972) Das Problem der Cl-Apatite. Lithos 5:315–324

Taylor HP, Forester RW (1979) An oxygen and hydrogen isotope study of the Skaergaard intrusion and its country rocks: a description of a 55-MY old fossil hydrothermal system. J Petrol 20:355–419

Thompson RN (1975) Is upper mantle phosphorus contained in sodic garnet? Earth Planet Sci Lett 26:417–424

Vainshtein EE, Tugarinov AI, Turanskaya NV (1956) Regularities in the distribution of rare earths in certain minerals. Geochem 1:159–178

Von Eckermann H (1948) The alkaline district of Alno Island. Sver Geol Unders Arsb Ser, p 36

Wass SY (1980) Geochemistry and origin of xenolith-bearing and related alkali basaltic rocks from the southern highlands, NSW, Australia. Am J Sci A280:639–666

Wass SY, Henderson P, Elliott CJ (1980) Chemical heterogeneity and metasomatism in the upper mantle: Evidence from rare earth and other elements in apatite-rich xenoliths in basaltic rocks from eastern Australia. Phil Trans R Soc Lond A297:333–346

Watkinson DH, Mainwaring PR (1975) The Kulyk Lake monazite deposit, northern Saskatchewan. Can J Earth Sci 13:470–475

Watson EB (1979) Apatite saturation in basic intermediate magmas. Geophys Res Lett 6:937–940

Watson EB (1980) Apatite and phosphorus in mantle source regions: An experimental study of apatite/melt equilibria at pressures to 25 kbar. Earth Planet Sci Lett 51:322–335

Watson EB, Capobianco CJ (1981) Phosphorus and the rare earth elements in felsic magmas: An assessment of the role of apatite. Geochim Cosmochim Acta 45:2349–2358

Watson EB, Green TH (1981) Apatite/liquid partition coefficients for the rare earth elements and strontium. Earth Planet Sci Lett 56:405–421

White DE, Waring GA (1963) Data of geochemistry, 6th edn. Volcanic emanations. US Geol Surv Prof Paper 440-K:1–29

Wylie AW (1950) Composition of some Australian monazites. Austr J Appl Sci 1:164–171

Zielinski RA (1975) Trace element evaluation of a suite of rocks from Reunion Island, Indian Ocean. Geochim Cosmochim Acta 39:713–734

Zielinski RA, Frey FA (1970) Gough Island: Evaluation of a fractional crystallization model. Contrib Miner Petrol 29:242–254

Chapter 7 Spatial and Temporal Controls on the Formation of Phosphate Deposits – A Review

PETER J. COOK

Introduction

Phosphorus is present in most rocks in minor to trace quantities ranging from an average P_2O_5 content of 0.04% in sandstone to 0.4% in intermediate igneous rocks (McKelvey 1973). It is only in phosphate rock that the P_2O_5 content is high enough for it to constitute a phosphate ore, attaining values as high as 40% P_2O_5[1] in some rocks. The lower limit of what is designated a phosphate rock depends on the nature of the rock and the available technology but rocks with phosphate contents as low as 4% P_2O_5 may constitute a phosphate rock in some circumstances. Phosphate rock occurs in deposits ranging in size from a few tons to many billions of tons. The phosphate is almost invariably in the form of apatite – generally fluorapatite $Ca_5(PO_4)_3F$, or carbonate fluorapatite which Altschuler (1973) represents by the approximate formula $Ca_{10}(PO_4)_{6-x}(CO_3)_x(F, OH)_{2+x}$. Less commonly it occurs as minerals such as crandallite ($CaAl_3(PO_4)_2(OH)_5 \cdot H_2O$), vivianite ($Fe_3(PO_4)_2 \cdot 8H_2O$), brushite ($CaHPO_4 \cdot 2H_2O$), and whitlockite ($Ca_3(PO_4)_2$). Most of these and other similar minerals form as secondary weathering products (Altschuler 1973) and seldom constitute an economic-phosphate deposit in their own right. The weathering of phosphate deposits is not only scientifically interesting, it can also be economically important, for many deposits are markedly upgraded by weathering. In some cases the phosphate is remobilized by weathering, then reprecipitated to form a deposit of secondary phosphate or phoscrete. In other instances the upgrading of the deposits results from the leaching of more soluble components, especially carbonates, to leave a residual phosphate. However, this paper will not be concerned with the nature and distribution of these residual or weathered deposits to any extent but will focus on the primary deposits of which there are three main types – igneous, guano, and sedimentary. This paper will examine the spatial, and where possible the temporal distribution of these three types. The sedimentary deposits will be examined in greater detail, not only because of their complexity, wide distribution and great scientific importance, but also because they provide the majority of current world phosphate rock production and most of the world phosphate rock reserves.

1 The phosphate content of phosphate rock is generally quoted at $\%P_2O_5$. Other forms quoted include %BPL (bone phosphate of lime) and %P. Conversion factors are: $\%P_2O_5 \times 2.1853 = \%BPL$; $\%P_2O_5 \times 0.4364 = \%P$

Research School of Earth Sciences, Australian National University, Canberra ACT, Australia;
Address now: Division of Continental Geology, Bureau of Mineral Resoures, Canberra, Australia

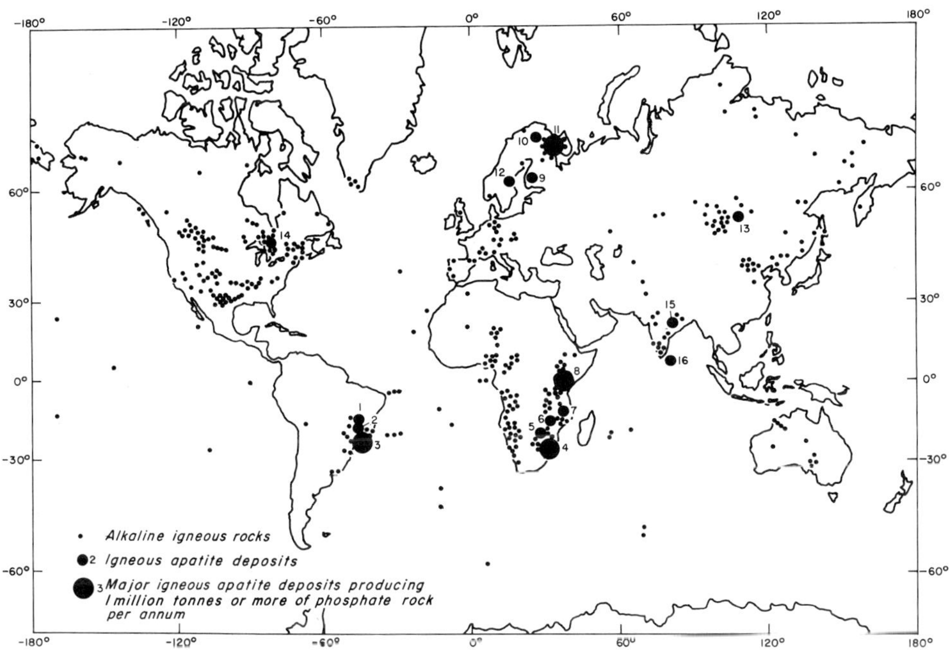

Fig. 1. Distribution of alkaline igneous rocks and igneous apatite deposits. Due to the abundance of alkalic complexes in some areas not all can be shown. Distribution compiled from Tuttle and Gittins (1966), Deans (1968), Sorensen (1974), Notholt (1979a, b), and Erdosh (1979). Important deposits are numbered as follows: *1* Araxa, Brazil; *2* Tapira, Brazil; *3* Jacupiranga, Brazil; *4* Palabora, South Africa; *5* Glendower, South Africa; *6* Dorowa, Zimbabwe; *7* Tundulu, Malawi; *8* Sukulu, Uganda; *9* Siilinjarvi, Finland; *10* Sokli, Finland; *11* Khibiny and Kovdor, USSR; *12* Grangesberg, Sweden; *13* Oshurkov, USSR; *14* Cargill, Canada; *15* Singhbhum, India; *16* Eppawalen, Sri Lanka

Igneous Deposits

Igneous phosphate deposits are geographically fairly widespread and range in age from Precambrian to Tertiary. They are associated with alkaline intrusive plutonic rocks, generally nepheline syenites, or carbonatites, or alkaline ultrabasic complexes, occurring mainly in plug-like bodies and ring complexes. Most associated phosphate deposits are small and non-commercial. However, a number are mined as a source of phosphate rock (Fig. 1), and provide approximately 16% of total world phosphate rock production (Howard 1979). Notholt (1979) estimates total world resources of igneous phosphate rock at 6,000 m tons averaging 4% P_2O_5 or more. By far the largest of these igneous deposits, with reserves of 2,700 m tons of ore, is that of Khibiny in the northern part of the Soviet Union (Fig. 1), with an annual production of about 16 m tons of phosphate rock concentrate with an average grade of 18% P_2O_5. The apatite is associated mainly with a zone of layered ijolites. The Palabora deposit of South Africa, which has reserves of 600 m tonnes is currently providing about 3 m tonnes of phosphate rock concentrate per annum, with an average grade of around 36% P_2O_5. This deposit is associated with a carbonatite plug which grades into the main ore zones of magnetite-olivine-apatite rock and apatite-forsterite-magnetite rock, with average phosphate contents of 6%–10% P_2O_5. Reserves are estimated at about 600 m tons. A third type of deposit, termed by Notholt (1979) primary apatitic carbonatite, it also fairly common. Several deposits of this type are worked as a source of phosphate rock but only one group of deposits, those associated with the Jacupiranga deposits of Brazil, are a major source. The Brazilian deposits are currently producing about 3 m tons of concentrate with an average phosphate content of about 36% P_2O_5, from primary ore with an average content of only 5% P_2O_5.

The occurrence of igneous apatite deposits is clearly a reflection of the much wider distribution of alkaline igneous complexes, which occur in well-defined provinces (Fig. 1). This topic has already been discussed in considerable detail by Tuttle and Gittins (1966) and Sorensen (1974) and consequently it is unnecessary to deal with it here to any extent. However, the majority of such complexes are not apatite-rich. For example, as pointed out by Deans (1968), whilst the Khibiny complex of the Kola Peninsula is apatite-rich, the nearby complex of Lovozero which is geologically very similar, contains no significant apatite deposits. The occurrence of a well-defined zone of alkaline igneous rocks does not then guarantee that an igneous apatite body will occur within that province. Nevertheless it is reasonable to assume that any pattern in the spatial and temporal distribution of igneous apatite bodies will be similar to that of alkaline igneous rocks in general. A very clear geographic distribution is apparent in Fig. 1, with the majority of alkaline igneous complexes occurring in the continental shield and platform areas especially in or adjacent to rift valleys and other tensional environments (Martin and Piwinskii 1972). The string of igneous phosphate deposits extending from Palabora in the south to Sukulu in the north (Fig. 1) are clearly associated with the East African Rift system for example. Similarly the Oshurkov deposit of the U.S.S.R. is associated with the Baikal Rift system. Less commonly, alkaline igneous complexes occur in mid-ocean ridge systems such as the mid-Atlantic ridge (Fig. 1) though none of these are known to be apatite-rich. Other obvious groupings of igneous apatite deposits include those associated with the alkaline igneous complexes of the Fenno-Scandian, Brazilian and Indian shields. Conversely the extensive zones of alkaline igneous rocks along the west coast of Africa and the western United States contain no known apatite deposits.

If alkaline igneous complexes are spatially related to particular tectonic features, it follows that they must also be temporally related to the development of those same tectonic features. Marsh (1973) showed that the alkaline igneous complexes of western African and eastern South America lie along distinct lineaments which can be related to transform faults. He was able to show that there were two distinct age groupings. The oldest of these, 120–135 m.y., appears to coincide with the initial rifting of the African-South American plate and the formation of the proto-Atlantic. The second grouping of 51–83 m.y. represents reactivation of magmatism along the transform directions resulting from stresses associated with the shift of the Cretaceous pole of rotation. Erdosh (1979) similarly showed that the 50 known carbonatite complexes (several of them phosphate-rich) which occur in the Ontario carbonatite province are almost invariably associated with major tectonic features. The younger two groups, dated at 120 m.y. and 570 m.y., are restricted to the Ottawa Graben. The two older groups dated at 1,100 m.y. and 1,700 m.y. are situated along the Kapuskasing High and Albany York and Carlo structures. Gittins et al. (1967) have suggested that these carbonatite ages indicate an eastern migration of carbonatite magmatism or alternatively a relative westerly movement of the continent over a deep source of carbonatite magma generation. These data are inadequate to indicate whether some of these episodes were more likely to produce igneous phosphate deposits than others. Elsewhere, the ages of the alkaline igneous provinces are too imperfectly known to ever delineate the episodes of alkaline igneous activity.

One additional factor in the formation of many igneous phosphate deposits is weathering and especially the leaching of calcium carbonate, also accompanied in some cases by the dissolution and reprecipitation of apatite. Sandvik and Erdosh (1977) have pointed to the formation of a karst topography and the developments of a residual accumulation of apatite sands at the Cargill Complex in Canada. In places there is also a secondary precipitate of crandallite. The age of this karstic weathering is uncertain, but it clearly is totally unrelated to the present weathering cycle in this part of Canada and it may be as old as Late Cretaceous to Paleocene. Apatite-rich residual soils are associated with the Sukulu carbonatite complex of Uganda (Bloomfield 1973). This is probably a Cretaceous igneous body (Notholt 1980). The age of the weathering is uncertain, but in this instance, unlike the Cargill Complex of Canada, the weathering could be related to recent tropical weathering. This may also be the case for the apatite-rich soils associated with the Triassic Araxa and Tapira deposits of Brazil. Therefore, if deep tropical-type weathering favors the development of residual or secondary phosphate deposits associated with alkaline igneous rocks, then obviously periods of tropical weathering and low (tropical) paleolatitudes would have favored the development of such residual or secondary deposits.

Guano Deposits

Cave Deposits

A number of small phosphate deposits with phosphate rock reserves of a few thousand tons or less are known to occur in caves (Fig. 2). They are sometimes worked on a small scale by villagers to provide a local phosphate source, but whilst they may have local importance, they provide a negligible proportion of total world phosphate rock production. The deposits are primarily formed by the accumulation of bat droppings. Much less commonly, the phosphate is derived from bird droppings, or accumulations of small vertebrate remains.

There are few accounts available on the distribution of cave guano deposits. The most detailed account is that by Hutchinson (1950); a number of additional localities are documented by Lee (1980). Despite the paucity of information it is apparent from Fig. 2 that there are concentrations of deposits in particular areas. First, it is evident that the majority of deposits occur in relatively low latitudes, and certainly not beyond latitude 50°. Within this latitudinal belt it is apparent that there are three areas where deposits are relatively common – the Caribbean region extending into northern Mexico and the southern United States; southern Europe extending into the Maghreb district of North Africa; and southeast Asia. It could

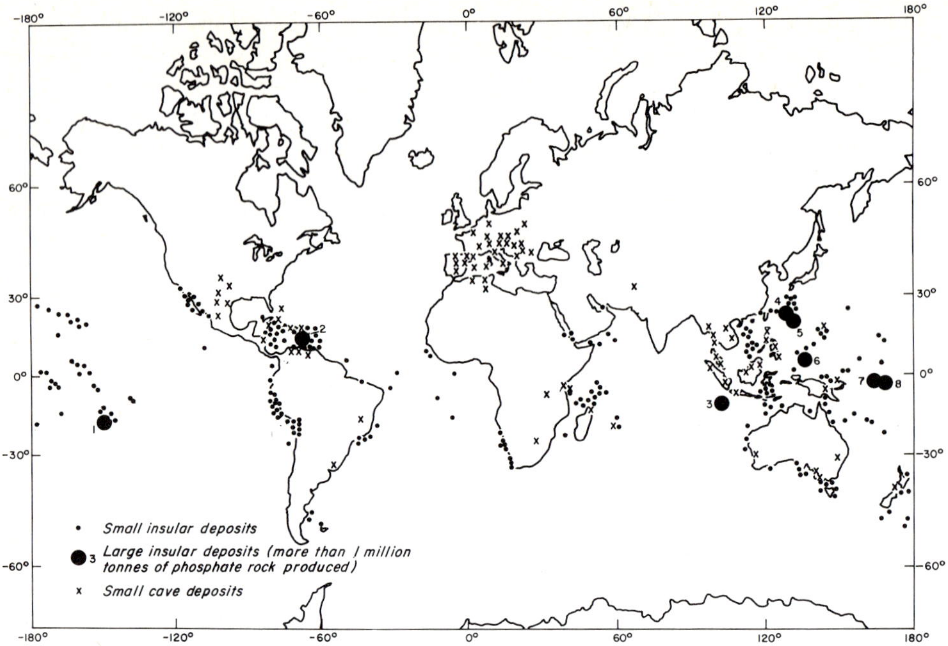

Fig. 2. Distribution of insular and cave guano deposits. Major deposits are numbered as follows: *1* Makatea, French Polynesia; *2* Curacao, Dutch Antilles; *3* Christmas Island, Indian Ocean; *4* Kita Daito Jima, NW Pacific; *5* Okino Daito Jima, NW Pacific; *6* Anguar, SW Pacific; *7* Nauru, central Pacific; *8* Ocean Island, central Pacific (now worked out)

be argued that the concentration in southern Europe is the consequence of European interest in caves. Similarly in southeast Asia, the apparent abundance of cave guano deposits may in part be a consequence of the intense agriculture of the region and the pressing need to find phosphate deposits there. Both of these "artifacts" may have influenced the distribution in Fig. 2 but nevertheless there are several important non-anthropogenic environmental factors which profoundly affect the distribution of such deposits. The first is that there must be bedrock appropriate for the development of large caves, i.e., limestone and dolomite. This is obviously not the only factor, for many areas, such as Western Canada, have extensive limestone caverns yet no cave guanos. This is essentially a thermal limitation, for large concentrations of bats are not found in cold climates. The third factor is aridity; this is important not only because the groundwater flow may be insufficient in very arid areas to produce large cave systems, but also because in such areas the insect populations and vegetation cover are insufficient to support a large bat population. Therefore the net result is that cave guano is found primarily in low-latitude humid areas with well-developed limestone or dolomite terrains. Temporally, cave guano is likely to have developed when these conditions were most likely to have been met. It is for this reason that most, if not all of the southern European cave guano was formed during interglacial periods. The other temporal factor is that bats did not appear until the early Tertiary. Consequently there are unlikely to be any major cave guano deposits greater than Tertiary in age.

Insular Guano

Insular phosphate deposits are formed directly or indirectly by the accumulation of bird droppings. The insular guano deposits contain a negligible proportion of the total world phosphate rock reserves and most have been worked out. However, because they are relatively easily mined and of high grade, they still provide approximately 2% of world phosphate rock production. Their geographic distribution is shown in Fig. 2. There are few modern accounts of insular deposits but they are documented in some detail by Hutchinson (1950) who describes two types, the low island and the high island deposits.

The deposits of the low islands (those rising only a few meters above sea level) are the result of the leaching of fresh nitrogenous guano to form thin accumulations, usually as cemented caps and crusts, on the low lying coral islands (Warin 1968). There is also some phosphatization of the coralline material. Most of the deposits are believed by Veeh (1979) to be as young as 3,000–5,000 years B.P. Deposits of this type occurred in the Pacific and Indian Oceans and in the Caribbean, but most were small and were worked out in the last century because they were both easy to recognize and of high grade (up to 40% P_2O_5). The deposits of the high islands (those rising 50 m or more above sea level) are much larger, though less common. They are also older than 800,000 years (Veeh 1979). Nauru, in the Pacific Ocean, with maximum reserves of about 40 m tons of high-grade phosphate rock and Christmas Island, in the Indian Ocean, with lower grade reserves of up to 200 million tons, are the only high island deposits currently being mined. These deposits have formed in part by the secondary phosphatization of the bedrock, which is generally composed of coralline material, to produce calcium phosphate minerals such as apatite or more rarely whitlockite. There are also some clayey deposits which may have a residual origin. In the case of Christmas Island there has been extensive phosphatization of volcanic rock to produce aluminum-bearing phosphate minerals such as crandallite, millisite and barrandite (Trueman 1965). The phosphatizing solutions are believed to be derived from accumulations of guano, which once overlay the islands. There are some apparent inconsistencies in this explanation. The lack of large colonies of nesting birds at the present day, the absence of modern guano and the paucity of fossil bird remains in the deposits, have been used as evidence in favor of a marine rather than avian origin for the Christmas Island phosphate. Bourrouilh (1980) has also questioned the validity of the guano hypothesis for the Nauruan deposits. Despite these reservations both types of insular deposits probably formed indirectly or less commonly directly as a result of the accumulation of bird droppings.

Insular deposits are most common in warm-arid or semi-arid areas with large bird populations either at the present day or in the fairly recent past. In his definitive volume on guano deposits Hutchinson (1950) states:

"Guano deposits are formed only when the organisms producing excreta are relatively large, feed over a wide tropophoric field, and return to a limited site for rest or reproduction. These, though necessary, are not sufficient conditions for the production of a deposit, for certain geophysical requirements must also be met. In general the geophysical conditions to be expected are that the substratum must be of a suitable form to retain the deposit and that the rainfall must not be so great that the deposit is entirely washed away. The biological and geophysical conditions are, however, not entirely independent for in the case of guano produced by marine birds, the same factors that make the regions of the richest guaneras arid also provide that the adjacent seas be unusually fertile, permitting an extraordinary quantitative development of the marine biota and extraordinary populations of marine birds."

These conditions of marine fertility to which Hutchinson alludes are primarily the result of the oceanic upwelling of deeper, colder, and more phosphorus-rich waters. This phenomenon occurs in quite specific geographic locations, most notably off the west coasts of continents (the result of entrainment of coastal waters by currents such as the Benguela and Peruvian Currents) and along the Equator (the result of divergence associated with the Equatorial Current System). In such areas the photic zone of the ocean is rich in nutrients and consequently is able to support a large biomass. Therefore, these are the areas where the major guano islands are located. Other obvious factors include the relative isolation of the island to ensure that the bird colonies feed over a wide trophophoric range, a lack of predators, and also the right species of bird, for not all colonial seabirds produce guano. Whether or not the deposits are preserved depends on the stability of the island. Relative sinking of the island will result in destruction of most guano deposits, unless the deposits are rapidly covered with a protective layer of sediment. Where the islands are undergoing uplift however, the guano deposits are removed from the zone of coastal erosion and stand a better chance of being preserved. Hutchinson (1950) reports emergent features on all of the high islands where there are large phosphate deposits.

The chronology of insular guano deposits is poorly known at the present time. There are at least two factors which will produce episodicity, or periodicity in the formation of the deposits – variations in the fertility of the surrounding seas, and changes of sea-level. Off the coast of Peru and nothern Chile, for example, there are marked perturbations of the marine fertility. One of the most important changes is the development of a warm south-flowing coastal counter current known as El Nino. This produces spectacular increases in the water temperature and associated mass mortalities not only of the marine faunas, but also of birds which feed upon them. Hutchinson (1950) documents seven such avian catastrophes between 1910 and 1943 which resulted from marked decreases in the anchovette population and in turn lead to a decline in the accumulation of guano. Variations in guano production on a longer time scale may similarly have resulted from more major and prolonged changes in the equatorial upwelling systems. Aharon (1982) has shown through an isotopic study of raised coral reefs of the Huon Peninsula that there have been marked changes in the productivity of the tropical surface waters of the Western Pacific, which can be related to glacial episodes. Unfortunately there are insufficient radiometric ages available to test this hypothesis directly on the guano deposits. Hutchinson (1950) found evidence of two or more periods of phosphate accumulation on the arid island deposits, notably Jarvis, Baker and Malden Islands. He suggests that this is the result of changes in the rate of accumulation of guano due to northward shift of the equatorial current system, and consequent climatic changes. Veeh and Burnett (1978) and Veeh (1979) reported ages of between 3,000 and 5,000 years B.P. for the apatite of Ebon Atoll and Fanning Islands in the Pacific and Tracey (1979) reports the probable time of phosphate accumulation at Enderbury Island at around 3,000–4,000 years B.P. These ages are consistent with the post-glacial rise of sea-level, and stabilization at around 6,000 years B.P. This does not necessarily preclude the accumulation of guano during low stands of sea-level. However, as pointed out earlier, deposits formed during high stands of sea-level are more likely to be preserved than those found at low stands because of the likelihood of submarine erosion of the guano following a post-glacial rise of sea level. Additionally, during periods of high sea level the number of islands is fewer and therefore the bird concentration higher. Dating of the high island deposits (Veeh 1979) revealed ages in excess of 800,000 years for Nauru, Makatea, and Christmas Island. Burnett (1980) supports this general age distribution, though he points out that there are some younger perhaps secondary deposits on Nauru. The reason for the apparent hiatus in the formation of insular guano deposits in the late Pleistocene is not known. It may be related to climatic and oceanographic changes during that time, but until more radiometric ages are obtained one can only speculate on the underlying causes.

Sedimentary Deposits

Sedimentary phosphate deposits – phosphorites – are by far the most important of the world's sources of phosphate rock. They currently provide about 82% of the total world phosphate production and 95% or more of the world's phosphate reserves (Howard, 1979). Reserves of phosphorites are probably in excess of 200 billion tons of phosphate rock, equivalent to about 30×10^9 tonnes of P_2O_5. The major part of the world's production is from the sedimentary deposits of Morocco and the United States. However, phosphorites occur on every continent (Fig. 3) and range in age from Precambrian to Recent, though almost all the commercially exploited deposits are of Phanerozoic age.

Most commonly the phosphorites occur as beds ranging in thickness for a few centimeters to tens of meters and are composed of grains (frequently termed pellets) of crypto-crystalline carbonate fluorapatite or collophane. Most of the grains are well rounded, up to about 1 mm in diameter. They contain included material such as quartz, clay, kerogen, and framboidal pyrite. Commonly the grains provide few obvious clues to their mode of origin, though in some deposits it is clear that many of them were originally grains of biochemical carbonate which were diagenetically phosphatized. Some were originally muds which were phosphatized. Oolitic grains are present in many deposits and may represent a primary precipitate but they constitute a very small proportion (about 3%) of the total phosphatic component in most deposits. Other forms of collophane include nodules, some of which may be up to tens of centimeters in diameter, and mudstone phosphorites (also termed microsphorite, collophane mudstone and phospholutite) which is composed of very fine-grained structureless collophane (de Keyser and Cook 1972; Ilyin and Ratnikova 1981). Although a few phosphorites have formed in lacustrine environments such as that of the Green River Formation (Love 1964) and the Glenns Ferry Formation (Swirydczuk et al. 1981) the overwhelming majority formed under marine conditions. Many contain marine fossils and other evidence of marine origin. Stromatolitic phosphorites, where the algal laminae are composed of collophane, have also been described. This type of phosphorite is especially common in the Precambrian (Bushinski 1969; Banerjee 1971) but also occurs in the Phanerozoic (Southgate 1980; Krajewski 1981).

It is not intended to review the origin of phosphorites thoroughly here, as this would require a separate chapter. Recent reviews on this topic include those of Cook (1976a), Baturin and Dubinchuk (1979), Cook and Shergold (1979), Sheldon and Burnett (1980), Bentor (1980), Slansky (1980), Kolodny (1981), and Sheldon (1981). For many years, the most popular theory for the formation of phosphorites was the so-called Upwelling Theory of Kazakov (1938). In essence, the theory invokes deep cold phosphorus-rich ocean water as the source of the phosphorus in phosphorites. The deep ocean water ascends into the shallow photic zone and in the original theory it was believed that apatite directly precipitated from the ascending water column and then accumulated as phosphate grains on the shelf. Since the early investigations of Murray and Renard (1891), phosphorites have been known to occur off the west coasts of continents (Fig. 3), notably in areas where there was strong oceanic upwelling at the present day, such as off Namibia. These occurrences were believed to represent modern examples of the upwelling mechanism at work, which could in turn then be extrapolated to explain many of the features of ancient deposits, such as the Phosphoria Formation (McKelvey et al. 1953; McKelvey 1967). The discovery by Kolodny (1969) that most offshore phosphorites are relict raised some doubts about the appropriateness of the upwelling model. Subsequent discovery of undoubted modern phosphorites off southwest Africa (Baturin et al. 1972) and Peru (Burnett and Veeh 1977; Burnett 1977) provide new support for the importance of upwelling, though not necessarily for the original Kazakov mechanism of inorganic precipitation of apatite. Particular stress has been placed by many writers on the west-coast phosphorites. However, phosphorites do also occur off several east coasts including Africa, North America, and Australia. Most of these may be relict, though insufficient dating has been undertaken so far to be certain of this. The east Australian phosphorites which were first described by von der Borch (1970), are of particular interest, for whilst they have a complex history with several phases of phosphatization (Marshall and Cook 1980), at least one of these phases has been shown by O'Brien and Veeh (1980) to be Holocene. The region is not one of strong upwelling or high productivity; the warm southward-moving East Australian Current is characterized by its low nutrient levels. However, south of about latitude 28 °S where the orientation of the coast changes to approximately south-southwest, a fairly well-defined gyral, known as Eddy J, develops off the continental margin. The phosphorites also occur in this region, suggesting the possibility of a genetic link between the gyral and the phosphate-rich sediments. The phosphorites off the east coast of the United States may similarly be associated with Gulf Stream gyrals, but at this stage far more work is required on these east coast occurrences. Investigation currently underway by S.R. Riggs should clarify the nature of the association.

So far the emphasis has been on modern phosphorites of the open shelf environment, occurring in water depths ranging from about 100 to 500 m. Most of these are nodular phosphorites or encrustated hardgrounds. Similarly, the phosphorites of the marginal plateaus such as the Blake Plateau are predominantly nod-

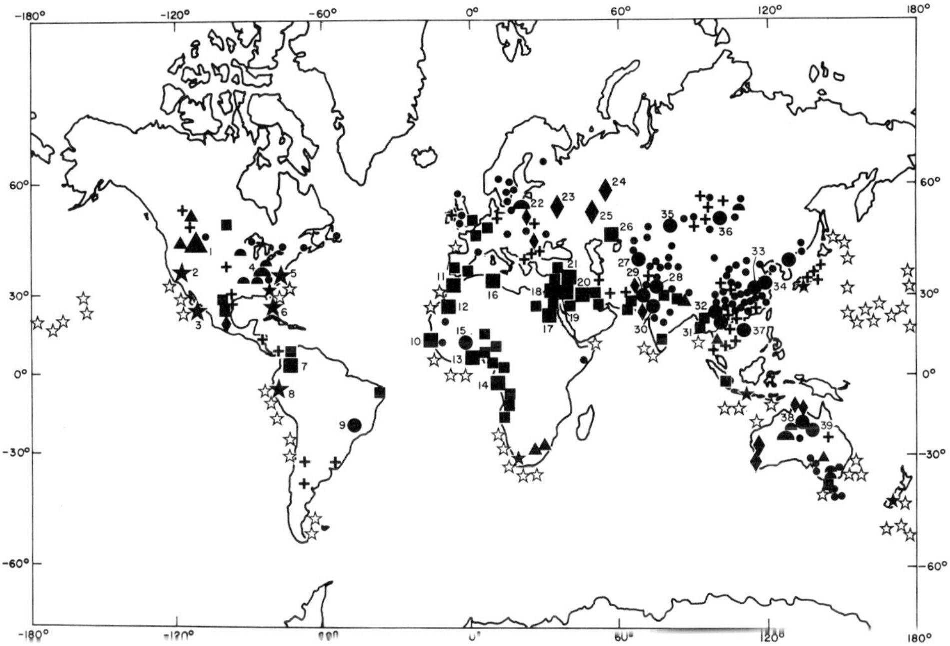

Fig. 3. Distribution of sedimentary phosphate deposits and occurrences. Major deposits are numbered as follows: *1* Phosphoria, USA; *2* Monterey, USA; *3* Baja California, Mexico; *4* Tennessee, USA; *5* N Carolina, USA; *6* Florida, USA; *7* Cordillera, Colombia; *8* Sechura, Peru; *9* Bambui, Brazil; *10* Taiba, Senegal; *11* Khourigba etc, Morocco; *12* Bu Craa, Morocco; *13* Hahotoe, Togo; *14* Cabinda; *15* Volta and Niger; *16* Gafsa, Tunisia and Djebel Onk, Algeria; *17* Abu Tatur, Egypt; *18* Machtesh area, Israel; *19* El Hassa, Jordan; *20* Iraq; *21* Palmyra, Syria; *22* Maadu, USSR; *23* Moscow Basin, USSR; *24* Yegorevsk, USSR; *25* Rudnichny, USSR; *26* Aktyubinsk, USSR; *27* Karatau, USSR; *28* Maldeota, India; *29* Hazara, Pakistan; *30* Rajasthan, India; *31* Laokoy, Vietnam; *32* Kunming, China; *33* Kaiyang, China; *34* Xingyang, China; *35* Abaka, USSR; *36* Hubsugul, Mongolian People's Republic; *37* Hainan, China; *38* Alexandria-Wonarah (NT), Australia; *39* Duchess etc (Qld), Australia

ular. Consequently, apart from the ancient nodular phosphorites associated with disconformity surfaces, these do not constitute a direct analog of ancient sedimentary phosphate deposit which are predominantly grainstone phosphorites, or more rarely, mudstone phosphorites. As stated by Bentor (1980) "It seems to follow that as far as phosphorite formation is concerned, a simplistic actualism does not work. Phosphorite deposition in the past probably took place in a number of ways, only one of which is operative at the present time". In fact there are probably several mechanisms operative even today. The deposits off Baja California may provide one of the best-known modern offshore analogs of an ancient deposit. In the region of Santo Domingo, black, well-rounded phosphate grains constitute up to 20% of the sands of the middle bay and adjacent lagoons. Whether these nearshore sites are merely zones of entrapments or also zones of genesis of the phosphate grains is not clear. Birch (1980) draws a number of comparisons between the Baja California deposits and those of southern Africa where phosphorites occur sporadically from the outer shelf to the inner coastal zone. The whole region is one of very high organic productivity as a consequence of upwelling not only on the outer shelf associated with the Benguela current, but also in the shallow nearshore zone due to the prevailing offshore winds. Birch (1980) proposes that on the outer shelf the prime process for the formation of phosphorites is the replacement of calcium carbonate by calcium phosphate. He also suggests that within the estuaries and lagoons of the adjacent coast the packstone phosphorites formed by the early diagenetic precipitation of apatite and the laminate mudstone phosphorites by direct precipitation from the water column.

Primarily as a result of studies of ancient phosphorites, Cook (1976a) proposed a six-stage sequence to produce a grainstone phosphorite.

"a) Influx of nutrient-rich water, generally by upwelling, into a shallow-marine region maximum water depth 500 m, but commonly considerably less with a slow rate of terrigenous deposition, and often a warm arid climate.

b) Development of a prolific biota.

c) Formation of anoxic organic-rich bottom sediments; loss of C, N, and H from dead organisms occurs either before or immediately after burial.

d) Interstitial waters rich in phosphorus from below the sediment/water interface by leaching of phosphate from organic remains as a consequence of low pH and high alkalinity in the sediments.

e) Localized patches of apatite develop by phosphatization of sediments in the presence of phosphate-rich pore waters or in some instances by direct precipitation of apatite from the pore waters. Diagenetic phosphatization will take place whether the sediments are clayey, siliceous, or calcareous. Phosphatized faecal pellets are a common feature of some deposits.

f) Reworking of sediments occurs in response to changes of current pattern or relative sea-level. The coarser patches of phosphatized sediment remain as a lag deposit and the finer matrix is winnowed out, resulting in extensive upgrading of the phosphatic sediments into a high-grade phosphorite.

Whilst this scheme still appears to be broadly valid, it is clear from recent investigations that some modifications are necessary. For example, it is becoming increasingly apparent that many major phosphate deposits accumulate in very shallow water – generally no more than a few tens of meters and commonly considerably less. Also, considerable difficulties are posed for the simple oceanic upwelling model by the fact that many old deposits were formed in shallow interior seas, far removed from the open ocean. In such cases any initial oceanic area of high organic productivity may have been far from the final site of accumulation of the phosphate grains. The possibility of a high productivity system being generated by other-than an oceanic upwelling system, such as estuarine-type circulation, also cannot be ignored. Sedimentological studies such as those undertaken by Elgueta (1981) on the Lady Annie deposits of the Georgina Basin have shown that many deposits form as accumulations of transported phosphate grains, and that in situ winnowing alone is not sufficient to produce a major deposit. However, the importance of transporting and trapping of phosphate grains is by no means universally accepted. For example, Sheldon (1980) states the alternative viewpoint quite clearly: "The evidence from the rocks themselves shows that most ancient phosphorite formed directly on the sea-floor". It is reassuring to see that controversy still flourishes in phosphate geology! Additionally, it is clear that stromatolitic phosphorites formed by a somewhat different mechanism possibly involving direct biochemical precipitation and/or fixation of phosphate by cyanobacteria, with little or no mechanical reworking.

Utilizing the Georgina Basin as an example Cook and Shergold (1980) empirically developed a set of prerequisites for phosphate deposits including: An age of between 650 and 540 m.y.; a low paleolatitude; a broad shallow downwarp flanking a major seaway; high productivity conditions in the general region; shallow marine sedimentation with minimal input of terrigenous sediments; a well-defined marine transgression; and a suitable trap such as a bay, an estuary, or a carbonate bank complex. Thus it is evident that to produce a sedimentary phosphate deposit it is necessary to have an interplay of a large number of factors, many of them related, some of them quite distinct. It is also necessary to have various scales of factors ranging from a very broad geographic scale or time interval giving a potentially favorable large-scale phosphogenic system, to the detailed geography of a basin or a particular and relatively short-lived event to produce the actual deposit.

The ideas of Kazakov (1938) represented one of the first attempts to explain the spatial distribution of phosphorites. In essence this theory suggested that because of the link with oceanic upwelling, phosphorites were most likely to be found where such events occurred, i.e., at low latitudes off the west coasts of continents. In one of the earliest attempts to apply the new techniques of paleomagnetism to a mineral deposit, Sheldon (1964) showed that phosphorites preferentially formed at low palaeolatitudes, i.e., within 40° of the equator. Freas and Eckstrom (1968) developed this further to show how paleolatitudes and paleo-oceanography can be used to explain the location of many phosphate deposits. It was left to Cook and McElhinny (1979) to attempt to fit phosphogenesis into a plate tectonic model. Using a large data base they were able to show that the preference for low latitude locations suggested earlier by Sheldon (1964) was indeed correct, though there were some variations and at times there is a bimodal distribution of the paleolatitudes. Using this and the types of vertical successions shown by phosphatic sequences it was possible for Cook and McElhinny (1979) to develop idealized models for phosphogenesis which could be related to various phases of continental drift Their spatial model A was for a latitudinal seaway and comprised a series of events including:

1. Continental sedimentation.
2. Low latitude rifting along an east-west axis.
3. Development of marginal evaporites during the phase of incomplete separation of the plates.
4. Complete separation of the plates resulting in a narrow east-west seaway, a strong westerly-directed current system, with associated dynamic upwelling, and the initiation of a high productivity system.

5. As spreading continues the seaway becomes broader, dynamic upwelling weaker, the nutrient supply diminished. Eventually the region will cease to be one of active phosphogenesis.

Spatial Model B is for a longitudinal ocean similarly relates phases of phosphogenesis to phases of seafloor spreading. If the longitudinal ocean is too narrow for an oceanic gyral to develop then west-coast upwelling will not occur and there will be no well-defined high productivity system. However, as sea-floor spreading proceeds and the ocean becomes wider, so oceanic upwelling eventually takes place. A minimum separation of the two plates of about 3,000 km is necessary before phosphogenesis commences. Ziegler et al. (1979) invoke monsoonal upwelling on diagonal east-facing coasts from 10° to 30° in association with monsoon circulation, as the mechanism for the establishment of a third model for phosphogenesis though at this stage no major deposits are known to be associated with such a monsoonal upwelling system. All of these models tend to stress the role of oceanic upwelling in the development of organic-rich high productivity systems. Such a bias is not unreasonable, given that a high percentage of the ocean biomass occurs in upwelling areas. Whilst specialized non-upwelling conditions such as fumarolic input (Riggs 1980) can be invoked to explain a single deposit, it is difficult to do this for a large number of deposits occurring over a wide area. In fact, oceanic upwelling appears to be one of the few potential mechanisms for generating organic-rich phosphogenic conditions over a wide area. The major phosphogenic provinces are apparent from the present-day distribution of phosphorites (Fig. 3), however in order to establish their original extent it is necessary to replot their locations on global reconstructions. Examples of just how large an area these ancient phosphogenic systems cover is provided by the two examples in Fig. 4. The Cretaceous-Eocene phosphogenic province extended for several thousand kilometers along the Tethys and provides the best example of Model A, the latitudinal seaway. However, the continuation of deposits of about the same age along the west coast of Africa is less readily explicable by such a model and it may be necessary to invoke a longitudinal model for that region. The paleogeography for the Cambrian period is still imperfectly known and a variety of reconstructions are available (Morel and Irving 1978; Scotese et al. 1979; Smith et al. 1981). In fact, all three reconstructions are consistent with low-latitude locations for the major Cambrian deposits. However, one of the difficulties with any Cambrian reconstruction is the extent of Cambrian time and the relatively rapid polar wander and continental drift during that time. If a latitudinal model is applied to the Cambrian data, then any coastal entrainment and upwelling may have been in response to westerly-directed, near-equatorial currents immediately to the north. Alternatively, as suggested by the data of Scotese et al. (1979), phosphogenesis may have occurred within the equatorial current system on the margins of continental fragments.

It should therefore be apparent that the development of seaways, the resulting ocean-continent configuration and the latitude of the coasts are of very considerable importance in establishing high organic productivity conditions in general and phosphogenic systems in particular. However, this is only one component in the spatial disposition of the actual deposits. Many ancient deposits were not formed in open shelf or upper slope locations but in epicontinental seas, trans-cratonic seaways, and coastal embayments. It is apparent that the major Cambro-Ordovician

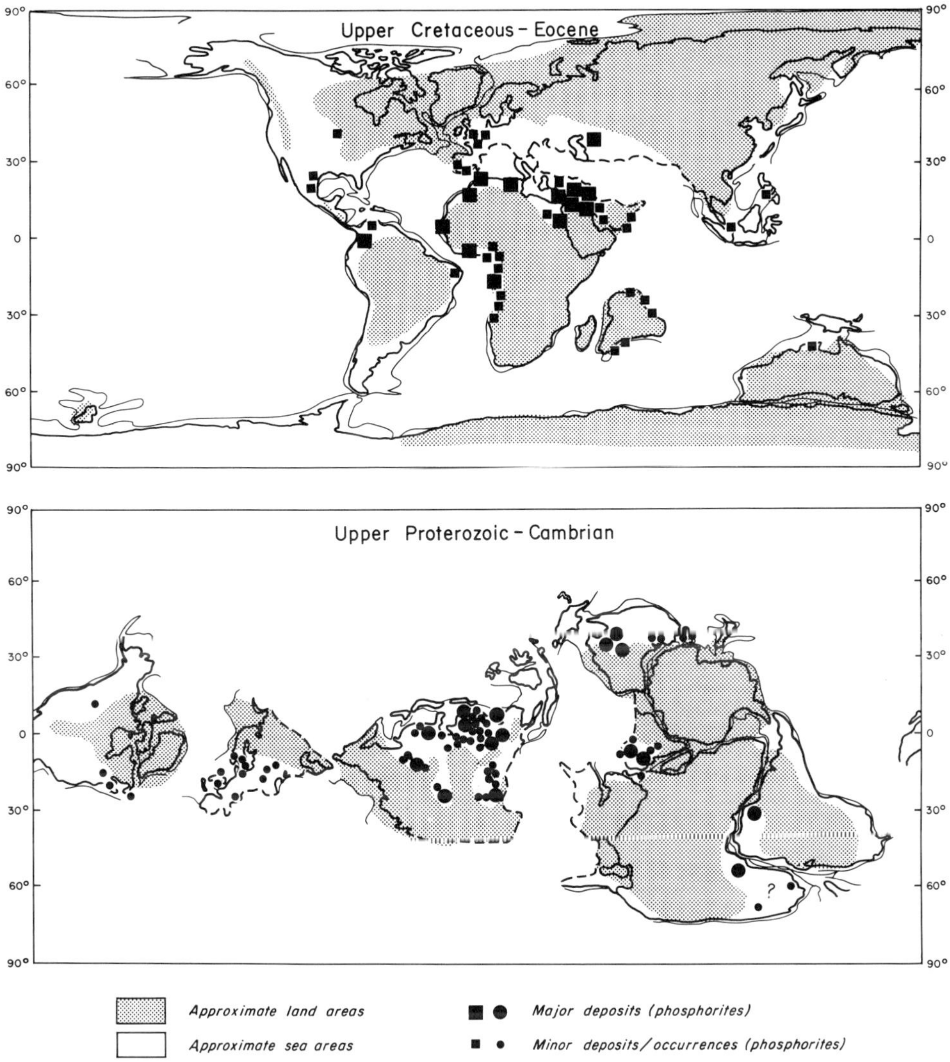

Fig. 4. Distribution of phosphorites. The reconstruction for the upper Cretaceous-Eocene utilizes a Santonian reconstruction, the Upper Proterozoic-Cambrian utilizes an Aldian reconstruction. Both reconstructions are after Smith et al. (1981) and are based on cylindrical equidistant projection

phosphate deposits of Australia for example (Fig. 5), were formed not on the outer shelf but in the epicontinental sea, and with a preference for times when there was a well-developed trans-cratonic seaway. This may be in part a result of the associated circulation system, the likely development of strong currents through the seaway and the formation of high productivity systems. It could equally well be related to the relative rise in sea-level which produced this inundation of the continent. This possible association with sea-level changes will be discussed later. If the precise location of the Australian deposits within the Middle Cambrian seaway is

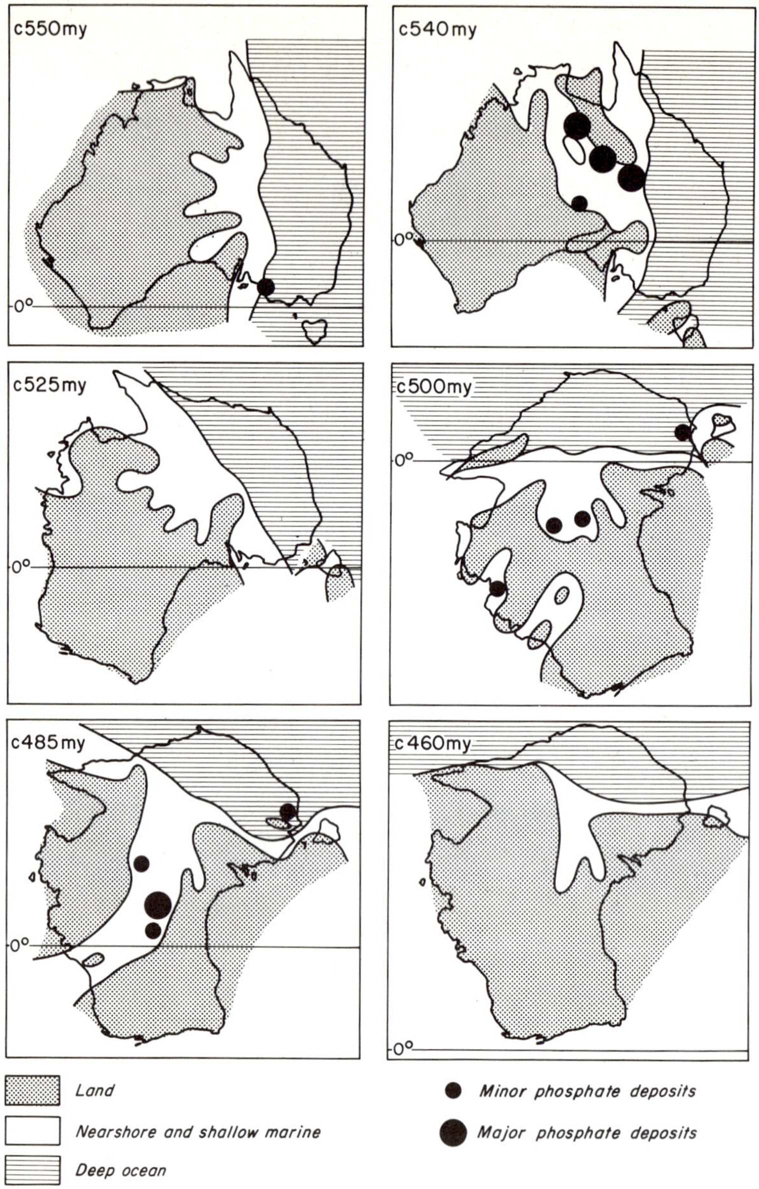

Fig. 5. Relationship between Australian Cambro-Ordovician palaeogeography and the location of phosphorites. (After Cook 1982)

considered (Fig. 6) it is immediately apparent that the deposits of the Georgina Basin are clustered around the margin of the present-day basin. Cook (1976b), Howard and Cooney (1976), and Howard and Hough (1979), consider that the present edge of the basin was close to the Cambrian margin. Shergold and Druce (1980) on the other hand consider that the Cambrian sea extended well beyond the present

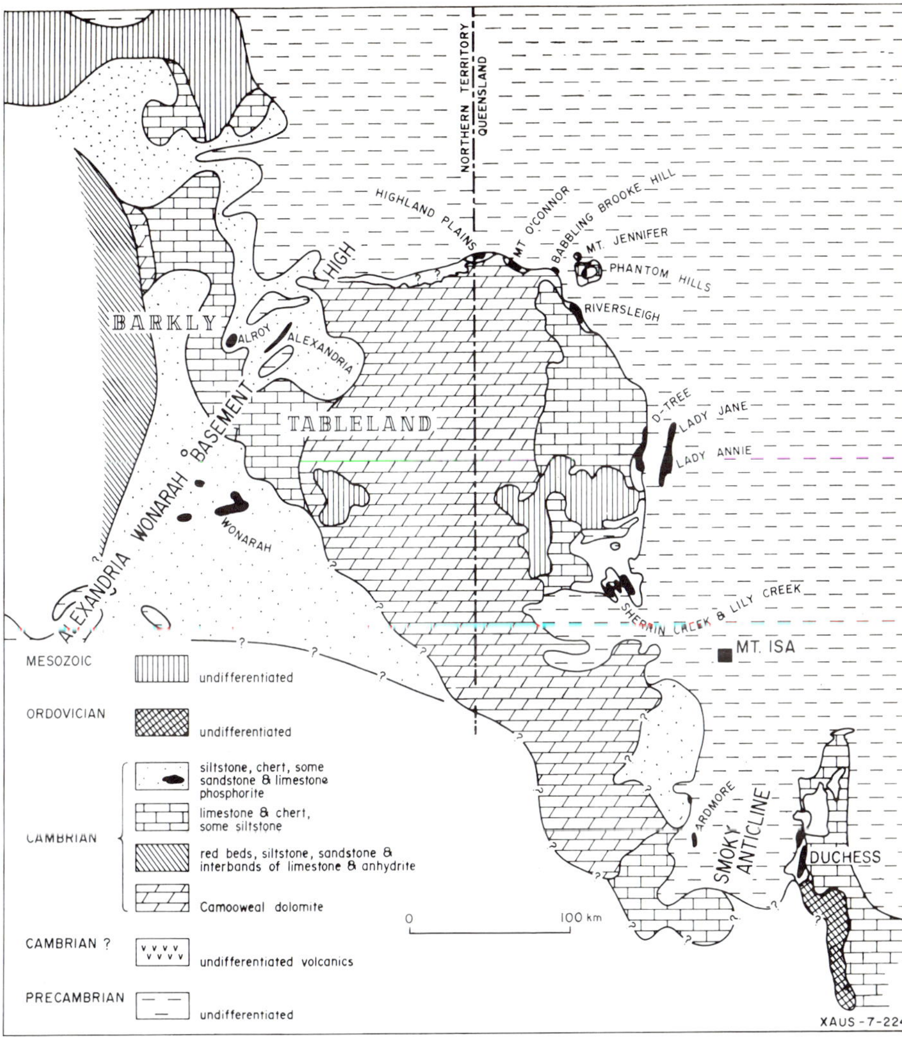

Fig. 6. Location of the major phosphate deposits of the Georgina Basin around the margins of the basin or adjacent to the Alexandria-Wonarah basement high. (After Howard 1972)

margins. However, all are agreed that, whether it was sub-aerial or sub-aqueous, the present Mount Isa block constituted a Cambrian topographic high, and that the phosphate deposits accumulated around and on the flanks of that high. As indicated earlier, it is suspected that grains in many grainstone phosphorites were transported from a site of formation to a site of accumulation. The studies of Elgueta (1981) in particular clearly demonstrate that the deposits of the western zone of Lady Annie formed as a result of the trapping of phosphatic allochems in a semi-protected microtidal tectonic depression marginal to the Georgina Basin epeiric sea. Elgueta (1981) also considers that the deposits (though not necessarily the grains making up that deposit) were formed in water depths of only a few meters,

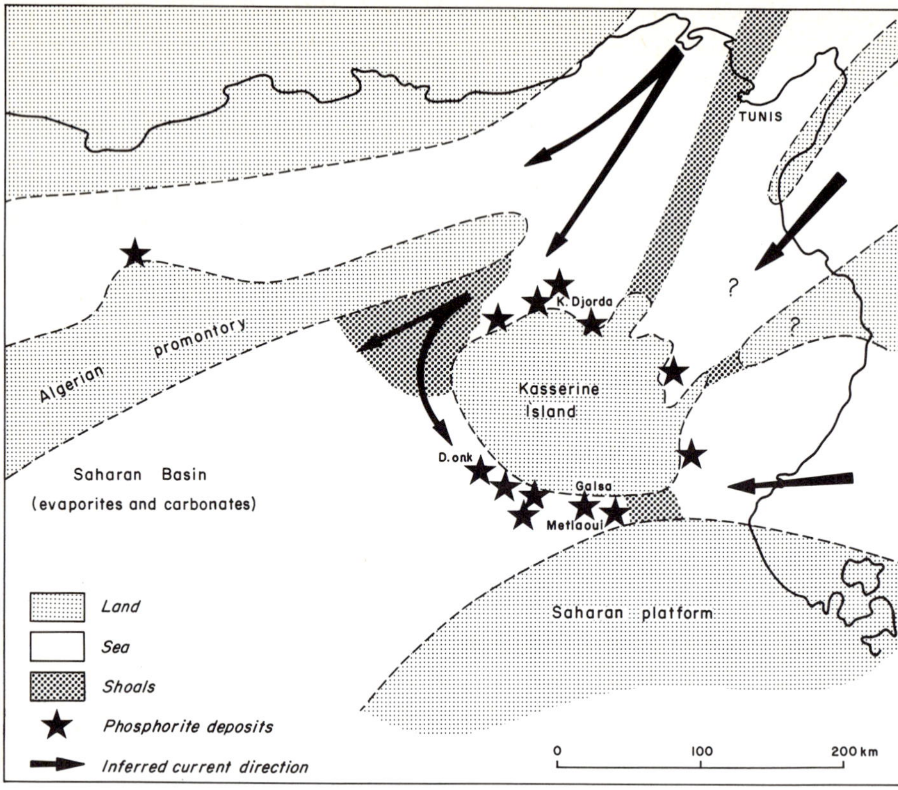

Fig. 7. Distribution of sedimentary phosphate deposits around the Eocene Kasserine Island of Tunisia-Eastern Algeria. (After Sassi 1980; Slansky 1980)

probably in an intertidal-subtidal system. Perhaps an even better example of a nearshore location is provided by the Tunisian and Algerian deposits which have been shown by the work of Sassi (1980) to be preferentially located adjacent to an Eocene island, Kasserine Island (Fig. 7). All the deposits appear to have been formed in small restricted basins and embayments. South and west of Kasserine Island lay the Saharan Basin, with abundant evaporites. At times the evaporitic conditions extended north and east to Kasserine Island and the phosphorites were interbedded with evaporites. Further to the north and east lay the open ocean. Sassi (1980) proposes that strong westerly-directed currents brought nutrient-rich conditions to the area and that abundant organic matter (from which the phosphorus making up the phosphorites was ultimately derived) was trapped in the areas flanking the Island. It seems probable that, as in the Georgina Basin, the phosphate grains making up the phosphorites were formed at one location then swept into the areas flanking the Island where they were trapped. The grains may have been transported for distances ranging from a few meters to tens of kilometers. In his review of the phosphate deposits of the southeastern United States, Riggs (1979a, b) also considers that the topographic or structural setting was a crucial component in the location of the deposits. He proposes that the structural framework which controlled the formation and deposition of the phosphorites of the Florida Miocene was a series of arches or topographic high associated with the major Ocala Upland and Sanford High. Extensive coastal, shallow nearshore shelf, and platform environments occurred around the highs and were the sites of major phosphorite sedimentation (referred to as the "phosphate machines") and associated entrapment basins. The perimeter phosphogenic belt of microsphorite-intraclast sedimentation is situated around the highs. The outer phosphogenic belt which occurs downdip, was characterized by pelletal phosphorite sedimentation. Phosphorite precipitation took place as cold phosphate-high waters, moved across the shallow platforms and into the coastal environments. Riggs suggests that the biologically stressed shallow-water environments received bacterially precipitated microcrystalline phosphorite mud, or microsphorite, as well as the other biologically produced phosphate grains. The benthic community populating these muds, injested and excreted the mud as fecal pellets, whilst other biological and physical processes produced intraclasts. Very locally, under high-energy conditions, some of the mud was aggregated to produce pseudo-oolites. The resulting phosphorite allochems (phosphorite gravel, sand, and clay) were then transported as clastic particles off the shoals by periodic high-energy conditions until they were deposited and accumulated in the adjacent entrapment basins and on the flanks of the structural highs. In other words the mechanism proposed by Riggs (1979a, b) is similar to that proposed here for the Georgina Basin and the Tunisia-Algeria deposits, in that trapping of phosphate grains in coastal locations is believed to be a crucial mechanism for the formation of the deposits. However, there is no evidence in the Georgina deposits to support the suggestion that most of the phosphate grains are formed from mudstone phosphorites which accumulated in intertidal-supratidal locations on the various highs and which were then broken up to form the rounded intraclasts which make up the grainstone phosphorite deposits. Instead, it is proposed that the phosphate grains are formed in comparable or greater depths than the eventual site of accumulation and that they were swept shorewards by currents or by marine transgressions resulting from relative rises of sea level. In Fig. 8 the implication is that the phosphate grains are formed well out on the shelf by the authigenic and diagenetic processes discussed earlier and then swept shorewards in response to a relative rise of sea-level. This mechanism may work for the Duchess deposits for example, which may have been relatively close to the open shelf. There are obvious problems invoking an outer shelf location for primary genesis of phosphate grains where the deposits, such as those of the Alexandria-Wonarah region of the Georgina Basin (Howard 1972), are situated well within the craton, though even in such cases there is still commonly an association of deposits with topographic highs. An additional problem is that many of the grains contain abundant organic matter and pyrite and show little evidence of transport under oxic conditions. As in many other respects, the Phosphoria Formation does not fit the normal picture for a phosphate deposit, and no clear relationship has yet been demonstrated between many of the deposits and topographic highs. However, the deposits probably were formed within fairly well-defined bathymetric limits (Sheldon et al. 1967), possibly in response to a favorable energy regime rather than in coastal traps. As a consequence of this, the main phophorite horizons are more widespread in the Western Phosphate Field than they are in most other phosphogenic basins,

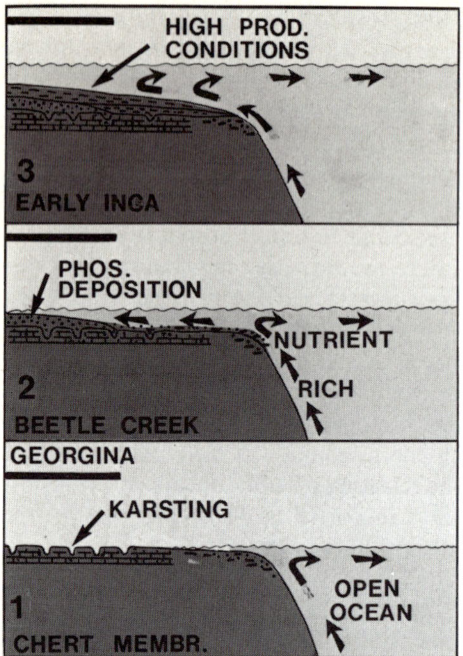

Fig. 8. Possible model for phosphate deposition in the eastern Georgina Basin and adjacent areas during Middle Cambrian time. Initially (*1*) much of the eastern Georgina Basin (approximate extent shown by the bar scale) was subjected to sub-aerial exposure at times during and immediately following the deposition of the Thorntonia Limestone. During this time only minor phosphate was deposited in the basin; phosphatic sediments may have been deposited further out on the shelf. Subsequently (*2*) a major transgression resulted in reworking of phosphatic shelf sediments and the shoreward transport of phosphate grains to form major deposits (the Beetle Creek Formation) such as Duchess, in the coastal zone and marginal embayments. Finally (*3*) deeper, quieter water conditions extended over much of the basin resulting in the deposition of organic-rich muds (the Inca Formation)

and it is possible that the depths of accumulation were no more than tens rather than the hundreds of meters suggested by McKelvey et al. (1953).

Turning now to the question of the temporal distribution of sedimentary phosphate deposits, again it is possible to discern variable time scales in the episodicity or periodicity of deposits. On the broadest scale it can be seen that phosphorites become progressively more abundant through the geological column: They are essentially unknown in the early Archean, rare in the late Archean–earliest Proterozoic, more abundant in the late Proterozoic and relatively common in the Phanerozoic. This could be constructed as merely a reflection of an increasingly complete geologic record in the younger rocks. However, the pattern is not regular and within this broad trend there do appear to be preferred times for phosphogenesis. An early attempt to examine the world distribution of phosphate deposits with time was made by Gimmel'Farb (1958), and also by Strakhov (1960). They identified a number of peaks of which the Cambrian, Permian, and Late Cretaceous-Eocene were the most prominent. Bushinski (1969) considers that phosphogenesis has been a relatively continuous process throughout the geologic record, but acknowledges that there is a certain regularity in the distribution of phosphorites which may be related to transgressions. Bushinski (1969) also explores the possibility of a relationship between glaciations and phosphogenesis but concludes that no such relationship exists. In contrast, many years earlier Mansfield (1927) suggested that it was possible to correlate the main phosphorite horizons of the Phosphoria Formation with the various glacial intervals of the Permian. The most recent com-

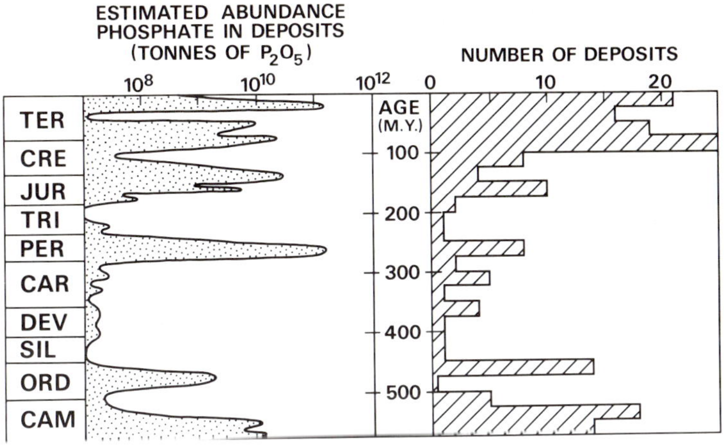

Fig. 9. Temporal distribution of phosphate deposits during the Phanerozoic on the basis of estimated abundance of phosphate in deposits and occurrences and the number of deposits and occurrence. (After Cook and McElhinny 1979)

pilation of the distribution of world phosphate deposits with time (Fig. 9) is after Cook and McElhinny (1979). They attempted to provide a more quantitative evaluation of the abundance of total phosphorus in the major phosphatic intervals and were able to delineate a number of quite distinct peaks in the Phanerozoic plus several less well-defined peaks in the Precambrian (Table 1). More recent work has not modified this distribution significantly though a recent compilation by Lee (1980) of phosphorites of the Asian-Pacific region can be interpreted as suggesting that there may possibly be a Devonian peak. Cook and McElhinny (1979) consider that plate tectonics played a major role in whether or not major phosphate deposits were formed. In their view phosphorites formed sporadically during the Proterozoic and throughout the Phanerozoic as continents drifted into low-latitude locations and seaways formed, though with a major period of phosphogenesis from about 800 m.y. to the present. They consider that phosphogenesis is not directly related to phases of volcanism, orogenesis, or evaporites, but all are associated through plate tectonic and the processes of plate separation and convergence. Volcanism and orogenesis are involved in the initiation of rifting; evaporites in the "rift-valley" stage and phosphorites are formed once the rift valley is breached and a seaway established. Using these features they developed the models discussed earlier to explain features (the nature of the sediments and their order of superposition) of east-west and north-south seaways. However, they also proposed that the chemistry of the oceans is of major significance to the formation of phosphorites. The distribution of iron in Precambrian seas especially around 2,200 m.y. was of great importance to the distribution of phosphorus in the hydrosphere at that time. The two are linked because upwelling was important to the formation of both iron ores and phosphorites. They also proposed that glacial periods may have been important because of the intensification of upwelling during glacial periods, and perhaps also because the onset of glaciation may have initiated oceanic overturn. Sheldon (1980) further developed this model drawing also on a hypothesis by Fischer

Table 1. Major phosphogenic episodes. (After Cook and McElhinny 1979)

Episode	Mean age (m.y.)	Locations
Miocene-Pliocene	14	Continental margins; *SE USA*[a], California, *Sechura*, Venezuela, Japan, Philippines
Upper Cretaceous-Eocene	65 {54[b], 77[b]}	*W Africa, N Africa, Mid East*, W Europe, N South America, Mexico, N India, Ukraine, Kazakhstan, W Australia, Pakistan, Texas, Manitoba
Jurassic	149	*Russia*, Mexico, N South America, W Europe, W Australia, N India, Pakistan, W Canada
Permian	250	*W USA*, Urals, Szechwan, Indochina, N India
Ordovician	464	*Tennessee*, Ala., Ken., Central and SE Australia, *Baltic area*, Iowa, E Canada, Bolivia, Baikal
Cambrian	542	*Karatau, Russia, Georgina Basin, Central and SE China, Vietnam, Hubsugul*, S Australia, Tien Shan, Sayan, Central Kazakhstan, Tennessee, E Canada, Mauritania
Upper Proterozoic I	620	*Volta area*, Mauritania, China(?)
Upper Proterozoic II	700– 800	*China*, Central Siberia, India(?), Brazil, Central Australia
Middle Proterozoic(?)	1,200–1,600	Yenisei, *Rajasthan*(?)
Lower Proterozoic	1,800–2,200	Rum Jungle, Broken Hill, Australia(?), Michigan, Finland

[a] Regions with very large deposits are italicized
[b] Secondary peaks may be present at about these times

and Arthur (1977) to explain secular variation of various features in pelagic sediments. In essence, Sheldon (1980) hypothesizes that:

1. The deep ocean has constituted a major geochemical sink or reservoir for phosphorus since about 2,022 m.y. before present.

2. The phosphorus content of this sink builds up during periods of slow vertical mixing and declines during periods of rapid vertical mixing.

3. Vertical circulation cycles in the ocean correlate with cycles of high and low stands of sea level.

4. Upward vertical mixing into shallow levels occurs primarily be equatorial upwelling and trade-wind-belt upwelling.

From these premises Sheldon (1980) considers that:

1. Episodes of phosphogenesis occur at the onset of episodes of vertical mixing after oceanic stability, during which the phosphorus content of the deep-ocean sink has built up to high levels.

2. The major phosphogenic episodes of the Cretaceous to early Tertiary are due to phosphorus withdrawal from the deep-ocean phosphorus sink primarily by equatorial upwelling at the time of high-level, warm seas.

3. The major phosphogenic episodes of the Cambrian, Ordovician, Permian, and Miocene periods are due to phosphorus withdrawal from the deep-ocean phosphorus sink primarily by trade-wind-belt upwelling at the time of transition (related to glacial episodes) from the high-level, warm oceans to low-level, cold oceans.

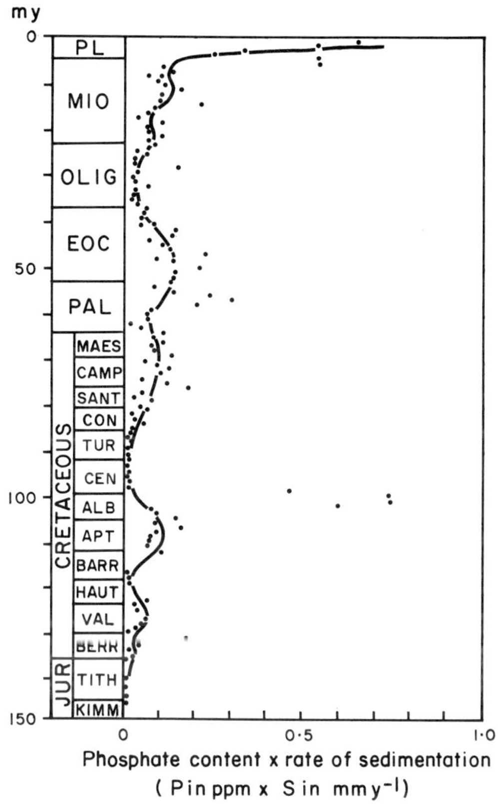

Fig. 10. Variation in the phosphate content of Indian Ocean deep sea sediments with time. Curve compiled from a total of about 3,000 analyses of DSDP samples from all Indian Ocean drill holes and then averaged over 1 m.y. intervals. To normalize phosphate contribution per unit time, the phosphate content (expressed as parts per million P) was multiplied by the rate of sedimentation (expressed in millimeters per year) given in the initial DSDP reports for Legs 22–28. The high Pleistocene values are probably a function of their lack of compaction (and hence an apparently rapid rate of sedimentation). No corrections are applied for compaction. It is assumed that from about the Miocene the sediments are uniformly compacted

This general hypothesis does appear to explain the distribution of phosphate deposits with time. Also, data recently obtained for the Indian Ocean (Fig. 10) do suggest variations in the phosphorus content of the pelagic realm. However, some aspects of the hypothesis are challenged by Arthur and Jenkyns (1981). They too consider that phosphorite genesis correlates in a general way with elevated sea level and warm climate and is therefore influenced by plate tectonics and continental drift. However, they also correlate phosphorite genesis with transgressions, resulting in the formation of nutrient-rich shallow transgressive seas. They point out that oceanic anoxic events (OAE's – times when there was widespread deposition and preservation of organic carbon), generally do not coincide with major phosphogenic episodes. Figure 11 demonstrates this lack of correlation between phosphogenesis and OAE's and lends support to the view of Arthur and Jenkyns (1981) that during OAE's the phosphorus is fixed in the deep sea sediments along with the organic carbon. Given that organic matter has an average atomic C : P ratio of about 106 : 1 (the Redfield Ratio) then extensive deep sea organic-rich facies, such as those documented from the Cretaceous (Schlanger and Jenkyns 1976; Summerhayes 1981), should indeed potentially act as phosphorus sinks. Consequently, the apparent negative correlation between phosphogenic episodes and OAE's is to be expected. In the opinion of Arthur and Jenkyns (1981) "The 'phosphorite giants' are commonly associated with 'black shales' in a given depositional setting (e.g.,

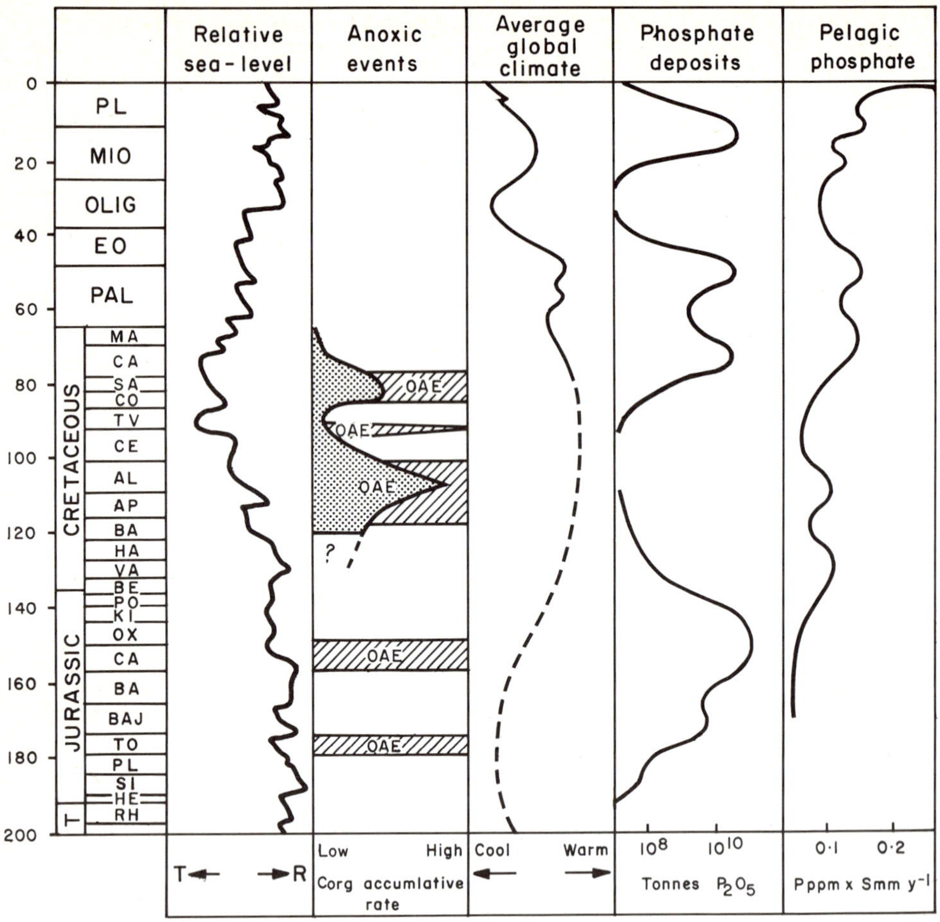

Fig. 11. Late Mesozoic and Cenozoic paleoceanographic parameters. Curves compiled mainly from Figs. 2 and 3 of Arthur and Jenkyns (1981) who used the data of Vail et al. (1977) for the sea-level curves; Arthur (1981), Jenkyns (1980), and Schlanger and Jenkyns (1976) for the anoxic events; Savin (1977) and Frakes (1979) for average global climate and Cook and McElhinny (1979) for phosphate deposits. Additionally, the Indian Ocean data for accumulation of phosphate in deep sea sediments (see Fig. 10) is also summarized

Piper and Codispoti 1975). However, widespread organic carbon-rich sediments formed during OAE's were not always associated in time or space with major phosphorite deposition. 'Phosphorite giants', in fact, tend to occur near the beginning or end of an OAE". However, recent data do not appear to wholly support this persuasively simple picture, for if there is a negative correlation between OAE's and phosphogenic episodes, then the rate of accumulation of phosphorus in deep sea sediments should similarly show a negative correlation. Currently data are only available for the Indian Ocean (Fig. 10) but when this is compared with the distribution of phosphogenic episodes for the period Late Cretaceous to Recent (Fig. 11) there appears to be a positive relationship (with the notable exception of the

Aptian-Albian OAE). In other words, during the phosphogenic episodes, when more phosphorus was reaching the shallow seas in the form of phosphorites, more phosphorus was also being incorporated into deep sea sediments. There is therefore no evidence from the Indian Ocean data to support the view that the phosphorus content of the shallow seas was increased by depleting the phosphorus content of the deep ocean. Clearly, this relationship needs to be more thoroughly investigated, particularly in the North Atlantic, where there are well defined black scale facies resulting from OAE's. Therefore the present data do suggest that during the formation of "phosphorite giants" there may have been phosphorus enrichment of the world ocean and not just of the shallow shelf seas. However, the possibility that the phosphorus content of pelagic sediments is primarily a function of the productivity of the overlying photic zone rather than the phosphorus concentration of the overlying deep water column, cannot be ignored. One further construction which can be put onto the apparent correlation is that it lends support to a further mechanism proposed by Arthur and Jenkyns (1981). They suggest that an increase in the rate of chemical weathering on land produced an increase in the supply of phosphorus to the ocean during times of globally warm climates, which in turn would have corresponded to times of relative highstands of sea level. But if this is the case, why then does not the mid-Cretaceous climatic optimum evident in Fig. 11 closely correspond with a phosphogenic episode? The answer may in part be that it is not one factor but a complex interplay of several which is responsible for the formation of a phosphate deposit. For example, evidence presented earlier (Fig. 5) showed a clear correlation in the Cambro-Ordovician of Australia between major transgressions and the formation of major phosphate deposits. However, as indicated in the model for the Georgina Basin (Fig. 8), this could be more a reflection of mechanical processes associated with a transgression, such as winnowing and transport of grains, than a major change in ocean chemistry. At the other end of the Phanerozoic, Burnett and Veeh (1977) have suggested from uranium series dating that phosphorite formation off South America was episodic rather than continuous, and that phosphorite formation occurred mainly during highstands of sea level though their data are sparse. As the South American phosphate nodules are mainly from depths greater than 200 m, the lack of phosphorite formation during glacial periods cannot be attributed merely to sub-aerial exposure of the seafloor at this time. Similarly, the radiometric ages obtained by Veeh et al. (1974) for phosphorites on the southwest African shelf also suggest episodic formation of phosphorites. Conversely, the Phosphoria Formation, one of the world's greatest repositories of phosphorus was formed during a period characterized by global cooling and a low stand of sea level. For this Arthur and Jenkyns (1981) propose that during such a low stand of sea level there were few sites suitable for the accumulation of phosphorites and that in Permian time the Phosphoria sea of the western United States was one of the few such sites. This would certainly seem to be entirely consistent with the apparent abundance of phosphorus in Permian phosphorites and especially the Phosphoria, compared with the scarcity of deposits of this age, as evident in Fig. 9.

So far, this consideration of importance of sea-level changes to phosphogenesis has focused on sea-level rises and falls associated with glaciations. Because of their rapidity these may indeed be especially important for the formation of phosphate deposits particularly, if as appears likely, many (perhaps most) deposits are detrital accumulations of phosphate grains. However, as pointed out by Cook and McElhinny (1979), the major periods of phosphogenesis in the Mesozoic and early Cenozoic appear to be totally unrelated to glacial episodes. Pitman (1978) has noted that there are a number of other ways in which relative rises or falls of sea level may occur. These include the flooding of a small ocean basin, desiccation, changes in the rate of subsidence of the seaward edge of the platform, and changes in the volume of the mid-ocean ridge system. Therefore, whilst plate tectonics only has an indirect link with sea-level changes associated with glaciation (the drift of continents to high latitude locations) some of the other mechanisms proposed by Pitman undoubtedly do have a more direct link with such plate tectonic processes as the initiation of rifting, the development of a mid-ocean ridge or a change in the rate of sea-floor spreading. These produce relatively slow changes in sea level and are unlikely to be responsible for other than the very broadest trends, i.e., changes at a scale of 10^7 years or greater. Therefore whilst such events may have been important in the development of a broad peak in phosphogenesis, such as that experienced around the Precambrian–Cambrian boundary, it is unlikely to have been of real significance to a relatively short-lived peak such as the Miocene peak.

Discussion

It has been previously demonstrated in this paper that phosphate deposits are not uniformly, or randomly, distributed throughout the globe, or through geologic time. They clearly were preferentially formed in certain regions and at particular times. Some of the more important features, responsible for directly influencing the distribution of phosphate deposits may be summarized as follows:

Spatial Features

Latitude (10^6–10^7 km^2)[1]

Igneous:	Associated secondary and residual deposits preferentially formed at low latitudes.
Cave:	Common in low latitudes (more food available).
Insular:	Mainly near-equatorial (upwelling) locations.
Sedimentary:	Restricted to low latitudes (0°–40°).

Climatic Zone (10^6–10^7 km^2)

Igneous:	Associated secondary and residual deposits preferentially formed under warm humid conditions.
Cave:	Warm humid zone (more food available).

1 Numbers in brackets indicate the spatial or temporal magnitude of the feature

Insular: Warm arid zone (associated with upwelling).
Sedimentary: Warm arid zone (frequently associated with upwelling).

Geographic Location (10^6–10^7 km^2)

Igneous: Continental (cratonic).
Cave: Continental.
Insular: Isolated oceanic islands.
Sedimentary: Oceanic (seamounts); continental shelf; epicontinental seas; coastal embayments (10^1–10^5 km^2).

Structural Location (10^2–$10^3 \times 10^1$–10^3 km)

Igneous: Rift valleys, volcanic belts, and possibly transform faults.
Cave: Only association is with regions of carbonate rocks.
Insular: Oceanic islands resulting primarily from volcanism, less commonly from transforms.
Sedimentary: Associated with early rifting phase (E–W oceans) or late oceanic phase (N–S oceans); transforms can affect coastal configuration and also produce sea-mounts; "basement" highs important; coastal embayments.

Temporal Features

Major Sea-floor Spreading Episodes (10^7–10^8 y)

Igneous: Enhanced activity along associated transforms likely to produce more alkalic igneous bodies.
Cave: No association.
Insular: Could result in more islands, producing more but smaller deposits; chances of preservation enhanced by uplift.
Sedimentary: Could result in new phosphogenic seaways; enhanced P input from more fumaroles, but could be balanced by P absorption by ocean-floor basalts.

Orogenic Episodes (10^7–10^8 y)

Igneous: Most commonly intruded during major tensional phases.
Cave: No association.
Insular: No genetic association but chances of preservation enhanced by uplift.
Sedimentary: Enhanced phosphate input during orogenic episodes suggested, but no evidence.

Volcanic Episodes (10^6–10^8 y)

Igneous: Strong association with episodes of alkalic volcanism.
Cave: No association.
Insular: Could result in more islands, producing more but smaller deposits.
Sedimentary: Increased P input from fumarolic activity or ash-induced mass mortalities suggested as link but no real evidence; major phase of oceanic volcanism can produce significant sea-level rise.

Global Warming/Cooling (10^4–10^8 y)

Igneous: Associated secondary and residual deposits would preferentially form during warm humid periods.
Cave: More abundant during warm humid periods.
Insular: Greater deposition of guano may occur during glacial periods due to greater marine organic productivity.
Sedimentary: Intensified upwelling during cold periods could produce enhanced phosphogenesis; major marine transgression during warming phases have important link with phosphogenesis.

Changes in Oceanic Circulation (10^4–10^7 y)

Igneous: No association.
Cave: No association.
Insular: Marked changes in rate of formation due to shift in major currents.
Sedimentary: Phosphorite more readily formed during periods of enhanced circulation; major phosphorus sinks in deep ocean during periods of decreased circulation.

Sea Level Changes (10^4–10^6 y)

Igneous: No association.
Cave: No association.
Insular: Could result in destruction of deposits by erosion. There are probably more islands at times of low sea-level than there are at times of high sea-level.
Sedimentary: Important reworking processes, especially during a sea-level rise, produce deposits.

Short-term Perturbation (10^1–10^2 y)

Igneous: None.
Cave: Association likely.

Insular: Decreased guano due to changes in current patterns (e.g., El Nino) and associated climatic effects.

Sedimentary: Increased rate of deposition of phosphate-bearing organic-rich sediments during El Nino-type events.

There are clearly a large number of factors operating over a range of scales which together are responsible for whether or not a phosphate deposit is formed at a particular place and time. The importance of considering multiple factors and their relevance to the formation of a phosphate deposit is perhaps best illustrated by sedimentary deposits. Though by no means the only possible set of factors necessary to produce a small sedimentary phosphate deposit with an area of around 10^2 km^2, spatial considerations could conceivably include: A low latitude location ($\sim 10^7$ km^2), aridity ($\sim 10^6$ km^2), upwelling ($\sim 10^6$ km^2), shallow seas ($\sim 10^5$ km^2), the nearshore zone ($\sim 10^4$ km^2), structural or basement features ($\sim 10^3$ km^2), and coastal embayments or lagoons ($\sim 10^2$ km^2). In addition there are a number of important temporal considerations including: the appropriate phase of plate tectonics (10^7–10^8 y), the appropriate phase of the global climate (10^7–10^8 y), enhanced oceanic circulation (10^4–10^7 y), and short-term sea level changes (10^4–10^6 y). In fact few deposits would have formed in response to all of these factors, but the more these factors coincide in time and space, the more likely it is that a phosphate deposit will indeed be formed. Conversely, if few of these factors coincide a deposit may be less likely.

A further consideration is whether there are spatial and temporal links between the various types of phosphate deposits? The obvious link between sedimentary and insular deposits has already been discussed, with the link provided primarily by oceanic upwelling. Insular, cave and sedimentary deposits, as well as the secondary and residual igneous deposits are all concentrated at low latitudes (cf. Figs. 1 3). The link again is basically climatic. It is conceivable that some igneous deposits and phosphorites could be spatially linked if the igneous deposit was in fact a metamorphosed sedimentary deposit. Apparent links can be misleading for in some instances the control is of primary significance whereas in others it is secondary. For example, Riggs (1980) has suggested that the alkaline igneous rocks could provide a direct source of phosphorus for phosphorites through hydrothermal systems; he has also pointed to apparent spatial relationships between zone of volcanism and major sedimentary phosphate deposits. However, it is dangerous to infer a direct genetic relationship just because two types of deposits are in juxtaposition. This can be illustrated by considering the coast of Southwest Africa. Comparison of Figs. 1–3, reveals that not only are the insular and sedimentary deposits juxtaposed there, but additionally there is a large concentration of alkaline igneous bodies on the immediately adjacent continent. Does this then mean that these three types of deposits must be genetically linked through, for example, hydrothermal systems? The short answer is no! There is absolutely no evidence of any genetic link between these igneous and sedimentary or igneous and guano deposits, though there is a strong link between insular guano deposits and sedimentary deposits, again with upwelling and associated phenomena providing that link. However, this and other similar examples should not then be taken as necessarily indicating that upwelling is therefore the single most important mechanism for phosphorites. For some deposits it may not have been operative at all as a significant mechanism. It is just one of a number of ingredients, both spatial and temporal which are required to produce a sedimentary deposit or an insular deposit. There is no unique mix of all the ingredients.

The patterns which are evident in the spatial and temporal distribution of deposits do nevertheless point to a limited number of key ingredients which cannot alone produce a phosphate deposit but which do have a particular responsibility for the periodic or episodic occurrence of deposits. These include tensional zones and volcanic episodes in the case of igneous deposits; warm humid climates for cave guano; warm arid conditions onshore, sea-level stillstands and fertile marine conditions for insular guano; low latitude locations, high organic productivity, increased rates of oceanic circulation, high phosphorus contents of the oceans, and major marine transgressions. Small-scale periodicities of tens to thousands of years can be identified in the guano deposits. Larger-scale periodicities may be present in the igneous deposits but the data are inadequate at present to be sure of this. However, large-scale patterns of 10^7–10^8 y are certainly readily identifiable in sedimentary deposits with a repetition in phosphogenesis ranging from about 50–150 m.y. The data are still too uncertain to be sure whether this represents a fairly irregular episodic pattern or a regular, though as yet ill-defined, periodicity. As mentioned previously, recent data for Asia and the Pacific by Lee (1980) suggests the possibility of a previously unsuspected peak of phosphogenesis in the Devonian. If this eventually proves to be the case then this combined with the evidence given in Fig. 9, provides support for a regular periodicity of about 10^8 y in phosphogenesis throughout much of the Phanerozoic. However, the new data presented in Figs. 10 and 11 suggest a shorter term periodicity of around 2–3×10^7 y during the Cretaceous and Cenozoic.

In conclusion then, it is clear that a sensible pattern in the spatial and temporal distribution of phosphate deposits is starting to emerge though more data are required. The underlying causes of these patterns are similarly becoming better understood but again there is more work to be done. An understanding of these patterns is essential if we are to fully understand the origin of phosphate deposits. However, the need to acquire a better understanding of these extends far beyond understanding phosphogenesis sensu stricto.

An understanding of alkalic igneous rocks and their associated phosphate deposits will provide a better understanding of the lithospheric processes and composition. Additional information on guano deposits such as their geochronology, and a better understanding of these deposits will enable us to better understand climatic and oceanic changes during the last million years. Similarly, more information on sedimentary deposits and phosphorites is required, such as more detailed biochronology, and radiometric dating, additional geochemistry (particularly stable isotopes, the lanthanides and organic geochemistry), and the development of more clearly defined depositional models. All of these will help us not only to better understand the formation of phosphorites, but also long-term changes in the hydrosphere, and may provide new insights into the evolution of the biosphere over the past two billion years.

References

Aharon P (1982) $^{13}C/^{12}C$ isotope ratio variations over the last 10^5 yr in a New Guinea coral-reef environment: Implications for the fertility shifts of the tropical ocean. In: Galbally IW, Freney J (eds) The cycling of carbon nitrogen sulfur and phosphorus in the terrestrial and aquatic ecosystems. Austral Acad Sci, Canberra, p 133–142

Altschuler ZS (1973) The weathering of phosphate deposits – geochemical and environmental aspects. In: Griffith EJ, Beeton A, Spencer JM, Mitchell DT (eds) Environmental phosphorus handbook. Wiley, New York, p 33–96

Arthur MA (1981) The carbon cycle: Controls on atmospheric CO_2 and climate in the geological past. In: Berger WH, Crowell JC (eds) Pre-Pleistocene climates. Nat Res Council, Washington

Arthur MA, Jenkyns HC (1981) Phosphorites and paleoceanography. Oceanol Acta, Proc 26th Int Geol Congr, Paris, p 83–96

Banerjee DM (1971) Precambrian stromatolitic phosphorite of Udaipur. Rajsthan, India, Geol Soc Am Bull 82:2319–2330

Baturin GN, Dubinchuk VT (1979) Microstructure of oceanic phosphorites. Nauka, Moscow, p 198 (in Russian)

Baturin GN, Merkulova AV, Chalov PI (1972) Radiometric evidence for recent formation of phosphatic nodules in marine shelf sediments. Marine Geol 13:M37–M41

Bentor YK (1980) Marine phosphorites. Soc Econom Palaeontol Mineral, Tulsa, Spec Pub, p 29

Birch GF (1980) A model of penecontemporaneous phosphatization by diagenetic and authigenic mechanisms from the western margin of Southern Africa. Soc Econom Paleontol Mineral, Tulsa, Spec Pub 29:79–100

Bloomfield K (1973) Economic aspects of Uganda carbonatite complexes. Overseas Geol Miner Res 41:139–167

Bourrouilh FG (1980) Phosphates, sols bauxitiques et karsts dolomitiques du Centre et SW Pacifique Comparisons sédimentologiques et géochimiques. BRGM, p 113–128

Burnett WC (1977) Geochemistry and origin of phosphorite deposits from off Peru and Chile. Geol Soc Am Bull 88:813–823

Burnett WC (1980) Oceanic phosphate deposits. In: Sheldon RP, Burnett WC (eds) Fertilizer mineral potential in Asia and the Pacific. Resour Syst Inst, Honolulu, p 119–144

Burnett WC, Veeh HH (1977) Uranium-series disequilibrium studies on phosphorite nodules from the west coast of S America. Geochim Cosmochim Acta 41:755–764

Bushinski GI (1969) Old phosphorites of Asia and their genesis. Israel Progr Sci Transl, Jerusalem, p 266

Cook PJ (1976a) Sedimentary phosphate deposits. In: Wolf KH (ed) Handbook of stratabound and stratiform ore deposits. Elsevier, Amsterdam 7:505–535

Cook PJ (1976b) Georgina Basin phosphatic province, Queensland and Northern Territory – Regional geology. In: Knight CL (ed) Economic geology of Australia and Papua New Guinea. Aust Inst Min Met Mon 5 (4):245–250

Cook PJ (1982) Cambrian palaeogeography of Australia and opportunities for petroleum exploration. APEA J 22 (1):42–64

Cook PJ, McElhinny MW (1979) A re-evaluation of the spatial and temporal distribution of sedimentary phosphate deposits in the light of plate tectonics. Econ Geol 74:315–330

Cook PJ, Shergold JH (eds) (1979) Proterozoic-Cambrian phosphorites. ANU, Canberra, p 106

Cook PJ, Shergold JH (1980) Proterozoic and Cambrian phosphorites of Asia and Australia – a progress report. In: Sheldon RP, Burnett WC (eds) Fertilizer mineral potential in Asia and the Pacific. Res Syst Inst, Honolulu, p 207–223

D'Anglejan BF (1967) Origin of marine phosphorites off Baja California, Mexico. Mar Geol 5:15–44

Deans T (1968) Exploration for apatite deposits associated with carbonatites and pyroxenites. Miner Res Develop Ser, UN, New York 32:109–119

De Keyser F, Cook PJ (1972) Geology of the Middle Cambrian phosphorites and associated sediments of northwestern Queensland. Bur Min Res Aust Bull 138:79

Elgueta AS (1981) Sedimentological study of the Western Zone of the Lady Annie phosphate deposit, Queensland, Australia. M Sc Thesis, Australian National University, p 238

Erdosh G (1979) The Ontario carbonatite province and its phosphate potential. Econ Geol 74:331–338

Fischer AG, Arthur MA (1977) Secular variations in the pelagic realm. Soc Econ Paleontol Mineral, Spec Pub 25:19–50
Frakes LA (1979) Climates throughout geologic time. Elsevier, Amsterdam, p 310
Freas DH, Eckstrom CL (1968) Areas of potential upwelling and phosphorite deposition during Tertiary, Mesozoic and late Palaeozoic time. UN Miner Res Develop Ser, UN, New York 32:228–238
Gimmel'Farb BM (1958) Regularity of the tectonic distribution of phosphorite deposits in the USSR. Izdetel'stvo, Moscow, 1 (in Russian)
Gittins J, Macintyre RM, York D (1967) The ages of carbonatite complexes in eastern Canada. Can J Earth Sci 4:651–655
Gulbrandsen RA (1969) Physical and chemical factors in the formation of marine apatite. Econ Geol 64:365–382
Howard PF (1972) Exploration for phosphorite in Australia – a case history. Econ Geol 64:365–382
Howard PF (1979) Phosphate. Econ Geol 74:192–194
Howard PF, Cooney AM (1976) D Tree phosphate deposit, Georgina Basin, Queensland. In: Knight CL (ed) Economic geology of Australia and Papua New Guinea. Aust Inst Min, Metal Mon (5) 4:265–273
Howard PF, Hough MJ (1979) On the geochemistry and origin of the D Tree, Wonarah, and Sherrin Creek phosphorite deposits of the Georgina Basin, Northern Territory. Econ Geol 74:260–284
Hutchinson GE (1950) The biogeochemistry of vertebrate excretion. Am Mus Nat Hist Bull 96:554
Ilyin AV, Ratnikova GI (1981) Primary, bedded, structureless phosphorite of the Khubsugul Basin, Mongolia. J Sed Petrol 51 (4):1215–1222
Jenkyns HC (1980) Cretaceous anoxic events: From continents to oceans. J Geol Soc (Lond) 137:171–188
Kazakov AV (1938) The phosphorite facies and the genesis of phosphorites. Int Geol Congr 17th Session, Moscow, p 95–113
Kolodny Y (1969) Are marine phosphorites forming today? Nature 224:1017–1019
Kolodny Y (1981) Phosphorites. In: Emiliani C (ed) The sea. Wiley, New York 7:981–1023
Krajewski KP (1981) Phosphate microstromatolites in the High-Tatric Albian limestones in the Polish Tatra Mts. Bull R'Acad Pol Sci 29 (2):175
Lee AIN (1980) Fertilizer mineral occurrences in the Asia-Pacific region. East West Centre, Honolulu, p 156
Love JD (1964) Uraniferous phosphatic lake beds of Eocene age in intermontaine basins of Wyoming and Utah. US Geol Surv Prof Paper 474-E:1–66
McKelvey VE (1967) Phosphate deposits. US Geol Surv Bull 1252-D:21
McKelvey VE (1973) Abundance and distribution of phosphorus in the lithosphere. In: Griffith EJ, Beeton A, Spencer JM, Mitchell DT (eds) Environmental phosphorus handbook, Wiley, New York, p 13–32
McKelvey VE, Swanson RW, Sheldon RP (1953) The Permian phosphorite deposits of the Western United States. Int Geol Congr 19th Session, Algiers 11:45–65
Mansfield GR (1927) Geography, geology and mineral resources of part of southeast Idaho. US Geol Surv, Prof Paper 152:453
Marsh JS (1973) Relationships between transform directions and alkaline igneous rock lineaments in Africa and South America. Earth Planet Sci Let 18:317–323
Marshall JF, Cook PJ (1980) Petrology of iron- and phosphorus-rich nodules from the E Australian continental shelf. J Geol Soc (Lond) 137:765–771
Martin RF, Piwinskii AJ (1972) Magmatism and tectonic settings. J Geophys Res 77:4966–4975
Morel P, Irving E (1978) Tentative paleocontinental maps for the early Phanerozoic and Proterozoic. J Geol 86:535–561
Murray J, Renard AF (1891) Scientific results, HMS Challenger, Deep Sea Deposits, p 391–400
Notholt AJG (1979) The economic geology and development of igneous phosphate deposits in Europe and the USSR. Econ Geol 74 (2):339–350
Notholt AJG (1980) Igneous apatite deposits. Mode of occurrence economic development and world resources. In: Sheldon RP, Burnett WC (eds) Fertilizer mineral potential in Asia and the Pacific. Res Syst Inst, Honolulu, p 263–285
O'Brien GW, Veeh HH (1980) Holocene phosphorite on the East Australia continental margin. Nature 288:690–692
Piper DZ, Codispoti LA (1975) Marine phosphate deposits and the nitrogen cycle. Science 188:15–18

Pitman WC III (1978) Relationship between eustacy and stratigraphic sequences of passive margins. Geol Soc Am Bull 89:1389–1403

Riggs SR (1979a) Petrology of the Tertiary phosphorite system of Florida. Econ Geol 74:195–220

Riggs SR (1979b) Phosphorite sedimentation in Florida – a model phosphogenic system. Econ Geol 74:285–314

Riggs SR (1980) Tectonic model of phosphate genesis. In: Sheldon RP, Burnett WC (eds) Fertilizer mineral potential in Asia and the Pacific. Res Syst Inst, Honolulu, p 159–190

Sandvik PO, Erdosh G (1977) Geology of the Cargill phosphate deposit in northern Ontario. Can Inst Min Metal Bull 69:90–96

Sassi S (1980) Contexte paléogéographique des dépots phosphates en Tunisie. Géologie comparée des gisements de phosphate et du pétrole. BRGM Mem, p 116

Savin SM (1977) The history of the earth's surface temperature during the last 100 millions years. In: Donath FA, Stehli FG, Wetherill GW (eds) Ann Rev Earth Planet Sci 5:319–355

Schlanger SO, Jenkyns HC (1976) Cretaceous anoxic events: Causes and consequences. Geol Minjbouw 55:179–184

Scotese CR, Bambach RK, Barton C, Van der Voo R, Ziegler AM (1979) Paleozoic base maps. J Geol 87 (3):217–277

Sheldon RP (1964) Paleolatitudinal and palaeogeographic distribution of phosphorites. US Geol Surv, Prof Paper 501C:106–113

Sheldon RP (1980) Episodicity of phosphate deposition and deep ocean circulation – a hypothesis. Soc Econ Paleontol Mineral, Spec Pub 29:239–247

Sheldon RP (1981) Ancient marine phosphorites. Ann Rev Earth Planet Sci 9:251–284

Sheldon RP, Burnett WC (1980) Fertilizer mineral potential in Asia and the Pacific. Res Syst Inst, Honolulu, p 481

Sheldon RP, Maughan EK, Cressman ER (1967) Sedimentation of rocks of Leonard (Permian) age in Wyoming and adjacent states. In: Hale LA (ed) Anatomy of the western phosphate field. 15th Ann Field Conf, Intermount Assoc Geol, Salt Lake City, p 1–12

Shergold, JH, Druce EC (1980) Upper Proterozoic and lower Palaeozoic rocks of the Georgina Basin. In: Henderson RA, Stephenson PJ (eds) The geology and geophysics of northeastern Australia. Geological Society of Australia. Queensland Division, Brisbane, p 149–175

Slansky M (1980) Geologie des phosphates sedimentaires. Bur Rech Geol Min 114:92

Slansky M (1980) Ancient upwelling models. Upper Cretaceous and Eocene phosphorite deposits around West Africa. In: Sheldon RP, Burnett WC (eds) Fertilizer mineral potential in Asia and the Pacific. Res Syst Inst, Honolulu, p 145–158

Smith AG, Hurley AM, Briden JC (1981) Phanerozoic paleocontinental world maps. University Press, Cambridge, p 102

Sorensen H (1974) The alkaline rocks. Wiley, London, p 622

Southgate PN (1980) Cambrian stromatolitic phosphorites from the Georgina Basin, Australia. Nature 285:395–397

Strakhov NM (1960) Climate and phosphate accumulation. Geol Rudnykh Mestor-Denii, 1 (in Russian)

Summerhayes CP (1981) Organic facies of Middle Cretaceous black shales in deep North Atlantic. Am Assoc Petrol Geol 65 (11):2364–2380

Swirydczuk K, Wilkinson BH, Smith GR (1981) Synsedimentary lacustrine phosphorites from the Pliocene Glenns Ferry Formation of southwestern Idaho. J Sed Petrol 51 (4).1205–1214

Tracey JI (1979) Quaternary episodes of insular phosphatization. In: Sheldon RP, Burnett WC (eds) Fertilizer mineral potential in Asia and the Pacific. Res Syst Inst, Honolulu, p 247–261

Trueman NA (1965) The phosphate, volcanic and carbonate rocks of Christmas Island (Indian Ocean). J Geol Soc Aust 12:261–283

Tuttle OF, Gittins J (1966) Carbonatites. Wiley, New York, p 591

Vail PR, Mitchum RM Jr, Thompson S (1977) Seismic stratigraphy and global changes of sea-level, Part 4: Global cycles of relative changes of sea-level. Am Assoc Petrol Geol Mem 26:83–97

Veeh HH (1979) Uranium-series ages of insular phosphorites. In: Burnett WC, Sheldon RP (eds) Report on the marine phosphatic sediments workshop. Res Syst Inst, Honolulu, p 16–17

Veeh HH, Burnett WC (1978) Uranium-series dating of insular phosphate from Ebon Atoll, Micronesia. Nature 274:460–462

Veeh HH, Calvert SE, Price NB (1974) Accumulation of uranium in sediments and phosphorites on the South West African Shelf. Marine Chem 2:189–202

Von der Borch CC (1970) Phosphatic concretions and nodules from the upper continental slope, Northern New South Wales. J Geol Soc Australia 16:755–759

Warin ON (1968) Deposits of phosphate rock in Oceania. Miner Res Develop Ser, UN, New York 32:124–132

Ziegler AM, Scotese CR, McKerrow WS, Johnson ME, Bambach RK (1979) Paleozoic palaeogeography. Ann Rev Earth Planet Sci 7:473–502

Chapter 8 The Mineralogy and Geochemistry of Phosphorites

YAACOV NATHAN

Introduction

Phosphorites are defined as sedimentary rocks which contain more than 18%–20% P_2O_5. Most phosphorites are of marine origin, and of these, francolite (carbonate-fluorapatite) is the main phosphate mineral.

Fluorapatite is the prototype of the apatite family, and the formula for pure fluorapatite is $Ca_{10}(PO_4)_6F_2$. The determination of its crystal structure was made by Naray-Szabo (1930) and refined by Beevers and McIntyre (1946). Many substitutions are possible for both anions and cations in the fluorapatite structure; some of these are summarized in Table 1.

Table 1. Some possible substitutions in the apatite structure

Constituent ion	Substituting ion
Ca^{2+}	Na^+, K^+, Ag^+
	Sr^{2+}, Mn^{2+}, Mg^{2+}, Zn^{2+}, Cd^{2+}, Ba^{2+}
	Sc^{3+}, Y^{3+}, R. E.$^{3+}$ (rare earths), Bi^{3+}
	U^{4+}
PO_4^{3-}	CO_3^{2-}, SO_4^{2-}, CrO_4^{2-}
	AsO_4^{3-}, VO_4^{3-}, $CO_3 \cdot F^{3-}$, $CO_3 \cdot OH^{3-}$
	SiO_4^{4-}
F^{1-}	OH^{1-}, Cl^{1-}, Br^{1-}
	O^{2-}

Francolite

In spite of the abundance of francolite and its economic importance, its composition and properties are still not known with certitude. This is mainly due to two reasons:

1. Francolite occurs as minute crystals (crystallites), which are only tens or hundreds of Ångstrom long. We also know that the composition of the crystallites is not homogeneous. Therefore every analytical result is only the mean of many compositions.

Geochemistry Division, Geological Survey of Israel, 30, Malchei Israel St., 95501 Jerusalem, Israel

2. The (apatite) lattice is an "open lattice" which allows a great number of diadochic substitutions. This increases the complexity of francolite composition and is the main cause of the heterogeneity of the crystallites.

One of the principal sources of mineralogical data on marine francolites is the studies made at the T.V.A. (Tennessee Valley Authority) laboratories in the United States (Lehr et al. 1968; McClellan and Lehr 1969). Francolite exhibits systematic isomorphous substitutions for both cations and anions (Table 1). However, the most characteristic and important of these for francolite is the carbonate substitution. This has also been the most controversial substitution. The first problem was: Is the carbonate present as a separate phase (Thewlis et al. 1939; Machatschki 1939; Brasseur et al. 1946) adsorbed on the surface of apatite crystallites (Neuman and Neuman 1953), or is the carbonate part of the apatite structure (Gruner and McConnell 1937; De Villiers 1942)? Proof of the presence of carbonate in the (apatite) lattice was first presented by Altschuler et al. (1952), and today there is near unanimity that the carbonate is part of the apatite structure. The second problem was, how does the carbonate enter the apatite structure and into which site? Two of the sites proposed are considered here, the F (or OH) site on the hexad axis (Wallaeys 1952) and the PO_4^{-3} site (McConnell 1938). McClellan (1980, p 676, Table 2) summarizes most of the proposals that have been made for the structural location of CO_3^{-2} in francolite, including support and opposition brought forward for these proposals.

Today most researchers in this field think that CO_3^{-2} substitutes for PO_4^{-3} in natural francolites and that the substitution for F^- is possible only in synthetic material prepared at high temperatures. Yet Brassens (1975) argues for the existence of two sites for CO_3^{-2} occurring in a natural Moroccan phosphorite.

Nevertheless, the carbonate substitution is an unusual one; since an essentially planar carbonate anion takes the place of a tetrahedral anion, the question arises: What is the exact structural position of the carbonate substitution? Moreover, since carbonate is a bivalent ion while phosphate is trivalent, another question arises: How does the charge balance occur?

One possible answer was suggested by Borneman-Starynkevitch and Belov (1940, 1953) who postulated that a hydroxyl (or fluorine) ion accompanies the carbonate and that the groups $(CO_3 \cdot OH)^{-3}$ or $(CO_3 \cdot F)^{-3}$, which probably form a distorted tetrahedron, replace the PO_4^{-3} group. Their proposal was based on chemical analyses of natural francolites, which usually show an excess of F over the stochiometry, and on crystallochemical considerations. Smith and Lehr (1966), Trueman (1966), Gulbrandsen (1966), and Elliot (1969), among others, support the $(CO_3 \cdot F)^{-3}$ substitution for phosphate in natural francolites with chemical and mineralogical data. Recently the association $(CO_3F)^{-3}$ and its introduction into the apatite lattice has been confirmed by ESR (Bacquet et al. 1980). Furthermore, heating experiments (see below) confirm the presence of the $(CO_3F)^{-3}$ in francolite.

However, the amount of F in excess of the stochiometry cannot in any case account for all the carbonate substitution. On the basis of 110 analyses of apatite fractions from various phosphorites, McClellan and Lehr (1969) showed that while one carbonate anion substituted for one phosphate anion, only 0.4 (on the average) of the carbonate groups were associated with F. Though the presence of $(CO_3OH)^{-3}$ groups cannot be excluded, and has been proposed among others by

McClellan (1980) to achieve a charge balance (electroneutrality) the presence of appreciable amounts of OH has not been confirmed in most sedimentary francolites. This means that a further substitution is needed to account for the charge imbalance. The ubiquitious occurrence of sodium in francolite and its substitution for calcium may be the substitution looked for since it acts to decrease the positive charges and thus helps balance the decrease in negative charges that is caused by the substitution of carbonate for phosphate. This hypothesis is supported by experiments made in the laboratory. The first of these was carried out by LeGeros (1967) who prepared, in aqueous systems at relatively low temperatures, carbonate-apatites in which carbonate ions replace phosphate ions. The extent of carbonate incorporation depends upon the presence of alkali metals such as Na. Still the amount of Na (calculated as Na_2O) found in natural francolites is generally less than 1%. Furthermore the substitution of sulphate $(SO_4)^{-2}$ for phosphate is frequent enough to complicate the situation. Although according to McClellan and Lehr (1969) this substitution is generally much less important than the carbonate substitution, it exists, and at times is considerable (Gulbrandsen 1966); Sassi (1974) showed that Tunisian apatites from the Gafsa basin contain up to 3.55% (by weight) of SO_3 for 28% of P_2O_5.

The sulphate ion is similar in size to the phosphate ion and has a similar tetrahedral configuration so as not to cause stress upon the structure as the carbonate ion does; but like the carbonate ion, it is divalent and therefore causes a further charge imbalance that must be compensated for. Sodium substitution for calcium appears to be one way of satisfying this requirement. This latter substitution already contributes to the balancing of the carbonate substitution and there is obviously not enough sodium for both purposes. One other possibility which can help to balance the charges is the silicate $(SiO_4)^{-4}$ substitution for phosphate. This substitution is possible and is known to be coupled with the sulphate substitution in wilkeite $Ca_{10}(PO_4, SiO_4, SO_4)_6(F \cdot OH)_2$ but to date, this has not been confirmed in natural francolites. The charge imbalance remains since the molar sum of carbonate and sulphate is usually more than the molar sum of the sodium and excess fluorine. (The other substitutions, though important from a geochemical point of view, are negligible as far as charge balance is concerned.) A possible solution to this problem is a Ca statistical deficiency. This has already been proposed for biological apatites (carbonate hydroxyapatites) by Posner and Perloff (1957) and Berry (1967a, b). Montel et al. (1981) who give a very thorough and detailed review of biological apatites propose a double deficiency of Ca and OH combining to balance one carbonate substitution. A Ca deficiency has also been proposed for francolite by Gulbrandsen et al. (1966).

In conclusion, the general formula for francolite can be idealized as follows:

$$Ca_{10-a-b-c}Na_aMg_b(PO_4)_{6-x}(CO_3)_{x-y-z}(CO_3 \cdot F)_y(SO_4)_zF_2 ,$$

where $x = y + a + 2c$ (c is the number of Ca vacancies).

Extent and Significance of the Carbonate Substitution

Synthetic carbonate-hydroxyapatites (dahllite) were prepared with up to 16% CO_2 (LeGeros et al. 1967). Natural francolites have in general a much lower substitution. One natural occurrence of submarine phosphorite (phosphorites occurring on the present sea-floor) from the Moroccan continental margin described by Elderfield et al. (1972) is reported to contain 16.1% CO_2 in the apatite lattice. While this number appears to be unusually high and therefore doubtful, there is no doubt that this is an area with a particularly high carbonate substitution. McArthur (1978, Table 4, p 94) reports values between 8.1% and 8.9% CO_2 in the francolite lattice for Moroccan sea-floor Fe-poor phosphorites. It should be noticed that these values and others given below were obtained by indirect methods – I.R. and X.R.D. – which are not always accurate. In a recent comprehensive and up-to-date review of submarine phosphorites Kolodny (1981) shows that the CO_2 values vary over a wide range; most of these fall between 4% and 7% (Table 4, Kolodny 1981). Lucas et al. (1978) found 6% in the marine phosphorites from the continental margin off northern Spain. The same 6% value is considered to be the convergent value for the carbonate substitution in on-land non-weathered phosphorites by Lucas et al. (1980a); this value is based mainly on Moroccan samples (Wadjiny 1979; Prévôt personal communication) and Tunisian samples (Chaabani 1978; Lucas et al. 1979a). In other areas different values were found; in Israel an average value of 4% was found for the non-weathered phosphorites of the Negev (Nathan et al. 1979) and the same value was found by Lucas et al. (1980b) in the phosphorites of Turkey. McClellan and Lehr (1969) found relatively high CO_2 values for the North and South Carolina phosphorites – 5% to 6%. Other phosphorites gave lower values; in particular the francolites from the Western Fields, United States, gave an average value of 2% (McClellan and Lehr 1969; Gulbrandsen 1970). These values were determined chemically on more or less pure apatite fractions after selective dissolution of calcite.

Gulbrandsen (1970) also gives the results of CO_2 content of apatites in 368 samples of phosphorites and phosphatic rocks from the Phosphoria Formation from Idaho and Wyoming (Western Fields). He shows a clear relationship between lithology and CO_2 content. The samples from the eastern localities (south-central Idaho) are mainly mudstone, chert and phosphorite and have low CO_2 values, 1.2%–1.7%, while the samples from the western localities (central Wyoming) are mainly carbonate and phosphorite and have relatively high CO_2 values, 3.2%–3.4%. Gulbrandsen suggests a difference in temperature and, more generally, a difference in the environment of deposition as the cause of the variation in the CO_2 content. A different interpretation for the variation in CO_2 content of francolites (together with other substituents such as Na, Sr, and SO_4) was given by McArthur (1978). He proposed two hypotheses. (1) The constant-composition hypothesis: upon formation all marine francolites have a similar composition. (2) The weathering hypothesis: weathering removes a majority of CO_2 (together with Na, Sr, and SO_4) in a systematic manner. Neither of the above hypotheses is entirely correct or entirely false.

The hypothesis which is the most consistent with known observations is the weathering hypothesis, with two reservations: (1) It should be expanded to the systematic alteration hypothesis as proposed by McClellan (1980) which includes diagenesis and metamorphism as well as weathering: e.g., Axelrod et al. (1980) showed systematic variations in CO_2 contents of peloids and bone fragments in the Negev phosphorites that are due to diagenesis. Matthews and Nathan (1977) showed that thermal metamorphism in the "Mottled Zone" accounted for the loss of structural CO_2 in francolite. (2) Alteration usually causes a lowering of the substituents and while this is almost always true for CO_2, this is not the case for Sr, which may at times be higher in weathered francolites than in unaltered samples. What happens is equilibrium between the weathering (diagenesis) solutions and the mineral.

The constant composition hypothesis is less clear-cut. To begin with, some recent phosphorites show relatively low CO_2 values and low F/P_2O_5 ratios (Romankevich and Baturin 1972; Price and Calvert 1978), showing that there is an enrichment in CO_2 and fluorine during their diagenesis. Furthermore, there are many variations which cannot be explained by alterations. The argument of McArthur (1978) that the samples which Gulbrandsen (1970) discussed are from two different members of the same formation is unconvincing if the constant composition hypothesis is correct. Nevertheless, there is a strong argument in favour of a relatively narrow range of CO_2 substitution (between 5% and 8%) for unaltered francolites in equilibrium with seawater (or interstitial water).

In conclusion, the amount of CO_2 substitution (and other substituents) in francolite is an important indicator of environment, but it cannot be interpreted mechanically; it reflects the whole history of the mineral, from the environment of deposition to supergene alteration through diagenesis (or metamorphism).

Influence of the Carbonate Substitution on Physical Properties

There is a direct relationship between the amount of CO_2 substitution and the birefringence (see Table 2 after Carlstrom 1968, cit. in McConnell 1973). There is also a non-linear decrease of the *a* parameter with increasing substitution of carbonate (Table 2). McClellan and Lehr (1969) show that in their (110) samples that the *a* values ranged from 9.392 Å (for sample 222 from Sweden with 0.11% CO_2 – adjusted weight % –) to 9.322 Å (for sample 22 from South Carolina with 5.91% CO_2). The *c* values for these two samples only varied from 6.896 to 6.900 Å; there were variations in the *c* parameter, but obviously the influence of the carbonate substitution on this parameter is negligible.

It should be stressed that other substitutions also influence the *a* parameter; e.g., the OH^- substitution for F^- causes an increase – the *a* parameter for a dahllite (carbonate-hydroxyapatite) with 3.51% CO_2 is 9.419 Å (Brophy and Nash 1968, Table 1, p 447).

McClellan and Lehr (1969, p 1382) proposed some mathematical models (their Table 3, p 1382) to correlate the amount of CO_2 and F with the cell parameters. McClellan (1980) further refined these models using new data (their Table 3, p 135).

Gulbrandsen (1970) (using the fact that the carbonate substitution causes changes mainly in the *a* parameter proposed a simple method to estimate the CO_2 content of francolite). He used the peak-pair difference method thus eliminating the necessity of an internal standard between the (410) and (004) peaks. Using the data on 74 samples from McClellan and Lehr (1969), he found the following empirical equation $y = 23.6341 - 14.7369\,x$, where y is the CO_2 weight percent and $x = \Delta 2\theta(004\text{–}410)$, 2θCuKα with a standard error of estimate of 0.5580.

Table 2. Changes in unit-cell dimensions and birefringence with carbonate substitution. (After Carlström 1968, cit. in McConnell 1973, Table 7.3 p. 44)

Sample	CO_2 %	F %	a (Å)	Birefringence
Syn Fap	–	3.78	9.371	0.0034
NRM 28259	1.26	3.78	9.361	0.0051
NRM 28260	1.32	3.71	9.360	0.0055
NRM 28258	2.01	3.83	9.352	0.0062
USNM 97339	2.57	3.94	9.350	0.0070
NRM 38648	3.18	3.83	9.349	0.0082
NRM 38745	4.22	4.06	9.342	0.0098
USNM 105532	4.39	4.59	9.334	0.0125

A second pair of peaks also proposed by Gulbrandsen (1970) is (002–300), which can be used when the pair (004–410) is too weak or too poorly defined to be measured. In this case the measurement is less accurate.

This method, though very popular and widely used, has a few drawbacks. It cannot be used in samples rich in dolomite since dolomite interferes; the dolomite peaks (116–009) are of medium intensity and are found around 51.2° and therefore interfere with francolite peak (410); the dolomite peak (006) which is around 33.5° interferes with francolite peak (300) (Chaabani 1978). The results obtained by this (Gulbrandsen's) method are sometimes systematically higher than those determined chemically (Menor 1975, Table 20, p 128).

The infra-red absorption spectrum is very sensitive to carbonate substitutions – even minute amounts of carbonate can be detected. It should be noted that examination of the middle infra-red spectrum (4,000–400 cm^{-1}) only may sometimes lead to incorrect interpretations. This has been shown to be the case for the PO_4^{-3} vibrations in apatites by studies of laser Raman spectra and for infra-red spectra (Kravitz et al. 1969; Levitt and Condrate 1970). The infra-red spectrum of the carbonate ions in carbonate-apatites is characterized by a splitting of the ν_3 vibration between 1,470 and 1,410 cm^{-1} (Gulbrandsen et al. 1966; Brophy and Nash 1968) and a splitting of the ν_3 vibration (less clear than the above) between 890 and 850 cm^{-1} (LeGeros et al. 1970). She and co-workers studied dahllite, but the same splitting (anomalous doublet) is observed in francolite with a slight shift in position towards lower values. The splitting of these vibration has been interpreted as indicating the CO_3^{-2} to be present in two different environments (Emerson and Fisher 1962; White 1974). LeGeros et al. (1970) list different explanations for this interpretation together with other alternatives.

The infra-red spectrum has also been used to estimate the amount of CO_2 within francolite. Lehr et al. (1968) proposed a "CO_2" index to estimate this amount, which is based on the relative intensities of the C–O and P–O absorptions. The C–O absorption is determined as the average of the ν_3 doublet measured from the background base at 1,800 cm^{-1}. The P–O absorption is measured as the intensity of the ν_4 vibration band between 605–600 cm^{-1}. Theoretically this should be a better way to estimate the amount of CO_2 than the X-ray (unit cell measurements) but in practice silicate, carbonate, sulphate minerals, and water have spectral contributions which interfere. Therefore these kinds of I.R. measurements give only a rough approximation. Recently Scheib et al. (in Press) have used Fourier transform infrared spectroscopy for this purpose. They show that through stripping of interfering spectra, a virtually interference free CO_2 index can be determined. The C–O absorption was measured by integrating the area from 1,550 to 1,375 cm^{-1} and the P–O absorption was measured by integrating the area from 690 to 535 cm^{-1}; the formula the authors calculated from 65 measurements on different phosphorites was: CO_2 index = 0.0342 + 0.226% CO_2.

LeGeros et al. (1967) showed that the carbonate substitution in synthetic apatites also has an influence on the size and shape of the apatite crystallites. When present it limits their size and makes them more equiaxial than needle-like. Lehr et al. (1968) observed the same changes in natural francolites. Furthermore, the carbonate substitution, since it weakens the structure, increases the solubility of the apatite. This property, which is considered to be nefarious in dental apatites, since it increases the susceptibility to caries, is an advantage in francolites; it allows the use of such a phosphorite by "direct application" or, in other words, the use of this phosphorite as a fertilizer without any chemical pre-treatment.

Changes Which Occur on Heating Phosphorite

When phosphorite is heated between 500° and 800 °C, the adsorbed moisture is expelled, organic matter (a very common constituent of phosphorites) is destroyed and francolite alters progressively to fluorapatite (Smith and Lehr 1966). The francolite loses its constituent CO_2 and some of its fluorine (Lehr et al. 1968). The loss of the structural CO_2 is gradual and can be monitored by the Gulbrandsen peak-pair method (Matthews and Nathan 1977). It is attended by an increase in crystalline size and the appearance of CaO and CaF_2 (Lehr et al. 1968; Axelrod 1980).

The reactivity of francolites depends to a large extent upon crystallite size. Decarbonation which leads to the annealing of crystals and recrystallization causes a decrease in reactivity, i.e., dissolution in citric acid – and other acids (Lehr et al. 1968; Champetier et al. 1980). Careful reading of the results of Lehr et al. (1968, Table VII, p 42–43) and unpublished results from our laboratory (Geological Survey of Israel) show that the loss of structural CO_2 occurs in two phases: (a) between 500° and 700 °C – this loss is constant for a given francolite; it ranges from 20% to 80% of the structural CO_2; samples with high CO_2 substitution lose a higher percentage of their CO_2 than those with a low substitution; little if any fluorine is expelled. (b) between 700° and 1,000 °C – all the remaining CO_2 is expelled; a proportional (equivalent) amount of fluorine being expelled at the same time. These results strengthen the assumption that CO_2 occurs in two different configurations in francolite – as $CO_3^{\ 2}$ and as $CO_3 \cdot F^{-3}$.

Heating at higher temperatures (above 1,000 °C) is not dealt with here since there is no more francolite at this temperature, but only fluorapatite. A voluminous literature exists on this subject (Reynolds et al. 1935, 1936; Van Wazer 1966, p 529).

Other Minerals

Other Phosphate Minerals. Although at least 300 distinct species of phosphate minerals are known (Moore 1980), practically the only phosphate mineral that occurs in unweathered marine phosphorites is francolite. The situation is different in weathered phosphorites where usually a number of calcium aluminum and aluminum phosphate minerals occur, as in the Thies area, Senegal (Capedecomme 1952). A thermodynamic approach to determine the environment of formation of calcium and aluminum phosphates was attempted by Nriagu (1976). Less frequently iron phosphates occur in weathered phosphorites, e.g., as in the Negev, Israel (Lucas et al. 1980a). A recent review of the mineralogy and geochemistry of weathered phosphate deposits has been given by Altschuler (1973).

Other Non-Phosphatic Minerals. Quartz is an ubiquitous mineral in sedimentary rocks and phosphorites almost always contain quartz. In the reworked deposit of the Pliocene Bone Valley Formation, this is the main gangue mineral (Altschuler et al 1964). Quartz occurs as a detrital mineral and also as an authigenic mineral, as chert or in silicified phosphorite. The association of cherts, carbonates and black shales with phosphorites is well known (McKelvey 1967; Cook 1976). Silica also occurs as opal C–T, which is a common constituent of the younger phosphorites, as, for example, in the Upper-Cretaceous Eocene Tethys province (Lucas et al. 1979a, b; Prévôt et al. 1979a; Nathan et al. 1979). Carbonates, calcite, and dolomite are very often associated with phosphorites of all ages (Emigh 1967; Bushinsky 1969; Prévôt et al. 1979b). The abundance of dolomite in phosphorites is somewhat puzzling in view of the fact that Mg ions inhibit apatite precipitation (Bachra et al. 1965; Martens and Harris 1970). Two explanations may be offered (and both may be correct) for the paradox: (a) dolomitization occurs in the late diagenesis, much after the apatite precipitation (or formation); (b) even minerals which belong to the same geological paragenesis, may form at what chemically is considered to be completely different times. Furthermore, dolomitization may be one of the ways to reduce the Mg/Ca ratio of the interstitial waters. Glauconite is also well known for its association with phosphorite (Collet 1908), nevertheless, it is more common in low-grade phosphatic rocks than in rich phosphorites and in many cases does not occur in these at all. The reason may be that glauconization and phosphatization occur in neighboring (but different) stages in the evolution of sedimentary basins (Odin and Letolle 1980). Burnett (1980) describes sediments off Peru and Chili where a general sequence, collophane mudstone – pelletal phosphorite – glauconized pelletal phosphorite, occurs. This sequence appears to represent an evolution from the most favorable conditions for apatite formation (well-developed oxygen minimum, high primary production, etc.) to those least favorable (high content of dissolved oxygen – low primary productivity). Illite (apart from glauconite) has been found in a number of phosphorites almost always as a minor component of the clay fraction. Mainly on the basis of a study of the Tertiary of the Atlantic Coast, U.S.A., Weaver and Wampler (1972) claim that illite is the only clay mineral present within the phosphate grains. They found (from apparent K–Ar ages) that most of the K-silicates were detrital in origin, though some illite and glauconite may be authigenic. Montmorillonite is the predominant clay mineral in many phosphorite deposits.

In many cases palygorskite occurs together with montmorillonite as in the Eocene unweathered calcium rocks of Senegal (Capedecomme 1952) or in the uppermost Miocene Hawthorne phosphate bearing limestones of Florida (Altschuler et al. 1964).

Clinoptilolite, a zeolite of the heulandite-clinoptilolite family, is also present in many phosphorites: the North Carolina Miocene phosphorites (Rooney and Kerr 1967), the Tunisian Upper Cretaceous phosphorites (Sassi and Jacob 1972), the Florida Miocene phosphorites (Riggs 1979), the Eocene and Oligocene phosphatic sequences of Senegal (Sommer 1972) and the Israeli Upper Cretaceous phosphorites (Nathan and Soudry 1982). Many of the silicates mentioned above may well be authigenic (Nathan and Soudry 1982).

Organic Matter

Organic matter is a characteristic constituent of phosphorites (Gulbrandsen 1966; Bentor 1980). The study of organic matter in phosphorites has become fashionable because of the present interest in phosphorites as source-rocks for petroleum (Powell et al. 1975; Maughan 1980; Belayouni and Trichet 1980; Bein et al. 1980; Sandstrom 1980). There appears to be an intimate relationship between the factors which generate phosphorite deposits and those which generate hydrocarbon source beds, both are probably coincident with the area of optimum trophism and biologic productivity (Maughan 1980). This does not contradict the usual inverse relationship between organic carbon and P_2O_5 content (Belayouni and Trichet 1980) found in many phosphorite deposits. This is because phosphorus is linerated only after the destruction of organic matter, a process which can be followed in recent phosphorites (Romankevich and Baturin 1972).

A direct proof of the possibility that organic phosphorus may be easily recycled to form inorganic apatite was given by Lucas and Prévôt (1981) who succeeded in synthesizing francolite, starting from calcite and phosphorus in organic matter (ribonucleic acid).

In general, organic matter in phosphorites appears to be in the main the result of anoxic microbial degradation of marine planton. Organic values (0.6%) appear to be lower than the average shale value (1.0%) but higher than the average carbonate value (0.3%) (Powell et al. 1975). Sulfur values are relatively very high, on the average 8% of the organic matter; not only sulfur, but also nitrogen and oxygen values of organic matter in phosphorites tend to be higher than those in other sediments (Powell et al. 1975). In many cases, since phosphorites were deposited in shallow basins and were never deeply buried, the deposits have not undergone extensive diagenesis and even in a Cambrian phosphorite the organic matter is thermally immature. This can be seen from the saturate to aromatic ratios (less than 1), the H/C ratios (1.10 to 1.27) and the relatively high oxygen content (Sandstrom 1980). This property (being thermally immature) appears to be characteristic of organic matter in phosphorites. Belayouni and Trichet (1980), reach this conclusion for the Tunisian phosphorites from their hydrogen and oxygen indices, and from chromatographs of their saturated products, and similar observations have been made for Israeli phosphorites (Bein et al. 1980). This property of organic matter may be related to the structural position of the areas favorable to phosphorite deposition and to the well-known fact that phosphorites are condensed series.

Table 3. Chemical analysis of phosphorites of different ages. (After Slansky 1980, Table 8, p. 34)

Locality	Age	P_2O_5	SiO_2	Al_2O_3	Fe_2O_3	CaO	MgO	Na_2O
Peru-Chili (off shore)	Holocene	22.61	22.13	5.15	2.85	33.93	1.07	0.85
Florida ("pebble")	Pliocene	32.07	9.31	1.29	1.57	46.98	0.19	0.21
Venezuela (Riecito)	Miocene	34.28	5.05	1.00	0.69	48.05	0.23	0.70
Benin	Mid. Eocene	28.15	13.15	5.40	0.17	40.94	0.52	0.30
Morocco (Khouribga)	Lower Eocene	34.26	0.03	0.37	0.26	52.78	0.48	0.84
Tunisia (Metlaoui)	Lower Eocene	26.09	8.90	1.53	0.60	42.85	0.50	1.45
Egypt (Quseir)	Upper Cretaceous	25.20	12.50	0.84	1.86	40.66	1.75	0.68
India (Mussoorie)	Permian	22.50	7.05	0.66	2.56	40.55	6.00	0.18
U.S.A. (Rocky Mts.)	Permian	30.50	11.90	1.70	1.10	44.00	0.30	0.60
Australia (Lady Annie)	Cambrian	35.00	10.30	1.59	0.14	48.20	0.15	0.04
China	Lower Cambrian	23.41	19.74	3.12	1.90	41.98	0.21	0.54
India (Udaipur)	Precambrian	35.46	10.56	1.18		47.75	0.28	

Major Elements

Major elements do not vary greatly in phosphorites; this is due in part to the definition of phosphorites. It is also due, according to Slansky (1980) to the permanency of the conditions necessary for formation of phosphorites. Slansky (1980) arrives at this conclusion from comparison between the chemical compositions of phosphorites from the entire geological column (Table 8, p 34, Slansky 1980). It can be seen from the data, which is in part reproduced here as Table 3, that there is no relation between age and chemical composition of phosphorites. The differences, when found, are mainly due to different concentrations of the so-called gangue minerals.

Trace Elements

Trace elements are considered to be enriched in phosphorites (Krauskopf 1955; Tooms et al. 1969). Table 4 summarizes some data on trace elements in phosphorites and compares them to the average concentration found in seawater, the crust and the average shale. It shows that a group of certain elements is enriched in phosphorites and this group is essentially the same, regardless of the way the average (for phosphorite) was found or of the basis for comparison, crustal abundances (Tooms et al. 1969) or the average shale (Altschuler 1980). This group is composed of at least Ag, Cd, Mo, Se, Sr, U, Y, Zn, and the R.E.E. (excluding Ce). This group appears to be characteristic of phosphorites and is probably indicative of their genesis. A few facts stand out, namely: (1) A parallelism between elements enriched in seawater and in phosphorites (Tooms et al. 1969). The similarity is even more striking when only the Y and R.E.E. group are considered, the group as a whole is enriched, Y is enriched relative to the group and Ce is depleted (Altschuler 1980). (2) There is in general no clear correlation between most of the enriched elements and P, even within the same phosphorite (Prévôt et al. 1979a). (3) A paral-

2	SO_3	F^-	Cl^-	H_2O^-	H_2O^+	C org.	$\frac{CaO}{P_2O_5}$	$\frac{F}{P_2O_5}$	$\frac{CO_2}{P_2O_5}$	$\frac{SO_3}{P_2O_5}$
		2.22					1.50	0.098		
7	0.59	3.68	0.013		1.88	0.053	1.46	0.115	0.09	0.018
0	0.70	2.23	0.08	0.42	1.81	0.09	1.40	0.065	0.08	0.02
9		2.65	0.48	1.77	3.45	0.07	1.48	0.094	0.06	
9	1.59	3.05	0.03		2.34		1.54	0.089	0.10	0.04
2	3.90	2.98	0.09	3.16	3.00	0.93	1.64	0.114	0.17	0.149
8		2.57	1.06	2.11	2.19	1.20	1.61	0.101	0.22	
0	1.35	2.15	0.06	0.24	0.75	0.76	1.80	0.095	0.67	0.06
0	1.80	3.10		0.60	2.10	2.10	1.44	1.101	0.07	0.059
0	0.07	3.16		0.53	0.70	0.60	1.37	0.090	0.03	0.002
2		2.72	0.67	0.33	0.24	0.62	1.79	0.116	0.23	
4	0.21	1.60			1.66		1.34	0.045	0.02	0.005

lelism between trace elements enriched in phosphorites and those concentrated in organic matter (Prévôt and Lucas 1979a). (4) Despite their similarity trace element concentrations vary considerably between different deposits (Table 4). To conclude, a characteristic group of trace elements is enriched in phosphorites; the composition of this group is due to: the marine origin of phosphorite; their paragenetic association with organic matter; and to the crystallochemical properties of francolite. Each deposit has a slightly different pattern, probably characteristic of its geological history. An even more promising way of typifying phosphorites has been proposed by Altschuler (1980), using geochemical ratios for coherent elements such as Zn/Cd, Th/U or Y/RE (see Table 6, p 26 in Altschuler 1980). Even in the Negev phosphorites which are anomalously rich in Zn (521 ppm) the Zn/Cd ratio (29) is still an order of magnitude lower than that found in the average shale (300).

Stable Isotopes

Many studies have been made on stable isotopes in phosphorites. Kolodny and Kaplan (1970) have analysed the carbon and oxygen isotopes within francolite CO_3^{2-} and in co-existing calcite. They found that both C and O in francolite are enriched in the respective light isotopes relative to calcite. The carbon and oxygen isotopic composition of the structural carbonate in francolite was also studied by Al-Bassam (1980) and McArthur et al. (1980). In one important point, the three works concur: samples which have ^{18}O depletion are also depleted in ^{13}C. As the authors point out, this may represent the effect of terrestrial ground water. Other data also indicate that a schizohaline environment is favorable for phosphate deposition (Kolodny 1980). The structural carbonate in francolite was also used for ^{14}C dating purposes (Haynes 1968).

The phosphate anion in francolite also contains oxygen and it was Urey et al. (1951) who first suggested that the study of phosphate oxygen isotopic composi-

Table 4. Trace element abundance data in seawater, Earth's crust average shale and phosphorites. All values in ppm except seawater, which is in ppb

	I	II	III	IV	V	VI	VII
Ag	0.07	0.28	0.07	3	2	1–50	
As	1.8	2.6	13	40	7	20.5	2[illegible]
B	10	4.45	100	13	–	16	1[illegible]
Ba	425	21	580	100	338	1–1,000	35[illegible]
Be	0.0006	2.8	3	–	–	1–10	
Cd	0.11	0.2	0.3	40	24	1–10	1[illegible]
Ce	0.0012	60	91	–	6.35	–	1[illegible]
Co	0.39	25	19	–	5	3.3	
Cr	0.2	100	90	1,000	227	285	12[illegible]
Cu	0.9	55	45	100	20	21.7	7[illegible]
Ga	0.03	15	19	–	–	–	
Hg ppb	0.15	80	400	–	50	–	5[illegible]
La	0.0034	30	40	300	8–55	–	14[illegible]
Li	170	20	66	–	1–2	1–10	
Mn	0.4	950	850	30	40	428	1,2[illegible]
Mo	10	1.5	2.6	30	40	18.7	
Ni	6.6	75	68	100	89	12.5	[illegible]
Pb	0.03	13	20	–	5	0–100	[illegible]
Sc	0.004	22	13	10	1.5–3	10–50	[illegible]
Se	0.09	0.05	0.06	10	0–15	2.7	
Sn	0.81	2	6	–	–	10–15	
Sr	8,100	375	300	1,000	2,300	1,900	7[illegible]
Ti	1	4,400	4,600	1,000	246	476	6[illegible]
U	3.3	1.8	3.7	90	111	190	1[illegible]
V	1.9	135	130	300	155	167	1[illegible]
Y	0.013	33	26	300	73	0–50	2[illegible]
Yb	0.00082	3	2.6	10	3–15	–	[illegible]
Zn	5	70	95	300	521	90	[illegible]

I Seawater abundances (Turekian 1969, Table 10–7, p. 309–311)
II Crustal abundances (Taylor 1964, Table III, p. 1,280–1,281)
III Average Shale (Turekian and Wedepohl 1961, Table 2, after p. 186)
IV Phosphoria Formation, modal values (Gulbrandsen 1966, Table III, p. 774)
V Negev, Israel, averages (Nathan, 1981 Table 1, p. 32) with some additional new values
VI Phosphorite averages (Tooms et al. 1969, Table 1, p. 55)
VII Average Phosphorite (Altschuler 1980, Table 5, p. 24)

tion could lead to the establishment of a thermometer which, when combined with the Ca-carbonate oxygen isotopic thermometer, would be independent of the isotopic composition of ocean water.

Tudge (1960) developed a reliable technique for analysis of $\delta^{18}O$ in PO_4^{-3}. This technique has been used by Longinelli (1965, 1966) and his coworkers (Longinelli and Sondi 1965; Longinelli and Nuti 1968, 1973a, b; Longinelli et al. 1976) to successfully analyse phosphates in carbonates, phosphorites, biogenic apatites, marine organism organic matter, and dissolved oceanic phosphate. Longinelli's pioneering studies were unfortunately marred by his own discovery of some systematic error in his analytical procedure (Longinelli and Nuti 1973a). The major result of these studies was the availability of a basic set of data on $\delta^{18}O$-temperature relationship

$t°(c) = 111.4 - 4.3(\delta_p - \delta_w + 0.5)$ where $\delta_p = \delta^{18}O$ of phosphate, $\delta_w = \delta^{18}O$ of water, on the SMOW scale. The slope of this line is parallel to the carbonate-water line. Thus the difference in δ values of carbonate and phosphate in equilibrium with ocean water is constant and independent of water temperature.

Longinelli was also the first to analyse $\delta^{18}O$ in bones and teeth of warm-bodied organisms (1974). This approach could lead to the establishment of a continental thermometer by obtaining environmental water $\delta^{18}O$ from $\delta^{18}O$ of constant-T, warm-blooded animals, and environmental temperatures from coeval bones of fish. Kolodny et al. (1981, 1982) followed this approach as well. Their major rationale for analysing oxygen in apatite is the fact that phosphate ions practically do not exchange their oxygen with water in inorganic exchange reactions, but their exchange is very fast in enzymatically catalyzed reactions. Thus the oxygen isotopic composition of apatite phosphates may be at equilibrium with water during an organism's life, and would be perfectly preserved after its death. Kolodny and Luz (in preparation) show that apatite formed in skeletons of both freshwater and marine fish obeys the Longinelli temperature relationship. It should be stressed that while at least part of the structural carbonate is exchangeable post-mortem (Hassan and Ortner 1977): recent results by Kolodny and Luz et al. (personal communication) indicate that post-mortem exchange of biogenic apatite PO_4 oxygen with its environment is indeed minimal.

S isotope studies have also been made in phosphorites by Bliskovskii et al (1978), and Nathan and Nielsen (1980). They determined the $\delta^{34}S$ of the structural sulfate in francolite from phosphorites of different ages and compared it to that of co-existing gypsum in some of the samples. The results support the theory that many phosphorites are formed in an early diagenetic reducing environment during bacterial sulfate reduction.

References

Al-Bassam KS (1980) Carbon and oxygen isotopic composition of some marine sedimentary apatites from Iraq. Econ Geol 75:1231–1233

Altschuler ZS (1973) The weathering of phosphate deposits – geochemical and environmental aspects. In: Griffith et al. (eds) Environmental Phosphates Handbook. Wiley & Sons, New York, p 33–96

Altschuler ZS (1980) The geochemistry of trace elements in marine phosphorites. Part I. Characteristic abundances and enrichment. In: Bentor YK (ed) SEPM Spec Publ 29:19–30

Altschuler ZS, Cisney EA, Barlow IH (1952) X-ray evidence of the nature of carbonate-apatite. Geol Soc Am Bull 63.1230–1231

Altschuler ZS, Cathcart JB, Young EJ (1964) The geology and geochemistry of the Bone Valley Formation and its phosphate deposits, West Central Florida. Guidebook Field Trip 6. Geol Soc Am Convention, p 68

Axelrod S (1980) The influence of the temperature on the physical chemical properties of natural apatite of Israel. D Sci Thesis, Technion, Haifa, p 101 (in hebrew)

Axelrod S, Rohrlich V (1982) Chemical variability in francolites from Israeli phosphorite macrograins. Mineralium Deposita 17:1–15

Axelrod S, Metzer A, Rohrlich V (1980) The petrography of Israeli phosphorites as related to their beneficiation. In: Bentor YK (ed) SEPM Spec Publ 29:153–165

Bachra BN, Trautz OR, Simon SL (1965) The effect of magnesium and fluoride ions on the spontaneous precipitation of calcium carbonates and phosphates. Arch Oral Biol 10:731–738

Bacquet G, Vo Quang Truong, Bonel G, Vignoles M (1980) Résonance paramagnétique électronique du centre F dans les fluorapatites carbonatées de type B. J Solid State Chem 33:189–195

Beevers CA, McIntyre DB (1946) The atomic structure of fluorapatite and its relation to that of tooth and bone material. Miner Mag 27:254–257

Bein A, Amit O, Nathan Y (1980) Organic matter in phosphorites. Isr Geol Surv Rep MN 4/80:16 (in hebrew)

Belayouni H, Trichet J (1980) Contribution à la connaisance de la matière organique du bassin phosphate de Gafsa. Informations fournies par l'analyse du potentiel pétrolier. Géol comp gisem phosph pétrol (Fr) Bur Rech Géol Minièr Mem 116:37–59

Bentor YK (1980) Phosphorites – The unsolved problems in marine phosphorites. In: Bentor YK (ed) SEPM Spec Publ 29:3–18

Berry EE (1967a) The structure and composition of some calcium-deficient apatites – I. J Inorg Nucl Chem 29:317–327

Berry EE (1967b) The structure and composition of some calcium-deficient apatites – II. J Inorg Nucl Chem 29:1585–1590

Bliskovskii VZ, Grinenko VA, Migdasov AA, Savina LI (1978) Sulfur isotopic composition of the minerals of phosphorites. Geochem Int 8:148–155

Borneman-Starynkevitch ID, Belov NV (1940) Isomorphic substitutions in carbonate-apatite. Compt Rend (Doklady) Acad Sci URSS XXVI 18:804–806

Borneman-Starynkevitch ID, Belov NV (1953) On carbonate-apatites. Dokl Akad Nauk SSR 90/1:89–92 (in Russian)

Brassens A (1975) Mise en évidence de deux types d'eau et de deux localisations des ions CO_2 dans la matière phosphatée d'un phosphate Marocain. CR Acad Sci Ser D, Pris, p 280

Brasseur H, Dallemagne MJ, Melon J (1946) Chemical nature of salts from bones and teeth and of tricalcium phosphate precipitates. Nature 157:453

Brophy GP, Nash TJ (1968) Compositional, infrared and X-ray analysis of fossil bone. Am Miner 53:445–454

Burnett WC (1980) Apatite-glauconite associations off Peru and palaeooceanographic implications. J Geol Soc (Lond) 137:757–764

Bushinski GI (1969) Old Phosphorites of Asia and their genesis. Jerusalem, Isr Program Sci Transl, p 266

Capdecomme L (1952) Etude minéralogique des gites de phosphates alumineux de la région de Thies (Senegal). Congr Geol Int Alger XI:103–107

Chaabani F (1978) Les phosphorites de la couche type de Foum-Selja (Metlaoui, Tunisie). Une série sédimentaire séquentielle à évaporites du Paleogène. Thèse 3ème cycle, Université Louis Pasteur. Strasbourg, p 113

Champetier Y, Blazy P, Joussemet R (1980) Enrichissement des phosphates carbonatés. Caracterisation gitologique et pétrographique. Comportement au traitement thermique. In: Proc 2nd Int Congr Phosphorus Compounds, Boston 1980, p 283–312, Institut Mondial du Phosphate, Paris

Collet LW (1908) Les Dépôts Marins. Doin, Paris, p 325

Cook PJ (1976) Sedimentary phosphate deposits. In: Wolf KH (ed) Handbook of Strata-Bound and Stratiform Ore Deposits. Elsevier, Amsterdam, p 505–535

De Keyser F, Cook PJ (1972) The geology of the Middle Cambrian phosphorites and associated sediments of north-west Queensland, Australia. Bur Miner Resour Geol Geophys Bull 138:79 p

Elderfield H, Holmefjord T, Summerhayes CP (1972) Enhanced CO_2 substitution in carbonate-apatite from the Moroccan continental margin. Univ Leeds, Inst African Geol Res 16th Ann Rep, p 51–52

Elliott JC (1969) Recent progress in the chemistry, crystal chemistry and structure of the apatites. Calcif Tissue Res 3:293–307

Emerson WH, Fischer EE (1962) The infra-red absorption spectra of carbonate in calcified tissues. Arch Oral Biol 7:671–683

Emigh DE (1967) Petrology and origin of phosphorites. Anatomy of the Western Phosphate Field – Intermountain Assoc Geol 15th Ann Field Conf, Proc, p 103–114

Gruner JW, McConnell D (1937) The problem of the carbonate apatites. The structure of francolite. Z Kristallogr A27:208–215

Gulbrandsen RA (1966) Chemical composition of phosphorites of the Phosphoria Formation. Geochim Cosmochim Acta 30:769–778

Gulbrandsen RA (1970) Relation of carbon dioxide content of apatite of the Phosphoria Formation to regional facies. US Geol Surv Prof, Paper 700-B:B9–B13

Gulbrandsen RA, Kramer JR, Besty LB, Mays RE (1966) Carbonate-bearing apatite from Faraday Township, Ontario, Canada. Am Miner 51:819–824

Hassan AA, Ortner DJ (1977) Inclusions in bone material as a source of error in radiocarbon dating. Archeometry 19:131–135

Haynes CV Jr (1968) Radiocarbon analyses of inorganic carbon of fossil bones and enamel. Science 161:687–688

Kolodny Y (1980) Carbon isotopes and depositional environment of a high productivity sedimentary sequence – the case of the Mishash-Ghareb formations, Israel. Isr J Earth-Sci 29:147–156

Kolodny Y (1981) Phosphorites. In: Emiliani C (ed) The Sea. Wiley & Sons, New York 7:981–1023

Kolodny Y, Kaplan I (1970) Carbon and oxygen isotopes in apatite CO and co-existing calcite from sedimentary phosphorite. J Sediment Petrol 40:954–959

Kolodny Y, Luz B (1982) Isotopic composition of phosphate oxygen as a tool in Pleistocene geochemistry. XI. INQUA Congr, Moscow 2:127

Kolodny Y, Luz B, Navon O (1981) Oxygen isotopes in phosphates. Rechecking the rules of the game. 7th Europ Colloq Geochron Cosmochron Isotope Geology, Jerusalem 1981

Krauskopf KB (1955) Sedimentary deposits of rare metals. Econ Geol 50:411–463

Kravitz LC, Kingsley JD, Elkin EL (1969) Raman and IR studies of coupled phosphate vibrations. J Chem Phys 49:4600–4610

LeGeros RZ (1967) Crystallographic studies on the carbonate substitution in the apatite structure. PhD Thesis, New York University

LeGeros RZ, Trautz OR, LeGeros JP, Klein E, Shirra WP (1967) Apatite crystallites: Effects of carbonate on morphology. Science 155:1409–1411

LeGeros RZ, LeGeros JP, Trautz OR, Klein E (1970) Spectral properties of carbonate in carbonate-containing apatites. Develop Appl Spectrosc 7B:3–12

Lehr JR, McClellan GH, Smith JP, Frasier AW (1968) Characterization of apatites in commercial phosphate rocks. In: Coll Int Phosphates Mineraux Solides, Toulouse 1967 2:29–44, Paris, Masson, 1448

Levitt SR, Condrate RA (1970) The vibrational spectra of lead apatites. Am Miner 55:1562–1575

Longinelli A (1965) Oxygen isotopic composition of orthophosphate from shells of living marine organisms. Nature 207:716–719

Longinelli A (1966) Ratio of oxygen-18; oxygen-16 in phosphate and carbonate from living and fossil marine organisms. Nature 211:923–927

Longinelli A (1974) Preliminary oxygen-isotope measurements of phosphate from mammal teeth and bones. CNRS Colloq Int 219:267–271

Longinelli A, Nuti S (1968) Oxygen isotopic composition of phosphorites from marine formations. Earth Planet Sci Lett 5:13–16

Longinelli A, Nuti S (1973a) Revised phosphate-water isotopic temperature scale. Earth Planet Sci Lett 19:373–376

Longinelli A, Nuti S (1973b) Oxygen isotope measurements of phosphate from fish teeth and bones. Earth Planet Sci Lett 20:337–340

Longinelli A, Sondi M (1966) Oxygen isotopic composition of phosphate from shells of some living crustaceans. Nature 211:727–728

Longinelli A, Bartelloni M, Cortecci G (1976) The isotopic cycle of oceanic phosphate, I. Earth Planet Sci Lett 32:389–392

Lucas J, Prévôt L (1981) Synthèse d'apatite à partir de matière organique phosphorée (ARN) et de calcite par voie bacterienne. CR Acad Sci (Paris) Ser II 292:1203–1208

Lucas J, Prévôt L, Lamboy M (1978) Les Phosphorites de la marge nord de l'Espagne. Chimie, Minéralogie, Génèse. Oceanol Acta 1:55–72

Lucas J, Chaabani F, Prévôt L (1979a) Phosphorites et évaporites: Deux formations de milieux sédimentaires voisins étudiés dans la coupe du Paléogène de Foum Selja (Metlaoui, Tunisie). Sci Geol Bull 32:7–19

Lucas J, Prévôt L, Ataman G, Gundogdu N (1979b) Etude minéralogique et géochemique de la serie phosphatée du sud-est de la Turquie (Mazidagl-Mardin). Sci Geol Bull 32:59–68

Lucas J, Flicoteaux R, Nathan Y, Prévôt L, Shahar Y (1980a) Different aspects of phosphorite weathering. In: Bentor YK (ed) SEPM Spec Publ 29:41–51

Lucas J, Prévôt L, Ataman G, Gundogdu N (1980b) Mineralogical and geochemical studies of the phosphatic formations in southeastern Turkey (Mazidagl-Mardin). In: Bentor YK (ed) SEPM Spec Publ 29:149–152

McArthur JM (1978) Systematic variations in the contents of Na, Sr, CO_2, and SO_4 in marine carbonate-fluorapatite and their relation to weathering. Chem Geol 21:41–52

McArthur JM, Coleman ML, Bremmer JM (1980) Carbon and oxygen isotopic composition of structural carbonate in sedimentary francolite. J Geol Soc (Lond) 137:669–673

McClellan GH (1980) Mineralogy of carbonate fluorapatites. J Geol Soc (Lond) 137:657–681

McClellan GH, Lehr JR (1969) Crystal chemical investigation of natural apatites. Am Miner 54:1374–1391

McConnell D (1938) A structural investigation of the isomorphism of the apatite group. Am Miner 23:1–19

McConnell D (1973) Apatite – Its crystal chemistry, mineralogy, utilization, and geologic and biologic occurrences. Springer, Wien Heidelberg New York, p 111

McKelvey VE (1967) Phosphorite deposits. US Geol Surv Bull 1252-D:21

Machatschki F (1939) Sind Abukumalit und Britholit Glieder der Apatitreihe? Zentralbl Miner A:161–164

Martens CS, Harriss DC (1970) Inhibition of apatite precipitation in the marine environment by magnesium ions. Geochim Cosmochim Acta 34:621–625

Matthews A, Nathan Y (1977) The decarbonation of carbonate fluorapatite (francolite). Am Miner 62:565–573

Maughan EK (1980) Relation of phosphorite, organic carbon, and hydrocarbons in the Permian Phosphoria Formation (western U.S.A.). Fr Bur Rech Geol Minieres Mem 116:63–91

Menor E (1975) La sedimentation phosphates, petrographie, mineralogie et geochimie des gisements de Taiba (Senegal) et d'Olinda (Bresil). These Doct – Ing, Université Louis Pasteur, Strasbourg, p 153

Montel G, Bonel G, Heughebaert JG, Trombe JC, Rey C (1981) New concepts in the composition crystallization and growth of the mineral component of calcified tissues. J Cryst Growth 53:74–99

Moore PB (1980) The natural phosphate minerals: Crystal chemistry. In: Proc 2nd Int Congr Phosphorus Compounds, Boston 1980, p 105–130, Institut Mondial du Phosphate, Paris

Naray-Szabo S (1930) The structure of apatite $(CaF)Ca_4(PO_4)_3$ Z Kristallogr 75:387–388

Nathan Y (1981) Structural position of trace elements in apatites – Implications for their recovery. Isr J Earth Sci 30:31–34

Nathan Y, Nielsen H (1980) Sulfur isotopes in phosphorites. In: Bentor YK (ed) SEPM Spec Publ 29:19–30

Nathan Y, Soudry D (1982) Authigenic silicate minerals in phosphorites of the Negev, Southern Israel. Clay Miner 17:249–254

Nathan Y, Shiloni Y, Roded R, Gal I, Deutsch Y (1979) The Geochemistry of the Northern and Central Negev phosphorites. Isr Geol Surv Bull 73:41

Neuman WF, Neuman MW (1953) Nature of the mineral phase of bone. Chem Rev 53:1–38

Nriagu JO (1976) Phosphate-clay mineral relations in soils and sediments. Can J Earth Sci 13:717–736

Odin GS, Letolle R (1980) Glauconitization and phosphatization environments: A tentative comparison. In: Bentor YK (ed) SEPM Spec Publ 29:227–237

Posner AS, Perloff A (1957) Apatites deficient in divalent cations. Nat Bur Stand J Res 58:279–286

Powell TG, Cook PJ, McKisdy DM (1975) Organic geochemistry of phosphorites: Relevance to petroleum genesis. Am Assoc Pet Geol Bull 59:618–633

Prévôt L, Lucas J (1980) Behaviour of some trace elements in phosphatic sedimentary formations. In: Bentor YK (ed) SEPM Spec Publ 29:31–39

Prévôt L, Lucas J, Doubinger J (1979b) Une correspondance entre le contenu palynologique et la composition minéralogique et chimique d'une série phosphatée sédimentaire (Gantour, Maroc). Sci Geol Bull 32:69–90

Prévôt L, Lucas J, Nathan Y, Shiloni Y (1979a) Répartition des éléments traces dans les phosphorites marines. In: Ahrens LH (ed) Origin and Distribution of the Elements. Pergamon, Oxford, p 293–304

Price NB, Calvert SE (1978) The geochemistry of phosphorites from the Namibian Shelf. Chem Geol 93:151–170
Reynolds DS, Jacob KD, Marshall HL, Rader LF Jr (1935) Phosphate fertilizers by calcination process. Ind Eng Chem 27:87–91
Reynolds DS, Marshall HL, Jacob KD, Rader LF Jr (1936) Phosphate fertilizers by calcination process. Ind Eng Chem 28:678–682
Riggs SR (1979) Phosphorite sedimentation in Florida – A model phosphogenetic system. Econ Geol 74:285–314
Romankevich YA, Baturin GN (1972) Composition of the organic matter in phosphorites from the continental shelf of Southwest Africa. Geochem Int 6:464–470
Rooney TP, Kerr PF (1967) Mineralogic nature and origin of phosphorite, Beaufort Co, North Carolina. Geol Soc Am Bull 7B:731–748
Sandstrom MW (1980) Organic geochemistry of some Cambrian phosphorites. Phys Chem Earth 12:123–131
Sassi S (1974) La sédimentation phosphatée au Paléocène dans le sud et le centre ouest de la Tunisie. Thèse Doct Sci, Université Paris-Sud, Orsay, p 252
Sassi S, Jacob G (1972) Decouverte de clinoptilolite dans le bassin phosphate de Metlaoui (Tunisie). CR Acad Sci (Paris) Ser D, 274:1128–1131
Scheib MR, Thrasher RD, Lehr RJ Chemical composition determination of francolite apatites by Fourier Transform Infrared Spectroscopy (FTIR). Int Confer Fourier Transform Infrared Spectroscopy, Univer South Carolina, Columbia, South Carolina, June 8–13, 1981 (in Press)
Slansky M (1980) Géologie des Phosphates Sedimentaires. Fr Bur Rech Geol Minières Mem 114:92
Smith JP, Lehr JR (1966) An X-ray investigation of carbonate apatite. J Agric Food Chem 14:342–349
Sommer F (1972) Néogenèse de clinoptilolite dans l'Eocène et l'Oligocène de Casamance (Senegal). Sci Geol Bull 25:251–258
Taylor SR (1964) Abundance of chemical elements in the continental crust: A new table. Geochim Cosmochim Acta 28:1273–1285
Thewlis J, Glock GE, Murray MM (1939) Chemical and X-Ray analysis of dental, mineral and synthetic apatites. Trans Faraday Soc 35:358–363
Tooms JS, Summerhayes CP, Cronan DS (1969) Geochemistry of marine phosphate and manganese deposits. Oceanogr Mar Biol Ann Rev 1:49–100
Trueman NA (1966) Substitutions for phosphate ions in apatite. Nature 210:937–938
Tudge AP (1960) A method of analysis of oxygen isotopes in orthophosphates – its use in measurements of paleotemperatures. Geochim Cosmochim Acta 18:81–93
Turekian KK (1969) The oceans, streams, and atmosphere. In: Wedepohl KH (ed) Handbook of Geochemistry. Springer, Heidelberg Berlin 1:297–323
Turekian KK, Wedepohl KH (1961) Distribution of elements in some major units of the earth's crust. Geol Soc Am Bull 72:175–192
Urey HC, Lowenstam HA, Epstein S, McKinney CR (1951) Measurements of paleotemperatures and temperatures of the Upper Cretaceous of England, Denmark, and the Southeastern United States. Geol Soc Am Bull 62:399–416
Villiers JE de (1942) The carbonate-apatites: Francolite from the Richtersveld, South Africa. Am J Sci 240:443–447
Wadjinny A (1979) Milieu de sédimentation et mécanismes de dépôt des couches inférieures de la série phosphatée de Ben Guerir (Gantour, Maroc); une etude sequentielle. These 3eme Cycle, Université Louis Pasteur, Strasbourg, p 100
Wallaeys R (1952) Contribution à l'étude des apatites phosphocalciques. Ann Chim (Paris) 7:808–848
Van Wazer JR (1966) Phosphorus and its compounds, V.1 Chemistry. Interscience, New York, p 954
Weaver CE, Wampler JM (1972) The illite-phosphate association. Geochim Cosmochim Acta 36:1–13
White WB (1974) The carbonate minerals. In: Farmer VC (ed) The infrared spectra of minerals. Miner Soc (Lond), p 227–284

Chapter 9 Weathering of Phosphate Minerals

René Flicoteaux[1] and Jacques Lucas[2]

Evolution and Genesis of Phosphate Minerals During Weathering – A Survey

The known natural phosphate minerals exceed 250 in number (Fisher 1973). However, the more common ones are small in number and belong to the apatite family. This family constitutes the main group of primary phosphates of the lithosphere. The majority of the other minerals result from a supergene alteration by groundwaters, or bat- and bird-guanos. This alteration has been well described in detail by Altschuler (1973), in a fundamental synthesis based on three main topics: the weathering of marine phosphorites, the weathering of carbonatites and the weathering caused by acid guanos. The weathering of phosphatic igneous and metamorphic rocks is not mentioned, probably because it is little known, except in the northeast of Brazil (Lima Da Costa, verbal comm.). Since Altschuler's synthesis, many studies have been published, some related to the weathering of phosphorites, others to the genesis of new minerals during the weathering process. Most of the mineralogical results obtained from these works are shown in Table 1. Consideration of Table 1 permits the findings of Altschuler (1973, p 78–80) to be extended and allows some general conclusions on the conditions of phosphate minerals weathering and on their behavior during weathering.

1. The guano reaction in limestones leads to the formation of calcium phosphates, in the order: brushite [$CaHPO_4 \cdot 2H_2O$], whitlockite [$Ca_3(PO_4)_2$] and hydroxylapatite [$Ca_5(PO_4)_3(OH)$]. This mineral sequence is characterized by a decrease of organic compounds and an increase of the Ca/P ratio.

2. The guano reaction on igneous rocks and soils, as well as the alteration by meteoric waters of rocks which are guanitized, phosphate deposits of marine sediments, karsts and caves, and soils, leads to the formation of aluminum and iron phosphates. One specific mineral paragenesis corresponds to each type of action. The guanos give leucophosphite [$K(Fe^{2+}, Al)_2(PO_4)_2(OH) \cdot 2H_2O$] and variscite-strengite isomorphous series, of which the most common is barrandite [$(Fe, Al)PO_4 \cdot 2H_2O$]. The groundwaters, after decarbonatation and in presence of clays, give millisite [$(Na, K)CaAl_6(PO_4)_4(OH)_9 \cdot 3H_2O$], crandallite [$CaAl_3(PO_4)_2(OH)_5 \cdot H_2O$] and its products of substitution by strontium, barium and rare earths, wavellite [$Al_3(PO_4)_2(OH)_3 \cdot 5H_2O$] and perhaps augelite [$Al_2PO_4(OH)_3$]. They give, in the presence of sulfates and iron sulfates, fairfieldite [$Ca_2(Mn, Fe)(PO_4)_2 \cdot$

1 Universite d'Aix-Marseille, Faculté des Sciences et Techniques de Saint-Jérôme, Laboratoire de Sédimentologie, 13397 Marseille Cedex 13, France
2 Laboratoire de Géologie de l'Université Louis Pasteur, 1 Rue Blessig, 67000 Strasbourg, France

Table 1. Paragenetic associations of aluminum and iron phosphates during weathering

Parent material	Original mineral	Agent	Products of alteration
Volcanic rocks[1,2]	Silicates (feldspar, peridot, pyroxene . . .)	Guano	Barrandite, leucophosphite, newberyite, phosphosiderite, strengite, variscite
Granitic rocks[1–3]	Silicates (feldspar. . .)	Guano	Barrandite, variscite
Magnetite - Hematite iron ore in gneiss[1–4]	Iron oxides	Guano	Leucophosphite, phosphosiderite, rockbridgeite, strengite
Limestones[1,2,5]	Calcite, aragonite	Guano	Brushite-monetite, hydroxylapatite, whitlockite
Decalcification clays in caves[1–5]	clay-minerals	Guano	Leucophosphite, taranakite
Guanitized rocks[1,2]	Barrandite	Groundwater	Crandallite, millisite, montgomeryite, wavellite
Banded calcium phosphates in caves[5]	Hydroxylapatite, Dahllite	Groundwater	Crandallite, montgomeryite
Argillaceous red coatings in banded calcium phosphates of karst[6]	Kaolinite	Groundwater	Crandallite, perhamite
Bauxitic soils, red soils, earthy fillings in reef karst, terra-rossas[1,2,7–9]	Clay-minerals?	Groundwater	Crandallite, gorceixite
Marine phosphorites with argillaceous, calcareous, quartzous, siliceous beds or matrix[1–10] to 27	Francolite and attapulgite, illite, mixed-layers or montmorillonites	Groundwater	Augelite, beraunite, cacoxenite orandallite, englishite, gordonite, goyazite, calci-millisite, turquoise, variscite, vivianite, wardite, wavellite
Glauconitic rocks with calcium phosphate beds or nodules[28–30]	Apatite, clays, glauconite	Groundwater	Crandallite, gorceixite, ferriferous variscite
Gypsiferous phosphatic and pyritic layers[23]	Fluor-apatite, francolite, pyrite?	Groundwater	Fairfieldite, lipscombite, phosphosiderite, strengite
Iron ore in metamorphic andesitic volcanism[31]	Aluminous and Phosphorous magnetite	Groundwater	Augelite, crandallite, wavellite, senegalite, turquoise (+ strengite, variscite?)
Carbonatites[32]	Apatite and silicates	Groundwater	Gorceixite, goyazite, monazite
Argillaceous and ferruginous coatings or matrix in lateritized phosphatic sediments[25]	Kaolinite, goethite, aluminous or calcium phosphates	Groundwater	Ferriferous-millisite
Phosphatic ferricrust[25]	Al-goethite and aluminous phosphates	Groundwater	Vivianite
Aluminous phosphates or phosphatic ferricrusts[25]	Aluminous phosphates and hematite or Al-goethite	Groundwater	Augelite

[1] Altschuler (1973), [2] Trueman (1965), [3] Du Toit (1917), [4] Axelrod et al. (1952), [5] Péneaud (1978), [6] Billaud (1982), [7] Sweatman (1961), [8] De Weisse (1970), [9] Tercinier (1972), [10] Larsen and Shannon (1930), [11] Capdecomme (1952, 1953), [12] Visse (1952), [13,14] Altschuler et al. (1956, 1958), [15] Frondel (1958), [16] Owens et al. (1960), [17] Slansky et al. (1964), [18] Blanchard and Donahan (1968), [19] Figueiredo Gomes (1968), [20] Zanin (1968), [21] Russel and Trueman (1971), [22] Flicoteaux et al. (1977), [23,24] Lucas et al. (1980 a and c), [25] Flicoteaux (1980), [26] Slansky (1980), [27] Lucas et al. (1981), [28] Charpy and Nahon (1978), [29] Milton et al. (1958), [30] Parron and Nahon (1980), [31] Johan (1976), [32] McKie (1962)

$2H_2O$], lipscombite [(Fe^{2+}, Mn), $Fe^{2+}(PO_4)_2(OH)_2$], phosphosiderite and strengite [$FePO_4 \cdot 2H_2O$]. The minerals so formed one another in a specific order, for example:

1. guanos + silicates → leucophosphite → barrandite
2. groundwaters + apatite + clays → millisite → crandallite → → wavellite → augelite.

Four main trends are observed in these paragenetic sequences: alkalis and hydration decrease, Al–Fe/P ratios increase and, in the case of iron phosphates, oxidation tends to a higher degree (Altschuler 1973; Poullen 1982).

3. Slight differences may occur in the order of appearance and disappearance of particular minerals in the sequences outlined above. In addition certain minerals observed in one paragenesis might be absent in another (Table 1). Finally, retrograde evolutions are locally observed. This can all be explained in three ways.

Firstly, the geochemical nature of the parent rock. This corresponds with what Paquet (1969) described as the geochemical particularities of the parent rock which conditions the nature of weathering minerals. Differences of composition of Florida and Senegal millisites and, also, low amount of Ca-millisite in numerous profiles (Flicoteaux 1980), are explained that way. We can then better understand the formation of gorceixite, rich in strontium and rare earths (Milton et al. 1958) and variscite rich in iron (Parron pers commun) in glauconitic beds; also the formation of minerals rich in barium, strontium and rare earths of the plumbogummite series by the weathering of carbonatite (Altschuler 1973). The association of montgomeryite, aluminous phosphate very rich in magnesium and rich in calcium, with crandallite, in weathered guanos of caves and limy outcrops (Trueman 1965; Peneaud 1978), is therefore better understood. Finally, the difference in nature of minerals resulting from guano action on silicates (Altschuler 1973) is also explained in this way.

Secondly, parent rock porosity, which controls the rate of leaching. This factor can explain how wavellite constitutes the most frequent mineral in the Florida phosphorites, in which calcium phosphate nodules are scattered in a quartzous material, while crandallite dominates in the deposits of Togo and western Senegal in which quartzous layers are accidents of the phosphatic sedimentation.

Thirdly, the topographic relief. This regulates drainage and surface slope changes, with passing time, and may cause changes in the nature or concentration of percolating solutions. We can explain thus the absence of evolution of (a) leucophosphite and taranakite on cave floors (Altschuler 1973); (b) the retrograde evolutions of variscite nodules from Fairfield (Larsen and Shannon 1930; Frondel 1958; Altschuler 1973); and (c) the augelites from the oxidized cap of Kouroudiako (Johan 1976) or the wavellites from Lam Lam leached facies (Flicoteaux et al. 1977; Flicoteaux 1980) towards crandallites and millisites.

4. Iron phosphates are less frequent than aluminum phosphates. The reaction of acid-guanos on igneous rocks, for example, usually leads to barrandite, an aluminum and iron phosphate. Likewise, acid meteoric waters lead to the generation of iron phosphates only in the absence of clayey material, or in very reducing conditions. These observations agree with the conclusions of Haseman et al. (1951), Kittrick and Jackson (1955a, b), Gastuche et al. (1963), Bromfield (1965) in Se-

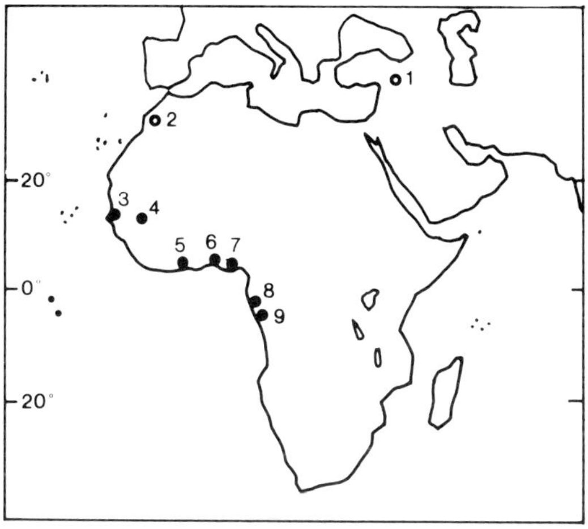

Fig. 1. Weathering of Upper Cretaceous and Eocene phosphorites. *1* Mazidagi; *2* Sidi Daoui; *3* Lam Lam, Thiès plateau, Taîba, Tobene; *4* Senegal valley; *5* Eboînda; *6* Hahotoe; *7* Oshosun; *8* Holle; *9* Cabinda. (*open circle* only calcite dissolution and apatite decarbonation; *full circle* alumino-phosphates appearance and development)

galen (1973), Taylor et al. (1964), Taylor and Gurney (1965) that phosphate ion reactions with iron hydroxides occur much more slowly than with aluminum hydroxides.

Weathering of Marine Phosphorites

Apatite Evolution During Weathering at Low Intensity

The most widespread deposits of phosphorite are those of marine origin. They consist of a phosphatic material, cement or grains of various aspects, associated with a non-phosphatic phase, which can be argillaceous, marly, carbonated, quartzous or cherty, more seldom ferruginous or gypsiferous. This phase is rich in organic matter, mainly when it is argillaceous. When it becomes the dominant component, it passes into real intercalations of clays, limestone or dolomite, chert or sandstone. In such phosphorites, the more frequent phosphate mineral is a carbonate-fluorapatite, with high CO_2 content, close to francolite. The CO_2 content is about 4.7%; but can reach or exceed 6%, as suggested by the successive observations of Parker and Siesser (1972) on Agulas Bank; Sassi (1974) and Chaabani (1978) in Tunisia; McArthur (1978) off Morocco; Birch (1979) on the South African continental plateau; Prévôt et al. (1979) at Ortegal Cape; and Jarvis (1980) in the North of France. This mineral probably represents the kind in equilibrium with the genesis environment of marine phosphorites (Lucas and Prévôt 1975).

The main transformations observed during weathering of these deposits are oxidization of organic matter, leaching of carbonate, evolution of clay and apatite mineral phases, appearance of free iron hydroxides and genesis of new phosphate and sulfate minerals, rich in iron and aluminum. The transformation intensity seems to depend greatly on climatic conditions. For example, only Eocene phosphorites, presently located in the tropical zone (Fig. 1), include secondary aluminous or ferruginous phosphate minerals in their paragenesis.

Table 2. Some characteristics of the apatites of Upper Cretaceous-Eocene deposits in West-African coastal basins. Non – weathered and weathered – primary apatites, secondary apatites compared with some typical minerals

Origin	F/P_2O_5	CaO/P_2O_5	% CO_2 Calc.	% CO_2 Meas.	a (Å)	c (Å)	c/a	References
Fluorapatite from Durango	0.087	1.32		0.05	9.391	6.878	0.7324	Young et al. (1969)
Synthetic fluorapatite					9.368	6.890	0.7355	ASTM–15 · 875
Synthetic carbonate fluorapatite					9.357–6	6.895–4	0.7348	Bonel and Montel (1966)
Francolite from Staffel	0.109	1.455		3.36	9.360	6.880	0.7350	Gruner and McConnell (1937)
Francolite from Staffel	0.109	1.460		3.36	9.340	6.880	0.7366	McConnell (1938)
Francolite from Wheal Franco	0.097	1.410		3.40	9.340	6.890	0.7377	Sandell et al. (1939)
Carbonate-fluorapatite from Foum Selja	0.110	1.550–1.570	4.96	4.50–6.30	9.324–9.329	6.890–6.896	0.7386–0.7396	Chaabani (1978)
Carbonate-fluorapatite from Ben Guerir	0.120–0.130	1.370–1.580	3.50–6.00	4.99–6.44	9.336	6.896	0.7380	Wadjinny (1979)
Sidi-Daoui: Non-weathered facies	0.120–0.150	1.540	3.40		9.323–9.348	6.880–6.900	0.7363–0.7388	El Mountassir (1977), Lucas et al. (1979, 1980a)
Sidi-Daoui: Weathered facies	0.080–0.120	1.290	2.10		9.331–9.392	6.870–6.890	0.7328–0.7358	
Taïba: Facies with carbonate	0.092	1.370	4.02		9.353	6.888		Menor (1975), Lucas et al. (1979, 1980a)
Taïba: Facies without carbonate	0.097	1.330	1.82	1.40	9.361	6.888		
Phosphorite from Tobene	0.107	1.415	2.36		9.339–9.384		0.7342–0.7393	Flicoteaux, Lucas and Prévôt (in preparation)
Phosphorite from Lam Lam	0.104	1.326	1.61		9.346–9.377		0.7360–0.7391	Flicoteaux (1980 and unpublished analysis)
Phosphorite from Hahotoe	0.111	1.327–1.509	1.85		9.333–9.368		0.7358–0.7397	Flicoteaux (idem)
Phosphorite from Thiès'Plateau	0.113	1.490	2.81		9.345–9.354		0.7375–0.7394	Flicoteaux (idem)
Phosphorite from Senegal Valley			3.63		9.324–9.334		0.7382–0.7401	Flicoteaux et Pascal (in preparation)
Carbonate-fluorapatite from Langebaanweg-Saldanhaarea	0.090–0.130	1.150–1.460		2.04–5.14	9.342–9.363	6.876–6.888	0.7349–0.7369	Tankard (1974b)

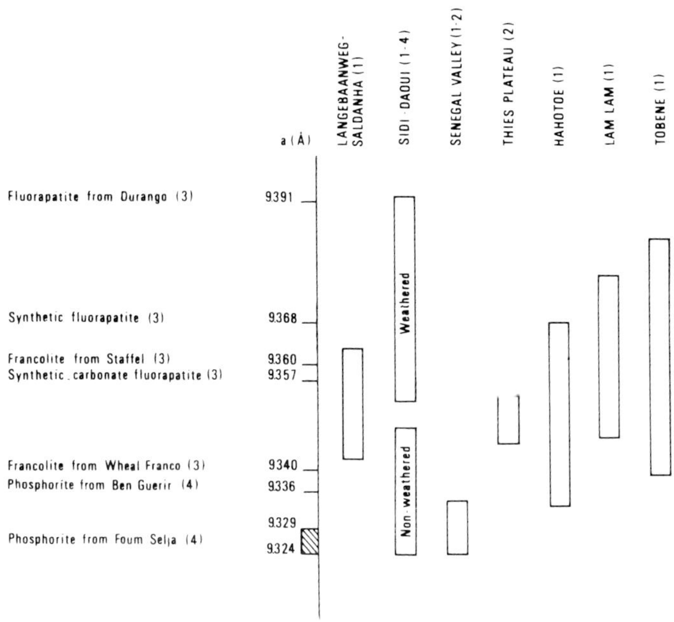

Fig. 2. Comparison of lattice constant a (Å) of weathered primary apatites (*1*) and precipitated secondary apatites (*2*) with some standard minerals (*3*) and non-weathered primary apatites (*4*). See Table 2 for references

The Sidi Daoui Deposit, an Example

The evolution of apatite composition under the influence of weathering has been studied in all details by El Mountassir (1977) and Lucas et al. (1979 b, 1980 a) in the Sidi Daoui deposit of Morocco, and can be used as a model. The unweathered part of the deposit consists of apatite, calcite and dolomite, which are often silicified, with minor amounts (about 1% to 5%) of clay. The apatite is a carbonate-fluorapatite with 1.2% to 5.5% CO_2 in the lattice; the CaO/P_2O_5 ratio is 1.56 and the F/P_2O_5 ratio is higher than 0.12 (Table 2). The clay minerals are smectite, Mg- and Fe-montmorillonites and some palygorskite (attapulgite).

The weathered zone is directly located under a topographic paleosurface which cuts the deposit. On weathering, the phosphatic series turned red and were enriched in clay. Calcite is partially dissolved, as can be seen under the microscope, and X-ray analysis reveals a poor crystallinity. Dolomite, depleted in magnesium, shows a trend to dedolomitization. Attapulgite disappears and kaolinite appears. The CO_2 in apatite decreases to between 0.5% and 3.5%; the CaO/P_2O_5 ratio decreases to 1.29 while the F/P_2O_5 ratio is lower than 0.12 (Table 2).

This evolution of chemical characters is accompanied by a modification of lattice constants [a] and [c], computed from X-ray diagrams, and the c/a ratio (Table II). The parameter [a] (Fig. 2) is between 9.323 and 9.348 Å in the unweathered phosphorite and increases to between 9.351 and 9.392 Å in the weathered phosphorite. Thus, it reaches dimensions clearly higher than in natural carbonate-fluorapatites or in the francolite of Wheal Franco. It is placed between a synthetic carbonate-fluorapatite and a fluorapatite.

Other Examples in Eocene Deposits

The Senegal and Togo deposits of Taîba (Slansky et al. 1964; Menor 1975; Lucas et al. 1979c, 1980a; Flicoteaux 1980), Lam Lam (Tessier et al. 1976; Flicoteaux 1980; Lucas et al. 1980a), Tobene and Hahotoe (under study) belong, as does the Sidi Daoui deposit, to the Upper Cretaceous-Eocene phosphorite deposits of the Atlantic continental margin of Africa. This suggest that the primary apatites must have been alike, indeed probably identical. These similarities thus facilitate comparison of the secondary characters due to weathering.

Except for the Taîba and Tobene deposits where the center is limy, these deposits consist of an argillaceous and phosphatic facies, practically free from calcite, with frequent interstratifications of chert and sand. Iron oxi-hydroxide products and aluminous phosphorites develop in their borders and upper parts. Thus, the weathering is more advanced than in the Sidi Daoui deposit. The numerous chemical analyses recorded (Table 2), the CO_2-contents, which were either calculated by the Gulbrandsen method or measured directly (Fig. 3) and the lattice constants also computed, show that:

Firstly, the apatite of the central Taîba deposit is still close to the unweathered apatite of Sidi Daoui, with 1% to 6% CO_2, a CaO/P_2O_5 ratio of 1.37 and a lattice constant [a] of 9.353 Å. This supports our hypothesis of the similarity of the primary apatites.

Secondly, the apatite of the Taîba periphery and of the Tobene, Lam Lam and Hahotoe deposits have undergone modifications of chemical and cristallographic characters, similar to the weathered apatite of Sidi Daoui. The 0% to 4.5% CO_2 and the lattice constant [a] of 9.333 to 9.348 Å show a trend towards fluorapatite.

General Discussion of Apatite Evolution During Weathering at Low Intensity

The weathering of marine primary apatites leads, as shown by the various examples of eocene phosphorites studied, to a loss of CO_2 from the lattice and to a concomitant increase of the lattice constant [a]. The phosphate mineral, initially close to carbonate-fluorapatite, is transformed into and develops the characteristics of fluorapatite. This transformation, by loss of CO_2, occurs in most cases; indeed, it has been recognized in the Senonian phosphate of Mazidagi, in Turkey (Lucas et al. 1979d, 1980b) and in the upper Precambrian phosphates of Volta, in West Africa (Lucas et al. 1980c; Lucas and Prévôt 1981). Consequently the same transformation is proposed in the case of the Pliocene phosphorites of the Langebaanweg-Saldanha area, in South Africa (Tankard 1974b). The CO_2-content (Table 2) and the lattice constant [a] (Fig. 2) are intermediate between carbonate-fluorapatites

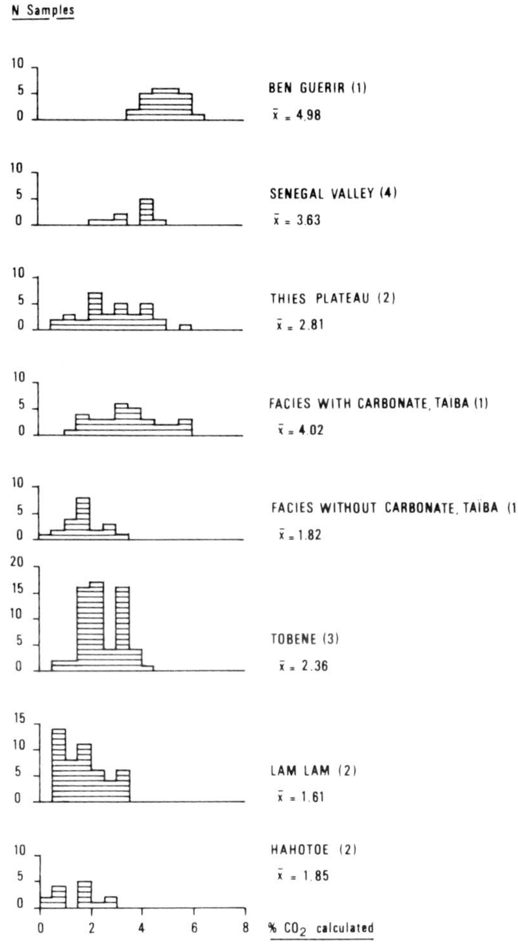

Fig. 3. Histograms of CO_2 content in percent, calculated by Gulbrandsen's method. *1* Lucas et al. (1979 b); *2* Flicoteaux (1980, modified); *3* Flicoteaux, Lucas and Prévôt (under study); *4* Flicoteaux and Pascal (under study)

and fluorapatites; besides aluminum phosphates are known in the same region (Du Toit 1917; Frankel 1943; Tankard 1974 a).

The decarbonation and the increase of [a] lattice constant are associated with the removal of many elements. The apatite lattice becomes depleted in calcium and fluorine (Table 2) and also in sodium and strontium (Prévôt et al. 1979; Lucas et al. 1980 a). In spite of some inconsistent data difficult to interpret (Lucas et al. 1980 b, c; Lucas and Prévôt 1981), this removal seems generally admitted (McArthur 1980). Fluorine is lost first, before calcium (Gusev et al. 1976), during the weathering process and it seems to us that strontium will be attracted by clay minerals (Lucas et al. 1979 b).

As long as carbonates are present in the rock they protect the associated apatite, stopping the evolution from going further than the fluorapatite stage. It is only after dissolution of the carbonates that apatite begins to disappear (Altschuler 1973; Lucas et al. 1980 a; Slansky 1980). The environment, depleted in calcium, becomes more acid, so that the calcium is totally removed from apatite, which is dissolved. In this intense weathering stage new minerals appear in the top and the bottom, or in the substratum of phosphorite deposits. In the top iron and aluminum phosphates develop, whose genesis will be discussed later. In the bottom and in the substratum secondary apatites and calcites develop. Such minerals are known in the lower part of the Florida, Georgia and Tennessee phosphorites (Sellard 1913; Espenhade and Spencer 1963; Altschuler 1965, 1973). They have been observed in the lower part of Lam Lam and Taîba (Flicoteaux et al. 1977; Flicoteaux 1980), in the Eocene calcarous substratum of aluminum phosphates in the Thies plateau (Tessier 1950; Flicoteaux 1980), close to Lam Lam, and in the Eocene argillaceous, carbonated and phosphatic complex from the Senegal valley (under study). Analysis of the apatites of the Senegal valley has revealed the characters of a carbonate-fluorapatite (Figs. 2 and 3; Table 2). The substratum apatites of the Thiès plateau show, on the contrary, a very clear evolution towards fluorapatites. Indeed, field observations show that the substratum is karstic and has been reached by deep weathering.

In summary, two important points can be made:

Firstly, an impoverishment in the CO_2 content of carbonate-fluorapatite emphasizes a disequilibrium between this mineral and the environment. The impoverishment can result either from a modification of the genetic conditions, as proposed by Gulbrandsen (1970) for the western part of the Phosphoria Formation, or from a transformation by weathering, after settlement, as shown by the preceding examples. Study of the various accompanying minerals will direct the choice towards one of the two possible interpretations. The decrease in the CO_2 content and the increase in lattice constant [a] which results from it, are good indicators of the importance of weathering.

Secondly, the weathering process of marine carbonate-fluorapatites includes two stages. The first stage is slight; it leads to decarbonated minerals of which the lattice constant [a] is close to that of fluorapatite. The second stage is more intense and leads to dissolution of the minerals transformed by stage one. It might occur at the bottom of a section and in a carbonated environment, accompanied by a precipitation of secondary apatite, characteristically related with the primary phosphates. So starts an iterative cycle of dissolutions and precipitations of the carbonate-fluorapatite.

Genesis of Aluminous Phosphates by Strong Meteoric Weathering

The intense weathering of marine phosphorites leads, after leaching of the carbonate fraction and decarbonation of apatite, to the appearance and growth of aluminous and ferruginous phosphates. Dissolution of apatite and etching of associated minerals occur. Then two pathways are possible (Altschuler 1973): Either direct fixing of phosphorus and calcium, released from apatite on other minerals, or combination of both ions with the aluminum and the iron leached from the lattice of these minerals. In the presence of clays, aluminous phosphates form as shown by the successive observations of Russ and Andrew (1924) in Nigeria; Capdecomme (1952, 1953); Slansky et al. (1964) in Senegal; Altschuler et al. (1956, 1958) in Flori-

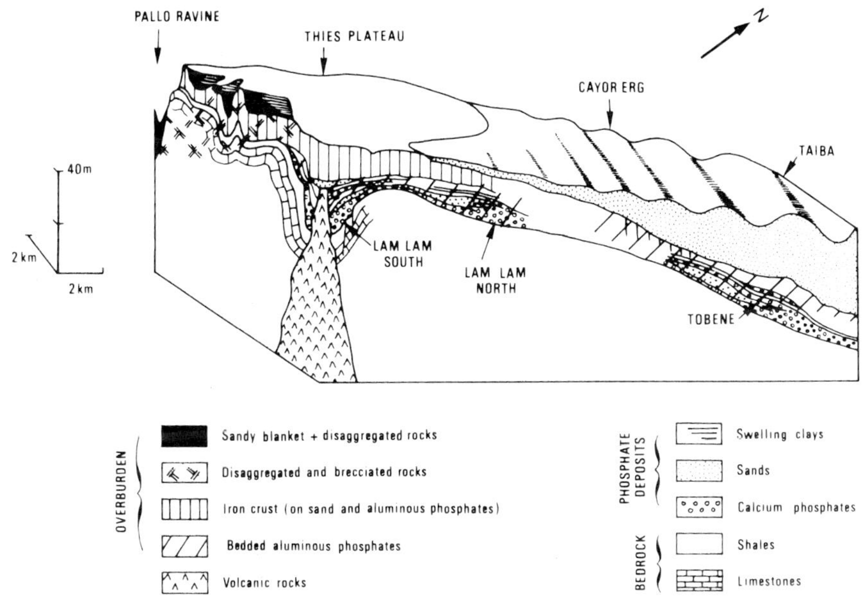

Fig. 4. Lithology and structure of weathered phosphorites in Western Senegal

da; Riedel (1959) in Congo; Bushinskii (1966) in South China and Vietnam; Zanin (1968) in Siberia; Russel and Trueman (1971) in Australia; Charpy and Nahon (1978), Parron and Nahon (1980) on the Ivory Coast; and Lucas et al. (1980 b), Lucas and Prévôt (1981) in the Upper Volta.

The Formation of Aluminous Phosphates in Western Senegal, an Example

Studies have been conducted on both the decarbonation of apatite and the genesis of aluminous phosphates in the various ores of Western Senegal at Lam Lam, Thiès plateau near Lam Lam, Taîba (Menor 1975; Flicoteaux et al. 1977; Lucas et al. 1979 c, 1980 a; Flicoteaux 1980) and also Tobene (Flicoteaux, Lucas and Prévôt, under study) and will be used as an example. This genesis is made at the expense of a sedimentary, Eocene and Oligocene to Miocene series, including from bottom to top: an argillaceous and calcareous substratum, phosphatic deposits (already mentioned), in which more or less argillaceous phosphorites, cherts, limestones and sandstones alternate, and a sandy cover. Clay minerals of the substratum are predominantly attapulgite and small amounts of montomorillonites. In the phosphorites they are montmorillonites, with traces of illite and mixed-layer clays.

The aluminophosphate genesis is polyphased (Fig. 4); it results from a succession and a superimposition of paragenetic associations and structural organizations:

The first stage corresponds to an intensification of the initial leaching of phosphorites. This intensification is shown by the appearance of kaolinite, goethite and also a paragenetic sequence of phosphate minerals consisting of Ca-millisite, crandallites and wavellite. Throughout these mineralogical transformations, the primary sedimentary and textural features in the parent rocks are maintained and the porosity is increased. In the field this evolution leads to a bedded aluminum phosphate layer.

The second stage corresponds to successive accumulation of kaolinite, Fe-millisite, Al-goethite and hematite in the voids that resulted from the first stage. This accumulation develops with preservation of primary sedimentary structures until the formation of pseudo-breccias and pseudo-pisolites, which destroy them. It leads to an induration, which results in a real phosphatic and iron crust.

The third stage corresponds to a deep transformation of the iron crust and the underlying horizons and finally to their disaggregation. The transformation is mainly characterized by a removal of iron, an individualization of augelite and a new generation of kaolinite. The disaggregation results from destruction of bedded or pseudo-pisolitic structures. This evolution leads to superficial and eluvial horizons rich in fragments and nodules.

The main characters of the phosphatic paragenetic associations developed in the first two stages are presented below.

Ca-Millisite, Crandallites and Wavellite Sequence, Composition, Texture, and Genesis. A whole suite of aluminophosphatic minerals appear at the expense of apatite in the first stage. These minerals develop in the profile (Fig. 5) according to a given order. First, Ca-millisite appears; second, a mixture of Ca-millisite and needle-like crandallite (1); then single needle-like crandallite (1); and then, sheaf-like crandallite (2) and arrow-head crandallite (3); and finally wavellite.

The Ca-millisite and needle-like crandallite result from weathering of clays and apatite by solutions that percolate through the profiles. This association appears at the borders of calcium phosphate beds and within argillaceous beds, while the main calcium phosphate beds do not show any modification. Within these same argillaceous beds Ca-millisite remains for the longest time while rising in the profiles. The Ca-millisite occurs in equant, tabular, small crystals (Fig. 6), with parallelepipedic sections; these crystals can stick together and show a spiral coil under optical and electron microscopes. The Ca-millisite and needle-like crandallite form a cement that progressively replaces the whole of the original apatitic cement. Once this calcium phosphate has disappeared, the process continues and needle-like crandallite replaces the Ca-millisite in its turn, to become the main constituent of aluminophosphatic cement.

Then, the sheaf-like and arrow-head crandallites succeed the needle-like crandallite; they are distinguished by their different habitus. The sheaf-like crandallite shows radiating prismatic crystals, which tend to overgrow the whole of the cement, but will retain the original texture. That represents a recrystallization in situ of needle-like crandallite. The arrow-head crystals of crandallite line the voids (Fig. 6), but they also develop within the cement. We must admit that they derive from needle-like or sheaf-like crandallite by dissolution and recrystallization in situ. Wavellite results from sheaf-like and arrow-head like crandallite in the same way.

The sheaf-like crandallite can also form directly at the expense of apatite, within calcium phosphate beds, as demonstrated by new observations. The arrow-head

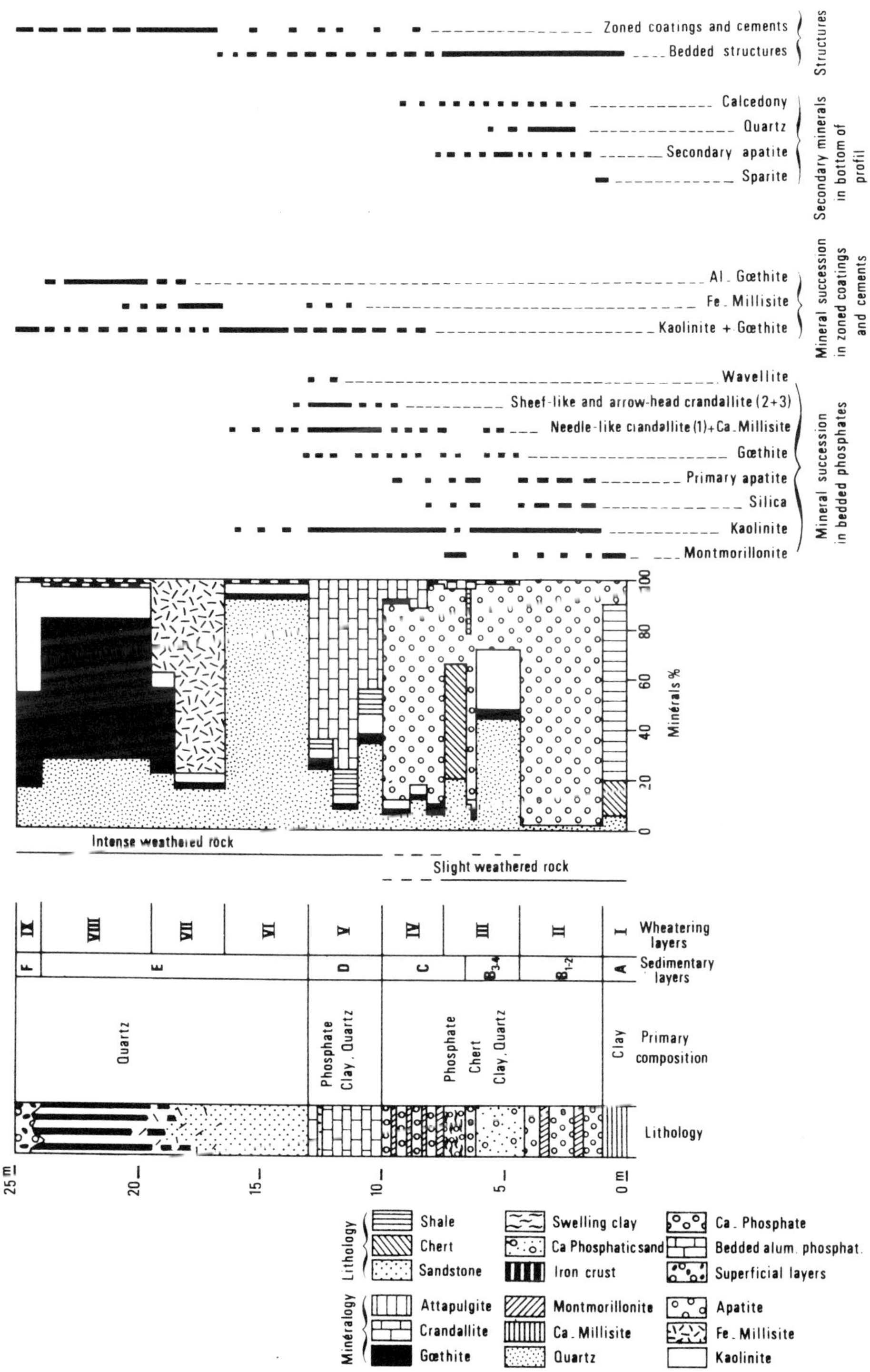

Fig. 5. Mineral succession in the two first stages of aluminous phosphate genesis in Western Senegal

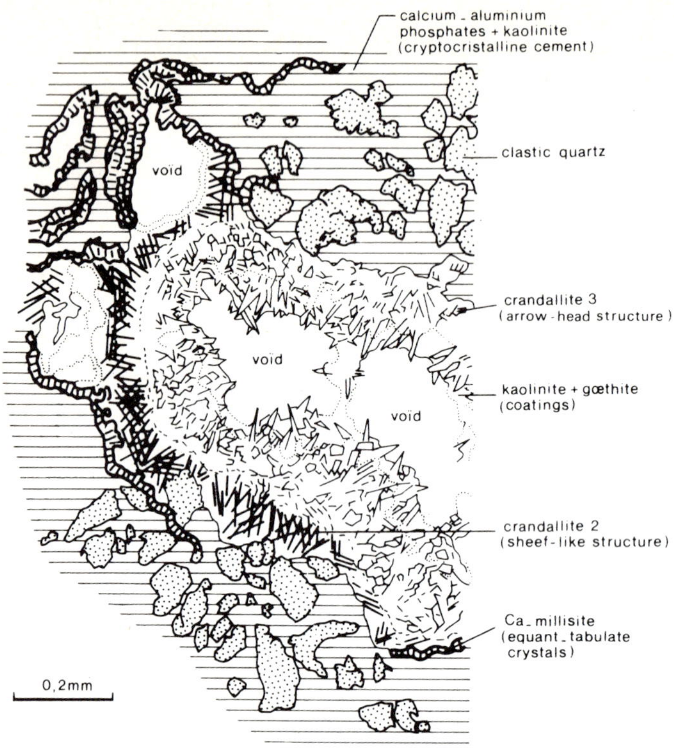

Fig. 6. Aluminum-calcium phosphate sequence on a void-edge of bedded alumino-phosphates

crandallite results always from sheaf-like crandallite by dissolution and recrystallization.

Likewise, the chemical analyses of pure mineral phases (Fig. 7 and 8); Tables 3 und 4) and the computation of lattice constants values (Fig. 9, Table 4) allow us to distinguish the Ca-millisite from the different kinds of crandallite. Ca-millisite is characterized by Al, Ca, Fe, Na. The Fe and Na are depleted relative to their content in Fe-millisite of the second stage, but Al and Ca are enriched. Ca-millisite has thus a Ca content intermediate between Fe-millisite and crandallites. Among the former, the arrow-head crandallite is less rich in Sr than the needle-like crandallite and also less than the sheaf-like crandallite. Moreover it appears that decrease in Sr content comes with decrease in the lattice constant [c].

The phosphate mineral sequence:

apatite →Ca-millisite→[Ca-millisite + crandallite 1]→crandallite 1
→crandallite 2→crandallite 3→wavellite,

in which the aluminous members appear during the first stage of intense weathering, shows, therefore, a genetic relationship. It is reflected by mineral form and composition. The sequence shows that aluminum concentrates, relative to the environment, by the leaching away of alkalis. This concentration is progressive, with the transitory formation of aluminum-calcium phosphates, they are schemati-

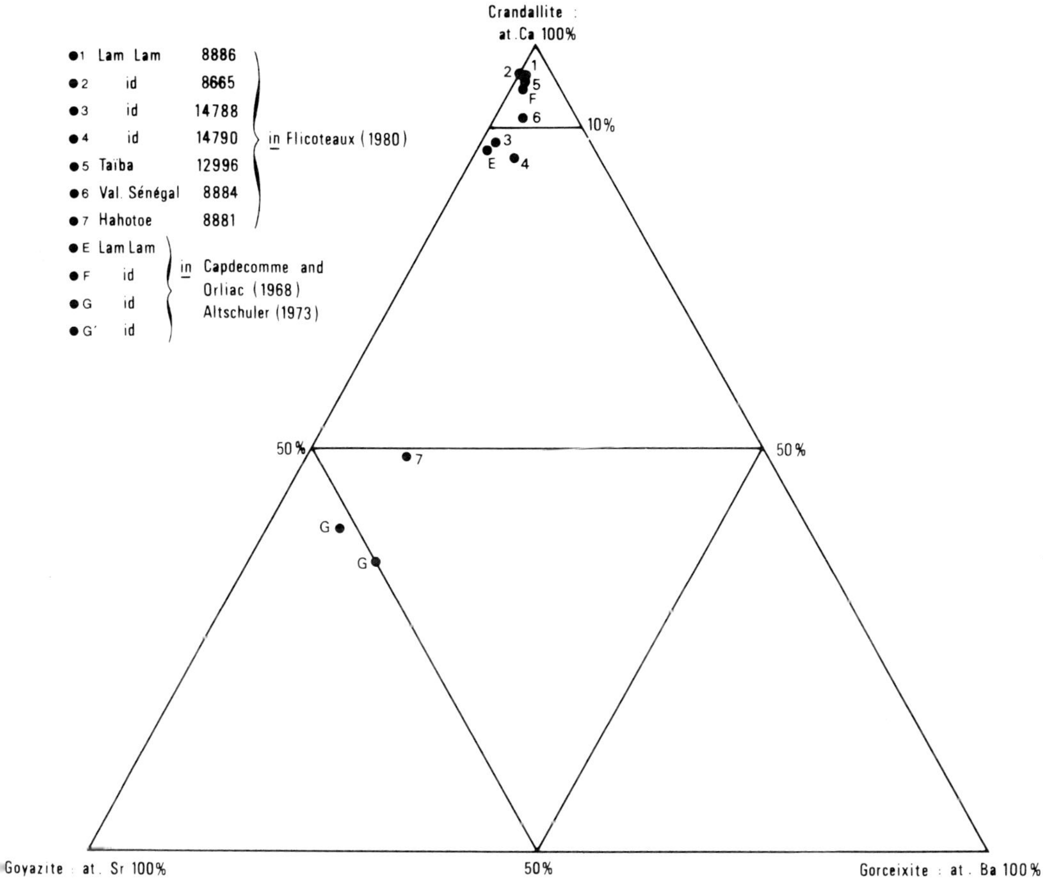

Fig. 7. Analysis of millisites from Senegal. Comparison with Florida and Utah

cally richer in Fe and Na in argillaceous beds and richer in Sr in calcium phosphate beds. Subsequently these elements are removed, Na seemingly more mobile than Sr, itself more mobile than Ca.

Fe-Millisite. Genesis and Transformation. Kaolinite, then Fe-millisite and finally Al-goethite succeed one another in the second stage and join minerals that resulted from the first stage (Fig. 10). The process takes place in two phases. During the first phase there is accumulation in cavities, also derived from the previous stage and epigenesis of kaolinite in Fe-millisite. The epigenesis also alters the minerals formed in the first stage, while they evolve in a retrograde manner. During the second phase there is transformation of Fe-millisite into Al-goethite.

1. Accumulation and Epigenesis of Kaolinite in Fe-millisite. The accumulation of kaolinite occurs as zoned coatings either by precipitation or by mechanical deposition. Those coatings line the voids of rocks; they can show a secondary red col-

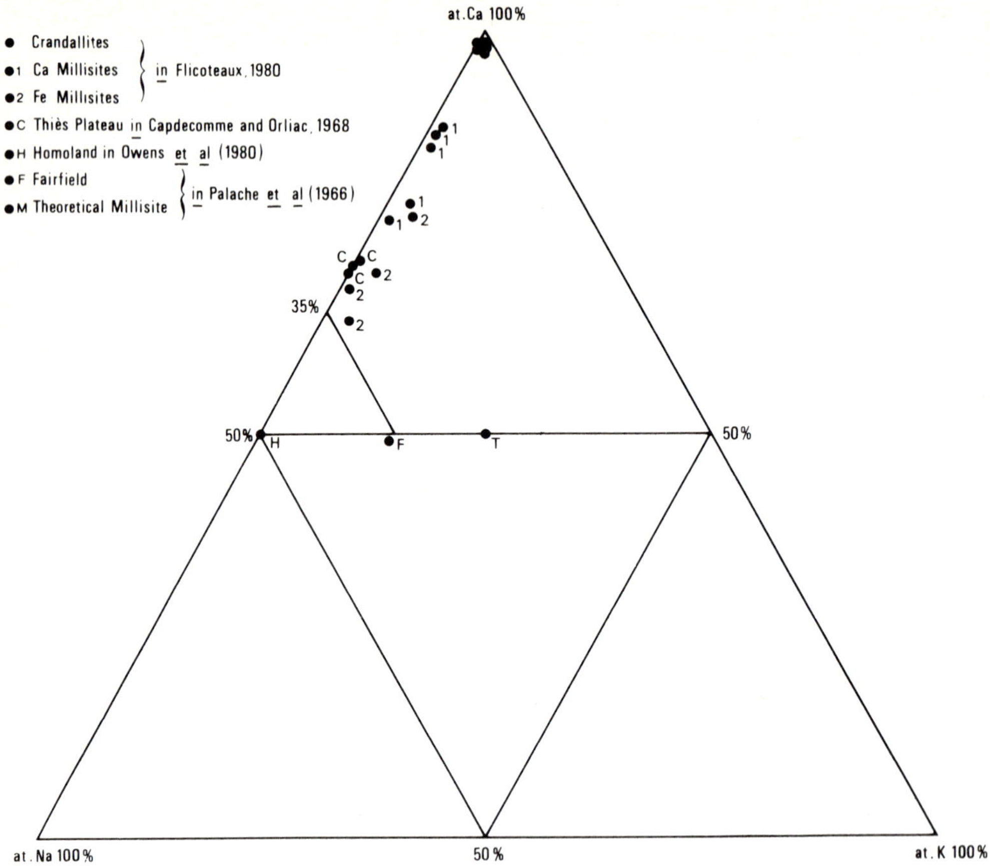

Fig. 8. Analysis of crandallites from Senegal, Mauretania and Togo

Table 3. Middle Composition of some Ca and Fe-Millisites from Western Senegal. (Cf. Flicoteaux, 1980, p. 31)

Mineral		P_2O_5	Al_2O_3	Fe_2O_3	L.O.I	CaO	Na_2O	K_2O
	5							
Ca-Millisite	$\bar{x}$ %	32.19	31.30	2.38	14.74	9.52	0.98	0.16
(n = 5)	6 %	1.92	1.26	0.58	0.65	0.62	0.33	0.07
Fe-Millisite	$\bar{x}$ %	32.29	27.90	10.95	15.63	8.08	1.79	0.21
(n = 4)	6 %	1.33	2.41	3.75	0.25	0.71	0.40	0.11

oration due to goethite. The kaolinic zoned coatings show lateral and vertical gradation to zoned coatings of millisite rich in iron and sodium. The transition is progressive from the unaltered kaolinite coating through a kaolinitic coating turning to millisite and the millisite coating with small traces of kaolinite. Moreover, the millisite coating preserves a zoned aspect and, compared to the kaolinitic coating, is thicker in the lower part of voids. Finally, it is cryptocrystalline. According

Table 4. Some characteristics of the crandallites from weathered Eocene deposits in West Africa compared with some typical minerals

Origin of crandallite	$\frac{Al_2O_3}{P_2O_5}$	$\frac{CaO}{P_2O_5}$	$\frac{SrO}{P_2O_5}$	$\frac{SrO}{CaO}$	SrO %	a (Å)	c (Å)	c/a	Ref.
1 Amberg, Bavaria	0.936	0.560				6.960	16.086	2.311	Van Tassel (1956)
2 Blaton, Belgium	1.090	0.399				7.040	16.200–260	2.301–2.310	
3 Homeland, Florida						6.980	16.200	2.321	Owens et al. (1960)
4 Fairfield, Utah						7.004	16.200	2.313	
5 Gore, Virginia					1–3.5	7.000	16.110	2.301	Mitchell and Knowlton (1971)
6 Alachua County, Florida	1.230	0.242	0.207	0.853	4.51	7.017	16.252	2.316	Blanchard (1972)
7 Fairfield, Utah						7.013	16.196	2.309	
8 Walpole Island	1.101	0.306	0.144	0.468	4.08	7.000	16.200	2.314	Tercinier (1972)
9 Fairfield, Utah	1.286	0.640	0.006	0.009	0.08	7.005	16.192	2.311	Blount (1974)
10 Synthetic crândallite						6.989	16.159	2.312	Slade (1974)
11 Caune de l'Arago, France	1.179	0.506				7.013	16.179	2.307	Peneaud (1978)
12 Taïba (12,996), Senegal	1.097	0.398	0.029	0.069	0.91	7.000	16.158	2.308	Flicoteaux (1980)
LamLam (8,665), Senegal	1.229	0.386	0.02	0.056	0.51	7.000	16.152	2.307	
LamLam (14,790), Senegal	0.994	0.331	0.068	0.207	2.15	7.004	16.176	2.310	
Kaedi (8,884), Mauretania.	1.[illegible]84	0.320	0.045	0.142	1.11	7.016	16.260	2.318	
Hahotoe (8,881), Togo	1.278	0.180	0.270	1.500	6.60	7.012	16.320	2.325	
13 Paraiba, Brazil						7.000	16.140	2.306	ASTM S–0615

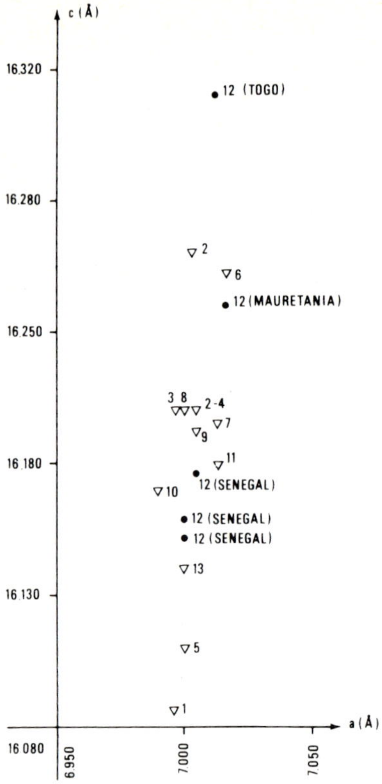

Fig. 9. Lattice constants of crandallites from Senegal, Mauretania and Togo. Comparison with published values. See Table 3 for references

to the observational data, the Fe-millisite coatings can be interpreted as the result of an epigenesis of kaolinite.

The internal mechanisms of epigenesis can be inferred from laboratory experiments on reactions between clays and phosphates. Two types of reactions are proposed. The first type is an adsorption of phosphorus on or in clay minerals; the second one, slower and more continuous, is a chemical etching of the clay lattice with precipitation of aluminum phosphates. Those two reactions are linked in time.

The adsorption of phosphorus is a complex and debated mechanism. The isomorphous replacement of silica tetrahedra by phosphoric acid tetrahedra proposed by Low and Black (1950) and by Hendrickx (1954), is rejected after the experiments of Baldwin and Tyener (1954), in Kittrick and Jackson (1956). The adsorption would be made at the level of hydroxides (Kittrick and Jackson 1956), or aluminum (Wey 1953, 1956). Muljadi et al. (1966a–c) propose, in addition, more complex penetrations into amorphous zones of the crystal surface (occlusions).

The etching of the clay itself is not well known yet. It might as well include diffusion of phosphorus within the lattice (Haseman et al. 1951) and release of aluminum in solution. It is what Low and Black (1948) called the "phosphatolysis". It leads to precipitation of a phosphate whose type varies with the nature and the concentration of the solutions used and to dissolution of the silica of clay minerals.

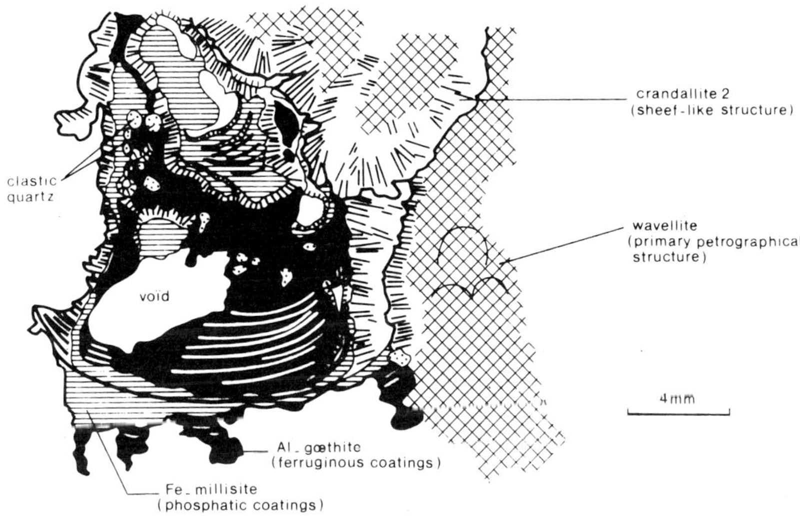

Fig. 10. Millisite coating superimposed in a cavity of bedded aluminophosphates

All this results in either an incongruent dissolution of clays and a precipitation of aluminous phosphates, or an incongruent reaction in the solid state. This also explains not only the kaolinite epigenesis in Fe-millisite, but also that of sedimentary clays (attapulgite, montmorillonite, illite) in Ca-millisite. The epigenesis corresponds to a relative accumulation of aluminum by leaching of silica and to an absolute accumulation of phosphorus, iron, calcium, and sodium. Compared with the first stage, this involves a modification in the nature or composition of the removing solutions. If the conditions that controlled the first stage reappear, Fe-millisite might evolve, like Ca-millisite, towards crandallites and/or wavellite.

2. Transformation of Fe-millisite into Al-goethite. The zoned coatings of Fe-millisite and the cement which derives from them by successive incorporations and homogeneizations, are in their turn replaced laterally and vertically by coatings and opaque cement of goethite. Here, the replacement is still progressive. Goethite formed in this way is aluminous. In Lam Lam and on the Thiès plateau, the calculated substitution amounts vary from between 12% and 16% to between 20% and 24% mol Al. This transformation is interpreted as the result of incongruent dissolution of phosphate and precipitation of a solid solution between ferruginous and aluminous oxi-hydroxide forms, or as the result of incongruent reaction in the solid state. It leads to a relative accumulation of iron and in part of aluminum and a release of phosphorus and alkalis. All this implies a modification of composition of the groundwaters.

Distribution and Behavior of Some Elements in the Two First Parageneses. The chemical analysis of samples, selected in function of their relative purity, constitutes the first way of characterizing millisites and crandallites, by the content of the most important elements. However, this concerns only a few samples. A second way of characterizing millisites and crandallites is by analysing large numbers of unselected samples, from alternances of clays and calcium phosphates up to bedded aluminous phosphates and phosphatic crusts. Then, the results of chemical and mineral analysis are treated to obtain the correlation matrix, correlation diagrams and groups of positive intercorrelations of elements and minerals, used by Besnus (1975), Besnus and Mosser (1976). This method gives the probabilities of distribution of various elements in the different minerals.

The second method has been applied to the Northern Lam Lam and Tobene deposits. In Lam Lam, sections going from clay and decalcified phosphatic facies to bedded aluminous phosphatic facies are studied. All those facies, of which the average composition is given in Table 5, result from moderate to intense leaching by groundwaters. At Tobene, sections identical with the precedent ones and accumulation horizons of Fe-millisite in the sandy cover, are studied. The accumulation horizons can be explained by the morphological and structural position of Tobene, presently located down dip of Lam Lam and the Thiès plateau (Fig. 4). Their presence above the leached horizons implies either a modification of topographic position of Tobene or a modification of drainage or groundwater composition during geological times.

The distribution of some elements between the various mineral phases of Lam Lam and Tobene is shown in Table 6. We see that in Tobene no element shows a positive correlation with the weak amounts of kaolinite, nor with the traces of goethite (Table 5).

The major elements correlate identically with the four other phases in both deposits, except Si, Ti, and Fe. P and Ca are strongly concentrated in apatite, Mg and K in montmorillonites, Al, L.O.I. (Pf) and Na in aluminum phosphates and Si in quartz. Some pertinent remarks are:

Si does not show positive correlation with the montmorillonites in Tobene. This may result from variations in the quartz and montmorillonite contents in the facies (Table 5). The affinity of Si with quartz is in proportion with the quartz content; the opposite is true for montmorillonites.

Fe shows a positive correlation with goethite in Lam Lam und thus is clearly located outside of the clays. It is divided between clays and aluminum phosphates in Tobene, where goethite is not truly expressed (Table 5).

Ti exists in kaolinite and is associated with quartz in Lam Lam; it is linked with aluminum phosphates in Tobene.

The suite of trace elements of the montmorillonites of Lam Lam and Tobene are also very similar, except Ga. We record, in both cases, an association of elements, V, Zn, Ni, Cu, and Cr, which are also correlated with kaolinite in Lam Lam and partially with apatite in Tobene. It is known that such elements have a common affinity for the apatite of sediments and phosphate deposits, but less than that for clays of the same deposits. Like Sr, they can be lost to the apatite lattice during slight weathering. Thus, this might explain the absence of correlations with apatite in Lam Lam, where the original facies are more decalcified and more decarbonated than in Tobene and thus more leached. As for the kaolinite, it derives, with goethite, from the leaching of montmorillonites, from which they trap some elements and from a beginning of secondary accumulation in voids. In the same way, Ca-millisite could trap Cd freed together from apatite and montmorillonites.

Table 5. Average chemical and mineralogical composition of weathered phosphate facies from LamLam and Tobène (Western Senegal)

	LamLam, average analysis of 51 samples	Tobène, average analysis of 80 samples
Apatite	20.62%	29.41%
Montmorillonites	6.76	14.97
Quartz	15.26	40.09
Kaolinite	12.49	2.64
Goethite	4.67	–
Crandallite	27.16	6.51
Millisite 1	11.32	–
Millisite 2	–	7.49
SiO_2	25.00%	50.15%
Al_2O_3	19.60	7.09
MgO	0.37	0.57
CaO	15.42	17.88
Fe_2O_3	6.95	3.02
Mn_3O_4	0.11	–
TiO_2	1.02	0.41
Na_2O	0.89	0.14
K_2O	0.22	0.23
L.O.I.	9,81	4.00
P_2O_5	20.87	15.18
Sr	2,119 p.p.m.	1,744 p.p.m.
Ba	956	237
V	576	448
Ni	205	96
Co	27	–
Cr	661	400
B	31	–
Zn	401	326
Ga	23	12
Cu	88	54
Pb	25	36
Sn	11	–
Cd	36	29
CaO/P_2O_5		1.07
Al_2O_3/P_2O_5		0.76

Other good correlations are observed with aluminum calcium phosphates of both deposits. In Lam Lam, Sr, Ba, and Pb show a close relationship with crandallite resulting from successive dissolutions and recrystallizations. The association Sr, Ba, Pb is made of elements which mutually replace one another during isomorphous substitutions that affect the Ca of crandallites. In Tobene, the crandallites are subordinate to the accumulation Fe-millisites (Table 5). This paragenesis always owns a close relationship with Sr and Ba and also with Ga, Cr, and Cu. The correlation with Pb has disappeared and again, the effect of abundance of quartz must be marked. Sr, which characterizes more particularly the crandallites, is the only element of the group that is not directly linked to Fe. Cu, intercorrelated with Cr and Fe to Al, can characterize the substitutions with Al; some turquoise has

Table 6. Occurrence of Elements in the different mineralogical phases in LamLam and Tobène

LamLam	Calcium							
(samples=51)	phosphate	P Ca						APA
	Aluminous					PbBaSr		CRA
	phosphates		AlPf			PbBa		CRA
			AlPf		Ga	Pb		
			AlPfNa					MiL
			Na				Cd	MiL
	Montmorillonites			MgK	VZn Cu			
				MgKSi	VZn			
				MgKSi	V			SME
				KSi	VZnGa Ni			
				KSi	VZnGa Cr			
				K	VZnGaCuNi			
				K	VZnGaCu			
	Kaolinite			SiTi	Ga			KAO
				SiTi	VZnGa Ni			
				SiTi	VZnGa Cr			
				Ti	VZnGaCuNi			
				Ti	VZnGaCu Cr			
				Ti	GaCuNi	Pb		
	Quartz			SiTi				QUA
	Goethite			Fe	Ni			GOE
Tobéne	Calcium	P			Cu		Cd	
(samples=80)	phosphate	Ca			Zn Ni		Cd	SME
	+Montmorillonites	P Ca			Zn		Cd	
		P Ca					Cd	PCA
	Aluminous		AlPfNa	TiFe	Ga	Ba		CRA–M
	phosphates		AlPfNa	Ti	Ga	BaSr		CRA–M
					Ga Cr	Sr		
					Ga Cr	Ba		
			Al	TiFe	Ga	Ba		CRA–M
			AlPf	TiFe	Ga Cr			
			Al	Fe	Cu Cr			
				FeVZn	CuNi			
	Montmorillonites				Zn Ni			SME
					VZn CuNiCr		Cd	SME
			Pf	K	Cu Cr			
				MgK	VZn Ni			SME
				MgK		Pb		
				Si		Pb		
				Si				QUA

been identified in traces. Finally, we note that the group of intercorrelated elements Ti, Ga, Cr, present in the Fe-millisites, is also found in the kaolinite of Lam Lam.

From this study we can conclude that the suite of major and trace elements can be used to characterize the weathering minerals of the argillaceous and phosphatic alternances and of their sandy cover: kaolinite, goethite and aluminous phosphate. Seemingly characteristic are the associations:

- Al – L.O.I. (Pf) – Na – Cd for the Ca-millisites;
- Al – L.O.I. (Pf) – Sr – Ba – Pb for the crandallites, particularly the most evolved,

- Al – L.O.I. (Pf) – Na – Ti – Fe – Ga – Ba – Cr – Sr – for the Fe-millisite and the subordinate crandallites (Table 5).

In addition, some elements linked to aluminous phosphates are also included in sedimentary clay minerals and in kaolinite and goethite resulting from leaching of these clays. That is particulary true in the case of Ti, Fe, Ga, Cr which are intercorrelated in the Fe-millisite. It is also seen that Fe-millisite has formed by epigenesis of kaolinite associated with goethite. Thus, the distribution of elements in aluminous phosphates depends on the weathering history and in particular on the mechanisms that control the various stages.

Crandallite, Wavellite and Millisite Succession and Superimposition Other Examples

Minerals of composition and form very close to the aluminous phosphates of Western Senegal have been described many times, in other examples of deeply weathered phosphorites. In the Ivory Coast (Parron and Nahon 1980), in the Senegal valley (Flicoteaux et al. 1979) and in Togo (under study), the phosphatic beds and nodules, interstratified within argillaceous beds, are transformed into crandallite and wavellite. In the field the original volumes and structures of the parent rocks are preserved. The chemical and optical analyses show that crandallite has sheaf-like forms, rich in strontium and acicular forms. We know from the studies of Altschuler and his colleagues of the U.S. Geological Survey (1956, 1958, 1965, 1973), that the phosphatic nodules, dispersed in an argillaceous and sandy matrix, are transformed into millisite, crandallite and wavellite. According to published documents, it seems that the crandallites have mainly radiant, sheaf-like forms and the millisites are characterized by an X-ray diffraction diagram and a morphology identical to those of the Fe-millisite of accumulation.

General Conclusions

In all the studied examples, the isovolumic relationships with successive development of Ca-millisite or crandallite and then of wavellite are independent of topography. They appear due to leaching by permanent groundwater, continuously replenished by the rainfall of a wet, tropical climate. On the contrary, the millisite-coatings and the zoned aluminocalcic phosphate cement occur close to a topographic surface. They must result from a superficial weathering which accumulates elements in suspensions or particles in a non-permanent groundwater. This implies an evolution towards a drier climate, which concentrated the groundwaters.

Not only the alumino-phosphate successions, but also the relationships of forms, dimensions and compositions of the various phosphates, enable the geochemical environment and the mechanisms associated with the different stages of weathering to be characterized. They constitute an equal number of criteria to follow its history during geological times.

References

Altschuler ZS (1965) Precipitation and recycling of phosphate in the Florida Land-Pebble phosphate deposits. US Geol Surv Prof Paper 525-B: 91–95

Altschuler ZS (1973) The weathering of phosphate deposits. Geochemical and environmental aspects. In: Griffith EJ, Beeton A, Spencer JM, Mitchel DT (eds) Environmental Phosphorus handbook, Wiley, New York, p 33–96

Altschuler ZS, Clarke RS, Young EG (1958) Geochemistry of uranium in apatite and phosphorite. US Geol Surv Prof Paper 314-D:45–90

Altschuler ZS, Jaffe ER, Cuttitta F (1956) The aluminum phosphate zone of the Bone Valley Formation, Florida, and its uranium deposits. US Geol Surv Prof Paper 300:495–504

Axelrod JM, Carron MK, Milton C, Thayer JP (1952) Phosphate mineralization at Bomi Hills and Bambuta, Liberia. Am Miner 37:883–909

Besnus Y (1975) Etude géochimique comparative de quelques gisements supergènes de fer. Thèse Sci Universitè Strasbourg, p 184

Besnus Y, Mosser C (1976) Les groupes à intercorrélations positives; une méthode statistique pour déterminer la répartition des éléments chimiques entre les phases constitutives d'une roche. CR Acad Sci, Paris, D 283:587–590

Billaud Y (to be published) Les paragnèses phosphatée du paléokarst des phosphorites du Quercy: Minéralogie et évolution. 9th RAST, Paris, 17–19 March

Birch GF (1979) Phosphorite pellets and rock from the western continental margin and adjacent costal terrace of South Africa. Marine Geol 33:91–116

Blanchard FN (1972) Physical and chemical data for crandallite from Alachua County, Florida, Am Miner 57:473–484

Blanchard FN, Denahan SA (1968) Cacoxenite and beraunite from Florida. Am Miner 53:2096–2101

Blount AM (1974) The crystal structure of crandallite. Am Miner 59:41–47

Buchinskii GI (1969) Old phosphorites of Asia and their genesis. PST, Jerusalem, p 266

Capdecomme L (1952) Sur les phosphates alumineux de la région de Thiès (Sénégal). CR Acad Sci, Paris, 235:187–189

Capdecomme L (1953) Etude minéralogique des gîtes phosphatés de la région de Thiès (Sénégal). 19th Inter Geol Congr, Algiers, 1952. CR, Sect 11, fasc 11:103–117

Capdecomme L, Orliac M (1968) Sur les caractères chimiques et thermiques des phosphates alumineux de la région de Thiès (Sénégal). Colloq Int Phosphates Miner Solides, Toulouse, 167. 2:45–55

Chaabani F (1978) Les phosphorites de la coupe-type de Foum Selja (Metlaoui, Tunisie). Une série sédimentaire séquentielle à évaporites du Paléogène. Thèse 3ème cycle, Univ L Pasteur, Strasbourg, p 113

Charpy DN, Nahon D (1978) Contribution à l'étude lithostratigraphique et chronostratigraphique du Tertiaire du Bassin de Côte d'Ivoire. Trav Lab Sci Terre, Univ Abidjan, 18, Multigr, p 33

Cowgill UM, Hutchinson GE, Joensuu O (1963) An apparently triclinic dimorph of crandallite from a tropical swamp sediment in El-Petèn, Guatemala. Am Miner 48:1144–1153

De Weisse G (1970) Bauxite sur un atoll du Pacifique. Mineralium Deposita 5:181–183

Du Toit AL (1917) The phosphate of Saldanha Bay. South Afr Geol Surv Memoir 10:34

El Mountassir M (1977) La zone rubéfiée de Sidi Daoui, altération météorique du phosphate de chaux des Ouled Abdoun (Maroc). Thèse de 3ème cycle, Univ L Pasteur, Strasbourg, p 126

Espenhade GH, Spencer C (1963) Geology of phosphate deposits of Northern Peninsular Florida. Bull US Geol Surv 1118:115

Figueiredo Gomes C de S (1968) Mineralogia da parte fosfatada de fosforitos de Cabinda (Angola). Serv Geol Min Angola Luanda 18:17–24

Fisher DJ (1973) Geochemistry of minerals containing phosphorus. In: Griffith EJ, Beeton A, Spencer JM, Mitchell DT, Environmental Phosphorus Handbook, Wiley, New York, p 1412–152

Flicoteaux R (1980) Genèse des phosphates alumineux du Sénégal occidental. Etapes et guides de l'altération. Thèse Sci Univ Aix-Marseille III, France, Multigr, p 229 and Mem Sci Geol 67:229

Flicoteaux R, Nahon D, Paquet H (1977) Genèse des phosphates alumineux à partir des sédiments argilo-phosphatés du Tertiaire de Lam Lam (Sénégal) Suite minéralogique. Permanence et changements de structures. Sci Géol Bull 30:153–174

Flicoteaux R, Lappartient JR, Parron C, Popoff M (1979) Les formations tertiaires du fleuve Sénégal. Influence des phénomènes diagénétiques et d'alteration sur les caractères lithostratigraphiques, 10e Coll Geol Afr Montpellier, 25–27 April 1979:153–154

Frankel JJ (1943) The constitution of the Langebaan phosphate rock. South Afr J Sci 39:102–108
Frondel C (1958) Geochemical scavenging of strontium. Science 128:1623–1624
Gastuche MC, Fripiat JJ, Sokolski J (1963) Fixation du phosphore par les hydroxydes de fer et d'aluminium amorphes et cristallisés. Pédol Gand 13:155–180
Gruner JW, McConnell D (1937) The problem of the carbonate-apatites. The structure of Francolite. Z Kristallogr 97:208–215
Gulbrandsen RA (1970) Relation of carbon dioxide content of apatite of Phosphoria formation to regional facies. US Geol Surv Prof Pap 700-B:9–13
Gusev GM, Zanin YN, Krivolutskaya LM, Lemina NM, Yusupov TS (1976) Transformation of apatite during weathering and leaching. Dokl Akad Nauk SSSR 229:971–973
Jarvis I (1980) Geochemistry of phosphatic chalks and hardgrounds from the Santonian to early Campanian (Cretaceous) of Northern France. J Geol Soc (Lond) 137:705–721
Johan Z (1976) La sénégalite, $Al_2(PO_4)(OH)_3 \cdot H_2O$, un nouveau minéral. Lithos 9:165–171
Haseman JF, Brown EH, Whitt DC (1951) Some reactions of phosphate with clays and hydrous oxides of iron and aluminium. Soil Sci 70:257–271
Hendrickx SB (1954) Base exchange of crystalline silicates. Ind Eng Chem 37:625–630
Kittrick JA, Jackson ML (1955a) Rate of phosphate reactions with soil minerals and electron microscope observations on the reaction mechanism. Soil Sci Soc Am Proc 19:292–295
Kittrick JA, Jackson ML (1955b) Application of solubility products to the variscite – kaolinite system. Soil Sci Soc Am Proc 19:455–457
Kittrick JA, Jackson ML (1956) Electron microscope observations of the reaction of phosphate with minerals, leading to a unified theory of phosphate fixation in soils. J Soil Sci 7:81–88
Larsen ES, Shannon EV (1930) The minerals of the phosphate nodules from near Fairfield, Utah. Am Miner 15:307–337
Low PF, Black CA (1948) Phosphate induced decomposition of kaolinite. Soil Sci Soc Am Proc 12:180–184
Low PF, Black CA (1950) Reactions of phosphate with kaolinite. Soil Sci 70:273–290
Lucas J, Prévôt L (1975) Les marges continentals pièges géochimques; l'exemple de la marge atlantique de l'Afrique à la limite Crétacé-Tertiaire. Bull Soc Géol Fr 496–501
Lucas J, Chaabani F, Prévôt L (1979a) Phosphorites et évaporites: Deux formations de milieux sédimentaires voisins étudiés dans la coupe du Paléogène de Foum Selja (Metlaoui, Tunisie). Sci Géol Bull 32:7–19
Lucas J, Menor E, Prévôt L (1979b) Le gisement de phosphate de chaux de Taïba (Sénégal). Un exemple d'enrichissement par altération. Sci Géol Bul 32:35–37
Lucas J, Prévôt L, El Mountassir M (1979c) Les phosphorites rubéfiées de Sidi Daoui, transformation météorique locale du gisement de phosphate des Ouled Abdoun (Maroc). Sci Géol Bull 32:21–37
Lucas J, Prévôt L, Ataman G, Gündogdu N (1979d) Etude minéralogique et géochimique de la série phosphatée du Sud-Est de la Turquie (Mazidagi-Mardin). Sci Géol Bull 32:59–68
Lucas J, Flicoteaux R, Nathan Y, Prévôt L, Shahar Y (1980a) Different aspects of phosphorite weathering. SEPM Spec Pub 29:41–51
Lucas J, Prévôt L, Ataman G, Gündogdu N (1980b) Mineralogical and geochemical studies of the phosphatic formations in Southeastern Turkey (Mazidagi-Mardin). SEPM Spec Pub 29:149–152
Lucas J, Prévôt L, Trompette R (1980c) Petrology, mineralogy and geochemistry of the late Precambrian phosphate deposits of Upper Volta (W. Africa). J Geol Soc (Lond) 137:787–792
Lucas J, Prévôt L (1981) Caractères pétrographiques et microchimiques de phosphorites précambriennes du bassin des Volta (Afrique de l'Ouest). Considérations génétiques. Bull Soc Géol Fr (7), 23:515–524
McArthur JM (1978) Systematic variation in the contents of Na, Sr, CO_2 and SO_4 in marine carbonate-fluorapatite and their relationship to weathering. Chem Geol 21:89–112
McArthur JM (1980) Postdepositional alteration of the carbonate-fluorapatite phase of Moroccan phosphates. SEPM Sp Pub 29:53–60
McConnell D (1938) A structural investigation of the isomorphism of the apatite group. Am Miner 23:1–19
McKie D (1962) Goyazite and florencite from two African carbonatites. Miner Mag 33:281–297
Menor E (1975) La sédimentation phosphatée, pétrographie, minéralogie et géochimie des gisements de Taïba (Sénégal) et d'Olinda (Brésil). Thèse Doc Ing Univ L Pasteur, Strasbourg, p 153
Milton C, Axelrod JM, Carron MK, Mac Neil FS (1958) Gorceixite from Daleco, Alabama. Am Miner 43:688–694

Mitchell RS, Knowlton SM (1971) Crandallite from Gore, Frederick County, Virginia. Miner Rec, p 223–224

Muljadi D, Posner AM, Quik JP (1966a) The mechanism of phosphate adsorption by kaolinite, gibbsite and pseudo-boehmite. Part. I. The isotherm and the effect of Ph on adsorption. J Soil Sci 17:212–229

Muljadi D, Posner AM, Quik JP (1966b) The mechanism of phosphate adsorption by kaolinite, gibbsite and pseudo-boehmite. Part. II. The location of the adsorption sites. J Soil Sci 17:230–237

Muljadi D, Posner AM, Quik JP (1966c) The mechanism of phosphate adsorption by kaolinite, gibbsite and pseudo-boehmite. Part. III. The effect of temperature and adsorption. J Soil Sci 17:238–247

Owens JM, Altschuler ZS, Berman R (1960) Millisite in phosphorite from Homeland, Florida. Am Miner 45:547–561

Paquet H (1969) Evolution géochimique des minéraux argileux dans les altérations et les sols des climats méditerranéens et tropicaux à saisons contrastées. Thèse Sci Strasbourg Mém Serv Carte Géol Als Lorr 30:212

Parker RJ, Siesser WG (1972) Petrology and origin of some phosphorites from the South African continental margin. J Sed Petr 42:423–440

Parron C, Nahon D (1980) Red bed genesis by laterite weathering of glauconitic sediments. J Geol Soc (Lond) 137:689–693

Péneaud D (1978) La paragenèse phosphatée de la grotte de la Caune de l'Arago (Pyrénées Orientales). Thèse 3ème cycle, Univ Paris VII, p 162

Poullen JF (1982) Etude de la séquence d'oxydation de la vivianite. Thèse Sci Univ Paris VI, France, Multigr, p 215

Prévôt L, Lucas J, Nathan Y, Shiloni Y (1979) Répartition des éléments traces dans les phosphorites marines. Sec Symp Orig distrib elements, UNESCO, Paris, 1977, Pergamon, Oxford, New York, p 293–304

Riedel L (1959) Contributo alla conoscenza della serie phosphatica di Holle (Republica del Congo). Boll Soc Adria Sci Nat Trieste 2:69–83

Russ W, Andrew CW (1924) The phosphate deposits of Abeokuta Province. Nigeria Geol Surv Bull 7:9–38

Russel RT, Trueman NA (1971) The geology of the Duchess phosphate deposits, Northwestern Queensland, Australia. Econ Geol 66:1188–1814

Sandell E, Hey M, McConnell D (1939) The composition of francolite. Miner Mag 25:395–401

Sassi S (1974) La sédimentation phosphatée au Paléocène dans le Sud et le Centre-Ouest de la Tunisie. Thèse Sci Paris-Sud Multigr, p 292

Segalen P (1973) L'aluminium dans les sols. Init Doc Techn ORSTOM 22:281

Sellards EH (1913) Origin of the hard rock phosphate of Florida. Ann Rep Fl Geol Surv 5:27–80

Slade PG (1974) The synthesis of crandallite $CaAl_3H(OH)_6/PO_4)_2$ N Jahrb Miner Mh Stuttgart H1:22–27

Slansky M (1974) Géologie des phosphates sédimentaires. Bur Rech Géol Min Mém 114:92

Slansky M, Lallemand A, Millot G (1964) La sédimentation et l'altération latéritique des formations phosphatées du gisement de Taïba (République du Sénégal). Bull Serv Carte Géol Als Lorr 17:311–324

Slansky M (1980) Géologie des phosphates sédimentaires. Mém BRGM 114:1–92

Sweatman TR (1961) Gorceixite, a naturally occuring soil phosphate mineral. Aust J Sci 24:242–243

Tankard AJ (1974a) Petrology and origin of the phosphorite and aluminium phosphate rock of the Langebaanweg – Saldanha area, Southwestern Cape Province. Ann S Afr Mus 65:217–249

Tankard AJ (1974b) Chemical composition of the phosphorites from the Langebaanweg-Saldanha area, Cape Province. Trans Geol Soc S Afr 77:185–190

Taylor AW, Gurney EL (1965) Precipitation of phosphate by iron oxide and aluminium hydroxide from solution containing calcium and potassium. Soil Sci Soc Am Proc 29:18–22

Taylor AW, Gurney EL, Moreno EC (1964) Precipitation of phosphate from calcium phosphate solution by iron oxide and aluminium hydroxide. Soil Sci Soc Am Proc 28:49–52

Tercinier G (1972) La crandallite, phosphate naturel répandu dans les sols et souvent abondant dans les produits de remplissage des karst. CR Acad Sci Paris 274:1445–1448

Tessier F (1950) Contribution à la stratigraphie et à la paléontologie de la partie ouest du Sénégal (Crétacé et Tertiaire), lère partie: Historique et Stratigraphie. Thèse Sci Marseille et Bull Dir Fed Min Géol AOF 14, p 267

Tessier F, Flicoteaux R, Lappartient J (1976) Etude géologique de la carrière de Lam Lam et de ses abords méridionaux (République du Sénégal). Trav Lab Sci Terre St Jérôme, Marseille, A, 9:122

Trueman NA (1965) The phosphate, volcanic and carbonate rocks of Christmas Island, Indian Ocean. J Geol Soc Aust 12:261–284

Van Tassel R (1956) Découverte de crandallite en Belgique. Bull Inst Sci Nat Belg 32, 33:1–10

Visse LD (1952) Pseudowavellite et millisite dans les minerais phosphatés dits latéritoïdes blancs de la région de Thiès (Sénégal). CR Acad Sci Paris 234:1377–1378

Wadjinny A (1979) Milieu de sédimentation et mécanisme de dépôt des couches inférieures de la série phosphatée de Ben Guerir (Ganntour, Maroc); une étude séquentielle. Thèse 3ème cycle, Univ L Pasteur, Strasbourg

Wey R (1953) Sur l'adsorption de l'anion phosphorique par la montmorillontie. CR Acad Sci Paris 236:1298–1300

Wey R (1955) Etude de la rétention des anions phosphoriques par les argiles montmorillonite et kaolinite. Thèse Sci Strasbourg Ann Agron 1:1–62

Young EJ, Myers AT, Munson EL, Conklin NM (1969) Mineralogy and geochemistry of fluorapatite from Cerro de Mercado, Durango, Mexico. US Geol Surv Prof Pap 680-D;89–93

Zanin YN (1968) Zones of lateritic weathering of secondary phosphorites of Altay-Sayan region. Int Geol Rev 10:1119–1127

Chapter 10 Formation and Stability of Base Metal Phosphates in Soils and Sediments

JEROME O. NRIAGU

Introduction

Most textbooks and journal articles contend that the stable forms in which the base metals are likely to occur in the aerobic environments are as the oxides and the carbonates. Since the base metal carbonates and oxides are fairly soluble (see Garrels and Christ 1965; Stumm and Morgan 1970), the conclusion is inescapable that, when in a dispersed form, the base metals are sorbed onto the sediment and soil particles. The dearth of pertinent thermochemical data apparently has forestalled any attempt to consider the supergene formation and stability of the base metal phosphates. Using the available (Nriagu 1974) or predictive thermochemical data, this report explores the role of the formation of base metal phosphates in the dispersion and fixation of the base metals in soils and sediments. Evidence which implicates the formation of insoluble lead phosphates in roadside soils is also considered in detail.

Stabilities of Base Metal Phosphates in Different Environments

Base metal phosphates are found widely as secondary minerals in the oxidized zones of sulfide ore deposits as well as in gossans and halos around ore bodies. They are also encountered as minor constituents in phosphatic beds, in soils and sediments. Although the occurrence and mineralogical characteristics of these minerals have received considerable attention (see e.g., Palache et al. 1951; Gmelins 1965) there has hitherto been little or no discussion of the chemical constraints which predicate the environments of their formation. The following discussion utilizes thermodynamic equilibrium models to predict and assess the phase relations for base metal phosphates under earth surface conditions and the geochemical behaviour of these minerals in subaqueous environments.

The thermochemical data used in the calculations are given in Table 1. Where the experimental information is not available, the free energy of formation (ΔG°) of the phosphate mineral has been approximated by the method outlined in Nriagu and Dell (1974) and Nriagu (1975). The predictive model is conceptually similar to that used by Tardy and Garrels (1974) to estimate the ΔG° data for layer silicates; the free energy data for the hydroxide and phosphate components *within* the mixed phases are used to calculate the ΔG° of the phosphate minerals. The calcu-

National Water Research Institute. Burlington, Ontario L7R 4A6, Canada

Table 1. Thermochemical data for base metal phosphates[a]

Mineral name	Chemical formula	ΔG_f^0, kcal mol^{-1}	log K_{sp}[b]
Plumbogummite	$PbAl_3(PO_4)_2(OH)_5 \cdot H_2O$	−1,221	− 99.3
Pyromorphite	$Pb_5(PO_4)_3Cl$	− 906.2	− 84.4
Parsonsite	$Pb_2UO_2(PO_4)_2 \cdot 2H_2O$	− 911.0	− 45.8
Tsumebite	$Pb_2Cu(PO_4)(OH)_3 \cdot 3H_2O$	− 592.4	− 51.3
Dumontite	$Pb_2(UO_2)_2(PO_4)_2(OH)_4 \cdot 3H_2O$	−1,653	− 91.4
Corkite	$PbFe_3^{3+}(PO_4)(OH)_6 \cdot SO_4$	− 809.8	−112.6
Hinsdalite	$PbAl_3(PO_4)(OH)_6SO_4$	−1,368	− 99.1
Dewindtite	$Pb(UO_2)_4(PO_4)_2(OH)_4 \cdot 8H_2O$	−2,169	− 92.6
Renardite	$Pb(UO_2)_4(PO_4)_2(OH)_4 \cdot 7H_2O$	−2,113	− 93.7
Przhevalskite	$Pb(UO_2)_2(PO_4)_2 \cdot 4H_2O$	−1,168	− 47.4
Cornetite	$Cu_3(PO_4)(OH)_3$	− 374.7	− 48.0
Libenthenite	$Cu_2(PO_4)OH$	− 288.0	− 28.0
Torbernite	$Cu(UO_2)_2(PO_4)_2 \cdot 10H_2O$	−1,567	− 41.0
Metatorbernite	$Cu(UO_2)_2(PO_4)_2 \cdot 8H_2O$	−1,454	− 41.3
Nissonite	$CuMg(PO_4)(OH) \cdot 2 \cdot 5H_2O$	− 548.6	− 23.6
Pseudomalachite	$Cu_5(PO_4)_2(OH)_4$	− 662.5	− 75.8
Tagilite	$Cu_2(PO_4)OH \cdot H_2O$	− 344.5	− 27.9
Pyromorphite	$Cu_5(PO_4)_3OH$	− 779.3	− 65.6
Turquois	$CuAl_6(PO_4)_4(OH_8 \cdot 4H_2O$	−2,426	−179.0
Chalcosiderite	$CuFe_6^{3+}(PO_4)_4(OH)_8 \cdot 4H_2O$	−1,773	205.7
Veszelyite	$CuZn_2PO_4(OH)_3 \cdot 2H_2O$	− 586.3	− 45.8
Hopeite	$Zn_3(PO_4)_2 \cdot 4H_2O$	− 867.3	− 35.3
Phosphophyllite	$Zn_2Fe^{2+}(PO_4)_2 \cdot 4H_2O$	− 851.8	− 35.8
Scholzite	$Zn_2Ca(PO_4)_2 \cdot 2H_2O$	− 849.3	− 34.1
Spencerite	$Zn_4(PO_4)_2(OH)_2 \cdot 3H_2O$	− 944.8	− 52.8
Tarbuttite	$Zn_2(PO_4)OH$	− 387.6	− 26.6
Zinc rockbridgeite	$ZnFe_4^{3+}(PO_4)_3(OH)_5$	−1,147	−138.6
Zinc pyromorphite	$Zn_5(PO_4)_3OH$	−1,030	− 63.1
Faustite	$ZnAl_6(PO_4)_4(OH)_8 \cdot 4H_2O$	−2,475	−177.7
Anglesite	$PbSO_4$	− 194.36	− 7.7[c]
Cerussite	$PbCO_3$	− 149.5	− 12.8[c]
Smithsonite	$ZnCO_3$	− 174.85	− 9.9[c]

[a] The data for the phosphate minerals are taken from Nriagu (1974a). The pertinent thermochemical constants for the dissolved species are from Wagman et al. (1968, 1969)

[b] These constants pertain to the stoichiometric ionization of the minerals

[c] Data from Wagman et al. (1968)

lative strategy, however, involves a corrective term which is of the form (see Chen 1975)

$$\Delta G^{\circ}_{correction} = (RT)A\Sigma ln BX_i$$

where A and B are empirical constants and X_i is an independent variable pertaining to the mole fractions. Unlike Chen's (1975) model, the assumption is made that the description of the formation reactions in terms of the simplest hydroxide and phosphate components gives the minimum dispersion between the experimental and predicted data. Thus A und B have been set equal to unity and X_i assigned the value of the reaction coefficient. The following discussions are centred around the *relative* stabilities of the solid phases. The strategy used in calculating the phase rela-

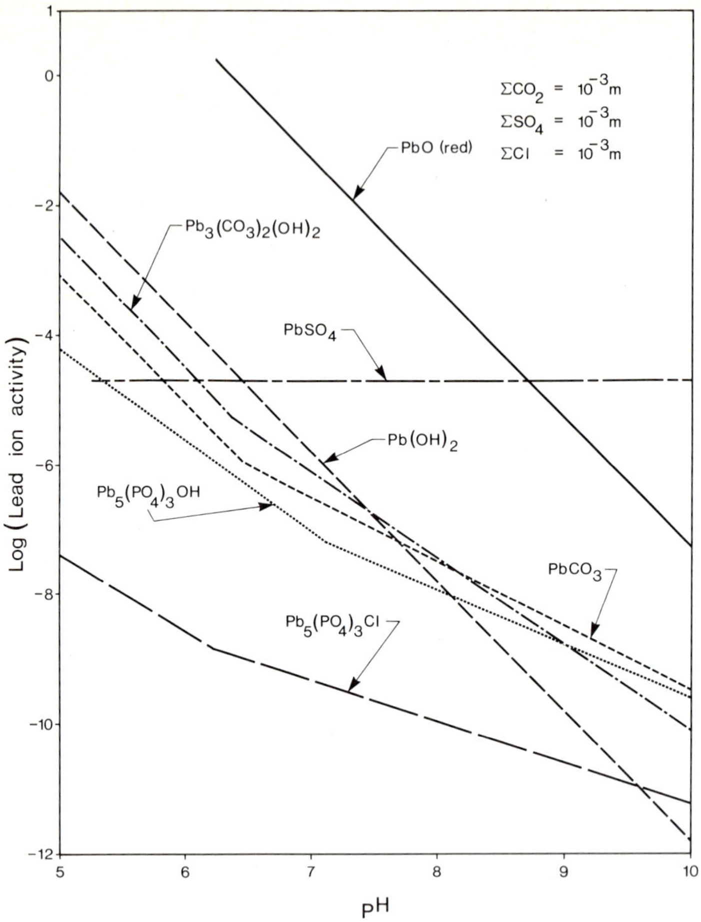

Fig. 1. Relative solubilities of low temperature lead compounds. The pertinent thermochemical data for the ionic species are from Wagman et al. (1968) and those for the solid phases are from Nriagu (1974). The total P concentration assumed to be 10^{-6} molar

tionships between two or more minerals clearly minimizes the effects of errors in the individual ΔG_f° values.

The method employed in calculating the phase relations in multicomponent systems follows closely that outlined in Stumm and Morgan (1970) and Nriagu (1974). Other models for estimating the free energies of formation of phosphate minerals are described by Vieillard and Tardy (This Vol.) The stabilities of the phosphate minerals in the presence of reduced sulfur species have not been considered; these minerals are found mostly in oxidized environments. The calculations assume HPO_4^{2-} as the predominant phosphate species. Alternate calculations based on $H_2PO_4^-$ or PO_4^{3-} as the principal ligand results in different topological projections but the essential geochemical inferences remain essentially the same.

Figure 1 shows the relative solubilities of the simple lead compounds at the fixed concentrations of the anions. This figure is used merely to demonstrate the

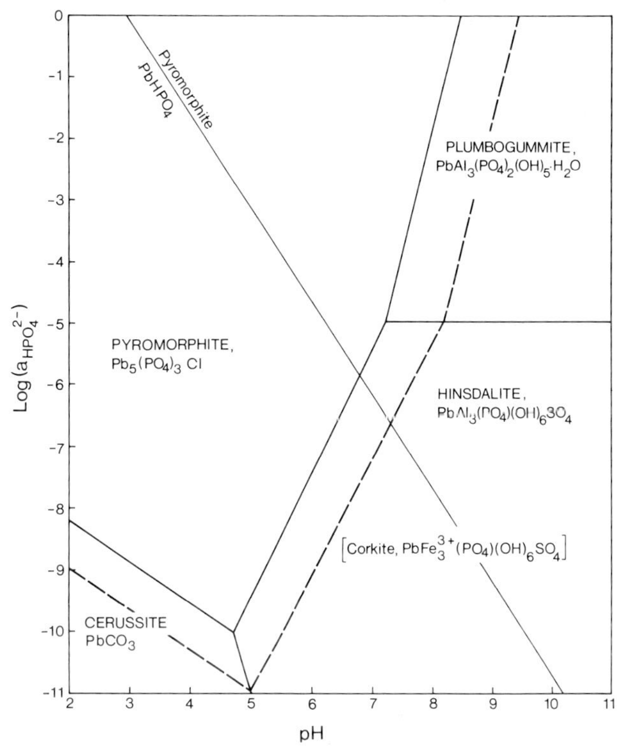

Fig. 2. Stability fields of secondary lead phosphates. The constraints on the ionic activities are: $a_{SO_4^{2-}} = a_{HCO_3^-} = 10^{-3}$; $a_{Al^{3+}} = 10^{-6}$; $a_{Pb^{2+}} = 10^{-6}$ (*solid lines*) and $a_{Pb^{2+}} = 10^{-5}$ (*broken lines*). The broken lines show the direction of change of "solubility" with increasing lead concentration. Minerals with closely related compositions and stabilities are assigned the same stability fields

greater stability of the phosphate minerals at earth surface conditions compared to the oxides, hydroxides, sulfates and carbonates. Previous considerations of the lead mineral stabilities in soils and sediments (see e.g., Krauskopf 1956) have missed this important fact. The stability relations between galena and pyromorphite have been considered in Nriagu (1974); it was shown that the pyromorphites remain stable even in moderately reducing environments.

The phase relationships for the important low-temperature lead phosphate minerals are shown in Fig. 2. It is immediately apparent that pyromorphite is the most stable mineral in neutral to acid environments, for the phosphate and lead concentrations likely to be encountered in most supergene environments. The large stability field for pyromorphite is consistent with its relatively widespread geological occurrence (see Palache et al. 1951). A decrease in Pb concentration favors the formation of hindsdalite, $PbAl_3(PO_4)(OH)_6SO_4$, corkite, $PbFe_3^{3+}(PO_4)(OH)_6$–SO_4 or plumbogummite, $PbAl_3(PO_4)_2(OH)_5 \cdot H_2O$ relative to pyromorphite. Corkite and hindsdalite form the stable association in alkaline environments and particularly at low P concentrations. Notice that the formation of plumbogummite also requires an alkaline environment. The formation of plumbogummite in soils and

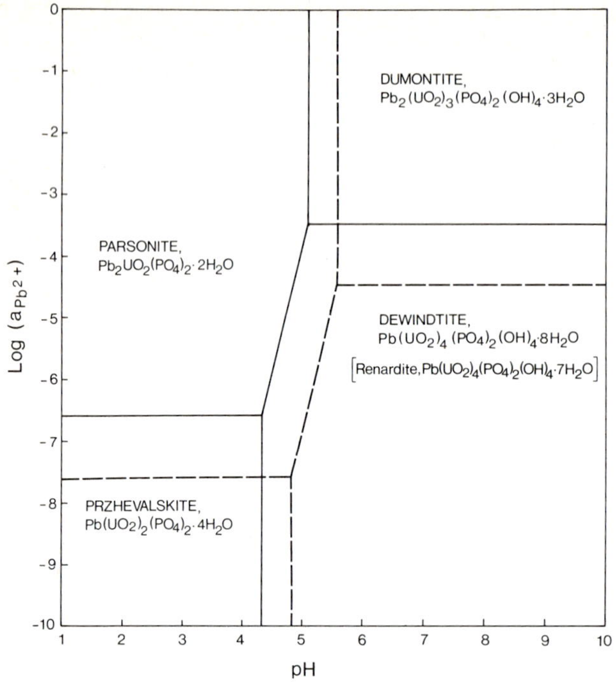

Fig. 3. Phase relations for Uranium-bearing lead phosphates assuming $a_{UO_2^{2+}} = 10^{-6}$. The activity of the HPO_4^{2-} ions is maintained at 10^{-6} (*solid lines*) or 10^{-5} (*broken lines*)

sediments is favored by high concentrations of Al which may be expected in weathering environments where Al is a mobile component.

In nature, pyromorphite is characteristically associated with iron oxides. Both pyromorphite and plumbogummite are also commonly found in association with cerussite, $PbCO_3$ and anglesite, $PbSO_4$. Figure 2, with no stable field for anglesite for the chemical constraints used, in fact, shows that the pyromorphite in association with cerussite and anglesite may not represent an equilibrium assemblage. Such an assemblage of generically incompatible phosphates, sulfates and carbonates suggests that the availability of dissolved P may be the prime limiting factor with regard to the supergene formation of base metal phosphate minerals.

Tsumebite, $Pb_2CuPO_4(OH)_3 \cdot 3H_2O$ is unstable relative to pyromorphite, plumbogummite and corkite under most supergene conditions. The formation of this mineral requires high Cu concentrations and is commonly encountered as one of the products of weathering primary Cu minerals. The positions of the phase boundaries for the following reactions

$$\text{Cerussite} + Pb^{2+} + Cu^{2+} + HPO_4^{2-} + 6H_2O \rightarrow \text{Tsumebite} + 3H^+ + HCO_3^- \quad (1)$$

$$K_1 = 10^{-5.5}$$

$$\text{Anglesite} + Pb^{2+} + Cu^{2+} + HPO_4^{2-} + 6H_2O \rightarrow \text{Tsumebite} + 4H^+ + SO_4^{2-} \quad (2)$$

$$K_2 = 10^{-7.7}$$

may explain the common coexistence of tsumebite-cerussite and tsumebite-anglesite in the oxidized zones of several sulfide ore bodies.

Figure 3 depicts the phase relations for uranium-bearing lead phosphates. The positions of the field boundaries are consistent with the well-known fact that two or more of these minerals generally occur together in most localities. The figure provides additional insight into the geochemical behavior of lead-uranyl phosphates. In alkaline environments, dewindtite, $Pb(UO_4)(PO_4)_2(OH)_4 \cdot 8H_2O$, or renardite, $Pb(UO_4)_4(PO_4)_2(OH)_4 \cdot 7H_2O$, is stable at low Pb concentrations whereas dumontite, $Pb_2(UO_2)_3(PO_4)_2(OH)_4 \cdot 3H_2O$ is the stable phase in Pb-rich environments. In acid environments, parsonsite, $Pb_2UO_2(PO_4)_2 \cdot 2H_2O$ (high Pb concentrations), and przhevalskite, $Pb(UO_2)_2(PO_4)_2 \cdot 4H_2O$ (low Pb concentrations), are the stable minerals. As may be expected, an increase in lead concentration favors the formation of parsonsite and dumontite.

The following reactions typically describe the transformation of UO_2-bearing phosphates (Fig. 3) into the hydroxyphosphates shown in Fig. 2:

$$\text{Parsonsite} + 3Pb^{2+} + HPO_4^{2-} + Cl^- \rightarrow \text{Pyromorphite} + H^+ + UO_2^{2+} + H_2O, \quad (3)$$

$K_3 = 10^{21.1}$

$$\text{Dumontite} + 3Al^{3+} \rightarrow \text{Plumbogummite} + H^+ + HPO_4^{2-} + Pb^{2+} + 3UO_2^{2+} + H_2O, \quad (4)$$

$K_4 = 10^{5.8}$

$$\text{Dewindtite} + 3Al^{3+} + SO_4^{2-} \rightarrow \text{Hinsdalite} + H^+ + 4UO_2^{2+} + HPO_4^{2-} + 6H_2O, \quad (5)$$

$K_5 = 10^{-9.5}$

$$\text{Przhevalskite} + 4Pb^{2+} + HPO_4^{2-} + Cl^- \rightarrow \text{Pyromorphite} + H^+ + 2UO_2^{2+} + 4H_2O. \quad (6)$$

$K_6 = 10^{24.7}$

Careful inspection of the above reactions reveals that under most natural environments, the mineral phases of Fig. 2 are unstable relative to the UO_2-bearing lead phosphates. In other words, the interaction of lead hydroxyphosphates with uranyl ions results in the formation of highly stable minerals. This inference is tempered by the fact that the UO_2^{2+} concentrations in natural waters are most often 10^{-7} to 10^{-8} *m*. It is significant that natural associations of the minerals in Fig. 2 with those in Fig. 3 are very rare (see e.g., Palache et al. 1951); the two groups of minerals are generically incompatible according to Eqs. (3)–(6). The great stabilities of UO_2-bearing phosphates have hitherto not been considered in the discussions of forms of uranium in soils and sediments. It is here suggested that the formation of these compounds may be one of the processes which demobilize uranium in sedimentary environments.

Figure 4 summarizes the phase relationships for copper phosphates. The great stabilities of UO_2-bearing phosphates are again evident. In nature, torbernite, $Cu(UO_2)_2(PO_4)_2 \cdot 10H_2O$, and metatorbernite, $Cu(UO_2)_2(PO_4)_2 \cdot 8H_2O$, are widely distributed, particularly in gossans of veins carrying Cu sulfides and uraninite (Palache et al. 1951). It should be noted that torbernite and metatorbernite are usually found with the other secondary uranium minerals rather than with the minerals in the figure. The copper phosphates to be expected in acidic environments include libethenite, $Cu_2(PO_4)OH$, tagilite, $Cu_2(PO_4)OH \cdot H_2O$ and nisso-

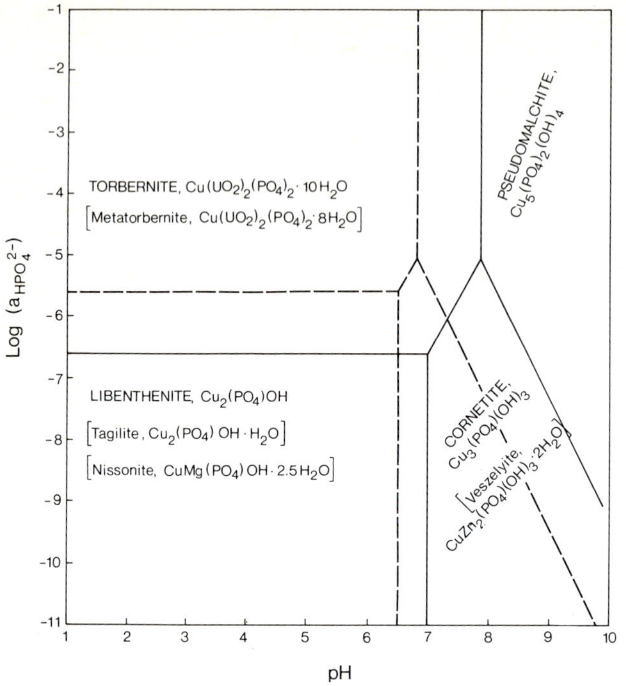

Fig. 4. Phase relations for low temperature copper phosphate minerals, assuming $a_{Mg^{2+}} = 10^{-3}$. For the *solid lines*, $a_{Cu^{2+}} = 10^{-6}$, $a_{UO_2^{2+}} = 10^{-7}$; for the *broken lines*, $a_{Cu^{2+}} = 10^{-4}$ and $a_{UO_2^{2+}} = 10^{-6.5}$. Minerals with closed related compositions have been assigned the same stability fields

nite, $CuMg(PO_4)OH \cdot 2\frac{1}{2}H_2O$. On the other hand, the stable Cu phosphates in alkaline environments include pseudomalachite, $Cu_5(PO_4)_2(OH)_4$, cornetite, $Cu_3(PO_4)(OH)_3$ and veszelyite, $CuZn_2(PO_4)(OH)_3 \cdot 2H_2O$. Notice (Table 1) that the replacement of Cu in a phosphate mineral lattice by Zn (as in veszelyite) increases the stability of the resultant phase whereas the substitution of Mg for Cu (as in nissonite) leads to a slight reduction in stability. Pseudomalachite and cornetite/veszelyite are commonly formed during the weathering of copper sulfides hosted by limestone formations (as in Mississippi-Valley type deposits). In such environments, alkaline condition necessary for the precipitation of these minerals is maintained by the carbonate equilibria.

Turquoise $CuAl_6(PO_4)_4(OH)_8 \cdot 4H_2O$, is the best known of the copper hydroxyphosphates. Natural turquoise invariably consists of components along the turquoise-chalcosiderite, $CuAl_6(PO_4)_4(OH)_8 \cdot 4H_2O$–$CuFe_6^{3+}(PO_4)_4(OH)_8 \cdot 4H_2O$ solid solution series. Although chalcosiderite is rare compared to turquois Table 1, however, suggests that it is thermodynamically more stable. It may be noted in passing that the so-called bone-turquoise or odontolite consists of microcrystalline apatite colored by vivianite, $Fe_3(PO_4)_2 \cdot 8H_2O$.

The formation of turquoise requires relatively high concentrations of Al^{3+} (or Fe^{3+}) and PO_4^{3-}. It is therefore most often formed in arid regions by the action of surface waters on aluminous igneous or sedimentary rocks. Turquoise found in

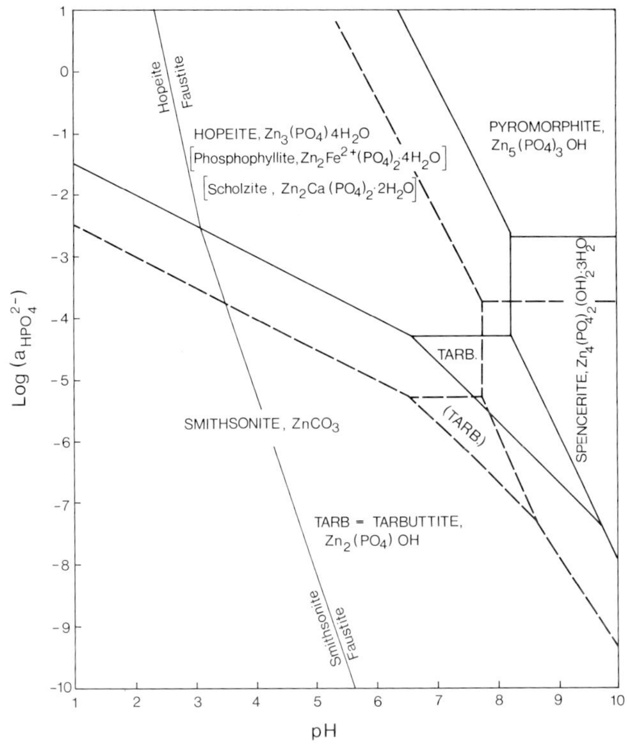

Fig. 5. Stability relationships for secondary zinc phosphate minerals, assuming $a_{Zn^{2+}} = a_{Fe^{2+}} = 10^{-6}$; $a_{Ca^{2+}} = a_{HCO_3^-} = 10^{-3}$ for the *solid lines*. For the *broken lines* the activity of Zn^{2+} ions has been assigned a value of 10^{-5}. Minerals with similar compositions and stabilities occupy the same domain in the diagram

aphanitic rocks (such as trachyte) is probably derived as the alteration products of the accessory apatite and disseminated Cu sulfides.

The following reactions illustrate the transformations of turquoise into the other Cu phosphates:

$$\text{Turquoise} + 1OH^+ + 2UO_2^{2+} \rightarrow \text{Torbernite} + 6Al^{3+} + 2HPO_4^{2-} + 2H_2O \quad (7)$$

$$K_7 = 10^{-1.3}$$

$$\text{Turquoise} + 10H^+ + Cu^{2+} \rightarrow \text{Libethenite} + 6Al^{3+} + 3HPO_4^{2-} + H_2O \quad (8)$$

$$K_8 = 10^{-15.7}$$

These equations serve to show that the other Cu phosphates are generally unstable relative to turquoise, in most cases, the field boundary lies outside the chemical regime of most supergene environments. In most environments however, the formation of turquoise would be limited by the low Al (or Fe^{3+}) activity. In nature, turquoise is rarely found in association with the other Cu hydroxyphosphates.

The stability fields of zinc phosphates are shown in Fig. 5. The widespread supergene and diagenetic occurrences of smithsonite, $ZnCO_3$, and the hopeite-phosphophyllite-scholzite, $Zn_3(PO_4)_2 \cdot 4H_2O$–$Zn_2Fe^{2+}(PO_4)_2 \cdot 4H_2O$–$Zn_2Ca(PO_4)_2 \cdot 2H_2O$, complex are consistent with the large stability fields of these minerals. According to Table 1, hopeite is slightly more stable than phosphophyllite but less stable than scholzite; the stability fields for the two minerals coincide essentially with that of hopeite in Fig. 5. In alkaline environments, hopeite may be altered to zinc pyromorphite, $Zn_5(PO_4)_3OH$ (high P concentration), tarbuttite, $Zn_2(PO_4)OH$ or spencerite, $Zn_4(PO_4)_2(OH)_2 \cdot 3H_2O$. The restricted field of stability of tarbuttite is consistent with its rare geological occurrence. It is noteworthy that all the phases in Fig. 5 are unstable relative to faustite, $ZnAl_6(PO_4)_4(OH)_8 \cdot 4H_2O$, the rare zinc analog of turquoise.

Inspection of Table 1 shows that for a given mineral composition, the stability sequence is Pb > Cu > Zn. Also, the inability of Zn phosphates to interact with uranyl ions is inexplicable. Cadmium and nickel phosphates are remarkably sparse in the geological record. The ionic sizes of these two metals suggest that Ni and Cd should be able to form several phosphates with compositions analogous to those of zinc and calcium respectively. It may be suspected that the dearth of Ni and Cd phosphate minerals is related to the low ionic activities of these metals in subaqueous, supergene environments. In this regard, it should be noted that many calcium and aluminum phosphates show significant concentrations of Cd and Ni (see Palache et al. 1951).

The calculations presented demonstrate the role of phosphate ions in the immobilization of base metals in low-temperature subaqueous environments. Such interactions of P with trace metals have several important environmental and economic consequences:

1. The localization of P in gossans and halos around ore bodies can be used as a prospecting guide. From the preceding discussions, it is clear that areas of high P anomaly are likely to indicate halos or weathered remnants of ore deposits. Since the base metal phosphates are unstable in the presence of sulfide ions, the formation of these minerals will likely be associated with the last phase of the (sulfide) mineralization process. Thus areas of high P anomaly may also define the margins of a metalliferous province.

2. Cultural activities now contribute tremendous quantities of both P and trace metals to the environment. It is therefore reasonable to suspect that the pollutant P and trace metals may immobilize each other as insoluble compounds. Such a self-purifying process may be significant in regulating both the rates and amounts of trace metals which can be loaded into the natural water systems. Furthermore, the trace metal phosphates are relatively insoluble (Table 1) and their formation (in the sediments) may be one of the buffer mechanisms regulating the levels of the trace metals in some natural bodies of water. The evidence for the interaction of pollutant lead with phosphate ions in the environment is considered in detail.

Lead Phosphates in Roadside Ecosystems

Many soils in the vicinity of roads and highways are severely contaminated with lead from automobiles that burn leaded gasoline (e.g., see Snyder et al. 1971; Hepple 1972). These soils thus provide a good setting for studying the biogeochemical transformations of pollutant lead in the environment. The mode of fixation (sorbed, complexed, organically or inorganically bound, etc.) of the lead in the roadside soils is of considerable ecological importance.

The reactivity of the automobile-emitted lead in the soil environments is largely controlled by the form in which the lead gets into the roadside soils. It is now widely recognized (see Bayard and Ter Haar 1971) that over 80% of the lead-containing compounds in the automobile exhausts is made of lead halides ($PbCl_2$, PbClBr, $PbBr_2$) and the double salts of lead and ammonium halides ($NH_4ClPbClBr$). Since most of the lead from the automobiles is deposited within 100 ft. of the roads and highways (Motto et al. 1970; Hepple 1972), the exhausts must have a short residence time in the atmosphere. Thus it may be inferred that the lead is deposited in the roadside soils as the halides and the double salts and the question arises as to what happens to these soluble compounds in the soil environments.

It has been suggested (Nriagu 1974) that the soluble lead compounds react with the phosphates in the soils to form plumbogummite and pyromorphites (see Fig. 6). The reaction scheme may be envisaged as,

$$\begin{aligned} & NH_4X \cdot PbX_2 + 5AlPO_4 \cdot 2H_2O \\ & \rightarrow PbAl_3(PO_4)_2(OH)_5 \cdot H_2O(\text{Plumbogummite}) \\ & + NH_4Al_2(PO_4)_2(OH) \cdot 2H_2O(\text{Leucophosphite}) \\ & + HPO_4^{2-} + 5H^+ + 3X^- + H_2O \end{aligned} \tag{9}$$

$$5PbX_2 + 3HPO_4^{2-} \rightarrow Pb_5(PO_4)_3X + 9X^- + 3H^+, \tag{10}$$

$$PbX_2 + 3Al(OH)_3 + 2H^+ + 2HPO_4^{2-} \rightarrow PbAl_3(PO_4)_2(OH)_5 \cdot H_2O + 2X^- + 3H_2O, \tag{11}$$

where X is a halide ion. Expectably, the precipitated phase will not be the above end members but components of the plumbogummite-crandallite and pyromorphite-apatite solid solutions.

The evidence (both direct and indirect) in support of the above suggestion is discussed in Nriagu (1974). Since then, other supportive evidence has been reported or realized: (1) Lead phosphate nodules have been identified (see Laitanen 1974) on the root surfaces and inside the stem and leaves of maize plants grown in phosphate rich hydroponic cultures. Furthermore, topical applications of lead acetate solution to plant cells have been shown to result in the formation of tiny crystals of lead pyromorphite within these cells (Malone et al. 1974). The phosphate nodules and crystals are believed to inhibit the uptake and metabolism of lead (Laitanen 1974) and phosphorus (Keeton 1973). These observations are consistent with the low solubilities of the lead phosphates (Fig. 1) and emphasize the strong affinity of lead for phosphate ions even under physiological conditions. (2) Studies of the interaction of lead and cadmium with several non-calcareous soils (using the equilibrium batch technique) indeed show that the lead concentrations are controlled by the formation of the insoluble lead phosphates Santillan-Medrano and Jurinak 1975). Even the concentrations of Cd were believed to be controlled by the formation of $Cd_3(PO_4)_2$. These studies are also significant in demonstrating that many soils contain phosphorus in forms which are reactive towards Pb^{2+} ions. (3) Plumbogummite, shown above to be a possible phase in the roadside soils has been identified in several soils adjacent to highways in Ontario, Canada.

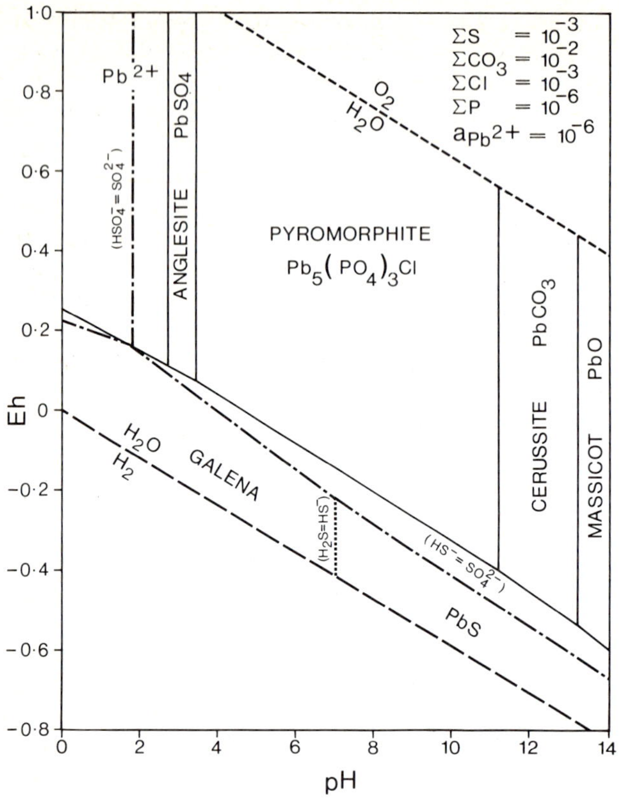

Fig. 6. Possible stability relationships for lead in roadside ecosystems. (Nriagu 1974)

It is believed that a detailed geochemical investigation of the forms of lead in soils and sediments close to the highways will likely show that most of the lead is tied up with the phosphorus. Two important implications of the lead-phosphate interactions deserve to be noted: (a) The fixation of lead as insoluble phosphates in these soils apparently regulates the quantity of lead that cycles annually through the various aquatic ecosystems. Additionally, the formation of the lead phosphates may be one of the buffer mechanisms regulating the concentrations of lead in natural water systems. (b) As noted above, lead bound to the phosphates is unavailable to plants. Thus the interaction has the beneficial effect of reducing the dietary intake of lead by man and other herbivorous animals.

Synergistic Effects of Ferromanganese Oxides

Ferromanganese oxyhydroxides promote the interaction of the base metals with phosphorus by serving as an adsorbent and a retaining matrix for both P and the base metals. It is known that a large fraction of the base metal burden of natural waters is sorbed into particles of ferromanganese oxides. When the particles settle out, they are immediately exposed to pore waters with elevated concentrations of orthophosphate ions (see Nriagu and Dell 1974) and the interaction of the metals with the orthophosphate ions may be expected to occur. The characteristic association of base metal phosphates with suites of ferromanganese oxides in recent sediments and older sedimentary rocks is evidence that the sorbed ions indeed are reconstituted into more stable phosphate phases. In sulfur-deficient sulfureta, the reduction of the ferromanganese oxides in response to changes in Eh will liberate the sorbed ions. There will thus be elevated concentrations of the metal ions and phosphate ions in the envelope surrounding the dissolving particles and these ions may be precipitated as phosphates. Also, Palache et al. (1951) have documented instances where the base metal phosphates have resulted from the (subsequent) phosphatization of host "limonite" beds. Clearly then, the synergetic effects of the ferromanganese oxyhydroxides will be to promote the formation of the base metal phosphates in a wide range of geological environments.

References

Bayard M, Ter Haar GL (1971) Composition of airborne lead particles. Nature 232:553–554

Chen C-HS (1975) A method of estimation of standard free energies of formation of silicate minerals at 298.15 °C. Am J Sci 275:801–817

Garrels RM and Christ CL (1965) Solutions, Minerals and Equilibria. Harper & Row, New York

Hepple P (1972) Lead in the Environment. Applied Science Publishers

Keeton CM (1973) Influence of lead compounds on the growth of barley. Soil Sci 43:401–411

Koeppe DE (1977) Environmental contamination by lead and other heavy metals, Vol 4: Soil-Water-Air-Plant Studies. Final Report, Inst. of Environmental Studies, Univ of Illinois, Urbana-Champaign, Illinois

Krauskopf KB (1957) Separation of manganese from iron in sedimentary processes. Geochim Cosmochim Acta 12:61–84

Malone CP, Koeppe DE, Miller RJ (1974) Localization of lead accumulated by corn plants. Plant Physiol 53:388–394

Motto HL, Dines RH, Chilko DM, Motto CK (1970) Lead in soils and plants: its relationship to traffic volume and proximity to highways. Environ Sci Technol 4:231–237

Nriagu JO (1974) Lead Orthophosphates. IV – Formation and stability in the environment. Geochim Cosmochim Acta 38:887–898

Nriagu JO (1975) Phosphate-clay mineral relations in soils and sediments. Can J Earth Sci 13:717–736

Nriagu JO, Dell CI (1974) Diagenetic formation of iron phosphates in recent lake sediments. Am Miner 59:934–946

Palache C, Berman H, Frondel C (1951) Dana's system of mineralogy, vol 2. Wiley & Sons, New York, p 654–1016

Santillan-Medrano J, Jurinak JJ (1975) The chemistry of lead and cadmium in soil: solid phase formation. Soil Sci Soc Amer Proc 39:851–856

Snyder RB, Wuebbles DJ, Pearson JE, Ewing BB (1971) A study of environmental pollution by lead. Nat Techn Inf Serv Rep LLEQ 71–7

Stumm W, Morgan JJ (1970) Aquatic chemistry. Wiley & Sons, New York, p 583

Tardy Y, Garrels RM (1974) A method for estimating the Gibbs energies of formation of layer silicates. Geochim Cosmochim Acta 38:1101–1116

Wagman DD, Evans WH, Halow I, Parker VB, Bailey SM, Schumm RH (1968, 1969) Selected values of chemical thermodynamic properties. NBS Techn Note 270–3, 270–4

Chapter 11 Chemistry and Structure of Precipitated Hydroxyapatites

AARON S. POSNER, NORMAN C. BLUMENTHAL, and FOSTER BETTS

Introduction

Our interest in the phosphate minerals, particularly in hydroxyapatite, comes from the fact that the latter is the prototype of the mineral in bone and other calcified tissues. To understand the structure and formation of the biological apatites it was necessary first to characterize hydroxyapatites which have been precipitated from solutions resembling body fluids. There is also a broad interest in hydroxyapatite throughout the scientific world. Workers in biological mineralization have profited from studies on this material which has taken place in a broad variety of fields.

The original igneous apatite mineral has been extensively leached by the Earth's waters, providing a resource for biological calcification which may occur in the form of apatite or, in many species, such as calcium carbonate. At death those calcium phosphate and calcium carbonate skeletons of marine species, which are not recycled by ingestion into the carnivorous sea life cycle, settle to the bottom of the Earth's waters (oceans, lakes, etc.) to form mineral deposits. Deep water apatite deposits may also form by direct precipitation from sea water (Posner 1969). Some of the skeletal calcium carbonate deposits are converted to apatite by reaction with dissolved phosphate. Most sedimentary or metamorphic calcium phosphate deposits have passed through a biological stage.

Precipitated hydroxyapatite are finely divided and differ in properties from the larger crystals of igneous, or hydrothermally formed hydroxyapatites. The submicroscopic crystals which are biologically or chemically precipitated are usually non-stoichiometric, carbonate-containing hydroxyapatites which are characterized by internal crystal disorder and a large and reactive specific surface. We will discuss the preparation and properties of the precipitated hydroxyapatites after first giving some background on the structure of well-crystallized hydroxyapatite.

Well-Crystallized Hydroxyapatite

The three-dimensional arrangement of the constituent ions in hydroxyapatites, as well as in a number of related apatites, has been delineated by X-ray and neutron diffraction (Elliott 1973; Young 1975). The unit cell contents of hydroxyapatite can be written $Ca_{10}(PO_4)_6(OH)_2$, but we often see isomorphous substitution in mineral and biological hydroxyapatites, e.g., F^-, or Cl^-, for OH^- and Sr^{2+}. It is difficult

The Hospital for Special Surgery Cornell University Medical College, New York, N.Y. 10021, U.S.A.

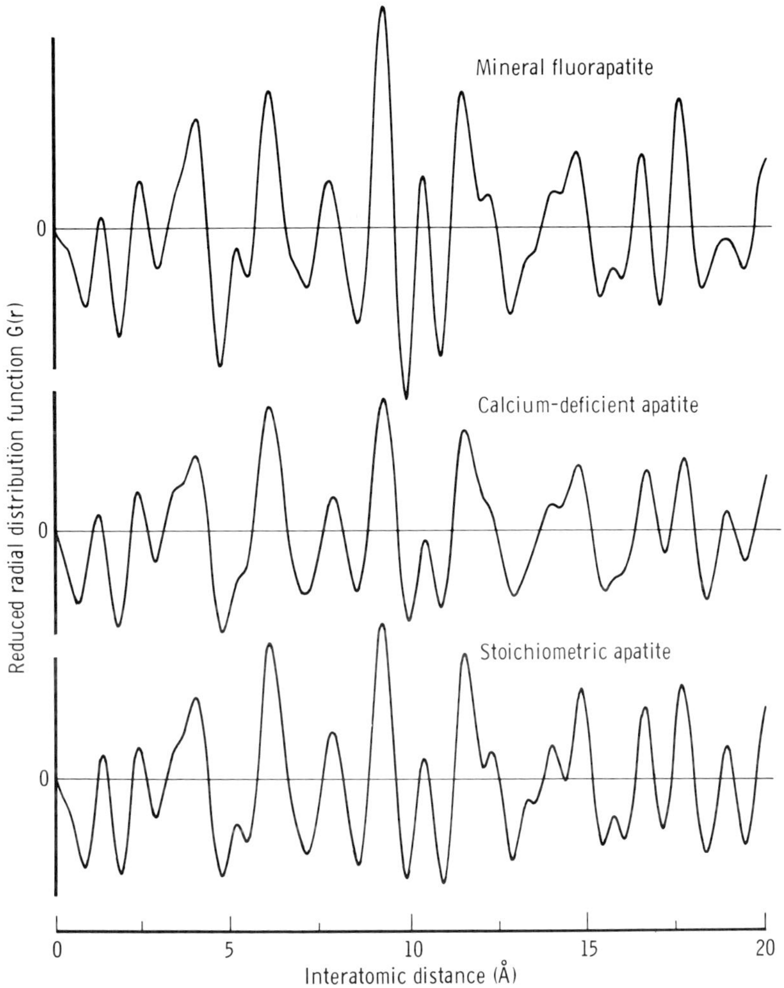

Fig. 1 Comparison of the X-ray radial distribution functions of well crystallized mineral fluorapatite, precipitated calcium-deficient and precipitated stoichiometric hydroxyapatite. The atomic structures of all three materials are essentially the same

to determine the atomic structure of the submicroscopic, precipitated hydroxyapatites because of the poorly resolved X-ray diffraction pattern they yield. However, it is reasonable to assume that the atomic arrangement in chemically precipitated (submicroscopic) hydroxyapatite is close to that of the well-crystallized hydroxyapatites which have been studied by standard, single crystal diffraction methods. This view is supported by the general resemblance of the X-ray radial distribution functions of well-crystallized mineral apatite to poorly crystallized, stoichiometric and poorly crystallized non-stoichiometric hydroxyapatites (Fig. 1). A radial distribution function (RDF) calculated from an X-ray diffraction powder pattern is a plot of atomic density vs. atomic separation which affords information about atomic arrangement in a material (Betts and Posner 1974).

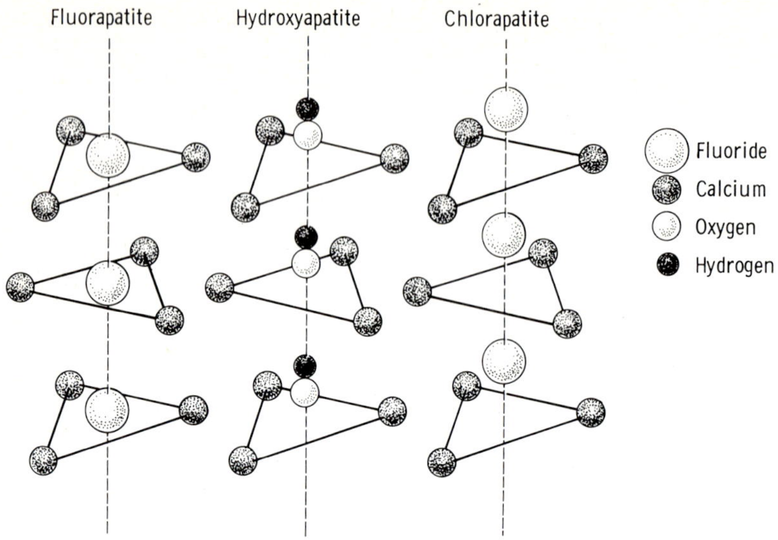

Fig. 2. Exercepts from the atomic structures of fluorapatite, $Ca_{10}(PO_4)_6F_2$, hydroxyapatite, Ca_{10}-$(PO_4)_6(OH)_2$, and chlorapatite, $Ca_{10}(PO_4)_6Cl_2$, to illustrate the F, OH, and Cl positions. Shown here are the columns of Ca triangles from each structure, the remainder of the structures are deleted. The F ions are closer and, thus, more tightly bound to the Ca triangles than the OH or the Cl ions

Determining the atomic structure of poorly crystallized, precipitated hydroxyapatites, synthetic and biological, has been difficult because, as noted, it is only possible to obtain poorly resolved powder diffraction patterns from these materials. Recently, Mackie and Young (1980) have begun to apply the method of Rietveld (1969) for the analysis of the neutron and X-ray diffraction powder patterns of dental enamel. Although definitive structures have not been worked out to date, this method promises to yield valuable structural information from powder diffraction patterns which have a number of well-resolved peaks, as is the case with dental enamel patterns.

The question of the substitution of carbonate in well-crystallized and poorly crystallized hydroxyapatite has received a great deal of attention (Posner 1969). It has been suggested, from infrared spectroscopy (Montel 1968), that the planar CO_3^{2-} ion substitutes in some random way, not yet fully elucidated, for the tetrahedral PO_4^{3-} group in hydroxyapatites prepared at body-to-room temperatures. In contrast, it was suggested that in high temperature (ca. 1,000 °C) hydroxyapatite preparations the CO_3^{2-} substitutes for OH^- with a specific orientation (Elliott 1965). The substitution and surface adsorption of carbonate on high specific surface, precipitated apatites will be discussed below.

Pure, defect-free $Ca_{10}(PO_4)_6(OH)_2$ is monoclinic in structure (Elliott et al. 1973) while the very similar fluorapatite, $Ca_{10}(PO_4)_6F_2$, is hexagonal. The difference in the structures lies only in the mirror planes, perpendicular to the c axis of $Ca_{10}(PO_4)_6F_2$, in which the F^- ions are situated surrounded in these planes by triangles of Ca^{2+} ions. The OH^- ion is located slightly above or below these mirror planes and because of this fact the symmetry of a stoichiometric hydroxyapatite, free of defects, degenerates into the monoclinic system because of preferential OH^- orientation (Elliott et al. 1973). Except for the OH^- positions, all other atomic positions in $Ca_{10}(PO_4)_6(OH)_2$ are essentially the same as in $Ca_{10}(PO_4)_6F_2$. In hydroxyapatite minerals and chemically prepared hydroxyapatites, where a small number of OH^- positions are randomly unoccupied and there is often some F^- substitution for OH^- (especially in minerals), the crystal system is apparently hexagonal. Here, as in fluorapatite, the OH^- ions create the mirror plane because of their placement randomly above and below the plane cancelling out the specific OH^- ion orientation found in pure monoclinic hydroxyapatite. Monoclinic hydroxyapatite is a laboratory rarity and in effect we must consider most hydroxyapatites to be isomorphous with fluorapatite, always remembering that the OH^- and F^- positions are slightly different. A very thorough discussion of this point is given by Young (1980) in a recent publication. Finally, Cl^-, when substituted in $Ca_{10}(PO_4)_6(OH)_2$ for the OH^-, sits further from the mirror plane than does OH^-. The important thing to remember about these subtle changes in structural position is that F^- is closer to the triangles of Ca^{2+} ions in the mirror plane to which it is bound electrostatically, than are OH^- and Cl^-, which lie progressively further from the same plane of calciums. Thus, the electrostatic bonds between the Ca^{2+} ions and F^- ions are stronger than the Ca–OH bond, which is stronger in turn than the Ca–Cl bond. This is illustrated in Fig. 2.

Precipitated Hydroxyapatite

When hydroxyapatite is precipitated from a highly supersaturated solution of calcium phosphate, an unstable amorphous calcium phosphate precursor solid phase is observed (Eanes et al. 1965a). This non-crystalline material, which has a Ca/P molar ratio of 3:2, rather than the value of 5:3 for hydroxyapatite, converts autocatalytically in water to a finely divided, non-stoichiometric hydroxyapatite. The latter tends with time in solution to improve slowly in stoichiometry and crystal size. On the other hand, when hydroxyapatite is precipitated from solutions of low supersaturation the amorphous precursor phase is not seen and fine dot-like crystals of hydroxyapatite, smaller than the high supersaturation crystals, are precipitated initially (Boskey and Posner 1976). No amorphous precursor phase was observed for calcium phosphate supersaturations (relative to well-crystallized hydroxyapatite) ranging from 10^5 to 10^9; an amorphous precursor is seen above a supersaturation of 10^9 (Boskey and Posner 1976). Later in this chapter we will treat the problem of solubility of finely divided compounds in relation to the equilibrium solubility product of well-crystallized hydroxyapatite.

Chemically precipitated hydroxyapatite differs from high temperature hydrothermal and igneous apatite preparations because of its submicroscopic crystal size, structural distortion and non-stoichiometry. Precipitated hydroxyapatites are usually calcium- and hydroxyl-deficient and contain internal hydrogen bonds between oxygens of adjacent orthophosphate groups not present in the stoichiometric compound. The existence of hydrogen bonds between oxygens of adjacent orthophosphates was presumed from infrared studies (Stutman et al. 1962). A general formula suggested for (carbonate-free) non-stoichiometric hydroxyapatites is:

$Ca_{10-x}H_x(PO_4)_6(OH)_{2-x}$, with x varying from zero, for stoichiometric hydroxyapatite, to almost 2. This formula was based, in part, on the amount of pyrophosphate produced when a series of calcium-deficient hydroxyapatites of differing Ca/P were heated to 600 °C (Berry 1967). When any solid orthophosphate containing hydrogen bonds between oxygens of adjacent PO_4^{3-} groups (e.g., $CaHPO_4$) is heated to an appropriate temperature there is a loss of water accompanied by pyrophosphate formation, with the amount of pyrophosphate formed related to the number of hydrogen bonds present in the compound. The OH^- ions were assumed to be missing, as shown in the formula above, to preserve electrical neutrality (Berry 1967). More recent work (Meyer 1979) has shown by a titration method that calcium-deficient apatites are in fact hydroxyl-deficient. Finally, an earlier crystallographic study (Posner and Perloff 1957) suggested that the Ca^{2+} ions were missing only from the columnar 3-fold axis positions and not from the 6-fold screw axis positions in the apatite structure.

Calcium-deficient hydroxyapatites are less stable thermally than stoichiometric hydroxyapatites. For example, the formation of pyrophosphate at about 600 °C occurs only in calcium-deficient apatites and not in perfect apatites. When heated to above 800 °C, a temperature which does not affect stoichiometric hydroxyapatite, calcium-deficient hydroxyapatite dehydrates to form β-$Ca_3(PO_4)_2$ (whitlockite) plus some residual apatite; if the Ca/P is equal to 1.5 there is no residual apatite seen after this pyrolysis. An interesting use of the phenomenon was made by Herrmann (1977). He showed that cremated fossil (Neanderthal) bone apatite contained β-$Ca_3(PO_4)_2$, while uncremated bone from the same era showed only bone apatite to be present by X-ray diffraction. He was thus able to determine which Neanderthal social groups cremated their dead.

The steady-state uptake of Ca^{2+} from solution by hydroxyapatites is inversely proportional to the Ca/P ratio of precipitated hydroxyapatites (Pak and Skinner 1968). Thus, calcium-deficient apatites tend to anneal in contact with calcium solutions. Synthetic apatites decreased in solubility in weak lactic acid as their Ca/P ratios were raised by such calcium treatment (Likins et al. 1958).

Finely divided, Ca-deficient apatites are used as catalysts in the dehydration and dehydrogenation of primary alcohols to aldehydes and ketones (Bowman and Piasecky 1964); butadiene, for example, is prepared from 2-butanol in this way. The rate of this dehydration was increased as the calcium deficiency of the catalytic hydroxyapatite increased (Bett et al. 1967). These workers suggested a catalytic dehydration mechanism which depended upon the presence of hydrogens in the structure of the calcium-deficient apatite catalyst.

Carbonate is always found in chemically precipitated hydroxyapatites unless stringent means are taken to exclude it. X-ray diffraction and infrared studies have shown that the finely divided apatites contain structural distortions which decrease their chemical stability. The primary source of structural distortion in precipitated hydroxyapatites is the substitution of the larger, planar CO_3^{2-} ions in phosphate positions. In preparations, with no precautions taken, we obtain a precipitated hydroxyapatite with about 1% CO_3^{2-} by weight. Biological apatites precipitated from body fluids are at least 4% in CO_3^{2-} content. Figure 3 illustrates the effect of CO_3^{2-} substitution for PO_4^{3-} on the regularity of interatomic distances (i.e., distortion) in three hydroxyapatites with different carbonate content. In this figure (Posner et

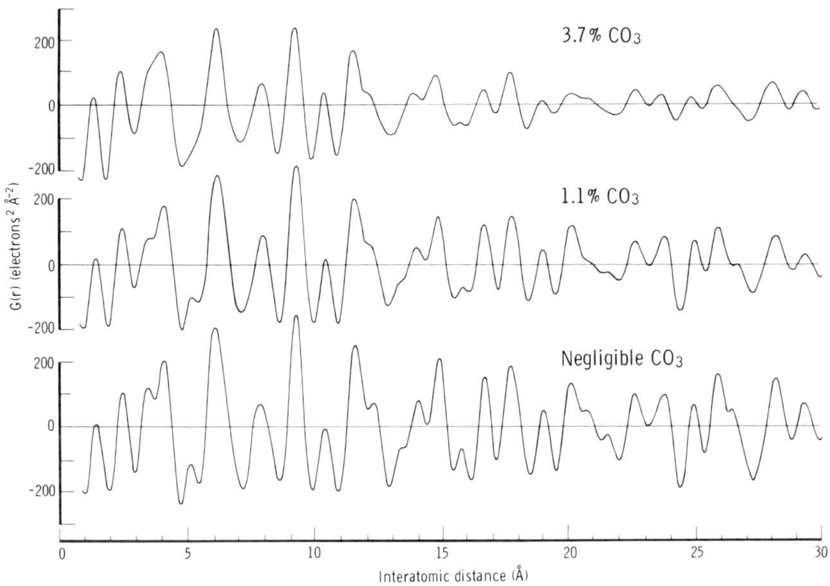

Fig. 3. Comparison of X-ray radial distribution functions of hydroxyapatites with different amounts of CO_3. The reduction of G(r) amplitude at greater interatomic distance (r) results from the structural distortion coming from the entry of CO_3 into the apatite structure

al. 1979) the X-ray radial distribution function decreases in amplitude at larger interatomic spacings with increasing carbonate content resulting from a reduced regularity of atomic positions, consistent with the strain induced by the entry of the CO_3^{2-} ion into the PO_4^{3-} position. Infrared absorption spectrophotometry also shows an increase in lattice distortion with increasing CO_3^{2-} content in hydroxyapatite (LeGeros et al. 1968). Blumenthal et al. (1981) showed that calcium-deficiencies cause very little structural distortion in precipitated apatites based on X-ray radial distribution analysis.

Amorphous Calcium Phosphate (ACP)

The unstable amorphous precursor phase seen in the precipitation of hydroxyapatite at high supersaturations can be stabilized by the presence of sufficient concentration of each of a number of chemical species in the transformation solution (Posner et al. 1978). Examples of such stabilizers are: Mg^{2+}, adenosine triphosphate (ATP), certain diphosphonates, pyrophosphate, proteoglycans, and glycoproteins. When magnesium and ATP are combined in the transformation solution there is a resulting synergistic stabilization effect on the amorphous phase which is much larger than the sum of their individual effects (Blumenthal et al. 1977). This result has significance in cell biology where it has been suggested that the amorphous calcium phosphate nodules seen in cell mitochondria are stabilized by a combination of magnesium and ATP in the intramitochondral fluid (Posner et al.

1977). Finally, some workers have suggested that the high Mg/Ca molar ratio (ca. 5) in sea water tends to stabilize the amorphous calcium phosphate in marine organisms and slow down the formation of poorly crystallized hydroxyfluorapatite (Lowenstam 1972; McConnell and Ward 1978).

The kinetics of the transformation of amorphous calcium phosphate to hydroxyapatite and the effect of stabilizers have been subjects of extensive study to elucidate the mechanism of normal biological mineralization and to search for drugs which will prevent pathological calcification in humans (Francis et al. 1973). Certain chemical species which prevent hydroxyapatite formation will slow down, or even stop, pathological apatite formation in soft body tissues. In addition, chemical moieties which stabilize the amorphous precursor also decrease the apparent solubility of already formed hydroxyapatite. An example of such a substance is the diphosphonic acid, ethane-1-hydroxy-1,1-diphosphonic acid (EHDP). Given orally, in proper dosage, EHDP stabilizes bone mineral in Paget's disease, a degenerative bone condition, thereby slowing down bone dissolution and decreasing the effect of the disease process (Fleisch et al. 1973). In this regard, X-ray diffraction studies in our laboratory have shown that oral EHDP treatment results in an improved crystallinity and presumably an improved stability of the bone mineral of Paget's patients.

The amorphous calcium phosphate precursor to precipitated hydroxyapatite can also be stabilized by dehydration in a freeze-dry apparatus. In fact, ACP can be heated in vacuum to as much as 500 °C without crystallizing. There was some question as to whether the amorphous precursor to hydroxyapatite was a single stoichiometric unit, for it was possible to prepare it with a Ca/P ratio ranging from 1.3 to 1.67, depending on solution conditions and whether or not a wash was used. X-ray radial distribution studies on a series of amorphous calcium phosphates prepared in different ways, with differing Ca/P ratios, showed that the basic structural unit is a spherelike cluster of about 9.5 Å in diameter with a stoichiometry of $Ca_9(PO_4)_6$ (Betts and Posner 1974). A number of these clusters are close-packed randomly to form the 300–1,000 Å spheres of amorphous calcium phosphate seen in the electron microscope (Fig. 4). The 15%–20% of water found in the freeze-dried amorphous solid is not an essential part of the $Ca_9(PO_4)_6$ structure, but is assumed to lie in the interstices between the clusters. The lack of 3/2 stoichiometry in Ca/P ratio has been shown to be due to adsorbed PO_4 ions (lower Ca/P) or occluded Ca ions (higher Ca/P).

In transforming to microcrystalline hydroxyapatite, the amorphous phase solubilizes and renucleates as apatite (Blumenthal et al. 1977). It is postulated that certain stabilizing species, such as ATP, diphosphonates and pyrophosphate, interfere with transformation not by acting on the amorphous phase itself but by a surface poisoning of the embryonic hydroxyapatite nuclei. This results in the prevention of further maturation of the highly soluble apatite embryos, thus, inducing their dissolution. The exact role of Mg^{2+} in preventing the amorphous to crystalline transformation remains unsolved. Our preliminary data suggest that, rather than poisoning surface growth in hydroxyapatite critical nuclei, Mg^{2+} enters the prenuclei structures disorting them and preventing further hydroxyapatite growth by creating a mismatch in epitaxial development. When present in a Mg/Ca molar ratio of greater than 0.2, the Mg can effectively prevent the transformation of

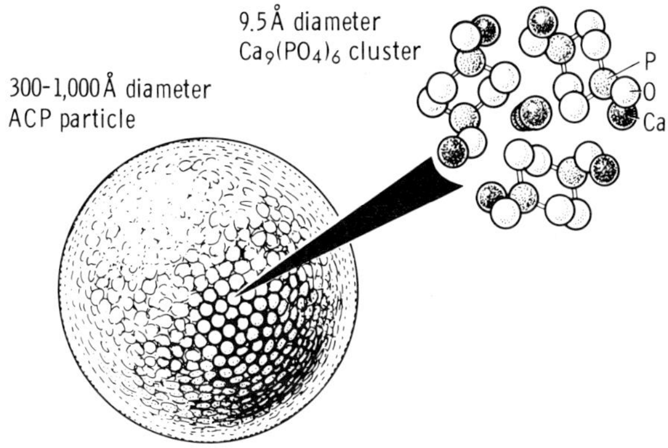

Fig. 4. Spherical amorphous calcium phosphate particle as seen in the electron microscope has been shown by X-ray radial distribution analysis to be composed of randomly close-packed $Ca_9(PO_4)_6$ clusters. The 15%–20% water found in dried ACP is situated in the interstices between the particles

amorphous calcium phosphate to crystalline hydroxyapatite, or the direct formation of HA from a low supersaturation solution (Boskey and Posner 1974).

Amorphous calcium phosphate was transformed into crystalline hydroxyapatite in a series of experiments (Blumenthal et al. 1981a) over a wide range of aqueous slurry concentrations. The higher slurries produced a more calcium-deficient hydroxyapatite (lower Ca/P ratio; larger amount of heat-produced pyrophosphate) from the mass action of excess phosphate in the higher slurry transformations. These experiments have biological relevance because it is believed that the apatite of bone is calcium-deficient and may be formed from a precursor amorphous phase. The stoichiometry of bone apatite could then be determined by regulating the ACP slurry density in the microvolume where bone apatite is formed.

Surface Chemistry

Precipitated hydroxyapatites have long been of interest to the surface chemist. The use of heat-treated bone apatite (bone char) in sugar refining, of apatite catalysts in the polymer industry (mentioned above), and of apatite chromatographic columns in protein chemistry are examples of the importance of the surface chemistry of apatites. Precipitated hydroxyapatites not only have a high specific surface but also have crystal surfaces which are reactive to specific molecules. In fact, it is the combination of these two factors which assures that bone mineral is in metabolic interrelationship with the rest of the body through the body fluids.

The specific surface of hydroxyapatites precipitated at or below body temperature, as measured by low temperature nitrogen adsorption, ranges from 100–200 $m^2\ g^{-1}$, the same as deproteinated bone mineral. Hydroxyapatite preparations at high temperatures result in a decreased specific surface. The amorphous calcium phosphate specific surface is generally lower than that of the hydroxyapatite which results from its conversion. It has been shown with nitrogen adsorption measurements that the amorphous phase can vary from 25 to 60 $m^2\ g^{-1}$, depending on the preparation conditions (Blumenthal et al. 1972).

Studies of the heat of adsorption of water vapor on a clean hydroxyapatite surface (heated to 500 °C), in vacuum showed that water molecules were only strongly

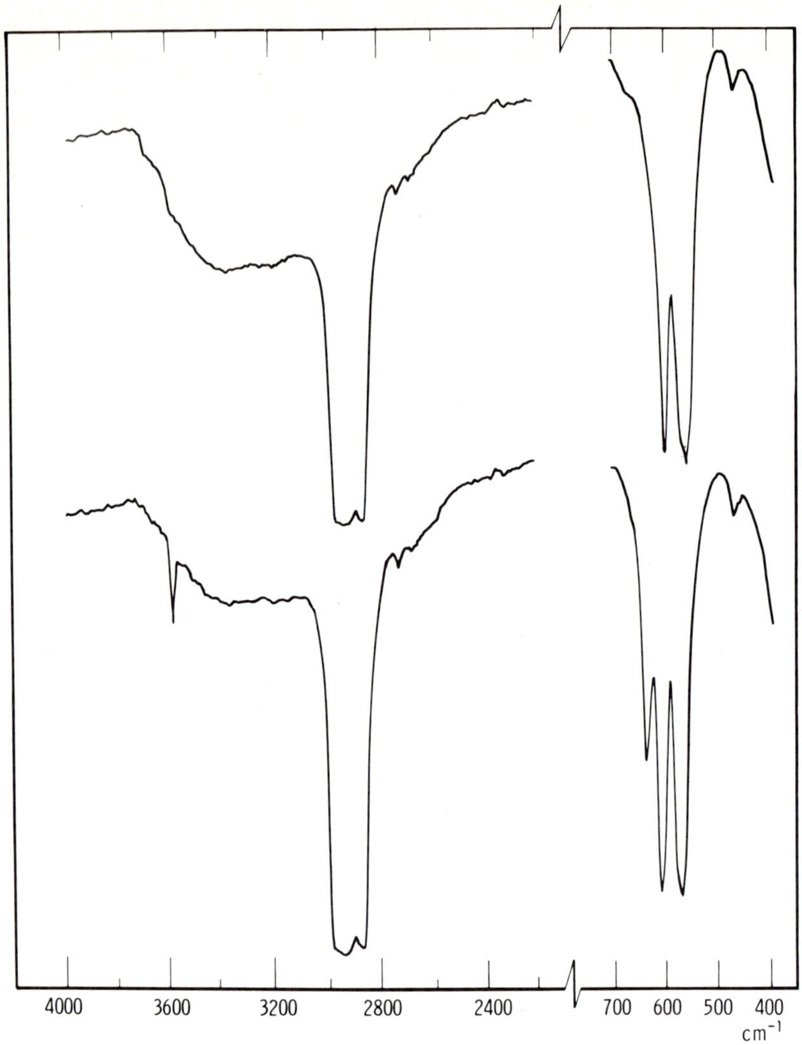

Fig. 5. Portions of the infrared spectra of two hydroxyapatite samples showing the regions of OH stretching mode (3572 cm^{-1}) and OH librational mode (630 cm^{-1}). The top pattern, representative of high specific surface hydroxyapatites, shows no evidence of OH bands. The bottom pattern was measured on a better crystallized hydroxyapatite prepared at 100 °C. The intense IR band at 2,900 cm^{-1} is due to the Nugol used in the sample preparation

adsorbed for two monolayers (Dry and Beebe 1960). The heat of adsorption of the initial water coverage on the precipitated apatite surface is 23 kcal mol^{-1}; the heat of adsorption approaches the heat of vaporization of water (11 kcal mol^{-}) as more coverage occurs. Thus, in a water slurry, after a coverage of two monolayers the water is not held specifically to the apatite surface but is equally attracted to the solution water.

It was suggested by earlier workers that the missing infrared hydroxyl adsorption modes reflected the fact that finely-divided apatites were hydroxyl-free (Termine 1972). Later work attributes the missing OH^- modes in the infrared patterns to the perturbation of hydroxyl stretching and bending modes by the hydrogen bonding of water on the apatite surface to surface OH^- ions (Blumenthal and Posner 1973). Beebe and Frankel (1968) showed that a high percentage of the OH^- ions lie on the surface of these high specific surface apatites. The infrared absorption modes due to unhydrated OH^- ions located centrally within the crystals are unperturbed by water. As the hydroxyapatite increases in size and the surface-to-weight ratio decreases, the water perturbation of the surface OH^- ions, though present, becomes unimportant and the infrared pattern exhibits only the unperturbed OH^- modes. Figure 5 illustrates this effect. Finally, a more recent work (LeGeros et al. 1978) showed that removing water from the apatite surface resulted in a sharpening of the OH^- bands in the infrared patterns.

In a study of the uptake of stearic acid ($C_{17}H_{35}COOH$) from cyclohexane solution on the surface of precipitated hydroxyapatite, it was shown that these long-chain molecules lined up in parallel bundles perpendicular to the surface (Beebe and Posner 1973). A strong polar bond between the carboxyl end group of this long chain molecule and the apatite surface was the reason advanced for this oriented uptake. A number of studies have been made on the uptake of biological macromolecules in aqueous solution on hydroxyapatite surfaces. Differential adsorption on apatite surfaces in chromatographic columns has been used to separate and collect specific polynucleotides and proteins. Native, double-stranded DNA is adsorbed strongly on the hydroxyapatite surface, while disordered (random coil) DNA is not, yielding a means of separating these species (Bernardi 1969).

The high specific surface of precipitated apatites is a major factor in the chemical and physical properties of these materials. Details of the methods and further results of studies on the apatite surfaces are given in a review paper by Posner and Beebe (1975). It should be noted that at least half of the carbonate present in finely divided hydroxyapatites is found adsorbed on the surface. This is discussed below in the section on bone apatite.

Solubility

The most important factor about the solubility of the precipitated hydroxyapatites is that their crystal sizes are so small that surface effects increase their solubilities. The classical Gibbs-Kelvin equation, which deals with the relationship between the crystal size and solubility, shows that below a certain particle size, peculiar to the compound in question, the equilibrium solubility rises sharply with decrease in particle size. Smales (1971) estimated this particle size for microcrystalline apatites and showed that, in substantial part, the greater solubility of biological apatites relative to well-crystallized apatites is due to the Gibbs-Kelvin effect.

Hydroxyapatites precipitated at or near room temperature are usually needle-like crystals, about 200 Å × 50 Å × 50 Å in size. These dimensions are apparently well below the point at which solubility is independent of crystal size, and so a small increase in size will result in a measurable decrease in solubility. The lack of stoichiometric and crystallographic perfection in precipitated hydroxyapatites also contributes to an apparent solubility which is higher than the solubility product determined on well-crystallized, stoichiometric hydroxyapatite. A major difficulty in determining the solubility of finely divided, poorly crystallized hydroxyapatite is the fact that this material reacts with water and slowly increases in crystal size and crystal perfection. Thus, we must realize that any solubility observed for such materials is not a true equilibrium measurement, for the solid being studied is changing in character with time. The same is true for amorphous calcium phosphate which converts to hydroxyapatite in water, with time. Fortunately, these changes in water, of both poorly crystallized hydroxyapatite and amorphous calcium phosphate, are slow enough for estimates to be made of what we can term an apparent solubility (Boskey and Posner 1973). The apparent solubility of bone mineral (a calcium-deficient, poorly crystalline, carbonate-containing hydroxyapatite) was measured to be 10^4 times greater than the solubility of well-crystallized hydroxyapatite; the K_{sp} for stoichiometric, well-crystallized hydroxyapatite is 10^{-59} (Brown and Chow 1976).

Bone Hydroxyapatite

The calcium phosphate in the bones of all vertebrates comprises a large part of this phosphate mineral in the Earth's biosphere. On a dry weight basis, mature bone is 60% to 70% calcium phosphate while the remaining organic fraction is composed largely (85%–90%) of the fibrous protein collagen. In the formation of bone, the organic matrix is first deposited extracellularly and then is infused with bone apatite. The possible appearance of an unstable precursor mineral phase is discussed below. The most distinguishing feature of bone apatite is its small crystal size and its lack of crystal and chemical perfection. This makes for a reactive material which, as noted earlier, satisfies the teleological need for normal biological resorption and redeposition of bone. It is the combination of mineral and protein fractions which yields the biomechanical strength required for the skeletal support role which bone performs as a structural component of the body.

As noted, one of the unsolved problems in the formation of bone apatite is the question of a possible precursor phase. Electron diffraction showed that a noncrystalline phase was the earliest mineral deposited in embryonic chick bone (Landis and Glimcher 1978), developing fish fin (Landis et al. 1981) and mineralizing bacteria (Landis and Boyan-Salyers 1981). These results seem to parallel the in vitro appearance of an unstable amorphous calcium phosphate precursor to precipitated hydroxyapatite. Studies on the embryonic chick suggested that a $CaHPO_4 \cdot 2H_2O$ (brushite) phase was the earliest mineral seen (Roufosse et al. 1979). This conclusion was based on the separation of the ground embryonic bone into a series of fractions of increasing density followed by X-ray diffraction analysis of each fraction. The fact that brushite appeared only in the lowest density (least mineralized) fraction led to the speculation that it was a precursor to the bone apatite found in all of the denser fractions. Later density fractionation of the actively calcifying region (epiphyseal growth plate) of fetal calf bone by other workers showed brushite to be present, but not in the lowest density fractions (Betts et al. 1979). The latter result was interpreted to mean that brushite might be present as an artifact of specimen preparation, or as an occurrence quite apart from bone formation. The role of brushite, which will convert in basic solution to hydroxyapatite, is still in debate.

There is another viewpoint which proposes that octacalcium phosphate, $Ca_8H_2(PO_4)_6 \cdot 5H_2O$, is the precursor phase to bone apatite in the tissue mineralization process (Brown and Chow 1976). There is no question that octacalcium phosphate under certain conditions can transform to hydroxyapatite (Brown et al. 1980). There is, to date, no diffraction evidence showing this material to be present in early bone formation. Obviously more work is needed before the matter of a bone apatite precursor can be adjudicated. Deciding about the precursor is important because the nature of the precursor will tell us something about the chemical milieu surrounding the earliest bone mineral deposition and throw light on the process itself. If an amorphous precursor appears, then we can feel certain that the first bone mineral precipitates from a solution highly supersaturated with respect to hydroxyapatite. If octacalcium phosphate or brushite appear first, then the mineralizing solution is probably slightly acidic or more acidic, respectively. If bone apatite precipitates directly, with no precursor, the body fluid from which the mineral deposits may be at low supersaturation. Obviously more work is needed to resolve this question.

The size and shape of bone apatite crystals have been measured by direct observation using the electron microscope and by indirect calculations employing X-ray diffraction line broadening. The literature on this subject is somewhat confusing because these parameters change with species, age and disease state. Bone mineral crystals are generally reported as plate-like, with the largest dimension having an average value of 250–350 Å, while the thickness is in the range of 25–50 Å. In each species the average crystal size is smallest at formation, increasing in average size to maturity, at which time there is a levelling off of this growth processes. It was shown that both animal and human bone mineral size increases with fluoride intake, presumably as a result of increased crystal growth rate (Eanes et al. 1965b). As noted, we also have shown that the diphosphonate EHDP will tend to improve the average crystal size when ingested by patients under treatment for a degenerative bone disease. With the increase of crystal size there is a concomitant improvement in chemical perfection with growth. The earliest bone crystal is calcium and hydroxyl deficient and hydrogen-bond containing. With maturation, the bone crystal approaches, but never reaches, the perfect hydroxyapatite formula (Menczel et al. 1965). As with synthetic calcium-deficient hydroxyapatites, bone produces β-$Ca_3(PO_4)_2$ when heated above 700 °C.

For many years it was debated whether or not the carbonate in bone mineral was substituted in the apatite structure or just adsorbed on the mineral surface. It is now clear that both statements, in a sense, are correct. As in the case of the synthetic hydroxyapatites, discussed above, bone apatite can be considered a carbonate-substituted material. In addition, a large part (about half) of the carbonate in bone mineral is on the crystal surface and is easily available for surface exchange. It is maintained by many workers that the readily exchangeable surface carbonate in bone is needed to help maintain a constant blood serum pH in conditions such as acidosis. Recent studies have shown that there is an interchange of carbonate between bone mineral, serum and the atmospheric carbon dioxide which enters the body through the lungs (Poyard et al. 1975). Although carbonate in bone does not

Table 1. Comparison of heats of adsorption. (Beebe and Posner 1973)

Adsorbent	Argon	Nitrogen	Carbon Monoxide
Carbon black	2.8	2.9	2.9
TiO_2	3.0	3.9	4.4
ThO_2	–	7.0	7.0
Bone Apatite	2.9	5.8	7.0
Bone Apatite with monolayer H_2O		3.5	
Bone Apatite monolayer CH_3OH	2.4	3.0	

Values expressed in kcal mol^{-1} gas adsorbed

play a major role in buffering extracellular fluids in acute acid-base abnormalities, it is safe to say that bone carbonate, at the very lest, has a minor, but significant, input into this process. Another important contribution carbonate makes is the increased reactivity it imparts to bone mineral by the structural disorder which result from its substitution within the crystal (Fig. 3).

The high surface reactivity of bone mineral was demonstrated by a study of the heats of adsorption of certain small molecular species on bone mineral surface. Table 1 illustrates a comparison of the heats of adsorption of argon, nitrogen, and carbon monoxide on several solid adsorbents including bone mineral, at low temperatures.
Table 1 shows that a polar (CO) or a polarizable molecule (N_2) will bond more strongly to polarizing surfaces (TiO_2, ThO_2, bone apatite) than to surfaces such as carbon black which do not polarize surface-bound molecules. Thus, bone apatite surfaces will bond polar groups more strongly than non-polar or non-polarizable groups. Table 1 also shows that a monolayer of water or methanol reduces the strength of the bond between bone apatite and a polarizable molecule such as nitrogen. Thus, it is apparent that a bond between some polar groups and the mineral apatite surface would be stronger if not made through an intermediate water layer. It is then likely that collagen in bone bonds directly to bone apatite surfaces and not through some intermediate hydrous bond system.

Bone Mineralization

There is at present no unified theory of the mechanism of bone mineralization. There are many, seemingly diverse, experiments which show that a number of individual chemical processes accompany the deposition of mineral in the organic matrix of bone. The earlier work on bone formation ignored the role of the cell and concentrated on extracellular processes, while the role of the bone cells has been considered more carefully in recent years. Excellent reviews of the biochemistry of bone mineralization are given by Howell and Pita (1976) and Irving (1973).

The mechanisms proposed for the infusion of the organic matrix of hard tissues with apatite mineral fall into three general categories: (a) raising the supersaturation in localized volumes to levels that would cause spontaneous precipitation, (b) providing substances which create nucleating sites or remove barriers to these sites and (c) removing or neutralizing bone mineral inhibitors. All these mechanisms may take place separately or simultaneously, as well as intracellularly and/or extracellularly. For example, in an early study Robinson (1923) proposed that alkaline phosphatase, present at calcification sites, hydrolyzed phosphate esters, thus producing an excess of free inorganic phosphate and elevating the local supersaturation of the body fluids. While alkaline phosphatase has an important role in bone mineralization, it is clear that this enzyme has multiple functions, for it is found throughout the body in non-calcifying as well as calcifying regions.

Most of the studies on the control of bone mineral formation have been on the role of certain molecular species, both nucleators and inhibitors, found in the regions of active calcification. The following have been proposed as nucleation substrates for hard tissue hydroxyapatite formation: collagen (Glimcher 1976), γ-carboxyglutamate containing proteins (Gallop et al. 1980), phosphoproteins (Veis 1978), Glycoproteins (Eastoe 1979), calcium-acidic phospholipid-phosphate complexes (Boskey and Posner 1977) and proteolipids (Boyan-Salyers 1980). All of these materials are present in bone and/or other normal and pathological mineral tissue. Some of these have been shown to nucleate hydroxyapatite in vivo, while others have only been shown to bind calcium ions.

The initial mineral deposits in the organic matrix of bone are shown by electron microscopy to be at discrete sites in, or on, the collagen fibrils (Glimcher 1976). The close association of bone mineral with collagen has led to extensive in vitro studies on the nucleation of apatite by this fibrous protein. Native collagen can nucleate hydroxyapatite from metastable (supersaturated) calcium phosphate solution (Glimcher 1976). This study also showed that modifying the native collagen by dehydration, by solution and redeposition into different tertiary structures, as well as by blocking the carboxylate group of aspartate and glutamate, will negate the apatite nucleating action of collagen. Most of the in vitro collagen studies have been carried out with skin and not bone collagen, both of which differ in some aspects of their intramolecular cross-linking. Yet it is difficult not to infer some close relationship between collagen and bone mineral. At a minimum, the in vivo apatite crystal orientation and crystal size seems to be under the control of collagen.

The phosphoproteins may play a structural and/or regulatory role in the calcifying organic matrix by providing an epitaxial substrate for apatite nucleation (Veis et al. 1977). It has been suggested that the steric arrangement of the phosphate groups in these phosphoproteins is optimal for the binding of calcium and, presumably, for the subsequent heterogeneous formation of apatite crystals (Fig. 6). It was shown that phosphoproteins will induce direct hydroxyapatite formation in highly supersaturated calcium phosphate solutions from which amorphous calcium phosphate would form in the absence of the phosphoprotein (Nawrot et al. 1976). No work has yet been done to show that phosphoproteins can induce apatite formation from more physiologically relevant (low supersaturation) calcium phosphate solutions.

It has been suggested that there are a variety of phosphoproteins in calcifying tissue which may have multiple roles including that of providing a substrate for apatite nucleation and controlling the size, shape and orientation of these crystals. Phosphoprotein bound to collagen in collagenous matrices may provide the mineral nucleating substrate, while in non-collagenous matrices (e.g., enamel and cal-

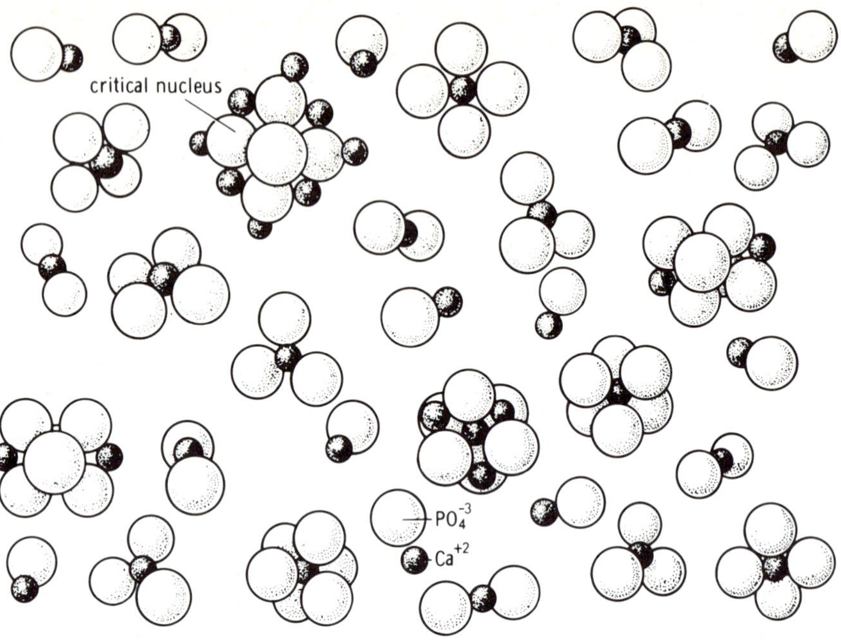

Fig. 6. Schematic of formation of hydroxyapatite critical nuclei from supersaturated calcium phosphate solutions. These events can occur (a) randomly at a rate as proportional to the supersaturation in homogeneous nucleation or (b) through the intervention of a heterogeneous substrate

cified carcinomas) the phosphoproteins are the principal matrix macromolecules. The exact role of the phosphoproteins found in calcifying tissues remains to be elucidated. The many papers reporting the presence of these macromolecules in calcifying tissue, such as bone, dentin, cementum, enamel, as well as calcified aorta, make it clear that this problem should be worked on further. The high concentration of these phosphoproteins at the mineralization front in calcified tissue suggests that this material must be involved, in a way yet to be determined, in the process of mineral deposition.

Although a great deal of work has been done on the γ-carboxyglutamic acid (GLA)-containing proteins in a variety of hard tissues, the role of this material is not clear. Some workers report that GLA-protein appears at the first deposition of bone mineral (Hauschka and Gallop 1979) while others find that this protein first appears after the initial mineral is deposited (Price et al. 1980). The fact that GLA-proteins are more abundant in the most highly mineralized (highest density) fraction of calcifying tissue (Glimcher et al. 1979) seems to support the latter observation, for, were the GLA-proteins initiators of mineralization, they should predominate in the region of earliest mineral deposit (lightest density fraction). Circumstantial evidence certainly links the GLA-proteins to a variety of mineralized tissues other than bone, for they are found in such tissues as calcified cartilage, dental cementum, dentin, and a variety of pathological calcification.

All of the collagenous hard tissue matrices contain a GLA-protein but only some of the non-collagenous hard tissue matrices contain these chemical moieties

(Boskey 1981). Not all of the GLA-proteins extracted from hard tissue are identical. For example, dentin has been shown to contain two non-identical GLA-proteins (Linde et al. 1980). One point seems to be clear about GLA-proteins and that is that their biosynthesis required the presence of Vitamin K (Gallop et al. 1980). Hoever, there are conflicting reports as to whether Vitamin K deficiency prevents in vivo mineralization.

It has been suggested that GLA-proteins when attached to a collagenous matrix are involved in orienting apatite crystals (King 1978). However, there has been no attempt to see if any of these proteins nucleate hydroxyapatite from in vitro calcium phosphate solutions. Much remains to be understood about the role of GLA-proteins in hard tissue.

Glycoproteins have been found in a variety of calcified tissues. The fact that glycoproteins in enamel are the first macromolecules made by the ameloblast (dental enamel cell) prior to calcification (Eastoe 1979) suggested that they serve as nucleating substrates for enamel apatite. There is, as before, no in vitro data supporting this hypothesis. A recent in vitro experiment (Blumenthal et al. 1981b) showed that glycoproteins prevented apatite deposition from solution by binding calcium and reducing the solution supersaturation. An earlier study of saliva stone formation cites glycoprotein as an apatite formation inhibitor (Van Dyke et al. 1979). Since glycoprotein is present in calcifying systems, it seems probable, given its affinity for calcium, that it is involved in some way in the deposition of mineral.

Certain lipids, too, have been shown to be connected with the initiation of tissue mineralization. Ca-acidic phospholipid-phosphate (Ca-APL-PO_4) complexes, isolated from a variety of hard tissues, are higher in concentration at sites of active apatite deposit (Boskey 1978). In contrast, no Ca-APL-PO_4 complexes are found in non-calcified tissues and in non-apatitic, pathological calcification, such as kidney stones (Boskey et al. 1978). A number of Ca-APL-PO_4 complexes isolated from hard tissue have been shown to induce hydroxyapatite precipitation from metastable calcium phosphate solution (Boskey 1978). The acidic phospholipids in the complexes extracted from bone are principally phosphatidyl serine and phosphatidyl inositol. In fact, placing either of these acidic phospholipids in a metastable calcium phosphate solution will result in the formation of a Ca-APL-PO_4 complex with subsequent hydroxyapatite precipitation (Boskey and Posner 1977). These workers showed in vitro that certain enzymes can alter the structure and thus the nucleating properties of the acidic phospholipid complexes (Boskey and Posner 1977).

These complexes are present in the membranes of extracellular vesicles and the membranes of cells in calcifying tissue. Proteolipids which are membrane components of most cells, have been shown to be the origin of the Ca-APL-PO_4 complexes in calcifying bacteria (Boyan-Salyers 1980). Not all proteolipids induce calcification, but proteolipids extracted from the following tissues have been shown to cause the in vitro precipitation of apatite: chick growth plate, marmoset bone, certain bacteria and calcified aortas (Boskey 1981).

The information gathered on the calcification by proteolipids parallels the results on studies of the Ca-APL-PO_4 complexes. It is possible that the proteolipids contain acidic phospholipids and the Ca-APL-PO_4 complexes are formed on the proteolipid surfaces; the apatite nucleation activity is primarily due to the complex.

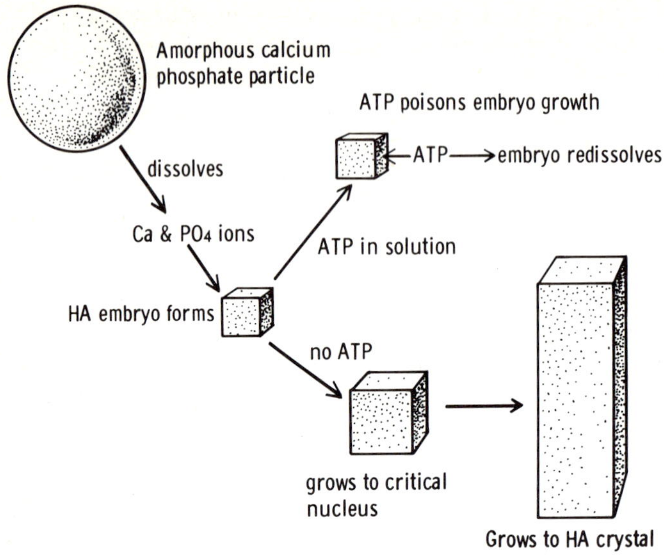

Fig. 7. Schematic for the mechanism of the stabilization of amorphous calcium phosphate by ATP

Our knowledge of the relationship between the Ca-APL-PO_4 complex and the proteolipids is incomplete at this point.

A number of biologically relevant chemical species slow down or inhibit direct hydroxyapatite precipitation, or the transformation of amorphous calcium phosphate to hydroxyapatite. Already mentioned is the fact that Mg^{+2} in high enough concentration (Mg^{+2}/Ca molar ratio >0.2) will prevent apatite formation. The pyrophosphate ion is a well known inhibitor of apatite as are its biologically relevant analogs ATP and ADP (Fig. 7). A number of synthetic analogs of diphosphonic acid also prevent apatite formation (Francis et al. 1973).

The very large biological macromolecules, the proteoglycans, are believed to be involved in the regulation of cartilage and bone mineralization, by acting as inhibitors of mineral deposition. The elastic properties of cartilage result from its composition of collagen infused with a proteoglycan gel (Maroudas 1974). Cartilage is of the order of 25–60 mg proteoglycan per ml of gel. Past work (Maroudas 1974) has indicated that proteoglycan must be removed or modified for cartilage to mineralize. As little as 10 mg ml^{-1} of bovine nasal proteoglycan aggregate can prevent the in vitro precipitation of apatite from a metastable solution from which precipitation normally takes place in about 30 min (Blumenthal et al. 1979). Smaller concentrations of this proteoglycan will delay but not prevent apatite formation. The proteoglycan subunit (monomer) is less effective than the aggregate but is still a potent apatite inhibitor. Both aggregate and subunit retard but do not prevent the transformation of amorphous calcium phosphate to hydroxyapatite. Parallel experiments have shown that dextrans with molecular weights in the same range as the proteoglycan subunits will prevent apatite precipitation. All of these enormous, extended molecules (synthetic and biological) exert influence over a large portion of the volume of the nucleating solution, reducing the probability of critical nucleus formation probably be reducing the rate at which ions, clusters and pre-critical embryos diffuse together in solution. In support of this view are the observation that enzymatic cleavage of the protein core and enzymatic reduction of the molecular configuration result in diminution of the inhibitory effect of proteoglycans (Blumenthal et al. 1980).

Summary

This has been a review of the chemistry and structure of precipitated hydroxyapatites, synthetic and biological. The most important features of these materials are their large and reactive surfaces as well as their crystal imperfection and nonstoichiometry which combine to make for a more reactive material than the well-crystallized, stoichiometric hydroxyapatite. It was indicated that precipitated hydroxyapatites, including bone mineral, are in the size range where their solubility decreases rapidly with a small increment of crystal growth. A discussion was given of the structure and chemistry of the amorphous calcium phosphate precursor to hydroxyapatite formed under high supersaturation conditions. Finally, nucleators and inhibitors of biological hydroxyapatite formation were discussed. It is clear that the body contains a number of nucleating and inhibiting mechanisms which seem to work in concert providing redundant pathways to the mineralization of tissue.

Acknowledgements. Most of the experimental work referred to in this chapter was supported by National Institutes of Health Grant AM-18412. This is publication number 144 of the Laboratory of Ultrastructural Biochemistry.

References

Beebe RA, Frankel SA (1968) The surface hydroxyl population in bone mineral. Koll Z Z Poly 222:56–61

Beebe RA, Posner AS (1973) Surface studies on bone mineral and related calcium phosphates. Phys-Chim Cristallogr Apatites D'Int Biol, 230 CMRS, Paris, p 275–281

Bernardi G (1969) Chromatography of nuclei acids on hydroxyapatite. I. Chromatography of DNA. Biochem Biophys Acta 174:423–434

Berry EE (1967) The structure and composition of some calcium-deficient apatites. J Inorg Nucl Chem 29:317–327

Bett HAS, Christner LG, Hall WK (1967) Studies of hydrogen held by solids. XII. Hydroxyapatite catalysts. J Am Chem Soc 89:5535–5541

Betts F, Posner AS (1974) A structural model for amorphous calcium phosphate. Trans Am Cryst Assoc 10:73–84

Betts F, Trotta R, Goldberg MR, Posner AS (1979) Non-apatite mineral in actively calcifying tissue. Orthop Trans 3:201

Blumenthal NC, Posner AS (1973) Hydroxyapatite: Mechanism of formation and properties. Calcif Tissue Res 13:235–243

Blumenthal NC, Betts F, Posner AS (1975) Effect of carbonate and biological macromolecules on formation and properties of hydroxyapatite. Calcif Tissue Res 18:81–90

Blumenthal NC, Betts F, Posner AS (1977) Stabilization of amorphous calcium phosphate by Mg and ATP. Calcif Tissue Res 23:245–250

Blumenthal NC, Betts F, Posner AS (1981) Formation and structure of calcium-deficient hydroxyapatite. Calcif Tissue Int 33:111–117

Blumenthal NC, Posner AS, Holmes JM (1972) Effect of preparation conditions on the properties and transformation of amorphous calcium phosphate. Mater Res Bull 7:1181–1190
Blumenthal NC, Posner AS, Silverman L, Rosenberg LC (1979) Effect of proteoglycans on in vitro hydroxyapatite formation. Calcif Tissue Int 27:75–82
Blumenthal NC, Posner AS, Rosenberg (1980) Further studies on the inhibition of hydroxyapatite formation by proteoglycans. Trans Orthop Res Soc 5:10
Blumenthal NC, Posner AS, Triffitt JT (1981) Inhibition of hydroxyapatite by a human glycoprotein. Trans Orthop Res Soc 6:56
Boskey AL (1978) The role of Ca-PL-PO_4 complexes in tissue mineralization. Metab Bone Dis Rel Res 1:137–142
Boskey AL (1981) Current concepts of the physiology and biochemistry of calcification. Clin Orthop 167:165–196
Boskey AL, Posner AS (1973) Conversion of amorphous calcium phosphate to microcrystalline hydroxyapatite: A pH-dependent, solution-mediated, solid-solid conversion. J Phys Chem 77:2313–2317
Boskey AL, Posner AS (1974) Magnesium stabilization of amorphous calcium phosphate: A kinetic study. Mater Res Bull 9:907–916
Boskey AL, Posner AS (1976) Formation of hydroxyapatite at low supersaturation. J Phys Chem 80:40–45
Boskey AL, Posner AS (1977) The role of synthetic and bone extracted Ca-phospholipid-phosphate complexes in hydroxyapatite formation. Calcif Tissue Res 23:251–258
Boskey AL, Goldberg MR, Posner AS (1978) Calcium-phospholipid-phosphate complexes in mineralizing tissues. Proc Soc Exp Biol Med 157:590–593
Bowman RS, Piasecky LJ (1964) Calcium phosphate catalysts and method of production. US Patent 3.149.082
Boyan-Salyers BD (1980) Proteolipid and calcification of cartilage. Trans Orthop Res Soc 5:9
Brown WE, Chow LC (1976) Chemical properties of bone mineral. Ann Rev Mater Sci 6:213–236
Brown WE, Tung MS, Chow LC (1980) Role of octacalcium phosphate in the incorporation of impurities into apatites. Proc 2nd Int Congr Phosph Comp. Inst Mond Phos, Paris, p 59–72
Dry ME, Beebe RA (1960) Adsorption studies on bone mineral and synthetic hydroxyapatite. J Phys Chem 64:1300–1304
Eanes ED, Gillessen IH, Posner AS (1965a) Intermediate stages in the precipitation of hydroxyapatite. Nature 208:365–367
Eanes ED, Zipkin I, Harper RA, Posner AS (1965b) Small angle X-ray diffraction studies on the effect of fluoride on human bone apatite. Arch Oral Biol 10:161–173
Eastoe JE (1979) Enamel protein chemistry-past, present and future. J Dent Res 58(B):753–764
Elliott JC (1965) The interpretation of the infrared absorption spectra of some carbonate-containing apatites. In: Stack MV, Fearnhead RW (eds) Tooth Enamel. Wright, Bristol, p 20–22
Elliott JC (1973) The problems of the composition and structure of the mineral components of the hard tissues. Clin Orthop 93:313–345
Elliott JC, Mackie PE, Young RA (1973) Monoclinic hydroxyapatite. Science 180:1055–1057
Fleisch H, Russell RGG, Bisaz S, Bonjour JP (1973) The effects of pyrophosphate and diphosphonates on calcium metabolism. In: Sognnaes RF, Vaughn J (eds) Hard tissue growth, repair and remineralization. Elsevier, London, p 331–358
Francis MC, Briner WS, Gray JA (1973) Chemical agents in the control of calcification processes in biological systems. In: Sognnaes RF, Vaughn J (eds) Hard tissue growth, repair and remineralization. Elsevier, London, p 57–90
Gallop PM, Lian JB, Hauschka PV (1980) Carboxylated calcium-binding proteins and vitamin K. N Engl J Med 302:1460–1466
Glimcher MJ (1976) Composition, structure and organization of bone and other mineralized tissues and the mechanism of calcification. In: Greep RO, Astwood EB (eds) Williams & Wilkins, Baltimore, Maryland (Handbook of physiology-endocrinology, vol 7, p 25–116)
Glimcher MJ, Kossiva D, Roufosse A (1979) Identification of phosphopeptides and γ-carboxyglutamic acid-containing peptides in epiphyseal growth plate cartilage. Calcif Tissue Int 27:187–191
Hauschka PV, Gallop PM (1979) Purification and calcium binding properties of osteocalcin and the γ-carboxyglutamate containing protein of bone in vitamin K metabolism and vitamin K-dependent proteins. In: Wasserman RH (ed) Calcium binding proteins and calcium function. Elsevier North Holland, New York, p 338–347

Herrman B (1977) Über die Reste des postcranielen Skelettes des Neanderthalers von Le Moustier. Z Morphol Anthropol 68:129–149

Howell DS, Pita JC (1976) Calcification of growth plate cartilage with special reference to studies of micropuncture fluids. Clin Orthop 118:208–229

Irving JT (1973) Theories of mineralization of bone. Clin Orthop 97:225–236

King K (1978) Distribution of γ-carboxyglutamic acid in calcified tissues. Biochim Biophys Acta 542:542–546

Landis WJ, Glimcher MJ (1978) Electron diffraction and electron probe microanalysis of the mineral phase of bone tissue prepared by anhydrous techniques. J Ultrastruct Res 63:188–223

Landis WJ, Boyan-Salyers BD (1981) Characterization of mineral deposits in *Bacterionema matruchotii* using anhydrous methods of specimen preparation. Trans Orthop Res Soc 6:325

Landis WJ, Geraudie J, Paine MC, Neuringer JR, Glimcher MJ (1981) An electron optical and analytical study of mineral deposition in the developing fin of trout, *Salmo gairdneri*. Trans Orthop Res Soc 6:271

LeGeros RC, Trautz O, LeGeros JP, Klein E (1968) Carbonate substitution in apatite structure. Bull Soc Chim Fr 1712–1718

LeGeros RZ, Bonel G, LeGeros R (1978) Types of "H_2O" in human Enamel and in precipitated Apatites. Calcif Tissue Res 26:111–118

Likins RC, Posner AS, Steere AC (1958) Effect of Ca treatment on solubility of synthetic hydroxyapatite and rat molar enamel. J Am Dent Assoc 57:335–339

Linde A, Brown M, Butler WT (1980) Non-collagenous proteins of dentin, a re-examination of proteins from rat incisor dentin utilizing techniques to avoid artifacts. J Biol Chem 255:5931–5942

Lowenstam HA (1972) Phosphate hard tissues of marine invertebrates: Their nature and mechanical function and some fossil implications. Chem Geol 9:153–166

Mackie PE, Young RA (1980) Crystallography of human tooth enemel: Initial structure refinement. Mater Res Bull 15:17–29

Maroudas A (1974) Physico-chemical properties of articular cartilage. In: Freeman MAR (ed) Adult articular cartilage. Grune & Stratton, New York, p 131–170

Martens CS, Harris RC (1970) Inhibition of apatite precipitation in the marine environment by magnesium ions. Geochem Cosmochem Acta 5:621–624

McConnell D, Ward P (1978) Nautical uroliths composed of phosphatic hydrogels. Science 199:208–209

Menczel J, Posner AS, Harper RA (1965) Age changes in the crystallinity of rat bone apatite. Isr J Med Sci 1:251–252

Meyer JL (1979) Hydroxyl content of solution precipitated calcium phosphates. Calcif Tissue Int 27:153–160

Montel G (1968) Conceptions nouvelle sur la physicochimie des phosphates de structure apatitique. Bull Soc Chim Fr 1693–1700

Nawrot CG, Campbell DJ, Schroeder JK, van Valkenburg M (1976) Dental phosphoprotein-induced formation of hydroxyapatite during in vitro synthesis of amorphous calcium phosphate. Biochem 15:3445–3449

Pak CYC, Skinner HCW (1968) Ionic interaction with bone mineral. IV. Varying affinity of synthetic calcium phosphates for Ca^{2+}. Biochim Biophys Acta 165:274–282

Posner AS (1969) The crystal chemistry of bone mineral. Physiol Rev 49:760–792

Posner AS, Perloff A (1957) Apatites deficient in divalent cations. J Res Nat Bur Stds 58:279–286

Posner AS, Beebe RA (1975) The surface chemistry of bone mineral and related calcium phosphates. Sem Arthritis Rheum 4:267–291

Posner AS, Betts F, Blumenthal NC (1977) Role of ATP and Mg in the stabilization of biological and synthetic amorphous calcium phosphates. Calcif Tissue Res 22:208–212

Posner AS, Betts F, Blumenthal NC (1978) Properties of nucleating systems. Metab Bone Dis Rel Res 1:179–183

Posner AS, Betts F, Blumenthal NC (1979) Bone mineral composition and structure. In: Simmons DJ, Kunin AS (eds) Skeletal research. Academic, New York, p 167–192

Poyard CF, Freminet A, Bursaux E (1975) The exchange of bone CO_2 in vivo. Respir Physiol 25:101–107

Price PA, Lothringer JW, Nishimoto SK (1980) Absence of the vitamin K-dependent bone protein in fetal rat mineral. Evidence for another γ-carboxyglutamic acid-containing component in bone. J Biol Chem 255:2938–2942

Rietveld HM (1969) A profile refinement for nuclear and magnetic structures. J Appl Cryst 2:65–71
Robinson R (1923) The possible significance of hexophosphoric esters in ossification. Biochem J 17:286–293
Roufosse AH, Landis WJ, Sabine WK, Glimcher MJ (1979) Identification of brushite in newly deposited bone mineral from embryonic chicks. J Ultrastruct Res 68:235–255
Smales FC (1971) The Kelvin effect and crystal growth of hydroxyapatite. In: Fearnhead RW, Stack MV (eds) Tooth Enamel II. Wright, Bristol, p 187–196
Stutman JM, Posner AS, Lippincott ER (1962) Hydrogen bonding in the calcium phosphates. Nature 193:368–370
Termine JD (1972) Biophysical properties of connective tissues. In: Slavkin H (ed) The comparative molecular biology of extracellular matrices. Academic, New York, p 443–450
Van Dyke TE, Levine MJ, Herzberg MC, Ellison SA, Hay DI (1979) Isolation of low molecular weight glycoprotein inhibitor of calcium phosphate precipitation from the extra parotid saliva of macaque monkies. Arch Oral Biol 24:85–89
Veis A (1978) The role of acidic proteins in biological mineralizations. In: Evrett DH, Vincent B (eds) Ions in macromolecular and biological systems. Colston Paper No 29, Scientechnia, Bristol, p 259–272
Veis A, Sharkey M, Dickson I (1977) Non-collagenous proteins of bone and dentin extracellular matrix and their role in organized mineral deposition, in calcium binding proteins and calcium function. In: Wasserman RH (ed) Calcium binding proteins and calcium function. Elsevier, North Holland, New York, p 408–418
Young RA (1975) Biological apatite vs. hydroxyapatite at the atomic level. Clin Orthop 113:249–262
Young RA (1980) Large effects from small structural differences in apatites. Proc 2nd Int Congr Phosph Comp, Inst Mond Phos, Paris, p 73–88

Chapter 12 Phosphate Minerals in Human Tissues

RACQUEL Z. LEGEROS and JOHN P. LEGEROS

Introduction

The mineralized or calcified tissues in biological systems are composed of two phases: organic and inorganic or mineral phases. In the invertebrates (e.g., echinoderms, mollusks, arthropods, etc.), the inorganic phase is usually calcium carbonate, $CaCO_3$, predominantly in the form of either calcite or aragonite or both. In the invertebrates, the inorganic phase consists of one or more types of phosphate minerals (predominantly calcium phosphates) depending on the nature of calcification, i.e., normal (e.g., bones and teeth) or abnormal or pathological (e.g., dental calculi, salivary and urinary stones, soft tissue calcifications, etc.). In several pathologically calcified tissues, the mineral is non-phosphatic, such as calcium oxalates (whewellite and weddellite), sodium urates, uric acid, cysteine.

The main purpose of this chapter is to review the occurrences and types of phosphate minerals in human tissues. Considered in this review are: (a) factors influencing the formation of the phosphate minerals in vivo and in vitro; (b) the co-existence of one type of phosphate mineral with another phosphate or non-phosphate mineral(s); and (c) their in vivo and in vitro transformations from one type of phosphate mineral to another. The background material was obtained from earlier reviews (21, 22, 40, 46, 83, 84, 95, 100–103, 105, 107, 108, 113, 117). Much of the illustrated experimental data are from past and current work of the authors (47–78, 116).

Occurrences of Phosphate Minerals in Human Tissues

Being the main inorganic constituent of bones and teeth, the most abundant mineral in human tissues is a basic calcium phosphate idealized as calcium hydroxyapatite, $Ca_{10}(PO_4)_6(OH)_2$. Other calcium phosphates [brushite or DCPD, $CaHPO_4 \cdot 2H_2O$; octacalcium phosphate, $Ca_8H_2(PO_4)_6 \cdot 5H_2O$; whitlockite or tricalcium phosphate, β-TCP, β-$Ca_3(PO_4)_2$; calcium pyrophosphate dihydrate (mono- and tri-clinic forms), CPPD, $Ca_2P_2O_7$; amorphous calcium phosphates, ACP]; and magnesium phosphates [struvite, $MgNH_4PO_4 \cdot 6H_2O$; newberyite, $MgHPO_4 \cdot 3H_2O$; amorphous calcium magnesium pyrophosphate, or calcium magnesium phosphate, ACMP] have been identified with or without association with apatite,

Department of Dental Materials Science New York University Dental Center, New York, N.Y. 10010

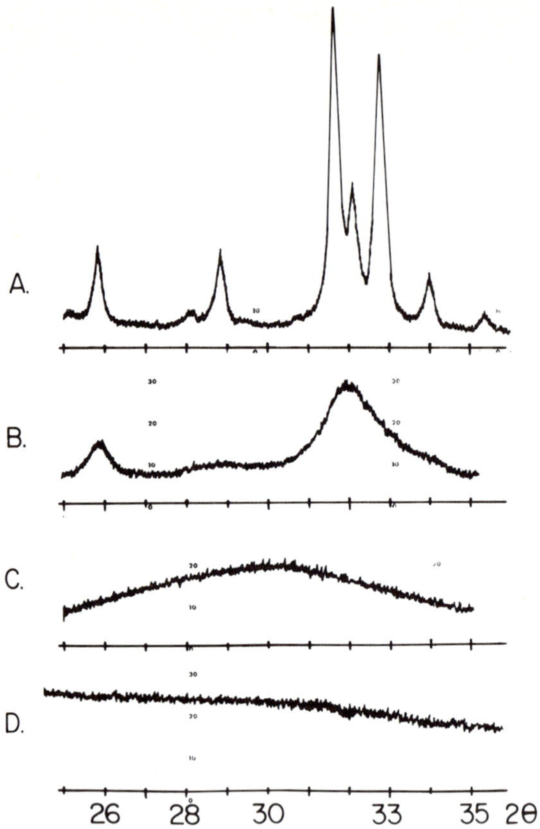

Fig. 1. X-ray diffraction patterns of some phosphate minerals in human tissues. *A* Tooth enamel; *B* dentine; *C* amorphous calcium phosphate (ACP) containing Mg^{2+} plus CO_3^{2-}; *D* ACP containing Mg^{2+} plus $P_2O_7^{4-}$. *A* and *B* are from normal calcifications, *C* and *D*, from pathological soft tissue calcifications. The X-ray diffraction pattern of bone is similar to that of dentine, *B*

as constituents of pathological calcifications (1, 13, 14, 16, 22, 24, 30, 78, 91–99, 105).

Apatites in normal calcified tissues of teeth and bone have been postulated to form either directly or indirectly by way of precursor calcium phosphates such as ACP, DCPD, OCP or TCP (1, 13, 14, 30, 39, 93, 99, 105). The presence of any of the possible precursors has not been detected in the X-ray diffraction patterns of tooth (enamel, dentine) or of bone (Figs. 1A, B). In contrast, several types of phosphate minerals co-exist in some pathological tissue calcifications (e.g., DCPD, OCP, β-TCP, and apatite in dental calculi, urinary and salivary stones) as shown in Figs. 2B, C, and 3C). In some cases of pathological tissue calcifications, the phosphate minerals co-exist with non-phosphate minerals (e.g., calcium and magnesium phosphates with calcium oxalates in urinary stones; calcium phosphate dihydrate, CPPD, with sodium urates in pseudodepositions). "Apatitic" and amorphous calcium phosphate deposits have been observed in articular, tumoral (Figs. 3A, B) and non-visceral calcifications in uremic patients (17, 59), in aortic valves (43) and pulmonary, ectopic and other extra-osseous calcifications (1, 12, 38, 106, 112).

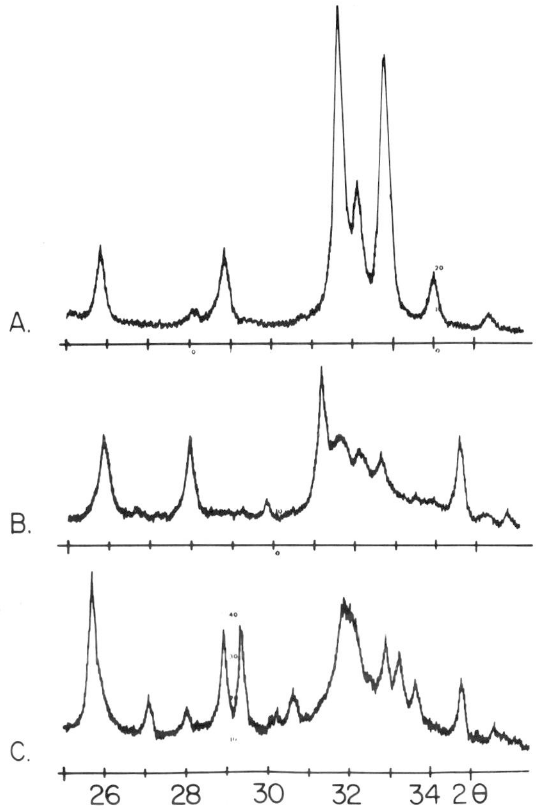

Fig. 2. X-ray diffraction patterns of phosphate minerals from normal and pathological tissue calcifications. *A* Tooth enamel; *B* dental calculi sample; *C* urinary calculi sample. *B* shows the mixed phases of β-TCP (Mg-containing) and apatite; *C* shows the mixed phases of struvite and apatite

Fig. 3a

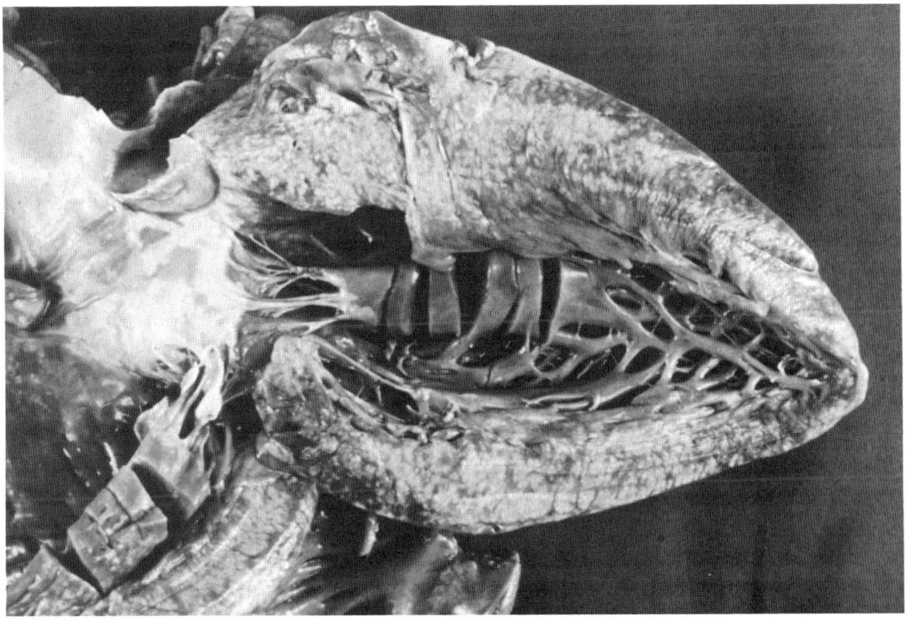

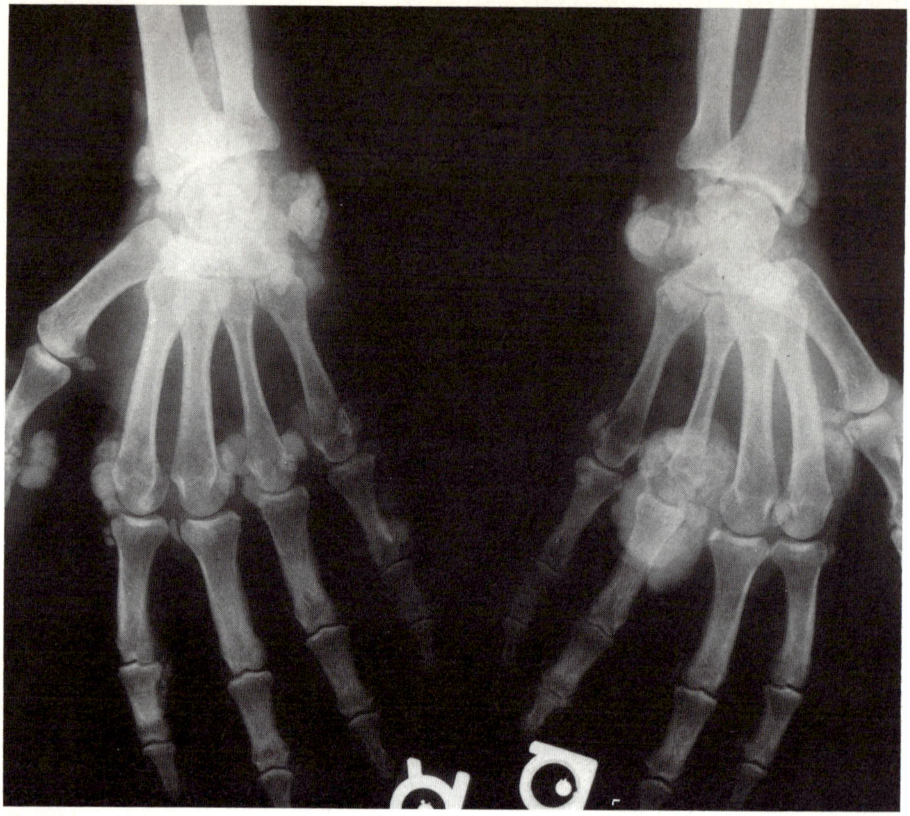

Fig. 3b

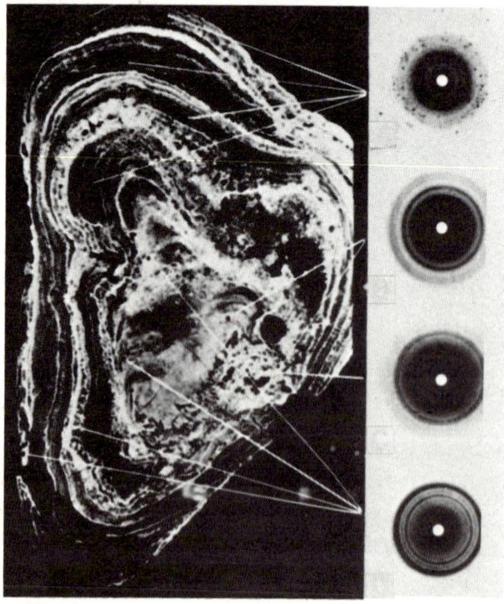

Fig. 3c

Fig. 3. a–c. Examples of pathological tissue calcifications. *a* calcified human heart; *b* tumoral calcifications; *c* thin section of a urinary stone showing the co-existence of different types of calcium phosphates: apatite, whitlockite, octacalcium phosphate, struvite. The calcium deposits of *a* and *b* are ACP and apatitic calcium phosphates, respectively (17, 59)

Table 1. Occurrences of phosphate minerals in human tissues

Phosphate minerals	Chemical formula	Occurrences
Apatite or "apatitic" calcium phosphates	$(Ca, Na, Z)_{10}$ $(PO_4, CO_3, Y)_6(OH, X)_2$[a]	Enamel[c], dentine[c], bone
		Salivary stones, dental calculi, urinary stones, soft-tissue calcifications (tumoral, periarticular)
Whitlockite, L–TCP	$(Ca, Mg)_9(PO_4)_6$[b]	Salivary stones, dental calculi, tubercular deposits, pulmonary calcifications, calcified cartilage (invertebral disc, abdominal, tracheal)
Octacalcium phosphate OCP	$Ca_8H_2(PO_4)_6 \cdot 5H_2O$	Dental and urinary calculi
Brushite, DCPD, calcium phosphate dihydrate	$CaHPO_4 \cdot 2H_2O$	Dental calculi, concretions in old bones, chondrocalcinosis
Calcium pyrophosphate dihydrate, CPPD (mono- and tri-clinic)	$Ca_2P_2O_7 \cdot 2H_2O$	Pseudo-gout deposits, in synovium fluid
Struvite	$MgNH_4PO_4 \cdot 6H_2O$	Urinary stones
Newberyite	$MgHPO_4 \cdot 3H_2O$	Urinary stones (rare)
Amorphous calcium phosphate, ACP (containing Mg^{2+} and/or $P_2O_7^{4-}$)	Variable composition	Non-visceral calcifications associated with uremia; aortic valves

[a] Where Z = cationic substituents, e.g., Sr^{2+}, Pb^{2+}, K^+, etc; Y = HPO_4^{2-}; X = F^-, Cl^-
[b] Modified by Dickens and Brown (18) to $Ca_7Mg_9(Ca, Mg)_2(PO_4)_{12}$
[c] Normal tissue calcifications
All other occurrences are pathological or abnormal tissue calcifications.
References: 2–5, 12, 17, 20–27, 33, 34, 36, 38, 40, 43, 44–46, 48–51, 58–60, 62–63, 69, 73, 77, 82, 83, 84, 91, 94–97, 100–103, 105–111, 113, 117

The general occurrence and co-existence of phosphate minerals in human tissues are summarized in Table 1 and are dicussed in detail in the following sections. Structural considerations have been described elsewhere (7, 8, 13, 18, 81, 86, 114, 115).

Amorphous Phosphate Minerals

Amorphous calcium phosphates, amorphous calcium-magnesium phosphates, and amorphous calcium-magnesium pyrophosphate have been identified in vivo and in vitro (1, 9, 10, 15, 28, 43, 47, 51, 56, 59, 61, 74, 91, 105). "Amorphous" synthetic and biological phosphate minerals are characterized by: (a) absence of definite peaks in their X-ray diffraction patterns (Figs. 1C, D); (b) lack of resolution in their infrared absorption spectra (e.g., Fig. 4C); (c) spheroidal, holey morphology when examined with electron microscope (61, 90) as shown in Fig. 5B; and (d) short- order range in their radial distribution patterns (9).

ACP is believed to be a constituent of bone mineral, the ACP/apatite ratio, still a debatable point (9, 95, 105). The ACP in bone is believed to transform to apatite,

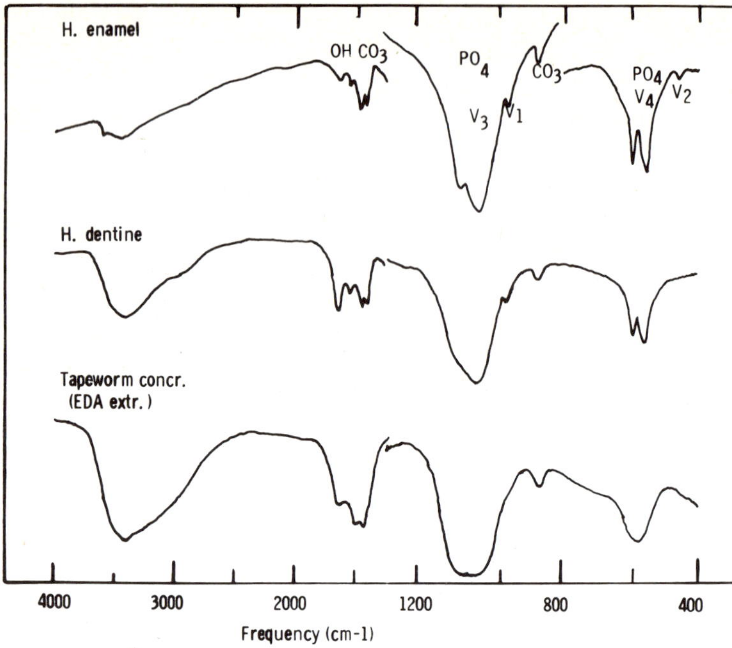

Fig. 4. Infrared spectra of normal human *A, B* and non-human calcifications. The X-ray diffractions corresponding to *A* and *B* are shown in Fig. 1A, B; to *C*, to Fig. 1C. The difference in the resolution of the absorption bands reflects the difference in crystallinity also observed in their X-ray diffraction patterns

thus acting as a transitory precursor of biological apatites just as they are supposed to be in some synthetic preparations (9, 10, 91, 95, 105). Stable ACP has been identified in non-human tissues, e.g. tapeworm concretions (Fig. 4C) and in human pathological soft tissue calcifications (17, 43, 59, 61, 112). These stable biological ACP contain magnesium plus carbonate (in the case of tapeworm concretions) and magnesium plus pyrophosphate (in the case of soft-tissue calcifications).

In synthetic systems (precipitations at 37 °C), the presence of carbonate or magnesium or pyrophosphate promotes the formation of ACP (47, 61). The simultaneous presence of carbonate and magnesium, carbonate and pyrophosphate, magnesium and pyrophosphate were even more effective in promoting the formation of ACP (47, 61). ACP formed in the presence of CO_3^{2-} transformed to apatite upon standing in the mother liquor or when hydrolyzed in water at 100 °C (47, 61). However, the ACP formed in the presence of Mg^{2+} or ACP hydrolyzed in the presence of Mg^{2+} or $P_2O_7^{4-}$ resisted hydrolysis to apatite (10, 28, 47, 61, 107). The comparative efficiency in promoting the formation of ACP parallels the efficiency in stabilizing the amorphous phase (47, 61) as abbreviated below:

$$[Mg^{2+} + P_2O_7^{4-}] \gg [P_2O_7^{4-}] \gg [Mg^{2+} + CO_3^{2-}] \gg [Mg^{2+}] \gg [CO_3^{2-}].$$

Since the presence of Mg^{2+} has been shown to stabilize the amorphous phase in synthetic systems (10, 47, 59, 61) and the presence of pyrophosphate was shown

a

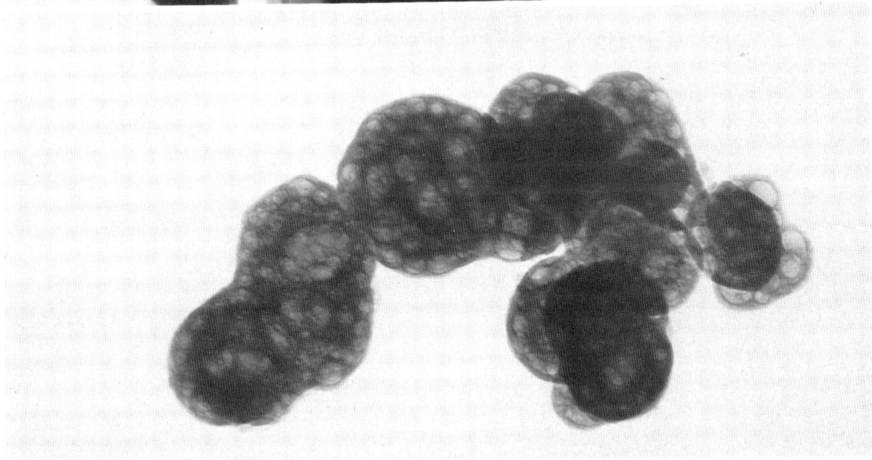

b

Fig. 5a, b. Electron micrographs of *a* human dentine and *b* ACP from tapeworm concretions. The EM shown in *b* is characteristic of all ACP regardless of composition (47, 61, 91)

to have even a greater stabilizing effect (28, 47, 61, 100); then the stability of ACP in human pathological tissue calcifications can be attributed to their magnesium and/or pyrophosphate contents (1, 17, 59).

Characterization of ACP by X-ray diffraction is not very helpful as seen in Fig. 1C, D. IR analyses give information on the involvement of other groups such as CO_3^{2-} or $P_2O_7^{4-}$ without chemical analyses (59, 61). A useful indirect method is by pyrolysis of the material. Such method causes the formation of stable phases from which the approximate composition of the milieu (and, indirectly, of the amorphous mineral) can be deduced (59, 61). For example, X-ray diffraction pat-

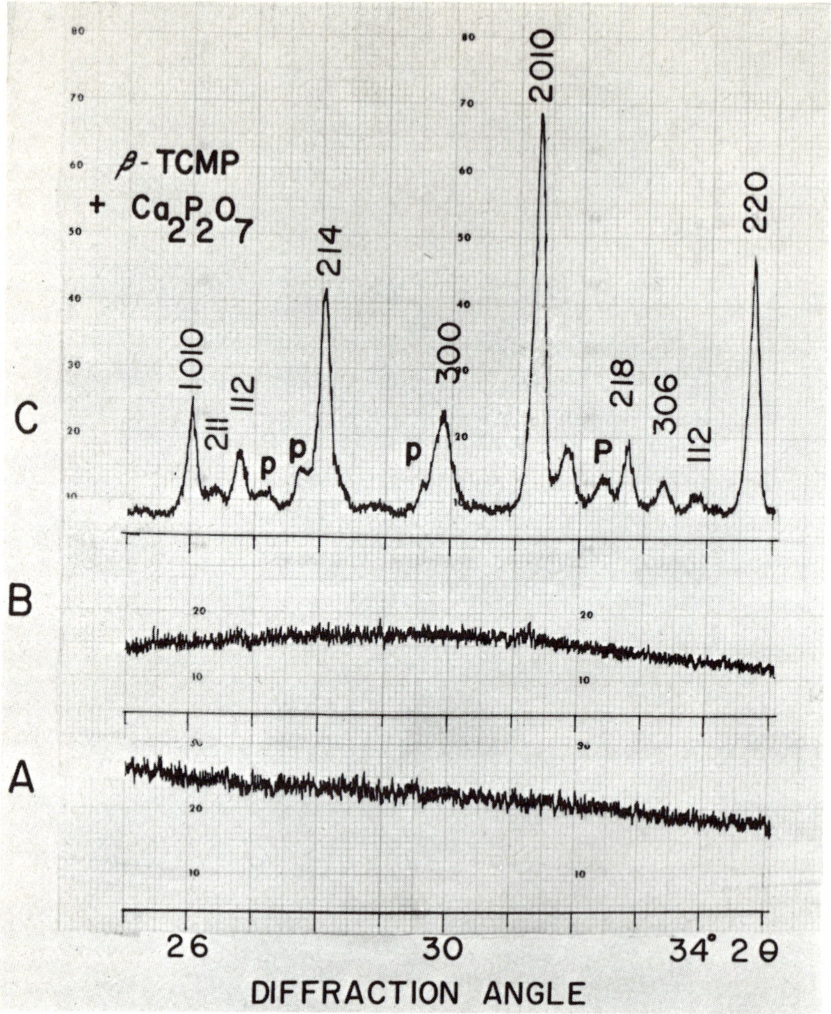

Fig. 6. Effect of pyrolysis on ACP. *A* Before pyrolysis; *B* after pyrolysis at 400 °C; *C* after pyrolysis at 700 °C showing the phases of β-TCP (Mg-containing) and γ-$Ca_2P_2O_7$. From the products of pyrolysis, one can deduce that the amorphous calcium phosphate mineral formed from milieu containing Ca^{2+}, Mg^{2+} and $P_2O_7^{4-}$ and HPO_4^{2-}

terns of the pyrolysis products from an amorphous mineral found in soft-tissue calcification (59) showed β-TCP (Mg-substituted) and γ-$Ca_2P_2O_7$ (Fig. 6C). Such information, combined with IR absorption and chemical analyses indicated that this mineral was formed from a milieu containing Ca^{2+}, Mg^{2+} and $P_2O_7^{4-}$.

Calcium Phosphate Dihydrate, DCPD, Brushite, $CaHPO_4 \cdot 2H_2O$

In vivo, DCPD crystals have been identified in "young" (3-day old) dental calculus samples (101), in "old" (800-yr old) human bones (84), and in several types of

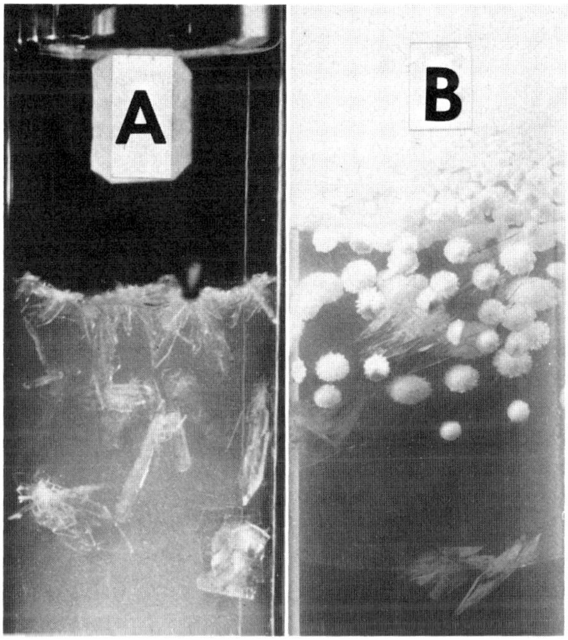

Fig. 7 A, B. Growth of calcium phosphates in gel media. *A* only DCPD, *B* DCPD co-existing with OCP (spherules). Starting pH 5 and 7, respectively, maintained at 37 °C (57, 58, 78)

pathological calcifications (26, 27, 48, 78, 85, 101). In vitro, DCPD crystals have been grown from solutions (7, 24, 47, 57, 66, 78, 92); from gel (silica, agar, collagen) systems (57, 58, 65, 78); from urine (24, 92); from unstimulated human saliva (70). DCPD crystals have also been grown on enamel and dentine surfaces of human and shark's teeth (57, 73, 78); on synthetic OH-apatite (7); on calcite crystal surface (75); and on calcium oxalate crystal surfaces (85). Because of the close relation of DCPD to apatite, DCPD has been proposed to be: (a) the nidus for the formation of some phosphatic and oxalic urinary stones (24, 92); (b) precursor in the formation of bone and tooth apatites (30, 98); initial extra-osseous calcium phosphate deposit in non-visceral calcifications in patients with hypercalcemia (1).

The formation of DCPD in synthetic (e.g., gel, solutions) and in biological systems (urine, saliva) is governed by factors such as the nature of the media, pH, temperature and presence of other "trace elements" (e.g., F^-, Mg^{2+}, $Sr^{2+} \cdot Cd^{2+}$, Mn^{2+}, $P_2O_7^{4-}$), and to a much lesser extent, the Ca/P ratio of the milieu (57, 58, 65, 66, 78). The properties affected by these factors are size and shape and crystal growth. Under similar conditions of Ca/P, pH and temperature, DCPD will grow in tabular habit in solution and in silica gel systems (Fig. 7A) and grows in both tabular and "clustered rods" from agar gel systems (Fig. 8). The growth of DCPD as rods originating from a common center was seen in silica gel and solution systems when low concentrations of $P_2O_7^{4-}$ was present (57, 58). The pH dependence of DCPD was observed – from solutions, nucleation took significantly longer time at ph 4 than at pH 6; in both gel and solution systems, crystals grown at pH 3 were larger than those at pH 6. DCPD crystals can be grown in gel and solutions at 5, 25, and 37 °C but not at 60 °C and higher temperatures. Growth of DCPD crystals

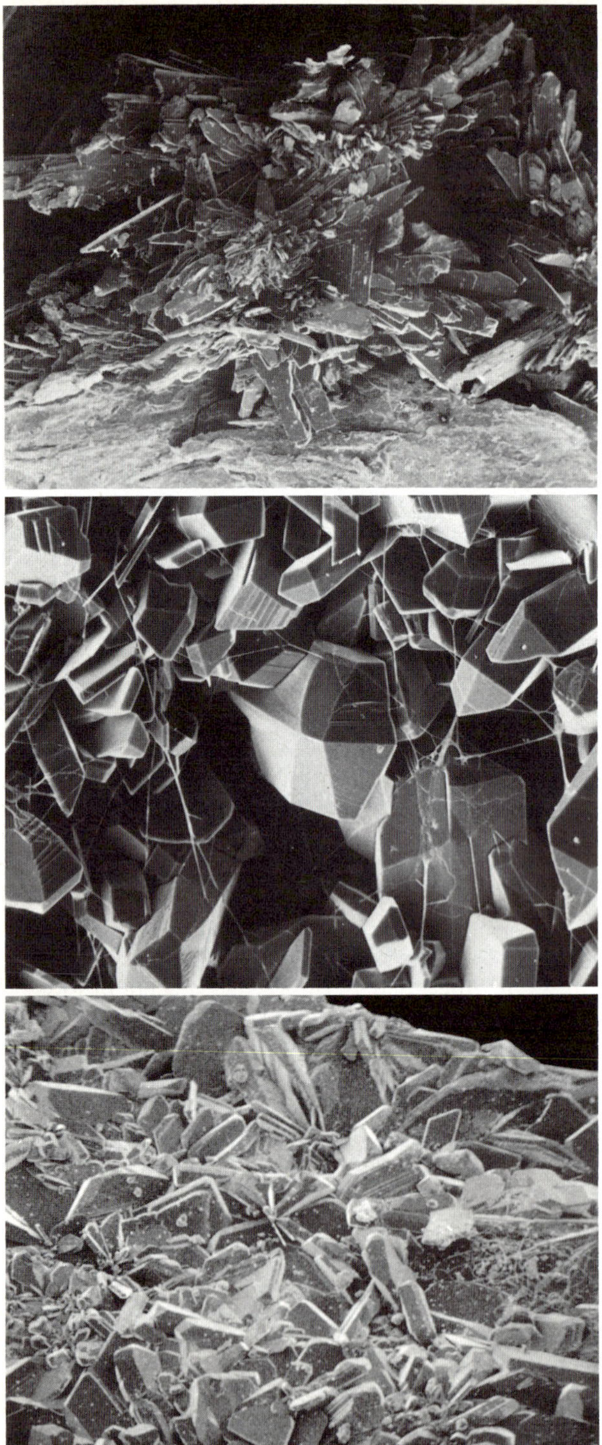

a

b

c

Fig. 8 a–c. Morphology of DCPD crystals growing on different Ca-bearing substrates: **a** on human bone, **b** on human enamel, **c** on calcite. **b** and **c** were grown in vitro at 37 °C, pH 5 when human tooth and calcite crystal were suspended separately in acidic phosphate solutions (57, 58, 75, 78). **a** were found in human bone from a 800-yr-old grave (84). In all cases, DCPD crystal grew by dissolution of apatite in **a** and **b** or of calcite in *c* and recrystallization of the more stable DCPD phase under acid pH. SEM magnifications × 100

Fig. 8 d–e

from different calcium-bearing surfaces (Fig. 9) also show differences in morphology.

Under conditions of changing pH, DCPD can co-exist with other calcium phosphates: with OCP when the starting pH is 7 (Fig. 7B); with apatite and OCP when the starting pH is 9 (58,62). Presence of other ions have been shown to suppress the crystal growth of DCPD and favor the growth of other calcium phosphates, e.g., in the presence of F, apatite is favored, in the presence of Cd^{2+}, Pb^{2+}, OCP is favored (62, 66, 78).

Formation of DCPD in vivo can be either by direct precipitation or by dissolution of apatite and reprecipitation of the more stable DCPD crystals under acid conditions. The different morphologies of DCPD in pathological calcifications may be due to the presence of other ions, e.g., $P_2O_7^{4-}$, Mg^{2+}, etc., as demonstrated in synthetic systems. X-ray diffraction patterns and infrared absorption spectra of DCPD crystals found in human tissues are identical to those of pure synthetic materials (Figs. 10a, 11a).

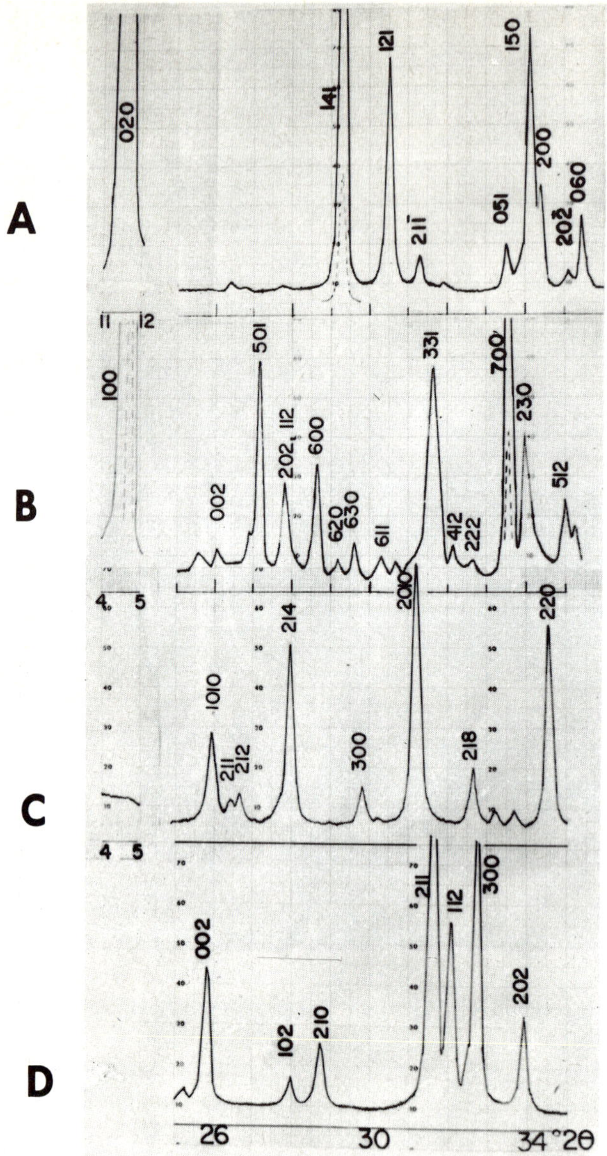

Fig. 9. X-ray diffraction patterns of calcium phosphate obtained from gel systems. *A* DCPD; *B* OCP; *C* β-TCP (Mg-containing); *D* OH-apatite.

Octacalcium Phosphate, OCP, $Ca_8H_2(PO_4)_6 \cdot 5H_2O$

OCP is one of the major crystalline components of dental calculi, usually co-existing with whitlockite and/or apatite (48, 101). The striking structural similarities of OCP and calcium OH-apatite (13, 114) prompted the theory that OCP is a precursor for biological apatite (13, 14, 16). The fact that OCP has not been detected as a constituent of enamel, dentine or bone mineral (Fig. 1A, B) has been explained

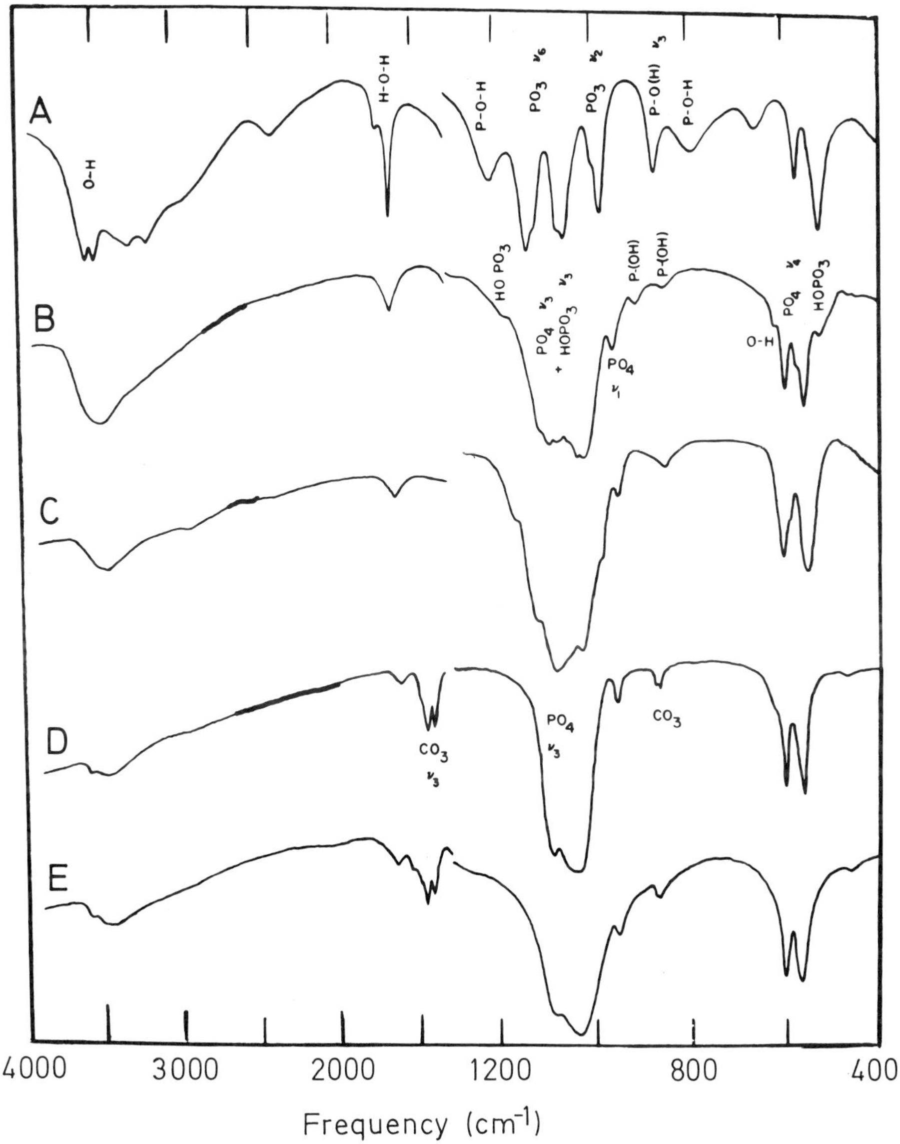

Fig. 10. Infrared absorption spectra of different calcium phosphates obtained from gel systems: *A* DCPD; *B* OCP; *C* β-TCP (Mg-containing); and *D* CO_3-containing OH-apatite; compared to the IR spectrum of human *enamel, E*

as being due to the rapid transformation of OCP to apatite in biological systems (14).

The formation of OCP in gel systems is reproducible while its formation from solutions (by precipitation) has been unpredictable (57, 58, 62). The most common method of preparing OCP is by the hydrolysis of DCPD in acetate buffer (13). OCP formation in synthetic systems is sensitive to changes of pH and temperature and

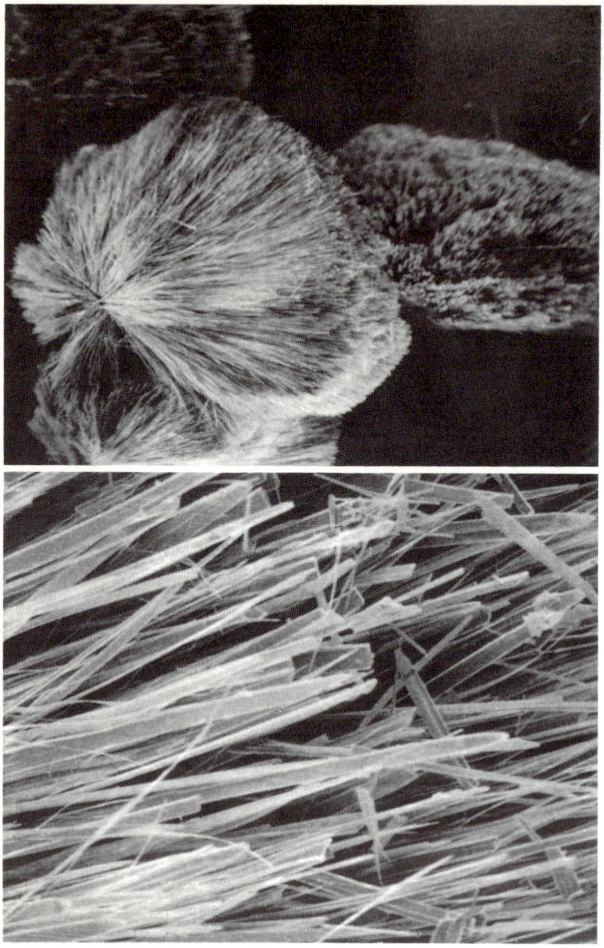

a
b

Fig. 11 a, b. SEM of OCP crystals grown in vitro by precipitation at 55 °C, pH 6. **a** OCP spherules, × 50; **b** individual OCP crystals comprising the spherules, × 500

to the presence of "trace elements". In the gel system (37 °C) where pH is not controlled, OCP will co-exist with DCPD when the starting pH is 7 (Fig. 7B); with apatite when the starting pH is 9 (58); and with β-TCP in the presence of Mg^{2+} in the media (58). In synthetic systems, OCP will form with or without apatite at 37 to 60 °C, but will not form at 100 °C (47, 58, 62). Once formed, OCP is stable in contact with gel (for more than 10 yrs) or in the air. OCP in dental calculus samples stored for more than 10 yrs also remained stable (48).

OCP usually forms as spherules or clusters consisting of individual needle-like crystals growing from a common center (Fig. 11). The OCP crystals are morphologically similar to apatite needles. OCP in human tissues and in synthetic systems vary in crystallite size. Since some of the diffraction peaks of OCP are close to those of apatite (Fig. 9), the presence of small amount of micro-crystals of OCP may escape detection by the conventional methods of analyses (e.g., X-ray diffraction, Infrared absorption). The formation of γ–$Ca_2P_2O_7$ upon pyrolysis of OCP at 700 °C is an indirect method for detecting the presence of OCP mixed with apatite (29).

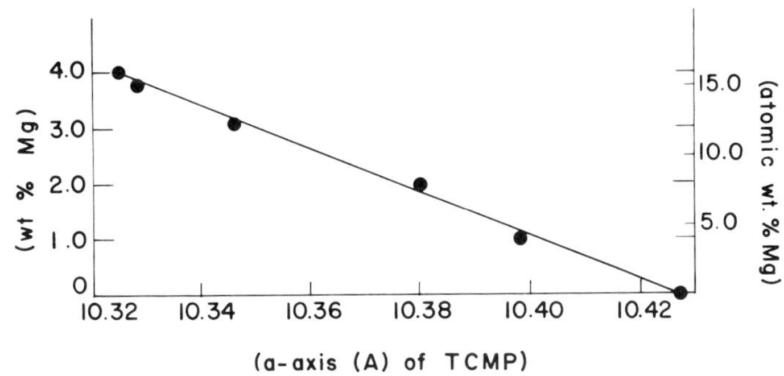

Fig. 12. Effect of Mg^{2+} incorporation in the β-TCP structure on the a-axis dimension. Maximum substitution in $Ca_9(PO_4)_6$ is 16 at wt.-% Mg (of 4 wt.-%). The Mg-for-Ca substitution in the whitlockite structure causes a contraction in the a-axis (47, 48, 100)

Whitlockite, β-TCP (or β-TCMP), (Ca, Mg)$_9$(PO$_4$)$_6$

Biological whitlockites are always magnesium-substituted, ~(Ca, Mg)$_9$(PO$_4$)$_6$, with varying Mg^{2+} concentrations (45, 47, 48, 100). They are not detected as constituents of normal tissue calcifications (teeth and bone mineral) but their presence in several pathological tissue calcifications and in human dental carious lesions has been reported (31, 45, 47, 48, 60, 69, 99, 101, 106, 108, 111). Whitlockite was found in greatest abundance in salivary stones and in dental calculi and rarely in urinary stones. It has been detected in several pathological tissue calcifications (abdominal aorta, appendix, lymph nodes of the mesentery) and in certain cases of abnormally calcified cartilage (invertebral disk, tracheal cartilage, nasal septum) and tubercular deposits. β-TCMP also appears as a separate phase when tooth enamel or dentine or apatitic and 'amorphous' calcium phosphate or amorphous calcium-magnesium-phosphate from pathological tissue calcifications are ignited at 700 °C (47, 51, 59, 61, 107).

Pure whitlockite, β-TCP, $Ca_9(PO_4)_6$, is not found in biological systems nor can it be prepared from aqueous systems. Pure whitlockite can only be prepared at high temperatures (900 °C) by solid state reactions. Biological whitlockites are characterized by shorter *a*-axis dimensions compared to pure β-TCP due to the partial Mg-for-Ca substitution (Fig. 12). The presence of Mg^{2+} in biological and in synthetic systems allows the formation of β-TCP at low temperatures (37 to 100 °C) and from aqueous systems (47, 48, 60, 74, 100). The presence of Mg^{2+} in biological and synthetic whitlockites stabilizes them from being hydrolyzed to apatite (47, 60, 74). Other ions, e.g., Mn^{2+}, Fe^{2+}, Co^{2+}, Ni^{2+} also promotes the formation of β-TCP (at the expense of apatite) from aqueous systems (74, 89), however, in biological systems, the influence of Mg^{2+} is the most significant. The Mg/Ca ratio in the biological and in synthetic milieu is the major determining factor in the formation of the whitlockite phase at the expense of apatite (47, 48, 51, 60, 61). However, in the presence of either CO_3^{2-} or F^- ions, the effect of the Mg/Ca ratio is modified or offset, such that the formation of the apatite phase rather than the whitlockite phase is favored (Fig. 13). In the gel systems, an increase in the Mg/Ca ratio causes the formation of TCP and OCP instead of apatite and OCP (58).

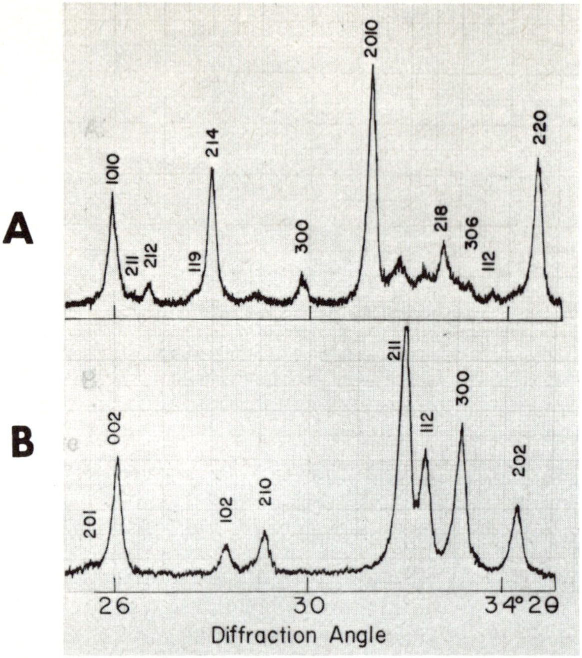

Fig. 13. Effect of fluoride on the formation of whitlockite. *A* formed without fluoride, mostly *β*-TCP (mg-containing) mixed with apatite (unlabelled diffration peaks); *B* formed in the presence of F^-, only (F, OH)-apatite (60)

Calcium Pyrophosphate Dihydrate, CPPD, $Ca_2P_2O_7 \cdot 2H_2O$

The triclinic and monoclinic forms of calcium pyrophosphate dihydrate [CPPD(T) and CPPD(M), respectively], crystals occur in synovium in joints characteristic of pseudo-gout diseases (15, 82, 97). CPPD in three forms (monoclinic, triclinic and orthorhombic), other hydrated forms of calcium pyrophosphates and mixed calcium-sodium pyrophosphate have been obtained from solutions or gel systems (15, 81, 97). The presence of Mg^{2+} in synthetic systems suppressed the formation of CPPD(T) and only allowed the formation of CPPD(M) mixed with what is believed to be (Ca, Mg)P_2O_7 (15). In other synthetic systems, the simultaneous presence of Mg^{2+} and $P_2O_7^{4-}$ caused the formation of amorphous calcium phosphate (20, 47, 61). In biological systems, the presence of amorphous calcium-magnesium pyrophosphate has been observed in soft-tissue calcifications in uremic patients (17, 59).

Struvite, $MgNH_4PO_4 \cdot 6H_2O$

Struvite is commonly found associated with apatite as a constituent of urinary stones which form as a consequence of bacterial infection (22, 34). In vitro experiments showed that struvite crystal formation was induced by bacterial (*Proteus*) or by alkalinization using NH_4OH (34, 80). In gel systems, struvite crystals grew very easily in systems containing Mg^{2+}, HPO_4^{2-}, PO_4^{3-}, $NH_4{}^+$; and the simultaneous presence of $P_2O_7^{4-}$ caused reduction in crystallite size, habit modifica-

a

b

Fig. 14 a, b. Struvite, $MgNH_4PO_4 \cdot 6H_2O$ grown in vivo **(a)** and in vitro **(b)**. Struvite is sometimes found on the outermost layer of a urinary stone **(a)**; and mixed with newberyite **(b)** from gel systems (58, 62)

tion and twinning of the struvite crystals (58, 62). The simultaneous presence of Ca^{2+} with Mg^{2+} affected the formation of struvite. For example, in systems where the Mg/Ca ratio in solution was 1/0 or 1/1, the main product was struvite whose crystallite size decreased with increasing Ca^{2+} (or decreasing Mg/Ca) concentrations in the solution; when the Mg/Ca in solution approached 1/1.5, the main product was DCPD, and when Mg/Ca was 1/4, the main product was whitlockite (62). Struvite crystals found in vivo (e.g., surface of urinary calculi, Fig. 14A) have similar morphology to those obtained in vitro (Fig. 14B).

Table 2. Comparative composition and some physical properties of human enamel, dentine and bone

Composition[a]	Enamel	Dentine	Bone
Calcium, Ca^{2+}	36.5	27.5	24.5
Phosphorus, as P	17.7	13.0	11.5
(Ca/P) by weight	2.06	2.12	2.33
(Ca/P) molar	1.58	1.62	1.80
Sodium, Na^+	0.3	0.6	0.7
Potassium, K^+	0.08	0.05	0.02
Magnesium, Mg^{2+}	0.34	0.81	0.55
Carbonate, CO_3^{2-}	3.5	5.0	6.0
Fluoride, F^-	0.01	0.02	0.02
Chloride, Cl^-	0.30	0.01	0.10
Pyrophosphate, $P_2O_7^{4-}$	0.02	0.08	0.05
ash (total inorganic)	97.0	70.0	65.0
total organic	1.0	20.0	25.0
adsorbed H_2O	2.0	10.0	10.0
Trace elements: Sr^{2+}, Pb^{2+}, Ba^{2+}, Fe^{2+}, Zn^{2+}, etc.			
Crystallographic properties			
Lattice parameters (± 0.003 Å)			
a-axis	9.441	9.419	9.417
c-axis	6.882	6.880	6.882
"Crystallinity index"	70–75	33–37	33–37
Crystallite size (Å)	1,300 × 300	200 × 40	200 × 40
Morphology	Needles	Rods	Rods
Ignition products (950° C)	Apatite + *B*–TCP[b]	Apatite + β–TCP[b]	Apatite + CaO

[a] Composition is expressed as wt.% of dried tissue. If expressed as wt.% ashed (500° C), the differences in calcium and phosphorus contents will be less significant and those of the minor constituents more significant

[b] β–TCP from biological systems are always Mg-containing

References: 4, 47, 49, 51, 52, 64, 67, 83, 107, 117

Newberyite, $MgHPO_4 \cdot 3H_2O$

Newberyite is a rare component of renal stone (102, 112). In the gel systems, neuberyite sometime formed with struvite (Fig. 14B), its formation being affected by pH and easily inhibited by the presence of other ions such as Sr^{2+}, Cu^{2+}, Mn^{2+}, and $P_2O_7^{4-}$ (58, 62).

Apatites, $\sim Ca_{10}(PO_4)_6(OH)_2$

Biological apatites are found as the principal inorganic phase of normal tissue calcifications (enamel, dentine, cementum, bone) and found associated with other phosphate- and non-phosphate minerals in pathological calcifications (2, 4, 12, 14,

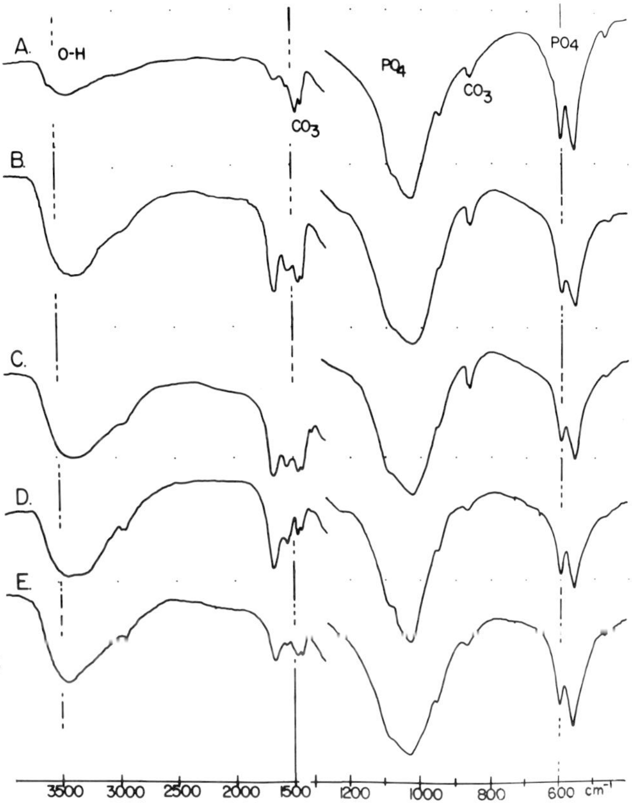

Fig. 15. Biological apatites from normal and pathological tissue calcifications. *A* Tooth enamel; *B* tooth dentine; *C* bone; *D* salivary duct stone; *E* innermost layer of a urinary stone. Note that all biological apatites contain carbonate (C–O absorption bands) in varying amounts

17, 19, 22–24, 31, 33, 34, 36, 40, 44–48, 51, 58, 59, 62, 79, 83, 93, 95, 102, 103, 106, 109, 112).

The apatites of human teeth and bones have been idealized as calcium hydroxy-apatite, approximating $Ca_{10}(PO_4)_6(OH)_2$ (8, 21, 36). However, differences in composition and in other properties make the biological apatites different from pure calcium hydroxyapatite and from each other (Table 2) and should be more appropriately referred to as "impure hydroxyapatites" (14, 21, 22, 47, 51, 83, 115). The biological apatites are microcrystalline (compared to the macro dimensions of mineral apatites), of variable composition and therefore heterogeneous even within each category and even within each calcified tissue (enamel, dentine, bone), thus making an accurate crystallochemical analysis and definition a difficult, if not an impossible undertaking. Some of the vexing problems concerning the biological apatites are: (a) non-stoichiometry (i.e., Ca/P of biological apatites) ranges from 1.54 to 1.73 compared to 1.67 for pure Ca-OH-apatite; (b) involvement (structural or surface) of some "impurities" e.g., Na^+, Mg^{2+}, F^-, HPO_4^{2-}, CO_3^{2-}, $P_2O_7^{4-}$, etc.; (c) co-existence or preexistence of possible precursors such as DCPD, OCP, TCP, ACP. Because of these difficulties, many studies on biological apatites have been made indirectly by studying apatites prepared from aqueous and non-aqueous systems. Such studies, however, cannot always be directly extrapolated to biological apatites (50, 51).

The biological apatites are uniquely similar in that they all contain carbonate in varying amounts as a substituent for phosphate in the apatite structure, as shown in Fig. 15 (33, 47, 50, 54, 55, 83). They differ in crystallite sizes, shapes and other physico-chemical properties such as chemical (susceptibility to acid dissolution) and thermal stabilities (14, 21, 23, 42, 47, 51). Apatite crystallites of enamel are much larger than those of dentine or bone – a difference reflected in their X-ray diffraction patterns (Fig. 1) and their infrared absorption spectra (Fig. 4) and confirmed by electron microscopic investigations. Apatites from pathological tissue calcifications (e.g., salivary stones, dental calculi) show greater crystallinity (reflecting greater crystallite size) than those of bone or dentine but less than those of enamel, except in the case of enamel pearls or enameloma in which the crystallinity of the apatite is more similar to that of enamel (2). Similarities and differences among biological apatites are demonstrated in Figs. 1, 2, 4, and 15.

Based on studies in synthetic systems (14, 22, 47, 50, 51, 54–56, 63, 64, 68, 72, 74, 75, 115), some of the "impurities" are presumed associated with biological apatites in the following abbreviated manner:

$Ca_{10}(PO_4)_6(OH)_2$

$\nearrow$	$\uparrow$	$\nwarrow$
Sr, Na,	CO_3,	Cl, F
K, Mg*,	HPO_4	H_2O
Z	P_2O_7*	CO_3

[*may also be surface-limited; Z = Pb, Ba, Zn, Fe, other cationic trace elements]

Considering that biological apatite is basically a calcium hydroxyapatite, other "impurities" can substitute for the Ca^{2+}, PO_4^{3-}, or OH^-, while others may be predominantly surface-bound. Again, based on observations from in vitro systems, some of these "impurities" affect the "crystallinity" (reflecting crystallite size and/or strain), morphology, lattice parameters of the biological apatites (4, 5, 47, 51) and as a consequence, impart stability or instability to the tissues involved. Some of these "impurities" are associated with the susceptibility or resistance of human tooth enamel to acid dissolution (caries) and of bone to resorption processes (5, 51, 53, 73, 76, 77). The effects of some of these "impurities" on the formation and stability of apatite are briefly discussed in the following sections.

Carbonate. The incorporation of carbonate in synthetic (aqueous) and biological apatites is coupled with the incorporation of Na^+ (47, 51). In synthetic non-aqueous (1,000 °C) systems, the mode of carbonate incorporation is CO_3-for-OH (type A); in aqueous systems (25 to 100 °C), CO_3-for-PO_4 (type B); and in biological apatites, predominantly CO_3-for-PO_4 (10, 23, 47, 50–56, 115). The extent of carbonate incorporation is influenced by the presence of other ions such as Na^+, K^+, Sr^{2+}. Na^+, and to a lesser extent, K^+ favors greater incorporation of carbonate into the apatite structure, Sr^{2+} limits the extent of incorporation.

The incorporation of CO_3^{2-} in apatite affects its properties in the following manner: (a) causes reduction in crystallite size; (b) causes changes in morphology from needles to rods to equi-axial crystals; (c) causes contraction in the *a*-axis and expansion in the *c*-axis dimensions when compared to unsubstituted calcium hydroxyapatite; (d) causes internal strain; (e) causes chemical instability (i.e., increases solubility; and decreases thermal stability). These effects are observed in

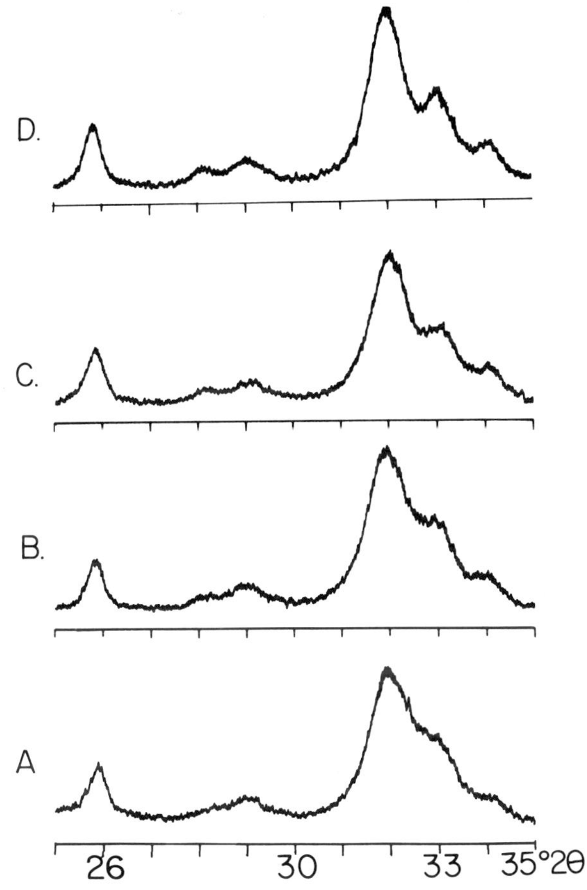

Fig. 16. Effect of fluoride incorporation on the "crystallinity" of the bone mineral. The fluoride contents (wt.-% in bones ashed at 500 °C) are: *A* 0.06; *B* 1.7; *C* 1.8; *D* 2.2. Fluoride incorporation causes increase in crystallinity reflecting increase in crystallite size and/or decrease in strain (77)

CO_3-containing apatites prepared by precipitation or by hydrolysis of monetite ($CaHPO_4$) or DCPD or OCP in CO_3-containing solutions maintained at 37, 60, 80 or 100 °C (16, 47, 51–55, 73, 76, 83, 89, 116).

Considering the effect of carbonate alone, the difference in crystallinity between enamel, dentine, bone (Figs. 1A, B, 4, 15) may be explained in part by the difference in their carbonate contents. The apatite with the least carbonate (enamel) has greater crystallinity (greater crystallite size and/or less strain) than the apatite with more carbonate (bone or dentine) showing less crystallinity (smaller crystallite size and/or greater strain). The differences in their solubilities in acid buffers (enamel being less soluble than either dentine or bone mineral) may also be explained in the same manner.

Fluoride. F-apatites, idealized as $Ca_{10}(PO_4)_6F_2$, are the most abundant kind of mineral apatites, containing carbonate in most cases, e.g. staffellite, marine phosphorites (47, 75, 83). In human tissues, partially substituted or (F, OH)-apatites in enamel and bone result from reactions of ingested or topically applied fluoride with enamel or bone apatites (5, 21, 51, 77).

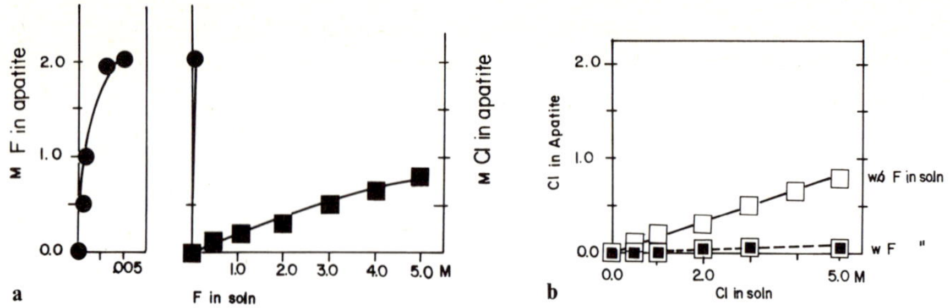

Fig. 17 a, b. Discrimination of synthetic apatite for the incorporation of fluoride and against the incorporation of chloride **(a)**. The incorporation of Cl^- in synthetic (aqueous) and biological apatites is very limited, especially in the simultaneous presence of F^- in the solution **(b)**

The observed effects of fluoride incorporation on the properties of biological and synthetic apatites are: (a) increase in crystallite size; (b) decrease in strain due to increased stability of the apatite structure; (c) contraction in the a-axis and no significant change in the c-axis when compared to unsubstituted calcium hydroxyapatite; (d) increase in chemical stability (decrease in solubility); and (e) increase in thermal stability (5, 6, 47, 51, 73, 76, 77, 87, 114). In vivo effect of ingested fluoride on the "crystallinity" of bone apatite whose fluoride content varied linearly with the concentration of administered fluoride is shown on Fig. 16.

The stabilizing effect of F^- as a substituent for OH^- in the apatite structure causes the above effects on the properties of synthetic and biological apatites. These observations explain the benefical effects of ingested or topically applied fluoride in the management of dental caries and of fluoride therapy in the management of osteoporosis.

Chloride, Cl^-. The maximum chloride incorporation in apatites obtained from non-aqueous systems is 100 mol.-% substitution, i.e., 2Cl for 2OH (6.7 wt.-% Cl) while in apatites from aqueous systems, the maximum incorporation is only 40 mol.-% (49, 51). The average chloride content of human enamel apatite is about 0.25 wt.-%, of bone or dentine, less than 0.01 wt.-% (47, 49, 117). Considering the significant concentrations of Cl^- ions in biological fluids, the small amounts of Cl^- incorporated in biological apatites and in apatites prepared from aqueous systems indicate that the forming apatites in these systems strongly discriminate against the incorporation of Cl (49). This is in marked contrast to the discrimination in favor of F^- incorporation into the apatite even under conditions of very low F^- concentrations (Fig. 17) in the milieu.

The substitution of Cl-for-OH in apatite [to form $Ca_{10}(PO_4)_6Cl_2$ or $Ca_{10}(PO_4)_6(OH, Cl)_2$ from non-aqueous and aqueous systems, respectively] results in the expansion of the a-axis and contraction of the c-axis dimensions (49, 114). No significant effect on the "crystallinity" of apatites have so far been observed as a consequence of chloride incorporation.

The atomic arrangement in Cl-apatite compared to OH- and F-apatites (114) and its large unit cell volume (Table 3) suggest that chloride ions contribute to the instability of the apatite structure and to its properties. The effect of Cl on the solubility of Cl-containing apatites has not been investigated. However, preliminary studies demonstrated that Cl-containing apatites are less stable thermally (and perhaps also chemically) than either OH- or F-apatites (50).

Acid Phosphate, HPO_3^{2-}. The non-stoichiometry of biological apatites, especially of human enamel apatite, has been partially attributed to the presence of HPO_4^{2-} ions, either surface-bound, latticebound (partially substituting for PO_4^{3-}), or as constituent of OCP interlayered with apatite (14, 22, 49, 51, 64). The HPO_4^{2-} contents of sound and carious human dental enamel (approximately 5 and 15 wt.-%, respectively) were determined on materials pyrolyzed at 400 and 500 °C using infrared methods (3). In biological and in synthetic apatites from aqueous systems, the original presence of HPO_4^{2-} is deduced from the formation of pyrophosphate upon ignition at temperatures 300 to 500 °C or formation of β-TCP upon ignition at 700 to 950 °C (32, 41, 47, 49, 64, 67). On this basis, enamel and dentine apatites contain much more HPO_4^{2-} than bone apatites (47, 51, 67).

The incorporation of HPO_4-for-PO_4 in apatite causes an expansion in the a-axis and slight contraction in the c-axis dimensions. This effect can partially explain the observed expanded a-axis of human dental enamel compared to synthetic (from non-aqueous systems) or mineral calcium OH-apatite (47, 49, 64). The effect of HPO_4^{2-} incorporation on the properties of apatite has not been directly investigated but can be predicted to cause instability causing the apatite to be more susceptible to acid-dissolution.

Pyrophosphate, $P_2O_7^{4-}$. The presence of pyrophosphate in solutions from which apatite forms cause suppression of crystal growth (28, 47, 51, 78, 110). Pyrophosphate concentrations in human enamel, dentine and bone differ, showing the least concentration in enamel (117). The presence of pyrophosphate in biological fluids, urine and saliva, has been associated with the inhibition of dental and urinary calculi formation (28, 110). The involvement of $P_2O_7^{4-}$ with the apatite structure, i.e., whether it is lattice- or surface-bound has not been determined. Its presence as a crystal surface poison has been implied (28, 62, 95).

Magnesium, Mg^{2+}. The presence of Mg^{2+} ions in solutions suppresses the crystal growth of apatite and promotes the formation of β-TCP at the expense of apatite (47, 51, 60, 74). At higher Mg/Ca ratio in the solutions, amorphous calcium magnesium phosphates are formed (47, 61).

The limited amount of Mg^{2+} (less than 1 wt.-%) in mineral, synthetic and biological apatites and the observed preferential release of Mg^{2+} during the initial dissolution of enamel, skeletal and synthetic apatites, were taken to indicate that Mg^{2+} in biological apatite is surface-limited and not lattice-bound (37, 47, 41, 60, 104).

Since the presence of Mg^{2+} causes disturbance in the crystallization of apatite, the lower amounts of Mg^{2+} in enamel compared to that in dentine or bone (Table 2) may partially explain the larger crystallite sizes of enamel apatite compared to those in dentine or bone.

Strontium, Sr^{2+}. Strontium subtitutes for calcium in synthetic apatites causing an expansion in both a- and c-axis dimensions as shown Table 3 (47, 51, 68, 74). In

Table 3. Effects of some substituents on the lattice parameter of synthetic apatites

Cationic substituents	Ionic rad (Å)	a-axis (Å)	c-axis (Å)
Strontium	1.12	9.768	7.214
Barium	1.33	10.150	7.795
Lead	1.20	9.880	7.410
(Calcium)	0.99	9.436	6.882
		9.422[a]	6.880[a]
Anionic substituents			
Fluoride	1.36	9.383	6.882
		9.375[a]	6.880[a]
Chloride	1.81	9.514	6.869
		9.634[a]	6.783[a]
(Hydroxyl)	1.53	9.436	6.882
		9.422[a]	6.880[a]
Substituents for PO_4^{3-}			
Carbonate	1.29	9.311	6.903
(Phosphate)	1.55	9.436	6.882
		9.422[a]	6.880[a]

a- and c-axis dimensions are ±0.003 Å

[a] Values for apatites prepared at high temperatures (1,000 °C), ref. 114. All other apatites in this table were prepared from aqueous systems at 100 °C (47, 49–51, 54, 68, 74)

CO_3-for-OH substitution is observed in apatites prepared at 100 °C. Such substitution causes an expansion of the a- and contraction of the c-axes when compared to CO_3-free apatites (10, 22, 50)

bone and tooth minerals, strontium is found as one of the minor constituents or "impurities" (5, 47, 51, 67, 95, 117). The incorporation of strontium in bone mineral was correlated with changes in the a-axis dimensions of the bone apatite (5), but in enamel apatite, the effect of Sr-for-Ca substitution on the *a*-axis was obscured by the simultaneous F-for-OH substitution which causes the opposite effect on the a-axis (63). In the bone mineral, the presence of strontium was associated with lower solubility (5), while in enamel, reports on the strontium effect have so far been contradictory.

Other "Impurities". Cations larger than Ca^{2+}, such as Sr^{2+}, Pb^{2+}, Ba^{2+} can be incorporated in the apatite structure to a much greater extent than cations whose ionic radii is similar to or smaller than Ca^{2+}, such as Na^+, Li^+, Mg^{2+}, Zn^{2+} (47, 54, 74). An exception to this is the incorporation of K^+ whose size belongs to the first category (i.e. larger radii than Ca^{2+}) but whose incorporation is similar to the ions in the second category (i.e. radii smaller then Ca^{2+}). Substitutions of the larger cations for Ca^{2+} cause the expected expansion in the lattice parameters (Table 3).

Anions, such as SO_4^{2-}, MnO_4^-, CrO_4^{2-} can substitute for PO_4^{3-} (83). However, such substitutions do not occur to a significant extent in biological apatites.

"Lattice H_2O". Pyrolysis below 200 °C of enamel apatites and apatites prepared from aqueous systems causes loss of weight without any significant change in their

lattice parameters; pyrolysis at temperatures between 300 and 400 °C causes a contraction in the a-axis dimensions concomitant with weight loss (41, 47, 49, 64, 67). This latter effect has been explained as being due to either loss of "lattice H_2O" or loss of H_2O from reactions between HPO_4^{2-} groups or reactions between HPO_4^{2-} and CO_3^{2-} (47, 64). Either H_2O-for-OH or HPO_4-for-PO_4 substitution in the apatite structure will cause an expansion in the lattice parameters and the loss of H_2O (from "lattice H_2O" or as a reaction product of HPO_4^{2-}) will cause the observed reduction in the a-axis resulting from the reordering of the apatite lattice after pyrolysis (64).

Concerted or Coupled Substitutions in the Apatite. The incorporation of Na^+ (and to a lesser extent, of K^+) in synthetic (aqueous) or biological apatites is coupled with the incorporation of CO_3^{2-} and/or HPO_4^{2-}, i.e., Na^+ and CO_3^{2-} for Ca^{2+} and PO_4^{3-} (47, 57, 64). Biological apatites with higher CO_3^{2-} contents also contain higher Na^+ contents (e.g., enamel apatite vs. dentine or bone).

The simultaneous or concerted substitutions of different ions in biological apatites (e.g., Sr^{2+}, Ba^{2+}, Pb^{2+}, K^+, Na^+ for Ca^{2+}; HPO_4^{2}, CO_3^{2-} for PO_4^{3-}; Cl^-, F^- or H_2O for OH^-) cause additive or conflicting effects on the physicochemical propertites of the apatite. Such propertites include: lattice parameters, "crystallinity" (expressing crystallite size and/or strain), chemical and thermal stabilitics. In some cases, the incorporation of one impurity suppresses the incorporation of another, e.g., the incorporation of F^- suppresses the incorporation of Cl^- or HPO_4^{2-}.

Thus, the observed expanded a-axis of human tooth enamel apatite compared to mineral or synthetic (non-aqueous) calcium hydroxyapatite (9.44 vs. 9.42 Å) despite its CO_3^{2-} content (known to cause contraction of the a-axis due to the CO_3-for-PO_4 substitutions in apatite from aqueous systems) may be explained by substitutions of HPO_4-for-PO_4, H_2O-for-OH, and larger cations-for-Ca. Also, based on observations on synthetic (aqueous) apatites, the presence of greater amounts of CO_3^{2-}, Mg^{2+} and $P_2O_7^{4-}$ in bone or dentine compared to human enamel, may account for the difference in their crystallinities.

The presence of some of the "impurities" contribute stability or instability to the apatite which is translated in terms of susceptibility to acid dissolution of the teeth and bone mineral. The presence of CO_3^{2-} causes instability (47, 53, 55, 56, 73, 76, 77, 89); of F^-, stability (5, 21, 47, 73, 76, 77, 87). When CO_3^{2-} and F^- are simultaneously present, the stabilizing effect of F^- becomes much more important than the opposite effect of CO_3^{2-} (73, 76).

Interrelationships of the Phosphate Minerals in Human Tissues

The formation of the different types of phosphate minerals in synthetic and biological systems is outlined in Fig. 18. Such a scheme takes into consideration the effects of pH, temperature and composition of the milieu on the type(s) of calcium or magnesium phosphate compounds formed. The factors of pH, temperature and composition are interdependent; the compositional factors having a more significant effect than the pH. For example, at pH 5, 37 °C, DCPD is the most stable

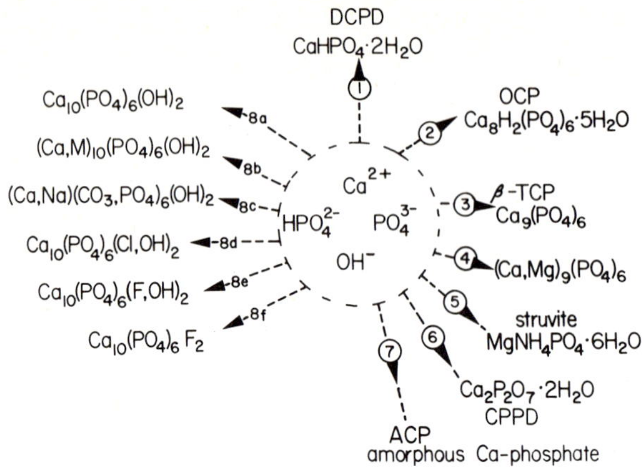

Fig. 18. Factors affecting the formation of different types of phosphate mineral in vitro. The environmental conditions: pH, temperature, composition of the milieu greatly affect the type(s) of phosphate mineral formed. In solutions containing only calcium and phosphate (HPO_4^{2-} and PO_4^{3-}) ions: *1* only DCPD crystals are formed when starting pH values are 4 to 6, temperature of the media, 5 to 50 °C; *2* OCP spherules at pH 6 and 6.5, 37 to 60 °C; *3* pure -TCP does not form in aqueous system; *4* Mg-containing-β-TCP forms between pH 6 to 9, 25 to 100 °C when Mg/Ca in solution exceeds 5/95; *5* struvite forms at pH 7 to 9, 37 °C, Mg^{2+} and NH_4^+ ions present with Ca^{2+} and PO_4^{3-} ions; *6* 37 °C, pH 7.4, $P_2O_7^{4-}$ present in elevated concentrations; *7* ACP forms at pH 7 to 11, 25 to 100 °C in the presence of Mg^{2+} and $P_2O_7^{4-}$; 25 to 37 °C in the presence of high concentrations of CO_3^{2-} and/or Mg^{2+}; *8a* OH-apatite forms at starting pHs 5 to 12, 70 to 100 °C; pH 7 to 10, 25 to 100 °C; the OH-apatite obtained is always Ca-deficient and contains varying amounts of HPO_4^{2-} or "lattice H_2O"; *8b* formed in the presence of cationic substituents, e.g., Sr^{2+}, Pb^{2+}, Ba^{2+}, K^+, Na^+; *8c* formed in the presence of Na^+ or K^+ and CO_3^{2-}; *8d* formed in the presence of Cl^-; *8e* formed in the presence of F^- ions, 25 to 100 °C, pH 5 to 12. (13, 18, 22, 28, 47–66, 68, 70, 72, 74, 78, 80, 90, 97, 104, 105, 107, 116)

phase. However, in the presence of a critical concentration of F^- ions in the solution, (F, OH)-or-F-apatites will form (66, 78); in the presence of low concentrations of $P_2O_7^{4-}$, the growth of DCPD crystals are suppressed and their morphology drastically modified (57, 58) and at higher concentrations, amorphous calcium phosphate forms (66, 78). Even more remarkable is the effect of Mg^{2+} which allows the formation of β-TCP (Mg-containing) from aqueous and biological systems (45, 47, 48, 60, 74, 90, 100, 102). [Pure β-TCP, does not form from aqueous systems (47, 60) and its presence in biological systems is always associated with magnesium (47, 48, 60, 100, 102).] The Mg/Ca ratio in the solutions determines which phosphate mineral (DCPD, β-TCP, struvite, or amorphous calcium-magnesium-phosphates) will form (62). Stability of amorphous calcium phosphates (i.e., resistance to hydrolysis to other types of calcium phosphates) depends on their composition, whether or not they contain CO_3^{2-}, Mg^{2+}, $P_2O_7^{4-}$, singly or in combination (11, 47, 61). The composition of biological apatites vary with the type(s) and concentrations of the "impurity" ions present in the solution. Most of these ions ultimately get incorporated in the apatite.

The transformation of one type of phosphate mineral to another in vivo or in vitro can proceed by hydrolysis or by dissolution and recrystallization of the more stable phosphate compound under conditions of fluctuating pH, temperature and

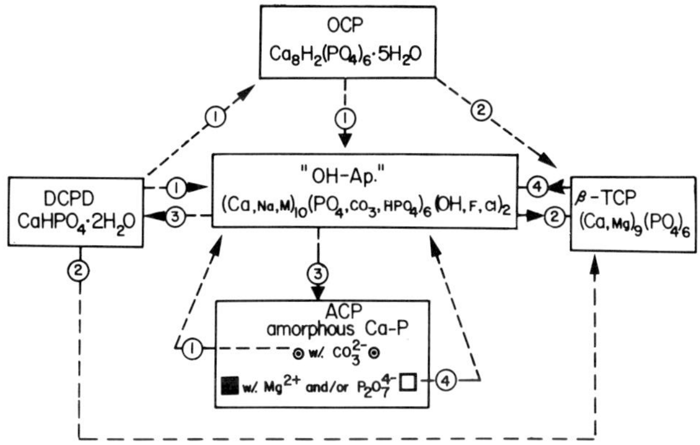

Fig. 19. Transformation of one type of phosphate mineral to another in vivo and vitro. Transformation occurs by hydrolysis (1); hydrolysis in the presence of Mg^{2+} (2); dissolution and recrystallization (3). Transformation *4* does not occur, e.g., Mg-containing TCP does not hydrolyze to apatite and ACP containing high concentrations of Mg^{2+} and/or $P_2O_7^{4-}$ does not hydrolyze to apatite. Such transformations had been demonstrated in vitro and speculated upon in vivo (1, 6, 11, 13, 16, 30, 47, 56, 59, 61, 75, 78, 99, 100, 105)

Table 4. Stoichiometry (Ca/P) of apatites and related calcium phosphates

Calcium phosphates	Ca/P (molar)
DCPD, $CaHPO_4 \cdot 2H_2O$	1.00
CPPD, $Ca_2P_2O_7 \cdot 2H_2O$	1.00
OCP, $Ca_8H_2(PO_4)_6 \cdot 5H_2O$	1.33
TCP, $Ca_9(PO_4)_6$	1.50
OH-Apatite, $Ca_{10}(PO_4)_6(OH)_2$	1.67
Biological apatites	
Enamel (human)	1.57 to 1.64
Dentine	1.62 to 1.68
Bone	1.72 to 1.80

composition (Fig. 19). DCPD, OCP or ACP (CO_3-containing) can hydrolyze to apatite. However, β-TCP (Mg-containing) and amorphous calcium-magnesium phosphate or amorphous calcium-magnesium-pyrophosphate are stable and do not hydrolyze to apatite (11, 47, 61, 111). DCPD and OCP can hydrolyze to β-TCP when Mg^{2+} is present in the solution. Biological apatites can undergo dissolution and recrystallization of DCPD can occur under acid conditions (Fig. 8A, B). Dissolution of biological apatites can also result in the formation of ACP when CO_3^{2-}, Mg^{2+} or $P_2O_7^{4-}$ ions are present; or in the formation of CPPD crystals when P_2O_7 concentrations are abnormally elevated (15, 47, 59, 61, 82). The dissolution of synthetic and biological apatite and reprecipitation of a "purer" or more stable apatite occurs specially in the presence of F^- ions (76).

Besides the transformation of one type of phosphate mineral to another, nonphosphatic mineral can also transform to phosphatic mineral by the process of dissolution and recrystallization. For example, $CaCO_3$ or calcium oxalate can dis-

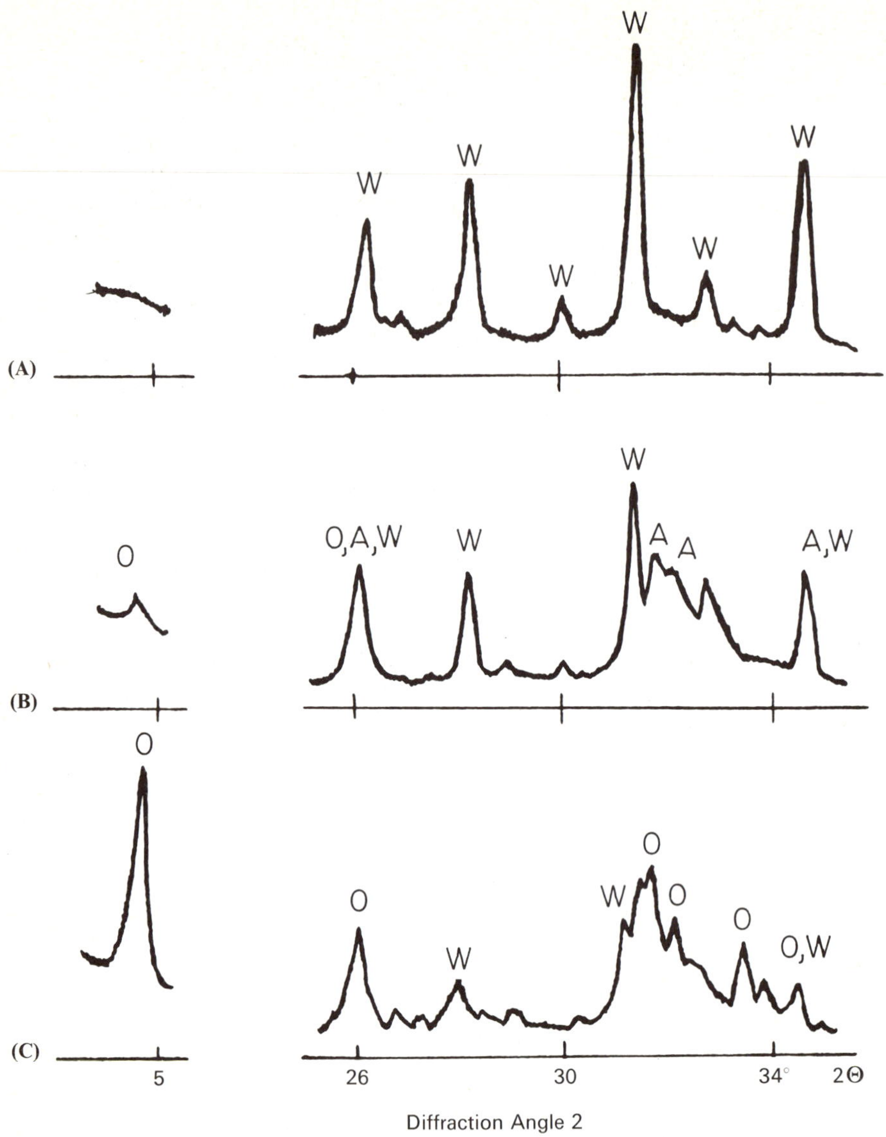

Fig. 20. Co-existence of different types of calcium phosphates in human dental calculi from sample to sample and from one layer to the next in the same sample. *A* only whitlockite in Mg-containing -TCP (W); *B* OCP, apatite and whitlockite (W); *C* OCP (O), and whitlockite (W). (*A*) and (*C*) are from the outermost and innermost layers of the same dental calculus sample (48)

solve and recrystallize to DCPD in acidic solutions containing phosphate ions (75, 78). By the same token, phosphate mineral can dissolve and recrystallize as non-phosphatic mineral, e.g., DCPD to calcium oxalate (78, 85).

The X-ray diffraction patterns and IR spectra of the phosphate minerals occurring in human tissues, specially DCPD and OCP, are similar to those of pure syn-

a

b

c

Fig. 21a–d. Samples of urinary stones **(a)** showing layered growth **(b)** and **(c)**. In many cases, variation in composition from layer to layer is observed **(d)**. **d** shows the presence of struvite (*Aa*) in the outer layer and of apatite (*Bb*) in the nidus of one urinary stone; and the presence of apatite in both inner and outer layers of another (*Ba, Bb*)

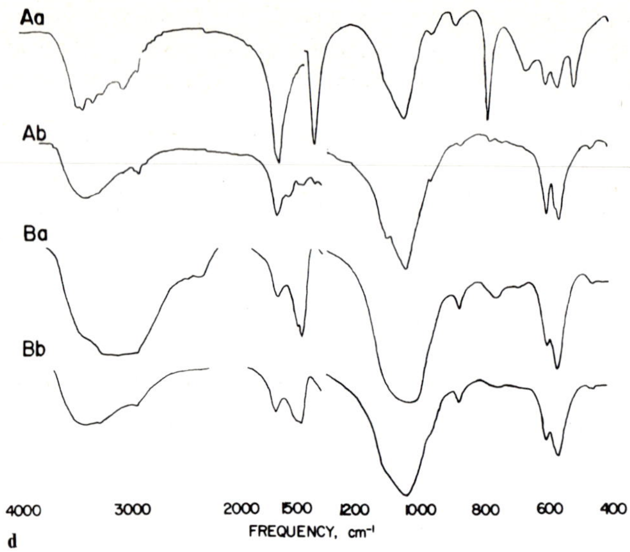

d

Fig. 21d. Legend see p. 379

thetic compounds (Figs. 9 and 10) although their "crystallinities" and morphology may vary from tissue to tissue (27, 28, 78, 102). The patterns and spectra of biologically occuring β-TCP are modified by the varying amounts of but persistent presence of Mg^{2+} in this phase (Figs. 2, 3, 9, 10, 12, 13, and 20); that of biological apatites are modified by the presence of HPO_4^{2-}, F^-, but specially by the ever-present CO_3^{2-} in these apatites (Figs. 4, 9, 14, 15, and 16). The comparative stoichiometry (Ca/P) of different calcium phosphates compared to that of biological apatite (human enamel) is shown in Table 4.

The co-existence of different types of phosphate mineral in pathological tissue calcifications, for example in dental calculi (Figs. 3C, 20) and in urinary calculi (Figs. 3C, and 21C, 21D) may be due to the independent formation of the individual crystalline components as defined by specific sets of conditions (pH, temperature, composition of the milieu) or to the transformation (by dissolution and recrystallization) of one type to another (Figs. 18 and 19). The layered growth of dental calculi (48) and of urinary stones (Figs. 3C and 21A, 1B) and the different compositions among layers (Figs. 3C, 20, 21D) suggest fluctuations in pH and in the composition of the milieu under pathological conditions.

The formation of apatites as the principal constituent of normal calcifications (enamel, dentine, bone) has been associated with specific types of organic matrices and presence or absence of inhibitors (20, 25, 28, 31, 96, 98). The organic matrices may act to locally concentrate the Ca^{2+} or PO_4^{3-} ions and cause initial nucleus formation of the phosphate mineral. The subtle compositional variations within a calcified tissue (bone or enamel) or the more significant compositional variations among calcified tissues (e.g., enamel vs. bone or dentine) must reflect the differences in the composition of their immediate environments. The dissolution and recrystallization reactions of biological apatites are especially important in considering the dissolution or demineralization and remineralization processes in human dental enamel (caries mechanisms) or in the mineralization and resorption of bone.

Acknowledgements. The authors gratefully acknowledge the valuable technical assistance of G. B. Quirolgico, P. Go and D. J. LeGeros; the professional collaboration of W. P. Shirra (AMR Corporation), Drs. Penugonda, M. Pines and T. K. Vaidjynathan (NYU Dept. of Dental Materials) and Dr. D. Lee (Cornell University) for the SEM investigations; and the support of the National Institutes of Health Research Grant DE-04123.

References

1. Alfrey AC, Solomons CC, Ciricillo J, Miller NL (1976) Extraosseous calcifications. Evidence for abnormal pyrophosphate metabolism in uremia. J Clin Invest 57:692–699
2. Amler MH, LeGeros RZ (1979) Infrared absorption analysis of enameloma. NY State J 45:502–503
3. Arends J, Davidson CL (1975) HPO_4 content in enamel and in artificial caries lesions. Calcif Tissue Res 18:65–75
4. Baud CA, Very JM (1975) Ionic substitutions in vivo in bone and tooth apatite crystals. Physico-Chemie et Crystallographie des Apatites D'Interet Biologique. Colloque Internationaux CNRS (Paris) 230:405–410
5. Baud CA, Bangs S, Very JM (1977) Minor elements in bone mineral and their effects in its solubility. J Biol Buccale 5:105–202
6. Barone JP, Nancollas GH, Tomson M (1976) The seeded growth of calcium phosphates. The kinetics of growth of dicalcium phosphate dihydrate on hydroxyapatite. Calcif Tissue Res 21:171–182
7. Beevers CA (1958) The crystal structure of dicalcium phosphate dihydrate, $CaHPO_4 \cdot 2H_2O$. Acta Cryst 11:273–277
8. Beevers CA, MacIntyre DB (1947) The atomic structure of fluorapatite and its relation to that of tooth and bone minerals. Miner Mag 27:254–259
9. Betts F, Posner AS (1974) An X-ray radial distribution study of amorphous calcium phosphate. Mater Res Bull 9:907–914
10. Bonel G (1972) Contribution de l'etude de la carbonatation des apatites. I. Synthese et etude des proprietes des apatites carbonatees de type A. Ann Chim 7:65–88
11. Boskey AL, Posner AS (1974) Mg stabilization of amorphous calcium phosphates: a kinetic study. Mater Res Bull 9:907–914
12. Bonner G, Jacob ET, Pevzner S, Jungmann A (1971) Diffuse calcification of lungs in patient on a maintenance hemodialysis. Isr J Med Sci 7:1182–1187
13. Brown WE, Smith JP, Lehr JR, Frazier AW (1962) Crystallographic and chemical relations between octacalcium phosphate and hydroxyapatite Nature 196:1048–1055
14. Brown WE, Schroeder LW, Ferris JS (1979) Interlayering of crystalline octacalcium phosphate and hydroxylapatite. J Phys Chem 83:1385–1388
15. Cheng PT, Pritzker, KPH, Adams ME, Nyburg SC, Omar SA (1980) Calcium pyrophosphate crystal formation in aqueous solutions. J Rheumatol 7:609–616
16. Chickerur NS, Tung MS, Brown WE (1980) A mechanism for incorporation of carbonate into apatite. Calcif Tissue Int 32:55–62
17. Contiguglia SR, Alfred AC, Miller NL, Runnells DE, LeGeros RZ (1973) Nature of soft tissue calcifications in uremia. Kidney Int 4:229–235
18. Dickens B, Brown WE (1971) The crystal structure of $Ca_7Mg_9(Ca, Mg)_2(PO_4)_{12}$. TMPM Tschermarks miner Petrol Mitt 16:79–104
19. de Jong WF (1926) La substance minerale dans les os. Rec Trav Chim 45:445–458
20. Dieppe P (1981) Inhibition of crystal formation in vivo and in vitro. J. Rheum. (in press)
21. Eanes ED (1979) Enamel apatite: chemistry, structure and properties. J Dent Res 58:829–834
22. Elliott JC (1973) The problems of the composition and structure of the mineral components of hard tissues. Clin Orthop 93:313–33

23. Elliott JS (1973) Structure and composition of urinary calculi. J Urol 109:82
24. Elliott JS, Quaide WL, Sharp RF, Lewis L (1968) Mineralogical studies of urine: the relationship of apatite, brushite and struvite to urinary pH. J Urol 80:269–271
25. Ennever J, Vogel J, Boyan-Salyers B, Riggan LJ (1979) Characterization of calculus matrix calcification nucleator. J Dent Res 58:619–623
26. Faure G, Netter P, Malaman B, Steinmetz J (1977) Monocrystalline calcium hydrogen phosphate dihydrate in destructive arthropathies of chondrocalcinosis. Lancet, 142–143
27. Faure B, Netter P, Malaman B, Steinmetz J, Duherille J, Gaucher A (1980) Scanning electron microscopic study of microcrystals implicated in human rheumatic diseases. Scanning Electron Microsc 1980, 111:163–176
28. Fleisch H, Russell RGG (1972) A review of the physiological and pharmacological effects of pyrophosphates on bones and teeth. J Dent Res 51:324–332
29. Fowler BO, Moreno EC, Brown WE (1966) Infrared spectra of hydroxy apatite, OCP and pyrolyzed OCP. Arch Oral Biol 11:477–487
30. Francis MD, Webb NC (1971) Hydroxyapatite formation from a hydrated calcium monohydrogen phosphate precursor. Calcif Tissue Res 6:335–342
31. Gabbiani G, Tuchweber B, Selye H (1973) Experimental ectopic calcification (calciphylaxis and calcergy). In: Zipkin Z (ed) Biological mineralization Wiley & Sons, New York, p 548–586
32. Gee A, Deitz FR (1965) Pyrophosphate formation upon ignition of precipitated calcium phosphates. J Am Chem Soc 77:2692–2696
33. Gonzalez-Diaz PF, Garcia-Ramos JV, Santos M (1979) Composition of apatites in human urinary calculi. Calcif Tissue Int 28:215–225
34. Griffith DP, Musher DM (1976) Pathogenesis of infection stones. In: Finlayson B, Thomas WC (eds) Colloquium on renal lithiasis. Univ Press Florida, p 54–65
35. Gron P (1973) Saturation of human saliva with calcium phosphates. Arch Oral Biol 15:613–617
36. Gross R (1926) Die kristalline Struktur von Dentine und Zahnschmelz. Festschr Zahn Inst Univ Greifswald, s 59–69
37. Hallsworth AS, Robinson C, Weatherell JA (1972) Mineral and magnesium distribution within the approximal carious lesion of dental enamel. Caries Res 7:345–354
38. Heath D, Robertson AJ (1977). Pulmonary calcinosis. Thorax 32:606–611
39. Heughbaert JC (1977) Contribution a l'etude de l'evolution des orthophosphates de calcium precipites en orthophosphates apatitiques. PhD These L'Int Nat Polytech Toulouse
40. Hodgkinson A, Marshall RW (1975) Changes in the composition of urinary tract stones. Invest Urol 13:131–135
41. Holcomb DW, Young RA (1980) Thermal decomposition of human tooth enamel. Calcif Tissue Int 31:189–201
42. Kerebel B, Daculsi G, Vervaere A (1976) High-resolution electron microscopy and crystallography study of some biological apatites. J Ultrastruct Res 57:266–276
43. Kim KM, Trump BF (1975) Amorphous calcium precipitations in human aortic valve. Calcif Tissue Res 18:155–160
44. Kohn NN, Hughes RE, McCarthy DJ (1962). The significance of calcium phosphate crystals in the synovial fluid of arthritic patients: the 'pseudogout syndrome'. II. Identification of crystals. Ann Intern Med 56:738–745
45. Knuutila M, Lappalinen R, Konturi-Nahrhi V (1980) Effect of Zn and Mg on the formation of whitlockite in human subgingival calculus. Scad J Dent Res 88:513–516
46. Lagergren C (1956) Biophysical investigations of urinary calculi. An X-ray crystallographic and microradiographic study. Acta Radiol 133 (Sp. Suppl)
47. LeGeros RZ (1967) Crystallographic studies of the carbonate substitution in the apatite structure. PhD Thesis, New York Univ
48. LeGeros RZ (1974) Variations in the crystalline components of human dental calculi. I. Crystallographic and spectroscopic analyses. J Dent Res 53:45–50
49. LeGeros RZ (1974) The unit-cell dimensions of human enamel apatite: Effect of chloride incorporation. Arch Oral Biol 20:63–71
50. LeGeros RZ (1977) Apatites from aqueous and non-aqueous systems: Relation to biological apatites. Proc 1st Int Congr Phosphorus Compounds, Rabat, Inst Mondial Phosphate IMPHOS (Paris), p 347–360
51. LeGeros RZ (1981) Apatites in biological systems. Prog Crystal Growth Charact 4:1–45

52. LeGeros RZ, LeGeros JP, Trautz OR (1965) A computer-diffractometer method of assessing crystallinity. Trans Am Cryst Assn: 40
53. LeGeros RZ, LeGeros JP, Trautz OR, Shirra WP (1967) Apatite crystallites: Effect of carbonate on morphology. Science 155:1409–1411
54. LeGeros RZ, LeGeros JP, Trautz OR, Klein E (1969) Two types of carbonate substition in the apatite structure. Experienta 24:5–9
55. LeGeros RZ, LeGeros JP, Trautz OR, Klein E (1970) Spectral properties of carbonate in carbonate-containing apatites. Dev Appl Spectrosc 7:3–10
56. LeGeros RZ, LeGeros JP, Trautz OR, Shirra WP (1971) Conversion of $CaHPO_4$ to apatites: Effect of carbonate on crystallinity and morphology. Adv X-ray Anal 14:57–66
57. LeGeros RZ, LeGeros JP (1972) Brushite crystals grown by diffusion. J Cryst Growth 13:476–480
58. LeGeros RZ, Morales P (1973) Renal stone crystals grown in gel systems. J Invest Urol 11:12–20
59. LeGeros RZ, Contiguglia SR, Alfrey AC (1973) Pathological calcifications associated with uremia: Two types of calcium phosphate deposits. Calcif Tissue Res 13:173–185
60. LeGeros RZ, Miravite MA, Klein I (1974) Biological whitlockites, its formation and occurences. J Dent Res 53:117
61. LeGeros RZ, Shirra WP, Miravite MA, LeGeros JP (1975) Amorphous calcium phosphates: Synthetic and biological. Physico-Chemie et Crystallographie des Apatites d'Interet Biologique. Colloque Internationaux CNRS (Paris) 230:105–115
62. LeGeros RZ, Miravite MA, Quirolgico G, Morales P (1976) Factors influencing crystal growth and morphology of urinary stone crystallites. In: Finlayson B, Thomas WC Jr (eds) Colloquium in renal lithiasis. Univ Florida Press, p 187–206
63. LeGeros RZ, Miravite MA, Quirolgico GB, Curzon MEJ (1977) The effect of some trace elements on the lattice parameters of human and synthetic apatites. Calcif Tissue Res [Suppl] 22:362–267
64. LeGeros RZ, Legros R, Bonel G (1978) Types of 'H_2O' in human enamel and in precipitated apaties. Calcif Tissue Res 26:111–118
65. LeGeros RZ, Kazimirof J, Nevins AJ (1978) The nature of the calcified material induced by collagen-calcium phosphate gel in tooth. J Dent Res A57:206
66. LeGeros RZ, Quirolgico GB, Go P (1978) $CaHPO_4 \cdot 2H_2O$ (DCPD): Effect of some trace elements on its crystal growth. J Dent Res A57:89
67. LeGeros RZ, Pentel L, Shirra WP (1979) Thermal stability of human enamel. Caries Res 13:96
68. LeGeros RZ, Quirolgico G, LeGeros JP (1979) Incorporation of strontium in apatite: Effect of pH and temperature. J Dent Res A58:168–169
69. LeGeros RZ, Shannon IL (1979) The crystalline components of dental calculi: Human vs. dog. J Dent Res 58:2371–2377
70. LeGeros RZ, Rankine C, Taheri H, Shannon I (1980) Calcium salts precipitated from human saliva. J Dent Res A59:366
71. LeGeros RZ, Suga S (1980) The crystallographic nature of fluoride in fish enameloids. Calcif Tissue Int 32:169–174
72. LeGeros RZ, Tung M (1980) Factors affecting fluoride incorporation in apatites. J Dent Res B59:973
73. LeGeros RZ, Kerebel LM, Silverstone L (1980) Comparative properties of fluoridated and non-fluoridated biological apatites. J Dent Res A59:523
74. LeGeros RZ, Taheri MH, Quirolgico GB, LeGeros JP (1980) Formation and stability of apatites: Effects of some cationic substituents. Proc 2nd Int Congr Phosphorus Comp, Boston, IMPHOS (Paris), p 89–103
75. LeGeros RZ, Vandemaele KH, Go P, Quirolgico GB, LeGeros DJ (1980) Transformation of calcium carbonates and calcium phosphates to carbonate apatites: Possible mechanisms for phosphorite formation. ibid, p 41–57
76. LeGeros RZ, Tung M (1982) Dissolution properties of CO_3-containing OH and F-apatites. J Dent Res 61:232
77. LeGeros RZ, Singer L, Ophaug RH, Thein A, LeGeros JP (1982) The effect of fluoride on the stability of synthetic and biological (bone mineral) apatites. In: Menczel J, Robin GC, Makin M (eds). Osteoporosis. J. Wiley, p 327–341
78. LeGeros RZ, Lee DD, Quirolgico GB, Shirra WP, Reich L (1982) In vitro formation of dicalcium phosphate dihydrate, $CaHPO_4 \cdot 2H_2O$ (DCPD). Scan Electron Microscop (II) in press

79. Leonard F, Boke JW, Ruderman RJ, Hegyeli AF (1972) Initiation and inhibition of subcutaneous calcification. Calcif Tissue Res 10:269–272
80. Lynch KL (1958) An evaluation of factors involved in the production of magnesium ammonium phosphate crystals of bacterial origin. MS Thesis Marquette Univ
81. Mandel NS (1975) The crystal structure or calcium pyrophosphate dihydrate. Acta Cryst B31:1730–1734
82. McCarthy DJ (1976) Calcium pyrophosphate dihydrate crystal deposition disease. Arthritis Rheum 19:275–285
83. McConnell D (1973) Apatite. Its crystal chemistry, mineralogy, utilization and biologic occurrences. Springer, Vienna New York, p 68–80
84. Mierau HD von, Trautz OR, Vahl J (1971) Kristallchemische Untersuchungen an menschlichen Skeletteilen und Zähnen. Dtsch Zahnarostl Z 26:37–45
85. Meyer JL, Bergert JH, Smith LH (1977). Epitaxial relationships in urolithiasis: The brushite-whewellite system. Clin Sci Mol Med 52:143–148
86. Moore PB (1984) Crystallochemistry and structures of phosphate minerals. In: Nriagu JO, Moore PB (eds) Phosphate minerals. Springer-Verlag, p 155–170
87. Moreno EC, Kresak M, Zahradnik RG (1977). Physico-chemical aspects of fluoride-apatite systems relevant to the study of dental caries. Caries Res [Suppl] 11:142–160
88. Myers HM (1965) Trapped water of dental enamel. Nature 206:713–716
89. Nelson D (1981) The influence of carbonate on the atomic structure and reactivity of hydroxyapatite. J Dent Res 60:1621–1629
90. Newesely H (1966) Changes in crystal types of low solubility calcium phosphates in the presence of accompanying ions. Arch Oral Biol 6:174–189
91. Nylen MU, Eanes ED, Termine JD (1972) Molecular and ultrastructural studies of non-crystalline calcium phosphates. Calcif Tissue Res 9:95–108
92. Pak CYC, Eanes ED, Ruskin B (1971) Spontaneous precipitation of brushite in urine: evidence that brushite is the nidus of renal stones originating as calcium phosphate. Proc Natl Acad Sci USA 68:1456–1460
93. Parfitt MA (1969) Soft-tissue calcification in uremia. Arch Intern Med 124:544–553
94. Peterson WC, Carlson CH, Armstrong WD (1963) Analysis of ectopic bone in osteoma cutis. Arch Dermatol 88:540–545
95. Posner AS (1969) Crystal chemistry of bone mineral. Physiol Rev 49:760–792
96. Prien EL (1949) Studies on urolithiasis. II. Relationships between pathogenesis, structure and composition of calculi. J Urol 61:821–836
97. Pritzker KPH, Phillips H, Luk SC, Koven IH, Kiss A, Houpt JB (1976) Pseudo-tumor of temporomandibular joint: destructive calcium pyrophosphate dihydrate arthropathy. J Rheum 3:70–81
98. Roufosse AH, Landis WJ, Sabine WK, Glimbher MJ (1979) Identification of brushite in newly deposited bone mineral from embryonic chicks. J Ultrastruct Res 68:235–255
99. Rowles SL (1968) The precipitation of whitlockite from aqueous solutions. In: Colloque International Sur Les Phosphates Mineraux Solides. Bull Soc Chim Fr pp. 151–156
100. Russell G, Robertson WG, Fleisch H (1973) Inhibitors of mineralization. In: Zipkin I (ed) Biological Mineralization. J. Wiley and Sons, New York, pp 807–825
101. Schroeder H (1969) Formation and inhibition of dental calculus. Hans Hubert Publ, Vienna
102. Sutor D, Wooley SE, Ellingsworth JJ (1974) Some aspects of the urinary stone problem in Great Britain and Northern Ireland. Br J Urol 46:275–288
103. Takasaki E (1971) An observation on the analysis of urinary calculi by infrared spectroscopy. Calcil Tissue Res 7:232–240
104. Taves DR, Neuman WF (1964) Factors controlling calcification in vitro: fluoride and magnesium. Arch Biochem Biophys 108:390–397
105. Termine JD (1972) Mineral chemistry and skeletal biology. Clin Orthop 85:207–232
106. Tao L-C (1978) Microliths in sputum specimens and their relationship to pulmonary alveolar microlithiasis. Am J Clin Pathol 69:482–485
107. Trautz OR (1967) The crystalline organization of dental mineral. In: Miles AEW (ed) Structural and Chemical Organization of Teeth. Academic Press, New York, pp 165–200
108. Trautz OR, Zapanta-LeGeros R, LeGeros JP (1964) Effect of magnesium on calcium phosphates. J Dent Res 43:751 (abstract)

109. Vahl von J, Hohling HJ, Frank RM (1964) Elektronenstrahlbeugung an rhomboedrisch aussehenden Mineralbildungen in kariösem Dentin. Arch Oral Biol 10:315–320
110. Vogel JJ, Amdur BH (1967) Inorganic pyrophosphate in parotid sailva and its relation to calculus formation. Arch Oral Biol 12:159–163
111. von Brand T, Scott DB, Nylen MU, Pugh MH (1965) Variations in the mineralogical composition of cestode calcareous corpuscles. Exp Parasitol 16:382–390
112. Wang N-S, Steele AA (1979) Pulmonary calcification. Arch Pathol Lab Med 103:151–157
113. Young RA (1975) Biological apatite vs. hydroxyapatite at the atomic level. Clin Orthop 113:249–260
115. Young RA, Elliott JC (1966) Atomic scale bases for several properties of apatites. Arch Oral Biol 11:699–707
116. Zapanta-LeGeros R (1965) Effect of carbonate on the lattice parameters of apatite. Nature 206:403–405
117. Zipkin I (1970) The inorganic composition of bones and teeth. In: Schraer H (ed) Biological calcification. Appleton-Century Crofts, New York, pp 69–99

Chapter 13 Precipitation of Phosphate Minerals in Waste Water Treatment Systems

MASON B. TOMSON and LAINE VIGNONA

Introduction

An approximate formula for algae or plant matter is (Garrels et al. 1975, p 106):

$C_{280}H_{560}O_{280}N_{19}P_1$

and although phosphorus constitutes only ~0.1 mol.-%, it is an essential nutrient of life, contained in protein and ATP. In many lakes and streams, phosphorus is the limiting nutrient to algal and plant growth. Algae require from 0.001 to 0.01 mg l^{-1} phosphorus for growth to take place. Above about 0.5 mg l^{-1} phosphorus is not a limitation to plant growth and algae, which along with other water plants, will often grow at an alarming rate and completely dominate a body of water. Consequently, the lower portions of the plant matter will die and while decaying at the bottom of the water, will use up most of the oxygen in the water. This process is referred to as eutrophication and results in fish kills and a general degradation of water quality. The natural level of dissolved phosphorus in fresh waters is probably about 0.01 mg l^{-1} P (Garrels et al. 1975) to 0.02 mg l^{-1} (Horne 1978, p 262) and about 0.07 mg l^{-1} P in the ocean (Horne 1978, p 322).

This discussion will focus on municipal waste water treatment. If a lake or stream receives waste water from a treatment plant, the phosphate concentration in the waste water may be quite high. The average total-P in domestic raw waste water is about 10 mg l^{-1} P, ~35% from feces, urine, and waste-food disposal and ~65% from synthetic detergent phosphate builders. If industry in the area uses phosphate for corrosion control, the waste water phosphate may increase by up to 20% (US-EPA 1976).

To protect local receiving waters from eutrophication many states and towns require phosphorus removal from the waste water before discharge to the ambient environment. Some laws require a fixed percent removal of phosphorus, such as 80%–95%, and some require removal to a fixed level, generally 1 mg l^{-1} as P. It should be noted that eutrophication appears to be a problem of local streams and lakes as receiving bodies, in that there is no evidence that the P-level in the world's waterways is being increased (Garrels et al. 1975).

The primary method of phosphorus removal from domestic waste water is by precipitation with Ca^{2+}, Fe^{2+}, Fe^{3+}, or Al^{3+} salts; the most common method is lime, $Ca(OH)_2$, treatment. These metal salts can be added to primary, secondary, or tertiary water treatment processes. Often, a polymer flocculation aid will be add-

National Center for Ground Water Research, Rice University, Houston, Texas 77251 ed downstream of the precipitant addition. Specifics of each of these processes will be covered below, along with the relevant water chemistry background. Whether in adding to an existing plant or in building a new plant, the choice of phosphate removal agent can generally be made so that the overall processes of biochemical oxygen demand (BOD) and suspended solids (SS) removal are aided. If the plant is already operating, full-scale tests of chemical addition at different locations in the plant can generally be performed efficiently. If pilot plant or laboratory tests are to be performed, they should be conducted with the actual waste water to be treated for as many day/night and seasonal cycles as possible.

Background

Phosphate Forms in Waste Water

Phosphate enters a waste water treatment plant (approximate concentration in mg l^{-1} as P) as either orthophosphate (5 mg l^{-1}), condensed polyphosphates ($H_5P_3O_{10}$ ~ 3 mg l^{-1} and $H_4P_2O_7$ ~ 3 mg l^{-1}), or organic phosphates (< 1 mg l^{-1}) (Snoeyink and Jenkins 1980). The phosphate form most readily usable by bacteria and in precipitation reactions is orthophosphate. Representative hydrolysis reactions of polyphosphates and organic phosphates might be:

$$\underset{\text{pyrophosphate}}{P_2O_7^{4-} + H_2O} \xrightarrow{k_1} \underset{\text{hydrogen-orthophosphate}}{2HPO_4^{2-}},$$

$$\underset{\text{tripolyphosphate}}{P_3O_{10}^{5-} + H_2O} \xrightarrow{k_2} HPO_4^{2-} + HP_2O_7^{3-},$$

$$RO(PO_3)_3^{4-} + H_2O \xrightarrow{k_3} ROPO_3H^{1-} + HP_2O_7^{3-}.$$

In the absence of catalysts at 7.0 pH, $k_1 \approx 1.5 \times 10^{-5}$ min^{-1}, and $k_2 \approx 8.0 \times 10^{-5}$ min^{-1} at 60 °C. The activation energy is about 25 kcal mol^{-1} which at room temperature corresponds to $t_{1/2}$ of 7.3 yrs and 1.4 yrs, respectively (Shen and Morgan 1973). The hydrolysis reactions are catalyzed by H^+, most metals, and enzymes. Hydrolysis reaction rates can be up to 10^6 times faster in the presence of specific enzymes. It is probably the high microbial population in the secondary treatment which is responsible for the degradation of condensed phosphates. There is a surprising lack of detailed kinetic data on the rate and mechanism of the various hydrolysis reactions as they occur in natural waters or waste water.

Equilibrium Chemistry

The variation of total soluble phosphate, T_P, vs. pH for concentrations of Fe^{3+}, Fe^{2+}, Al^{3+}, and Ca^{2+} relevant to waste water treatment is plotted in Fig. 1. Assumed initial phosphate (10 mg l^{-1} as P or $10^{-3.5}$ M) along with 90% ($10^{-4.5}$ M) and 99% ($10^{-5.5}$ M) removal are marked on each graph for comparison. For iron

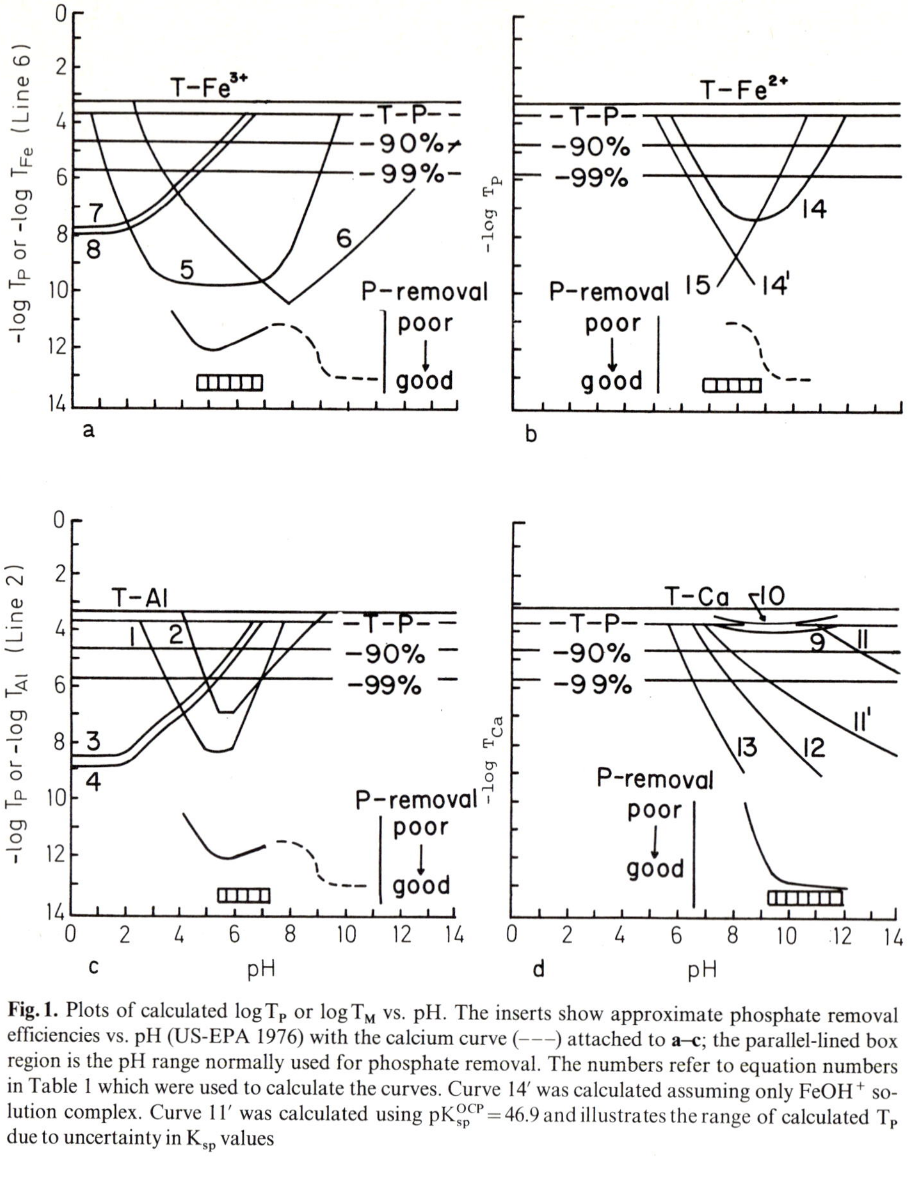

Fig. 1. Plots of calculated log T_P or log T_M vs. pH. The inserts show approximate phosphate removal efficiencies vs. pH (US-EPA 1976) with the calcium curve (---) attached to **a–c**; the parallel-lined box region is the pH range normally used for phosphate removal. The numbers refer to equation numbers in Table 1 which were used to calculate the curves. Curve 14′ was calculated assuming only $FeOH^+$ solution complex. Curve 11′ was calculated using $pK_{sp}^{OCP} = 46.9$ and illustrates the range of calculated T_P due to uncertainty in K_{sp} values

and aluminum total molar metal was assumed to be two times the total phosphate, i.e., $T_M \approx 10^{-3.2}$ M, which has been shown to be reasonable by field practice (see below). For calcium in Fig. 1d a constant 10^{-3} M T_{Ca} was assumed.

Table 1 contains a summary of the equations used to produce the correspondingly numbered curves in Fig. 1. Only acid/base or hydrolysis reactions were included in the equations due to both numerical uncertainty in the equilibrium constants and uncertainty about the actual precipitation process. The metal-phosphate solubility Eqs. (1), (5), (9)–(14), in Table 1 were derived from the correspond-

Table 1. Summary of solubility equations used in Fig. 1

From $T_P = P\alpha_P$; $T_{Al} = Al\,\alpha_{Al}$, etc. define:

$$\alpha_P = (1 + H/K_3 + H^2/K_2 \cdot K_3 + H^3/K_1 \cdot K_2 \cdot K_3)$$
$$\alpha_{Al} = (1 + K_{11}^{AlOH}(OH) + \beta_{12}^{Al(OH)_2}(OH)^2 + \beta_{14}^{Al(OH)_4}(OH)^4)$$
$$\alpha_{Fe^{2+}} = (1 + K_{11}^{FeOH}(OH) + \beta_{12}^{Fe(OH)_2}(OH)^2)$$
$$\alpha_{Fe^{3+}} = (1 + K_{11}^{FeOH}(OH) + \beta_{12}^{Fe(OH)_2}(OH)^2 + \beta_{13}^{Fe(OH)_3}(OH)^3)$$

Define

$f(x) = [x/(3-3x)]^{[3/(3-2x)^2]}$ for $(MP)_x[M(OH)_3]_{1-x}$ solid solution
$M = Al, Fe(III)$

Equation	Solid Phases
1. $T_P = K_{sp}^{Alp}\ \alpha_{Al}\alpha_P/T_{Al}$	AlP
2. $T_{Al} = K_{sp}^{Al(OH)_3}\ \alpha_{Al}/(OH)^3$	$Al(OH)_3$
3. $T_P = K_{sp}^{Alp}\ \alpha_P(OH)^3/K_{sp}^{Al(OH)_3}$	$AlP + Al(OH)_3$
4. $T_P = K_{sp}^{Alp}\ \alpha_P(OH)^3\ f(x)/K_{sp}^{Al(OH)_3}$	$(AlP)_x(Al(OH)_3)_{1-x}$ solid solution
5. $T_P = K_{sp}^{Alp}\ \alpha_{Fe}\alpha_P/T_{Fe}$	FeP
6. $T_{Fe} = K_{sp}^{Fe(OH)_3}\ \alpha_{Fe/(OH)^3}$	$Fe(OH)_3$
7. $T_P = K_{sp}^{FeP}\ \alpha_P(OH)^3/K_{sp}^{Fe(OH)_3}$	$FeP + Fe(OH)_3$
8. $T_P = K_{sp}^{FeP}\ \alpha_P(OH)^3\ f(x)/K_{sp}^{Fe(OH)_3}$	$(FeP)_x(Fe(OH)_3)_{1-x}$ solid solution
9. $T_P = K_{sp}^{DCPD}\alpha_P K_3/T_{Ca} \cdot H$	$CaHP \cdot 2H_2O$
10. $T_P = K_{sp}^{DCP}\alpha_P K_3/T_{Ca} \cdot H$	CaHP
11. $T_P = \alpha_P(K_{sp}^{OCP}/T_{Ca}^4 \cdot H)^{1/3}$	$CaHP \cdot 2.5\ H_2O$
12. $T_P = \alpha_P(K_{sp}^{TCP}/T_{Ca}^3)^{1/2}$	Ca_3P_2
13. $T_P = \alpha_P(K_{sp}^{HAP}/T_{Ca}^5 \cdot OH)^{1/3}$	Ca_5P_3OH
14. $T_P = \alpha_P(K_{sp}^{Fe_3P_2}\ \alpha_{Fe}^3/T_{Fe}^3)^{1/2}$	$Fe_3P_2 \cdot nH_2O$
15. $T_P = \alpha_P(K_{sp}^{Fe_3P_2}(OH)^6/K_{sp}^{Fe(OH)_2})^{1/2}$	$Fe_3P_2 \cdot nH_2O + Fe(OH)_2$

ing solubility products, assuming that the total metal is constant and that no other solid phases are present. Aluminum phosphate, (AlP) where $P = PO_4^{3-}$, will be used for illustration. The solubility product is:

$$K_{sp}^{AlP} = (Al^{3+})(P^{3-}).$$

From the equations for total aluminum and phosphate in solution, with charges omitted for clarity:

$$\begin{aligned} T_{Al} &= (Al) + (AlOH) + Al(OH)_2 + Al(OH)_4 \\ &= Al[1 + K_{11}^{Al(OH)} \cdot (OH) + \beta_{12}^{Al(OH)_2} \cdot (OH)^2 + \beta_{14}^{Al(OH)_4} \cdot (OH)^4] \\ &= Al \cdot \alpha_{Al}, \\ T_P &= P + HP + H_2P + H_3P \\ &= P(1 + H/K_3 + H^2/K_2 \cdot K_3 + H^3/K_1 \cdot K_2 \cdot K_3) \\ &= P \cdot \alpha_P. \end{aligned}$$

Then:

$$K_{sp}^{AlP} = (T_{Al}/\alpha_{Al}) \cdot (T_P/\alpha_P),$$
$$\text{or } T_P = K_{sp}^{AlP} \cdot \alpha_{Al} \cdot \alpha_P/T_{Al}. \qquad \text{[Eq. (1), Table 1].}$$

For completeness, the metal solubility of $Fe(OH)_3$ and $Al(OH)_3$ are included [Eq. (2) and (6)]. It is often suggested that both the metal hydroxide and the metal phosphate are present and are at equilibrium in solution at the same time; in this case the T_P and T_M in solution are fixed by the pH alone, according to Eqs. (3), (7), and (15). This can be illustrated for aluminum as follows:

$$K_{sp}^{AlP} = (Al)\,(P),$$

$$K_{sp}^{Al(OH)_3} = (Al)\,(OH)^3,$$

$$K_{sp}^{AlP} = (K_{sp}^{Al(OH)_3}/(OH)^3)\cdot(T_P/\alpha_P)$$

or

$$T_P = K_{sp}^{AlP}\cdot\alpha_P\cdot(OH)^3/K_{sp}^{Al(OH)_3} \qquad \text{[Eq. (3), Table 1].}$$

Instead of forming separate solids, the metal phosphate and hydroxide may form a single solid solution, e.g., for aluminum:

$$(AlP)_x[Al(OH)_3]_{1-x},$$

where x is the mole fraction of the phosphate in the solid phase. If it is assumed that the solid phases mix to form a solid solution, the excess free energy of mixing can be assumed, as a first approximation, to be a consequence of the configurational entropy of mixing only. With this assumption the derivation by Berndt and Stearns (1973) was extended to include the present solid solutions. Corresponding to their Eqs. (7) and (10) respectively, we obtain for aluminum:

$$\frac{K_{sp}^{Al(OH)_3}}{K_{sp}^{AlP}}\underbrace{\left[\frac{3(1-x)}{x}\right]^{\left(\frac{3}{(3-2x)^2}\right)}}_{f(x)} = \frac{(OH)^3}{(P)}, \qquad (16)$$

$$(K_{sp}^{AlP})^x\,[K_{sp}^{Al(OH)_3}]^{1-x}\underbrace{\left(\frac{x}{3-2x}\right)^{\left(\frac{x}{3-2x}\right)}\left(\frac{3-3x}{3-2x}\right)^{\left(\frac{3-3x}{3-2x}\right)}}_{g(x)} \qquad (17)$$

$$= (Al)\,(P)^x(OH)^{3(1-x)}.$$

Equation 16 corresponds to Eqs. (4) and (8) of Table 1. With the exception of the f(x) and g(x) terms, Eqs. (16) and (17) can be seen to follow from the quotient and products of the solubility products. The values of f(x) and g(x) generally may vary over a small range of 0.3 to 3.0, e.g., at $x=0.5$, $g(x)=0.57$, and $f(x)=2.28$, which lowers the predicted solubility of phosphate by 0.4 log units from that which would have been obtained for separate solid phases (see Fig. 1a, c).

One of the major problems with applying equilibrium calculations to real problem of water chemistry is the large variation in the values of reported equilibrium constants for the same reaction. In Table 2 are compiled some representative solubility product values for the constants in Table 1 (see also Vieillard and Tardy, this Vol., Chap. 4). The values used to construct the lines in Fig. 1 are underlined. In no way should these underlined values be interpreted as "best" or recommended

Table 2. Table of solubility product values at 25 °C ($-\log K_{sp}$)

$Al(OH)_3$(am)	31.7 (4)	$Ca_5(PO_4)_3OH$	55.9 (2)
	32.4 (1)		57.0 (3)
	33.0 (2)		*58.0* (6)
		Fe(II)	
$AlPO_4$	18.2 (1)	$Fe(OH)_2$	14.0 (1)
	21.0 (3)		14.5 (2)
			15.1 (4)
$CaHPO_4$	6.6 (3)		
	6.7 (2)		
	6.89 (9)	$Fe_3(PO_4)_2$	33.0 (2)
			36.0 (5)
$CaHPO_4 \cdot 2H_2O$	*6.59* (8)		
		Fe(III)	
$Ca_3(PO_4)_2$	24.2 (2)	$Fe(OH)_3$ (am)	37.4 (4)
	25.0 (1)		*38.0* (2)
	26.0 (2)		38.7 (3)
	28.7 (4)		39.4 (1)
	28.9 (7)		
		$FePO_4$	17.9 (2)
$Ca_4H(PO_4)_3$	*37.8* (10)		21.9 (1)
	46.9 (11)		*23.0* (3)

(1) Pagenkopf (1978)
(2) Snoeyink and Jenkins (1980)
(3) Stumm and Morgan (1981)
(4) Frankenthal (1963)
(5) Nriagu (1972)
(6) Moreno et al. (1968)
(7) Gregory et al. (1974)
(8) Gregory et al. (1970)
(9) McDowell et al. (1971)
(10) Boulet and Marier (1961)
(11) Moreno and Brown (1960)

values. The effect of typical variation in K_{sp}'s is illustrated for OCP (octacalcium phosphate) in Fig. 1d. Considerable work is needed in the area of establishing reliable equilibrium constants for field conditions. The iron and aluminum systems are particularly poorly understood from the point of view of solution chemistry. Until better equilibrium constants are available, diagrams such as those in Fig. 1 will be usable only as predictors of overall solubility trends.

The observed phosphate removal from waste water vs. pH for each system is sketched as an insert in Fig. 1a–d, along with the pH range normally used (dashed area). Once the pH is above about 9.5 pH, phosphate removal is good in each system; this is probably due to calcium phosphate precipitation. All of the calcium phosphate phases shown in Fig. 1d are used by different researchers as model phases for phosphate removal using lime. Moreno, Brown and Osborn (1960) and Zoltek (1976), and others have presented convincing evidence that OCP is probably the best model calcium phosphate phase to use. Either OCP, TCP (tricalcium phosphate) or HAP (hydroxyapatite) properly predict the pH trend (see Fig. 1d insert), but the lower OCP solubility curve appears to fit the field data best. Precipitated particles from these systems are so small that they do not produce good X-ray patterns nor lend themselves readily to stoichiometric analysis. Therefore, pH-dependent solubility studies such as those of Zoltek (1976) and Singer (1972) or "constant composition" techniques of Tomson and Nancollas (1978) may be the most reliable methods for phase identification in real systems.

Both AlP and FeP exhibit minima in their calculated solubility in the same pH range in which optimal phosphate removal is observed (Fig. 1). In neither case does the observed phosphate level go as low as the predicted value. This is probably a consequence of both an inadequate equilibrium model and real kinetic limitations. Both Al^{3+} and Fe^{3+} are subject to extensive hydrolysis and both form polynuclear complexes which are kinetically slow, but even less is known about the kinetics than is known about the equilibria involved.

Finally, it is generally observed that Fe^{2+} phosphate removal is effective at higher pH values than either Al^{3+} or Fe^{3+}, as is suggested by Fig. 1b. The importance of Fe^{2+} to phosphate removal is not well known. When Fe^{2+}, pickle liquor, is added to waste water in the secondary stage it may be oxidized to Fe^{3+} before it has time to react with phosphate. Alternatively, when Fe^{3+}, as either iron in solution or as precipitated iron comes in contact with organic matter under anaerobic conditions, it may be reduced to Fe^{2+} and precipitate as vivianite, Fe_2P_3, solid (Nriagu 1972). Vivianite has been identified in waste water and sludge, but its origin and extent of occurrence are not clear and are subjects of ongoing research.

Waste Water Treatment

A typical waste water treatment scheme is shown in Fig. 2. Also noted in Fig. 2 are points where chemicals for phosphate removal may be added. There are four phases in most waste water treatment plants: primary, secondary, and tertiary treatment, and anaerobic digestion. Primary treatment refers to simple settling of the heavy solids from the influent raw sewage to produce sludge. The primary sludge is generally pumped to an anaerobic sludge digester. During secondary treatment, the lighter and smaller solids in the liquid from the primary settler are microbially metabolized in the presence of excess air. Trickling filters and forced air-activated sludge treatment plants are the most common types of secondary treatment. In a trickling filter, primary effluent is sprayed downward over 3- to 5-inch rocks from nozzles in the bottom of a rotating pipe. As the liquid percolates downward through the rocks, attached slime on the rocks sorb and metabolize suspended and dissolved solids in the liquid. Some attached slime is sloughed off into the liquid phase. The effluent goes to a secondary settler. In an activated sludge unit, liquid from the primary settler is vigorously aerated as it flows slowly (4–6 h) from one end of a tank to the other. As with the trickling filter, the bacteria actively metabolize the dissolved and suspended solids and the effluent goes to a secondary settler. Sludge from the secondary settler is either recycled for additional biological treatment or is combined with the primary sludge and pumped to an anaerobic sludge digester.

Two functions of the anaerobic sludge digester are first, to further metabolize suspended solids and second, to concentrate the remaining suspended solids into a minimal volume with least percent water. Disposal by incineration, land application, etc. of the undigested solids from the anaerobic sludge digester is often a major fraction of the operating cost of a waste water treatment plant. Therefore, an

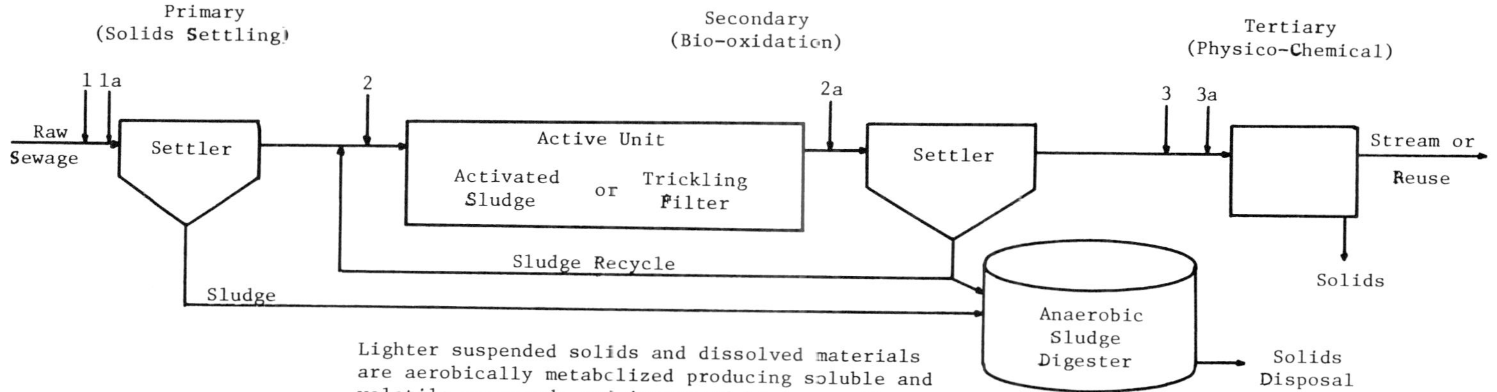

Bulky suspended solids are removed from the raw sewage by simple sedimentation.

Lighter suspended solids and dissolved materials are aerobically metabolized producing soluble and volatile compounds and increased biomass. There are two major types of secondary treatment, "activated sludge" with forced aeration and "trickling filter" down over rocks. The increased biomass and remaining suspended solids are flocculated, settled and removed as sludge to the anaerobic sludge digester.

Tertiary processes include: chlorination, activated carbon adsorption, chemical precipitation, ion exchange, electrodialysis.

Time:	Settler 1–2 h	Active Unit 4–6 h (AS) ½–1 h (TF)	Settler 1–2 h	Several days (Sludge)	Tertiary Few minutes

	Raw Sewage	Primary %-Removal	Secondary %-Removal	
Suspended solids (mg l^{-1})	275	35–65%	70–90% TF 80–95% AS	
BOD (mg l^{-1})	200	25–40%	65–95% TF 85–95% AS	Effluent of almost any quality can be produced by sufficient physico-chemical treatment.
Bacteria no./100 ml	10^{11}	50–60%	70–95% TF 90–95% AS	

Fig. 2. Typical waste water treatment scheme. Phosphate removal chemicals may be added at points *1*, *2*, or *3* and flocculant at points *1a*, *2a*, or *3a*

Table 3. Normal range of percent BOD, SS, and P removal during waste water treatment with and without chemical addition. (US–EPA 1976)

Treatment	Phosphorus		SS		BOD	
	Without	With	Without	With	Without	With
Primary	5–10	70–90	40–70	60–75	25–40	40–50
Secondary						
Trickling filter	10–20	80–95	70–92	85–95	80–90	80–95
Activated sludge	10–20	80–95	85–95	85–95	85–95	85–95

Table 4. Summary of phosphate removal from waste water

Primary[a]	Secondary	Tertiary
1.[b] $Al_2(SO_4)_3$, Alum. Easily used. Lowers pH and alkalinity. Adds sulfate, which may aid in flocculation. Lowered pH does not affect secondary treatment	$Al_2(SO_4)_3$. See Primary. May be added toward end of secondary treatment unit. Normally, ~1.5 to 2.5 Al/P required. Little carryover of Al into effluent. Normal pH range is 5.5 to 6.5	$Al_2(SO_4)_3$. See primary
2. $NaAl(OH)_4$, Sodium aluminate. Reacts similarly to alum but does not consume alkalinity. Also, no sulfate is added to effluent	$NaAl(OH)_4$. Similar to alum, see Primary	$NaAl(OH)_4$. Similar to alum; see Primary
3.[b] $FeCl_3$, Ferric chloride. Reacts similarly to aluminum. Easily handled and metered. Often produces carryover of dissolved iron. Most effective at ~1/2 to 1 pH unit lower than aluminum	$FeCl_3$. See Primary. Both iron and phosphate may be released when sludge becomes anaerobic	$FeCl_3$. Iron carryover in effluent normally precludes use of iron in tertiary treatments
4.[b] $FeCl_2/HCl$, Pickle liquor. Generally, chosen because of availability from local steel industry. The excess acid may require neutralization. Reacts similarly to $FeCl_3$	$FeCl_2/HCl$. See Primary and $FeCl_3$	$FeCl_2/HCl$. See $FeCl_3$, Tertiary
5. $Ca(OH)_2$, Lime. 9.5–11.5 pH required. Increased pH is neutralized either by CO_2 addition to the primary effluent or by CO_2 produced in the activated sludge unit	$Ca(OH)_2$. Not used	$Ca(OH)_2$. Preferred when high quality discharge is required. Operationally complicated. 75% of lime can be recovered

[a] Often a significant percentage of phosphate in the primary clarifier is as condensed phosphates, which do not precipitate readily

[b] Particularly with iron and aluminum and to a lesser extent with lime, addition of a polymer (0.2 to 1 ppm) flocculation aid will normally greatly improve phosphate and suspended solids removal at little added cost

objective of phosphate removal is generally to reduce, or minimally increase, the volume of solids produced from the overall plant. By proper choice of chemical(s), concentration, and point of addition (discussed below) it is generally possible to reduce the overall volume of solids or settled material by improving the flocculation or agglomeration properties of the suspended solids.

Tertiary treatment refers to any advanced physicochemical method used. Water of almost any quality can be produced if sufficient physical and chemical treatment is used. A few such methods which have been tried are listed in Fig. 2. Most of these methods (many developed in the early 1970's) add considerably to the cost of water treatment and have become less popular in the past few years. Chemical precipitation is the one of primary interest to phosphate removal. Finally, most waste water is disinfected by chlorine before being discharged into a stream, lake, or ocean.

Phosphate Removal

Phosphate removal will be discussed in terms of where chemical addition takes place. Approximate points of chemical addition are noted in Fig. 2. An overall summary of phosphate, suspended solids, and BOD removal which can be expected with and without chemical addition, is presented in Table 3. A general summary of phosphate removal through primary, secondary, and tertiary treatment is given in Table 4. For a 10 million gallon per day (mgd) plant in 1978, total waste water treatment cost was about 40 cents/1,000 gal (Rice and Gilbert 1978). The addition of chemicals for phosphate removal generally adds about 5 to 10 cents/1,000 gal including capital, labor, and chemicals (US-EPA 1976).

Primary Settler

The advantage to adding chemicals before the primary settler is to provide increased reaction time and to provide decreased solids and BOD loading in the secondary treatment stage. Often a polymer or flocculation aid is added (point 1a, Fig. 2) to further increase the solids removal in the primary clarifier. The major disadvantage to chemical addition before the primary settler is that often 40%–60% of the total phosphorus in solution is still as condensed polyphosphate which is less reactive than orthophosphate.

All precipitants in Fig. 3 are used to remove phosphate by addition before the primary settler (point 1 in Fig. 1). The molar ratio, metal:P, necessary for 70%–90% removal for Al, Fe(II), and Fe(III) are presented in Table 5. Particularly with iron and aluminum, some anionic flocculation polymer is often added at point 1a in Fig. 2 at 0.3–1.0 ppm; this adds about 0.5 to 1.5 cents/1,000 gal treated. Lime is also used before the primary clarifier for phosphate removal, but the key variable seems to be the solution pH obtained (US-EPA 1976):

CaO mg l^{-1}	289	400	400	500
pH	10.2	11.0	11.0	11.5
Flocculant ($FeCl_3$) mg l^{-1}	24	14	–	–
% P removed	93	91	75	90

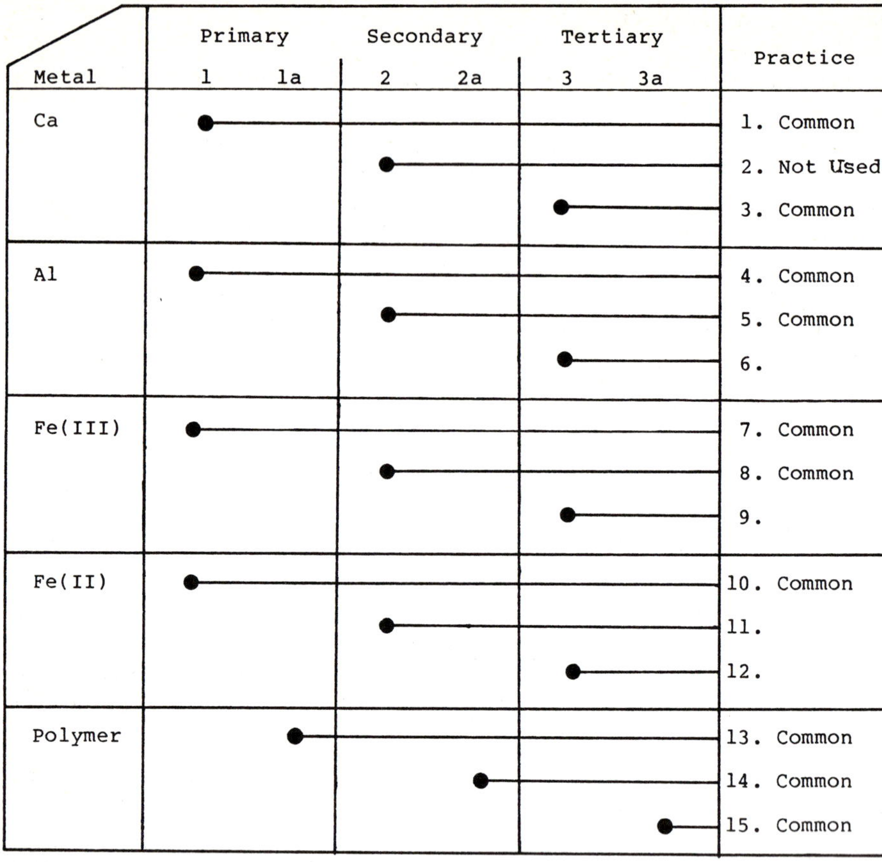

Fig. 3. Summary of chemical addition for phosphate removal options

Table 5. Molar ratio of metal/phosphorus for 70%–90% phosphorus removal from waste water. (US–EPA 1976)

Al/P	Fe(II)/P	Fe(III)/P
$Al_2(SO_4)_3$	Pickle liquor	$FeCl_3$
1.5 to 2.5	1.3 to 2.0	1.1 to 1.8

These data are consistent with Fig. 2 and the assumption that OCP is the stable solid phase which precipitates from solution, although most texts suggest that HAP is the model stable phase. Two concerns with lime addition at point 1 are the primary effluent pH increase and the increased $CaCO_3$ precipitation (increased solids production). If high lime dosage is used at point 1, CO_2 is often added at point 2 to decrease the pH before secondary treatment. Lime addition is particularly attractive when the influent raw sewage has a low alkalinity, requiring minimal lime to reach desired pH. The high pH "limestone-like" sludge is often calcined, slaked, and reused; this can reduce additional lime purchases by 40%–60%.

Trickling Filter

If a trickling filter is used for secondary treatment, normally either iron (II) or (III) or aluminum (as alum or sodium aluminate) is added for phosphate removal. Either iron or aluminum may be added at pont 2, Fig. 2, without adversely affecting the trickling filter BOD and SS removal, but generally it is not recommended to add minerals directly to the trickling filter unit. Rather, minerals are normally added at point 1, and polymer at point 1a in a trickling filter secondary treatment plant, resulting in 75%–85% of the phosphorus being removed in the primary clarifier and another 5%–10% removed in the trickling filter unit itself. When polymer is not added the phosphate removal drops by about 20%.

Activated Sludge

Activated sludge treatment is used for 70%–80% of waste water receiving secondary treatment. Phosphorus can be removed in a secondary treatment plant using activated sludge with similar efficiencies of 70%–95% as noted before (Table 3). Normally, either iron or aluminum is used. The following quote from the US-EPA Manual of Phosphorus Removal (1976) provides an excellent summary of potential phosphate removal in activated sludge secondary treatment plants:

"The mechanism of phosphorus removal by mineral salt addition, usually iron or aluminum compounds, in a biological system is through a combination of precipitation, adsorption, exchange, and agglomeration as influenced by the pH and ionic composition of the water. The phosphorus removal technique is operationally simple and is accomplished by direct mineral addition to an aeration tank. Treatment costs are largely a function of the required effluent phosphate residual. Through optimization techniques, any degree of phosphorus removal may be provided. The main liability is the introduction of dissolved solids."

By appropriate choice of metal, polymer, and addition point it is generally possible to remove phosphate and to shorten the treatment time in the aeration basin due to improved settleability of the floc. Therefore, in an overloaded activated sludge plant it could be desirable to add minerals to the aeration basin even if phosphate removal were not a requirement. If $Al_2(SO_4)_3$ or $FeCl_3$ is used, there is often a substantial drop in water alkalinity due to hydrolysis, especially during periods of high storm water. This drop in pH or alkalinity can be avoided by using sodium aluminate. Also, iron from $FeCl_3$ or pickle liquor generally leaks and appears in the final effluent more than aluminum from either alum or sodium aluminate. If low alkalinity water is to be treated in the aeration basins for phosphate removal, the metals can be added near the end of the basin, just before solids removal.

Tertiary Lime Treatment

For effective phosphate removal using lime, the pH generally needs to be greater than 8 or 9 (Fig. 2) and often pH 10 to 11 is used. Since activated sludge requires 6–8 pH for effective metabolism, lime treatment is not used in the activated sludge unit itself. The secondary effluent can be treated with lime in a tertiary treatment system. Although this tertiary treatment is more expensive to operate, it can often produce an effluent with lower total phosphorus, lower suspended solids, and even lower dissolved solids. Also, tertiary lime treatment is considered to be a more "reliable" treatment method than iron or aluminum addition (US-EPA 1976).

There are two lime methods in use: a high lime or two-stage method with lime recovery and a low lime or single-stage method. If the alkalinity is 150 to 200 mg l^{-1} $CaCO_3$ or greater, the simpler single stage system can be used. For single stage treatment 150 to 200 mg l^{-1} of lime (as CaO) is added to raise the pH to only 9.5 to 10. Time is allowed for $CaCO_3$, calcium phosphate, and other solids to precipitate, flocculate, and settle. Carbon dioxide may be added to reduce the pH before discharge. If the alkalinity is less than about 150 mg l^{-1} $CaCO_3$, a two-stage system is generally needed because there is not enough solids production to promote flocculation and settling. Thus, 300 to 400 mg l^{-1} (as CaO) lime is added which increases the pH to above 11. At this pH, $CaCO_3$, Ca-phosphate, and $Mg(OH)_2$ precipitate. Then, CO_2 is bubbled through the waste water to lower the pH to about 10 where more $CaCO_3$ precipitates and settles and removes the suspended solids. Additional CO_2 may be added to lower the pH before discharge. Up to 75% of the lime can be recovered, recalcined, slaked, and reused. With lime recovery the added cost of two-stage treatment is estimated at about U.S.$ 10/1,000 gal of water treated. Advantages of tertiary lime treatment are process reliability and high quality effluent water. High quality effluent water is often required if the receiving water body is used for fishing, swimming, or recreation.

Acknowledgements. The authors gratefully acknowledge financial support for this work from the Gas Research Institute, 5080-321-0317, U.S. Environmental Protection Agency National Center for Ground Water Research, R-805292-02 and CR-806931-03, and Maurine Lee for typing. In no way does this constitute an endorsement by GRI or EPA of any product or view contained herein.

References

Berndt AF, Stearns RI (1973) The equilibrium between a solid solution and an aqueous solution of its ions. J Chem Educ 50:415–417

Boulet M, Marier JR (1961) Precipitation of calcium phosphates from solutions at near physiological concentrations. Arch Biochem Biophys 93:157–165

Frankenthal RP (1963) Fundamental data: Solubility products. In: Mertes L (ed) McGraw-Hill, New York (Handbook of analytical chemistry, 1st ed, p 113–118)

Garrels RM, Mackenzie FT, Hunt C (1975) Chemical cycles and the global environment: Assessing human influences. Kaufmann, Los Altos, California, p 105–109, 157–162

Gregory TM, Moreno EC, Brown WE (1970) Solubility of $CaHPO_4 \cdot 2H_2O$ in the system $Ca(OH)_2$–H_3PO_4–H_2O at 5, 15, 25, and 37.5 °C. J Res Nat Bur Std, A Phys Chem A74:461–475

Gregory TM, Moreno EC, Patel JM, Brown WE (1974) Solubility of β-$Ca_3(PO_4)_2$ in the system $Ca(OH)_2$–H_3PO_4–H_2O at 5,15, 25, and 37 °C. J Res Nat Bur Std, A Phys Chem A78:667–673

Horne RA (1978) The chemistry of our environment. Wiley & Sons, New York, p 261–263

McDowell H, Brown WE, Sutter JR (1971) Solubility study of calcium hydrogen phosphate. Ion pair formation. Inorg Chem 10:1638–1643

Metcalf Eddy Inc (1979) Wastewater engineering: Treatment, disposal, reuse, 2nd ed. McGraw-Hill, New York, p 745–751

Moreno EC, Brown WE (1960) Solubility of dicalcium phosphate dihydrate in aqueous systems. Soil Sci Soc Am Proc 24:94–98

Moreno EC, Brown WE (1960) Stability of dicalcium phosphate dihydrate in aqueous solutions and solubility of octacalcium phosphate. Soil Sci Soc Am Proc 24:99–102

Moreno EC, Gregory TM, Brown WE (1968) Preparations and solubility of hydroxyapatite. J Res Nat Bur Std, A Phys Chem A72:773–782

Nesbitt JB (1973) Phosphorous in wastewater treatment. In: Griffith EJ, Belton A, Spencer JM, Mitchell DT (eds) Environmental phosphorous handbook. Wiley Intersci, New York, p 203–239

Nriagu J (1972) Stability of vivianite and ion-pair formation in the system iron (II) orthophosphate-phosphoric acid-water. Geochem Cosmochim Acta 36:459–470

Pagenkopf G (1978) Introduction to natural water chemistry, vol 3. Envir Sci Tech Ser. Dekker, New York, p 259

Rice RC, Gilbert RG (1978) Land treatment of primary sewage effluent: Water and energy conservation. Hydrology and water resources in Arizona and the Southwest. 8:33–36

Shen CY, Morgan JW (1973) Hydrolysis of phosphorous compounds. In: Griffith EJ, Beeton A, Spencer JM, Mitchell DT (eds) Environmental phosphorous handbook. Wiley Intersci, New York, p 244

Singer PC (1972) Anaerobic control of phosphate by ferrous iron. J Water Pollut Control Fed 2:663–669

Snoeyink VL, Jenkins D (1980) Water chemistry. Wiley & Sons, New York, p 298–312

Stumm W, Morgan JJ (1981) Aquatic chemistry, 2nd ed. Wiley Intersci, New York, p 238–249, 282–285

Tomson MB, Nancollas GH (1978) Mineralization kinetics: A constant composition approach. Science 200:1059–1060

US Environ Prot Agency (1976) Process design manual for phosphorous removal. EPA 625/1-76-001a.

Williams BG, Patrick WW Jr (1973) Dissolution of complex ferrous phosphates under controlled E_h and pH conditions. Soil Sci Am Proc 37:33–36

Zoltek J Jr (1976) Identification of orthophosphate solids formed by lime precipitation. J Water Poll Control Fed 48:179–182

Chapter 14 Phosphorus in the Environment

R. T. Oglesby[1] and D. R. Bouldin[2]

Introduction

As with all writing assignments, an initial decision had to be made relative to the comprehensiveness and depth to which the subject was to be treated. We have elected to constrain our material within the boundaries of, first, broadly defining the distribution and movement of phosphorus through the environment and, second, how human influence on distribution and movement of P has created water quality problems in some situations.

There are at least two other environmental aspects of P the reader might want to pursue if a comprehensive treatment of the subject is desired. One concerns the group of organophosphate compounds which possess anticholinesterase properties. These were first synthesized during World War II as chemical warfare agents, the so-called "nerve gases". Subsequently, they have found widespread use as a class of insecticides, Parathion and Malathion being the best-known examples. Inch (1978) provides a well-synthesized discussion of the organophosphates.

A second topic involves environmental problems produced by the release of elemental (e.g., red) phosphorus into the environment. An instructive case history of such an incident, one resulting in a substantial kill of marine life in Placentia Bay, Newfoundland as the result of industrial waste discharge, is provided by Vallentyne (1974, p 82–85).

Content and Cycling of Phosphorus in the Biosphere

Phosphorus is the tenth most abundant element in the planet Earth, with an average concentration in the crust of 0.1%. Continental sediments average about 0.07% P, soils 0.08%, igneous rock 0.10%, and marine sediments 0.12% (Brinck 1978). Since P is essential for all forms of life, any portion of the Earth's surface which contained no P would be incapable of autotrophic production. Although difficult to document, we know of no place on the Earth's surface where this is true.

However, P is frequently an element limiting autotrophic productivity as evidenced by the fact that:

a) Additions of soluble forms to many aquatic systems will stimulate production.

---

[1] Department of Natural Resources, Fernow Hall and [2] Department of Agronomy, Bradfield Hall, Cornell University, Ithaca, New York 14853

b) Phosphorus must be added to a large fraction of Earth's land surface before high yields of agricultural crops (e.g., cereal grain yields in excess of 1,500 to 2,000 kg ha^{-1}) can be obtained. Relatively little information is available on non-farmed terrestrial systems and hence whether or not P additions will stimulate production in "natural" terrestrial systems is largely unknown.

In short, P availability is a critical factor in determining autotrophic production in many ecosystems. Although not a universal regulator which everywhere sets limits on plant and, ultimately, animal productivity, it is a very important factor over large areas of the Earth's surface.

In terrestrial systems there is seldom a deficiency of P in the sense that all of that in the rooting medium is removed by the vegetation; in fact, usually there is a large amount of P in the rooting medium which the plants cannot get at because it is: (a) physically buried within a solid matrix, (b) in organic forms which cannot be assimilated by the plant or (c) transfer from the soil solid phase to the plant root is too slow to meet the needs of the growing plant. This inability of plants to remove all or even most of the total P in land-based systems has led to the use of the word "available" to describe that fraction of the total P which plants can assimilate. Although the concept of "available" P may seem quite clear, any quantitative definition is ambiguous, whether based on biological or chemical assay. There are numerous operational definitions based on either biological or chemical assays, but none is universally useful.

In aquatic systems there is often a correlation between total P in the photic zone and biomass. In the case of temperate latitude lakes, on the order of 1/2 of the total P will be in the algae. This is particularly true in deep lakes with a well-defined thermocline. However, if one considers all of the P in a water column 1 m square and includes the P in the sediment to a depth of a few centimeters, the relationship between biomass and total P will vary among lakes and only a relatively small fraction of the P in the system will be in the biomass (Golterman et al. 1969; Golterman 1976). Thus in aquatic systems all of the P is not available to the organisms unless the portion of the system being considered is restricted to the water column.

In many respects this review is an account of how different disciplines define and use the term "available" phosphorus, how this moves through and is recycled in the biosphere and how "available" phosphorus is, or can be, managed in terrestrial and aquatic systems.

A corollary to the above is the thesis that tracing of total P is of limited usefulness in assessing biological impact. Thus, although we will discuss total phosphorus, emphasis will be on biologically available forms and on understanding the biogeochemistry of this element.

Overall, we will proceed from the Earth's biosphere to smaller and more specific ecosystems with emphasis on cycling within the system and interchanges between ecosystems. We will discuss the "availability" of the P in the various compartment where appropriate.

Table 1 summarizes major P fluxes on a global scale. The recycling within the marine and terrestrial biota exceeds inputs from mining by a factor of about 80. Looked at on a global scale, it is not possible (nor useful) to detect deviations from a steady state. Changes in the size of the various compartments in Table 1 are too small relative to the uncertainties in the estimates of the various fluxes to be

Table 1. Summary of phosphorus fluxes on a global scale

Global Flux (10^8 m ton yr^{-1})	
Cycled between marine biota and ocean water	9.6
Cycled between terrestrial biota and crust (includes 0.12 input to land from mined phosphates)	2.3
Mining of phosphate minerals[b]	0.15
Transport from land to ocean	0.02[a]
Contribution to ocean sediments	0.02 ?
Human food[c]	0.02

[a] Corresponds to a P loading to oceans of 6 mg P $m^{-2} yr^{-1}$
[b] Brinck (1978)
[c] Based on per capita consumption of U.K. (Bowman 1978), food in U.S.A., 1959–1976 (USDA 1977) and world population of 4×10^9

meaningful on a global scale. The comparison of phosphate mined and that contained in human food should be noted. About 90% of the P mined is applied as fertilizer used in food production (McClelland and Hignett 1978). This exceeds the P in human food by a factor of about six to seven, which suggests P is accumulating in the soils used for food production or else it is being removed by some mechanism other than cropping. Most of this accumulation is occurring in the developed countries, and hence this is a very misleading comparison on a global scale.

Bowman (1978) developed a fairly detailed P balance sheet for the agriculture-human food system in the United Kingdom. Inputs into the system (in thousands of metric ton P yr^{-1}) were: fertilizers 210, detergents 36, and imported food and feed 77. Losses from the agricultural-human food chain were: sewage 54, refuse and animal excreta 26, and soil losses 20. Thus, annual inputs exceed losses by 223 thousand metric tons. Input of fertilizer P exceeded the food produced domestically by a factor of 8, but it was only 1.1 times the P of harvested plant material used for feed and food combined.

Essentially all of the excess of inputs over losses accumulated in the cultivated soils, which received the bulk of the fertilizers, manure and sewage sludges. Furthermore, most accumulated in the soils as inorganic P derived from fertilizer because (a) fertilizers are inorganic phosphates; (b) less than 10% of added fertilizer is taken up by plants in the year of application; and (c) since the U.K. soils have been cultivated for a long time under similar management, the organic matter and organic P content should be approaching a steady state where rate of decomposition equals rate of addition. The major increase in the various portions of the system would be in the inorganic compartment of the cultivated soils with perhaps a small increase in organic P and probably no important change in the size of the biomass compartments. Fluxes between compartments are probably constant or at least change slowly with time. Considerable discussion following Bowman's paper (Bowman 1978) was devoted to whether or not phosphorus fertilizer additions are needed to maintain present production levels; opinions on this question were divided. Recently Karlovsky (1980) made the following very clear statements:

a) "The central problem in the world phosphorus economy does not consist, as claimed by some participants of the CIBA Foundation Symposium 'largely of digging phosphorus at one place and stor-

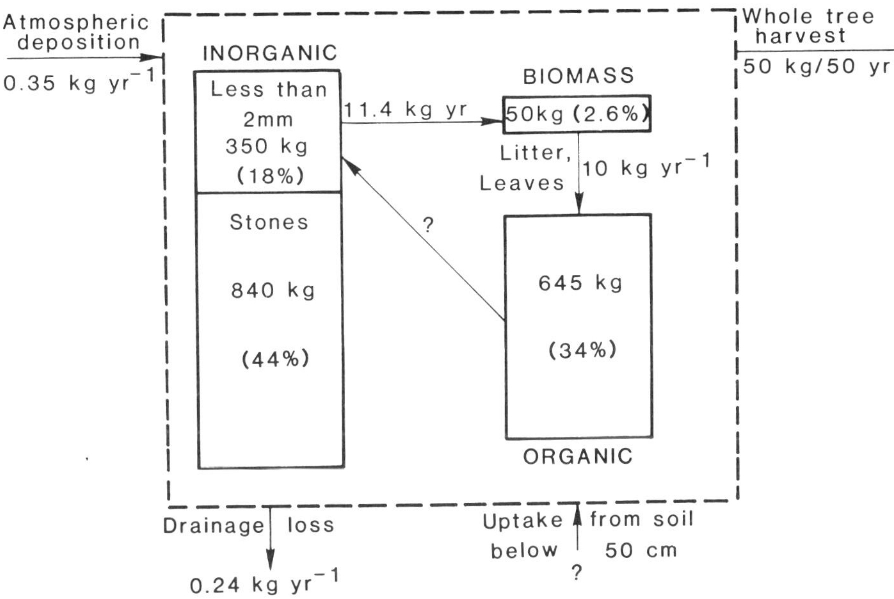

Fig. 1. Phosphorus pools and fluxes in a hectare of Meathop Wood (Harrison 1978)

Table 2. Range of biomass, and annual uptake, return and retention of overstory (Ovington 1962; Sukachev and Dylis 1964; Duvigneaud 1971; Tsutsumi 1971; Likens et al. 1977; Cole and Rapp 1981)

Biomass content		Uptake	Release	Retention
kg ha^{-1}			kg ha^{-1} yr^{-1}	
Aboveground 7–224	Roots 3–81	0.8–12	0.7–7	0.3–10

ing it in the soil at another, but is due to the failure of traditionally trained soil scientists to interpret the farm and experimental evidence in a logical way, taking into account all parts of the phosphorus cycle and

b) the efficiency of P utilization in the great majority of agroecosystems in very high and is as high or higher than the utilization of other macro-elements."

The phosphorus cycle in several wooded systems has been partially or completely estimated (Ovington 1962, 1964; Duvigneaud and Denaeyer-DeSmet 1971; Likens et al. 1977; Harrison 1978). Some of the important portions of the P content and cycling in Meathop Wood (Wales) as reported by Harrison (1978) are illustrated in Fig. 1. With respect to the pools and fluxes in Fig. 1, values for other wooded systems are as shown in Table 2.

These data illustrate the following generalizations:

1. Soil and litter contain very large amounts of P relative to biomass. A precise description of compartmentalization is difficult because the total content of the litter is subject to considerable uncertainty since more or less "soil" may be included.

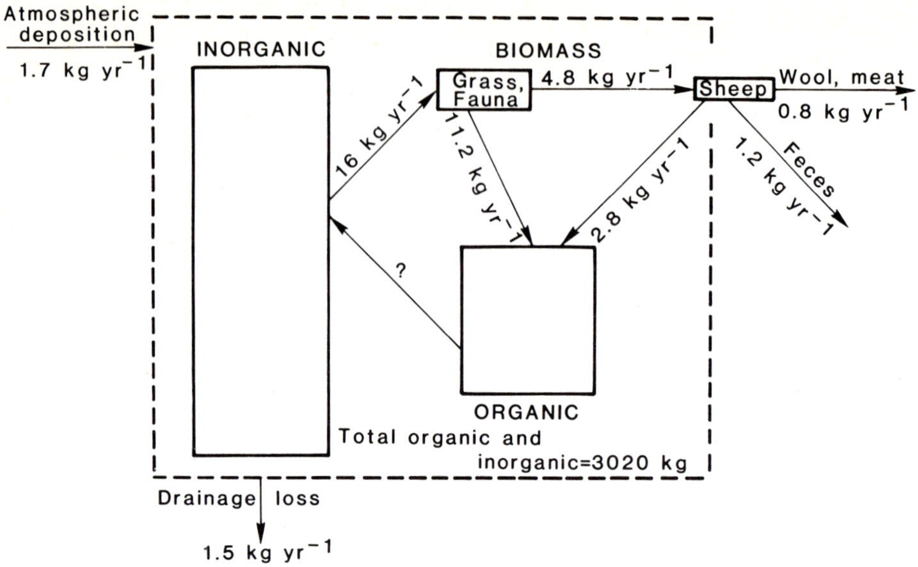

Fig. 2. Phosphorus pools and fluxes in a hectare of grassland under grazing by sheep (Harrison 1978)

Relatively few total P analyses of soil were found. More often values were for extractable P; but they were of limited usefulness for comparative purposes because the diversity of methods used precluded any useful comparison.

2. The fluxes between the compartments are small relative to the size of biomass, litter or soil pools. This means estimation of these fluxes cannot be based on changes in size of the various pools over short periods of time or when the system is at steady state.

3. The inputs in precipitation and outputs in soil drainage are small relative to flux between soil and biomass.

4. Data presented in Fig. 1 illustrate the importance of P occluded within particles; in the greater than 2 mm soil fraction there is 840 kg P ha^{-1}; clearly, this will become available to plants only over many centuries.

5. Harrison (1978) states that Meathop Wood soil is deficient in available P. A 700% increase in yield was obtained by adding P to samples of the soil in greenhouse cultures. Perhaps P was limiting growth of these and other forests, but this is not clear on the basis of any other data we found.

6. Removal of P in wood harvested periodically (every 20 to 60 years) is on the order of 0.1 to 1 kg P ha^{-1} yr^{-1} (Harrison 1978; Ovington 1962).

7. Forested watersheds act as sinks for P in the terrestrial environment.

Harrison (1978) presented a P balance for grassland grazed by sheep from April to October in the United Kingdom. The results are summarized in Fig. 2. He states that phosphorus limits plant production. Although the cumulative annual uptake of P by the plants is about twice that in the woodland systems (16 vs. 9 kg) it is still only a small fraction (0.5%) of the total P in the system. Again the balance of P is negative in that the inputs are smaller than losses by drainage, and removal

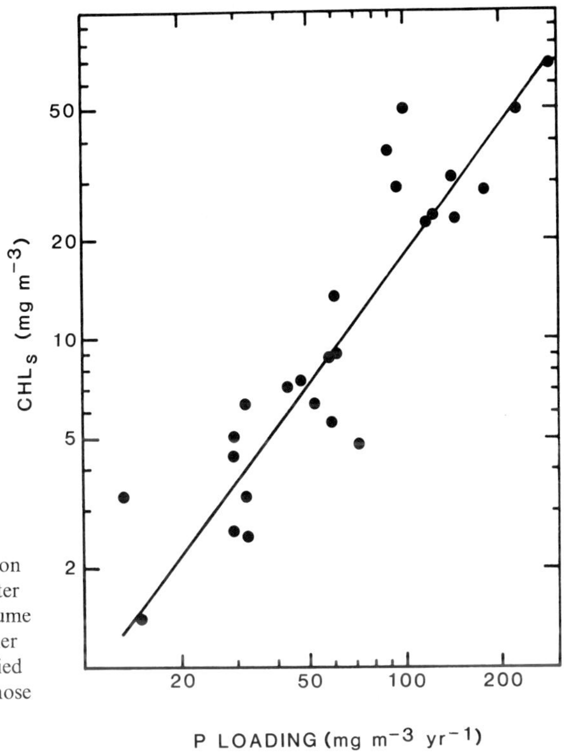

Fig. 3. Mean summer chlorophyll concentrations in 26 lakes as a function of "bio-available" P loading. The latter is calculated as annual input to a volume equivalent to that of the lake's summer mixed zone, i.e., epilimnion of stratified systems and whole lake volume for those that are mixed from top to bottom (Replotted from Oglesby 1977)

in animal products and feces (the animals spend the night off the pasture and hence there is a loss of P in manure from the grazed portion of the pasture).

Clark et al. (1980) summarized studies with native grasslands in the arid – semiarid regions of North America and found about 14 kg P ha^{-1} in the grass biomass (tops, crowns, roots). They estimated in one case the biomass of the soil microorganisms was 19.8 kg P ha^{-1}. Total P in the soil ranged from 1,000 to 2,000 kg ha^{-1}. Presumably the annual cycling of P from soil to grass is about equal to the P in the biomass (14 kg ha^{-1}), in these two cases. At the Canadian site (Halm et al. 1972), addition of P alone did not stimulate growth but that of H_2O, or N or $N+H_2O+P$ did. Perhaps the extreme in low amounts of P in grass biomass is afforded by the data of Bülow-Olsen (1980) where less than 1 kg P ha^{-1} was found in the living biomass of an unproductive grazing area in Denmark.

The data we found on non-intensively grazed or native grassland ecosystems indicate that P cycling is of the same order of magnitude as that in woodland except that the total storage in the biomass is about an order of magnitude less in the former. One interesting observation is the importance placed by many of the investigators in mineralization of organic P as a key variable in controlling cycling of P (Cole et al. 1977; Clark et al. 1980; Harrison 1978).

The retention of P by the soil/vegetation compartments of watersheds is illustrated in Fig. 3. These represent naturally vegetated systems, but this does not mean that human influences are necessarily lacking. For example, Taughannock

Creek one of the systems depicted in this figure, has almost 50% of its land area in active agricultural use. Values are for total P so land uses that increase erosion accelerate output. Direct input of household wastes also increases output above that attributable to "natural" leakage.

Karlovsky (1980) summarized P "inputs" and "consumable outputs" for 43 agro-ecosystems. He found the latter ranged from 0.1 to 53 kg P ha^{-1} yr^{-1}. Output/input ratios ranged from 0.04 to ∞, but most were less than 1. Karlovsky points out with considerable emphasis that where inputs exceed outputs by a considerable margin, the pool of available P is increased and hence crops in future years will benefit from this increased available P.

As Karlovsky (1980) points out, the enigma of P behavior which creates so much confusion among agronomists, farmers and ecologists is the fact that seldom does plant recovery of fertilizer P exceed 10% in the year of application, yet what remains is available to crops over a long period of time. Experimental quantification of the residual effect by field experimentation (the only way to get a "final" answer) requires on the order of 5 to 20 years. Even in highly weathered tropical soils characterized by large amounts of iron and aluminum oxides, the residual effects are very high and persist for several years (Lathwell 1979).

Transfer Among Ecosystems

In the preceding sections, major emphasis was placed on pools of P and recycling within terrestrial ecosystems. Now attention will be turned to transfers of P into and among ecosystems. Basically this is accomplished by one or more of the following processes:

a) Deliberate enrichment of soils by applications of fertilizer P, manures and other residues.

b) Discharge of sewage and industrial wastes containing phosphorus into water.

c) Particulate matter injected into the atmosphere by human activity (there are no important gaseous forms of P).

d) Transfer of dust by wind.

e) Transfer of particulate and dissolved material by water.

Major adverse effects of transfers of P from system to system are associated with enrichment of water with P to the extent that excessive algal growth occurs. We know of no major adverse, direct effects of P additions on terrestrial ecosystems except for the misuse or improper disposal of certain synthesized organic compounds.

Most estimates of the net inputs to Earth's surface from precipitation range from 0.04 to 0.6 kg P ha^{-1} yr^{-1} (Ingrahm 1950; Viro 1953; Miller 1963; Crisp 1966; Hobbie and Likens 1973; Likens 1974; Schindler et al. 1974; Smith 1979; Ahl 1980). These include more or less "dry deposition" (dust fallout). The majority of these estimates are for "total" P and hence include not only that in solution but also any associated with the solid phase. Thus there is no easy means to estimate the "availability" of these P inputs.

Similarly, the outputs in the streams draining what could be described as "undisturbed" watersheds (e.g., no major human disturbance or activity) range from 0.01 to 0.5 kg P ha^{-1} yr^{-1} (Viro 1953; Miller 1963; Crisp 1966; Vollenweider 1968; Hobbie and Likens 1973; Dillon and Kirchner 1974; Porter 1975; Ahl et al. 1977; Likens et al. 1977; Smith 1979; Kjensmo 1980). Again, these outputs are a combination of phosphorus dissolved in water and P associated with particulate matter suspended in the water and hence the "availability" of the P is a matter of considerable speculation.

Turning attention now to how human activity influences transfer of P by flowing water, we will discuss the Fall Creek watershed in the Northeastern U.S.A. (N.Y. State) in considerable detail since study results are available to illustrate most of the processes. Details are available in Porter (1975) and Johnson et al. (1976).

The Fall Creek watershed lies in the cool-temperature zone of the United States. Annual precipitation averages 85 cm with about equal distribution throughout the year. Approximately one-half of the precipitation flows out of the watershed into Cayuga Lake, a body of water where nutrient-phytoplankton interactions have been studied extensively.

The total area of the watershed is 331 km^2. Land use is variable, ranging from forested land which has been essentially undisturbed by human activity for the past 50 or so years to intensively used agricultural land largely devoted to dairy farming.

The population of the watershed is twelve thousand. There are several villages within the watershed one of which has a sewage treatment plant that discharges directly into a tributary stream. The remainder of the sewage is disposed of by on-lot systems.

The outflow of the watershed was intensively sampled (total of 2,500 samples collected) for about 3 years and subwatersheds with variable characteristics were sampled for shorter periods of time. The P in the water samples was separated into three fractions on the basis of operational definitions as follows: (a) particulate-defined as the total P in the solid phase separated by high speed centrifugation; (b) dissolved inorganic-based on the colorimetric analysis of the supernatant after reaction with molybdate and (c) total dissolved – defined by colorimetric analysis of the supernatant following persulfate oxidation. By definition dissolved organic is considered to be (c)–(b).

The flow-weighted mean concentrations were 110 µg P l^{-1} in particulate form and 32 µg P l^{-1} in the two dissolved forms. During the year, the relative and absolute concentrations of these two fractions differed; when flow was high the particulate matter increased by orders of magnitude and the dissolved forms increased by a factor of three to four.

Based on a number of arguments (see below) and available data, the important fraction in terms of impact on Cayuga Lake (of which Fall Creek is a major tributary) is the dissolved forms of P. Thus, major effort was devoted to sources and behavior of these.

During the year September 1972 through August 1973 about 6,600 kg dissolved P was carried out of the watershed, or, roughly, 0.2 kg ha^{-1} yr^{-1}. Based on concentrations in outflow from two forested watersheds devoid of human activity and samples from shallow wells, the expected dissolved phosphorus output with no human activity was estimated at 15 μg P l^{-1} or roughly 0.095 kg ha^{-1} yr^{-1}.

Origin of the remainder of the dissolved P output (~3,400 kg) was about equally divided between "diffuse" and "point" sources. The former were primarily associated with surface runoff from farmed land which is enriched with fertilizer P and receives frequent applications of manure. The point sources were sewage discharges; one sewage treatment plant plus the impact from the on-lot sewage disposal.

Based on a number of approximations, the estimated "leakage" of sewage P from on-lot disposal systems was 13% of sewage P leaving the watershed in dissolved form. The remainder was immobilized in the soil adjacent to the disposal field and some of the P which was put directly in streams was immobilized on sediments and carried out of the watershed in particulate form.

The estimated input of P in manure, fertilizer and precipitation over a 20-month period was 441,000 kg; most of the manure and fertilizer was applied to about 7,000 ha of intensively farmed land. Based on the analyses described above, about 0.5% of this was lost in dissolved form.

The relatively low losses of applied P can be explained as follows: (a) In Fall Creek the bulk of the manure and fertilizer were applied to the better drained soils in the watershed; on about 40% of the land which received manure and fertilizer, surface runoff was probably very low; (b) most of the water in excess of evapotranspiration percolated through the well-drained soils and the P concentration was reduced to the level of that from the forested land (~15 ppb); (c) thus the enrichment of the runoff in P was limited because of the relatively low portion of surface runoff associated with agricultural land. If all the manure and fertilizer had been applied to poorly drained soils where surface runoff was a substantial fraction of the excess, then much larger amounts of runoff enrichment would have occurred and the percentage of the manure and fertilizer P found in the dissolved forms in runoff would have been much higher.

Limited studies were made of the amount of P associated with the particulate matter which could be easily removed by leaching with water. The results indicated that on the order of 10% was easily desorbed. Presumably 90% of the particulate P was essentially no different from the geological forms associated with soils not enriched with fertilizer and manure.

Finally, one other aspect of the data requires comment. Very clear evidence of loss of dissolved P from the stream water was found during intervals of low flow downstream from the outfall of the sewage treatment plant. Presumably, this was a consequence of the reaction of the dissolved P with the sediments in the stream. During high flow events, the sediments in the stream were resuspended and some of the P redissolved while the remainder was carried out of the watershed as particulate P.

The following statements summarize some of the generalizations which can be made relative to runoff studies of watersheds.

1. A substantial portion of the total phosphorus delivered out of many watersheds is associated with the particulate fraction. Basically, the amount of particulate P exported is nearly proportional to erosion and hence varies by orders of magnitude both within and between watersheds. How "available" is the P carried on

the sediment? Some generalizations based on agricultural experience follow; other aspects will be addressed in the section on lake response.

As increasing amounts of P are reacted with soils, the concentration in water extracts increases, with the relationship between concentration and quantity added usually being described by an adsorption isotherm of some form. Similarly, as P is removed by successive extractions with water the concentration in water extracts decreases. Sooner or later the concentration in the water extract becomes on the order of 10 $\mu g\ l^{-1}$ or less and the desorption study is discontinued; the P content of the particulate matter is often several hundred $\mu g\ l^{-1}$ at this point and is referred to as "unavailable"; the summation of desorbed P is referred to as "available". Because of the tedious nature of numerous desorptions, extractants with acids, bases or complexing agents are often substituted as methods for estimating "available" P. Algal assays of some sort may also be used. Thus there are a wide range of methods used to estimate the "available" P on sediment.

There is no universal agreement about the interpretation of these measurements. The fraction of total P that is used by algae under conditions where the P could be obtained by the algae over a period of several weeks is in the range of 5% to 40% (Golterman 1976; Cowan and Lee 1976; Allan and Williams 1978; Green et al. 1978; Logan et al. 1979; Trautmann 1981). In general, the total P in sediment ranges from 500 to 2,000 $\mu g\ g^{-1}$ (0.05 to 0.2%) with most samples falling in the 1,000 to 1,500 range. "Available" P on sediment may therefore range from a few ppm to as much as 500 to 600 ppm on highly enriched samples from the Rhine River (Golterman 1976).

2. Dissolved P from point sources is reduced as it moves downstream during low flow episodes. Presumably some of this reacts with sediments and hence it converted to particulate P. Depending upon circumstances, some particulate P may be redissolved during high flow episodes (Taylor and Kunishi 1971; Porter 1975; Hill 1981; Verhoff and Baker 1978; Trautmann 1981). Another portion of the dissolved P is taken up by the stream biota during the growing season. Some of this is lost from the system in the form of emerging insects and the harvest of fish, some is exported in the stream as living and senescent biomass entrained in the flow and the remainder, represented by the mean annual standing crop, represents the extent to which a stream's biota acts as a P sink.

3. Dissolved P varies from 10–15 $\mu g\ p\ l^{-1}$ in water draining from pristine watersheds to 500 $\mu g\ l^{-1}$ or more in surface runoff from barnlots and heavily manured fields.

4. Most watersheds of lakes and reservoirs are a mosaic of sources. Conceptually, suppose the landscape can be aggregated into n units which meet the following specifications: (a) the area in each unit is to have approximately the same amount of surface runoff; and (b) concentration of P in the surface runoff is approximately the same (e.g., well-drained soils treated with manure in one unit, poorly drained soil treated with manure in another, forest land, etc.). The total amount of dissolved P in the *surface runoff* from these n units is then given by the following equation:

$$DP]\,{}^{\text{surface}}_{\text{diffuse}} = \sum_{i=1}^{i=n} S_i C_i A_i, \tag{1}$$

where $DP]^{surface}_{diffuse}$ is the dissolved phosphorus in the surface runoff from diffuse sources.

S_i = surface runoff per unit area
C_i = concentration of P in the surface runoff
A_i = Area of the unit
n = Number of units in the watershed.

The above equation only deals with surface runoff; there is also a throughflow component which infiltrates into the soil, flows through the soil profile/unconsolidated mantle and eventually finds its way to a stream. Since this water flows slowly for a long distance through unconsolidated material which has not been enriched by human activities, the concentration in solution by the time it reaches a stream will be reduced to the "geo-chemical" equilibrium level; therefore, the concentration will be everywhere the same and it will be approximately equal to that in outflow from forested watersheds at low flow (Porter 1975; Ryden et al. 1973). Modifying the preceding equation accordingly:

$$DP]^{subsurface}_{diffuse} = C_G \sum_{i=1}^{i=n} (P_i - ET_i - S_i)A_i. \tag{2}$$

C_G = concentration of P in the throughflow
P_i = precipitation
ET_i = evapotranspiration

The following approximations may be applied to the above equations which simplifies them further:

a) P − ET = F, a constant for all units (total outflow from each area is the same)
b) The surface runoff from land not enriched with P and the subsurface flow are equal in concentration and the same for all subunits ($=C_G$)
therefore

$$DP]_{diffuse} = FC_G \sum_{i=1}^{i=n} A_i + \sum_{i=1}^{i=M} (C_i - C_G)S_iA_i. \tag{3}$$

C_G = subsurface flow from all land and concentration of P in surface runoff from undisturbed land
M = subunits where surface runoff is enriched ($C_i > C_G$)

Equation (3) is satisfying in a conceptual sense since the first term on the right side is an estimate of DP if there were no human activity in the watershed and the second term on the right side is the impact of human activity on diffuse source dissolved P (note that point source contributions would still need to be added).

However, there is one serious difficulty with the above concept; during episodes of high flow (when surface runoff occurs), sediment concentrations are often high and hence there is interaction between the sediment and dissolved phosphorus due to the following: (a) The sediment is derived from the same areas that produced the runoff and hence is in equilibrium with it; (b) sediment from stream banks, road ditches, etc., by virtue of being unenriched with phosphorus, has a capacity to react with dissolved phosphorus in excess of C_G; and (c) sediments from stream beds receiving point source phosphorus are highly enriched with this element. During high flow episodes (when the bulk of the water, dissolved phosphorus and sediment are delivered out of the watershed), dissolved phosphorus inputs of highly variable concentrations are being mixed with sediments from different sources enriched to variable degrees with phosphorus. The result of all of these reactions does not lend itself to easy analysis both because of the complexity of the reactions involved and the large amount of input data needed.

Despite the difficulties cited in the preceding paragraph, Eq. (3) is still a useful approximation so long as: (a) point sources do not dominate the phosphorus inputs; (b) sediment with high capacity to either remove or dissolve large amounts of P is not put into the system (Porter 1975). Furthermore, the conversion of dissolved P to P sorbed on particles (or vice versa) may not be too important in terms of biological impact in shallow bodies of water (this is discussed further below).

5. Statistical correlation methods and regression analysis has been used fairly extensively in relating nutrients in outflow to human activities, land use, and soil types. Usually the different "independent" variables are correlated with each other and some are very highly correlated. Therefore, the statistical analysis cannot be used to establish cause and effect relationships; unfortunately this fact is ignored in many discussions of water quality management programs.

Lakes and Phosphorus

Development of Perceptions and Concepts

The eutrophication of lakes in North America constituted one of the more widely publicized environmental issues of the 1960's. While there is still much to be learned about the subject, limnologists and water quality management specialists generally agree that the problem has been defined, causal factors identified and considerable progress made toward a solution. The principal agent responsible for the accelerated eutrophication of temperate latitude lakes during the post WW II period is phosphorus.

Hutchinson (1969, 1973) has provided two insightful reviews of the early research from which the trophic state concept evolved. Insights from paleolimnological analyses of stratigraphic sequences observed in lake sediment cores converged with those obtained from studies of processes taking place in the water column to provide patterns of changes in lakes associated with alterations in nutrient loadings. An important adjunct was the recognition that responses were probably modified by lake morphology. The broad picture is one of lakes maturing from a low nutrient condition (oligotrophy), with consequent low levels of biological production, to one of high input of available nutrients (eutrophy) characterized by abundant growth of plants and animals. Stratigraphic records indicate that, for some lakes, this process has included periods of reversal towards oligotrophy as the result of changes in climate and watershed vegetation (Sanger and Gorham 1972; Whitehead and Crisman 1978) and from particular kinds of human influences, both unwitting (e.g., Deevey et al. 1979) and as a consequence of efforts

Table 3. Concentration of elements in fresh water relative to estimated demand. (Vallentyne 1974)

Element	Demand	Supply	Demand: Supply
	Plants %	Water %	Plant: Water (approx.)
Oxygen	80.5	89	1
Hydrogen	9.7	11	1
Carbon	6.5	0.0012	5,000
Silicon[a]	1.3	0.00065	2,000
Nitrogen	0.7	0.000023	30,000
Calcium	0.4	0.0015	< 1,000
Potassium	0.3	0.00023	1,300
Phosphorus	0.08	0.000001	80,000
Magnesium	0.07	0.0004	< 1,000
Sulfur	0.06	0.0004	< 1,000
Chlorine	0.06	0.0008	< 1,000
Sodium	0.04	0.0006	< 1,000
Iron	0.02	0.00007	< 1,000
Boron	0.001	0.00001	< 1,000
Manganese	0.0007	0.0000015	< 1,000
Zinc	0.0003	0.000001	< 1,000
Copper	0.0001	0.000001	< 1,000
Molybdenum[a]	0.00005	0.0000003	< 1,000
Cobalt	0.000002	0.000000005	< 1,000

[a] Known to be required only by certain plant groups

at control (e.g., Oglesby 1969; Funk and Gibbons 1979; Welch 1979; Edmondson and Lehman 1981).

By the late 1940's it was recognized that the process of eutrophication was being accelerated in many North American lakes. Increased societal affluence and a greater desire to enjoy the amenities offered by lakes occurred during the same period. Taken together, and reinforced during the 1960's by the environmental movement, these factors provided the impetus for research programs to establish causes and seek remedies.

Sawyer (1947) was responsible for the first major breakthrough in establishing a quantitative basis for relating nutrient regimes to trophic status. Seeking a solution to the eutrophication which had long plagued the lakes in and around Madison, Wisconsin, he did two things which provided a framework for management and also served as the basis for some of the research which followed two decades later. First, he evaluated nutrient (P and N) sources and assessed relative lake response. Second, he developed a tentative relationship between the concentrations of nutrients in the water column at the end of the biologically quiescent winter period and the presence/absence of nuisance algal blooms the following summer. In other words, it was implied that the vernal pool of nutrients was indicative of the lakes' nutritional status. Both of these concepts have been widely and usefully employed in subsequent research on lake eutrophication.

The next research development of importance occurred in 1967 when R. A. Vollenweider presented a paper at the international symposium *Eutrophication:*

Causes, Consequences, Correctives held at the University of Wisconsin (Madison). This work was subsequently enlarged and published in what has become the most cited paper (Vollenweider 1968) concerning the subject of lake eutrophication. In this he provided both a theoretical and a practical basis for relating nutrient loading to the response of lakes as modified by their morphometry and hydrology.

By this time limnologists were increasingly coming into agreement that P supply was the critical factor in controlling the net algal production of temperate latitude lakes. Despite the intervening distraction of arguments to the contrary, there is a general consensus to this effect today. The supporting evidence is of three kinds: that based on knowledge of P biogeochemistry, insight supplied by extensive observations on lakes and the results of experimentation both in the laboratory and *in situ*.

The dominant geochemical factors bearing on the criticality of P to lakes are lack of a gaseous phase in the P cycle, the tendency of this element to form insoluble compounds with such common elements as calcium, iron and aluminum, and the strong adsorption of phosphate to clay particles. Vallentyne (1974) has shown how the consequences of these properties is the establishment of an imbalance between supply and the demand for P, relative to other essential nutrients, by freshwater plants (Table 3). The imbalance is probably greater than that shown since data are for total P. Adding to Vallentyne's argument consideration of the uneven spatial distribution of P minerals, as discussed above, and the greater geochemical mobility of the other major plant nutrients provides sufficient evidence of itself to permit the deduction that, over much of the globe, P could be expected to act as a controlling factor in the production of aquatic plant communities.

Direct evidence for the controlling role played by P has been provided by extensive comparative studies of lakes, by bioassay studies and through whole lake experiments. The literature is much too voluminous to review comprehensively here. Examples of the first two kinds of evidence are referred to elsewhere in this chapter.

The most extensive and insightful whole lake experimental research has been that carried out by David Schindler and his co-workers with their Experimental Lakes Area (ELA) project in northwestern Ontario. This has been described in a special issue (Vol. 71, No. 1) of the *Journal of the Fisheries Research Board of Canada*. Selective additions of various nutrients, singly and in combination, over a period of years to this group of small, oligotrophic lakes has demonstrated the unique role of P in controlling their primary production (e.g., Schindler 1975).

Employing the concept of nutrient loading, i.e., input per unit surface area or volume of water per unit time, involves decisions as to whether this should be expressed areally or volumetrically and the time span the loading value is to cover. A good deal of attention has been paid to depth as a parameter influencing lake response to any specified P input (e.g., O'Melia 1972; Vollenweider 1975). Most commonly, this has been treated as a discrete parameter so that loading is expressed as input per unit surface area rather than volumetrically, the depth effect then being added as a modifier of lake response.

The time span chosen by Vollenweider (1968) was the year, a decision concurred in by most others who have worked on nutrient loading questions. For most temperature-subtropical systems this reflects a natural cycle of input and uptake

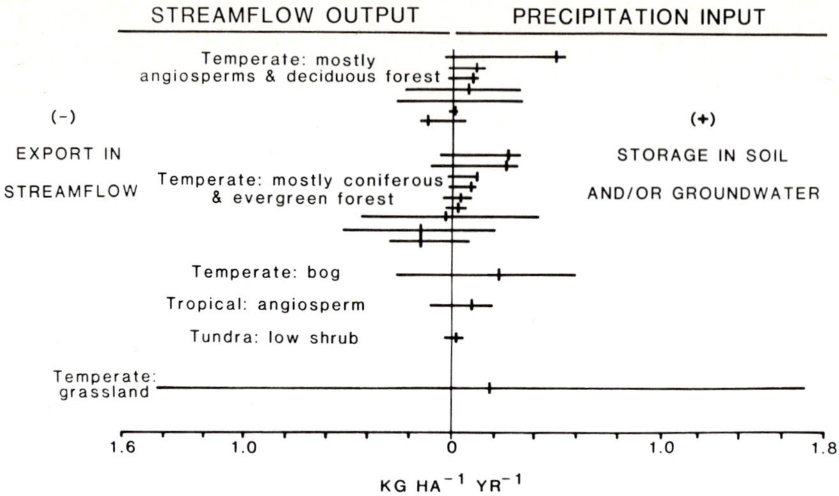

Fig. 4. Phosphorus input via precipitation and export in surface water outflow for 20 largely non-agricultural watersheds. Heavy vertical bars indicate net storage in soil and/or groundwater (*right*) or loss in outflow (*left*) (Data from Viets 1975; Likens et al. 1977; Harrison 1978)

by the biota and thus is a rational time frame. The exceptions to this involve lentic systems with a flushing time greater than once per year, e.g., many run-of-the-river impoundments. Vollenweider 1976; Schindler et al. 1978 and others have dealt with this by considering hydraulic retention time as a separate parameter, or expressions have been developed with boundaries which exclude the short retention time systems (e.g., Oglesby 1977).

The response of lakes to phosphorus inputs was first defined in terms of trophic status (Vollenweider 1975) with "permissible" and "dangerous" levels defined for management purposes. This caused considerable uncertainty in interpretation, since the trophic state continuum is not a quantitatively defined entity. Most recent definitions of lake response are in terms of chlorophyll concentration, usually averaged over the growing season for some specified depth of water column (e.g., Fig. 4). This provides a picture of phytoplankton abundance which then may be converted to an estimate of water transparency (Edmondson 1972; Bachman and Jones 1974; Oglesby and Schaffner 1975; Carlson 1977).

The response of lakes to P loadings has at present been defined only for these simple criteria. The effect of phosphorus input on littoral zone macrophytes, the interactions between inshore and limnetic production and the nature and extent to which grazing and other biotic parameters effect response (Shapiro 1979) are only just beginning to be investigated.

Since 1975, numerous models of varying type and sophistication have been developed to describe phosphorus input-lake response. The first generation of these (Vollenweider 1968, 1975, 1976; O'Melia 1972; Vollenweider and Dillon 1974; Larsen and Mercier 1975; Snodgrass and O'Melia 1975; Chapra and Tarapchak 1976; Schaffner and Oglesby 1978; Oglesby and Schaffner 1978; Uttormark and Hutchins 1978) were essentially mass balances which, by various means, compensated for the influence of major physical factors on biotic response. Specific relations were developed empirically using comparable data sets for temperate latitude lakes. For most, the scatter of data points about the curves describing these models is great enough to preclude their application for accurately predicting changes in the response of a particular lake to variations in P loading. Their principal utility has been in providing verification of the controlling nature of P supply and broad target loadings for management purposes.

Recently, another step in modeling sophistication has manifested itself in the form of four P loading–lake response dynamic mass balance models. They were developed in response to an intensive focus on managing the Laurentian Great lakes. These models have been reviewed and preliminarily evaluated by Bierman (1980) and Simons and Lam (1980). After an impressive review and analysis, the latter conclude (p 105) "... that the uncertainty surrounding the formulation of sedimentation and nutrient regeneration in conjunction with the sensitivity of models to assumptions regarding dynamic balance between lake concentrations and nutrient loadings, undermine the predictive capability of dynamic water quality models". Other modeling approaches that have been used include multivariate analysis (Shannon and Brezonik 1972), stepwise discriminant analysis (Yeasted and Morel 1978), simulation analysis (Chapra 1977), and systems theory (e.g., Halfon 1980).

Which approaches and methodologies may ultimately be most appropriate and useful in providing scientific understanding and for the purpose of lake management will be determined only after further testing and evaluation. Certainly, the very number and variety of these is an indication of the intense interest generated by the P loading–lake response question in recent years.

Unresolved Issues

Whichever approaches are taken towards managing the input of P to lakes, two major issues will have to be clarified before an efficient solution can be implemented. One is the role that "internal" P loading, i.e., that released from bottom sediments into the water column, plays in lake fertilization. A second involves the biological availability of various forms of P.

The first of these topics has received considerable attention beginning with studies in the 1940's of iron-phosphorus equilibria as affected by redox potential at the mud–water interface (Mortimer 1941). Many studies of the phosphate mineralogy of lake sediments have been performed since then, especially during the last decade. In summarizing the geochemical view Jones and Bowser (1978) state (p 227): "The phosphate mineralogy of lake sediments is dominated by authigenic processes. Though detrital apatite is little altered in the lacustrine environment, solute or organic phosphate is sorbed by iron oxides and eventually incorporated either into vivianite or more complex mixed valence phosphates in the sediment. This material may eventually be further transformed into diagenetic apatite and essentially removed from further water-sediment interaction. Regeneration of solute phosphorus by reduction of amorphous iron oxides in lake sediments is an important control on the rate at which sedimentation and mineral diagenesis remove phosphorus from lake water."

For temperate lakes the significance of interaction between sediment and overlying water under the anaerobic conditions which may pertain in those subject to density stratification depends upon the degree to which released P is subsequently available to the phytoplankton community. This is an exceedingly complex question involving the processes of turbulent and passive diffusion, the reaction kinetics between P and cations with which it forms compounds and the degree to which littoral zone macrophytes compete for the phosphorus. In reviewing seasonal distribution of phosphorus and chlorophyll for a number of lakes, a minor peak in phytoplankton production is often seen following autumn mixing. This is presumably a consequence of the vertical transport to the euphotic zone of nutrients which accumulated in the hypolimnion during summer stratification. The resultant algal "bloom" is seldom of much importance when considered from the water quality manager's viewpoint.

Physical transport of solubilized phosphorus across thermoclines and into the euphotic zone is probably maximized in shallow lakes with large epi- to hypolimnion ratios and in those where topography and/or climatic winds produce upwelling. Shagawa Lake, Minnesota is an example of such a system. Sewage discharges to the lake began receiving tertiary treatment for phosphorus removal in 1973 and scientists of the U.S. Environmental Protection Agency have been following the response of the lake since then. Recycling of phosphorus from bottom sediments (P 0.2%–0.5% of dry weight) amounted to an internal supply of 240–680 kg wk^{-1} over a 3- to 9-week interval. Studies of the sediments suggest that this internal source of phosphorus may be enough to keep Shagawa Lake in a eutrophic state for some years despite the reduction of external loading by about 80% (Larsen et al. 1981).

A very different mechanism operates to supply sediment-associated P to the water column when the bottom is stirred and fine particles suspended in response to the direct action of wind and waves. Sorption reactions which take place under such conditions would be exceedingly difficult to estimate. However, their importance may be inferred from the weight given to lake mean depth in the various P loading–lake response models.

As already noted above, the procedures used in analyzing the various forms of P do not define the availability to the biota of the forms determined. A conservative, if inefficient, approach to the control of lake P loading would employ total P data when comparing sources and estimating inputs. The temptation to express loading in this form is enhanced by the greater availability of data on total P and the relative unreliability of much of that for other forms of P due to the critical sample handling and storage procedures which the latter require if they are to be accurately measured.

The management of lakes requires, as one of the first steps, that sources of phosphorus be identified and evaluated. If any substantial portion of the loading is in a form unavailable to phytoplankton, this becomes an important criterion to apply when controlling sources if management is to be efficient.

Given the strong tendency for phosphorus to adsorb to clay particles, it is logical to question the biological availability (Viets 1975), and hence the practical significance, of phosphorus borne on such sediments when they enter lakes. Golterman et al. (1969) tested a Frisian clay for the availability of its phosphorus to phytoplankton and found that only about 4% of its total P content could be used by a test species of algae in batch culture. Cowan and Lee (1976) obtained similar results for clays from western New York State but values were higher (averaging about 30%) for sediments from some other watersheds in the Great Lakes' basin. J.P. Barlow (pers. commun.) has used continuous cultures of both selected test organisms and natural lake communities to study sediments of a central New York watershed. His results are comparable to those obtained by Golterman and for the New York system reported on by Cowan and Lee.

Peters (1981) studying the phosphorus in Lake Memphramagog (Vermont/Quebec) and its tributaries, used ^{32}P exchange and the enzymatic release of soluble from particulate P when stored with chloroform to assess bioavailability. He estimated that, while 83% of the lake total P was available, only 18%–57% of that in the tributaries was in this category. Proportions of available to total P showed large spatial and temporal variability. He concluded (p 1160): "If variations in the presence of these forms occur consistently among bodies of water, at least a five-fold variation in biological effect per unit of total P can be expected."

The logistical problems, finesse of technique and problems of interpretation associated with bioassay studies have encouraged attempts to develop physical and chemical methods to define biologically available phosphorus. As noted above these have included simple desorption procedures with sediments suspended in solution and various extractions (acids, bases, NTA, EDTA and ion exchange resins). The different methodologies have been reviewed and a number of them compared with bioassays by Trautmann (1981). From her results and the review of Lee et al. (1980) it is apparent that no one method has as yet reached a level of development where it is acceptable for general use.

After reviewing the question of biological availability, the International Joint Commission (1980) concluded (p 36):

"In view of this continued uncertainty, the concerns of the Water Quality Board, and testimony received during its public hearings on the crucial role of this factor in developing phosphorus management strategies, the Commission determined that further investigation was required before advising the Governments on the importance of biological availability and indeed on the entire question of phosphorus target loads and control strategies."

Discussion

Many investigators of environmental questions concerning specific elements or compounds have concluded that relatively simple mass balances are more useful for management and offer more insights into processes than any other kind of analysis. While the general concept of a mass balance is a straightforward one, numerous complexities emerge in practice. Nevertheless, most of the research concerning phosphorus in the environment has relied on this approach.

Most phosphorus-related phenomena of concern at the ecosystem or environment level are seasonal in nature. Agronomists, limnologists and forest scientists have made considerable progress in delineating, both deterministically and empirically, the processes manifested on this time scale.

However, management decisions should be made within a decadal time frame. Field studies of this length and of a scope necessary to adequately describe P-related parameters are extremely rare. Thus, little is known relative to "natural" levels of variance in the elements of a mass balance. Researchers working on lake eutrophication bypassed this problem through the use of comparative data sets for large numbers of systems. Such an approach is a valid one for developing general, static management models. These also provide support for cause-and-effect arguments but are severely limited in their ability to predict phenomena related to a particular lake over a specified time interval.

A second, and very critical, complicating factor in structuring phosphorus mass balances is to decide what forms of P should be included. Ideally, those which are or will become biologically available would be used. Since this is not well defined by existing analytical procedures, a rationale and methodologies for estimating bio-availability from data for forms which can be directly measured becomes necessary. As imprecise as such estimates may be, they are an absolutely necessary step in identifying sources of phosphorus that need to be controlled because of their impact on water quality.

The contribution of P used in agricultural systems to the loading of lakes and impoundments receiving cropland runoff is especially ambiguous. Soils having a high adsorptive capacity for phosphorus, e.g., glacial tills in eastern North America, act as a major sink for P. Even when P-bearing particles from fertilized soils of this type enter a watercourse through erosion, they tend to lose only a small fraction of their bound phosphorus. If they have not been influenced by cultural practice, they may even act to sorb dissolved phosphorus.

Most of the phosphorus from diffuse agricultural sources enters lakes during the more intense runoff events such as those following heavy of prolonged rain or rapid snow melt. This measurement of the diffuse agricultural component requires intensive sampling of these largely unpredictable events. Variation among events is large and related to condition of the vegetation, season of the year and farming practices. Finally, there is also large year-to-year change in total loading and since climate tends to be cyclic (several consecutive below average followed by several consecutive above average years) long term sampling is required. There are very few data of this nature.

Logan and Adams (1981) provide an interesting example of the difficulties of defining the impact of specific agricultural practices on algal activity in lakes. In a recent review they document that limited tillage reduced sediment loading, but increased dissolved P loading when compared to conventional tillage. Thus if reduction of total P loading is desired, limited tillage is the better practice; but if management of dissolved P loading is the objective, then conventional tillage is the better practice.

The above examples illustrate the complexities of phosphorus mass balances, especially when they are intended to identify specific sources needing control. In general, they demonstrate that: (1) Management of such P additions to control water quality problems will certainly continue to be an imprecise activity due to the number of complex variables involved, the aperiodic, unpredictable nature of events determining P loss and the changes in form that can occur during transport. (2) Assumptions about the significance of particular sources need careful substantiation and (3) Sampling to define P movement in the environment may often have to be both intensive and extensive during runoff events and continued over a long period of time to cover many diverse events.

Considering questions of phosphorus management as it relates to lakes, on the positive side many of the processes and the general magnitude of most source compartments are becoming increasingly well understood and defined. Problems facing scientists and resource managers have changed from ones such as "Can lake eutrophication be controlled?" to "What level of control is desired?" and "How can this goal be reached most efficiently?"

The story of lake eutrophication and P is not one free of mistakes and failures nor of challenge for the future. In the former category, the scientific community has been particularly lax in reviewing the assumptions which underlie some of the more widely used models. Another issue which has been a problem in the past and one that seems likely to extend well into the future is the lack of long term data sets. Government agencies show little sympathy towards acquiring such information, yet without it there is little hope of "proving" cause-effect relations. Nor is it likely that background levels of intralake variability, a feature against which the progress of remedial programs needs to be judged (Trautmann et al. 1982), will be defined.

In taking an overview, it becomes apparent that the public's concern with microscopic plants which grow in lakes, sometimes to excess, has had some rather profound consequences for our understanding of phosphorus in the environment. Scientists and engineers from many disciplines have focused on this element, its movements through the landscape, and its relationship to water quality. Many of their findings have been quickly adapted into management programs so that in the United States, Canada, Scandinavia and some parts of Europe eutrophication is a problem that is understood and under reasonable control

Before concluding this discussion, another perception of the phosphorus-lake eutrophication situation should be noted. Up to this point only the negative aspects of phosphorus-induced algal populations have been considered. But increased production of these microscopic plants also means larger inputs into the base of food webs with the consequence that fish production is increased (Melack 1976; Oglesby 1977; Liang et al. 1981). Where qualitative aspects of a fishery are of primary importance, increased production may or may not be appreciated since in some types of lakes and impoundments there is a tendency, under highly eutrophic conditions, to shift from the more desired species to those less valued. If production of cheap, high quality fish protein or maximizing the catch per unit fishing effort are viewed as optimal objectives, increased bioavailable P loading leading to eutrophication, at least up to moderately high levels, may be desirable. Management objectives conflicting with those focussed on traditional water quality considerations result – an issue generally not considered in the debates which have led to programs of eutrophication control.

As a consequence of concern about effects of phosphorus on the environment, research on the biogeochemistry of this element has been greatly expanded over the past two decades. The size of storage pools and flux rates between them are broadly defined as are biological response of some types of ecosystems. Strangely enough, given the relatively long history of phosphorus use in fertilizers, one of the least understood but most important concerns about P is the long term availability to plants of that stored in soil. What forms is this in? How is its motility affected by soil chemistry, climate and cultural practices? These are questions that need explication both for their importance to a better understanding of how phosphorus behaves in the environment and for more efficient management of phosphorus recources and the agricultural and forests systems where P is applied as a fertilizer.

References

Ahl T (1980) Eutrophication in relation to the load of pollution. Prog Water Technol 12:49–61

Ahl T, Haverinen A, Thorell L, Wartiovaara J (1977) Discharge of nitrogen phosphorus and organic matter into the Gulf of Bothnia. Ambio 6:273–275

Allan RJ, Williams JDH (1978) Trophic status related to sediment chemistry of Canadian Prairie Lakes. J Environ Qual 7:99–106

Bachman RW, Jones JR (1974) Phosphorus inputs and algal blooms in lakes. Iowa State J Res 49:155–160

Bierman VJ (1980) A comparison of models developed for phosphorus management in the Great Lakes. In: Loehr RC, Martin CS, Rast W (eds) Phosphorus management strategies for lakes. Ann Arbor Sci Publ, Ann Arbor, Michigan, p 235–255

Bowman JC (1978) Balance sheet for phosphorus in the UK 1978. In: Ciba foundation symposium 57. Phosphorus in the environment: Its chemistry and biochemistry. Elsevier, Amsterdam, p 243–251

Brinck JW (1978) World resources of phosphorus. In: Ciba foundation symposium 57. Phosphorus in the environment: Its chemistry and biochemistry. Elsevier, Amsterdam, p 23–48

Bülow-Olsen A (1980) Nutrient cycling in grassland dominated by *Deschampia flexuosa* Trin. and grazed by nursing cows. Agroecosyst 6:209–220

Carlson RE (1977) A trophic state index for lakes. Limnol Oceanogr 22:361–369

Chapra SC (1977) Total phosphorus model for the Great Lakes. Proc Am Soc Civil Eng 103, No EE2, p 147–161

Chapra SC, Tarapchak ST (1976) A chlorophyll *a* model and its relationship to phosphorus loading plots for lakes. Water Resour Res 12:1260–1264

Clark FE, Cole CV, Bowman RA (1980) Nutrient cycling. In: Breymeyer AI, VanDyne GM (eds) Grasslands, systems analysis and man. Inter Biol Progr 19. Cambridge Univ Press, NY, p 659–712

Cole CV, Innes GS, Stewart JWB (1977) Simulation of phosphorus cycling in semiarid grasslands. Ecology 59:1–15

Cole DW, Rapp M (1981) Elemental cycling in forest ecosystems. In: Reichl DE (ed) Dynamic properties of forest ecosystems. Cambridge Univ Press, NY, p 341–409

Cowan WF, Lee GF (1976) Algal nutrient availability and limitation in Lake Ontario during IFYGL, part I. US Envir Prot Agency Rep 600/3-76-094a

Crisp TD (1966) Input and output of minerals for one area of Pennine moorland: The importance of precipitation, drainage, peat erosion and animals. J Appl Ecol 3:327–348

Deevey ES, Rice DS, Rice PM, Vaughan HH, Brenner M, Flannery MS (1979) Mayan urbanism: Impact on a tropical korst environment. Science 206:298–306

Dillon PJ, Kirchner WB (1974) The effect of geology and land use on the export of phosphorus. Water Res 9:135–148

Duvigneaud P, Denaeyer-De Smet S (1971) Cycle des éléments biogènes dans les écosystèmes forestiers d'Europe (principalement forêts caducifoliées). In: Duvigneaud P (ed) Productivity of forest ecosystems. No 4, UNESCO, Paris, p 527–542

Edmondson WT (1972) Nutrients and phytoplankton in Lake Washington. Am Soc Limnol Oceanogr Spec Symp 1:172–188

Edmondson WT, Lehman JT (1981) The effect of changes in the nutrient income on the condition of Lake Washington. Limnol Oceanogr 26:1–29

Funk WH, Gibbons HL (1979) Lake restoration by nutrient inactivation. In: US Environ Prot Agency. Lake restoration: Proceedings of a national conference, August 22–24, 1978, Minneapolis, Minnesota. US Environ Prot Agency, Washington, DC, p 141–151

Golterman HL (1976) Sediment as a source of phosphate for algae growth. In: Golterman HL (ed) Interactions between sediment and fresh water. Junk, Hague, p 286–293

Golterman HL, Bakles CC, Jakobs-Mogelin J (1969). Availability of mud phosphates for the growth of algae. Verh Int Verein Limnol 17:467–479

Green DB, Logan TJ, Smeck NE (1978) Phosphate adsorption – desorption characteristics of suspended sediments in the Maumee River Basin of Ohio. J Environ Qual 7:208–212

Halfon E (1980) Mathematical modeling of phosphorus dynamics through integrations of experimental work and systems theory. In: Scavia D, Robertson A (eds) Perspectives on lake ecosystem modeling. Ann Arbor Sci, Ann Arbor, Michigan, p 77–85

Harrison AF (1978) Phosphorus cycles of forest and upland grassland ecosystems and some effects of land management practices. In: Ciba foundation symposium 57. Phosphorus in the environment: Its chemistry and biochemistry. Elsevier, Amsterdam, p 175–199

Hill R (1981) Stream phosphorus exports from watersheds with contrasting land uses in southern Ontario. Water Res Bull 17:627–634

Hobbie GE, Likens GE (1973) Output of phosphorus, dissolved inorganic carbon and fine particulate carbon from Hubbard Brook watershed. Limnol Oceanogr 18:734–742

Hutchinson GE (1969) Eutrophication: Past and present. In: Nat Acad Sci. Eutrophication: Causes, consequences, correctives. Proc Symp Madison, Wisconsin, June 11–15, 1969. Nat Acad Sci, Washington, DC, p 17–26

Hutchinson GE (1973) Eutrophication. Am Sci 61:269–279

Inch TD (1978) The biological importance of organophosphorus compounds containing a carbon-phosphorus bond. In: Ciba foundation symposium 57. Phosphorus in the environment: Its chemistry and biochemistry. Elsevier, Amsterdam, p 135–153

Ingrahm G (1950) Effect of materials absorbed from the atmosphere in maintaining soil fertility. Soil Sci 70:205–212

International Joint Commission (1980) Pollution in the Great Lakes basin from land use activities. An IJC report to the governments of the United States and Canada. Int Joint Comm, Windsor, Ontario, p 36

Johnson AH, Bouldin DR, Goyette EA, Hedges AM (1976) Phosphorus loss by stream transport from a rural watershed: Quantities, processes and sources. J Environ Qual 5:143–157

Jones BF, Bowser CJ (1978) The mineralogy and related chemistry of lake sediments. In: Lerman A (ed) Lakes: Chemistry, geology, physics. Springer, Berlin Heidelberg New York, p 179–235

Karlovsky J (1980) Cycling of nutrients and their utilization by plants in agricultural ecosystems. Agroecosyst 7:127–144

Kjensmo J (1980) Eutrophication of Norwegian freshwaters in relation to natural conditions. Prog Water Technol 12:39–47

Larsen DP, Mercier HT (1975) Lake phosphorus loading graphs: An alternative. US Environ Prot Agency Work Pap 174, p 1–26

Larsen DP, Schults DW, Malueg KW (1981) Summer internal phosphorus supplies in Shagawa Lake, Minnesota. Limnol Oceanogr 26:740–753

Lathwell DJ (1979) Phosphorus response on oxisols and ultisols. Cornell Int Agricult Bull 33. NY State Coll Agricult Life Sci, Cornell Univ, Ithaca, NY, p 1–40

Lee GF, Jones RA, Rast W (1980) Availability of phosphorus to phytoplankton and its implications for phosphorus management strategies. In: Loehr RC, Martin CS, Rast W (eds) Phosphorus management strategies for lakes. Ann Arbor Sci, Ann Arbor, Michigan, p 259–308

Liang Y, Melack JM, Wang J (1981) Primary production and fish yield in chinese ponds and lakes. Trans Am Fish Soc 110:346–350

Likens GE (1974) Water and nutrient budgets for Cayuga Lake, New York. Tech Rep 82. Cornell Univ, Water Resour Mar Sci Center, Ithaca, NY, p 1–91

Likens GE, Bormann FH, Pierce RS, Eaton JS, Johnson NM (1977) Biogeochemistry of a forested ecosystem. Springer, Berlin Heidelberg New York, p 101

Logan TJ, Verhoff FH, DePinto JV (1979) Biological availability of total phosphorus. Tech Rep Ser. Lake Erie Wastewater Management Study. US Army Eng District, Buffalo, NY

Logan TJ, Adams JR (1981) The effects of reduced tillage on phosphate transport from agricultural land. Tech Rep Ser. Lake Erie Wastewater Management Study, US Army Eng District, Buffalo, NY

McClelland GH, Hignett T (1978) Some economic and technical factors affecting use of phosphate raw materials. In: Ciba foundation symposium 57. Phosphorus in the environment: Its chemistry and biochemistry. Elsevier, Amsterdam, p 49–64

Melack JM (1976) Primary productivity and fish yields in tropical lakes. Trans Fish Soc 76:322–334

Miller RB (1963) Plant nutrients in hard beech. I. The immobilization of nutrients. II. Seasonal variation in leaf composition. III. The cycle of nutrients. NZ J Sci 6:365–413

Mortimer CH (1941) The exchange of dissolved substances between mud water in lakes. J Ecol 29:280–329

Oglesby RT (1969) Effects of controlled nutrient dilution on a eutrophic lake. In: Jenkins SH (ed) Advances in water pollution research. Proc 4th Int Conf, Prague, 1969. Pergamon, Oxford, p 747–757

Oglesby RT (1977) Phytoplankton summer standing crop and annual productivity as functions of phosphorus loading and various physical factors. J Fish Res Board Can 34:2255–2270

Oglesby RT (1977) Relationship of fish yield to lake phytoplankton standing crop, production and morphoedaphic factors. J Fish Res Board Can 34:2271–2279

Oglesby RT, Schaffner WR (1975) The response of lakes to phosphorus. In: Porter KS (ed) Nitrogen and phosphorus: Food production, waste and the environment. Ann Arbor Sci, Ann Arbor, Michigan, p 25–57

Oglesby RT, Schaffner WR (1978) Phosphorus loadings to lakes and some of their responses. Part 2. Regression models of summer phytoplankton standing crops, winter total P, and transparency of New York lakes with known phosphorus loadings. Limnol Oceanogr 23:135–145

O'Melia CR (1972) An approach to the modeling of lakes. Schweiz Z Hydrol 34:1–33

Ovington JD (1962) Quantitative ecology and the woodland ecosystem concept. Adv Ecol Res 1:102–197

Peters RH (1981) Phosphorus availability in Lake Memphremagoy and its tributaries. Limnol Oceanogr 26:1150–1161

Porter KS (1975) Nitrogen and phosphorus: Food production, waste and the environment. Ann Arbor Sci, Ann Arbor, Michigan, p 61–120

Ryden JC, Syers JK, Harris RF (1973) Phosphorus in runoff and streams. Adv Agron 25:1–41

Sanger JE, Gorham E (1972) Stratigraphy of fossil pigment as a guide to the postglacial history of Kirchner Marsh, Minnesota. Limnol Oceanogr 17:840–854

Sawyer CN (1947) Fertilization of lakes by agricultural and urban drainage. J N Engl Water Works Assoc 61:109–127

Schaffner WR, Oglesby RT (1978) Phosphorus loadings to lakes and some of their responses. Part 1. A new calculation of phosphorus loading and its application to 13 New York lakes. Limnol Oceanogr 23:120–134

Schindler DW (1975) Whole-lake eutrophication experiments with phosphorus, nitrogen and carbon. Verh Int Ver Limnol 19:3221–3231

Schindler DW, Welch HE, Kalff J, Brunskill GJ, Kritsch N (1974) Physical and chemical limnology of Char Lake, Cornwallis Island. J Fish Res Board Can 31:585–607

Schindler DW, Fee EJ, Ruszczynski T (1978) Phosphorus input and its consequences for phytoplankton standing crop and production in the Experimental Lakes Area and in similar lakes. J Fish Res Board Can 35:190–196

Shannon EE, Brezonik PL (1972) Eutrophication analysis: A multivariate approach. Proc Am Soc Civil Eng 98, No SA1, p 37–57

Shapiro J (1979) The need for more biology in lake restoration. In: US Environ Prot Agency. Lake restoration: Proceedings of a national conference, August 22–24, 1978, Minneapolis, Minnesota. US Environ Prot Agency, Washington, DC, p 161–167

Simons TJ, Lam DCL (1980) Some limitations of water quality model for large lakes: A case study of Lake Ontario. Water Resour Res 16:105–116

Smith RV (1979) Sources of phosphorus in The Lough Neagh system and their reduction. Prog Water Technol 11:209–217

Snodgrass WJ, O'Melia CR (1975) Predictive model for phosphorus in lakes. Environ Sci Technol 9:937–944

Taylor AW, Kunishi HM (1971) Phosphate equilibria on stream sediment and soil in a watershed draining an agricultural region. J Agric Food Chem 19:827–831

Trautmann NM (1981) Biological availability of wastewater phosphorus in the sediments of two New York streams. MS Thesis, Cornell Univ, Ithaca, New York, p 107

Trautmann NM, McCullock CE, Oglesby RT (1982) Statistical determination of data requirements for assessment of lake restoration programs. Can J Fish Aquat Sci 39:607–610

Tsutsumi T (1971) Accumulation and circulation of nutrient elements in forest ecosystems. In: Duvigneaud P (ed) Productivity of forest ecosystems. No 4. UNESCO, Paris, p 543–552

USDA (1977) Agricultural statistics. US Govern, Washington, DC, p 561

Uttormark PD, Hutchins ML (1978) Input/output models as decision criteria for lake restoration. Water resources center, Univ Wisconsin, Madison, Wisconsin. Tech Complet Rep C-7232:61

Vallentyne JR (1974) The algal bowl: Lakes and man. Fish Res Board Can Misc 22:82–85, 166

Verhoff FH, Baker DB (1978) Moment methods for analyzing river models with application to point source phosphorus. In: Verhoff FH, Melfi DA, Yaksich SM, Baker DB (eds) Phosphorus transport in rivers. Technical Report Series. Lake Erie Wastewater Management Study. US Army Corps of Engineers. Buffalo, NY, p 64–88

Viets FG (1975) The environmental impact of fertilizers. CRC Crit Rev Environ Control 5:423–453

Viro PJ (1953) Loss of nutrients and the natural nutrient balance of the soil in Finland. Comm Inst Forest Fenn 42:1–50

Vollenweider RA (1968) The scientific basis of land and stream eutrophication with particular reference to phosphorus and nitrogen as eutrophication factors. OECD (Organ Econ Coop Dev) Tech Rep DAS/CS1/68 27:1–82

Vollenweider RA (1975) Input-output models with special reference to the phosphorus loading concept in limnology. Schweiz Z Hydrol 37:53–83

Vollenweider RA (1976) Advances in defining critical loading levels for phosphorus in lake eutrophication. Mem Ist Ital Idrobiol 33:53–83

Vollenweider RA, Dillon PJ (1974) The application of the phosphorus loading concept to eutrophication research. Nat Res Counc Can, NRCC 13690:42

Welch EB (1979) Lake restoration by dilution. In: US Environ Prot Agency. Lake restoration: Proc Nat Conf, August 22–24, 1978, Minneapolis, Minnesota. US Environ Prot Agency, Washington, DC, p 133–139

Whitehead DR, Crisman TL (1978) Paleolimnological studies of small New England (USA) ponds. Part I. Late-glacial and post-glacial trophic oscillations. Pol Arch Hydrobiol 25:471–481

Yeasted JG, Morel FMM (1978) Empirical insights into lake response to nutrient loadings with application to models of phosphorus in lakes. Environ Sci Techn 12:195–201

Subject Index

Abukumalite, *see* britholite
Achondrites, phosphorus contents of 200
Acid/base hydrolysis reactions in waste waters 388–390
Actinide phosphate micaceous structures 168–170
Activated sludge, removal of phosphorus using 397
Adsorption of water by amorphous calcium phosphates 337–339
Aldermanite 6
Alforsite 6, 7
Algae, approximate chemical composition of 386
Algerian phosphate deposits 258
Alkaline igneous rocks, apatite deposits found with 243–245
Alluaudite series 6, 7
Althausite 7
Aluminium, interaction with phosphorus in waste waters 388–392
Amblygonite serie 7, 8
Amorphous calcium phosphates, *see* amorphous phosphate minerals
  reactivity in human tissues 376–381
  reactivity of 335–337
  solubility of 339, 340
  surface chemistry 337–339
Amorphous phosphate minerals, *see* amorphous calcium phosphates
Amorphous phosphate minerals
  in human tissues 355–358
  pyrolysis of 358
Anapaite 8
Ancient phosphate deposits, origins of 252, 253
Andrewsite 8
Anglesite in roadside ecosystems 327, 328
Anglesite, stability field of 322
Apatite
  adsorption of gases 342
  chemistry of 330–347
  extent of carbonate substitution into 176–181
  formation in bones 340–347
  in lunar rocks 210, 211
  occurrence in igneous rocks 217, 218
  solubility 339, 340
  structure of 330–347
  surface chemistry 337–339
  thermochemical data for 178, 184
Apatite series
  chemical composition 8, 9
  lattice constants 9
Apatites, *see* fancolite, *see* bone minerals
  analysis of 218–222
  analysis of metamorphic samples of 231, 232
  behavior during magmatic intrusion 221
  chemistry in igneous rock 217–226
  coupled substitutions in 375
  evolution during low intensity weathering 295–297
  fluid inclusions in 232, 233
  halogen contents of 226–229
  histogram of carbonate concentrations in 299
  igneous deposits of 243–245
  infra-red spectra 280
  minor elements in metamorphic phases of 231
  occurrence in human tissues 351–355, 368–381
  occurrence in metamorphic rocks 230–232
  physical properties of 279–281
  possible substitutions in 275
  rare earths in 221–226
  reactivity in human tissues 376–381
  solid inclusions in 232, 233
  trace elements in 218–221
  unit-cell dimensions 9, 10, 279, 296, 297
Arakawaite, *see* veszelyite
Archrite 10
Arctite 10, 11
Ardealite 11
Argon, adsorption by bone minerals 342
Arrojadite, structure of 164
Arrojadite series 11
Arsnocrandallite 11, 12
ATP, role in bone mineralization 16, 346
Attakolite (attacolite) 12
Auerlite, *see* thorite
Augelite
  paragenetic affiliations 293
  stability of 192
  thermochemical data for 185
Augelite 12, 13
Australia, phosphate deposits in 256, 257
Autunite 13
  thermochemical data for 184
Avelinoite, *see* cyrilovite
Average shale, elemental abundance in 286

Babefphite 13, 14
Baldaufite, *see* hureaulite
Barbosalite 14
Baricite 14, 15
Barium, incorporation into bon minerals 370
Barrandite, *see* aluminous strengite
  formation in guano deposits 292–294
Bassetite 15
Bastinite, *see* lithium hureaulite
Beckelite, *see* britholite
Belovite 15
Beraunite 16
Bergenite 16
Berlinite 16, 17
Bermanite 17
Bertossaite 17, 18
Beryllonite 18
Beta-lomonosovite 18, 19
Beusite series 19
Biochemical oxygen demand in waste waters 394
Biosphere, reservoirs and cycling of phosphorus in 400–406
Biotite, association with apatite in igneous rocks 228
Biphosphammite 19
Birefringence of apatites 280
Bjarebyite series 19, 20
Bobierrite 20
  thermochemical data for 176, 180
Boggildite 20
Bolivarite 21
Bone hydroxyapatite, chemistry and structure of 340–342
Bone mineral
  nucleation of 343, 344
  reactivity of 342
Bone mineralization 342–347
Bone minerals, *see* apatites
  composition of 368
  in human tissues 351–355, 368–381
  impurities in 370–375
  physical properties 368
Borickite 21
Bornemanite 21, 22
Bosphorite, *see* oxidized vivianite
Brabantite 22
Bradleyite 22
Brazilianite (brasilianite) 22, 23
Brianite 23
  electron analysis microprobe of 207
  in meteorites 200–208
  structure of 161, 163
Britholite, *see* apatite series
Britholite 23, 24
Brockite 24
Broggerite, *see* uraninite
Bromine in igneous apatites 229
Brushite 24, 25
  crystal morphology 360, 361
  factors regulating formation of 359–360
  formation in embryonic chick 340
  occurrence in human tissues 351, 355, 358–362
  paragenetic affiliations 292, 293
  reactivity in human tissues 376–381
  thermochemical data for 180
Buchwaldite (rhenanite) 25
  analysis of 207
  in meteorites 200–206
Burangaite 25
Burton-Cabrera-Frank theory of crystal growth 144

Cacoxenite 25, 26
Cacoxenite, paragenetic affiliations 293
Calcified human heart, phosphate minerals in 353
Calcioferrite 26
Calcium phosphates, important mineral phases 145
Calcium phosphates, solubilities of 146, 147
Calcium pyrophosphate dihydrate in human tissues 351, 355, 366
Calcium-deficient apatite, crystallographic functions of 331
Cambrian phsophate deposits during 261, 262
Cambro-Ordovician locations of phosphorite deposits 256
Carbon monoxide, adsorption by bone minerals 342
Carbonate, incorporation into bone minerals 370, 371
Carbonate, loss upon heating francolite 281
Carbonate apatite 26, 27
Carbonate substitution into francolite structure 276–281
Carbonates in apatites 334
Carbonatite complexes associated with apatite deposits 244
Carboxyglutamic acid, role in bone mineral formation 344
Cassidyite 27
Cathophorite, *see* brabantite
Cave guano phosphate deposits 245, 246
Cayuga Lake, phosphorus flux into 407–411
Cenozoic phosphate deposits 262–266
Cerite, structure of 161
Cerussite
  in roadside ecosystems 327, 328
  stability field of 320–323
Chalcolite, *see* torbernite
Chalcosiderite, *see* turquois
Chalcosiderite 27
Changes which occur on heating phosphorite 281
Chavesite 27, 28
Chemical additives for phosphorus removal from waste waters 394–396
Chemistry of hydroxyapatites 330–347
Cheralite (monazite group) 28
Chert associated with phosphorites 282
Childrenite 28, 29
Chinoite, *see* libenthenite
Chlorapatite
  atomic structure of 332
  electron microprobe analysis of 202
Chlorapatite in meteorites 200–208

Chloride, incorporation into bone minerals 372, 373
Chlorine in igneous apatites 226–229
Chlorophyll concentrations in lakes 414
Chondrites, phosphorus contents of 200
Churchite, *see* weinschenkite
Clay minerals associated with phosphorites 282
Clay-phosphate mineral paragenesis 192, 193
Climatic zones, influence on phosphate deposition 266
Clinophosinaite 29
Clinopyroxene, phosphorus contents of 225
Clinostrengite, *see* phosphosiderite
Cluster condensation for ions in phosphate minerals 159
Cluster hypothesis of phosphate mineral structures 155–160
Cluster types found among phosphate minerals 167, 168
Coconinoite 29, 30
Coeruleolactite, *see* turquois
Coeruleolactite 30
Coffinite, structure of 161
Collagene, role in bone mineral formation 344, 345
Collinsite 30, 31
Collophane, *see* carbonate apatite
Complex aqueous species, thermochemical data for 173, 174
Complex species distribution in sea water 190, 191
Composition of sea water 190, 191
Composition of weathered phosphate facies 311
Concentration of phosphorus in the biosphere 400–406
Concerted substitutions in apatite 375
Controls of bone mineral formation 343
Copper phosphates, stabilities of 323, 324
Corkite 31
  stability field of 321–323
Corner-sharing octahedral chain structures 157–167
Cornetite 31
  stability field of 324, 325
Coupled substitutions in apatite 375
Crandallit (pseudowavellite) 32
Crandallite
  composition of 302–313
  genesis of 302–313
  paragenesis of 167, 192, 302–313
  paragenetic affiliations 292 294
  stability of 192, 193
  texture of 302–313
Cretaceous phosphate deposits 262–266
Cretaceous-Eocene phosphate deposits 255
Critical cluster in nucleation of phosphate minerals 139
Criticality of phosphorus to lakes 411–417
Crustal abundance of the elements 286
Crystal growth, reaction kinetics 143–152
Crystal growth, theories 141–145
Cubic close-packed structures 156–167
Cuprouranite, *see* torbernite
Curetonite 32
Cycling of phosphorus in the biosphere 400–406
Cyrilovite (avelionite) 33

Dahllite, *see* carbonate apatite
Debye-Hückel coefficients for phosphate ions 189
Decarbonation of apatites 281, 299
Defects in apatites 331–333
Deltaite, *see* crandallite
Delvauxite 33, 34
Dental calculi
  calcium phosphates in 378
  phosphate minerals in 353–355
Dentine, phosphate minerals in 352–355
Dewindtite 34
  stability field of 322, 323
Diadochite 34
Dicalcium phosphate dihydrate, precipitation of 145–152
Dickensonite series 34, 35
Differentiated intrusions, apatite distribution in 221, 222
Discrdited apatitic names 9
Distribution of phosphate deposits 243–266
Dittmarite 35
DNA, adsorption by hydroxyapatite 339
Dolomite associated with phosphorites 282
Dorfmanite 35, 36
Drugmanite 36
Dry deposition of phosphorus 406
Dufrenite 36
Dumontite 36, 37
  stability field of 322, 323
Dynamic mass balance models for phosphorus in ecosystems 406–411

Ecosystems
  adverse effects of phosphorus in 411–417
  transfer of phosphorus among 407–411
Effects of substituents on lattice parameters of apatites 374
Egueiite 37
Electron microprobe analysis of phosphate minerals 202–204
Ellestadite 37
Embreyite 37, 38
Englishite 38
  paragenetic affiliation 293
Enthalpy of phosphate minerals 173–189
Entropy, *see* heat capacity
Environment, pollution with phosphorus 400–420
Eosphorite 38, 39
Epigenesis of kaolinite in millisite 305–309
Epitaxial growth on seed phase 142
Equilibrium chemistry of phosphorus in waste waters 387–392
Erikite, *see* rhabdophane
Ernstite 39
Eutrophication induced by phosphorus deficiency 401
Evansite 39

Evolution of apatite during weathering 295–313
Evolution of phosphate minerals during weathering 292–295
Extent of carbonate substitution into apatite lattice 275–281
Extraterrestial rocks, phosphate minerals in 199–212
Eylettersite 39–40

Factors affecting formation of phosphate minerals in vitro 376–378
Faheyite 40
Fairfieldite, paragenetic affiliation 293
Fairfieldite series 40, 41
Fall Creek Watershed, cycling of phosphorus in 407–411
Farringtonite 41
  electron microprobe analysis of 204
  in lunar rocks 210, 211
  in meteorites 200–208
  thermochemical data for 176, 180
Faustie, *see* turquios
Faustite, stability field of 325
Fenghuangite 41, 42
Ferri-sicklerite 42
Ferroalluaudite 6, 7
Ferromanganese oxides, synergistic effects 328, 329
Fillowite series 42
  structure of 163
Finchenite, *see* apatite
Finite cluster structures 157–168
Florencite 42, 43
Fluellite (kreuzbergite) 43, 44
Fluid inclusions in apatites 232, 233
Fluorapatite 44
  atomic structure of 332
  structure and chemistry 331–333
Fluoride
  effect on cell dimensions of apatites 280
  incorporation into bone minerals 371, 372
Fluorine in igneous apatites 226–229
Fluxes of phosphorus on a global scale 402
Foggite 44, 45
Forms of phosphate in waste waters 387
Foucherite 45
Francoanellite 45
Francolite, *see* carbonate apatite, *see* apatite
  birefringence of 280
  composition of 275–277
  effects of heating on 281
  major elements in 284, 285
  significance of carbonate substitution into 277–281
  stability of 192, 193
  stable isotope studies on 285–287
  trace elements in 284, 285
Francolite, effect of substituted carbonate on properties of 279
Fremontite, *see* natromontebrasite
Frequency of nucleation 140
Frondelite series 45
Fuastite 41
Fugacity of phosphorus at high temperature 230
Furongite 45, 46

Galena in roadside ecosystems 327, 328
Garnet, phosphorus contents of 225
Gatumbaite 46
General formula for francolite 277
Genesis of alumino-phosphates during weathering 300–309
Genesis of phosphate minerals during weathering 292–295
Geochronology of phosphorite deposits 285–287
Geographic locations of phosphate deposits 244–251, 255–258, 261
Gibbs free energy of formation for phosphate minerals 174–186
Giniite 46, 47
Glaserite derived structures 156–167
Glauconite in phosphorites 282
Global fluxes of phosphorus 402
Global warming as a factor in phosphate deposition 268
Glycoproteins, role in bone mineralization 35
Goedkenite 47
  isomorphism 169
Goethite association with millisite 309
Gorceixite, paragenetic affiliations 293
Gormanite 47
Goyazite 48
  paragenetic affiliation 293
Graftonite 48
Grainstone phosphorite, origins of 252, 253
Grassland, cycling of phosphorus in 404, 405
Grayite (brockite) 49
Griphite 49
  structure of 163
Grodnolite, *see* francolite
Guano phosphate deposits 245–249
Guano reaction in calcarious environments 292
Guano reaction in soils 293
Gutsevichite 49, 50

Hagendorfite 50
Hahotoe phosphate deposit, weathering of 296–313
Halogens in igneous apatites 226–229
Hannayite 50
HAP, *see* hydroxyapatite
  growth of 150–152
Heat capacity of phosphate minerals 173–189
Herderite 50, 51
Heterosite series 51
  thermochemical constants for 183
Hexagonal close-packed structures 156–167
High temperature phosphate structures 160–165
Hilgenstockite, thermochemical data for 185
Hinsdalite 52
  stability field of 321–323
Histogram of carbonate concentration in apatites 299
History of phosphate minerals 1–4

Holtedahlite (satterlyite) 52
Homogeneous nucleation of phosphate minerals 141
Hopeite
  stability field of 325
  thermochemical data for 176, 181
Hopeite (parahopeite) 52, 53
Huhnerkobelite 53
Human calcifications, phosphate minerals in 351–381
Human dentine
  brushite occurrence in 359
  composition of 368
  electron micrograph of 357
  physical properties 368
Human enamel
  brushite occurrence in 359
  composition of 368
  composition of apatite in 369
  physical properties 368
Human tissues, amorphous phosphate minerals in 355–358
Hureaulite 53, 54
Hurlbutite 54
Huttonite, isomorphism 169
Hydroxyapatite 54, 55
  atomic structure of 332
  in meteorites 200–208
  solubility of 145
Hydroxyapatite, *see* HAP

Igneous phosphate deposits 243–245
Igneous rocks
  apatites in 216, 217
  halogens in apatites from 226–229
  phosphate deposits associated with 243–245
  phosphorus contents of 216, 217
  rare earths in apatites from 219–226
Ijolite zones, association with apatite deposits 243
Illite associated with phosphorites 282
Ilmenite, phosphorus content of 225
Indian Ocean sediments, phosphorus contents 263
Infrared spectra of apatites 280
Infrared spectra of hydroxapatite 338
Infrared spectra of phosphate minerals 356, 363, 369, 380
Insular guano phosphate deposits 247–249
Interrelations of phosphate minerals in human tissues 375–381
Iron
  incorporation into bone minerals 370
  interaction with phosphorus in waste waters 387–389
Iron meteorites, phosphate minerals in 205–208
Isokite 55
Isomorphisms among phosphate minerals 161

Jagowerite 55
Jahnsite series 55, 56
Jezekite, *see* morinite
Jezekite 56
Johnsomervilleite 56
Jungite 56, 57
Jurassic phosphate deposits 262, 263

Kagome semi-regular net 163
Kaolinite, stability of 192
Kaolinite association with millisite 305–309
Karnasurtite (kozhanovite) 57
Keckite 57
Kelley power function for phosphate minerals 187
Kemmlitzite 58
Kidwellite 58
Kingite 58
Kingsmountite 59
Kipushite, *see* veselyite, *see* veszelyite
Kleemanite 59
Kolbeckite 59, 60
Koninckite 60
Kribergite 60
Kryzhanovskite 60, 61
  structure of 166
Kulanite 61
Kurskite, *see* francolite

Lacroixite 61
Lakes
  chlorophyll concentrations in 414
  concentrations of elements in 412
  effects of phosphorus pollution on 411–417
  management of 416, 417
  response to phosphorus inputs 414–417
Lal Lam phosphate deposits, mineralogical associations 312
Lam Lam phosphate deposits, weathering of 296–313
Landesite 62
Lanthanide phosphate structures 156, 168–170
Laplandite 62
Latitudinal distribution of phosphate deposits 266
Lattice water in bone minerals 374, 375
Laubmannite 62, 63
Laueite series 63
Lazulite series 63, 64
Lead, incorporation into bone minerals 370
Lead phosphates
  in roadside ecosystems 326–328
  relative solubilities of 320
Lehiite 64
Lermontovite 64
Lessingite, *see* britholite
Leucophosphite 65
  formation in guano deposits 292–294
  in roadside ecosystems 327
Libenthenite 65, 66
  stability field of 324, 325
Likasite 66
Lime treatment, removal of phosphorus by 394–398
Lingaitukuang, *see* brabantite
Lipids, role in bone mineral formation 355
Lipscombite 66
Lipsombite, paragenetic affiliation 293
Liroconite 66, 67

Lithiophilite series 67
Lithiophosphate 67, 68
Lithiophosphatite, solubility product of 179
Lithology of weathered phosphorites in western Senegal 301
Lomonosovite 68
Lucinite, *see* metavariscite
Ludlamite 69
Lunar rocks, phosphate minerals in 209–211
Luneburgite 69
Lusungite 70

Magnesium, incorporation into bone minerals 373
Magnetite, phosphorus content of 225
Maier power function for phosphate minerals 187
Major elements in igneous apatites 217, 218
Major elements in phosphorites 284, 285
Management of lakes 416, 417
Mantle derived rocks, apatites in 222, 224–226
Maricite 70
Marine phosphorites, weathering of 295–297
Massicot in roadside ecosystems 328
Meathop Wood, phosphorus pools and fluxes in 403, 404
Mechanisms of bone mineralization 342, 343
Mechanisms of mineral nucleation 139–141
Melkovite 71
Melonjosephite 71
Merwinite, structure of 161
Mesosiderites, phosphate minerals in 204, 205
Mesozoic phosphate deposits 262–266
Messelite 71, 72
Meta-ankoleite 72
Meta-autunite 72, 73
Meta-Uranocircite I and II 74
Metalaite 70, 71
Metalomonosovite 68, 69
Metamorphic rocks, apatites in 230–232
Metaschoderite 73
Metasomatic phases of phosphate minerals 165–167
Metastrengite, *see* phosphosiderite
Metatorbernite 73, 74
  stability field of 324
Metavariscite 74, 75
Metavauxite 75
Metavivianite, *see* oxykerchenite
Millisite 75
  composition of 302–313
  genesis of 302–313
  paragenesis of 167, 192, 302–313
  texture of 302–313
Mineral succession during phosphate genesis in Senegal 303
Mineralogical composition of weathered phosphate facies 311
Minyulite 75, 76
  thermochemical data for 178, 185
Miocene phosphate deposits 262, 263
Misconceptions about phosphorus behavior in lakes 411–415
Mitridatite series 76
Mixed valence phosphate compounds 165–167
Model for phosphate deposition in Cambrian time 260
Monazite 76, 77
  analysis of 235, 236
  in igneous and metamorphic rocks 233–235
  isomorphism of 169
Monetite 77
  paragenetic affiliation 293
  thermochemical data for 176, 180
Monite, *see* carbonate apatite
Montebrasite series 77, 78
Montgomeryite 78
  paragenetic affiliations 293
Moraesite 78, 79
Morinite 79
Mundite 79, 80
Munkforssite, *see* manganoan apatite

Nacaphite 80
Nagelschmidtite 80
Nahpoite 80, 81
Nastrophite 81
Natromontebrasite 81
Natrophilite 81, 82
Natrophosphate 82
Nauruite, *see* francolite
Neomesselite 71, 72
New mineral reports, history of 3
Newberyite 82, 83
  formation in guano deposits 13, 293
  occurrence in human tissues 17, 351, 355, 363
  reactivity in human tissues 17, 376–381
  thermochemical data for 176, 180
Ningyoite 83
Ningyoite, thermochemical data for 177, 184
Nissonite 83
  stability field of 324
Nitrogen, adsorption by bone minerals 342
Non-agricultural watersheds, cycling of phosphorus in 405, 406
Novacekite (autunite group member) 83, 84
Nucleation of phosphate minerals 139–141

Occurrence of
  guano phosphate deposits 245–249
  igneous phosphate deposits 243–245
  sedimentary phosphate deposits 249–252
Occurrence of phosphate minerals in human tissues 351–381
Octacalcium phosphate
  crystal morphology 364
  occurrence in human tissues 351–355, 362–364
  precipitation of 376–381
  reactivity in human tissues 376–381
Octahedral corner sharing sheet structures 157–167
Olgite 84
Olivine, phosphorus contents of 225
Olmsteadite 84, 85
Olympite 85
Ordovician phosphate deposits 262, 263
Organic matter in phosphorites 283, 285

Origin of phosphorite deposits 250–266
Ornithite, *see* carbonate apatite
Orogenic episodes as a factor in phosphate deposition 267
Orthopyroxene, phosphorus contents of 225
Overite 85
Overstory, cycling of phosphorus in 403
Oxyapatite, *see* voelckerite
Oxygen isotopic composition of phosphorites 285–287
Oxykerchenite 85

Packing efficiency in phosphate minerals 160–162
Palermoite series 85, 86
Pallasites, phosphate minerals in 203, 204
Palygorskite in phosphorites 282
Panasqueiraite 86
Panethite 86, 87
  analysis of 207
  in meteorites 200–206
Paragenesis of phosphate minerals during weathering 292–295
Parahopeite 87
Parascholzite 87
Paravauxite 87, 88
Parent rock composition, impact on phosphate weathering 294
Parsonite 88
  stability field of 322, 323
Pathological calcifications, phosphate minerals in 351–381
Penikisite series 88, 89
Perhamite 89
Periodicity in insular guano deposits 248
Perloffite 89
Permian phosphate deposits 262, 263
pH
  effect of phosphorus speciation in waste waters 387, 388
  effect on phosphate precipitation in vitro 376
Phanerozoic distribution of phosphate deposits 261
Phase relations, phosphate systems at high temperatures 229, 230
Phosinaite 89, 90
Phosphoferrite series 90
Phosphophyllite 90, 91
  stability field of 325
Phosphoproteins, role in bone mineral formation 343
Phosphoria Formation 265, 278
  modal values for elements in 186
Phosphorite
  average elemental abundance in 286
  analyses of 284
  carbonate distribution in 276–282
  distribution of 249–252, 255–259
  effects of heating on 281
  occurrences of 249–252
  origin of 250–266
  stable isotope studies on 285–287
  weathering of 293–313
Phosphorroesslerite 91
Phosphorus fugacity 230
Phosphosiderite 91, 92
  paragenetic affiliations 293
Phosphuranylite 92
Phuralumite 92, 93
Phurcalite 93
Physical properties of francolite 279–281
Plagioclase, phosphorus content of 225
Planerite, *see* turquois
Plant matter, approximate chemical composition of 386
Plumbogummite 93, 94
  in roadside ecosystems 327
  stability field of 321–323
Polyquartz 94
Power function coefficients for entropy 174
Precipitated hydroxyapatites, structure and chemistry 333–335
Precursor phases in precipitation of calcium phosphates 147–152
Primary apatites, characteristics of 296
Primary settlers, removal of phosphorus in waste waters by 395–397
Proteogylcan, role in bone mineral formation 346
Proterozoic phosphate deposits 255
Protolipids, role in bone mineralization 345, 346
Przhevalskite, stability field of 322, 323
Pseudo-gout deposits, phosphate minerals in 355
Pseudoautunite 94
Pseudolaueite 94
Pseudomalachite 94, 95
  stability field of 324, 325
Pseudowavellite, *see* crandallite
Pyroguanite, *see* phosphate rock
Pyromorphite series 95, 96
  in roadside ecosystems 326–328
  stability field of 321–323
  thermochemical data for 178, 185
Pyrophosphate, incorporation into bone minerals 373

Quartz, in phosphorites 282

Ranunculite 96
Rare earths in igneous apatites 219–226
Rare earths in phosphorites 286
Rate of crystal growth 143–152
Rate of nucleation of phosphate minerals 141
Reactivity of amorphous calcium phosphates 335–337
Reddingite series 96, 97
Reference aqueous ions, thermochemical constants for 171–173
Relative solubilities of lead compounds 320
Renardite 97
Rhabdophane 97, 98
Rhodophosphite, *see* manganoan apatite
Richellite 98
Roadside ecosystems, formation of lead phosphates in 326–328

Robertsite series 98
Rock porosity, influence on phosphate weathering 294
Rockbridgeite 98, 99
paragenetic affiliation 293
Rod structures in phosphate minerals 160–165
Roscherite 99, 100
Rosieresite 100
Rusakovite 100

Saamite, *see* strontium apatite
Sabugalite 100, 101
Saleeite 101
Salmonsite, *see* hureaulite + jahnsite
Sampleite 101, 102
Samuelsonite 102
Sarcopside 102, 103
electron microprobe analysis of 207
in meteorites 200–208
Saryarkite 103
Sasaite 103
Satterlyite 103, 104
Schertelite 104
Schoderite 104
Scholzite 104, 105
stability field of 325
Schoonerite 105
Schreibersite (rhabdite) 105, 106
Schreibersite in meteorites 199, 204
Scorzalite series 106
Sea level changes and phosphate deposition 268
Sea water, composition of 190, 191
Sea-floor spreading episodes of phosphate deposition 267
Seamanite 106, 107
Seawater abundance of the elements 286
Secondary apatites, characteristics of 296
Sedimentary phosphate deposits, *see* phosphorites
Sediments, role in phosphorus cycling in lakes 415, 416
Seeded growth of calcium phosphates 149–152
Segelerite 107
Senegal, weathering of phosphate deposits of 296–313
Senegalite 107
paragenetic affiliation 293
Shape of apatite crystallites 281
Shape of bone apatite crystals 341
Sicklerite 107, 108
Sidi Daoui phosphate deposit, weathering of 297, 298
Sidorenkite 108
Sigloite 108
Significance of carbonate substitution into apatite 276–281
Sincosite 109
Size of bone apatite minerals 341
Smithsonite, stability field of 325
Soft tissue calcifications, phosphate minerals in 355
Soils, accumulation of phosphorus in 402
Solid inclusions in apatites 232, 233
Solubility
aluminium compounds in waste waters 389
amorphous calcium phosphates 339, 340
iron compounds in waste waters 389
secondary lead compounds 320
Solubility equilibrium constants 176–178, 391
Solubility products of phosphate minerals 176–178
Solubility products, calcium phosphates 146
Sombrerite, *see* phosphate rock
Souzalite 109, 110
Spatial controls on the formation of phosphate deposits 242–270
Spencerite 110
stability field of 325, 326
Spodosite, thermochemical constants for 185
Stabilities of
copper phosphates 323, 324
lead phosphates 320–323
zinc phosphates 325, 326
Stabilization of amorphous calcium phosphate by ATP 346
Stable isotopic studies on phosphorites 285–287
Staffelite, *see* francolite
Stanfieldite 110
electron microprobe analysis of 204
in meteorites 200–206
Stearic acid, adsorption by hdroxyapatite 339
Steenstrupine, structure of 164
Stepwise growth of mineral phases 142–144
Stercorite 110, 111
Stewartite 111
Stone meteorites, phosphate minerals in 200–204
Strengite series 111, 112
paragenetic affiliations 293
Strong weathering of phosphates by meteoric waters 300–309
Strontium, incorporation into bone minerals 373, 374
Strontium phosphate, precipitation of 138
Strontium-apatite (saamite) 112
Structural controls on phosphate deposition 255–260
Structure of hydroxyapatites 330–347
Strunzite 112, 113
Struvite 113
occurrence in human tissues 351, 355, 366, 367
reactivity in human tissues 376–381
thermochemical data for 178, 184
Sulfate ion substitution into apatite lattice 277
Sulfur isotopic composition of francolite 287
Supersaturation, phosphate mineral systems 139–141
Surface chemistry of amorphous calcium phosphates 337–339
Surface runoff, tranfer of phosphorus by 409–411
Suspended solids in waste waters 394
Svanbergite 113, 114
Switzerite 114
Synergistic effects of ferromanganese oxides 328, 329

Tagilite, *see* pseudomalachite
stability field of 324
Taiba phosphate deposit, weathering of 296–313
Tangeite, *see* calciovolborthite
Tapeworm concretions, phosphate minerals in 357
Taranakite 114, 115
thermochemical data for 178, 184
Tarbuttite 115
stability field of 325, 326
Tavistockite, *see* fluorapatite
Tavorite 115, 116
Temperature, effects on phosphate precipitation in vitro 376
Temporal controls on the formation of phosphate deposits 242–270
Temporal distribution of phosphate deposits 260–266
Terrestial igneous rocks, *see* igneous rocks
Tertiary lime treatment, removal of phosphorus using 397–398
Thadeuite 116
Thermochemical data
base metal phosphates 318–320
complex aqueous species 173, 174
phosphate minerals 173–189
reference aqueous ions 171–173
Threadgoldite 116
Tinticite 117
Tobene phosphate deposit, weathering of 296–313
Tobene phosphate deposits, mineralogical associations 312
Tooth enamels, phosphate minerals in 352–355
Topographic relief, effect on phosphate weathering 294
Torbernite series 117
stability field of 324, 325
Trace elements in igneous apatites 218–221
Trace elements in phosphorites 284–286
Transfer of phosphorus among ecosystems 406–411
Tricalcium phosphate, precipitation of 145–152
Trickling filter, removal of phosphorus using 397
Trimeric face-sharing structures 156–167
Triphylite series 118
Triplite series 118, 119
Triploidite series 119
Trolleite 119
Tsumebite 119, 120
stability field of 322, 323
Tubercular deposits, phosphate minerals in 355
Tumoral calcification, phosphate minerals in 354, 355
Tunisian phosphate deposits 258
Turquois group
history of 1–3
paragenetic affiliation 293
stability field of 324–326

Unit-cell dimensions for apatites 9, 10, 279, 296, 297
Upalite 120, 121
Uralolite 121
Uramphite 121
Uranite 121
Uranocircite 121, 122
Uranospathite 122
Urinary stone
layered growth in 379
phosphate minerals in 354, 355

Variscite series 123
paragenetic affiliation 293
stability of 192
thermochemical data for 177, 182
Varulite 123
Vashegyite 123
Vauquelinite (laxmannite) 123, 124
Vauxite 124
Vayrynenite series 124, 125
Veszelyite 125
stability field of 324, 325
Vibertite, *see* bassanite
Viitaniemiite 125
Vilateite, *see* metastrengite
Viseite 125, 126
Vitusite 126
Vivianite 126, 127
paragenetic affiliation 293
thermochemical constants for 182
Voelckerite 127
Vuonnemite 127

Wagnerite 127, 128
thermochemical constants for 185
Wardite 128
paragenetic affiliation 293
Waste waters
acid/base hydrolysis reactions in 388–390
equilibrium chemistry of phosphorus in 387–392
forms of phosphate in 387
phosphorus removal processes in 394
removal of phosphorus using
activated sludge 397
settlers 395, 396
tertiary lime treatment 397, 398
trickling filter 397
solubility of phosphate compounds in 389
treatment schemes to remove phosphorus from 392–395
Wavellite 128, 129
composition of 302–313
genesis of 302–313
paragenesis of 167, 192, 302–313
paragenetic affiliation 293
stability of 192
texture of 302–313
Waylandite 129
Weathered phosphate facies, composition of 311
Weathering of phosphate minerals 292–314
Weinschenkite 129, 130

Well-crystallized apatite, *see* hydroxyapatite
Well-crystallixed apatite, chemistry and structure 330–333
Wenzelite, *see* hureaulite
Whiteite 130
Whitlockite 130, 131
analysis of 236
electron microprobe analysis of 202
in igneous and metamorphic rocks 235–237
isomorphism 169
in lunar rocks 210, 211
in meteorites 200–208
occurrence in human tissues 351–355, 365
paragenetic affiliation 292–295
reactivity in human tissues 376–381
structure of 161
thermochemical data for 176, 180
Whitmoreite 13
Wicksite 131
Wilkeite 131, 132
Wolfeite 132
Wooded ecosystems, cycling of phosphorus in 403, 404
Woodhouseite 132, 133
Wyllieite 133

X-ray diffraction patterns for apatites 330–332, 335
X-ray diffraction patterns for P minerals 352, 353, 358, 362, 366, 371
Xanthoxenite 133, 134
Xenotime 134
analysis of 235, 236
in igneous and metamorphic rocks 235–237
isomorphism 169
structure of 161
Xiangjiangite 134

Yoshimuraite 134, 135

Zairite 135
Zapatalite 135, 136
Zeolite associated with phosphorite 283
Zinc, incorporation into bone minerals 370
Zinc phosphats, stabilities of 325, 326
Zwieselite series 136

Author Index

Aballain M 4, 136
Abbona F 138
Acree SF 146
Adams JR 418
Adams ME 355, 366, 377
Adib D 201, 202
Afans'ev YA 186
Agasyan PK 174
Agrell SO 201, 211
Aharon P 248
Ahl T 406, 407
Ahrens LH 200
Aia MA 178
Al-Bassam KS 201, 285
Albee AL 210, 237
Aleksandrov YaI 186
Aleshchkina AE 174
Alfred AC 352, 354–357, 366, 369
Alfrey AC 352, 354–359, 365, 366, 369, 376, 377
Allan RJ 409
Allen RO 228, 229
Allulli S 186
Altschuler ZS 167, 242, 276, 282, 284–286, 292–294, 299, 300, 305–307, 313
Alyamovskaya KV 178, 179
Amit O 283
Amjad Z 150
Amler MH 355, 368, 370
Amli R 225, 235–237
Andon JL 185
Andrew CW 300
Aoki K 215, 228
Appleman DE 169
Araki T 161, 162, 164, 166
Arends J 355, 373
Armstrong WD 352, 355
Arthur MA 262–265
Ashcroft SJ 188
Ashuiko VA 186
Aspden JA 232, 233
Ataman G 278, 282, 298, 299, 301
Axelrod JM 293, 294
Axelrod S 278, 281
Axon HJ 211
Avnimelech Y 179
Bachman RW 414
Bachra BN 282
Bacon CR 222
Bacquet G 276
Bailey SM 172–174, 185, 188, 319, 320
Baker DB 409
Baker WE 179
Bakles CC 401, 416
Balducci G 186
Baldwin WG 178
Ball JW 178
Bambach RK 254
Bandy MC 167
Banerjee DM 250
Bangs S 355, 370–372, 374
Barlow IH 276
Barner HE 172, 174
Barone JP 372, 377
Bartelloni M 286
Barton C 254
Bass JD 186
Bates RG 146
Batkibekova M 179
Baturin GN 250, 279, 283
Baud CA 355, 368, 370–372, 374
Bayard M 327
Becker R 142
Bedell RL 204
Beebe RA 338, 339, 342
Beevers CA 275, 355, 359, 369
Behne W 228
Bein A 283
Belayonni H 283
Belov NV 276
Belyaevskaya TK 188
Bentor YK 250, 251, 283
Bergert JH 359, 378
Bergman SC 224
Berman H 4, 136, 318, 321, 323, 326, 329
Berman R 167, 293, 306, 307
Bernardi G 339
Berndt AF 390
Berry EE 277, 334
Besnus Y 310
Besty LB 277, 280
Bett HAS 334
Betts F 147, 331, 335–337, 340, 355, 356
Beswick AE 224, 225, 237
Bevan AWR 201
Bierman VJ 415
Bild RW 205, 207
Billaud Y 293
Birch GF 252
Birch WD 217
Bisaz S 336
Bishop FC 222, 225
Black CA 179, 308
Blanchard FN 293, 307
Blazy P 281
Bliskovskii VZ 287
Bloomfield K 245
Blount AM 307
Blumenthal NC 147, 335–337, 339, 345–347
Bobrownicki W 186
Boistelle R 138
Boke JW 369
Bondar LA 186
Bonel G 152, 276, 277, 339, 355, 356, 386, 370, 373–376
Bonjour JP 336
Bonner G 352, 355, 368
Borisov MS 174
Bormann FH 404, 405, 407
Borneman-Starynkevitch ID 276
Borosik M 228
Borr AB 228
Boskey AL 333, 337, 339, 343, 345, 376, 377
Bouldin DR 407
Boulet M 391
Bourrouilh FG 247
Bousquet J 188
Bowman JC 402
Bowman RA 405
Bowman RS 334
Bowser CJ 415
Boyan-Salyers BD 340, 343, 345, 355, 380
Brandt ELM 222, 231
Brassens A 276
Brasseur H 276
Brecevic Lj 147

Bremmer JM 285
Brenner M 411
Brett R 211
Brezonik PL 415
Briden JC 254, 255
Brinck JW 400, 402
Briner WS 336, 346
Brixner LH 221
Brodsky A 178
Brophy GP 279, 280
Brown EH 294, 208
Brown GM 221, 222, 227, 228
Brown M 345
Brown WE 137, 146–148, 150, 152, 161, 178, 179, 188, 189, 191, 340, 341, 352, 355, 362–364, 368–370, 373, 376, 377, 391
Brunfelt AO 222
Brunskill GJ 406
Buchwald VF 199, 200, 205–207
Bull RK 205
Bülow-Olsen A 405
Bunch TE 200, 202, 205, 207, 209, 210
Burnett WC 248–250, 265, 282
Burns RG 166
Burri C 231
Bursaux E 341
Burton JC 207
Burton WK 144
Buseck PR 201–204
Bushinski GI 250, 260, 282
Butler WT 345

Cabrera N 144
Calvert SE 265, 279
Calvo C 165, 201
Campbell DJ 343
Capdecomme L 282, 293, 300, 305, 306
Capobianco CJ 229, 230
Carlson CH 352, 355
Carlson RE 414
Carmichael ISE 222, 224–226, 228, 230, 237
Carron MK 234, 293, 294
Carswell DA 231
Cathcart JB 282
Chaabani F 278, 280, 282, 295, 296
Chaigneau M 232, 233
Chalov PI 250
Chambolle P 4, 136
Champetier Y 281
Chang SC 179
Chapra SC 414, 415
Charitonova WJ 201
Charpy DN 293, 301
Chen C-Hs 319
Cheng PT 355, 366, 377
Chester AH 136
Chickerur NS 152, 352, 362, 371, 377
Chien SH 179
Chilko DM 327
Chodos AA 210
Chodos SA 237
Chow LC 340, 341
Christ CL 186, 318
Christiansen JAC 140
Christner LG 334
Chughtai A 146, 174
Chukhlantsev VG 178, 179
Churney KL 172, 174, 185, 188
Ciricillo J 352, 355, 357, 359, 377
Cisney EA 218
Clanton JS 211
Clark FE 405
Clark JS 179
Clarke RS Jr 203, 293, 300, 313
Codispoti LA 264
Coffy G 185, 188
Cohen D 178
Cole CV 405
Cole DW 403
Coleman ML 285
Collett LW 282
Collin RL 138
Condrate RA 280
Conklin NM 296
Contiguglia SR 352, 354–358, 365, 366, 369, 376, 377
Cook PJ 249, 250, 252, 253, 256, 261, 262, 264, 266, 282, 283
Cooney AM 256
Cordfunke EHP 185
Counsell JF 185
Cowan WF 409, 417
Cressman ER 259
Crisman TL 411
Crisp TD 406, 407
Cronan DS 284, 286
Cruft EF 215, 217, 231, 233
Curzon MEJ 152, 355, 370, 374, 376
Cuttitta F 293, 300, 313

Dalculsi G 370
Dallemagne MJ 276
Davidson CL 355, 373
Davies CW 144, 146, 148
Davies KA 218
Davis AM 204
Dawson JB 222, 224, 225, 228, 229
Deans T 243, 244
De Bruyn PL 148, 151
Deer WA 215, 217, 218
Deevey ES 411
Deitz FR 373
De Keyser F 249
Delaney JS 204, 224, 228
Dell CI 318, 329
De Maria G 186
Denaeyer-De Smet S 403
Denotkina RG 179
De Pinto JV 409
Derec-Poussier F 4, 136
Despotovic R 148
Deutsch Y 278, 282
De Vries T 178
De Weisse G 293
Dickens B 161, 355, 376
Dickson I 343
Dieppe P 355, 366, 380
Dillon PJ 407, 414
Dines RH 327
Diot M 188
Donahan SA 293
Dongarra G 189
Doubinger J 282
Dowty E 209, 210
Druce EC 256
Dry ME 338
Dubinchuk VT 250
Duff EJ 185, 186, 191, 226
Duffield WA 222
Du Fresne ER 200, 203
Duherille J 355, 359, 380
Dunn PJ 232
Durrani SA 205
Duschene JC 220, 222
Du Toit AL 293, 299
Duvigneaud P 403
Dyakonova MI 201

Eakle AS 231
Eanes ED 147, 148, 333, 341, 351, 352, 355–357, 359, 369–371
Eastman J 137
Eastoe JE 343, 345
Easton AJ 200, 201
Eaton JS 403, 405, 407
Eby GN 222
Eckstrom CL 253
Edmondson WT 412, 414
Egan EP 185, 186, 188
Eichholz DE 2
Elderfield H 278
Elesin AA 174
Elgueta AS 252, 257
Elkin EC 280
Ellingsworth JJ 351, 355, 368, 369, 376, 380
Elliott CJ 200, 222, 228
Elliot JC 152, 276, 330, 332, 333, 351, 352, 355, 369, 370, 373, 374, 376
Elliott JS 352, 355, 359, 366, 369
Elliott RB 216
Ellison SA 345
El Mountassir M 296–298, 301
Embrey PG 1, 136

Emerson WH 280
Emigh DE 282
Emmermann R 219, 222, 231
Ennever J 355, 380
EPA (United States Environmental Protection Agency) 386, 388, 394, 395, 397, 398
Epstein S 285
Erdosh G 243–245
Erlank AJ 200
Erlichman J 200, 207
Esbensen K 206
Espenhade GH 300
Essene EJ 231, 232
Evans HT Jr 169
Evans WH 172–174, 185, 186, 188, 191, 319, 320
Evensen NM 223
Everett AJ 178
Ewald PP 159
Exley RA 220

Fakeev AA 186
Faure B 355, 359, 380
Faure G 355, 359
Faust GT 231
Fee EJ 414
Feenstra TP 138, 148, 151
Ferguson JF 137
Ferris JS 148, 352, 362, 363, 368–370, 373
Figueiredo Gomes C de S 293
Filipovic-Vincekovic N 148
Filomonov VT 174
Fischer AG 262
Fisher DJ 292
Flannery MS 411
Fleisch H 336, 351, 355–357, 365, 373, 376, 377, 380
Fleischer M 136
Fletcher MH 218
Flicoteaux R 278, 282
Folinsbee RE 202
Forester RW 229
Fowler BO 364
Frakes LA 264
Francis MC 336, 346
Francis MD 147, 352, 359, 377
Frank FC 142, 144
Frank RM 355, 369
Frankel JJ 299
Frankel SA 339
Frankenthal RP 391
Frazier AW 174, 178, 179, 276, 280, 281, 293, 294, 296–301, 305–307, 313, 352, 355, 362, 363, 376, 377
Freas DH 253
Fredriksson K 200, 204, 205, 207
Freminet A 341
French BM 209
Frey FA 216, 222, 225, 228
Friel JJ 211
Fripiat JJ 294
Fritz B 191
Frondel C 4, 136, 200, 293, 294, 318, 321, 323, 326, 329
Frondel JW 209
Fuchs LH 199–205, 207, 211
Fuge R 228, 229
Fuger J 172, 174, 186
Fuller JP 1, 136
Funk WH 412
Furedi-Milhofer H 147, 148

Gabbiani G 365, 369, 380
Gal I 278, 282
Gallop PM 343–345
Gancarz AJ 210, 237
Garcia-Ramos JV 355, 369, 370
Garrels RM 186, 318, 386
Gastuche MC 294
Gebert E 203, 204
Gerassimenko SV 186
Geraudie J 340
Gibbons HL 412
Gigli G 186
Gilbert RG 395
Gil[illegible]on IH 147, 148, 333
Gimmel'Farb BM 260
Girault J 232, 233
Gittins J 243–245
Glass JJ 234
Glimcher MJ 340, 343, 344, 352, 359, 380
Glock GE 276
Glybin VP 186
Go P 359–361, 370, 371, 376–378
Goeminne A 179
Goldberg MR 340, 345
Goldschmidt VM 160
Goldstein JI 199, 211
Golterman HL 401, 409, 416
Gomez M 203
Gonzales-Diaz PF 355, 369, 370
Gopal R 201
Gorham E 411
Goyette EA 407
Graham AL 201
Gramaccioli CM 235
Graziani G 231
Green DB 409
Green DH 222, 225, 228
Green TH 217, 220, 229
Gregory TM 146, 178, 188, 189, 191, 391
Griffin WL 225, 231, 236, 237
Griffith DP 355, 366, 369
Grinenko VA 287
Gross R 355, 369
Gruner JW 231, 276, 296
Guidi G 231
Guido M 186
Guillemin C 4, 136
Gulbrandsen RA 276–280, 283, 286, 300
Gulson BL 236
Gündogdu N 278, 282, 296, 298
Gurney EL 174, 178, 179, 295
Gusev GM 299
Gvelesiani GG 188

Halfon E 415
Hall WK 334
Hall-Beyer M 202
Hallsworth AS 373
Halow I 172–174, 185, 188, 319, 320
Hamilton PJ 223
Handley CD 226, 228
Hanson GN 225
Harlow GE 204
Harper RA 341
Harris JW 236
Harris RF 410
Harrison AF 403–405
Harriss DC 282
Haseman JF 294, 308
Haskin LA 220
Hassan AA 287
Hauschka PV 343–345
Hausel WD 225, 230, 236, 237
Haverinen A 407
Hawthorne FC 156
Hay DI 345
Haynes CV Jr 285
Heath D 352, 355
Hedge CE 230
Hedges AM 407
Hegyeli AF 369
Heier KS 225, 236, 237
Heinrich KJF 209
Henderson EP 200, 202, 205, 207
Henderson P 220, 222, 225, 228
Hendricks SB 231
Hendrickx SB 308
Hepple P 326, 327
Herrmann B 334
Hervig RL 203, 222, 224, 228
Herzberg MC 345
Heughebaert JC 145, 277, 352
Hey MH 200, 296
Hignett T 402
Hildreth EW 216, 220, 222, 226–228
Hill R 409
Hillig WB 144
Hobbie GE 406, 407
Hodgkinson A 351, 355, 369
Hohling HJ 355, 369
Holcomb DW 373, 375
Holdsworth E 201–204
Holland HD 228
Holmefjord T 278

Horne RA 386
Hough MJ 256
Houpt JB 352, 355, 366, 376
Howard PF 243, 249, 256, 257, 259
Howie RA 215, 217, 218
Hubbard NJ 211
Hughes RE 355, 369
Hunt C 386
Huntington HD 221, 227, 228
Hurley AM 254, 255
Huss GI 202, 205
Hutchins ML 414
Hutchinson GE 245, 247, 248, 411
Hutchison R 201
Hyde JR 203

Ilyin AV 249
IMPHOS 155
Inch TD 400
Ingamells CO 215, 217
Ingrahm G 406
Innes GS 405
Inouye TS 178
Irving AJ 220, 222, 228
Irving E 254
Irving JT 342
Irving RJ 185
Ivanov VI 210

Jackson ML 179, 294, 308
Jacob ET 352, 355, 368
Jacob G 283
Jacob J 231
Jacob KD 281
Jacobs DC 228
Jaffe ER 293, 300, 313
Jakobs-Mogelin J 401, 416
Jarmakowicz J 186
Jarosewich E 203
Jarvis I 295
Jenkins D 147, 387, 391
Jenkyns HC 263–265
Jensen KJ 203, 204
Jérome DY 201
Johan Z 293, 294
Johnson AH 407
Johnson ME 254
Johnson NM 403, 405, 407
Jones AL 144, 148
Jones BF 174, 179, 186, 189, 415
Jones JR 414
Jones RA 417
Jong WF de 369
Joussemet R 281
Javonovic S 209
Jowett M 179
Jungmann A 352, 355, 368
Jurinak JJ 178, 327

Kalff J 406
Kanizawa S 228
Kaplan I 285
Karlovsky J 402, 406
Karpov VI 179
Kazakov AV 250, 253
Kazimirof J 359, 376
Keen E 188
Keeton CM 327
Keil K 202, 205, 209, 210
Kelley KK 173, 188
Kerebel B 370
Kerebel LM 355, 359, 370–372, 375
Kerr PF 283
Kharaka YK 179
Khermann A 178
Khodakovsky IL 179, 185, 186, 188
Kielland J 189
Kim KM 352, 355, 356
Kind A 215, 217
King EA 200
King K 345
Kingsley JD 280
Kipfer A 136
Kirchner WB 407
Kirova OA 201
Kiss A 352, 355, 366, 376
Kittrick JA 294, 308
Kjensmo J 407
Kleeman JD 222
Klein C 204
Klein E 277, 280, 281, 335, 370, 371, 374–376
Klein I 355, 365, 373, 376
Klygin AE 179
Knowlton SM 307
Knutson J 217, 229
Knuutila M 355, 365, 369, 376
Koeppe DE 327
Kogarko LN 179, 186, 188, 229
Kohn NN 355, 369
Kolodny Y 250, 278, 285, 287
Konturi-Nahrhi V 355, 365, 369, 376
Koritnig S 4, 215–217
Kossel W 142
Kossiva D 344
Koster von Groos AF 226
Koutsoukos PG 150, 152
Koven IH 352, 355, 366, 376
Kracher A 205
Krajewski KP 250
Kramer JR 277, 280
Krauskopf KB 284, 321
Kravitz LC 280
Kresak M 372
Krigman LD 229
Kritsch N 406
Krivolutskaya LM 299
Krogh TE 236
Kunishi HM 409
Kunz GF 1, 3
Kurat G 205
Kuznetsova VK 179
Kydynov MK 179

Ladle GH 211
Lagergren C 351, 355, 369
Lahiri SC 174
Lallemand A 293, 298, 300
Lam DCL 415
Lamboy M 278
Landis WJ 340, 352, 359, 380
Langmuir D 189
Langmuir O 172, 174, 186
Lappartient JR 298, 313
Larsen DP 414, 416
Larsen ES 201, 218, 293, 294
Larson LM 218–220
Lathwell DJ 406
Laufer B 2
Lavrov AV 186
Le Bas MJ 226, 228, 232, 233
Lebedev IA 174
Lee AIN 245, 261, 270
Lee DD 352, 359, 360, 361, 373, 376, 377, 380
Lee DE 222, 231
Lee GF 409, 417
Leeman WP 216
Le Geros DJ 359, 360, 370, 371, 377, 378
Le Geros JP 152, 277, 280, 281, 335, 354–357, 359, 360, 362, 365, 366, 368, 370, 371, 373–377
Le Geros RZ 152, 277, 280, 281, 335, 339, 352, 354–362, 364–380
Legros R 339, 386, 370, 373, 375, 376
Lehman JT 412
Lehr JR 276–281, 352, 355, 362, 363, 367, 377
Lehr RJ 280
Lemina NM 299
Leonard F 369
Letolle R 282
Levine MJ 345
Levitt SR 280
Levskij LK 201
Lewis L 352, 355, 359, 369
Lian JB 343, 345
Liang Y 419
Liebau F 4, 164
Likens GE 403, 405–407
Linde A 345
Lindsay WL 178, 179
Liou JG 201, 202
Lippincott ER 333
Litkins RC 334
Logan TJ 409, 418
Logsden MJ 228
Loh SE 186
Loiacono GM 221
Longinelli A 286, 287
Lothringer JW 344

Love JD 249
Lovering JF 201, 203, 222
Low PF 308
Lowell WR 231
Lowenstam HA 285, 336
Lucas J 278, 282–285, 293, 295–299, 301
Ludington S 226
Luff BB 185, 186, 188
Luk SC 352, 355, 366, 376
Luz B 287
Lynch KL 366, 376

Maaløe S 215
MacCann HG 179
MacDonald RA 188
MacDowell H 146, 178
Machatschki F 276
MacIntyre DB 355, 369
Macintyre RM 245
Mackenzie FT 386
Mackie PE 332, 333
MacNeil FS 293, 294
Maier CG 173
Mainwaring PR 235
Makarov ES 210
Malaman B 355, 359, 380
Malissa H 202, 204, 205
Malone CP 327
Malueg KW 416
Mandarino JA 3
Mandel NS 355, 366
Mansfield GR 260
Marc R 144
Marcus V 174
Marier JR 391
Marinova LA 179, 186
Maroudas A 346
Marsh JS 244
Marshall HL 281
Marshall JF 250
Marshall R 146, 174
Marshall RR 205
Marshall RW 351, 355, 369
Martell AE 174, 178, 179
Martens CS 282
Martin JF 185
Martin RF 244
Marvin UB 204
Masalovich VM 174
Mash CJ 185
Mason B 200, 201, 203, 204
Massucci A 186
Mathers JE 178
Mathew M 150, 161
Mathez EA 209, 210
Matthews A 278, 281
Maughan EK 259, 283
Mayer SW 174
Mays RE 277, 280
McArthur JM 278, 279, 285, 299
McCallum IS 209, 210
McCammon CH 166
McCandless T 222
McCarthy DJ 355, 366, 369, 377
McClellan GH 276–281, 402
McConnell D 5, 136, 215, 218, 231, 276, 279, 280, 296, 336, 351, 355, 368–371, 374
McCullock CE 419
McDowell H 391
McElhinny MW 253, 261, 262, 264, 266
McIntyre DB 275
McKay DS 211
McKelvey VE 242, 250, 260, 282
McKerrell H 185
McKerrow WS 254
McKie D 293
McKinney CR 285
McKisdy DM 283
Melack JM 419
Mellini M 161, 164
Mellor JW 136
Melon J 276
Melson WG 200
Menczel J 341
Menor E 280, 296–299, 301
Mercier HT 414
Merkulova AV 250
Merlino S 161, 164
Mertie JB Jr 233
Mertzman S 225
Metzer A 278
Meyer C 211
Meyer JL 147, 148, 334, 359, 378
Migdasov AA 287
Mignon R 4, 136
Mikheev HB 178
Miller NL 352, 354–357, 359, 366, 369, 377
Miller RB 406, 407
Miller RJ 327
Millot G 293, 298, 300
Mills AA 232
Mills KC 178
Milton C 293, 294
Miravite MA 152, 355–357, 361, 362, 364–369, 373, 376, 377
Misra KC 211
Mitchell L 231
Mitchell RS 136, 307
Mitchum RM Jr 264
Mohan MS 148
Montel G 152, 277, 332
Mooney RW 178
Moore JG 216
Moore PB 5, 136, 155, 156, 158, 161, 162, 164–167, 200, 207, 215, 281
Morales P 355, 359–362, 364, 365, 367–369, 373, 376
Morel FMM 415
Morel P 254
Moreno EC 146, 178, 179, 188, 189, 191, 295, 364, 372, 391
Morgan JJ 318, 320, 391
Morgan JW 387
Morrison DA 211
Mortimer CH 415
Mortimer CT 188
Moskvin AI 174, 179, 186
Moss AA 200
Mosser C 310
Motto CK 327
Motto HL 327
Muljadi D 308
Munson EL 296
Murata KJ 234
Murray J 250
Murray MM 276
Musher DM 355, 366, 369
Muto T 179, 186
Muysson J 215, 217
Myers AT 296

Nagasawa H 220, 222
Nahon D 192, 193, 293, 294, 300, 301, 313
Nancollas GH 144–146, 148–150, 152, 174, 372, 377, 392
Naray-Szabo S 275
Nasarov MA 210
Nash RH 185, 188
Nash TJ 279, 280
Nash WP 216, 217, 221, 222, 225–228, 230, 236, 237
Nassau K 221
Nathan Y 278, 281–284, 286, 287, 293–301
Naumov GB 185, 188
Navon O 287
Nawrot CG 343
Needham J 2
Nehru CE 204
Nelen J 203
Nelson D 365, 371, 375
Nesbitt RW 215
Netter P 355, 359, 380
Neuman MW 276
Neuman WF 276, 373, 376
Neuringer JR 340
Nevins AJ 359, 376
Newberry NG 231, 232
Newesley H 150, 151, 355, 372, 376, 377
Nielsen AE 140, 144
Nielsen H 287
Nikol'skaya NA 179
Nishimoto SK 344
Nixon PH 231
Nordstrom DK 178
Notholt AJG 243–245
Novikov GI 185, 186

Nriagu JO 174, 178, 179, 186, 192, 282, 318, 320, 321, 327–329, 391, 392
Nuti S 286
Nyburg SC 355, 366, 377
Nylen MU 352, 355–357, 365, 377

O'Brien GW 251
Odin GS 282
Oetting FL 172, 174, 186, 188
Oglesby RT 412, 414, 419
Okada A 201, 203
Olsen E 199, 200, 202–205, 207
Omar SA 355, 366, 377
O'Melia CR 413, 414
O'Nions RK 223
Ophaug RH 355, 370, 371, 375
Orliac M 305, 306
Orlovskii VP 188
Ortner DJ 287
Osborn G 146, 178, 179
Ousubaliev D 179
Ouweltjes W 185
Overstreet W 233, 234
Ovington JD 403, 404
Owens JM 293, 306, 307
Owens JP 167

Pagenkopf G 391
Paine MC 340
Pak CYC 334, 352, 359
Palache C 4, 136, 318, 321, 323, 326, 329
Paquet H 293, 294, 300, 301
Parfitt MA 352, 369
Parker R 231
Parker RJ 295
Parker VB 172–174, 185, 186, 188, 191, 319, 320
Parron C 293, 301, 313
Parry WT 228
Pastor TP 220
Patel JM 146, 178, 191, 391
Patel PR 146, 178, 179, 189
Peacor DR 231, 232
Pearson JE 326
Peck DL 216
Peckauskas RA 147, 151
Peckett A 221, 222, 227, 228
Peech M 179
Pellas P 203
Péneaud D 293, 294, 307
Pentel L 368, 374, 375
Perez L 145
Perloff A 277, 334
Perminov PS 174, 186
Peters RH 417
Peterson WC 352, 355
Petrova YeN 229
Pevsner S 352, 355, 368
Phillips H 352, 355, 366, 376
Phinney WC 210
Piasecky LJ 334
Pierce RS 407, 403, 405
Pierrot R 4, 136
Piper DZ 264
Piret P 179
Pita JC 342
Pitman WC III 266
Piwinskii AJ 244
Pliny 2
Popoff M 313
Porter KS 407, 409–411
Posner AM 308
Posner AS 147, 148, 151, 277, 330–337, 339–343, 345–347, 351, 352, 355, 356, 360, 373, 374, 376, 377
Poullen JF 294
Povarennykh AS 136
Powell BN 205
Powell TG 283
Poyard CF 341
Prévôt L 278, 282–285, 293, 295–299, 301
Prewitt CT 160, 169, 201, 221
Price HI 179
Price NB 265, 279
Price PA 344
Price RC 225, 230, 231
Prien EL 352, 355, 380
Prins P 218
Prinz M 204, 209, 210, 216
Pritzker KPH 352, 355, 366, 376, 377
Prophet H 186
Puchelt H 219, 222, 231
Pugh MH 355, 365, 377
Purdie N 146

Quaide WL 352, 355, 359, 369
Quik JP 308
Quirolgico GB 152, 352, 355, 359–362, 364, 365, 367–371, 373, 374, 376–378, 380
Quist AS 146

Racz GJ 178
Rader LF Jr 281
Rankin AH 232
Rankine C 359, 376
Rapp GR Jr 1, 136
Rapp M 403
Rast W 417
Rat'kovskii IA 186
Ratnikova GI 249
Reed GW 228, 229
Reed GW Jr 209
Reed RB 185, 186, 188
Reich L 352, 359–361, 373, 377, 380
Renard AF 250
Rep'ko VP 188
Rey C 277
Reynolds DS 231, 281
Ribbe PH 202
Rice DS 411
Rice PM 411
Rice RC 395
Riedel L 300
Rietveld HM 332
Riggan LJ 355, 380
Riggs SR 254, 258, 259, 269, 283
Roberts WL 1, 136
Robertson AJ 352, 355
Robertson WG 351, 355, 357, 365, 376, 377
Robinson C 373
Robinson R 343
Roded R 278, 282
Roden MF 225
Roegge JS 228
Roelandts I 220, 222
Rogers AF 231
Rohrlich V 278
Romankevich YA 279, 283
Romig AD 199
Rooney TP 283
Rose HJ Jr 222, 231, 234
Rosenberg LC 346, 347
Rosenquist IT 178
Ross M 169
Rothbard DR 201
Roufosse AH 340, 344, 352, 359, 380
Rowles SL 352, 356, 377
Roy SD 225
Roy SK 200, 203
Ruderman RJ 369
Rud'ko PK 179, 185, 186
Runnels DE 352, 354–357, 366, 369
Ruskin B 352, 359
Russ W 300
Russel RT 293, 301
Russell G 351, 355, 357, 365, 376, 377
Russell RGG 336, 355–357, 373, 376, 380
Ruszczynski T 414
Ryden JC 410
Ryzenko B 185, 188

Sabine WK 340, 352, 359, 380
Sandell E 296
Sandstrom MW 283
Sandvik PO 245
Sanger JE 411
Santillan-Medrano J 327
Santos M 355, 369, 370
Sarcia JA 4, 136
Sassi S 258, 277, 283, 295
Savin SM 264
Savina LI 287
Sawyer CN 412
Schaffner WR 414
Schauwecker DS 220
Scheib MR 280
Scheuerman RV 172, 174
Schindler DW 406, 413, 414

Schlanger SO 263, 264
Schroeder H 351, 355, 358, 359, 362, 365
Schroeder JK 343
Schroeder LW 148, 161, 352, 362, 363, 368–370, 373
Schults DW 416
Schumm RH 172–174, 185, 188, 319, 320
Schwartz SD 174
Sclar CB 186
Scotese CR 254
Scott DB 355, 365, 377
Sears DW 200
Segalen P 295
Segalstad TV 235
Sellards EH 300
Selye H 365, 369, 380
Sereda BP 174

Shahar Y 278, 282, 296
Shannon EE 415
Shannon EV 201, 293, 294
Shannon IL 355, 359, 365, 376
Shannon RD 160, 169, 221
Shapiro J 414
Sharp RF 352, 355, 359, 369
Sheldon RP 250, 253, 259–262
Shelyakina AM 174, 186
Shen CY 387
Shergold JH 250, 253, 256
Shevaleevsky ID 210
Shevchenko VB 179
Shiloni Y 278, 282, 284, 295, 299
Shima M 201, 203
Shirra WP 152, 277, 281, 352, 355–357, 359–361, 365, 366, 368, 370, 371, 373–377, 380
Shyu LJ 145
Siesser WG 295
Sillén LG 174, 179
Silverman L 346
Silverstone L 355, 359, 370–372, 375
Simmons EC 230
Simon SL 282
Simonds CH 210
Simons TJ 415
Singer L 355, 370, 371, 375
Singer PC 178
Singer PL 391
Skinner HCW 229, 334
Slade PG 307
Slansky M 250, 258, 284, 293, 298–300
Smales FC 339
Smeck NE 409
Smith AG 254, 255
Smith DGW 202
Smith LH 359, 378
Smith JP 174, 178, 276, 280, 281, 352, 355, 362, 363, 376, 377
Smith JV 203, 215, 222, 224, 225, 228, 229
Smith RM 174, 178
Smith RV 406, 407
Snodgrass WJ 414
Snoeyink VL 387, 391
Snyder RB 326
Sokolski J 294
Solomons CC 352, 355, 357, 359, 377
Solovova IP 229
Sommer F 283
Sondi M 286
Soper RJ 178
Sørensen H 243, 244
Soudry D 283
Southgate PN 250
Spencer C 300
Spitsyn VI 178
Starotin AD 186
Staveley LAK 185, 188
Stearns RI 390
Steele AA 352, 356, 368, 369
Steele IM 203
Steere AC 334
Steinmetz J 355, 359, 380
Stepanov SI 179
Stewart JWB 405
Stormer JC 226
Storzer D 203
Strakhov NM 260
Stranski IN 142
Strickland-Constable 144
Strunz H 136, 201, 212, 231
Stull DR 186
Stumm W 318, 320, 391
Stutman JM 333
Summerhayes CP 263, 278, 284, 286
Sun SS 215, 225
Sunner S 185, 188
Sutor D 351, 355, 368, 369, 376, 380
Sutter JR 178, 391
Swanson RW 250, 260
Sweatman TR 293
Syers JK 410

Taborsky FK 215, 231, 232
Taheri MH 152, 355, 359, 365, 370, 373, 374, 376
Takasaki E 351, 355, 369
Tankard AJ 296, 299
Tao L-C 352, 355, 365, 369
Tarasov LS 210
Tardy Y 192, 193, 318
Taves DR 373, 376
Taylor AW 174, 178, 179, 295, 409
Taylor HP 229
Taylor LA 207, 211
Taylor SR 200, 225, 230, 231, 286
Tercinier G 293, 307
Ter Haar GL 327
Termine JD 147, 151, 339, 351, 352, 355–357, 376, 377
Tessier F 298, 300
Thayer JP 293
Thein A 355, 370, 371, 375
Thewlis J 276
Thompson RN 222
Thompson S 264
Thorell L 407
Thrasher RD 280
Tomassini N 186
Tomazic BB 148, 149
Tomson MB 149, 150, 372, 377, 392
Tong MS 150, 152
Tooms JS 284, 286
Tracey JI 249
Tranquard A 188
Trautmann NM 409, 417, 419
Trautz OR 152, 277, 280–282, 335, 351, 355, 356, 358, 360, 365, 368, 370, 371, 374–377
Trichet J 283
Triffitt JT 345
Trombe JC 277
Trompette R 293, 298, 299
Trotta R 340
Trueman NA 247, 276, 293, 294, 301
Truesdell AH 174, 179, 186, 189
Trump BF 352, 355, 356
Tsagareishvili DS 188
Tsutsumi T 403
Tuchweber B 365, 369, 380
Tudge AP 286
Tugarinov AI 234
Tung MS 341, 352, 362, 370, 371, 375–377
Turanskaya NV 234
Turekian KK 191, 286
Turner FJ 230
Tuttle OF 243, 244

Urey HC 285
USDA (United States Department of Agriculture) 402
Uttormark PD 414

Vahl J 351, 355, 358, 360
Vail PR 264
Vainshtein EE 234
Vallentyne JR 400, 412, 413
Valyashko VM 179, 186, 188
Vandemaele KH 359, 360, 370, 371, 377, 378
van der Voo R 254
Vanderzee CE 146
van Dyke TE 345
van Loenen RE 222, 231
van Schmus WR 202
van Straten HA 151

van Tassel R 307
van Valkenburg M 343
van Wazer R 281
Vaughan HH 411
Veeh HH 247–251, 265
Veis A 343
Verhoff FH 409
Verhoogen J 230
Vervaere A 370
Very JM 355, 368, 370–372, 374
Vieillard P 189, 191–193
Viets FG 405, 416
Vignalou JR 188
Vignoles M 276
Villiers JE de 276
Viro PJ 406, 407
Visse LD 293
Vitaliano CJ 216
Vo QT 276
Vochten R 179
Vogel J 355, 380
Volkov AI 185, 186
Vollenweider RA 407, 413, 414
Volmer M 140
Von Brand T 355, 365, 377
Von der Borch CC 250
Von Eckermann H 218
Von Michaelis H 200
Von Mierau HD 351, 355, 358, 360
Von Vahl J 355, 369

Wadjinny A 278, 296
Wagman DD 172–174, 185, 186, 188, 191, 319, 320
Wakefield ZT 186, 188
Walker BM 211
Wallaeys R 276
Wilcox TS 209
Wampler JM 282
Wang J 419
Wang N-S 352, 356, 368, 369
Ward P 336
Warin ON 247
Waring GA 230
Warner JL 210
Wartiovaara J 407
Wass SY 224
Wasson JT 200
Waterfield CG 185, 188
Watkinson DH 235
Watson EB 220, 225, 229, 230
Weatherell JA 373
Weaver CE 282
Webb NC 147, 352, 359, 377
Webber MD 178
Weber J 1, 136
Wedepohl KH 286
Welch EB 412
Welch HE 406
Wey R 308
White DE 230
White WB 280
Whitehead DR 411
Whitt DC 294, 308
Wiik HB 200
Wilkinson BH 249
Williams JDH 409
Willis JP 200
Wise WS 186
Wooley SE 351, 355, 368, 369, 376, 380
Woolley AR 233
Wright TL 216
Wuebbles DR 326
Wyllie PJ 326

Yagi K 201, 203
Yaglov VN 179, 185, 186
Yakovlev GN 174
Yeasted JG 415
York D 245
Young EG 293, 300, 313
Young EJ 282, 296
Young HS 228
Young RA 150, 330, 332, 333, 351, 355, 369, 370, 373, 375
Yousoupov V 179
Yusupov TS 299

Zahradnik PG 372
Zanazzi PF 161, 164
Zanin YN 293, 299, 301
Zapanta-LeGeros R 351, 355, 365
Zavrazhnova DM 179
Zawacki S 145
Zharovskii FG 178
Ziegler AM 254
Zielinski RA 216, 225
Zindler A 225
Zipkin I 341, 351, 355, 368, 372–374
Zolotavin VL 179
Zoltek J Jr 391
Zussman J 215, 217, 218